BOOKS BY RICHARD RHODES

FICTION

Sons of Earth

The Last Safari

Holy Secrets

The Ungodly

VERITY

How to Write

Nuclear Renewal

Making Love

A Hole in the World

Farm

The Making of the Atomic Bomb

Looking for America

The Inland Ground

Richard Rhodes

DARK SUN

THE
MAKING
OF THE
HYDROGEN
BOMB

A TOUCHSTONE BOOK
Published by Simon & Schuster

TOUCHSTONE
Rockefeller Center
1230 Avenue of the Americas
New York, NY 10020

First Touchstone Edition 1996.

TOUCHSTONE and colophon are registered trademarks
of Simon & Schuster Inc.

Designed by Irving Perkins Associates, Inc.

Manufactured in the United States of America

1   3   5   7   9   10   8   6   4   2

Library of Congress Cataloging-in-Publication Data
Rhodes, Richard.
Dark sun : the making of the hydrogen bomb.
p.   cm.
1. Hydrogen bomb—History.   I. Title.
UG1282.A8R46   1995
623.4'5119—dc20   95-11070
CIP
ISBN 0-684-80400-X
ISBN 0-684-82414-0 (pbk)

The author is grateful for permission to reprint the following:

Arthur Lawrence Norberg interviews with Norris Bradbury, Darol Froman, John Manley, J. Carson Mark and Raemer Schreiber quoted by permission of The Bancroft Library.

Excerpts from *The Journals of David E. Lilienthal, Vol. 2: The Atomic Energy Years, 1945–1950* by David E. Lilienthal. Copyright © 1964 by David E. Lilienthal. Copyright renewed 1992. Reprinted by permission of HarperCollins Publishers, Inc.

Excerpts from *Mission with LeMay* by Curtis E. LeMay with MacKinlay Kantor. Copyright © 1965 by Curtis E. LeMay and MacKinlay Kantor. Used by permission of Doubleday, a division of Bantam Doubleday Dell Publishing Group, Inc.

Excerpts from Yuli Khariton and Yuri Smirnov, "The Khariton Version," *Bulletin of the Atomic Scientists,* v.93, reprinted by permission of Yuri Smirnov.

Excerpts from the unpublished papers of John Manley quoted by permission of Kathleen B. Manley; of Raemer Schreiber quoted by permission of Raemer Schreiber.

Excerpts from Ilya Ehrenburg and Konstantin Simonov, *In One Newspaper.* Trans. Anatol Kagan. Copyright © 1985 by Sphinx Press and reprinted by permission.

(continued at back of book)

FOR ARTHUR L. SINGER, JR.

The author acknowledges with gratitude the support of the John D. and Catherine T. MacArthur Foundation and the Alfred P. Sloan Foundation in the research and writing of this book.

THIS BOOK IS PUBLISHED AS PART OF AN
ALFRED P. SLOAN FOUNDATION PROGRAM

Contents

Part Three

Scorpions in a Bottle

Preface to the Sloan Technology Series

TECHNOLOGY IS THE APPLICATION of science, engineering and industrial organization to create a human-built world. It has led, in developed nations, to a standard of living inconceivable a hundred years ago. The process, however, is not free of stress; by its very nature, technology brings change in society and undermines convention. It affects virtually every aspect of human endeavor: private and public institutions, economic systems, communications networks, political structures, international affiliations, the organization of societies and the condition of human lives. The effects are not one-way; just as technology changes society, so too do societal structures, attitudes and mores affect technology. But perhaps because technology is so rapidly and completely assimilated, the profound interplay of technology and other social endeavors in modern history has not been sufficiently recognized.

The Sloan Foundation has had a long-standing interest in deepening public understanding about modern technology, its origins and its impact on our lives. The Sloan Technology Series, of which the present volume is a part, seeks to present to the general reader the stories of the development of critical twentieth-century technologies. The aim of the series is to convey both the technical and human dimensions of the subject: the invention and effort entailed in devising the technologies and the comforts and stresses they have introduced into contemporary life. As the century draws to an end, it is hoped that the Series will disclose a past that might provide perspective on the present and inform the future.

The Foundation has been guided in its development of the Sloan Technology Series by a distinguished advisory committee. We express deep gratitude to John Armstrong, S. Michael Bessie, Samuel Y. Gibbon, Thomas P. Hughes, Victor McElheny, Robert K. Merton, Elting E. Morison and Richard Rhodes. The Foundation has been represented on the committee by Ralph E. Gomory, Arthur L. Singer, Jr., Hirsh G. Cohen, Raphael G. Kasper and A. Frank Mayadas.

Alfred P. Sloan Foundation

Fundamentally, and in the long run, the problem which is posed by the release of atomic energy is a problem of the ability of the human race to govern itself without war.

A Report of a Panel of Consultants on Disarmament of the Secretary of State, January 1953

Much that follows is new, and some of it surprising. A discussion of sources appears ahead of the Notes beginning on page 591; the Notes are keyed to a Bibliography that begins on page 689.

Readers unfamiliar with Russian names may take comfort in knowing that they are transliterated phonetically from their original Cyrillic, an alphabet borrowed from Greek and Hebrew. Sounding them out aloud two or three times usually fixes them in memory. A Glossary of Names, with approximate pronunciations, begins on page 671.

Prologue: Deliveries

THE WAR WAS OVER. The troops were coming home. Sick of mud and olive drab, of saltwater showers and sweltering holds, twelve million American soldiers and sailors counted their service points to see how soon they could ship out for Brooklyn and Ukiah and St. Joe. Tens of thousands of warplanes, ships, tanks, artillery pieces sat abandoned, the full industrial output of a prosperous nation, the work the women and the older men had done, soon to be junked. The Second World War had been the most destructive war in history, obliterating fifty-five million human lives. The German invasion of the Soviet Union and the obdurate Soviet response had accounted for more than half those deaths; with them, in Germany and the Soviet Union both, had followed general ruination. In the end, out in the Pacific, two planes carrying two bombs had compelled the war's termination. The two atomic bombs, ferocious as minor suns, had given an emperor descended from a god reason to surrender. The war was over. It was hard to imagine that there might ever be another.

Luis Alvarez, an American experimental physicist, a tall, ruddy Californian with ice-blond hair, had understood the message of the bomb on his way back from Hiroshima. Alvarez collected adventures. He liked to be on hand when history was made. After he invented ground-controlled approach radar he had flown a prototype unit to wartime England and personally tested it talking down British bombers returning through fog. At the secret laboratory at Los Alamos in New Mexico where the atomic bombs were designed and built by hand, he had arranged to observe intensely radioactive test explosions up close in a lead-lined tank. He had invented a new electric detonation system for the Fat Man plutonium implosion bomb that fired its multiple detonators with microsecond simultaneity. As the time to deploy the revolutionary new weapons approached, Alvarez had found a way to justify flying the historic first mission.

The Hiroshima bomb, Little Boy, was a uranium gun. It used sixty-four kilograms of rare uranium 235, all of that dense, purple-black metal the United States had been able to accumulate up to the end of July 1945. The uranium gun was an extremely conservative design. "We were confident it would work," Alvarez writes, but it had not been tested. To determine its efficiency, Los Alamos had needed to know its explosive yield. So Alvarez had invented a device for measuring that yield, a set of parachute-deployable pressure gauges to be dropped ahead of the bomb that would radio their readings to a backup plane. Riding in that backup plane, a B-29 named the *Great Artiste,* Alvarez had seen the bright flash of the Hiroshima explosion, had watched its pressure pulses register on the oscilloscopes mounted in the rear compartment he occupied, had felt the two sharp slaps of direct and ground-reflected shock waves slamming the plane like flak explosions, had moved to the window then and searched below while the plane circled the rising mushroom cloud. "I looked in vain for the city that had been our target. The cloud seemed to be rising out of a wooded area devoid of population." On the intercom the pilot confirmed that the aiming had been excellent; Alvarez could not see the city because the city had been destroyed.

On the way back to Tinian, the island in the Marianas from which the atomic bombing had been staged, Alvarez had passed the time writing a letter to keep for his son Walter, then four years old. "This is the first grown-up letter I have ever written to you," the physicist began. He reminded his son that they had inspected a B-29 together in Albuquerque—"probably you will remember climbing thru the tunnel over the bomb bay," he teased him, "as that really impressed you at the time." Then Alvarez described "what has happened to aerial warfare" as a result of the *Enola Gay*'s mission that morning:

Last week the 20th Air Force . . . put over the biggest bombing raid in history, with 6,000 tons of bombs (about 3,000 tons of high explosives). Today, the lead plane in our formation dropped a single bomb which probably exploded with the force of 15,000 tons of high explosive. That means that the days of large bombing raids, with several hundred planes, are finished. A single plane disguised as a friendly transport can now wipe out a city. . . .

What regrets I have about being a party to killing and maiming thousands of Japanese civilians this morning are tempered with the hope that this terrible weapon we have created may bring the countries of the world together and prevent further wars. Alfred Nobel thought that his invention of high explosives would have that effect, by making wars too terrible, but unfortunately it had just the opposite reaction. Our new destructive force is so many thousands of times worse that it may realize Nobel's dreams.

A second atomic bomb exploded three days later over Nagasaki reinforced the point and on August 14, 1945, the Japanese had surrendered. After the surrender, Robert Serber, the theoretical physicist who had directed the design of the Little Boy bomb, a lean, gentle Philadelphian with a steel-trap mind, had walked the streets of the city his bomb had destroyed. With other scientists and physicians, Serber had been assigned to visit the two atomic-bombed cities to study the damage; from Tokyo his group had caught a ride down Honshu in the personal plane of Admiral Richard E. Byrd, the Antarctic explorer, who wanted to see the destruction at first hand. In Nagasaki and then Hiroshima, Serber and British hydrodynamicist William Penney had collected dented gas cans, concrete rubble, a charred crate, a beaverboard panel burned with the shadow of a window frame. They had talked to returning Australian and Dutch prisoners of war temporarily housed in Nagasaki, living skeletons whom the Japanese had brutally abused and starved. They had visited a Japanese civilian hospital and seen women and children ill with flash burns and radiation sickness, an experience Serber still characterized almost fifty years later as "really harrowing." It had been easy to leave the United States during wartime. Returning now that the war was over was more complicated. "We had a little trouble in San Francisco," Serber remembers. "Peacetime practices were now in effect. We had to go through Customs (squashed gas cans, hunks of concrete, charred crate) and Immigration and it turned out that Bill didn't have a passport. However, our other identifications so impressed the immigration official that he decided he could call Bill a British RAF [Royal Air Force] officer and let him in." To a nation weary of war, the scientists who built the atomic bombs were heroes.

Major General Curtis LeMay riddled a different oracle from the ashes of Hiroshima and Nagasaki. A swarthy, burly, taciturn thirty-eight-year-old Ohio-born engineer, LeMay commanded the B-29s that had firebombed Japan to destruction, lifting from the vast coral runways of Guam, Saipan and Tinian like the thousand silver throwing-stars of a warrior god. LeMay still remembered vividly—would remember all his life—how unprepared the United States had been at the beginning of the war. "We came into the war with practically nothing," he told an interviewer in 1943. To an audience of fellow Ohio State alumni later in 1945 he would insist starkly:

We tottered on the brink of defeat for two years before we could strike back. I know the feeling of our men [besieged] on Bataan and Corregidor because I commanded a bomb group in England in the early days of the war where we found the same situation—50 bombers against the entire German air forces. There came a time when we could see that at the existing loss rate with no reinforcements the last B-17 would take off to bomb Germany within 30 days. Fortunately, that unhappy day never arrived because the first trickle of help came just in time. It is quite an experience to see the reaction on people who have reconciled themselves to dying, [who] suddenly finish their combat tour and look forward to living again. I hope no American ever has to go through that experience in the future.

In England, LeMay had led his bombardment group's first combat mission. He had invented defensive formations that saved crew lives and bombing techniques that put twice and three times as many bombs on target as less imaginative commanding officers arranged. His byword was preparation. "Hit it right the first time," he taught his men, "and we won't have to go back." They called him Iron Ass because he trained them relentlessly, but they also called him "absolutely the best CO in the Army." From England in 1944 he had moved to India to attempt the thankless task of bombing the Japanese from bases in China supplied by air from India over the Himalayas, the infamous Hump. The B-29, the first intercontinental bomber, was just then coming into production and the leaders of the Air Forces, still a branch of the Army,* needed to prove the value of the investment. LeMay's B-29s had to haul their own gasoline over the Hump; it took a half-dozen Hump flights with bomb bays tanked with fuel to support one combat mission over Japan. Japan's weather moved in through north China, which Mao Zedong's army controlled. LeMay traded the Communist guerrilla leader medical supplies for crew rescues and weather reports.

The four-engine B-29, half the size of a football field, with electric control systems and two capacious bomb bays, was supposed to be a high-altitude precision bombing machine, aiming bombs down chimneys with the famous Norden bombsight from thirty thousand feet. But the force assembling in the Marianas while LeMay's crews labored from China had the bad luck to discover the jet stream. From one mission to the next it blew the planes off their targets. The Norden bombsight had not been designed to compensate for such furious drift. Once, when the B-29s were supposed to be bombing an aircraft factory ten miles north of Tokyo, they discovered their bombs had exploded in Tokyo Bay; the Japanese joked that the Americans were trying to drown them. LeMay was called in to fix the problem early in 1945. While he worked on improving precision, he and his staff studied strike photos and flak reports. They realized the Japanese had no night fighters and noticed that Japanese anti-aircraft fire clustered high. "We couldn't find any low-altitude defense," LeMay concludes.

Daylight precision bombing from low altitude would put LeMay's crews at risk. Advanced radar bombsights were not yet available for precision bombing at night. The USAAF wanted to end the war with air power before an Army and Navy invasion of Japan. LeMay worked out a radical change in strategy, ordered his B-29s stripped of armament to increase their carrying capacity, had 325 planes loaded with ten thousand pounds each of jellied-

* The US military air arm was called the Army Air Corps until June 1941, when its name was changed to the Army Air Forces (USAAF). In July 1947 the air arm separated from the Army; as an independent service it was and is designated the United States Air Force (USAF).

gasoline firebomb clusters and sent them over Tokyo on the night of March 10, 1945, staggered at from five to nine thousand feet, with pathfinder B-29s going ahead of them to mark out huge Xs in flame at their designated aiming points. LeMay's subsequent mission report emphasized that the object of the attack "was *not* to bomb indiscriminately civilian populations. The object *was* to destroy the *industrial and strategic targets* concentrated" in the Tokyo urban area. The firebombing successfully destroyed or damaged "twenty-two industrial target[s] . . . and many other unidentified industries." But the destruction that first windy night was in fact indiscriminate to the point of atrocity, as LeMay himself understood: 16.7 square miles of the Japanese capital burned to the ground, 100,000 people killed and hundreds of thousands injured in one night. "The physical destruction and loss of life at Tokyo," LeMay quotes from the official Air Force history of the Second World War, "exceeded that at Rome . . . or that of any of the great conflagrations of the western world—London, 1666 . . . Moscow, 1812 . . . Chicago, 1871 . . . San Francisco, 1906. . . . Only Japan itself, with the earthquake and fire of 1923 at Tokyo and Yokohama, had suffered so terrible a disaster. No other air attack of the war, either in Japan or Europe, was so destructive of life and property." With such compelling evidence that the new bombing strategy worked, LeMay laid on firebombings night after night against city after Japanese city until his supply depots ran out of bombs; resupplied, he pursued the firebombing campaign relentlessly through the spring and summer of 1945 until the end of the war, by which time sixty-three Japanese cities had been totally or partially burned out and hundreds of thousands of Japanese civilians killed, at a total cost to the Air Forces, as LeMay would lecture later, of "485 B-29s" and "approximately 3,000 combat crew personnel." Hiroshima and Nagasaki survived to be atomic-bombed only because Washington had removed them from Curtis LeMay's target list.

Long after the war, a dauntless cadet asked LeMay "how much moral considerations affected his decisions regarding the bombing of Japan." LeMay, as hard a man as Ulysses S. Grant, answered with his usual bluntness:

Killing Japanese didn't bother me very much at that time. It was getting the war over that bothered me. So I wasn't worried particularly about how many people we killed in getting the job done. I suppose if I had lost the war, I would have been tried as a war criminal. Fortunately, we were on the winning side. Incidentally, everybody bemoans the fact that we dropped the atomic bomb and killed a lot of people at Hiroshima and Nagasaki. That I guess is immoral; but nobody says anything about the incendiary attacks on every industrial city in Japan, and the first attack on Tokyo killed more people than the atomic bomb did. Apparently, that was all right. . . .

I guess the direct answer to your question is, yes, every soldier thinks something of the moral aspects of what he is doing. But all war is immoral, and if you let that bother you, you're not a good soldier.

At the Japanese surrender ceremonies on the battleship *Missouri* in Tokyo Bay on September 2, LeMay's B-29s, nearly five hundred of them, had roared overhead in salute while LeMay stood on the deck watching Douglas MacArthur stern at the table where the Japanese foreign minister grimly signed the surrender. LeMay was thinking of the boys who had died to get them there, he wrote later, thinking "that if I had done a better job we might have saved a few more crews." That was the overriding message Curtis Emerson LeMay took with him from the long, bloody war: preparation. "I think the main experience that I wouldn't want to repeat is the war experience that I had," he told the same cadets who heard his opinion of killing Japanese. "There is nothing worse that I've found in life than going into battle ill-prepared or not prepared at all." To the lesson of that elemental experience he would attribute the massive work he would accomplish postwar of building up a strategic air force.

"Like many other folks" at the end of the war, he writes, he was "pretty tired." He took time to fly up and down the Japanese coast to view the results of his firebombing, then returned to his headquarters on Guam. His aide-de-camp notes on September 3 that "General LeMay spent the night at General Spaatz's house—a last stand all night poker game. The game broke up at 0600 hours the morning of the fourth." Spaatz was LeMay's boss, Carl "Tooey" Spaatz, commanding general of the Strategic Air Force in the Pacific; who won the poker game, the aide doesn't record.

At the end of August, LeMay had heard through Spaatz that Washington had asked General James Doolittle, the air pioneer and Eighth Air Force commander, to lead a flight of three B-29s nonstop from Tokyo to Washington, and that Doolittle had recommended including LeMay. "Offhand," says LeMay, "I would guess that this flight was dreamed up to demonstrate and dramatize . . . the long-range capability of the [B-]29 to the American people and to the world at large." To make the long flight—nearly seven thousand miles—the bomb bays of the aircraft would need to be fitted with extra fuel tanks. Doolittle on Okinawa had studied the matter and concluded that six tanks would give the B-29s a gross takeoff weight of 142,800 pounds. "The trip can be made," Doolittle had messaged Spaatz by courier, provided they could find an airfield in Japan long enough and with enough bearing capacity to handle the load.

Spaatz replied on September 5 that "there are no fields in Japan suitable for take off at gross weight necessary. . . . Flight is not feasible." Never one to take no for an answer, Doolittle flew to Guam three days later to confer with LeMay. "We got together," writes LeMay, "and talked the thing over; we examined photographs and charts. The only field which might accommodate

the B-29's was Mizutani, up on the northern Japanese island of Hokkaido.
. . . Trouble was, we didn't have any troops in there as yet. . . . There was
nobody of whom we could make inquiry concerning the runways." LeMay
sent one of his commanders to scout Hokkaido in a B-17. The Japanese naval
officers at Mizutani had heard their emperor's surrender broadcast and
didn't shoot him. The runways, the man reported, would do.

LeMay ordered three B-29s stripped of spare equipment and outfitted with
bomb-bay tanks. In the meantime, Doolittle was called ahead to Washington.
Lieutenant General Barney Giles, commander of the Central Pacific Air
Forces, took over Doolittle's place in the lead plane; LeMay and Brigadier
General Emmett "Rosie" O'Donnell, Jr., would fly the other two. The three
B-29s left Guam on Sunday, September 16, fueled at Iwo Jima and flew to
Hokkaido, where they topped off their tanks with drum gasoline flown in
on C-54s. "That night we slept in a barracks with three thousand polite
Japanese sailors surrounding us," LeMay recalls. "No sweat." The trio of
generals with their eleven-man crews took off for North America at 0600
hours on Wednesday, September 19, flew a Great Circle route northeast,
crossed the International Date Line into the Western Hemisphere's Wednesday, made radio contact with Nome, reached their halfway point over
Whitehorse in the Yukon at nine A.M. Eastern War Time and approached the
northern Middle West late that afternoon. They had bucked headwinds most
of the way that slowed their average speed to less than 250 knots and ate up
their fuel. LeMay wanted to take a chance on making it to Washington, where
the weather was reported marginal, but Giles and O'Donnell opted to refuel
in Chicago. "I went on awhile," writes LeMay nonchalantly, "then received
another Washington report. This time the weather was *really* marginal, and
that didn't seem to make very good sense, with the small reserve of gas I'd
have. I turned around and went back." From Chicago they flew on to Washington the same night and landed at National Airport just before nine to the
clangor of a brass band the Air Forces had deployed for the occasion. Curtis
LeMay, too, had come home.

The *Chicago Tribune* thought "the only significance" of the intercontinental nonstop flight of three US heavy bombers was "that it is going to be
possible very soon to fly from here to Tokyo in 24 hours by commercial
airliner." The Army Air Forces saw further significance in intercontinental
flight. A document titled *A Strategic Chart of Certain Russian and Manchurian Urban Areas* had gone to Brigadier General Leslie R. Groves, the head
of the atomic-bomb project, already on August 30, 1945; the document
identified the important cities of the Soviet Union and Manchuria and
charted their area, population, industrial capacities and target priority. Thus
Moscow was estimated to have a population of four million, an area of 110
square miles, priorities of 1 for industry and 3 for oil and was estimated to
supply 13 percent of Soviet plane output, 43 percent of truck output, 2

percent of steel and 15 percent of copper, machine-building, oil refinery and ballbearing output. Baku produced 61 percent of the Soviet Union's oil, Gorki 45 percent of its guns, Chelyabinsk 44 percent of its zinc. The list descended to cities of only 26,000 population, but was then refined to selections of "15 key Soviet cities"—Moscow, Baku, Novosibirsk, Gorki, Sverdlovsk, Chelyabinsk, Omsk, Kuibyshev, Kazan, Saratov, Molotov, Magnitogorsk, Grozny, Stalinsk, Mishni Tagil—and "25 leading Soviet cities." An appendix estimated how many atomic bombs would be needed to destroy each city—six each for Moscow and Leningrad. A map centered on the North Pole accompanied the chart; around the world from bases in Nome; Adak, in the Aleutians; Stavanger, Norway; Bremen, Germany; Foggia, Italy; Crete; Lahore, India; and Okinawa, B-29 flight paths had been overlaid darkly like segments of radar sweeps to cover the USSR.

The plan was something of a wish list. LeMay, Giles and O'Donnell had flown one way intercontinentally and then only by loading their bomb bays with fuel tanks. The realistic range of a B-29 with a bomb load was three thousand miles. Nor were all those convenient bases available. Before the US would have a force capable realistically of striking the Soviet Union, it would need forward bases, aerial refueling or a longer-range bomber. In the autumn of 1945 none of those capabilities yet existed.

If the Soviet Union had been the United States's Second World War ally, it was also the only possible enemy to survive the general destruction with sufficient military power to challenge American hegemony. Its army occupied the eastern half of Europe. The United States believed it had a trump card in the atomic bomb, but even that advantage was a wasting asset. On September 19, while Curtis LeMay and his colleagues were en route from Hokkaido to Washington proving that atomic bombs could be delivered great distances by plane, physicist Klaus Fuchs, a member of the British Mission at Los Alamos, was finishing up delivering information about the atomic bomb by hand to Harry Gold, an American industrial chemist who was a courier for Soviet intelligence. Fuchs had been passing information on the atomic-bomb project to Soviet agents since 1941. In June he had delivered to Gold a complete description of the Fat Man plutonium implosion bomb, including detailed cross-sectional drawings, which had been sent along immediately to Moscow. Now, driving Gold up into the Santa Fe hills overlooking the New Mexican capital in the early evening, Fuchs reported on the rate of US production of U235 and plutonium and on advanced concepts for improved bomb designs. In October 1945, with Fuchs's information and information from other US and British spies, the head of Soviet foreign intelligence in Moscow was able to send to the commissar for state security newly appointed to direct the Soviet atomic-bomb program, Lavrenti Beria, a detailed plan of the plutonium implosion bomb for Soviet scientists to duplicate. The war was over. The atomic arms race had begun.

PART ONE

A Choice Between Worlds

His decision to become a Communist seems to the man who makes it as a choice between a world that is dying and a world that is coming to birth.

WHITTAKER CHAMBERS

1

'A Smell of Nuclear Powder'

EARLY IN JANUARY 1939, nine months before the outbreak of the Second World War, a letter from Paris alerted physicists in the Soviet Union to the startling news that German radiochemists had discovered a fundamental new nuclear reaction. Bombarding uranium with neutrons, French physicist Frédéric Joliot-Curie wrote his Leningrad colleague Abram Fedorovich Ioffe, caused that heaviest of natural elements to disintegrate into two or more fragments that repelled each other with prodigious energy. It was fitting that the first report of a discovery that would challenge the dominant political system of the world should reach the Soviet Union from France, a nation to which Czarist Russia had looked for culture and technology. Joliot-Curie's letter to the grand old man of Russian physics "got a frenzied going-over" in a seminar at Ioffe's institute in Leningrad, a protégé of one of the participants reports. "The first communications about the discovery of fission... astounded us," Soviet physicist Georgi Flerov remembered in old age. "... There was a smell of nuclear powder in the air."

Reports in the British scientific journal *Nature* soon confirmed the German discovery and research on nuclear fission started up everywhere. The news fell on fertile ground in the Soviet Union. Russian interest in radioactivity extended back to the time of its discovery at the turn of the century. Vladimir I. Vernadski, a Russian mineralogist, told the Russian Academy of Sciences in 1910 that radioactivity opened up "new sources of atomic energy ... exceeding by millions of times all the sources of energy that the human imagination has envisaged." Academy geologists located a rich vein of uranium ore in the Fergana Valley in Uzbekistan in 1910; a private company mined pitchblende there at Tiuia-Muiun ("Camel's Neck") until 1914. After the First World War, the Red Army seized the residues of the company's extraction of uranium and vanadium. The residues contained valuable radium, which transmutes naturally from uranium by radioactive decay. The

Soviet radiochemist Vitali Grigorievich Khlopin extracted several grams of radium for medical use in 1921.

There were only about a thousand physicists in the world in 1895. Work in the new scientific discipline was centered in Western Europe in the early years of the twentieth century. A number of Russian scientists studied there. Abram Ioffe's career preparation included research in Germany with Nobel laureate Wilhelm Roentgen, the discoverer of X rays; Vernadski worked at the Curie Institute in Paris. The outstanding Viennese theoretical physicist Paul Ehrenfest taught in St. Petersburg for five years before the First World War. In 1918, in the midst of the Russian Revolution, Ioffe founded a new Institute of Physics and Technology in Petrograd.* Despite difficult conditions—the chemist N. N. Semenov describes "hunger and ruin everywhere, no instruments or equipment" as late as 1921—"Fiztekh" quickly became a national center for physics research. "The Institute was the most attractive place of employment for all the young scientists looking to contribute to the new physics," Soviet physicist Sergei E. Frish recalls. ". . . Ioffe was known for his up-to-date ideas and tolerant views. He willingly took on, as staff members, beginning physicists whom he judged talented. . . . Dedication to science was all that mattered to him." The crew Ioffe assembled was so young and eager that older hands nicknamed Fiztekh "the kindergarten."

During its first decade, Fiztekh specialized in the study of high-voltage electrical effects, practical research to support the new Communist state's drive for national electrification—the success of socialism, Lenin had proclaimed more than once, would come through electrical power. After 1928, having ousted his rivals and consolidated his rule, Josef Stalin promulgated the first of a brutal series of Five-Year Plans that set ragged peasants on short rations building monumental hydroelectric dams to harness Russia's wild rivers. "Stalin's realism was harsh and unillusioned," comments C. P. Snow. "He said, after the first two years of industrialization, when people were pleading with him to go slower because the country couldn't stand it:

> To slacken the pace would mean to lag behind; and those who lag behind are beaten. We do not want to be beaten. No, we don't want to be. Old Russia was ceaselessly beaten for her backwardness. She was beaten by the Mongol khans, she was beaten by Turkish beys, she was beaten by the Swedish feudal lords, she was beaten by Polish-Lithuanian pans, she was beaten by Anglo-French capitalists, she was beaten by Japanese barons, she was beaten by all—for her backwardness. For military backwardness, for cultural backwardness, for agricultural backwardness. She was beaten because to beat her was profitable and went unpunished. You remember the

---

* St. Petersburg, renamed Petrograd by Czar Nicholas in 1914 and Leningrad by the new Soviet government in 1924.

words of the pre-revolutionary poet: "Thou art poor and thou art plentiful, thou art mighty and thou art helpless, Mother Russia."

We are fifty or a hundred years behind the advanced countries. We must make good the lag in ten years. Either we do it or they crush us.

Soviet scientists felt a special burden of responsibility in the midst of such desperate struggle; the heat and light that radioactive materials such as radium generate for centuries without stint mocked their positions of privilege. Vernadski, who founded the State Radium Institute in Petrograd in 1922, wrote hopefully that year that "it will not be long before man will receive atomic energy for his disposal, a source of energy which will make it possible for him to build his life as he pleases." World leaders such as England's Ernest Rutherford, who discovered the atomic nucleus, and Albert Einstein, who quantified the energy latent in matter in his formula $E = mc^2$, disputed such optimistic assessments. The nuclei of atoms held latent far more energy than all the falling water of the world, but the benchtop processes then known for releasing it consumed much more energy than they produced. Fiztekh had spun off provincial institutes in 1931, most notably at Kharkov and Sverdlovsk; in 1932, when the discovery of the neutron and of artificial radioactivity increased the pace of research into the secrets of the atomic nucleus, Ioffe decided to divert part of Fiztekh's effort specifically to nuclear physics. The government shared his enthusiasm. "I went to Sergei Ordzhonikidze," Ioffe wrote many years later, "who was chairman of the Supreme Council of National Economy, put the matter before him, and in literally ten minutes left his office with an order signed by him to assign the sum I had requested to the Institute."

To direct the new program, Ioffe chose Igor Vasilievich Kurchatov, an exceptional twenty-nine-year-old physicist, the son of a surveyor and a teacher, born in the pine-forested Chelyabinsk region of the southern Urals in 1903. Kurchatov was young for the job, but he was a natural leader, vigorous and self-confident. One of his contemporaries, Anatoli P. Alexandrov, remembers his characteristic tenacity:

I was always struck by his great sense of responsibility for whatever problem he was working on, whatever its dimensions may have been. A lot of us, after all, take a careless, haphazard attitude toward many aspects of life that seem secondary to us. There wasn't a bit of that attitude in Igor Vasilievich. . . . [He] would sink his teeth into us and drink our blood until we'd fulfilled [our obligations]. At the same time, there was nothing pedantic about him. He would throw himself into things with such evident joy and conviction that finally we, too, would get caught up in his energetic style. . . .

We'd already nicknamed him "General.". . .

Within a year, justifying Ioffe's confidence in him, Kurchatov had organized and headed the First All-Union (i.e., nationwide) Conference on Nuclear Physics, with international attendance. With Abram I. Alikhanov, he built a small cyclotron that became, in 1934, the first cyclotron operating outside the Berkeley, California, laboratory of the instrument's inventor, Ernest O. Lawrence. He directed research at Fiztekh in 1934 and 1935 that resulted in twenty-four published scientific papers.

Kurchatov was "the liveliest of men," Alexandrov comments, "witty, cheerful, always ready for a joke." He had been a "lanky stripling," his student and biographer Igor N. Golovin writes, but by the 1930s, after recovering from tuberculosis, he had developed "a powerful physique, broad shoulders and ever-rosy cheeks." "Such a nice soul," an Englishwoman who knew him wrote home, "like a teddy bear, no one could ever be cross with him." He was handsome, Sergei Frish says—"a young, clean-shaven man with a strong, resolute chin and dark hair standing straight up over his forehead." Golovin mentions lively black eyes as well, and notes that Kurchatov "worked harder than anyone else. . . . He never gave himself airs, never let his accomplishments go to his head."

When Igor was six, his father, a senior surveyor in government service, took a cut in pay to move west over the Urals from the rural Chelyabinsk area to Ulyanovsk, on the Volga, where the three Kurchatov children could attend a proper academic *gymnasium*. Three years later, in 1912, Igor's older sister Antonina sickened with tuberculosis. For her health the family moved again, to the balmier climate of Simferopol on the Crimean Peninsula. The relocation proved to be a forlorn hope; Antonina died within six months.

The two surviving Kurchatov children—Igor and his brother Boris, two years younger—thrived in the Crimea. Both boys did well in *gymnasium,* played soccer, traveled into the country with their father during the summer on surveying expeditions. Igor ran a steam threshing machine harvesting wheat the summer he was fourteen. Another summer he worked as a laborer on the railroad.

A chance encounter with Orso Corbino's *Accomplishments of Modern Engineering* encouraged the young *gymnasium* student to dream of becoming an engineer. The Italian physicist would influence Kurchatov's career again indirectly in the 1930s when Corbino sponsored Enrico Fermi's Rome group that explored the newly discovered phenomenon of artificial radioactivity. The discoveries of the Rome group would inspire and challenge Kurchatov's Fiztekh research.

The Great War impoverished the Kurchatov family. Igor added night vocational school to his heavy schedule, qualified as a machinist and worked part-time in a machine shop while taking nothing but 5's—straight A's—during his final two years of *gymnasium.*

After the Revolution, in 1920, when he was seventeen years old, Kurchatov matriculated in physics and mathematics at Crimean State, one of about seventy students at the struggling, recently nationalized university. None of the foreign physics literature in the university library dated past 1913 and there were no textbooks, but the rector of the school was a distinguished chemist and managed to bring in scientists of national reputation for courses of lectures, among them Abram Ioffe, theoretical physicist Yakov I. Frenkel and future physics Nobel laureate Igor E. Tamm.

In the wake of war and revolution there was barely enough to eat. After midday lectures, students at Crimean State got a free meal of fish soup thickened with barley so flinty they nicknamed it "shrapnel." The distinction of an assistantship in the physics laboratory in the summer of 1921 gratified Kurchatov in part because it won him an additional ration of 150 grams— about five ounces—of daily bread.

Kurchatov finished the four-year university course in three years. He chose to prepare a thesis in theoretical physics because the university laboratory was not adequately equipped for original experimental work; he defended his dissertation in the summer of 1923. His physics professor, who was leaving for work at an institute in Baku, invited the new graduate to join him. Drawn from childhood to ships and the sea, Kurchatov chose instead to enroll in a program in nautical engineering in Petrograd. He suffered through a winter short on resources in the bitter northern cold, eking out a living as a supervisor in the physics department of a weather station, sleeping on a table in the unheated instrument building in a huge black fur coat. "This is no life I'm living," he wrote a friend that winter, uncharacteristically depressed, "but a rusted-out tin can with a hole in it." But the station director gave him real problems to solve, including measuring the alpha-radioactivity of freshly fallen snow, and the work finally won him for physics. He returned to the Crimea in 1924 to help his family—his father had been sentenced to three years of internal exile—and later joined his former teacher in Baku.

In the meantime, one of Kurchatov's physics classmates, his future brother-in-law Kirill Sinelnikov, had caught Ioffe's eye and accepted his invitation to work at Fiztekh. Sinelnikov told the institute director about his talented friend. Off went another invitation. Kurchatov returned to Leningrad, this time to take up his life's work. (He married Sinelnikov's sister Marina in 1927.)

Kurchatov quickly impressed Ioffe. "It was almost routine to chase him out of the laboratory at midnight," the senior physicist recalls. In the interwar years Ioffe sent twenty of his protégés abroad "to the best foreign laboratories where [they] could meet new people and familiarize [themselves] with new scientific techniques." Like a young entrepreneur too busy to bother going to college, Kurchatov never found time for foreign study. "He kept putting off taking advantage of [this opportunity]," Ioffe adds.

"Everytime it was time to leave he was on an interesting experiment that he preferred to the trip."

Others left and won international reputations. Peter Kapitza explored cryogenics and strong magnetic fields at Cambridge University and became a favorite of Ernest Rutherford, the New Zealand–born Nobel laureate who directed the Cavendish Laboratory there; Kapitza would earn a Nobel in his turn. So would theoretician Lev Landau, who worked in Germany during this period with his young Hungarian counterpart Edward Teller. The German emigré physicist Rudolf Peierls remembers a walking tour of the Caucasus with Landau after Landau had returned home when the Soviet theoretician pointed out that a nuclear reaction that produced secondary neutrons, if it could be found, would make possible the release of atomic energy—"remarkably clear vision in 1934," comments Peierls, "just two years after the discovery of the neutron." Less conspicuously, but with more enduring influence on Soviet history, Yuli Borisovich Khariton, the youngest son of a St. Petersburg journalist and an actress in the Moscow Art Theater —"compact, ascetically slight and very sprightly," a friend describes him— worked at Fiztekh on chemical chain reactions with Semenov, their discoverer, before earning a doctorate in theoretical physics at the Cavendish in 1927. Alarmed by the growing mood of fascism he found in Germany on his return passage, Khariton at twenty-four organized an explosives laboratory in the new Institute of Physical Chemistry, a Fiztekh spinoff. These were only a few of Ioffe's talented protégés.

Their talents barely protected them from the Great Terror that began in the Soviet Union after the assassination of Central Committee member Sergei Mironovich Kirov in December 1934 as Stalin moved to eliminate all those in power whose authority preceded his imposition of one-man rule. "Stalin killed off the founders of the Soviet state," writes the high-level Soviet defector Victor Kravchenko. "This crime was only a small part of the larger blood-letting in which hundreds of thousands of innocent men and women perished." According to a Soviet official, the slaughter claimed not hundreds of thousands but millions: "From 1 January 1935 to 22 June 1941, 19,840,000 enemies of the people were arrested. Of these 7 million were shot in prison, and a majority of the others died in camp." Exiled Soviet geneticist Zhores Medvedev notes that "the full list of arrested scientists and technical experts certainly runs into many thousands." Kharkov, where Kirill Sinelnikov had moved to direct the high-voltage laboratory after studying at Cambridge, lost most of its leaders, though Kurchatov's brother-in-law himself was spared.

The British Royal Society had funded an expensive laboratory in its own dedicated building in the courtyard outside the Cavendish for Peter Kapitza. Perhaps suspecting that he intended to defect, the Soviet government detained him during a visit home in the summer of 1934 and barred him from returning abroad. His detention shocked the British, and for a time he was too depressed to work, but the Soviet government bought his Cambridge laboratory equipment and built a new institute for him in Moscow. (A frustrated Kapitza had to order such unavailable consumer goods as wall clocks, extension telephones and door locks from England.) Eventually he went back to work, as he wrote People's Commissar Vyacheslav Molotov, "for the glory of the USSR and for the use of all the people." Niels Bohr, the Danish physicist, after visiting him in Moscow in 1937, observed that "by his enthusiastic and powerful personality, Kapitza soon obtained the respect and confidence of Russian official circles, and from the first Stalin showed a warm personal interest for Kapitza's endeavors."

Kapitza's golden captivity was not yet terror, but he needed all his connections when Lev Landau was arrested in April 1938, convicted of being a "German spy" and sent to prison, where he languished for a year and became ill. Landau had been working at Kapitza's Institute for Physical Problems. Kapitza determined to save him, writes Medvedev:

> After a short meeting with Landau in prison, Kapitza took a desperate step. He presented Molotov and Stalin with an ultimatum: if Landau was not released immediately, he, Kapitza, would resign from all his positions and leave the institute. . . . It was clear that Kapitza meant business. After a short time Landau was cleared of all charges and released.

In old age, Edward Teller would cite his friend's arrest and imprisonment as one of three important early influences on his militant anti-Communism (the other two, Teller said, were the Great Terror itself and Arthur Koestler's novel *Darkness at Noon*): "Lev Landau, with whom I published a paper, was an ardent Communist. Shortly after he returned to Russia, he went to prison. After that he was no longer a Communist." Communist or not, Landau continued to work at Kapitza's institute in Moscow.

Not even Ioffe escaped the general harrowing. "Although the majority of [Soviet] scientists realized the importance of work in the field of nuclear physics," writes Alexandrov, "the leadership of the Soviet Academy of Sciences and of the Council of People's Commissars believed that this work had no practical value. Fiztekh and Ioffe himself were heavily criticized at the 1936 general assembly of the Academy of Sciences for 'loss of touch with practice.'" With the Great Terror destroying lives all around them, Soviet physicists understandably learned caution from such charges. "In those years," writes Stalin's daughter Svetlana Alliluyeva, "never a month went by in peace. Everything was in constant turmoil. People vanished like shadows in the night." Her father brooded over it all, reports the historian Robert Conquest: "Stalin personally ordered, inspired and organized the operation. He received weekly reports of . . . not only steel production and crop figures, but also of the numbers annihilated." Shot in the back of the head at Lubyanka prison, truckloads of bodies to the crematorium at the Donskoi Monastery, smoking ashes bulked into open pits and the pits paved over. That was the era when Osip Mandelstam suffered three years' exile and then five years in a gulag camp—five years that killed him—for writing a poem, "The Stalin Epigram," the most ferocious portrait of the dictator anyone ever devised:

> *Our lives no longer feel ground under them.*
> *At ten paces you can't hear our words.*
>
> *But whenever there's a snatch of talk*
> *it turns to the Kremlin mountaineer,*
>
> *the ten thick worms his fingers,*
> *his words like measures of weight,*
>
> *the huge laughing cockroaches on his top lip,*
> *the glitter of his boot-rims.*
>
> *Ringed with a scum of chicken-necked bosses*
> *he toys with the tributes of half-men.*
>
> *One whistles, another meows, a third snivels.*
> *He pokes out his finger and he alone goes boom.*
>
> *He forges decrees in a line like horseshoes,*
> *one for the groin, one the forehead, temple, eye.*
>
> *He rolls the executions on his tongue like berries.*
> *He wishes he could hug them like big friends from home.*

Igor Kurchatov organized the initial Soviet study of nuclear fission at Fiztekh in the early months of 1939, following Joliot-Curie's letter to Ioffe and confirmation of the discovery in scientific journals. Landau's remark to Peierls in 1934 about secondary neutrons points to one universal line of inquiry: examining whether the fission reaction, which a single neutron could initiate, would release not only hot fission fragments but additional neutrons as well. If so, then some of those secondary neutrons might go on to fission other uranium atoms, which might fission yet others in their turn. If there were enough secondary neutrons, the chain reaction might grow to be self-sustaining. Joliot-Curie's team in Paris set up an experiment to look for secondary neutrons in late February; in April the French reported 3.5 secondary neutrons per fission and predicted that uranium would probably chain-react. Enrico Fermi, now at Columbia University in flight from anti-Semitic persecution (his wife Laura was Jewish), and emigré Hungarian physicist Leo Szilard, also temporarily working at Columbia, soon independently confirmed fission's production of secondary neutrons. At a Fiztekh seminar in April, two young members of Kurchatov's Fiztekh team, Georgi Flerov and Lev Rusinov, reported similar results—between two and four secondary neutrons per fission. (In 1940, Flerov and Konstantin A. Petrzhak would make a world-class discovery, the spontaneous fission of uranium, a consequence of uranium's natural instability and a phenomenon that would prove crucial to regulating controlled chain reactions in nuclear reactors. Before the young Russians succeeded, the American radiochemist Willard F. Libby, later a Nobel laureate, had tried two different ways unsuccessfully to demonstrate spontaneous fission.)

Down the street at the Institute of Physical Chemistry, Yuli Khariton and an outstanding younger colleague, theoretician Yakov B. Zeldovich, began exploring fission theory. "Yuli Borisovich notes a curious detail," Zeldovich recalled: "we considered the work on the theory of uranium fission to be apart from the official plan of the Institute and we worked on it in the evenings, sometimes until very late." Zeldovich was a brilliant original— "not a university graduate," comments Andrei Sakharov; ". . . in a sense, self-educated"—who had earned a master's degree and a doctorate "without his ever bothering about a bachelor's degree." "We immediately made calculations of nuclear chain-reactions," Khariton remembers, "and we soon understood that on paper, at least, a chain-reaction was possible, a reaction which could release unlimited amounts of energy without burning coal or oil. Then we took it very seriously. We also understood that a bomb was possible." Khariton and Zeldovich reported their first calculations in a seminar at Fiztekh in the summer of 1939, describing the conditions necessary for a nuclear explosion and estimating its tremendous destructive capacity—one atomic bomb, they told their colleagues, could destroy Moscow.

Theoretical physicist J. Robert Oppenheimer at Berkeley, Fermi, Szilard, Peierls in England, all quickly came to similar conclusions. "These possibilities were immediately obvious to any good physicist," comments Robert Serber. But it was also soon obvious from work by Niels Bohr that a formidable obstacle stood in the way of making bombs: only one isotope of uranium, U235, would sustain a chain reaction, and U235 constituted only 0.7 percent of natural uranium; the other 99.3 percent, chemically identical, was U238, which captured secondary neutrons and effectively poisoned the reaction.* There were then two difficult technical questions that needed to be resolved by any nation that proposed to explore building an atomic bomb: whether it might be possible to achieve a controlled chain reaction —to build a nuclear reactor—using natural uranium in combination with some suitable moderator, or whether the U235 content of the uranium

---

* An atomic bomb and a nuclear reactor exploit different circumstances using significantly different arrangements: a bomb creates a chain reaction using fast neutrons, a nuclear reactor using slow neutrons.

would have to be laboriously enriched; and how to separate U235 from U238 on an industrial scale for bomb fuel when the only exploitable distinction between the two isotopes was a slight difference in mass. Enrichment and separation were essentially identical processes ("separated" bomb-grade uranium is natural uranium enriched to above 80 percent U235) and would use the same massive, expensive machinery that no one yet knew how to build; while a reactor fueled with natural uranium, if such would work, might be a straightforward enterprise.

Khariton and Zeldovich approached these questions from first principles, as it were, carefully calculating what was not possible as well as what might be. In the first of three pioneering papers they published in the Russian *Journal of Experimental and Theoretical Physics* in 1939 and 1940 (papers that went unnoticed outside the Soviet Union) they demonstrated that a fast-neutron chain reaction was not possible in natural uranium. Isotope separation would therefore be necessary to build a uranium bomb.

A second, longer paper, delivered a few weeks later on October 22, 1939, developed important basic principles of reactor physics. Khariton and Zeldovich correctly identified the crucial bottleneck that experimenters would have to bypass to build a natural-uranium reactor that worked. Visualize a stray neutron in a mass of natural uranium finding a U235 nucleus, entering it and causing it to fission. The two resulting fission fragments fly apart; a fraction of a second later they eject two or three secondary neutrons. If these fast secondary neutrons encounter other U235 nuclei they will continue and enlarge the chain of fissions. But there is much more U238 than U235 in the mass of natural uranium, making an encounter with a U238 nucleus more likely, and U238 tends to capture fast neutrons. It is particularly sensitive to neutrons moving at a critical energy, twenty-five electron volts (eV), a sensitivity which physicists call a "resonance." On the other hand, U238 is opaque to slow neutrons. To make a reactor, then, Khariton and Zeldovich realized, it would be necessary to slow the fast secondary neutrons from U235 fission quickly below U238's twenty-five eV resonance. The way to do that, they proposed, was to make the neutrons give up some of their energy by bouncing them off the nuclei of light atoms such as hydrogen. "In order to accomplish [a chain] reaction [in natural uranium]," they wrote, "strong slowing of the neutrons is necessary, which may be practically accomplished by the addition of a significant amount of hydrogen."

The simplest way to mix uranium with hydrogen would be to make a slurry—a homogeneous mixture—of natural uranium and ordinary water. But Khariton and Zeldovich demonstrated in this second paper that such a mixture would not sustain a chain reaction, because hydrogen and oxygen also capture slow neutrons, and in a reactor fueled with natural uranium such capture would subtract too many neutrons from the mix. Important consequences followed from this conclusion. One was that instead of hydrogen in ordinary water it would apparently be necessary to use heavy hydrogen—deuterium, H^2 or D, an isotope of hydrogen with a smaller appetite for neutrons than ordinary hydrogen—perhaps in the form of rare and expensive heavy water. (In a review article published in 1940, Khariton and Zeldovich proposed carbon and helium as other possible moderators, both materials that later proved to work.) Alternatively, wrote the two Soviet physicists, "another possibility lies in the enrichment of uranium with the isotope 235." They calculated that natural uranium enriched from 0.7 percent U235 to 1.3 percent U235 would work in a homogeneous solution with ordinary water.

In a third paper submitted in March 1940, Khariton and Zeldovich identified two natural processes that would make it easy and "completely safe" to initiate and control a chain reaction in a nuclear reactor. The fissioning process would heat the mass of uranium and cause it to expand, which in turn would increase the distance the neutrons would have to travel to cause additional fissioning and would therefore slow down the chain reaction, allowing the mass of uranium to cool and the chain reaction to accelerate. This natural oscillation could be controlled by increasing or decreasing the volume of uranium. Another natural process—delayed neutrons released in fission which would "significantly increase" the oscillation period—subsequently proved more significant for reactor control. (Apparently critics within the Soviet scientific community had made safety a point of attack; in this third paper Khariton and Zeldovich vigorously disputed what they called "hasty conclusions . . . on the extreme danger of experiments with large masses of uranium and the catastrophic consequences of such experiments." Because of the natural processes they had identified, they scoffed, such conclusions "do not correspond to reality.")

Khariton and Zeldovich summarized these early and remarkable insights in the introduction to their third paper:

It would appear (the lack of experimental data precludes any categorical assertions) that by applying some technique, creating a large mass of metallic uranium either by mixing uranium with substances possessing a small capture cross-section (e.g., with heavy water) or by enriching the uranium with the U^{235} isotope . . . it will be possible to establish conditions for the chain decay of uranium by branching chains in which an arbitrarily weak radiation by neutrons will lead to powerful development of a nuclear reaction and macroscopic effects. Such a process would be of much interest since the molar heat of the nuclear fission reaction of uranium exceeds by $5 \cdot 10^7$ [i.e., 5,000,000] times the heating capacity of coal. The abundance and cost of uranium would certainly allow the realization of some applications of uranium.

Therefore, despite the difficulties and unreliability of the directions indicated, we may expect in the near future attempts to realize the process.

At the annual All-Union Conference on Nuclear Physics, held in 1939 in November at Kharkov in the Ukraine, Khariton and Zeldovich reported their conclusion that carbon (graphite) and heavy water were possible neutron moderators. They also reported that a controlled chain reaction even with heavy water would be possible in a homogeneous reactor only with uranium enriched in U235. Since uranium enrichment was notoriously difficult, and would require the development of an entirely new industry, their conclusion made the possibility of building a working nuclear reactor within a reasonable period of time and for a reasonable amount of money appear remote. But there are other possible arrangements of natural uranium and graphite or heavy water that they overlooked, even though their second 1939 paper had offered an important clue. Why two such outstanding theoreticians should have overlooked more promising alternative arrangements is a question worth exploring.

The effectiveness of a moderator such as graphite or heavy water is limited crucially by its probability of capturing rather than reflecting neutrons. That probability, called a "cross section," can only be determined by experiment. Physicists quantify capture cross sections (and other such probabilities) in extremely small fractions of a square centimeter, as if a cross section were the surface area of a target the incoming neutron might hit. The two theoreticians had calculated that to achieve a chain reaction in a mixture of ordinary uranium and heavy water, the cross section of deuterium for neutron capture must not be larger than $3 \cdot 10^{-27}$ cm^2. They lacked the laboratory equipment they needed—a powerful cyclotron and a large quantity of heavy water—to measure the actual capture cross section of deuterium (the entire Soviet supply of heavy water at that time amounted to no more than two to three kilograms). For the 1939 All-Union Conference they must have offered an approximation drawn from the international physics literature.

Apparently they continued to search the literature to see if someone had determined a more accurate value for the deuterium capture cross section. They found an estimate in a letter to the editor of the American journal *Physical Review* published in April 1940. In that letter, University of Chicago physicists L. B. Borst and William D. Harkins noted a "quantitative estimate" of $3 \cdot 10^{-26}$ cm^2, a full order of magnitude too large (-26 rather than -27). "Thus," Igor Kurchatov would explain in 1943 in a top secret report, "we came to the conclusion that it is impossible to achieve a chain reaction in a mixture of [ordinary] uranium and heavy water." And if not in heavy water without investing expensively in isotope enrichment, then also not in carbon, where tolerances were even closer. "Contrary to the opinion of a small group of enthusiasts," Khariton would comment late in life, "the dominant opinion in our country was that a technical solution to the uranium problem was a matter for the remote future, and that success would require fifteen to twenty years." Khariton and Zeldovich's disappointing conclusion must certainly have contributed to that conservative assessment. But the "small group of enthusiasts," which included Khariton, Zeldovich, Kurchatov and Flerov, was not deterred. "In the case of a homogeneous reactor, the enterprise looked doomed," Khariton would note, "but there was still some hope that a loophole was possible. The cross sections were not very reliable and we felt that we had to dig through the material."

Believing that a nuclear reactor as well as a bomb would require increasing the U235 content of natural uranium, Kurchatov's group examined various methods of uranium enrichment. Gaseous diffusion—pumping a gaseous form of uranium against a porous barrier through which the lighter U235 isotope would diffuse faster than the heavier U238, selectively enriching the product—the physicists discounted as impractical. Instead they recommended separating U235 from U238 in gaseous form in a high-speed centrifuge, a method Khariton had studied in detail in 1937 but one for which the technology had not yet been developed.

These early discussions caught the attention of Leonid Kvasnikov, the head of the science and technology department of the state security organization, the People's Commissariat of Internal Affairs, known by its Russian initials NKVD. The NKVD, which had orchestrated the Great Terror (which then swallowed up some 28,000 of its own), had been headed since 1938 by Stalin's brutally efficient fellow Georgian Lavrenti Pavlovich Beria. It maintained a network of spies throughout the world run by NKVD *rezidents* stationed in Soviet consulates and embassies. One important field of *rezidency* work was industrial espionage—stealing industrial processes and formulas to save the Soviet Union the expense of licensing these technologies legitimately from their developers. The American industrial chemist Harry Gold, who began a long career of espionage for the Soviet Union in 1935, mentions among such information "the various industrial solvents used in the manufacture of lacquers and varnishes . . . , such specialized products as ethyl chloride (used as a local anesthetic) and in particular, absolute (100%) alcohol (used to blend, i.e., 'extend,' motor fuels)." These commonplace products, Gold understood, "would be a tremendous boon to a country [that was] back in the 18th century, industrially speaking (in spite of some localized advances)." They "could go toward making the harsh life of those who lived in the Soviet Union a little more bearable."

Early in 1940, Kvasnikov alerted the *rezidency* network to collect information on uranium research. According to Georgi Flerov, the early focus of Soviet concern was on German more than on Anglo-American work, just as it was in England and America:

It seemed to us that if someone could make a nuclear bomb, it would be neither Americans, English or French but Germans. The Germans had brilliant chemistry; they had technology for the production of metallic uranium; they were involved in experiments on the centrifugal separation of uranium isotopes. And, finally, the Germans possessed heavy water and reserves of uranium. Our first impression was that Germans were capable of making the thing. It was obvious what the consequences would be if they succeeded.

Espionage, then, accompanied the Soviet development of nuclear energy from its earliest days.

In the spring of 1940, George Vernadsky, who taught history at Yale University, sent his father, V. I. Vernadski, an article about atomic energy published in the *New York Times*. Vernadski wrote a letter to the Soviet Academy of Sciences about the article, following which the academy created a Special Committee for the Problem of Uranium. Khlopin, who had succeeded Vernadski as director of the State Radium Institute, was appointed to head the Uranium Committee, which also included Vernadski, Ioffe, the distinguished geologist A. Y. Fersman, Kapitza, Kurchatov and Khariton as well as a number of senior Soviet scientists. The committee was directed to prepare a scientific research program and assign it to the necessary institutes, to oversee the development of methods of isotope separation and to organize efforts toward achieving a controlled chain reaction—that is, building a nuclear reactor. The decree that established the committee also ordered the construction, completion or improvement of no fewer than three Soviet cyclotrons, two already at hand in Leningrad and one to be built in Moscow; set up a fund for the acquisition of uranium metal, which Soviet industry at that time did not have the technology to produce; and appointed Fersman to lead an expedition into Central Asia to prospect for uranium. ("Uranium has acquired significance as a source of atomic energy," Vernadski wrote a colleague in July. "With us uranium is a scarce metal; we extract radium from deep brine [pumped from oil wells], and any quantity can be obtained. There is no uranium in these waters.")

Kurchatov was disappointed with the committee's plan, which the Academy of Sciences approved in October 1940. He believed it to be unduly conservative. Despite the expectation that uranium would have to be enriched, he wanted to move directly to building a nuclear reactor. At the Fifth All-Union Conference on Nuclear Physics in Moscow in late November, he analyzed fission studies published throughout the world to demonstrate that a controlled chain reaction was possible and listed the equipment and materials he would need. Asked if a uranium bomb could be built, he said confidently that it could and estimated that a bomb program would cost about as much as the largest hydroelectric plant that had been built in the Soviet Union up to that time—an estimate low by several orders of magnitude, but comparable to one Rudolf Peierls and Austrian emigré physicist Otto Robert Frisch had prepared in England eight months earlier for the

British government. In any case, as Frisch commented later, the cost of a plant for separating U235 "would be insignificant compared with the cost of the war."

Golovin was an excited eyewitness to the November debate:

> The situation... during Kurchatov's talk was rather dramatic. The work-shop took place at the Communist Academy on Volkhonka Street, in a large hall with an amphitheater overcrowded by numerous participants. In the course of the presentation the excitement of the audience kept growing and by the end of it the general feeling was that we were on the eve of a great event. When Kurchatov finished his talk, and, together with the chairman of the meeting, Khlopin, went to the adjacent room from the rostrum, Ioffe, Semenov, [A. I.] Leipunski, Khariton and others started to move there one after the other. Meanwhile, the discussion over Kurchatov's talk was continued in the hall.... The break was delayed. Instead of the ordinary five or ten minutes between talks, the chairman, Khlopin, didn't return even in twenty minutes.... A noisy discussion was taking place [in the adjacent room].

The Great Terror had taught its survivors wary circumspection. In the fifteen months since the beginning of the Second World War on September 1, 1939, Germany had overrun Europe. To buy time, Stalin had concluded a nonaggression pact with Hitler, but the Soviet Union was gearing up for the war with Germany that Stalin understood was coming; in May 1941 he would tell his inner circle, "The conflict is inevitable, perhaps in May next year." The Soviet leadership had made clear its suspicion of "impractical" science, and Stalin had ordered the scientists in no uncertain terms to roll up their sleeves and get down to practical work. Nor had Khariton and Zeldovich's calculations encouraged optimism in an older generation still suspicious of the new physics. Surprisingly, even Ioffe was skeptical. He was not a nuclear physicist, and after the discovery of fission he had taken a long view of its potential, predicting that "if the mastering of rocket technology is a matter of the next fifty years, then the utilization of nuclear energy is a matter of the next century." All these factors would have influenced the noisy discussion going on in the adjacent room at the Communist Academy. Golovin:

> A quarter of an hour later, Khlopin returned to the rostrum and declared that he had come to the conclusion that it was too early to ask the government for large grants since the war was going on in Europe and the money was needed for other purposes. He said that it was necessary to work a year more and then make the decision whether there would be some grounds to involve the government.... The audience was disappointed.

The development of a capacity to build atomic bombs required a massive commitment of government funds, funds that would have to be diverted from the conventional prosecution of the war. If atomic bombs could be built in time they would be decisive, in which case no belligerent could afford *not* to pursue them. But making that judgment depended critically on how much scientists trusted their governments and how much governments trusted their scientists.

Trust would not be a defining issue later, after the secret, the one and only secret—that the weapon worked—became known. This first time around, however, it was crucial, as the Russian physicist Victor Adamsky emphasizes in a discussion of why Nazi Germany never developed an atomic bomb:

The tension [between scientists and their governments] stemmed from the fact that there existed no *a priori* certainty of the possibility of creating an atomic bomb, and merely for clarification of the matter it was necessary to get through an interim stage: to create a device (the nuclear reactor) in order to perform a controlled chain reaction instead of the explosive kind. But the implementation of this stage requires tremendous expenses, incomparable to any of those previously spared for the benefit of scientific research. And it was necessary to tell this straight to your government, making it clear that the expenses may turn out to be in vain—an atomic bomb may not result. . . . .

Scientists and their governments developed confidence and mutual understanding in England and the United States, Adamsky concludes, but not in Germany. At the end of 1940, such confidence and mutual understanding had not yet developed in the USSR.

The overwhelming German surprise attack along the entire western border of the Soviet Union at dawn on June 22, 1941, one month after Stalin's prediction that a shooting war would not begin for another year, mooted the issue of how large an effort should be devoted to what Soviet physicists called the "uranium problem." Stalin met with military and other leaders for eleven hours that first day and almost continuously for several days thereafter, Beria at his side. The Wehrmacht decimated the Soviet Air Force, rolled over Belorussia and the Ukraine and thrust up through the Baltic states toward Leningrad. Once the magnitude of the disaster sank in, says Stalin biographer and General of the Soviet Army Dmitri Volkogonov, the dictator "simply lost control of himself and went into deep psychological shock. Between 28 and 30 June, according to eyewitnesses, Stalin was so depressed and shaken that he ceased to be a leader. On 29 June, as he was leaving the defense commissariat with Molotov, [Kliment] Voroshilov,

[Andrei] Zhdanov and Beria, he burst out loudly, 'Lenin left us a great inheritance and we, his heirs, have fucked it all up!' " Stalin retreated to his dacha at Kuntsevo; it took a visit from the Politburo, led by Molotov, to mobilize him. "We got to Stalin's dacha," Anastas Mikoyan recalled in his memoirs. "We found him in an armchair in the small dining room. He looked up and said, 'What have you come for?' He had the strangest look on his face. . . ."

By the time the Soviet dictator rallied, the Germans were bombing Moscow. Volkogonov chronicles the debacle:

> Soviet losses were colossal. Something like thirty divisions had been virtually wiped out, while seventy had lost more than half of their numbers; nearly 3,500 planes had been destroyed, together with more than half the fuel and ammunition dumps. . . . Of course, the Germans too had paid a price, namely about 150,000 officers and men, more than 950 aircraft and several hundred tanks. . . . The [Red] army was fighting. It was retreating, but it was fighting.

Stalin finally rallied the Soviet people on July 3. Molotov and Mikoyan had written the speech and they almost had to drag Stalin to the microphone. The Soviet writer Konstantin Simonov, a front-line correspondent throughout the war, recalled the momentous occasion in his postwar novel *The Living and the Dead:*

> Stalin spoke in a toneless, slow voice, with a strong Georgian accent. Once or twice, during his speech, you could hear a glass click as he drank water. His voice was low and soft, and might have seemed perfectly calm, but for his heavy, tired breathing, and that water he kept drinking during the speech. . . .
> Stalin did not describe the situation as tragic; such a word would have been hard to imagine as coming from him; but the things of which he spoke—*opolcheniye* [i.e., civilian reserves], partisans, occupied territories, meant the end of illusions. . . . The truth he told was a bitter truth, but at last it was uttered, and people now at least knew where they stood. . . .

"It was an extraordinary performance," reports the Russian-born journalist and historian Alexander Werth, who covered the war in the USSR for the London *Times,* "and not the least impressive thing about it were these opening words: 'Comrades, citizens, brothers and sisters, fighters of our Army and Navy! I am speaking to you, my friends!' This was something new. Stalin had never spoken like this before."

But Stalin's secret police had surprises in store for any of his newfound "friends" whose loyalty might be suspect, particularly if their background was German. "In every village, town and city," notes Victor Kravchenko, "long blacklists were ready: hundreds of thousands would be taken into custody. . . . The liquidation of 'internal enemies' was, in sober fact, the only part of the war effort that worked quickly and efficiently in the first terrible phase of the struggle. It was a purge in the rear in accordance with an elaborate advance plan, as ordered by Stalin himself. . . ." Half a million people—the entire population of the Volga German Republic—were transported to internal exile in Siberia. "In Moscow alone thousands of citizens were shot under martial law in the first six months," Kravchenko concludes. ". . . The magnitude of the terror inside Russia cannot be overstated. It amounted to a war within the war."

In the course of his July 3 speech, Stalin announced the formation of a State Defense Committee (GKO), in which he vested "all the power and authority of the State." He appointed himself chairman of the five-man committee, Molotov deputy chairman, and as members Red Army Marshal Kliment Voroshilov ("an utterly mindless executive with no opinion of his own," scoffs Volkogonov), the assiduous bureaucrat Georgi Malenkov and Beria.

Thus Lavrenti Beria came into his own. Born in the Sukhumi district of Georgia in 1899, he had worked his way to power first as police chief and then party chief of Georgia and the Transcaucasus (where he had personally organized the terrible purges) and now at the center in Moscow. Stalin had summoned him from Georgia in 1938 to purge the NKVD itself. "By early 1939," according to a biographer, "Beria had succeeded in arresting most of the top and middle-level hierarchy of [his predecessor's] apparatus. . . ." He inherited a gulag slave-labor force of several million souls. "Camp dust," he liked to call them. "A magnificent modern specimen of the artful courtier," Svetlana Alliluyeva mocks; she blamed Beria for her father's excesses. The Yugoslavian diplomat Milovan Djilas met Beria in the course of the war: a short man, Djilas says, "somewhat plump, greenish pale, and with soft damp hands," with a "square-cut mouth and bulging eyes behind his pince-nez" and an expression of "a certain self-satisfaction and irony mingled with a clerk's obsequiousness and solicitude." Beria's brutality extended to casual rape—of teenage girls plucked off the street and delivered to his Lubyanka office—and official torture and murder. He was nevertheless an exceptional administrator. Stalin gave him huge responsibilities: for evacuating wartime industry eastward over the Urals, for mobilizing gulag labor, for overseeing industrial conversion and for moving troops and matériel to the front. "Beria was a most clever man," Molotov testified, "inhumanly energetic and industrious. He could work for a week without sleep." In the early months of the war he almost certainly did.

"Beria was no engineer," observes Victor Kravchenko, a factory manager in those days. "He was placed in control for the precise purpose of inspiring deadly fear. I often asked myself—as others assuredly did in their secret hearts—why Stalin had decided to take this step. I could find only one plausible answer. It was that he lacked faith in the patriotism and national honor of the Russian people and was therefore compelled to rely primarily on the whip. Beria was his whip."

According to Marshal K. S. Moskalenko, who told a group of senior military officers in 1957 that he heard it from Beria himself, Stalin colluded with Beria and Molotov in late July to offer a surrender, "agreeing to hand over to Hitler the Soviet Baltic republics, Moldavia, a large part of the Ukraine and Belorussia. They tried to make contact with Hitler through the Bulgarian ambassador. No Russian czar had ever done such a thing. It is interesting that the Bulgarian ambassador was of a higher caliber than these leaders and told them that Hitler would never beat the Russians and that Stalin shouldn't worry about it."

The war emptied out the Leningrad institutes. The scientists crated up their movable equipment and shipped it on tracks crowded with troop trains to the other side of the Urals, out of range of German bombers. Fiztekh went to Kazan, four hundred kilometers east of Moscow on the Volga. Whole factories moved east,* reports Sergei Kaftanov, minister of higher education and deputy for science and technology to the State Defense Committee:

How long would it take today to move a big industrial enterprise to a new site? Two years? Three years? During the war it took only months for plants that had been moved a thousand kilometers to start up again. The regular order of construction is: walls—roof—machines. We were doing it this way: machines—roof—walls. War pressed us for quick solutions.

Quick solutions meant solutions, including scientific solutions, that contributed immediately to the defense of the beleaguered country. In the late summer of 1941, Kurchatov and Alexandrov set up a laboratory together in the Crimean port of Sevastopol, on the Black Sea, organized a test site for demagnetizing ships to protect them against magnetic mines and trained Navy crews in the lifesaving technology until September, when the Germans began bombing Streletskaya Bay. Alexandrov went north then to work with the Northern Fleet; Kurchatov stayed on in Sevastopol demagnetizing submarines.

Boris Pasternak compacted the mood that terrible autumn into a shudder of dread:

---

* "Altogether, between July and November 1941 no fewer than 1,523 industrial enterprises, including 1,360 large war plants, had been moved to the east—226 to the Volga area, 667 to the Urals, 244 to Western Siberia, 78 to Eastern Siberia, 308 to Kazakhstan and Central Asia. The 'evacuation cargoes' amounted to a total of one and a half million railway wagon-loads." Werth (1964), p. 216.

Do you remember that dryness in your throat
When rattling their naked power of evil
They were barging ahead and bellowing
And autumn was advancing in steps of calamity?

In October there was panic in Moscow. The Germans had advanced to within a hundred kilometers of the city and it seemed they might succeed in seizing it. A young Red Army cipher clerk stationed in training nearby, Igor Gouzenko, had been given a pass into Moscow on October 16 and witnessed the debacle. "The street was crowded with people carrying bundles, sacks and suitcases," Gouzenko recalled after the war. "They were scurrying in all directions. No one seemed to know where they were fleeing. Everyone was just fleeing. Most astounding of all was the strange silence hanging over the scene. Only the stamp of hurrying feet created an undertone of frantic rhythm." Andrei Sakharov, who was then a young university student, remembered that "as office after office set fire to their files, clouds of soot swirled through streets clogged with trucks, carts, and people on foot carrying household possessions, baggage, and young children. . . . I went with a few others to the [university] Party committee office, where we found the Party secretary at his desk; when we asked whether there was anything useful we could do, he stared at us wildly and blurted out: 'It's every man for himself!' "

At the Scientific Research Institute where Igor Gouzenko's sister had been working, a notice had been posted on the door on the authority of the chairman of the Moscow Soviet: "The situation at the front is critical. All citizens of the City of Moscow, whose presence is not needed, are hereby ordered to leave the city. The enemy is at the gates."

Gouzenko thought the notice qualified as "the most panicky document of World War II." Warranted or not, Moscow emptied out; by the end of October, more than two million people had been evacuated officially and many more had simply fled. Stalin stayed. The counterattack outside Moscow, the first major Soviet offensive, began early in December and saved the city. "West of Moscow," observes Alexander Werth, ". . . miles and miles of road were littered with abandoned guns, lorries and tanks, deeply embedded in the snow. The comic 'Winter Fritz,' wrapped up in women's shawls and feather boas stolen from the local population, and with icicles hanging from his red nose, made his first appearance in Russian folklore." But the siege of Leningrad had begun, and that winter nearly half the population of the city, a million people, died of starvation.

Georgi Flerov had been drafted into the Soviet Air Force at the beginning of the war and assigned to the Air Force Academy in Ioshkar-Ola to train as an engineer. He was a stubborn man; he suspected that other nations, including the fascist enemy, were working on a uranium bomb; he believed passionately that his country should develop such a weapon first. He said as much in a letter to the State Defense Committee in November, but the letter went unanswered.

That month German bombs and artillery barrages finally drove the Soviet Navy from the Sevastopol harbor. Kurchatov left ruined Sevastopol then, evacuating first by boat to Poti, south of Sukhumi on the eastern shore of the Black Sea, then beginning the long journey by train to Kazan, seven hundred kilometers east of Moscow, to resume work at the temporary Fiztekh installation there. On his way, the Soviet physicist spent a night on a below-zero station platform and caught cold. Suzanne Rosenberg, a daughter of Canadian Communists who had returned to the Soviet Union to support the Revolution, describes a similar railroad ordeal evacuating Moscow during the October panic:

> So crammed with evacuees was the train that we spent the first twenty-four hours standing on the wind-swept platform between the carriages. Later we took brief turns sitting down on the benches inside. Our journey lasted nineteen days: normally it took forty-six or fifty hours. We learned to sleep standing up, like horses, to do without water and with little food for whole days. The German Messerschmitts were on our trail. Hearing their approach we would jump off the train, tumbling over one another, and scurry off in all directions. If there were woods we made a dash for their cover. If not, we fled into the open fields and stretched out in the frozen grass, faces buried in the icy ground.

In December Flerov won leave to present a seminar on the uranium problem to the Academy of Sciences, which, like Fiztekh, had been evacuated to nearby Kazan. He missed Kurchatov, who was still in transit, but wrote him a long letter in a school notebook that repeated the gist of his report. One of the participants remembers:

> Flerov's report was well-argued. As usual, he was vivid and enthusiastic. We listened to him attentively. Ioffe and Kapitza were present. . . . The seminar left the impression that everything was very serious and fundamental, that work on the uranium project should be renewed. But the war was going on. And I don't know what the outcome would have been if we'd had to decide whether to start work immediately or to delay beginning for another year or two.

Flerov was proposing work on a fast-neutron chain reaction: a bomb. He argued that an atomic bomb was possible and that 2.5 kilograms of pure U235 would yield 100,000 tons of TNT equivalent. "He suggested developing a 'cannon' design," reports Khariton, "that is, quickly driving together two hemispheres made of U235. He also expressed the important idea of the use of 'compression of the active material.' " The record is silent on how Flerov proposed to achieve such compression in a uranium gun, which assembles but does not compress. Flerov's 2.5 kilograms was at best a rough approximation, far below the minimum quantity of U235 necessary to sustain a chain reaction,* but it compares with the 1 kg that Rudolf Peierls and Otto Frisch in England had first roughly estimated and was probably derived similarly from the known cross section of uranium for neutron capture, the geometric cross section, 10^{-23} cm^2.

By the time Kurchatov arrived behind the Urals, at the end of December 1941, his cold had turned to pneumonia. He took to his bed. His wife Marina Dmitrievna joined him in Kazan and nursed him. Abram Ioffe nursed him. During his illness he chose not to shave. When he recovered, early in 1942, he emerged into Russian winter with a full-blown beard, "which," says Golovin, "he declared no scissors would touch till after victory." It was unusual in those days for a young Russian to wear a beard. Kurchatov would make his famous.

Khariton says Kurchatov cherished Flerov's report, saving it in his desk to the end of his life. Admiring Flerov's enthusiasm was not the same as trusting his judgment, however. "Kurchatov knew," comments Golovin, "that Flerov did not and indeed could not have proofs; he only had a passion for experimentation and would not back down from his ideas. . . . Cares of the day distracted Kurchatov. He was recalled to fleet duty and left for Murmansk."

"Scientific work which is not completed and produces no results during the war," Peter Kapitza explained in a lecture in 1943, "may even be harmful if it diverts our forces from work which is more urgently required." With ships to demagnetize, tank armor to harden and radar to invent, the Soviet scientific establishment concluded once again, that hard winter of 1941, that it would be imprudent to undertake expensive, problematic and long-term nuclear-fission research in the midst of war.

---

* Critical mass for a bare U235 sphere, 56 kg; for a U235 sphere surrounded by a thick uranium tamper, 15 kg. King (1979), p. 7.

2
Diffusion

"I THINK THAT THE WORLD in which we shall live these next thirty years will be a pretty restless and tormented place," Robert Oppenheimer wrote his younger brother Frank from Berkeley in 1931; "I do not think that there will be much of a compromise possible between being of it, and being not of it." Many thoughtful men and women felt that way in the decades between the two world wars, and for some of them, Communism seemed to promise what the *Time* essayist and Communist agent Whittaker Chambers called a "solution." "In the West," Chambers observed of that period, "all intellectuals [who] become Communists [do so] because they are seeking the answer to one of two problems: the problem of war or the problem of economic crises." Chambers explained:

> The same horror and havoc of the First World War, which made the Russian Revolution possible, recruited the ranks of the first Communist parties of the West. Secondary manifestations of crisis augmented them—the rise of fascism in Italy, Nazism in Germany and the Spanish Civil War. The economic crisis which reached the United States in 1929 swept thousands into the Communist Party or under its influence.

But commitment to Communism was also always personal, Chambers emphasized, the resolution of a crisis of faith; "his decision to become a Communist seems to the man who makes it as a choice between a world that is dying and a world that is coming to birth." Partisan observers then and since have ridiculed such commitment, judging it naive or even delusional, but it was no more so than any other religious conversion seen from outside the circle of faith.

For committed Communists it followed that the Soviet Union was the new world's vanguard. Some acknowledged its unparalleled violence, its rule by terror; some did not. "The Communist Party presents itself," Chambers noted, "as the one organization of the will to survive the crisis. . . . It is in the name of that will . . . that the Communist first justifies the use of terror and tyranny . . . which the whole tradition of the West specifically repudiates." "We were defending the first socialist country," insisted Ruth Kuczynski, a German Communist who lived in exile in England. "We didn't know—I didn't know—about Stalin's crimes," she told an interviewer late in life. "We knew how the capitalist West wanted to destroy the Soviet Union. It really seemed possible that they had managed to insert all these agents [who were purged during the Great Terror] into high places. . . . I believed Stalin."

Blindered or open-eyed, some among the faithful invested the raw, brutal, revolutionary new nation with their hopes of connection. Through its instrumentalities, they hoped that they could fight fascism, anti-Semitism, ignorance, inequality. Harry Gold believed he was attacking a universal and all-encompassing anti-Semitism:

In only the Soviet Union was anti-Semitism a crime against the State. . . . Here, too . . . was the one bulwark against the further encroachment of that monstrosity, Fascism. To me Nazism and Fascism and anti-Semitism were identical. This was the ages-old enemy of the Roman Arena, the ghetto, of the inquisition, of Pogroms, and now of concentration camps in Germany. Anything that was against anti-Semitism I was for, and so the chance to help strengthen the Soviet Union seemed like a wonderful opportunity.

Soviet intelligence networks made productive use of Communist Party members even though such volunteers were not trained agents and even though their Party affiliation made them suspect to their own governments; they were such people as money could not buy.

Recruiting usually followed a standard pattern. Committed Party members looked out for potential converts with useful skills or affiliations, made them welcome, proselytized them, obligated them with favors and gifts. Out of work in the depths of the Great Depression, Harry Gold got a job with the help of a Party recruiter, Tom Black. "That wonderful $30.00 every Saturday kept our family off relief. . . . I was grateful to Black, very much so." A 1946 Royal Commission investigating Soviet intelligence operations in Canada found that there were "numerous . . . groups where Communist philosophy and techniques were studied. . . . To outsiders these groups adopted various disguises, such as social gatherings, music-listening groups and groups for discussing international politics and economics. . . . These study groups were in fact 'cells' and were the recruiting centres for agents, and the medium of development of the necessary frame of mind which was a preliminary condition to eventual service of the Soviet Union in a more practical way." Besides commitment to the cause, the "necessary frame of mind" was secrecy:

This object is to accustom the young Canadian adherent gradually to an atmosphere and an ethic of conspiracy. The general effect on the young man or woman over a period of time of *secret* meetings, *secret* acquaintances, and *secret* objectives, plans and policies, can easily be imagined. The technique seems calculated to develop the psychology of a double life and double standards.

A candidate dropped out of Party activity when he agreed to become an agent, dividing and isolating him still further.

This theme of recruiting had significant variations. Morris Cohen, a native New Yorker born in 1910 to immigrant Russian parents and a high school football star, had joined the Communist Youth League in 1933 at the University of Illinois and subsequently volunteered to fight with the Abraham Lincoln International Brigade in the Spanish Civil War. While recovering from wounds in a hospital in Barcelona, Cohen was invited to attend the Republican Army's nearby Barcelona Intelligence School, which operated under the code name Construction. There he was recruited for US espionage by a Soviet intelligence officer. "In April 1938," Cohen wrote in his NKVD autobiography, "I was one of a group of various nationalities sent to a conspiratorial school in Barcelona. Our chief commissar and leaders were Soviets." Cohen completed his course of espionage training in February 1939 and returned to the United States to begin a productive career.

Ruth Kuczynski's older brother Jurgen was the political leader of the German Communist Party in England. Jurgen had escaped Nazi Germany in 1933 through Czechoslovakia and taken up teaching at the London School of Economics. Ruth, born in Berlin in 1907, came west by a different route; trained in Moscow as a clandestine radio operator, she had already worked out of Czechoslovakia, Trieste, Cairo, Bombay, Singapore, Hong Kong, Shanghai, Peking and Poland. By the time she settled in England in 1938 she was a major in Red Army intelligence (GRU as opposed to secret police intelligence, NKVD; the two entities maintained parallel and independent networks).

The most productive cell in the history of Soviet espionage developed at Cambridge University in the 1930s. While physicists at the Cavendish Laboratory probed the real world of the atomic nucleus with the new tool of neutron bombardment, a brilliant and fanatic group of young Cambridge intellectuals at Trinity College lauded the certainties of Marxian metaphysics. The majority of the group were homosexual or bisexual in a society that branded homosexual acts as felony crimes; sexual orientation contributed to affiliation even as it taught the young conspirators double standards and a double life. But Communism in any case was intensely fashionable at English universities between the world wars. Michael Straight, an American student at Cambridge at the time, estimates that "the Socialist Society had two hundred members when I went to Cambridge and six hundred when I left. About one in four of them belonged to Communist cells."

The nucleus of the Cambridge group was Guy Burgess, recruited in 1933 by a Russian agent who worked in London as a journalist under the alias Ernst Henri. Burgess, the handsome son of a well-married naval commander, took prizes at Eton and first-class honors in history at Cambridge. His brilliance and charm won him election to the Cambridge Conversazione Society, an elite secret society whose members were known as the Apostles. He enlisted at least two of the members of his cell by seduction. "At one time or another," wrote a don who adored him, "he went to bed with most of [his] friends, as he did with anyone who was willing and was not positively repulsive, and in doing so he released them from many of their frustrations and inhibitions." Of the four other men who came to be known as the Cambridge Five, Anthony Blunt and Donald Maclean certainly count among Burgess's sexual conquests. Kim Philby and John Cairncross were already dedicated Communists, but Cairncross at least acknowledged finding Burgess "fascinating, charming and utterly ruthless."

John Cairncross was a tall, rangy Scotsman from Glasgow, born in 1913. He studied at Glasgow University for two years beginning in 1930, when he was seventeen, took a year at the Sorbonne in Paris, then won a scholarship to Cambridge. Anthony Blunt was one of his Trinity supervisors there and directed him to Burgess, who recruited him for espionage in 1935. In the autumn of 1936, after he graduated from Cambridge with first-class honors in modern languages, Cairncross joined the British Foreign Office. Maclean, the tall, athletic namesake of the Liberal politician Sir Donald Maclean, was already on staff. "It's like being a lavatory attendant," Maclean would say later of espionage; "it stinks, but someone has to do it."

Though he worked at making friends, Cairncross was not a success in the Foreign Office. "Cairncross was always asking people out to lunch," one of his colleagues, John Colville, remembers. ". . . He ate very slowly, slower than anyone I've ever known." Colville judged him "a very intelligent, though sometimes incoherent, bore." In 1938, Cairncross transferred from the Foreign Office to the Treasury, probably at the request of the NKVD. Cairncross's real espionage breakthrough came in September 1940, a year into the European war, when Lord Hankey, minister without portfolio in Winston Churchill's War Cabinet, appointed him his private secretary. Hankey had full access to top secret War Cabinet papers and oversight of British intelligence. He also chaired the Scientific Advisory Committee.

It was probably John Cairncross who first passed information on Anglo-American atomic-bomb research to "Henry," the Cambridge Five's London NKVD control Anatoli Borisovich Gorsky, at the end of September 1941, when the Wehrmacht was besieging Leningrad and Igor Kurchatov was demagnetizing ships in Sevastopol. Gorsky—"a short, fattish man in his mid-thirties, with blond hair brushed straight back and glasses that failed to mask a pair of shrewd, cold eyes" according to one of his wartime agents—was "Vadim" to Moscow Center, the NKVD home office. Cairncross was probably "List." Vadim's report, "#6881/1065 of 25.IX.41 from London," summarized a meeting of the British Uranium Committee held on September 16. The information corresponds to information contained in the secret "Report by MAUD Committee on the Use of Uranium for a Bomb" prepared that summer for the British Cabinet and transmitted to the United States. At some time Moscow Center acquired a complete copy of the MAUD report.

"The uranium bomb may very well be developed within two years," Vadim's report began dramatically. Measurements of U235 cross sections would be accomplished by December. The British firm Metropolitan Vickers had been commissioned to develop a twenty-stage gaseous-diffusion pilot plant, a task which had "high priority," construction to begin "immediately." The government had contracted with Imperial Chemical Industries (ICI) for uranium hexafluoride, the gaseous form of uranium, which the Vickers plant would process.

Some of the information in this first transmission was garbled. A second transmission sent October 3 cleared up the confusion. "It is thought that the critical mass [of U235] falls within the range from 10 to 43 kg," the document reported. ICI had already produced three kilograms of uranium hexafluoride. "Production of U235 is realized by diffusion of uranium hexafluoride in a vaporized state through a number of membranes consisting of a grid of very fine wire." (This configuration was German emigré chemist Franz Simon's first approximation of a diffusion "membrane" or "barrier"—he had pounded out a kitchen strainer to demonstrate the idea to his Oxford staff.) In 1939, Yuli Khariton and Yakov Zeldovich had dismissed gaseous diffusion as an impractical method of separating U235; here was information that the British considered it superior. The document reported problems, however. "Development of the separation plant design is meeting with serious difficulties." Vadim enumerated the perverse physical characteristics that made "hex" hellish stuff—the heavy, corrosive gas destroyed lubricant, dissociated in the presence of water vapor and attacked equipment. A gaseous-diffusion plant would be huge, the British had calculated, 1,900 ten-stage units occupying a plant area of some twenty acres.

From gaseous diffusion the report then veered back to the bomb, echoing Peierls and Frisch's early realization that a weapon that derived its explosive force from nuclear fission would have unique characteristics: "It should be noted that besides the uranium bomb's tremendous destructive effect, the air at the site of the explosion will be saturated with radioactive particles capable of killing everything alive."

September 1941 was a banner month for Soviet nuclear espionage. While Vadim was reporting from London, Morris Cohen weighed in from New

York. Cohen had married a fellow Communist, Leontine Patka, known as Lona, the day Germany invaded the USSR. The invasion had depressed him, but after he had mulled it over for a few days he had revealed his affiliation to his wife and convinced her to join him in espionage work. Together they had already collected and passed along information from an engineer in Hartford on a new aircraft machine gun, even delivering a prototype of the machine gun to Morris's Soviet contact, the long barrel concealed in a bass viol case. Now Cohen reported a remarkable development. An American physicist whom he knew from Spanish Civil War days had contacted him for an introduction to Amtorg, the Soviet trading corporation in New York that clandestinely organized North American espionage. The physicist told Cohen he had been invited to work on a secret project to develop an American atomic bomb. Cohen wanted to know if he could recruit the man. Moscow Center approved.

Lavrenti Beria received these independent reports of Allied nuclear-research activity with his habitual cynicism. Anatoli Yatzkov, the NKVD's New York *rezident* during the Second World War, notes that "from the very beginning [Beria] suspected that these materials contained disinformation and thought that our adversaries [sic] were trying to drag us into tremendous expenditures and efforts on dead-end work. He gave them to a group of physicists for review. The scientists concluded that even if nuclear weapons were possible, they could only be built in the remote future."

Early in 1942, a new GRU volunteer began contributing to the volume of information reaching the Soviet Union. He was a refugee in England from Nazi Germany, a devoted Communist already gone underground and an exceptional young physicist and he worked for Rudolf Peierls:

> I . . . found many problems piling up on the theoretical side, and I could not deal with all of them fast enough. . . . I needed some regular help— someone with whom I would be able to discuss the theoretical technicalities. I looked around for a suitable person, and thought of Klaus Fuchs.

Born in 1911 in Rüsselsheim, in the Rhine Valley south of Frankfurt, Fuchs at thirty-one had already seen enough conflict and tragedy for a lifetime. He claimed later that he had "a very happy childhood," but it culminated with his mother's violent suicide—she drank hydrochloric acid—when he was nineteen. His elder sister Elizabeth would also be a suicide, though her act may have been protective: a Communist who was active politically against the Nazis, she jumped in front of a train when she was about to be arrested. Fuchs's father Emil was a politically contentious parson who left the Lutheran Church when Fuchs was fourteen and became a Quaker. "My father always told us that we had to go our own way," Fuchs remembered, "even

if he disagreed. He himself had many fights because he did what his conscience decreed, even if these [sic] were at variance with accepted convention." Klaus Fuchs would become his father's son, but he broke away from
his father's philosophy, he said, over pacifism.

Fuchs joined the Socialist Party at the University of Leipzig, where he
began studying physics and mathematics in 1930. After two politically active
years he went on to the University of Kiel. There he quit the Socialists over
the party's decision to support the presidency of Paul von Hindenburg, the
conservative field marshal who would pass the chancellorship of Germany
to Adolf Hitler. "At this point," Fuchs recalled, "I decided to oppose the
official policies openly, and I offered myself as a speaker in support of the
Communist candidate." He joined the Communist Party soon afterward and
worked actively on its behalf in student politics, his work culminating in a
strike which the Nazi leaders called in SA brownshirts to break. "In spite of
that I went there every day to show that I was not afraid of them. On one of
these occasions they tried to kill me and I escaped."

After the Reichstag fire early in 1933 that gave Hitler an excuse to invoke a
state of emergency and round up the opposition, Fuchs went underground:

> I was lucky because on the morning after the burning of the Reichstag I
> left my home very early to catch a train to Berlin for a conference of our
> student organization, and that is the only reason why I escaped arrest. I
> remember clearly when I opened the newspaper in the train I immediately
> realized the significance and I knew that the underground struggle had
> started. I took the badge of the hammer and sickle from my lapel. . . .

"I was ready to accept the philosophy that the Party is right," Fuchs continues, "and that in the coming struggle you could not permit yourself any
doubts after the Party had made a decision." Long afterward, Rudolf Peierls
would ask Fuchs how a scientist could accept Marxist orthodoxy and would
be shaken by the "arrogance and naiveté" of his answer. "You must remember what I went through under the Nazis," Peierls reports Fuchs answering.
"Besides, it was always my intention, when I had helped the Russians to take
over everything, to get up and tell them what is wrong with their system."

Fuchs remained underground until he left Germany for Paris in July 1933.
He was then twenty-one years old. "I was sent out by the Party, because they
said that I must finish my studies because after the revolution in Germany
people would be required with technical knowledge to take part in the
building up of the Communist Germany." To Harry Gold, who would meet
him later in America, Fuchs's dedication would always be "noble":

> Here: While Klaus was a mere boy of 18 he was head of the student chapter
> of the Communist Party at the University of Kiel . . . and Klaus, a frail, thin
> boy, led these boys in deadly street combat against the Nazi storm troopers

... and later, when the Nazis had put a price on his head, he barely managed to escape with his life to England. . . . For a man of such convictions who fought this horror of Fascism at the risk of his life, I cannot help but express my admiration.

Student friends helped Fuchs find his way to England, where a Bristol family with Communist connections took him in. Theoretical physicist Nevill Mott, a professor at Bristol University, gave him an assistantship. Mott thought Fuchs "shy and reserved," but saw another side at meetings of the Bristol branch of the Society for Cultural Relations with the Soviet Union, which sometimes staged dramatic readings of the texts of the purge trials then underway in Moscow. Fuchs chose to read the part of the prosecutor, shrill Andrei Vyshinsky, "accusing the defendants with a cold venom that I would never have suspected from so quiet and retiring a young man."

After four years at Bristol, Fuchs moved in 1937 to Edinburgh to work with Max Born, one of the pioneers of quantum mechanics and himself an emigré. In Edinburgh, says Peierls, Fuchs "did some excellent work in the electron theory of metals and other aspects of the theory of solids." Like Mott, Born also found the young German "a very nice, quiet fellow with sad eyes"; after Bristol Fuchs seems to have dissembled his political radicalism and swallowed his rage, although he did organize sending propaganda leaflets from Scotland to Germany.

He must have had trouble containing himself when he was interned as an enemy alien in May 1940 and sent to a camp on the Isle of Man. From there, jammed in with hundreds of other undesirables, he was deported by ship to internment in Canadian army camps that were short on latrines and running water. England was in a jingoist mood; by July, it had interned more than twenty-seven thousand Germans and Italians, many of them refugees from fascism, and would ship more than seven thousand abroad. Shattered by this second deportation, some of them committed suicide. A German U-boat torpedoed the *Arandora Star,* one of the passenger liners carrying the unlucky internees to exile; of 1,500 aboard, only 71 survived. Everyone's papers went down with the *Arandora Star* and for a time in Canada, as a result, Fuchs was billeted among Nazis. "I felt no bitterness by the internment," he claimed later, "because I could understand that it was necessary and that at the time England could not spare good people to look after the internees, but it did deprive me of the chance of learning more about the real character of the British people." How he assessed the British people in ignorance of their real character he chose not to say, but he did say, of his state of mind during the next several years, that he "had complete confidence in Russian policy and . . . believed that the Western Allies deliberately allowed Russia and Germany to fight each other to the death." No less a figure than Missouri Senator Harry S. Truman argued publicly for just such a policy when Germany invaded the USSR in June 1941. "If we see that Germany is winning we ought to help Russia and if Russia is winning we ought to help Germany and that way let them kill as many as possible," Truman told the Senate, "although I don't want to see Hitler victorious under any circumstances. Neither of them think anything of their pledged word." This early expression of Truman's hostility to the Soviet Union suggests that his move to a hard line after the war was a move from the Roosevelt policy of cooperation and accommodation back to long-standing conviction more than simply a response to Soviet intransigence.

After inquiries and the intercession of friends, Fuchs was returned to England and released from internment on December 17, 1940, twelve days before his twenty-ninth birthday. He went back to Edinburgh and Max Born and his chosen work of physics, a thin, pale, stoop-shouldered young man of average height with prominent forehead and Adam's apple, myopic brown eyes watchful behind thick glasses, a habit of swallowing hard, frequently and audibly, a chain-smoker with stained fingers. Someone eventually wrote a clerihew about him:

> *Fuchs*
> *Looks*
> *Like an ascetic*
> *Theoretic.*

Rudolf Peierls requisitioned Fuchs from Born sometime after the first of the year and took him in as a lodger; Peierls's wife Genia was exuberantly Russian and a great mother of young men, having previously taught Otto Frisch to shave daily and dry dishes faster than she could wash them. "[Fuchs] was a pleasant person to have around," Peierls recalls. "He was courteous and even-tempered. He was rather silent, unless one asked him a question, when he would give a full and articulate answer; for this Genia called him 'Penny-in-the-slot.' "

Since Fuchs was still an enemy alien, and was known to have been an active Communist in his homeland, clearance was delayed. The quiet young German started work on the atomic bomb at Birmingham in May 1941.

"When I learned the purpose of the work," Fuchs testified later, "I decided to inform Russia and I established contact through another member of the Communist Party." Fuchs went up to London in late 1941 and talked to Jurgen Kuczynski. "On his first contact with Kuczynski," an FBI report paraphrases his testimony, "he informed him of his desire to furnish information to the Soviet Union." Kuczynski put Fuchs in touch with a man he would come to know as "Alexander": Simon Davidovitch Kremer, secretary to the military attaché at the Soviet Embassy, who became his GRU control. In the next six months, Fuchs met with Alexander two or three times, once at the embassy, and gave him copies of the reports he was writing for Peierls. These included studies of isotope separation and calculations of critical mass as well as reviews of published German work in the field.

By early 1942, Lavrenti Beria's agents had bombarded him with so much information about British, French, German and American research toward an atomic bomb that he could no longer discount it. He ordered the British documents that the NKVD had received gathered together and a report prepared for Stalin. Copy No. 1 of that report, KZ-4, went to Stalin over Beria's signature in March 1942.

"Study of the question of military use of nuclear energy has begun in a number of capitalist countries," Beria began cautiously. Work on the development of new explosives using uranium was being carried out in an atmosphere of "strict secrecy" in France, England, Germany and the US. Top secret documents obtained by the NKVD from its agents in England revealed that the British War Office was intensely interested in the problem of military use because of concern that Germany might solve the problem first.

Drawing directly on the MAUD report, Beria noted that "well-known English physicist G. P. Thomson" was coordinating the work in England and that U235 was the explosive isotope involved, extracted from ores of which there were large reserves in Canada, the Belgian Congo, Sudetenland and Portugal. In a significant garble, Beria reported that the French scientists Hans Halban and Lew Kowarski had developed a method for extracting U235 using uranium oxide and heavy water; in fact, Halban and Kowarski (using most of the world's supply of heavy water, fifty gallons spirited out of France just ahead of the Germans in tin cans by car and boat) had determined that a controlled chain reaction was possible using such materials without enrichment, information Yuli Khariton and Yakov Zeldovich would benefit in the course of time from learning.

Beria went on to discuss gaseous diffusion, noting that the British hoped to cooperate in development with the United States. Then he took up the bomb itself.

Peierls, Beria reported, had determined that ten kilograms of U235 would form a critical mass. "Less than this amount is stable and absolutely safe, but a mass of U235 greater than 10 kilograms develops in itself a fission chain reaction, leading to an explosion of tremendous force." The British therefore proposed to design a bomb in which the "active part consists of two equal halves" and to drive them together at around six thousand feet per second. "Professor Taylor"—presumably Geoffrey Taylor, the English hydrodynamicist—"has calculated that the destructive action of 10 kg of U235 would correspond to 1,600 tons of TNT."

Imperial Chemicals had estimated that a plant to separate U235 "using Dr.

Simon's system" would cost £4.5 to £5 million, Beria went on. Then he offered a justification for bomb building that demonstrates how little anyone yet understood the revolutionary nature of the potential new explosive:

> Given production of 36 bombs per year by such a plant, the cost of one bomb would be £236,000 compared to the cost of 1,500 tons of TNT at £326,000.

Beria concluded that the British leadership considered the military application of uranium solved in principle and that the War Office was laying plans to produce uranium bombs. He recommended: (1) forming a special scientific committee attached to the State Defense Committee to coordinate Soviet work on atomic energy and (2) passing the espionage documents along to "prominent specialists and scientists" for assessment and use.

Coincidentally, the timing of Beria's report to Stalin matched within a few days a report US science czar Vannevar Bush sent to Franklin Roosevelt describing an American program that was then in the process of expanding from laboratory research to industrial development. "If every effort is made to expedite [research and production]," Bush concluded, an American bomb could be delivered in 1944. On March 11, 1942, Roosevelt responded enthusiastically, "I think the whole thing should be pushed. . . . Time is of the essence." In contrast, Stalin moved cautiously. He acted on Beria's second recommendation but not yet his first. The Soviet leader sent the file of espionage documents to Molotov with instructions to pass it for evaluation in turn to Mikhail Georgievich Pervukhin, the newly appointed People's Commissar of the Chemical Industry.

Molotov called him in, Pervukhin later told an interviewer, and expressed concern that other countries "might have achieved a major advance in the field, so that if we didn't restart our work we might seriously lag behind. . . . Then he said: 'You should talk to the scientists who know the field and then report on it.' That's what I did."

April 1942 brought further confirmation that the giant of nuclear fission was stirring. A Red Army colonel who commanded partisan detachments behind the German lines sent a captured document to Sergei Kaftanov, the State Defense Committee deputy for science. "Ukrainian partisans had brought him the notebook of a dead German officer," Kaftanov recalls. ". . . The notebook contained certain chemical formulae . . . [which] appeared to concern the nuclear transformations of uranium. The notes in general showed that the officer had a professional interest in nuclear energy. It seemed he'd come to the occupied territories specifically to look for uranium." Kaftanov gave a Russian translation of the German officer's notes to A. I. Leipunski, a senior Ukrainian physicist on the staff of the ill-fated institute at Kharkov. Leipunski responded with the safe and standard litany, says Kaftanov: "In three days the answer came. Leipunski believed that in the coming fifteen to twenty years the problem of developing nuclear energy would hardly be solved and that it wasn't worth spending money on it in the midst of war." Pervukhin heard much the same message.

Georgi Flerov had lost patience with timid Academicians and stodgy bureaucrats. He was a lieutenant in the Air Force now, assigned to a reconnaissance squadron in Voronezh, near the confluence of the Voronezh River and the Don some five hundred kilometers south of Moscow, but he was still strafing the government with letters and telegrams—no fewer than five telegrams to Kaftanov in recent months, with no response. Nor was official indifference to the cause of uranium research his only resentment. Although he and Konstantin Petrzhak had been nominated for a Stalin Prize for their 1940 discovery of spontaneous fission—an honor that customarily included tangible gifts—the nomination had not been confirmed because scientists in other countries had not welcomed the discovery in print or cited it in their publications. The university at Voronezh had been evacuated eastward, leaving behind its library. Flerov decided to check the scientific journals there to see if any new citations had turned up.

He found more missing from the foreign journals he consulted than merely references to his own work. Nuclear physics itself was missing; all the leading American nuclear physicists had stopped publishing. Flerov immediately understood that their work must have been classified. To Flerov that meant that the United States must be developing an atomic bomb. Twenty-nine years old and a mere lieutenant, but a physicist who understood the energy that matter might release if it were properly arranged, he notched his sights up then from assaulting the bureaucracy and in April 1942 appealed directly to Stalin:

Dear Josef Vassarionovich:

Ten months have already elapsed since the beginning of the war, and all the time I have felt like a man trying to break through a stone wall with his head.

Where did I go wrong?

Am I overestimating the significance of the "uranium problem"? No, I am not. What makes the uranium projects fantastic are the enormous prospects that will open up if a successful solution to the problem is found. . . . A veritable revolution will occur in military hardware. It may take place without our participation—due simply to the fact that now, as before, the scientific world is governed by sluggishness.

Do you know, Josef Vassarionovich, what main argument has been advanced against uranium? "It would be too good if the problem could be solved. Nature seldom proves favorable to man."

Perhaps, being at the front, I have lost all perspective. . . . I think we are making a big mistake. . . .

Flerov went on to propose a conference where he might state his case, with Stalin and a jury of ranking physicists present—he asked for Ioffe, V. G. Khlopin, Kapitza, Leipunski, Landau, Kurchatov, Khariton, Zeldovich and others. "I see this as the only means to prove that I am right," he argued, ". . . because other means . . . are simply being passed over in silence. . . . That is the wall of silence which I hope you will help me break through. . . ."

Stalin enjoyed springing traps. "To choose one's victim," he mused once, "to prepare one's plans minutely, to slake an implacable vengeance, and then to go to bed . . . there is nothing sweeter in the world." After he received Flerov's letter and consulted with Kaftanov he called in four of his Academicians—Ioffe, Kapitza, Khlopin and Vladimir I. Vernandski—and berated them, indignant that a young tyro like Flerov had recognized a danger to the country that they had ignored. Golovin says he "asked them bluntly how serious the information he had was concerning the possibility of developing the atom bomb in the next few years. . . . His guests unanimously confirmed the importance of this work."

The expense of building a new industry in the midst of war mobilization worried the Soviet dictator. Two of his advisers predicted that a bomb would cost as much again as the entire war effort. Kaftanov defended the expense:

I said that of course a degree of risk was involved. We would risk tens, perhaps hundreds of millions of rubles. In the first place, we would have to spend money on science anyway, and investment in a new field of science is always fruitful. But if we did not take the risk, a much greater risk would then emerge: that we might one day face an enemy possessing nuclear weapons while we ourselves were unarmed.

After some hesitation, adds Kaftanov, "Stalin said: 'We should do it.'"

It was then May 1942 and the Wehrmacht was still smashing its way across the western USSR. The possibility that Germany might develop an atomic bomb had strongly influenced the Anglo-American decision to go forward. The possibility that Germany was working on an atomic bomb and the certainty, confirmed by espionage, that England and the United States were, had now catalyzed the Soviet decision as well.

———

Deciding was one thing. Embodying the decision in difficult research and fantastic, extravagant technology would be quite another. "The Stalingrad victory was far ahead," write Golovin and Russian physicist Yuri Smirnov of the desperate spring and summer of 1942. ". . . Moscow was the front line and nearly depopulated. Anti-aircraft batteries stood on alert, the Kremlin stars had been covered with canvas, barrage balloons guarded the approaches, German and Soviet planes were dogfighting over the city. A curfew began at dusk and the streetlights had been shut off; automobiles found their way with headlights dimmed and narrowed to blue beams.... Food and goods were rationed. Many ministries and departments were still in evacuation." On a train ride from Murmansk to Moscow during the first week in June, Alexander Werth observed the results of wartime shortages and German successes:

> Civilians were badly underfed, and many suffered from scurvy; old women especially were tearful and pessimistic, and thought the Germans were terribly strong.... Morale among soldiers and officers was rather better.
> ... All the same, they were far from underrating the power of the Germans, and in their game of dominoes, they called the double-six "Hitler"—"because it's the most frightening of them all."

As of June 22, official Soviet combat casualties, probably underestimated, totaled 4.5 million; German totals approached 1.6 million. On July 28, Stalin issued his notorious Order No. 227 acknowledging the loss of the Ukraine, Belorussia and the Baltics to the German advance. "We now have fewer people and industrial plants, less bread and metal," Stalin declared. "... Any further retreat will be fatal for us and for the Motherland.... Not a step backward! At any cost, we must stop the enemy, push him back and defeat him!"

One tried, effective way to save time and expense was industrial espionage. A coded radio message went out from Moscow Center on June 14, 1942, to NKVD *rezidents* in Berlin, London and New York:

> Top secret.
>
> Reportedly the White House has decided to allocate a large sum to a secret atomic bomb development project. Relevant research and development is already in progress in Great Britain and Germany. In view of the above, please take whatever measures you think fit to obtain information on:
> —the theoretical and practical aspects of the atomic bomb projects, on the design of the atomic bomb, nuclear fuel components, and the trigger mechanism;
> —various methods of uranium isotope separation, with emphasis on the preferable ones;
> —transuranium elements, neutron physics, and nuclear physics;
> —the likely changes in the future policies of the USA, Britain, and Germany in connection with the development of the atomic bomb;
> —which government departments have been made responsible for co-

ordinating the atomic bomb development efforts, where this work is being done, and under whose leadership.

Morris Cohen was drafted into the US Army in July and left New York for basic training and service in Europe. It took Anatoli Yatzkov two months clandestinely to reestablish contact with Morris's wife Lona, but she agreed to replace her husband as a courier.

Fuchs's arrangements also changed that summer. Traveling to London was awkward in wartime; to deceive Genia Peierls, Fuchs had to fake illnesses and pretend to be visiting a physician. At his third meeting with "Alexander," the Russian proposed a more convenient link. Fuchs would not quite remember if Alexander also told him he was leaving England; in any case the new arrangement would give Fuchs a contact closer to Birmingham.

Fuchs's courier would be a woman this time. Her code name was "Sonia." He knew her as Ruth Kuczynski, the sister of the man whom he had first approached to propose espionage. She was living in Oxford under the name Ruth Brewer with her children and her English husband Len, a fellow spy, clandestinely broadcasting coded espionage information to Moscow using a shortwave radio she had built herself. She was tall, slender and attractive, and at their meetings in Banbury and in the countryside near Birmingham —Fuchs rode out on a bicycle—she offered Fuchs a welcome change from what he would later call the "controlled schizophrenia" of his double life. "It was a great relief for him to have someone he could talk to openly," she told an interviewer many years afterward. "He never met any comrades in Britain with whom he could talk about things." He was, she thought, "a good, decent man." For his part, Fuchs confessed, he had "no hesitation in giving all the information I had."

In Moscow, the search went forward for someone to direct the new project. According to Golovin, Stalin consulted with Beria. Beria suggested Ioffe or Kapitza. Stalin disagreed; they were world-famous scientists, he argued, they were already burdened and their disappearance into secret work would be noticed. "He said that it was necessary to promote a young, not well-known scientist," writes Golovin, "for whom such a post would be . . . his life work." Kaftanov describes a different, or perhaps a complementary, sequence:

I got the job of finding people, finding a place and organizing the necessary institutions. I began with Ioffe. The most important issue was who would head this extraordinary project. I suggested that he himself should head it. He said that he was already too old (he was then sixty-three), and that we needed a young, energetic scientist. He proposed a choice of two [physicists]: thirty-nine-year-old [Abram] Alikhanov and forty-year-old Kurchatov.

Yuli Khariton's wife Maria Nikolaevna encountered Kurchatov in Kazan that summer. "After the epic events in Sevastopol I saw Kurchatov with a beard. I asked him, 'Igor Vasilievich, what are you doing with that pre-Petrine ornamentation on your face?'* He answered with two lines of a popular song: 'First we're gonna beat back Fritz, then, when there's time, we'll all shave.' . . . The beard suited that tall and imposing man very well." Bearded Kurchatov traveled to Moscow for consultations. So, presumably, did Alikhanov.

"Alikhanov," Kaftanov explains, "was by that time quite famous. He was already a corresponding member of the Soviet Academy of Sciences and winner of a Stalin Prize. He was known for his discovery of positron-electron pairs and his work in the field of cosmic rays. Kurchatov was less well-known." But Kurchatov, Kaftanov continues, had worked with uranium and with nuclear fission. He had not only participated in this work but directed it. "It was also in his favor that he had joined the Navy, which showed that he was willing to work where he was most needed."

The government chose its man sometime in September 1942. A Kaftanov senior aide, S. A. Balezin, recalls Kurchatov's final interview:

> We invited Kurchatov to Moscow simply to meet him before rejecting his candidacy. But he entered the room and immediately impressed everyone with his modesty and charm: he had a very good smile. He also appeared to be a thorough man. I had shown him translations of the German officer's notebook and he had read them through. I didn't tell him that the decision to restart uranium work had already been made. I only asked him: if such work should start, would he accept the leadership? He became thoughtful for a while, smiled, patted his beard—it was a short one then—and said, "Yes."

Apparently the interview made the difference. "The outcome of any enterprise," says Kaftanov, "is finally determined by competence, energy, organizing skills and devotion to the cause." He offered Kurchatov the job. Kurchatov asked for a day to think it over. "On the next day he came and said, 'If it is necessary, I'm ready. This is a tremendously difficult task. But I hope that the government will help, and of course that you will help too.'"

One other version of how Kurchatov was chosen has surfaced. Molotov, who notes that he "was in charge" of atomic-bomb research, says he picked Kurchatov:

> I had to find a scientist who would be able to create an A-bomb. The [NKVD] gave me a list of names of trustworthy physicists. . . . I summoned Kapitsa, an Academician. He said we were not ready, that it was a matter

---

* In the eighteenth century, Peter the Great had made shaving compulsory as part of his program to Europeanize his subjects.—RR

for the future. We asked Ioffe. He too showed no clear interest. To make a long story short, I was left with the youngest and least-known scientist of the lot, Kurchatov; they had been holding him back. I summoned him, we chatted, and he impressed me.

Kurchatov returned to Kazan and told Alexandrov. "The work on nuclear physics will continue. There's information that the Americans and the Germans are making nuclear weapons." "How is it possible for us to develop a thing like that in wartime?" Alexandrov asked. "They said don't be shy," Kurchatov told him. "Order what you need and begin work immediately."

3

'Material of Immense Value'

VYACHESLAV MOLOTOV—"Stalin's shadow," says Dmitri Volkogonov, "a harsh man"—assumed overall direction of the Soviet atomic bomb program at its inception in autumn 1942. Molotov had earned Lenin's contempt in the early years of the new state for "generating the most shameful bureaucratism and the most stupid." "His leadership style," Yuli Khariton reports, "and correspondingly, its results, were not terribly effective." Born in northwestern Russia in 1890 and one of the few Old Bolsheviks to survive the purges, Molotov was square and dark, with close-cropped curly hair and a strip of black mustache pasted across his upper lip. Like Beria, he affected pince-nez; when he grimaced at Stalin in devotion, baring his teeth, he looked like Teddy Roosevelt, but a Russian poet who had occasion to work with him found him not exuberant but "modest, precise and thrifty," the kind of man who could not pass an empty room without turning off the lights.

If Molotov told Kurchatov not to be shy and to order all he needed, the vice-premier did not yet give the new project carte blanche. The atomic-bomb program in the United States, which the US Army Corps of Engineers was now administering and had code-named the Manhattan Engineer District, was awarded first priority for materials and personnel over any other program of the war. In the Soviet Union, to the contrary, atomic-bomb research began ad hoc, Kurchatov and his colleagues pulling together whatever resources they could find.

The vicissitudes of war partly determined the Soviet program's modest initial priority. Molotov had assigned chemical industry commissar Mikhail Pervukhin to work with Kurchatov and with Sergei Kaftanov of the State Defense Committee. "It was difficult to organize the works to the desired scale," Pervukhin recalls, "because the country was in the heaviest period of the war; the nation's full potential was already mobilized to defeat the enemy." Research institutes had been evacuated to the east, Pervukhin adds;

the cyclotron under construction in Leningrad had to be moved with its big magnet to Moscow; Kurchatov needed time to prepare a feasibility study.

But bureaucratic politics interfered as well. Kurchatov's lack of scientific rank, which Stalin had counted in his favor, worked against him in council. "Our suggestion to the State Defense Committee was to form an institute," says Pervukhin, "but we were told that we should start in a more modest way, with a laboratory, since Kurchatov had been only a laboratory director up to that time. Start with a laboratory, they said, and develop a program of works to be done."

Nor was it easy to corral the necessary organizations and personnel. "There were many difficulties in those years," Pervukhin continues:

> For instance, we had problems drawing institutes into our work. We asked Academician Ilia Iliich Chernyayev of the Institute for Inorganic Chemistry to develop some chemical methods for us, but he refused: "Why should we do it? It's not our work. We have our own job to do." We couldn't agree to that and we got a decision obliging the institute to do the work. Then . . . along came the deputy director of the institute and the secretary of its Party organization, complaining that we were interfering with their scientific programs and ruining the institute's specialization. We had to explain to these comrades that they were wrong.

Bureaucrats similarly resisted aiding the new enterprise. "It was very difficult to negotiate with Ministers," complains Pervukhin. "They said, 'You're taking our people from us when we have our own state plans to fulfill. We won't give our people away!'" Pervukhin had to invoke the State Defense Committee to enforce his requisitions. "Until 1945," Khariton confirms, "this program was carried out by only a few researchers who had scarce resources."

Everyone was preoccupied with the Battle of Stalingrad, which raged through the autumn and early winter. "Stalingrad was the key to the rest of the country still in Russian hands," comments Alexander Werth—"the whole of European Russia east of Moscow, the Urals and Siberia." Blocked in the north at Leningrad, stopped and pushed back before Moscow, the Germans had launched a major summer offensive up through the Crimea and eastward through central Russia southeast of Moscow intended to capture or destroy Stalingrad and then turn south to claim the vital oil areas of the Caucasus at Maikop, Grozny and Baku. Soviet industry had not yet revived sufficiently to supply the Red Army with the equipment it needed to match the German onslaught; "with 1,200 planes in this area of the front," writes a Soviet historian, "the enemy had great superiority in aircraft, as well as in guns and tanks."

On August 23, 1942, a raid of six hundred German bombers on Stalingrad killed forty thousand civilians. The Wehrmacht began a major ground assault on September 13. "Whole columns of tanks and motorized infantry were breaking into the center of the city," writes the commander of one of the defending Soviet armies. "The Nazis were now apparently convinced that the fate of Stalingrad was sealed, and they hurried towards the Volga. . . . Our soldiers—snipers, anti-tank gunners, artillerymen, lying in wait in houses, cellars and firing-points, could watch the drunken Nazis jumping off the trucks, playing mouth organs, bellowing and dancing on the pavements." Stalingrad with its suburbs and factories, war correspondent Konstantin Simonov wrote back from the front, was one "whole, huge, thirty-seven-mile-long strip along the Volga":

This city is no longer as we saw it from the Volga steamer [before the war]. It has no white buildings climbing the mountain in a merry throng, no little landing piers on the Volga, no quays with rows of baths, kiosks, and small buildings running along the river. At present this city is smoke-filled and grey and the fire dances about it and the soot whirls day and night. This is a soldier-city, scorched in battle, with strongholds of self-made bastions built from the stones of its heroic ruins. . . .

The Wehrmacht pushed the Soviets back east across the river—the Soviets were able to maintain only about twenty thousand troops on bridgeheads on the west bank—but "the other side of the Volga," says a Red Army lieutenant who fought there, "was a real ant-heap. It was there that all the supply services, the artillery, air force, etc., were concentrated. And it was *they* who made it hell for the Germans." Artillery shells and Katyusha rockets roared over the bridgeheads and smashed into the city. Fighting went on day to day and hand to hand. The Germans began another all-out offensive on October 14 that the Soviet Army commander characterizes as "a battle unequalled in its cruelty and ferocity throughout the whole of the Stalingrad fighting." The Germans wanted to make a hell out of the city, Simonov wrote: "The sky burns overhead, and the earth shudders underfoot." Wehrmacht forces drove their way to within four hundred yards of the Volga, close enough to rake the bridgeheads with machine-gun fire; the Soviets had to build stone walls under fire to protect their positions.

In November, the Red Army was able at last to mount a great counteroffensive. Forces from the Don and Northwest Fronts pushed down from the north while Stalingrad Front armies pushed up from the south; in four days they sealed off the Germans in what they named a "cauldron." It was cold by then and it got bitterly colder in December, as much as forty degrees below zero. Until too late the German high command had withheld winter clothing from its armies in Stalingrad, afraid the realization that they would have to fight through the winter would damage the soldiers' morale. A Luftwaffe attempt to airlift supplies foundered on bad weather and poor organization. But the starving Germans refused to surrender and in January 1943 Soviet forces liquidated the cauldron, barraging the ruined city from seven thousand mortars and guns, bombing, crashing in with tanks and infantry. They had encircled 330,000 men; they took fewer than 100,000 prisoners. They stacked up the frozen German bodies like cordwood. "Funny blokes," a boy told Werth. ". . . Coming to conquer Stalingrad, wearing patent-leather shoes." Werth heard that children in a nearby village were using one dead German for a sled.

One night after the liquidation of the cauldron, when it was minus forty-four degrees, Werth drove shivering toward Stalingrad in a van full of journalists through the victorious armies:

All the forces in Stalingrad were now being moved. . . . About midnight we got stuck in a traffic jam. And what a spectacle that road presented. . . . [There were] lorries, and horse sleighs and guns, and covered wagons, and even camels pulling sleighs. . . . Thousands of soldiers were . . . walking in large irregular crowds, to the west, through this cold deadly night. But they were cheerful and strangely happy, and they kept shouting about Stalingrad and the job they had done. . . . In their *valenki* [wool felt boots], and padded jackets, and fur caps with the earflaps hanging down, carrying tommy-guns, with watering eyes, and hoarfrost on their lips, they were going west. How much better it felt than going east!

Stalingrad was the turn of the tide.

Moving and other preliminaries kept Kurchatov busy until early 1943; in January the Navy even ordered him to Murmansk to work on German mines. The State Defense Committee (GKO) officially awarded him authority over the uranium project on February 11, 1943. "At that time it required special permission from the GKO to enter Moscow," Kaftanov recalls. "We obtained permission for approximately a hundred people and a respective number of apartments and began inviting the chosen specialists."

Working out of a room at the Moscow Hotel, on Marx Prospekt within sight of the Kremlin, Kurchatov assembled a core team of talents to prepare a feasibility study: theoretical physicists Georgi Flerov, Yuli Khariton and Yakov Zeldovich, experimentalists Isaak Konstantinovich Kikoin and Abram Alikhanov. Kikoin was a specialist in diffusion processes; Alikhanov, the young Academician and cosmic-ray expert, had competed with Kurchatov to head the project. Golovin:

In no hurry to expand his staff, [Kurchatov] tried to determine the main lines of attack and clearly formulate the scientific and engineering task ahead. He made numerous estimates and gave more detailed consideration to the possible ways of achieving a uranium fission chain reaction, carefully discussing them all. The group soon decided to build a [nuclear reactor] powered by [slow] neutron fission and simultaneously to work out means for separating large quantities of uranium isotopes.... Kurchatov did not settle for half measures but at once boldly got started on estimates for a uranium bomb whose explosive power would come from fast-neutron fission, though he did not yet have so much as a microgram of pure U-235 and though neither he nor the other members of the group had an inkling as to the possibility of producing ... plutonium. ...

At the beginning of 1943, that is, Igor Kurchatov and his colleagues in the Soviet Union were planning to build a nuclear reactor to prove that a chain reaction was possible in uranium and then to build a uranium bomb using U235 separated laboriously from natural uranium by physical means.

That would be a long, slow, expensive route to a bomb, one that Kurchatov certainly would not have chosen if he had known any shorter, faster and cheaper approach. The Soviet Union had only limited known reserves of uranium ore. It had only a few kilograms of heavy water and no facilities for making more, but a reactor moderated with heavy water would require several tons. It lacked the technology to make large quantities of pure graphite, an alternative to heavy water. It lacked the technology to make uranium metal or uranium hexafluoride. U235 had not yet been separated from U238 in the Soviet Union even at laboratory scale, and separating enough U235 for a bomb—tens of kilograms—would require developing a vast new industrial plant based on one or more new and difficult technologies. The gun bomb that Flerov had proposed and that Kurchatov had in mind would be prodigal of material, requiring several critical masses of U235 in its design.

The Soviet scientists had not yet appreciated that a reactor would transmute a portion of its larger inventory of U238 into a new man-made element heavier and less stable than uranium. Early in 1941, a team of American scientists at Berkeley led by radiochemist Glenn T. Seaborg had transmuted the first millionth of a gram of the new element in the big sixty-inch Berkeley cyclotron; the team had isolated the first sample on March 28, but the discovery was classified and would not be announced until after the war. In 1942, Seaborg had named the new element plutonium. By then, the Americans had determined what the Soviets did not yet know: that plutonium was even more fissionable than U235, with a fission cross section for fast neutrons 3.4 times as large as natural uranium. Since it could be separated chemically from the matrix of natural uranium in which it was bred, and since chemical separation was a far less difficult and therefore less costly process than physical separation, plutonium would probably be a shortcut to a bomb. So the leaders of the American program had come to believe. As a result, the Manhattan Project was now gearing up to breed plutonium in graphite and heavy-water reactors as well as to separate U235 using gaseous diffusion, thermal diffusion and electromagnetic means. A new secret laboratory that would open its doors on a mesa in the northern New Mexico wilderness in April 1943 would begin developing gun designs for both uranium and plutonium.

With its high priority and unlimited resources, the American program could afford to hedge its bets. At that early point in any case it would be prudent to explore alternatives, as Kurchatov also understood—when Ukrainian physicist Anatoli Petrovich Alexandrov asked him why he wanted thermal diffusion explored when there were better methods and it wouldn't be used, Kurchatov shot back, "The Devil knows what will be used. We have to try this way just in case." But given a choice, a country with fewer resources might do better to give priority to plutonium. As of early 1943, Igor Kurchatov was evidently not aware of the existence of such a choice.

Then he saw the accumulated NKVD espionage. "He said he still had a lot to clear up," Molotov remembers. "I decided then to provide him with our intelligence data. Our intelligence agents had done very important work. Kurchatov spent several days in my Kremlin office looking through this data. . . . I asked him, 'So what do you think of this?' I myself understood none of it, but I knew the material had come from good, reliable sources. He said, 'The materials are magnificent. They add exactly what we have been missing.' "

On March 7, 1943, Kurchatov finished drafting a fourteen-page review for Mikhail Pervukhin of the documents and transmissions that Moscow Center had collected. He only refers to British material—most of it probably passed by Klaus Fuchs—which almost certainly means that no American technical information had yet come in. But the British knew enough, and Kurchatov learned enough, to transform the Soviet program.

"Having reviewed the material," Kurchatov began directly, "I came to the conclusion that it is of immense value for our science and our country. Its value cannot be overestimated."

The material "shows what serious and intensive research and development work on the uranium problem has been undertaken in England," Kurchatov explained. It also, he wrote, "provides some quite important reference points for our research, informing us of new scientific and technical approaches and enabling us to skip labor-intensive phases of development."

Kurchatov judged at that point that the most valuable information in the espionage material dealt with isotope separation. The Anglo-American preference for gaseous diffusion as a means of separating U235 from U238 was unexpected, he explained; the Soviet scientists had believed the centrifuge approach to be much more promising. The espionage material "made us include diffusion experiments in our plans along with centrifuge."

Next Kurchatov went through the theoretical work on diffusion, "a very detailed study." Fuchs and Peierls had done that study and Fuchs later admitted passing a number of reports on diffusion theory to Alexander and to Sonia. The study provided a complete description of Franz Simon's proposed gaseous-diffusion unit. "Our theoreticians haven't yet checked this extensive work," Kurchatov reported to Pervukhin, "but as far as I can judge, it is the work of a group of prominent scientists who based their well-founded and laborious calculations on clear physical principles." The study was so complete, Kurchatov exulted, that it would enable his team "to skip the initial stage" and to move immediately to developing gaseous diffusion in the Soviet Union.

Kurchatov wanted further information on the machinery the British were developing for gaseous diffusion. He included five questions on the subject in this section of the report, clearly intending for Pervukhin to pass them to the NKVD and GRU to guide further espionage. The scientific director of the Soviet program to build an atomic bomb, that is, was not a passive recipient of espionage materials but an active participant in an extensive program of espionage directed against his country's wartime allies, Britain and the United States. On the other hand, they were allies that had decided to exclude his suffering country from a secret joint program to develop a decisive new weapon of war; he surely felt justified.

The material Kurchatov reviewed contained brief analyses of the usefulness of thermal diffusion, centrifuge and mass-spectrographic approaches to isotope separation. Thermal diffusion he discounted as "inefficient because of high energy consumption." It was, but it would save the American project in 1944 when problems in barrier development delayed start-up of the big gaseous-diffusion plant under construction at Oak Ridge, Tennessee. The British analysis dismissed the centrifuge approach because of the difficulty of making a centrifuge that would hold together at the high rate of rotation necessary for isotope separation. "This conclusion may be challenged," Kurchatov writes, defending the primary approach of the Soviet program so far. It would not be successfully challenged in the Soviet Union until long after the end of the war.

Kurchatov headed the second section of his March 7, 1943, report "The Problems of Nuclear Explosion and Chain Reaction." Here were more revelations from espionage, first of all "the statement that it is possible to realize a nuclear chain reaction in a mixture of regular uranium oxide (or metallic uranium) and heavy water. For Soviet scientists this conclusion is unexpected and contradicts the established point of view; we considered it to be proven that without isotope separation it is not possible to achieve a chain reaction with heavy water."

In 1940, misled by the cross-section estimate that Borst and Harkins had reported in their letter to the *Physical Review*, Yuli Khariton and Yakov Zeldovich had reached that pessimistic conclusion. Now Fuchs reported Hans Halban and Lew Kowarski's actual measurements of deuterium cross sections using, as Kurchatov said, "the entire world reserve" of heavy water; the report convinced Kurchatov that his theoreticians should review their conclusions once more. He also wanted more information, via espionage, about the French scientists' work:

It is mentioned in the [espionage] material that Halban and Kowarski intend to continue their experiments with larger amounts of heavy water in America, where, it is said, production of this substance is organized on a very large scale. . . . Therefore it is extremely important to find out if Halban and/or Kowarski went from Britain to America (in 1941–1942) and if they carried out [their] experiments. . . .

Kurchatov vigorously defended Khariton and Zeldovich's heavy-water calculations, stressing that the theoreticians could not have produced a more accurate cross-section estimate because they lacked the necessary experimental equipment. He understood that he and his colleagues were being watched. The only mistakes the Soviet leadership tolerated were its own; "wrecking," real and metaphorical, was a crime that crowded the camps of the gulag.

Then Kurchatov played a brilliant hunch. His hunch reveals that as of March 1943, word had not yet reached the Soviet Union of the construction and successful operation of the world's first man-made nuclear reactor in the United States on December 2, 1942—Enrico Fermi's uranium-graphite reactor, CP-1, stacked by hand in a doubles squash court under the west stands of the University of Chicago's Stagg Field. Kurchatov wrote: "All experiments with systems of uranium and moderator that have so far been conducted and published used homogeneous mixtures of these components." But a heterogeneous system might be better, he guessed, one where "the uranium is concentrated within the mass of [moderator] in spaced blocks of appropriate dimensions." That conclusion had come independently to Fermi and to Leo Szilard at Columbia in 1940; CP-1 was just such a three-dimensional lattice. "Kurchatov was an exceptional leader," Khariton and Smirnov comment, "who organized a strategically correct program from the very beginning." He had an "unerring ability to find correct ways of attaining goals . . . despite the scarce and incomplete initial scientific data." As soon as Kurchatov realized that a heterogeneous arrangement might be superior to the homogeneous systems he had been sponsoring—might make possible a reactor assembled from natural uranium without enrichment—he asked his team to study the two different arrangements theoretically and experimentally and asked Pervukhin to set Soviet intelligence to find out which kind of system the British and the Americans were studying. "In the end," Khariton and Smirnov conclude, referring to members of the Kurchatov team, "it was I. I. Gurevich and Isaak Pomeranchuk who successfully solved Kurchatov's problem, showing the decisive advantage of a heterogeneous reactor."

Part III of Kurchatov's report, "The Physics of the Fission Process," primarily discussed espionage information that confirmed what Soviet scientists had already worked out on their own, but it included more questions which Kurchatov hoped further espionage might resolve. Kurchatov made a point of emphasizing that Otto Frisch in England had "confirmed the phenomenon of the spontaneous fission of uranium, discovered by Soviet physicists G. N. Flerov and K. A. Petrzhak," that the fact of spontaneous fission made it necessary to keep a critical mass disassembled until the moment of explosion (so that a stray secondary neutron would not cause it to chain-react prematurely) and that Flerov's calculation of the necessary speed of assembly closely matched British estimates.

Finally, Kurchatov assessed the issue that had bothered Beria, and probably Stalin as well, since Vadim's first report had come in: was this harvest of espionage information or disinformation?

> Naturally, the question is raised whether the materials received reflect the real status of research and development in Britain rather than a legend aimed at misdirecting our science. This question is of special importance for us because in many important areas we are not in a position to test the data (because we lack the technical base necessary to do so).
>
> Based on a close examination of the material, I formed the conclusion that it reflects the real state of things.
>
> Certain conclusions, even some that refer to quite important parts of the work, seem dubious to me, some of them not well founded, but for this the British scientists are responsible rather than the reliability of the information.

The most crucially important information about atomic-bomb development that the Soviet Union would acquire through espionage—information that would accelerate the Soviet program by a full two years—took up only a brief paragraph in Kurchatov's March 7 report. There he noted that he would discuss it in more detail in a separate letter. He sent Pervukhin that separate seven-page letter two weeks later, on March 22, 1943.

"Fragmentary remarks" in the espionage materials he reviewed, Kurchatov wrote in his March 22 letter, referred to the possibility of using plentiful $U238$ as well as rare $U235$ to make a bomb. The documents contained "very important remarks on the use for bomb material of an element with mass

239, which should be produced in the 'uranium pile'* as the result of the absorption of neutrons by uranium 238."

The suggestion had sent Kurchatov to the library to look through the last papers that American scientists had published on transuranium elements in the *Physical Review* before wartime security clamped down. What Kurchatov found elated him; announcing his discovery in his handwritten report, he capitalized the key words and underlined them twice: "I was able to see a NEW direction to solving the entire uranium problem. . . . THE PROSPECTS OF THIS DIRECTION ARE EXTREMELY PROMISING."

Kurchatov proceeded to review the workings of a "uranium pile," pointing out that "it was always assumed that only the light isotope of uranium—U235, which constitutes only 1/140th part of regular uranium—would be useful in a 'pile.' The rest of the uranium—U238, constituting 139/140ths—would be useless, since it does not emit large amounts of energy or produce secondary neutrons when hit by a slow neutron. . . . This conclusion may be totally wrong."

Kurchatov was alluding to the transmutation of U238 under neutron bombardment that Edwin McMillan and Philip Abelson at Berkeley had successfully explored in 1940. McMillan and Abelson had found, in Kurchatov's words, that "the nucleus of U238, hit by a neutron, passes through certain changes and transforms itself into uranium-239. This element is unstable and in twenty minutes (on average) transforms spontaneously into element 93 (which doesn't exist on earth)—the element called eka-rhenium." McMillan and Abelson had a better name for the first artificial new element beyond uranium (and hence "transuranic"), but they had not published it in their *Physical Review* paper; they called element 93 neptunium.

Uranium was the heaviest element to occur naturally on earth because its nucleus, densely packed with positively charged protons that repelled each other electrically, was only marginally stable. It should follow that man-made transuranic elements like element 93, with even more protons packed into their nuclei, would be even less stable. The German physicist Carl Fredrich von Weizsäcker had independently worked out these consequences of U238 bombardment in the summer of 1940, had assumed that 93 might fission and chain-react and had reported the idea to his government, which failed to take up the suggestion. In fact, 93 was more like U238 than U235, and in any case its 2.3-day half-life made it unsuitable for use in a weapon. But since 93 was radioactive, spontaneously emitting beta electrons and gamma rays, it followed that it would quickly transmute itself further. And on theoretical grounds, element 94 ought to fission and chain-react even more vigorously than U235. McMillan and Anderson had men-

---

* "Pile" was Fermi's coinage; that much at least had leaked through (previously the Soviets had called a nuclear reactor a "boiler").—RR

tioned this "daughter product 94^{239}" in their paper on 93 and had measured the outside limits of several of its physical characteristics. (Glenn Seaborg and his colleagues had then taken over the research and discovered 94— plutonium—but by then the work was classified.)

Now, in March 1943, Kurchatov realized the same possibilities. "It appears," he wrote, "that although eka-rhenium [93] is somewhat more stable than uranium-239, it also possesses a short half-life . . . and by itself transforms into element 94—this element is called eka-osmium.* . . . According to all current theoretical ideas, the collision of a neutron with a nucleus of eka-osmium [94] will be accompanied by the release of a large amount of energy and the emittance of secondary neutrons, so in this respect it should be analogous to U235."

In Part Two of this March 22, 1943, letter, Kurchatov quickly explained the enormous import of his extrapolation:

> If eka-osmium [94] really possesses properties similar to those of U235, it could be extracted [chemically] from the "uranium pile" and used as material for an "eka-osmium" bomb. The bomb hence would be made of "unearthly" material, material which has disappeared from our planet.
>
> As one can see, given this solution to the entire problem, there is no more necessity to separate uranium isotopes. . . .

But having stated his conclusion at its most optimistic, Kurchatov then qualified it: "These unusual properties . . . are of course not yet proved in many respects. Their realization is only possible if it is true that eka-osmium-239 is analogous to U235 and also only if a 'uranium pile' may be built one way or another. . . . The scheme demands quantitative analysis of every detail." He had already assigned that work to Zeldovich, he wrote, but it would not be possible to study the properties of element 94 fully "earlier than mid-1944, when our cyclotrons will be restored and operating." He therefore asked Pervukhin to "request that the organs of intelligence find out what has been done in this direction in America" and appended a list of seven laboratories to be infiltrated, including the University of California's Berkeley Radiation Laboratory where he thought McMillan was working (in fact McMillan had gone off to MIT to work on radar), Yale University, the University of Michigan and Columbia. Kurchatov wanted to know if 94 fissioned under the action of fast or slow neutrons, what the relevant cross sections were and whether 94 fissioned spontaneously, all information of

---

* Radiochemists had coined these preliminary names for 93 and 94 in the 1930s because the anticipated new transuranic elements were assumed (incorrectly, as it turned out) to resemble the metals—rhenium, osmium and so on—that fell directly above their anticipated positions on the periodic table of the elements; "eka" means "beyond."—RR

great relevance to determining if 94—plutonium—could be used in a bomb.

Kurchatov had mentioned the new information about element 94 in the summary to his March 7 report and gave a glowing assessment there of the value to the Soviet Union of the espionage material:

CONCLUSION

The intelligence material . . . requires us to review many of our established opinions and introduces three directions of work new for Soviet physics:
1. Separation of U235 through [gaseous] diffusion.
2. Realization of nuclear burning in the mixture uranium–heavy water.
3. Exploration of the element eka-osmium 94^{239}.

In conclusion it is necessary to mention that the material as a whole shows that it is technically possible to solve the entire uranium problem in a much shorter period than our scientists believed before they were informed about developments in this field abroad.

Pervukhin, impressed, had underlined this last sentence.

"Don't tell anyone about this letter to you," Kurchatov closed his March 22 follow-up report cautiously. Long afterward, defending the contribution of Soviet scientists, Yuli Khariton would assert that "one should not overestimate the importance of [the] Soviet intelligence community in setting up the atomic program. . . ." Based on Kurchatov's own documented responses to the espionage material he reviewed, one should not *underestimate* the importance of espionage either. American scientists had been right to withhold their work on plutonium from publication; the possibility of transmuting U238 to a new fissionable element that could be separated chemically from uranium was the most important secret of the early years of the nuclear arms race. "The world learned about plutonium at Nagasaki," Glenn Seaborg remarks. Thanks to Klaus Fuchs, Soviet scientists learned about plutonium early in 1943.

In May, to facilitate and reward his work on the Anglo-American atomic bomb, Klaus Fuchs received the gift of British citizenship.

———

Between Igor Kurchatov's first two espionage reviews, on March 10, 1943, the Presidium of the Soviet Academy of Sciences confirmed his appointment as director of the Soviet atomic-bomb program. He was forty years old and not yet even a full member of the academy (that election came six months later, on September 29, 1943).

Wisely, Kurchatov did not abandon work on uranium isotope separation because espionage had revealed element 94 to be a possible alternative, any more than the United States had done. Pursuing multiple approaches to a bomb, however redundant and expensive, was the only way to hedge against failure in the days before it was certain which approach would work.

Kurchatov needed a home for his new secret laboratory. He was allowed to commandeer space temporarily in the old Seismological Institute about a kilometer southwest of the Kremlin within a meandering loop of the Moscow River, on Pyzhevski Lane in the Zamoskvorechie district, where Gogol, Tolstoy and Chekhov had lived in pre-revolutionary days. Kurchatov and his small staff, no more than twenty people, moved into the institute workshop—"a neat small three-story building surrounded by linden trees," Golovin and Smirnov describe it. Kurchatov named the operation the Laboratory for Thermal Engineering.

His staff grew slowly, paced by the lack of facilities and equipment, but people joined the adventure willingly from their far-flung assignments in the military or in industry, Golovin recalls:

> Most of the personnel came to Kurchatov with only the clothes on their backs and what had been thrown together in small suitcases. Their other belongings, including the books and manuscripts so important to scientists, would have been lost in evacuations or during air raids. Kurchatov's first concern was to feed and house the new arrivals. This was a great morale-builder for people who had suffered wartime privations.

"We used to lunch for coupons in the House of Scientists on Kropotkinskaya Street," Golovin writes. "We went to lunch with Kurchatov in a covered truck, the entire team. By the standards of that time these lunches were real feasts. We were very happy with the fresh salad which was grown near the House of Scientists during the summer instead of flowers."

Almost immediately they ran out of laboratory space and took over another evacuated building on Bolshaya Kaluzhskaya Street that belonged to the Institute for Inorganic Chemistry. "On Kaluzhskaya," says Golovin, "for the first time armed guards appeared at the entrances. . . ." Shortages caused delays that made everyone impatient. "We had to find our way through," Georgi Flerov recalled their embattled mood, "just like the soldiers fighting in the front lines. . . . We were poor at first; fortunately, we were authorized to scavenge voltmeters and other equipment from the army and the institutes of the Academy of Sciences. Sometimes, when we discussed what was most important . . . , it would seem that what was most important was whatever hadn't yet been done. And everything which had already been done might be spoiled if some minor thing went wrong."

While occupying these first, temporary facilities, Kurchatov began looking for permanent quarters. Kaftanov says his aide S. A. Balezin and Kurchatov "examined many buildings which had belonged to various institutes that had been evacuated from Moscow. . . . We were interested in an appropriate building in a suitable location which could be extended in the future—it was clear from the beginning that extension would be necessary." Pervukhin also sometimes accompanied Kurchatov on his real-estate rounds. "Igor Vasilievich and I examined the unfinished buildings of the All-Union Institute for Experimental Medicine in Pokrovskoye-Streshnevo," the exclusive suburb of Silver Woods in northwestern Moscow. "We decided to organize the main laboratory for nuclear physics in one of the buildings which already had a roof. . . . Kurchatov's laboratory took over the entire territory of the institute's compound." Five hectares of pine woods had been enclosed; a creek had been culverted underground. Golovin, who would serve as assistant director of the new laboratory, describes the setting:

> Kurchatov . . . decided on an unfinished three-story brick building beyond the belt-line railroad on the edge of a sprawling potato field a kilometer from the Moscow River. A few hundred meters from the building were two unfinished one-story stone cottages and a couple of warehouses, also roofless, and a half a kilometer farther off stood the two-story building of a small factory that made clinical X-ray machines. A pine grove, a few log cabins, and two railroad spurs across the field completed the picture. . . . Here on the edge of an area once called Khodynskoe Field, for many decades an artillery and machine-gun range, construction began on Laboratory No. 2 of the Academy of Sciences of the USSR. . . .

(In July 1943, when the Red Army pushed the Wehrmacht back across the Ukraine west to the Dnieper River, liberating Kharkov, Kirill Sinelnikov had immediately returned to his laboratories there and gone to work restoring the electrostatic generators so that he could begin measuring cross sections; that operation took the name Laboratory No. 1. Sinelnikov's English wife Eddie described the destruction her husband found in Kharkov in a December letter home: "Kira has been on an official visit to Kharkov and the Institute. He says in places you might think the streets had not been destroyed. The outline is the same, but when you walk down a street you find it is only the shell of the buildings that remain. . . . He found an absolutely empty and dirty flat. Our beautiful Steinway [piano] lying on the road near the garage having been used by the Germans as a platform for washing lorries. . . . We shall have to begin again from the beginning.")

That summer of 1943, the seventy-five-ton magnet of the Leningrad cyclotron arrived in Moscow (brought, remarkably, through the German blockade) and construction began in the basement of the laboratory at the Silver Woods site on what would be the most powerful particle accelerator in Europe—and a way of making a few micrograms of plutonium, work which Kurchatov assigned to his chemist brother Boris.

Sometime in spring 1943, Moscow Center passed Igor Kurchatov the first flood of espionage material from the United States. An unidentified person with access to the files of the reference committee of the Washington-based National Research Council that controlled publication of any research that had military significance—Morris Cohen's physicist friend, whose code name according to Yatzkov was "Perseus," or some other unidentified American spy—had either copied or summarized the contents of 286 classified scientific papers filed with the committee in lieu of publication.

Kurchatov completed his review of this extensive secret literature by early summer 1943. On July 3 he sent Pervukhin his analysis of 237 works he deemed relevant. Their content covered a full range, as a Russian historian notes:

> Of 237 analyzed works, twenty-nine were devoted to the separation of isotopes by [gaseous] diffusion, which Kurchatov believes to be the main method under development in the USA, eighteen to centrifuge separation, four to electromagnetic separation, six to thermal diffusion, five to general problems of isotope separation, ten to the design of a U235 bomb (his analysis of these works takes relatively much space). Thirty-two works concern uranium-heavy-water piles, twenty-nine concern uranium-graphite piles, fourteen concern transuranics (plutonium and neptunium), three concern the [rare] uranium isotopes U232 and U233, thirty concern general issues in the neutron physics of nuclear fission, fifty-five relate to the chemistry of uranium (production of metallic U, oxide, hexafluoride used for diffusion separation and other substances, including metallorganic substances with uranium) and three works concern the physiological action of uranium.

Kurchatov wrote in his July 3 report that the twenty-nine works that concerned the development of a uranium-fueled, graphite-moderated nuclear reactor constituted "the main results of American work on a uranium-graphite pile." By then he had decided that the Soviet Union should also take that route to achieving a controlled chain reaction in uranium; clearly his decision had been significantly influenced by the espionage information he had seen.

The twenty-nine summaries or abstracts gave "only a brief presentation of the general results of research," however, and did not include "important technical details" (which implies that the American espionage agent who assembled the collection was a secretary or a clerk operating within the National Research Council rather than a physicist, who would have realized that abstracts were inadequate). The fact that the summaries concerned technical problems such as "temperature of the walls of cooling tubes, diffusion of fission products in uranium under high temperatures, etc.," that were "characteristic of a technical project rather than an abstract physical scheme" gave further evidence of "the seriousness of the attempts the American scientists are making to realize a uranium-graphite pile in the nearest future." Kurchatov asked for more information: "It is extremely important to receive detailed technical material on this problem from America."

The reports on nuclear reactor development in the United States that Kurchatov was reviewing were a full year out of date. In fact, CP-1 had operated successfully at the University of Chicago beginning in December 1942;* its successor, CP-2, had been assembled and was operating with shielding at a site in the Argonne Forest outside Chicago; a vast tract of land had been purchased near Hanford, Washington, where industrial-scale reactors for plutonium production then being designed would be built; a one-thousand-kilowatt air-cooled reactor that would produce gram quantities of plutonium was under construction at Oak Ridge, Tennessee; and distilleries were going up in the United States and Canada that were planned to produce three tons of heavy water per month by October 1943 for a heavy-water-moderated reactor to be assembled at Argonne. Either the espionage documents in Kurchatov's hands had been passed at least a year before he saw them, which is unlikely (why would the NKVD have allowed him to review equally sensitive British files but withheld this cornucopia of American information from him during the previous several months when he and his team were formulating their first plans?); or they reflected some abrupt cutoff of information. A cutoff is probable. Morris Cohen had been drafted into the US Army in the summer of 1942, breaking the espionage connection with his contact or contacts until Yatzkov reestablished it with Lona Cohen several months later. Morris Cohen's contact may not have been able to pass the information he or she had collected until after reconnecting with Lona. Alternatively or additionally, whoever was supplying the Soviets with information may have lost access to the files.

Kurchatov next reviewed fourteen works that contained "detailed information on the physical properties of elements 93 [neptunium] and 94 [plutonium]." The cross section for fission of 94^{239} by slow neutrons, for example, had been reported in a classified May 29, 1941, paper by Berkeley physicists

---

* That Kurchatov was still ignorant in July 1943 of CP-1's successful operation is clear proof that NKVD officer Pavel Sudoplatov is lying when he claims he showed Kurchatov "a full report on the first nuclear chain reaction . . . in Chicago" in February. Kurchatov's ignorance also confirms that neither Fermi nor Bruno Pontecorvo had passed word of the pile's completion to the Soviets the previous January, as Sudoplatov further alleges (Pontecorvo in any case was working at that time for an oil drilling company in Oklahoma and had no access to US secret research). Cf. Sudoplatov and Sudoplatov (1994), p. 182.

and chemists J. W. Kennedy, Glenn Seaborg, Emilio Segrè and A. C. Wahl to be even larger than that of U235. Kurchatov wanted to see a further classified work by Seaborg and Segrè, for which he seems to have had at least a reference, "devoted to fission of 94—element eka-osmium—by fast neutrons." Kurchatov explained:

> In its response to the action of neutrons, this element is similar to U235, for which the action of fast neutrons hasn't yet been explored. Thus Seaborg's data on eka-osmium 94^{239} is of interest for the problem of realizing a U235 bomb. That's why we consider it especially important to receive the results of this work of Seaborg and Segrè.

The Soviet project director noted that all this work showed that the United States was involved in a major effort to build an atomic bomb. It also showed, he wrote, that Soviet research was being conducted "(although of course not in sufficient volume)" along the same lines as American research with two exceptions—a uranium-heavy-water pile and electromagnetic separation of uranium isotopes—for which work in the Soviet Union had not yet been started.

"I think that we ought to begin working in both of these directions," Kurchatov concluded, "and the first of them demands the most serious attention." A heavy-water pile would be more difficult to build than a graphite pile, he pointed out, because they would have to organize heavy-water production on a scale of tons. But it would resolve a serious problem: the Soviet project's lack of uranium. The realization of a heavy-water pile, Kurchatov revealed, "would demand not 50, but 1–2 tons of uranium, the amount we have at hand in 1943, while it remains unclear when our country will accumulate a stock of uranium of as much as 50 tons."

"It was . . . the results of earlier research by the Soviet scientists themselves," writes Yuli Khariton of himself and his colleagues, "that put them in a strong starting position when they embarked on a solution to the atomic problem." By the summer of 1943, an open flood of espionage from England and the United States had added significantly to that base. But even if the Soviets could have caught up with the Anglo-Americans technically, they could not yet have built a bomb; they lacked the necessary raw materials, and had not yet begun to develop the vast industrial enterprise they would need to process them.

4

A Russian Connection

LIKE MANY AMERICANS who spied for the Soviet Union, Harry Gold had a Russian connection. In 1881, following the murder of Czar Alexander II by revolutionaries and the subsequent repeal of reforms and outbreak of violent pogroms, a vast Jewish exodus began from Russia. Some 3.5 million Jews fled to the United States between 1882 and 1920. Harry Gold's parents, Sam and Celia Golodnitsky, left Russia around 1904. They stopped for a decade in Switzerland, where Sam found work as a cabinetmaker; Henrich —Harry—was born in Bern on December 12, 1910. The Golodnitskys continued on to America in 1914, were assigned the name Gold by an immigration officer on Ellis Island, landed temporarily with a relative in Little Rock, Arkansas, moved to Chicago and work in the stockyards and coalyards and finally settled in South Philadelphia in 1915. Harry's younger brother Yosef was born two years later.

South Philadelphia was a tough neighborhood. Harry Gold thought the "fertile soil" of his "earnest ... desire" to work with the Soviet Union lay there, in his early experience of anti-Semitism:

> When I was about twelve I made regular trips to the Public Library at Broad and Porter Streets, a distance of about two miles from my home. On returning from one such trip I was seized by a group [of] about 15 gentile boys at 12th and Shunk Streets and was badly beaten. . . .

Gangs of Neckers, kids who lived in the marshy Neck section of South Philadelphia near the city dump "under extremely primitive conditions and amid the mosquitoes and dirt," staged "brick throwing, window smashing, lightning forays" into Gold's neighborhood, "their special hatred ... directed at the Jews. . . ."

Gold's father, a hard and honest worker, was similarly harassed at the

Victor Talking Machine Company where he was employed sanding radio cabinets. Immigrant workers who "were crudely anti-Semitic . . . made Pop, one of the few Jewish workers, the object of their 'humor.'. . ." An Irish foreman in particular "who hated the Jews far more bitterly than anyone Pop had ever encountered" assigned Sam to sand alone on a fast production line:

> So Sam Gold would come home at night with his fingertips raw and with the skin partially rubbed off. This was no exaggeration. Mom would bathe the fingers and put ointment on them and Pop would go back to work the next morning. But he never quit, not Pop, and he never uttered one word of complaint to us boys.

"Many other such incidents could be described," Harry Gold summarizes. "This was a scheme to which I built up a tremendous resentment throughout the years and [a] desire to do something active to fight and to combat it. Something on a much wider scale than by combat of an individual anti-Semitic."

If Gold's resentment of anti-Semitism descended in part through his father, whom he idolized, his interest in socialism developed through his mother. Celia Gold was "fascinated" with Socialist Party founder and presidential candidate Eugene V. Debs, her son would testify. The Golds subscribed to the *Jewish Daily Forward,* which "also espoused the theory of Socialism." Beginning in high school, Harry "became a great admirer of Norman Thomas and thought him a very great man indeed." By contrast, for Harry at that time "Bolshevism or Communism was just a name for a wild and vaguely defined phenomenon going on in a primitive country thousands of miles away. . . . 'A Communist'—I was horrified."

A good student in high school, a member of the Latin and Science clubs, Harry went to work for the Pennsylvania Sugar Company after graduation and saved money for college. He started at the University of Pennsylvania in 1930 and managed to get through two years before his money ran out. Pennsylvania Sugar rehired him; a decade later he would tell his draft board that he began contributing to his mother's support—to family expenses, that is—in March 1932, the nadir of the Great Depression. One week before Christmas 1932 he was laid off. "Here . . . was [raised] the disgraceful spectacle and deep ignominy of charity. The first thing that followed my discharge was the necessity of returning a parlor suite (the first in 14 years) to Lit Brothers—that $50.00 refund was so necessary and loomed so large." His mother was opposed to charity, "violently so." (Celia Gold, according to one of Gold's later employers, was "somewhat of a tyrant in that she ruled the Gold household with an iron hand.") Gold looked for work "frantically for five weeks." That was when he met Tom Black, the Party recruiter, who was a fellow chemist.

A friend called Gold and told him about a job that Black had just vacated with a soap manufacturer in Jersey City. Gold needed to see Black that night for a briefing and a recommendation. "Mom hurriedly and anxiously packed a brown cardboard suitcase and I borrowed $6.00 ... as well as a jacket which closely matched my pants, and I was bundled on a Greyhound bus to Jersey City." Gold slogged through the snow to the address he was given. "Black was waiting for me downstairs. I can still see that huge, friendly, freckled face, the grin and the feel of the bear-like grip of his hand." They stayed up all night; Black briefed Gold on soap chemistry and told Gold frankly that he hoped to recruit him for the Communist Party. Most of the five hours the two men talked Black spent attacking capitalism and selling Communism. Gold got the job; the thirty dollars a week it paid kept his family off relief.

Black took Gold to Party meetings in Jersey City, gatherings of misfits where Gold felt "nothing was ever accomplished." At more sophisticated meetings in Greenwich Village they drank wine and ate spaghetti and oysters while their host read "incredibly funny" Thurber stories from *The New Yorker*. One night someone attacked the decadence of bourgeois family life and Gold erupted. "To me this was the worst sort of heresy and I hotly defended the concept of the happy and closely knit group of parents and children." Like his own; he was returning to Philadelphia the next day to begin working at the sugar company again, happy to be released from his obligation to Tom Black, happy to be home.

By then it was September 1933; that winter Gold started night school at the Drexel Institute of Technology, studying chemical engineering. But Black continued to come around. "My family was naturally very glad to greet the man who, in effect, had been our economic savior, and Tom with his bluff and hearty ways quickly endeared himself to them." The big chemist, "with his build and features a two-hundred-year throwback to those of a British peasant," began to propagandize Gold's parents "but then suddenly stopped." He also stopped urging Gold to join the Party. Gold understood later that the change in Black's behavior was significant. In fact Black had gone boldly to Amtorg, the Soviet trading company in New York, and volunteered to work as a scientist in the USSR. The Soviets had proposed instead that he undertake industrial espionage; for good measure they had assigned him undercover duty keeping track of Trotskyites.

Eventually, at the turn of the year between 1934 and 1935, Black told Harry about his new work and asked Gold to sign on. Gold was ready. "I said that I would think it over, but actually I had already made the decision. ... I was even to a certain extent eager to." Long afterward, reviewing his life, Gold would carefully, layer by layer, examine his reasons for becoming a Soviet agent: that he owed Black, that he genuinely wanted to help the people of the Soviet Union "to enjoy some of the better things of life," that acquiescing got Black off his neck. "But these were really surface circumstances." There were also underlying motives, "far more powerful" that he "did not realize at the time." Fighting anti-Semitism in all its disguises, including "Nazism and Fascism," was one of them. "It might be asked, why didn't I try to fight anti-Semitism here in the United States? Frankly, this seemed to me like a pretty hopeless business."

His "almost suicidal impulse to take drastic, and if need be, illegal action, when [he] believed a situation required it," was another motive he discerned. He also suspected that "there must have been in my makeup a certain basic lack of faith in democratic processes. . . . Unswervingly through all these years of work with the Russian agents I thought of myself as an American citizen working, outside the law, and underhandedly it is true, for the Soviet Union. . . . If I had thought that my actions might in any way harm the United States I would never have gone ahead." Gold understood later how absurd that rationalization must sound: "Here I was unwittingly fooling myself." He understood as well that he was "letting down the strong barriers against deceit, trickery and thieving, barriers which had been built up by my mother over so many years." Indeed, he was explicitly deceiving his mother with "the lies I had to tell at home and to my friends to explain my supposed whereabouts (Mom was certain that I was carrying on a series of clandestine love affairs)." Espionage, Gold's secret life, *was* a love affair of sorts for a shy, lonely, workaholic bachelor who lived at home with his parents and his bachelor brother, under his mother's thumb:

> The planning for a meeting with the Soviet agent; the careful preparations for obtaining data . . . ; the writing of reports; the filching of blueprints for copying and then returning them; the meeting with [agents] in New York or Cincinnati or Buffalo . . . ; the difficulty in raising money for the various trips; the weary hours of waiting on street corners in strange towns where I had no business to be and the killing of time in cheap movies . . . —all this became quite ingrained in me. It was drudgery, and I hated it; anyone who had an idea this work was glamorous and exciting was very wrong indeed—nothing could have been more dreary. But here is one curious fact:
>
> [After the war] when . . . my activity ceased, after a while I actually began to miss it. . . . Once, I discussed this with Black and he said that it was really a mistake that I had gotten into espionage work. . . . "But you know, Tom," I said, "in some funny manner I still long for that life which now seems dead. . . ." And Black replied, "It is peculiar, I do too, even though it has caused me so much grief and disaster in the last 14 years."

For most of 1935, Black served as Gold's contact. Harry was supposed to steal chemical processes from his boss at Pennsylvania Sugar, but neither he nor Black could afford the cost of photographing the documents. In those early years Gold struggled to find funds to pay for his espionage. Considering what they got in return, the Soviets were surprisingly stingy. Eventually Black convinced Amtorg to handle photocopying. All Harry had to do was to deliver the material to New York and return it. "Best of all, the man who was providing all of this service, a Russian engineer from Amtorg, was very anxious to meet me." In November 1935, young Harry Gold met "Paul."

Harry was twenty-five years old and evidently dazzled by his entrance into the underworld of international espionage, but he was nothing if not independent, and from the beginning he had doubts. The doubt that bothered him most of all was professional:

> It had to do with the Soviets' seeming lack of initiative in chemical engineering research, and [their] utter horror of any pioneering efforts in that field.
>
> From the very first, in 1935, Paul instructed me that what was wanted were processes already in successful operation in the United States; and Paul, and the others who followed him, continually said that they not only preferred, but absolutely insisted upon, only having the details of a plant already in successful and proven operation in America as compared to another which, though it might promise to be very superior, still was only in the experimental stage. On several occasions, when I made efforts to submit material which represented work not yet in full-scale production, I would have my knuckles smartly rapped. So, I desisted; but I wondered.
>
> When there is added to this their absolute veneration of American technological skill, I wondered again. . . . But I was told that the Soviet Union was so desperately in need of chemical processes that they could afford to take no chances on one which might not work. . . .

Soviet agents wanted only conservative, reliable, tested technology from America. Their bosses were managers, not engineers or scientists, and had no way to evaluate untried ideas. The agents knew the penalty for taking risks when mistakes counted as heinous crimes.

Gold worked with Paul, who had silver-blond hair and might have been Danish; with a huge man "with a heavyweight boxer's build"; with "a small, dark man with a mustache [who] was a fanatical martinet" and whom Harry hated. Amtorg found Harry reliable and soon shifted him to more responsible duty as a courier—a cut-out, a go-between whose knowledge of the chain of espionage agents was limited to his information sources on the one side and his immediate Soviet superior on the other, reducing the network's exposure. To increase his competence the Soviets paid his way to Xavier College in Cincinnati, where he graduated summa cum laude with a degree in chemistry in 1940. They kept Gold on a short leash, however; he had to earn his Amtorg scholarship by trying to bribe and then to blackmail a

Wright Field aeronautical engineer named Ben Smilg, but Smilg successfully resisted both attempts.

In 1940, Gold registered for the draft. The draft examiner found him at twenty-nine years of age to have brown hair, hazel eyes and a "brown complexion." He was short and broad: five feet, six inches tall and 180 pounds. He had developed what he called "a fabulous appetite" as a child at camp. It had "stayed with me yet," he would write proudly, quoting a friend of his who once said, "Harry will eat anything which will stand still long enough [and which] won't eat him first." He had a broad, Slavic face, heavily jowled; he looked older than his years, and he turned out to be hypertensive; his draft board classified him 4-F and exempted him from military service. People who worked with him found him pleasant—"a hard worker," a fellow chemist said, "conscientious, and a sincere individual." One woman at work remembered him as "nervous," around women at least. "When he talked," she would testify, "especially to a woman, his face would become flushed. . . . He was a quiet individual who would sometimes converse a bit with a man, but would only talk to a woman when he had a job for them to do."

Gold got a new control in the fall of 1940, his favorite, Semen N. Semenov, an MIT graduate engineer, a man he would know only as "Sam." Sam, Gold says, "had a swarthy complexion, almost Mexican-like in texture, black dancing eyes, and a really warm and friendly smile." Sam was the only Soviet whom Gold ever met who might have passed for an American because of the way he spoke, dressed and acted, "and especially in the way in which he wore his hat. For some reason foreigners never wear their hats as Americans do. . . ." The MIT engineer was contemptuous of paid agents; Harry never gave offense in that regard, asking for no reward, but beginning with Sam the Soviets fully reimbursed Gold's travel expenses, so that across his years of devoted espionage he at least broke even.

Sam sent Gold off abruptly to Buffalo to rendezvous with a man named Al Slack, who worked for Eastman Kodak and was passing Amtorg information on Kodachrome film manufacture. The work was routine and not especially productive; in the spring of 1941 Sam told Harry "I was not needed anymore."

But when Germany invaded the Soviet Union that June, Amtorg's priorities changed. In autumn 1941, "Sam called me up, I met him, and he told me that we had to begin an intensive campaign for obtaining information for the Soviet Union." Gold made a series of runs up to Syracuse, Rochester and Buffalo, collecting more material from Al Slack and three other men.

With Gold's next assignment, routine espionage descended to soap opera. An exasperated Soviet operative named Jacob Golos had turned over to Sam a difficult, mercurial chemist named Abraham Brothman, a Columbia University graduate with a penchant for missing meetings and making promises he failed to keep. Brothman's previous contact had been Elizabeth

Terrill Bentley, Golos's mistress, a Vassar graduate whose contacts expanded during the war years under Golos's direction to include Communists in Canada, in the US government in Washington and in the ranks of industrial espionage. Bentley was a specialist in Italian literature, not chemistry, and the Soviets had decided Brothman needed a contact with a technical background.

Sam told Harry that Brothman was "an important government official, an engineer." After several postponements, the two chemists connected. Gold slipped into Brothman's car in the Manhattan garment district on a Monday night in September 1941. While Gold was identifying himself, the Joe Louis–Lou Nova fight came on Brothman's car radio and they listened to it together in silence for the two or three rounds Louis needed to knock Nova out. In the car and later that night in a Bickford restaurant they talked for three hours.

Then began Gold's Sisyphean labor of trying to coax useful information out of Brothman, who was not a government official but worked for private industry under government contracts. The Columbia chemist seems to have been manipulating the Soviets in the hope that they would eventually set him up in business, Gold thought:

Starting in early 1942, . . . Brothman, on many occasions, I would say at least six, openly and directly asked me if I could obtain legitimate backing from the Soviet Union so that he could set up an enterprise and do work on chemical processes for the Soviets. When I first mentioned this to Sam, he laughed hilariously and said that he had never heard of such damned fool nonsense in his life. . . . By legitimate backing, Brothman meant sums ranging from $25,000 to $50,000.

In the meantime Brothman gave the Soviets only enough information to string them along, and Gold bore the brunt of Sam's frustration. When Harry finally confronted Brothman, the man counterattacked and called the Soviets "a bunch of fools." He told Harry he had already given them, in Gold's words, "a drawing of a turbine type of engine for aircraft, and also information on one of the earliest jeep models which had been designed by him." He promised to deliver the complete design of an explosives plant.

Brothman soon reneged on the explosives-plant design, but promised something far more desirable. "He told me he was in possession of complete information on the manufacture of Buna-S, a synthetic rubber. He also told me that not only was he in possession of complete information, but that he had the complete design material [for a synthetic rubber plant]. . . . When I told Sam about this, he was highly elated." Buna-S was one of the items on Semenov's wish list. Gold arranged to meet with Brothman on New Year's Day 1942 to pick up the Buna-S plans. Brothman came downstairs from his office two hours late, empty-handed. "I remember this occasion very clearly and distinctly," Gold testified bitterly, "because it was a cold morning and I waited outside the Exchange Bar, which unfortunately was closed, on New Year's morning."

Once during this ongoing negotiation Sam exploded at Brothman's callous disregard of Gold's misery:

> He said, "Look here, you fool, this scoundrel will not have the information on Sunday. He won't have it next Sunday or the Sunday after that. I bet you that it will be a month or two months before you will get it; then I doubt that it will be complete. He doesn't have it complete now; he doesn't have half of it complete; maybe it isn't even started on yet." . . .
>
> Then [Sam] became so enraged, actually not at me but at Brothman, that he was almost beside himself and actually stopped talking from the force of his anger. After he cooled down, he said, "Look, we are going to have a couple of double Scotches, and you are going to have something to eat. We will sit there and will talk of music and we will talk of opera, and we will not talk of that son-of-a-bitch Brothman."

But eventually Brothman came through. On a rainy evening in March 1942, he passed to Harry Gold a complete report on the manufacture of Buna-S synthetic rubber, including blueprints for a plant—several hundred single-spaced typewritten pages and a dozen blueprints. In April, Sam told Gold to congratulate and praise Brothman "because . . . the information he had turned over . . . had been received in the Soviet Union and had been hailed as a remarkable, extremely valuable piece of work. . . . The Soviets were immediately beginning to set up a plant for the manufacture of Buna-S."

Al Slack came through as well, though not before an accumulation of disappointments nearly led Gold to quit:

> Once, in the fall of 1942, I did waver. Things were going very badly. I had lost contact with Al Slack . . . and things were going very poorly with Brothman . . . and the whole business seemed futile. Also, at this time my increased absences from home had depressed my mother very much, and I was greatly concerned. To top it off, on that very evening in New York, the usually ebullient [Sam] had been very subdued regarding some failures of his own, and so, after I left him and went to Penn Station I came to the determination to be through with this work once and for all; I felt that I had done enough. I had some fifteen minutes for my train to Philadelphia and sat down in the smoking room of the station. Thereupon, I was approached by a swaying drunk who proceeded to vilify me as a "kike," a "sheeny bastard" and a "yellow draft dodger and money grabber" plus a series of far more horrible epithets.

Gold walked away. "But as I did so, so went my resolution to quit espionage work. It seemed all the more necessary to work with the utmost vigor, to fight any discouragement and to do everything possible to strengthen the Soviet Union, so that such incidents could not occur. To fight anti-Semitism here seemed so hopeless."

Gold reconnected with Slack by going to Kingsport, Tennessee, where Slack had moved, and looking him up in the phone book. Slack delivered information that autumn 1942 on the superior new high explosive RDX, including two one-pound rubber containers of the material itself in what Slack assured Gold was a nonexplosive form. Gold hoped so; "just before I turned the RDX over to [Sam], I had been narrowly missed by a speeding cab while crossing Sixth Avenue in New York, near the Gimbel liquor store." Subsequently Slack was transferred to Oak Ridge, Tennessee, where the Kodak subsidiary Tennessee Eastman had contracted to operate the electromagnetic isotope separation plant that the Army was building there to process uranium for the atomic bomb; at that point the Soviets severed him from Gold, telling Harry to forget about him.

At the end of 1942, Sam directed Gold to set up an elaborate charade for Abe Brothman's benefit. "The purpose of this meeting had been carefully discussed with Sam before I suggested it to Brothman, and was essentially to be in the nature of a pep talk. . . . I was to represent Sam as a visiting Soviet dignitary. . . . The whole idea of the meeting was to 'butter up' Brothman so that he would work on processes in which we were interested. . . . Brothman readily agreed to this meeting."

The conspirators assembled around nine o'clock one midwinter evening in a room Gold had rented at the Lincoln Hotel in Manhattan. "Sam was extremely genial and expansive during this meeting. . . . He called up and had some wine and some sandwiches sent up. We then proceeded to talk until one, possibly two o'clock in the morning." Sam praised Brothman at length. He also brought up a subject that Gold had not heard him mention before:

A good deal of conversation [concerned] mathematics and the application of mathematics to practical problems of engineering. . . . Sam very gently and extremely diplomatically hinted to Brothman . . . that Brothman should try to get work in fields . . . relating to military endeavor, or military equipment. . . . I believe . . . that here may have come the first hint . . . of the interest of the Soviets in Atomic Energy* . . . and also there may have been some

---

* In 1965 Gold would remember an even earlier incident: "One evening in New York City, about October–November 1942, Semenov asked me if I had heard anything of a military weapon involving a 'pressure wave' of hitherto unknown power. I was puzzled. Pressure wave? (I had a mental picture of some kind of advancing front, [such] as a storm formation.) So Semenov asked me to watch the technical literature very closely and also to see if any even small bit of information was let drop at scientific meetings or by one of my professional acquaintances." Gold (1965b), p. 47.—RR

conversation relating to Brothman's acquaintance with Dr. Harold Urey at Columbia University. I believe that here Brothman stated to Sam that he was a former pupil of Dr. Urey's. . . . I am emphasizing this because at this time, I had no idea that anything was going on in regard to Atomic Energy in the United States.

Harold Urey was a specialist in isotope separation who won the 1934 Nobel Prize in Chemistry for first isolating deuterium. At the end of 1942 he was a member of the government S-1 Committee that oversaw the Manhattan Project and was directing research at Columbia on gaseous diffusion. The Columbia team had just developed a workable barrier material made of compressed nickel powder.

Then it was Harry Gold's turn to be buttered up. Harry met with Sam on schedule in Manhattan in November 1943 and Sam told him they would conduct no business that evening; instead they were going to celebrate. The two men went to the bar in the Park Central Hotel. They took a table, as they always did in bars.

From time to time in the years of their relationship Sam had fretted that Harry's demanding courier work made it impossible for him to lead an ordinary life. "His greatest concern seemed to be over the fact that I had no wife and family of my own," Gold would write. " 'I realize that it is because of this work,' he said. 'But it's not natural or good. You are not ascetic and you have normal instincts and desires. We must find some solution to this problem. Obviously you cannot take on the responsibilities of marriage and still do this work (and do not think that our people fail to realize the sacrifice you are making).' " But the only solution Sam could propose was a fantasy:

And, Sam would continue: "The obtaining of information in this under-handed way will not always be necessary. You'll see. After the war is over there will come a great period of cooperation between all nations and people will be able to travel freely. . . . You will openly come to Moscow and will meet all of your old friends again—They will be so glad to see you—and we'll have a wonderful party and I'll show you all around the town. Oh, we'll have a great time."

Gold thought Sam was sincere, but he was never sure. "I am puzzled," he wrote in 1951, "even now, as to whether this was all part of a gigantic confidence scheme. . . . I just don't know." He knew that evening in the Park Central bar that there were "ulterior motives involved" in what Sam did next, that it was "carefully planned and staged," but he also thought it contained "the element of a genuine reward for work well done. . . ."

When they were comfortable together at their table, Sam announced that because of his outstanding work, Harry had been awarded the Order of the Red Star. The Soviet agent showed Harry the citation, "an affair in a rather gaudy red color," Harry recalled later, "and with a large seal." Sam apologized that he could not actually present Harry with the citation or the medal; security considerations obviously made that impossible. He told Harry about the privileges that came with the award. The privilege that amused Harry and stuck in his mind was "free trolley rides in the city of Moscow." But Gold was proud of the honor; he even told Tom Black and Abe Brothman about it.*

Gold soon learned why he had been singled out for special honor. At a meeting a month or two later, in December 1943 or January 1944, "I was told by Sam that there was an extremely important mission coming up for me and that before he could tell me about [it] he wanted to know, would I undertake it. I unhesitatingly agreed." It was to be "work of so critical a nature that I was to think twice and three times before I ever spoke a word concerning it to anyone, or before I made a move. . . ." Sam told Gold to drop completely his association with Abe Brothman and never to see him again. Following standard espionage practice, Semenov was disconnecting Gold to avoid cross-linking contacts, which could compromise them if either was exposed.

"Sam then told me that the mission was far more important than anything that I had ever done before, and concerned matters of not only immediate necessity but of world-shaking importance." The Soviet agent "didn't elaborate on what the nature of the work actually was" but simply gave Gold the details of an arrangement to meet a man.

Gold could not remember afterward if Sam told him the man's name. "In any event," he testified, he and his new contact met for the first time ". . . in, I believe, late February or early March of 1944 [at the Henry Street Settlement on the East Side of New York]. I introduced myself to him as Raymond. He never used the name. He knew it was a phony. He introduced himself to me as Klaus Fuchs."

Klaus Fuchs had come to America.

---

*To Gold's disgust, Brothman later bragged to his friends that it was *he* who had received a Red Star.

5

'Super Lend-Lease'

GREAT FALLS, MONTANA, is located about two hundred miles due north of Yellowstone National Park at the confluence of the Sun and Missouri Rivers. Gore Field, its airport during the Second World War, extended its ten-thousand-foot runway on a mesa of tableland three hundred feet above the city at 3,674 feet elevation. Montana weather in the winter is extremely cold and dry, and as a result Gore Field offered more than three hundred clear flying days a year.

In 1942, when German submarines made Allied efforts to ship aircraft to the Soviet Union through the North Sea hazardous and windblown sand damaged aircraft flown to Soviet Georgia across Africa, the United States proposed and the Soviet Union reluctantly agreed to open a trans-Alaskan route across Siberia. The staging point within the US for this air ferry route —the Alsib Pipeline, it came to be called—would be Gore Field in Great Falls.

The pipeline was one conduit of the program of Lend-Lease that Franklin Roosevelt proposed in January 1941 to help cash-strapped Britain and other allies defend themselves against Germany while the US maintained at least the appearance of neutrality. Roosevelt's proposal frightened isolationist senators such as Republican Arthur Vandenberg of Michigan, who correctly foresaw that it would carry the United States a long stride closer to war; when the Lend-Lease bill passed the Senate in March, Vandenberg wrote in his diary:

> I believe we have *promised* not only Britain but every other nation (including Russia) that joins Britain in this battle that *we* will see them through. I fear this means that we must actively engage in the war ourselves. I am *sure* it means *billions upon billions* added to the American public debt. . . . I do not believe we are rich enough to underwrite all the wars of the world.

In the course of the war, under the Lend-Lease Act, the United States delivered some $46 billion worth of equipment, supplies and services to Britain, China and other allies—the preponderance of it by sea, but the most urgent of it by air. Twenty-five percent of that total, $11 billion, went to the Soviet Union after the German invasion of the USSR, of which $1.5 billion paid for services. Of the remaining $9.5 billion, munitions accounted for about half the value of Soviet shipments, including thousands of B-25 bombers and other aircraft, more than 400,000 trucks ("Just imagine," Nikita Khrushchev would say later, "how we would have advanced from Stalingrad to Berlin without [American trucks]"), $814 million worth of ordnance and ammunition, thousands of tanks, a merchant fleet and 581 naval vessels. The other half, nonmunitions, included thirteen million pairs of winter boots, five million tons of food, two thousand locomotives, eleven thousand boxcars, 540,000 tons of rails and $111 million worth of petroleum products. Nonmunitions also, pointedly, included entire factories: "complete alcohol, synthetic rubber, and petroleum cracking plants," in the words of a postwar congressional report, "together with the requisite engineering drawings, operating and maintenance manuals, spare parts lists, and other pertinent documents." Harry Gold's collections from Abe Brothman gave the Soviet Union an early start; but by 1943 the United States was supplying directly the plans for synthetic rubber and other factories that Gold had shivered in the cold in 1942 to accumulate by espionage.

None of this largesse was contraband. It was tangible support. Until the Anglo-American invasion of Normandy on June 6, 1944, the Soviet Union fought Germany essentially alone on the European continent except for the Anglo-American strategic bombing campaign; had the USSR lost that fight, hundreds of German divisions bulwarked with Soviet resources would have been freed to turn west and challenge Britain and the United States. Averell Harriman, back from a mission to Moscow for Franklin Roosevelt in October 1941, made the point in a radio speech to the American people; "to put it bluntly," he said, "whatever it costs to keep this war away from our shores, that will be a small price to pay." The United States agreed to furnish Lend-Lease and the Soviets did not doubt that they had earned it—at Leningrad, at Stalingrad, in the monstrous enclosures in the western USSR where the Germans, as they advanced, confined Soviet prisoners of war completely exposed without water or food. At least 4.5 million Soviet civilians and combatants had been killed by 1943; at least three million combatants died in enclosures and camps throughout the war; at least 25 million Soviet civilians and combatants died before the eventual Allied victory. From the Soviet point of view, Lend-Lease was the least America could do when the Russian people were dying; anything the Soviets could grab, legally or illegally, must still have seemed less than a fair exchange. "We've lost millions of people," a Russian told Alexander Werth after the US ambassador, Admiral William H. Standley, complained at a Moscow press conference in March

1943 of the "ungracious" Soviet attitude toward Lend-Lease, "and they want us to crawl on our knees because they send us Spam." The point was to win the war. "One can bear anything," novelist and journalist Ilya Ehrenburg incited the men and women of the Red Army in August 1942: "the plague, and hunger and death. But one cannot bear the Germans. . . . Today there are no books; today there are no stars in the sky; today there is only one thought: Kill the Germans. Kill them all and dig them into the earth. Then we can go to sleep. Then we can think again of life, and books, and girls, and happiness."

But more than Lend-Lease aircraft loaded with urgent supplies staged from Gore Field. If the cold, windswept airport high and flat under the vast Montana sky was a pipeline for war matériel, it was also a tunnel under the border that directly connected the US to the USSR.

The American in charge of expediting deliveries through the Gore Field end of the Alsib Pipeline was a tall, rugged USAAF officer named George Racey Jordan who had served with Eddie Rickenbacker's First Pursuit Group during the First World War. Jordan, an older officer who was a businessman in peacetime, had begun working with the pipeline when it was based at Newark Airport and had learned there firsthand that it was sacrosanct. A taxiing American Airlines DC-3 had bumped a medium bomber consigned to the Soviets, a minor mishap in Racey Jordan's book. The Soviet head of mission, Colonel Anatoli N. Kotikov, taking offense, had called someone in Washington, and shortly afterward the Civil Aeronautics Board had suspended *all* civilian traffic through Newark, rerouting it to La Guardia across the Hudson in Queens. Jordan understood that Kotikov had a direct line to Harry Hopkins, the first administrator of the Lend-Lease Act and Roosevelt's personal emissary to the Soviet government.

Jordan got to know Kotikov better after the pipeline moved to Great Falls in November 1942. Kotikov was a Soviet hero, Jordan records, who "made the first seaplane flight from Moscow to Seattle along the Polar cap; Soviet newspapers of that time called him 'the Russian Lindbergh.'" Jordan liked him and the two officers worked well together. Kotikov noticed that Jordan was outranked by many of the other American officers assigned at Great Falls and arranged to improve his standing. "Capt. Jordan work any day here is always with the same people," Kotikov wrote Jordan's superior in his newly acquired English: ". . . Major Boaz . . . Major Lawrence . . . Major Taylor . . . Major O'Neill. . . . He is much hindered in his good work by under rank with these officers who he asks for things all time. I ask you to recommend him for equal rank to help Russian movement here." Jordan was promoted promptly from captain to major; at his promotion ceremony, Kotikov pinned on his new gold oak leaves.

But soon what Jordan called "the black suitcases" began to arrive, "the unusual number of black patent-leather suitcases, bound with white window-sash cord and sealed with red wax, which were coming through on the route to Moscow." They raised Jordan's suspicions. The first six, in charge of a Russian officer, Jordan passed as personal luggage. "But the units mounted to ten, twenty and thirty and at last to standard batches of fifty, which weighed almost two tons and consumed the cargo allotment of an entire plane. The officers were replaced by armed couriers, traveling in pairs, and the excuse for avoiding inspection was changed from 'personal luggage' to 'diplomatic immunity.'"

Jordan remonstrated with Kotikov that the black suitcases were not coming from the Soviet Embassy but from the Soviet Purchasing Commission in Washington. "Highest diplomatic character," Kotikov insisted. "I am sure he knew," Jordan writes, "that one of these days I would try to search the containers."

One afternoon in March 1943, Kotikov flashed a brace of vodka bottles at Jordan and invited the American officer to dine with the Soviet contingent at a restaurant in Great Falls. Since the Soviets always dined separately and seldom picked up bar tabs, Jordan assumed that free vodka and a dinner invitation meant they wanted something from him. He took the precaution of arranging to travel to the dinner in his own staff car. Before he left, he asked his maintenance chief if the Soviets were planning any flights that night. The maintenance chief "answered yes, they had a C-47 staged on the line, preparing to go." American pilots flew all Lend-Lease aircraft as far as Fairbanks, Alaska, where Soviet pilots took over for the trans-Siberian leg of the route; Jordan had authority to ground any plane at any time. He left word with the tower that no cargo plane should be cleared for the Soviet Union without his approval.

At the Carolina Pines restaurant in Great Falls, the five Soviets on hand for the occasion plied Jordan with vodka. They first toasted Stalin, then Red Army Air Forces commander-in-chief Novikov, then a Soviet ace named Pokryshkin with forty-eight German aircraft to his credit. Jordan proposed Franklin Roosevelt and then USAAF commander Hap Arnold. Thus warmed, the group sat down to dine.

Before Jordan had finished eating, a call came from the Gore Field tower. Jordan took the call at a pay phone downstairs from the second-floor restaurant. The C-47, reported the tower, was demanding clearance. The American officer threw on his coat and never looked back. It was twenty below zero outside. Jordan's driver raced the four miles to the field:

As we neared the Lend-Lease plane there loomed up, in its open door, the figure of a burly, barrel-chested Russian. . . . I clambered up and he tried to stop me by pushing hard with his stomach. I pushed back, ducked under his arm, and stood inside the cabin.

It was dimly lighted by a solitary electric bulb in the dome. Faintly visible was an expanse of black suitcases, with white ropes and seals of crimson wax. . . .

It had been no more than a guess that a fresh installment of suitcases might be due. My first thought was: "Another bunch of those damn things!" The second was that if I was ever going to open them up, now was as good a time as any.

The Soviet courier resisted. Jordan called in an armed GI. One of the couriers jumped off the plane and ran for a telephone.

Jordan carried a razorblade-loaded packing knife in his pocket. In the dim light of the C-47's cargo hold, he began cutting ropes and prying open suitcases, making notes on the backs of two envelopes of what he discovered. "Always just 50 black suitcases each load with 2 or 3 Couriers—usually 3 weeks apart," he noted to remind himself. In the suitcases he found tables listing railroad mileages between American cities, a load of road maps marked with American industrial plants, a full load of documents from Amtorg, a collection of Panama Canal maps, folders of naval and shipping intelligence, hundreds of commercial catalogues and scientific journals. Folders from the State Department included one labeled "From Hiss." "I had never heard of Alger Hiss," Jordan wrote in 1952, after Whittaker Chambers had accused the former special assistant to the US Secretary of State of spying for the Soviets and Hiss had been convicted of perjury and was serving a prison term, "and made the entry because the folder bearing his name happened to be second in the pile. It contained hundreds of photostats of what seemed to be military reports."*

Jordan continued opening black suitcases while his hands went numb with cold. He found voluminous copies of secret reports sent back to the State Department from American attachés in Moscow. He found other State Department documents with their edges trimmed, either to conserve space or, he suspected, to cut away classification stamps. He found a large map which bore a legend he recorded as "Oak Ridge—Manhattan Engineer Dept. or District I think it was," a place he had never heard of before. He wrote down words he did not recognize from other documents he skimmed: "Uranium 92—neutron—proton and deuteron—isotope—energy produced by fission or splitting—look up cyclotron. . . . Heavy-water hydrogen or deuterons."

It was eleven o'clock before Colonel Kotikov arrived; by then Jordan was nearly finished. He opened a few more suitcases in Kotikov's presence to

---

* Robert Lamphere, the FBI agent who pursued Soviet espionage during and after the Second World War, is "dubious" that Hiss's name appeared on the files Jordan saw. It would have been a remarkable coincidence. Robert Lamphere, personal communication, vi.94.

underscore his authority and cleared the C-47 for departure. He fully expected to be transferred even farther out into the boondocks for his temerity in bucking the Soviets. But the suitcases were on their way to Moscow and apparently Kotikov chose not to lodge a complaint.

In later shipments of black suitcases Jordan claimed he found blueprints of American factories, including the General Electric plant in Lynn, Massachusetts, where aircraft turbochargers were manufactured and the Electric Boat Company of Groton, Connecticut, which built submarines. Entire planeloads of copies of US patents went through Great Falls. A congressional committee determined after the war that the number of patents the Soviets thus legally acquired "runs into the hundreds of thousands."

"Another 'diplomatic' cargo which arrived at Great Falls," Jordan discovered, "was a planeload of films. . . . A letter from the State Department [authorized the Soviets] to visit any restricted plant, and to make motion pictures of intricate machinery and manufacturing processes. I looked over a half dozen of the hundreds of cans of films. That one plane carried a tremendous amount of America's technical know-how to Russia."

During his two years with the Alsib Pipeline, Jordan observed other Soviet Lend-Lease operations as well:

I began to realize an important fact: while we were a pipeline to Russia, Russia was also a pipeline to us. . . . The entry of Soviet personnel into the United States was completely uncontrolled. Planes were arriving regularly from Moscow with unidentified Russians aboard. I would see them jump off planes, hop over fences, and run for taxicabs. They seemed to know in advance exactly where they were headed, and how to get there.

From the beginning Jordan kept a record of every Soviet with whom he came in contact during the war, including in particular those who passed through Gore Field; by the end of the war he had a list of 418 names.

Jordan acquired copies of the Soviets' own itemized lists of Lend-Lease shipments and confirmed what he had recorded at Gore Field: that the Roosevelt administration shipped quantities of what he called "atomic materials" to the USSR as part of Lend-Lease. From the Soviet lists he extracted the relevant totals, including materials useful in constructing and controlling a nuclear reactor and a small quantity of heavy water (about 1.2 quarts):

Beryllium metals	9,681 lbs.
Cadmium alloys	72,535 lbs.
Cadmium metals	834,989 lbs.
Cobalt ore and concentrate	33,600 lbs.
Cobalt metal and scrap	806,941 lbs.
Uranium metal	1 kg.

Aluminum tubes	13,766,472 lbs.
Graphite, natural	7,384,282 lbs.
Graphite electrodes	21,131,124 lbs.
Deuterium oxide (heavy water)	1,100 grs.
Thorium salts and compounds	25,352 lbs.
Uranium nitrate	500 lbs.
Uranium nitrate (UO_2)	220 lbs.
Uranium oxide	500 lbs.
Uranium oxide (U_3O_8)	200 lbs.

The Soviet Purchasing Commission placed orders for uranium oxide and uranium nitrate in March 1943, just as Igor Kurchatov and his team were preparing their plan for atomic-bomb research and development. Brigadier General Leslie R. Groves, the head of the Manhattan Engineer District, authorized the shipments—under pressure from the Lend-Lease Administration, he testified later. "Where that influence came from," Groves told a congressional committee after the war, "you can guess as well as I can. It was certainly prevalent in Washington, and it was prevalent throughout the country, and the only spot I know of that was distinctly anti-Russian at an early period was the Manhattan Project. . . . There was never any doubt about [our attitude] from sometime along about October 1942."

The small amount of uranium metal on Racey Jordan's itemized list, one kilogram (2.2 pounds), represented Groves's grudging response to a Soviet Purchasing Commission request on January 29, 1943, for twenty-five pounds, which he authorized to be prepared only after the Soviets called the Lend-Lease Administration in March and threatened to arrange a black-market transaction. The kilogram of metal was not delivered until February 16, 1945, and Groves made sure it was an impure sample. According to Jordan, Lawrence C. Burman, the Manhattan Project expert on rare metals, "urged the [uranium metal production] firm to make sure that its product was of 'poor quality.' He did not explain why. But the metal, of which 4.5 pounds was made, turned out to be 87.5 per cent pure as against the stipulated 99 per cent."[*]

Racey Jordan's story of Soviet espionage shipments through Great Falls has never been corroborated in its entirety, but enough pieces of it have found independent confirmation to establish its general credibility. Air Force Major General Follette Bradley, who pioneered the Alsib Pipeline, would tell the *New York Times:*

---

[*] The Congressional Joint Committee on Atomic Energy slyly corroborated Jordan's story in a 1950 report, noting that the results of an assay of the metal "were considerably at variance with assays of uranium metal used by the Manhattan Engineer District . . ." JCAE (1951), p. 188.

Of my own personal knowledge I know that beginning early in 1942 Russian civilian and military agents were in our country in huge numbers. They were free to move about without restraint or check and, in order to visit our arsenals, depots, factories and proving grounds, they had only to make known their desires. Their authorized visits to military establishments numbered in the thousands.

I also personally know that scores of Russians were permitted to enter American territory in 1942 without visa. I believe that over the war years this number was augmented at least by hundreds.

In 1950, Victor Kravchenko, who had served as economic attaché of the Soviet Purchasing Commission from August 1943 to April 1944, described preparing a shipment of black suitcases during the war:

On the seventh floor of the Soviet Purchasing Commission, behind an iron door at 3355 Sixteenth Street, Washington, D.C. . . . there was a special department of the NKVD. . . . One day in February 1944, I don't remember the date, [Semen] Vasilenko, myself, and Vdovin got ready to fly to the Soviet Union six large bags, and Vasilenko took the six bags to the Soviet Union. I saw that material. Some of this material was about the production of planes and the new technological processes; some was about artillery; some was about new technological processes in metallurgy; some was about the possibilities of industrial development. . . . All departments of the Soviet Purchasing Commission—aviation, transportation, all of them— were working for this purpose [of gathering material]. We transferred to the Soviet Union not just this one package; we transferred to the Soviet Union dozens of tons of material, and not just by airplane. We also were using Soviet ships that came from Lend-Lease for the Soviet Union, and they called this material *Super Lend-Lease.* . . .

Jordan's wartime diary confirmed that Semen Vasilenko passed through Great Falls on February 17, 1944, en route to Moscow with what a postwar investigator called "diplomatic mail."

Igor Gouzenko, the Soviet cipher clerk who visited Moscow during the October 16 panic, characterized the Soviet espionage system from personal experience as "mass production." "There were thousands, yes thousands, of agents in the United States," he estimated; "thousands in Great Britain, and many other thousands spread elsewhere throughout the world." America and England were particularly well covered, Gouzenko reported. "When I worked in the Special Communications branch [in Moscow] the vast majority of the telegrams came from England and the United States. Telegrams from other countries were lost in the flood." The military attaché at the Soviet Embassy in Washington had five cipher clerks working for him, added

Gouzenko, "which gives some indication of the amount of information he alone sent."

> The persistence and patience of the [Soviet Intelligence] experts seldom failed to get the wanted information. . . . Often we would send out the same telegram to twenty or more addresses in various parts of the world. One "urgent" query of this nature asked for an item of information about some alleged scientific innovation in the United States. . . . Neither of two agents in the United States could enlighten the experts, but complete and identical information on the American development was received from agents in Canada and England.

In 1943, Gouzenko was posted to Canada. His superior there told him that with a population of fewer than thirteen million people, "This one country . . . has nine separate intelligence networks operating in direct contact with Moscow."

Elizabeth Bentley, the American Communist courier who handled Abe Brothman before passing him along to Harry Gold, independently confirmed the wholesale character of Soviet espionage:

> What the Russians wanted to know [from US agents] was practically limitless. They asked for information on Communists they were considering taking on as agents, on anti-Soviet elements in Washington, on the attitudes of high-up government officials in a position to help or hinder the Soviet Union. . . . They sought military data: production figures, performance tests on airplanes, troop strength and allocation, and new experimental developments such as RDX and the B-29. They were avid for so-called political information: secret deals between the Americans and the various governments in exile, secret negotiations between the United States and Great Britain, contemplated loans to foreign countries, and other similar material.

Bentley reported personally moving some forty rolls of microfilm, thirty-five exposures to the roll, from Washington to New York every two weeks, as well as knitting bags full of documents.

Racey Jordan's superior officer at Gore Field, Colonel Roy B. Gardner, summed up Soviet activity there simply and bluntly in a radio interview after the war. "I know nothing first-hand about the shipment of atomic materials," Gardner said. "I do know that, while I was in command at Great Falls and in charge of this operation, the Russians could and did move anything they wanted to without divulging what was in the consignment."

6

Rendezvous

KLAUS FUCHS ARRIVED at Newport News, Virginia, aboard the passenger ship *Andes* on December 3, 1943. He had accepted assignment among a group of fifteen British scientists, including Rudolf Peierls, Franz Simon and Otto Robert Frisch, to participate in gaseous-diffusion development in the United States with engineers of the Kellex Corporation and a team of physicists and chemists at Columbia University led by Harold Urey.

The *Andes* had zigzagged west across the Atlantic, unconvoyed. Its sparse company of scientists rattled around in its spacious staterooms, gaining weight after British rationing on hearty American breakfasts of bacon and eggs. The train from Newport News up to Washington stopped in Richmond, Virginia, where the unaccustomed luxury of bright lights at night shining on fruit stalls piled with oranges sent Otto Frisch into "hysterical laughter." In Washington, General Groves, having accepted British intelligence's warranty that the new arrivals were not security risks, lectured them on security. The British team traveled on from Washington, then to Manhattan, and lodged at the Taft Hotel. Fuchs disliked the Taft or wanted cover; within a few days he moved to less collegial lodging at the Barbizon Plaza off Central Park. On December 22, he and the other members of the British team attended an important meeting initiating a review of American progress on developing a suitable barrier material for filtering U235 from U238.

Fuchs's younger sister Kristel and her family lived in Cambridge, Massachusetts. After the barrier meeting in New York, Fuchs caught a train to Boston to spend Christmas there, arriving in Cambridge on December 23. When Fuchs's father Emil had been arrested in Germany in the spring of 1933, Kristel, then twenty years old, had fled to Zurich and begun her university studies. She had returned to Berlin in 1934, by which time Emil was free from Gestapo custody awaiting trial and had set up a car rental agency in Berlin as a cover for the Fuchs family's dangerous work of smuggling Jews and anti-Nazi Christians out of Germany. In 1936, Emil had arranged through American Quakers to enroll Kristel at Swarthmore College in Swarthmore, Pennsylvania, safe harbor. There Fuchs's sister had met Robert Heineman, a student from Wisconsin four years her junior who was a member of the American Communist Party active in the Swarthmore Young Communist League. A year later Kristel had dropped out of college; she and Heineman married in October 1938. Heineman had graduated from Swarthmore the following June and the couple had moved to Cambridge, where Robert had taken up graduate study at Harvard. A son had been born in 1940, a daughter in 1942. The marriage was troubled and intermittent; Robert had moved away to Philadelphia for a year beginning in 1942. By 1944 he was back in Cambridge working at the General Electric plant in Lynn.

Fuchs returned to New York after Christmas. The review of gaseous-diffusion technology then underway culminated in a stormy meeting with General Groves early in January 1944 when Groves won British endorsement of manufacturing a new and superior barrier material at Kellex that would supersede existing barrier production. Retooling for the new barrier would significantly delay starting up the big Oak Ridge gaseous-diffusion plant then under construction; Harold Urey, for one, understood the decision to mean that the United States was pursuing a postwar nuclear-weapons capability, not simply trying to beat the Germans to the bomb. Thereafter Fuchs concentrated on gaseous-diffusion theory as a consultant to Kellex, working first from offices at 43 Exchange Place, later out of the British Mission of Supply at 37 Wall Street. By February 1, 1944, he had settled into a furnished apartment in a brownstone at 128 West 77th Street passed along by a member of the British Mission who was returning to England.

Fuchs would recall later that he first met with the man he knew as "Raymond"—Harry Gold—"around Christmas 1943." But Fuchs had been in Cambridge at Christmas; Gold remembered more accurately meeting Fuchs for the first time in "late January or very early February 1944." It was standard Soviet practice to prearrange recognition signals between agents unknown to each other; Gold would testify that Sam had instructed him "to carry a pair of gloves in one hand, plus a green-covered book, and Dr. Fuchs was to carry a hand ball." Sonia had briefed Fuchs before he left England on recognition signals and meeting place, which was to be outside the Henry Street Settlement House on the Lower East Side of Manhattan.* Fuchs was apprehensive about this first American meeting; rather than risk asking

---

* Sonia had also, on instruction, passed Fuchs over from GRU military intelligence to Beria's NKVD, although Fuchs never knew the difference and perhaps never cared. Beria was maneuvering to control the most important source of information then available about the Anglo-American atomic-bomb program.

someone how to find Henry Street he had bought a map and worked out the subway route from the stop nearest the Barbizon Plaza. Fuchs remembered the initial contact differently from Gold, recalling that Gold "was wearing gloves and carried an additional pair of gloves in his hand and I had a tennis ball in my hand." (Gold recalled stopping on his way to the meeting to buy a pair of gloves, presumably an extra pair since it was winter.) Gold introduced himself as "Raymond"; Fuchs gave his real name. Raymond "indicated he had been expecting [me]," Fuchs reported, "and he stated definitely that he was pleased to have been selected for such an important assignment."

"We went for a brief walk," Gold recalled of that first rendezvous, "and then took a cab uptown to [Manny Wolfe's] restaurant around 3rd Avenue in the 50's, where we had dinner, but we did not speak much there. Afterwards we went for a walk, during which we completed arrangements for further meetings."

Fuchs reported the discussion in more detail in paraphrased testimony:

[Fuchs] told "Raymond," in answer to questions, where he was living and where he was working. They also arranged to hold another meeting in the immediately near future. He discussed with "Raymond" his plans. He also discussed with him orally some of the officials for whom he was working and told him where, in fact, he was working at the time. "Raymond" specifically suggested that at future meetings Fuchs make sure that he was not being followed. The attitude of "Raymond" at all times was that of an inferior. At this first meeting Fuchs believes that he made a statement to "Raymond" about atomic energy, and he knows that the words "atomic energy" and "atomic bomb" were both mentioned, and "Raymond" must have known about them as he did not ask any questions of interpretation or explanation. He also believes that the comparative strength of an atom bomb was also mentioned at this first meeting. . . .

But Fuchs remembered no dinner together that first time out; he thought the first meeting lasted only about twenty minutes, though he did remember having dinner with Gold at least once during their New York contacts, and agreed that it might have been then. Evidently Gold was dazzled to be working with a man who he believed to be "one of the world's foremost mathematical physicists." If Fuchs characterized their relationship as that of superior to inferior, Gold recalled it more generously: "I liked this tall, thin, somewhat austere man . . . with the huge horn-rimmed glasses . . . from the very first, and in his stuffy, repressed British manner he reciprocated." To Gold, Fuchs was no less than a "genius (a word I always use with caution)."

After he and Fuchs parted, Gold rendezvoused that same evening with Sam and reported what Fuchs had told him. Thus the connection between

Fuchs and Kurchatov's team in Moscow was renewed. "Intelligence information was channeled directly to [Kurchatov]," Anatoli Yatzkov, who was about to become Gold's control, wrote late in life. "Representatives of the Intelligence Service contacted him directly. He studied the materials, produced detailed reviews and compiled lists of questions, which were immediately sent to *rezidents.*"

On January 4, Eddie Sinelnikov wrote her sister in England from "Near Moscow" describing the conditions under which the Soviet scientists were living and working:

Our present abode is in rather nice surroundings and I begin to appreciate the beauty of the real Russian winter—not in town—but communication with Moscow is not all that could be desired—but we get pleasure from visits to Marina [Kurchatov] and the Kapitzas. Garry [i.e., Igor Kurchatov] is now an Academician and has grown a beard! We can't decide whether it was originally due to lack of razor blades or mufflers. Anyhow he looks very amusing and friendly with it and on New Year's Eve I measured it and discovered the said beard to have the drastic length of twelve centimeters! Jill [the Sinelnikovs' young daughter] and Garry are great friends.

Gee! Isn't the news from the front splendid? Every day fireworks—and such jolly ones. Bang! Bang! and up into the air, over Moscow sail hundreds of brightly coloured balls—like so many bouquets of flowers thrown on high.

"A turning point came in the war," Igor Golovin explains the fireworks. "Our armies drove back the foe relentlessly. In November, 1943, Kiev was liberated; in January of the new year, 1944, the siege of Leningrad was lifted. . . . Moscow hailed the victories with salvoes." By the end of 1943, the Red Army with increasing mastery had liberated two-thirds of Soviet territory. The Soviets called 1943 the *perelom* year: the year of the great turning point. The fireworks Eddie Sinelnikov enjoyed included 120-gun victory salutes that had begun on August 5 with the liberation of Kursk and continued throughout the rest of the war as towns and regions were liberated, more than three hundred salutes in all. Soldiers were still dying, an average of five thousand a day throughout the war. "None of the Russian offensives in 1944 were in the nature of a walkover," Werth reports, "and the nearer the Russians got to Germany, the more desperate became German resistance." But the Katyushas were rolling west. Soviets called the multiple-banked rocket launchers "Stalin Organs"; the Germans, on the receiving end, called them "The Black Death."

Kurchatov's colleagues had nicknamed him "the Beard." Some of them were puzzled at his stock of ideas and information. "The reason for selection of graphite as a moderator [for the first small nuclear reactor the Soviets were planning], by Kurchatov, immediately in the spring of 1943, remains unclear," writes Golovin; "one can only guess why he did so." Kurchatov evidently did so because he had learned that the United States had done so, successfully. When Kurchatov presented a laboratory group with two versions of calculations to compare, experimental physicist Lev Altshuler recalled, "the joke was that one version came from the 'ceiling'—meaning Beria—and the other came from the Beard." Altshuler understood that they were "testing that this [data derived from espionage] was correct information rather than disinformation."

Gold met once more with Sam before his second meeting with Fuchs. Sam had surprising news: he was passing Gold off to another control, a man Gold would know as "John." Subsequently Gold met John for the first time across the street from the Manhattan 34th Street bus terminal. "He was younger than I, and was taller by some inches; he had a shy, boyish grin and a lock of dark hair that kept falling over his right forehead, and this he would always brush back with a characteristic motion. . . ." John led Gold to a nearby bar—the Russian had a purposeful but duck-like walk, Gold noticed—where Sam joined them and they discussed the transfer of control. When Gold next met Klaus Fuchs, he would deliver his report to John.

"John" was Anatoli Antonovich Yatzkov, known during his years in the United States by an assumed name, Yakovlev. Born in 1911, trained like Semenov in engineering, Yakovlev had entered the United States in February 1941. Though Gold always assumed that John, like Sam, worked for Amtorg, his new control was officially a clerk at the Soviet Consulate in New York; as New York NKVD *rezident,* he also controlled the Cohens. "I failed to master English in the three-month term which I was allocated," Yatzkov remembered in old age, "but I took the risk and went to America. My luck was to talk to Americans, which made it easier to learn English, but I progressed slower than I would like." Now he could talk to Harry Gold as well.

An FBI informant who bumped into Yatzkov/Yakovlev at consulate receptions during this period remembered him complaining "about being continuously overworked and homesick." He was married, with twin children, Victoria and Pavel, born four months after his arrival in the US, and Gold would find him not complaining but optimistic; John and Sam, Gold remembered, "spoke with great pride of their wives and their children, and would elaborate on their great plans for the future of the young ones." When discussions among the Allies began in San Francisco "which led to the formation of [the United Nations], I can recall the enthusiasm with which Yakovlev discussed the affair. We both thought it was such a great thing."

Fuchs and Gold met again in February, on the northwest corner of 59th Street and Lexington Avenue. They walked east toward the Queensboro Bridge, "the intention in my mind," says Gold, "being that we would walk across the bridge and into Queens." They found the bridge closed to foot traffic and walked uptown along First or Second Avenue instead, possibly as far as 75th Street. It was "anything but an exclusive area," Fuchs remembered; Gold recalled "several passages on the dark deserted streets."

At this second meeting Fuchs told Gold about the Manhattan Project work on isotope separation. Gold was captivated. For years he had studied developing a process for recovering valuable compounds from industrial waste gases using thermal diffusion, the type of isotope separation with which Otto Frisch had experimented at Birmingham in 1939 that had led him to conclude that U235 could be separated from U238 to make an atomic bomb. "That is my baby, that is my dream," Gold exclaimed in 1950 when FBI agents asked him about his interest in thermal diffusion; he told them he had written a dissertation on the subject, a claim they later confirmed. So he was surprised when Fuchs seemed not to know about the process:

> Klaus knew of only two methods for the separation of the isotopes from uranium, that is, methods as were being pursued here in the United States, and . . . these methods were, (1) The gaseous diffusion process, (2) The electromagnetic separation process.

It was Harry's chance to impress a man he considered to be a "genius." He made bold to do so: "I . . . mentioned to Klaus the possibility of the use of thermal diffusion as a means of separating isotopes, but . . . Klaus . . . brushed this aside." Gold must have been crushed. Fuchs could be arrogant as well as insensitive; in fact, when problems with the gaseous-diffusion plant that Fuchs was helping design delayed its full operation, Groves would jury-rig a thermal-diffusion plant of physicist Philip Abelson's design which would process a significant portion of the uranium enriched for the Little Boy uranium gun bomb exploded over Hiroshima; without thermal diffusion there would have been no uranium bomb ready to use in August 1945.

Fuchs emphasized to Gold then and later that Manhattan Project scientists, as Gold recalled, "worked in extremely tight compartments, and that one group did not know what the other group was doing. This I can verify by the fact that he told me that he thought that there was [the] possibility of a large-scale installation for isotope separation projected for future development somewhere, he thought, down in Georgia or Alabama. This, of course, later turned out to be Oak Ridge." And was, of course, the very plant that Fuchs was helping design.

Gold "made good mental notes of such data," he would testify, and after the meeting ". . . at the first opportunity I put this material in writing, and later handed it over to John." John sent the information back to Moscow Center in coded cables; Fuchs's code name in the cables was "Rest." The cables went out over commercial telegraph lines, which made it possible to intercept them. After the Japanese surprise attack on Pearl Harbor on December 7, 1941, and US entry into the war, the US State Department had promoted a "drop copy" program whereby the cable companies held up message transmissions long enough to copy them, ostensibly for the US Office of Wartime Censorship. The copies went through the censorship office to the Army security agency, where the FBI had access to them. Soviet espionage cables were coded on one-time pads, however—five-unit ranks of random numbers on pads of paper, used only once, that matched pads kept in Moscow—so that without access to the code pads they were effectively indecipherable; thousands of such coded Soviet wartime cables piled up at Army security, Fuchs's ongoing disclosures among them.

Fuchs met Gold for a third time in March 1944 on Madison Avenue in the 70s. "It was still quite cold and we both wore overcoats," Gold recalled. ". . . We immediately turned into one of the dark deserted sidestreets toward 5th." For the first time, Fuchs passed Gold documents. To reduce the risk that both might be apprehended together, it was standard practice between Soviet agents to separate immediately after a document transfer. "The whole affair took possibly 30 seconds or one minute," Gold testified, "and I immediately walked ahead of Klaus and down 5th toward 57th Street and 6th Avenue, where approximately 15 minutes later I turned over the information to John."

Fuchs and his colleagues, particularly Rudolf Peierls, were working on a series of papers for Kellex, designated the MSN series, laying out gaseous-diffusion theory. During the period when Fuchs was based in New York, the British completed nineteen papers in the MSN series. Of those, Fuchs personally wrote thirteen. "Two or more MSN papers," Fuchs testified somewhat inaccurately in FBI paraphrase, "were passed to Raymond by him at each of the approximately 5 meetings held after the first meeting." To evade security, Fuchs simply took advantage of the trust the Manhattan Project accorded him:

> I, with other scientists, prepared certain highly confidential and classified documents . . . referred to as the MSN Series. . . . I would first prepare a draft. . . . [This draft] would be routed for duplication. . . . In all instances, when I prepared the draft a proof copy and the original draft would be returned to me. Each of the duplicated copies was numbered for control and security purposes, due to the highly confidential character of the contents. I would personally retain the original draft, which most of the time I had prepared in longhand, and I personally furnished all of the drafts of my own composition directly to the individual known to me as Raymond. . . . These documents were at times folded and at other times in package form and were delivered by me personally in groups of one or more at most of the . . . prearranged meetings, after the initial contact meeting which I had covertly with Raymond in New York City during 1944.

From this point on, Fuchs's and Gold's accounts of their meetings diverge. Gold remembered dinners together and personal confidences that represented, he said, at least "a deviation from the rules." Fuchs remembered strict compliance and businesslike formality; confronted, later, with Gold's testimony of bonhomie, the emigré physicist rejected testimony and eager witness both with withering contempt:

[Fuchs] advised that there would have been no occasion for any meeting except to deliver written information since the knowledge and background of Raymond was insufficient to enable him to understand technical details and his lack of scientific knowledge of the type necessary to understand the problems on which Fuchs was working would have made it very unlikely that [Fuchs] would have arranged any meeting with Raymond after the first for any purpose other than to deliver information in writing to him.

But the volume of related information that Fuchs testified he furnished Gold orally implies extended conversation: "the manpower employed by Kellex and the nature of the work being performed by the British Mission and all that he knew concerning personnel and general activities in the Manhattan Engineer District. . . . The identity of the officers and the high-ranking scientists. . . . He also discussed some of the personnel orally." Gold also reported confidences about Fuchs's family which the chemist could not easily have learned from any other source. Fuchs was a bachelor alone in a strange country, unable because of the double life he was leading to confide in colleagues, penny-in-the-slot. Under similar circumstances in England he had confided similarly in Sonia. He was both "stuffy" and "repressed," as Gold accurately characterized him. In repudiating Gold, Fuchs sounds like someone angered to hear his confidences betrayed and incensed that a mere industrial chemist, a bag man, would presume. (The question is important. Later, when Gold was exposed as a courier and testified for the US government against Americans accused of spying, there were attempts to discredit him as a fantasist, a lonely bachelor who invented tales and connections to thrust himself into the limelight. But in fact, allowing only a little for the vagaries of recollection across fifteen crowded years of espionage work, Gold's remarkably detailed memories of events almost always prove accurate wherever they can be checked.)

So at their fourth meeting, in the Bronx, in April, Gold recalled, "we went for a walk partly along the Grand Concourse . . . during which time we discussed the next meeting . . . at which a second transfer of information was to take place. . . . After this I took Klaus to dinner, it was a wet and somewhat chilled night for April, and as I recall, he had a bad cough, and I did not wish to expose him to the elements any more than was necessary. . . . We had a dinner at which we discussed a number of matters, including music and chess." Among other matters, they may have discussed dissatisfaction within the British Mission at the progress of its work in America, information that Fuchs is more likely to have passed orally than in writing; a cryptic note in J. Edgar Hoover's hand underlined in the file that the FBI opened on Fuchs in 1949 reports such a discussion at about this time and hints that the Soviet New York *rezident* may have raised the possibility with Moscow Center of having Fuchs arrange to be transferred back to England, which would have been a devastating mistake:

May 8, 1944. F[uchs] advised Russians [that the] work of [the] Brit[ish] Com[mission] on A[tomic] E[nergy] [was] meeting with no success in U.S. & [that there was] dissatisfaction. Russia proposed to send F[uchs] back to G[reat].B[ritain].

Then or later, according to Hoover's notes, Fuchs also advised the Soviets —presumably through Gold, the only contact Fuchs acknowledged in the US—that Britain and the US were slowing down research work on diffusion (they may have been; they were moving on toward industrial development), that the Americans had informed the British that construction of a diffusion plant in England would directly contradict the spirit of the agreement on atomic energy signed together with the Atlantic Charter, and that someone from England was in Washington "at that time looking into details of trans-ferring the work to G[reat] B[ritain]." All this information probably came to Hoover after the war from decoded intercepts. There is no further reference to it in the files that the FBI has declassified; it hints, however, as does much else in Fuchs's testimony, at more extensive communications between Fuchs and Gold than Fuchs chose to acknowledge.

At dinner that April evening, Gold recalled, he and Fuchs also concocted a cover story together, "should either of us ever be questioned," that they "had met at one of the New York Philharmonic's concerts . . . in Carnegie Hall; the idea was that we had had adjacent seats and had talked together in the lobby during the intermission." Gold agreed to look up the date and the program of such a concert so that they would both agree on when they attended and what they heard. After dinner, Gold and Fuchs shared a cab to a bar on Madison Avenue where they had further drinks. Then Gold put Fuchs in a cab to cross Central Park to his apartment on the West Side.

At the meeting they had scheduled the next month in Queens, Fuchs passed Gold "some 25 to 40 pages" of information. Gold could not resist sneaking a look. "After leaving the Elevated I was in the general area where I was to meet John. I still had about five minutes to wait and I recall stopping near a drug store and taking a glimpse of the information. . . . This was in a very small but distinctive writing; it was in ink, and consisted mainly of mathematical derivations. There was also further along in the report a good deal of descriptive detail." Two minutes of delicious snooping and Gold moved on to his rendezvous with Yatzkov.

In June, the two conspirators met in Brooklyn; Gold remembers Fuchs discussing a personal dilemma of the sort that Fuchs may later have resented Gold revealing:

> During this meeting I recall that Klaus Fuchs told me that there was some possibility that his sister who lived in Cambridge, Massachusetts—he did not give me her name, however—might come to New York. He explained to me that his sister was married and had two children, and that she was having great difficulty with her husband and that she was fully intending to leave her husband and come to New York. Should this occur, Klaus told me that he would like very much to be able to share an apartment with his sister. . . . He brought the matter up because he first wanted me to inquire of my superior whether such an action would be all right. I said that I would make the inquiry.

For this meeting, John had given Gold "several typewritten pieces of paper about three by nine inches, of irregular size, which had contained a number of questions relating to atomic energy. The phraseology of these questions was extremely poor, and I had great difficulty in making any sense out of them." Gold thought the questions had probably suffered in decoding or in translation from Russian. Here may be the origin of Fuchs's conviction that Harry Gold was technically illiterate when in fact he was a competent industrial chemist with a good working knowledge of at least one process of isotope separation. Gold:

> I did make what sense I could out of the message, and on this occasion . . . began to tell Klaus about what further information was desired. I did not get very far along this course because Klaus seemed to take offense at being instructed and said very briefly that he had already covered all of such matters very thoroughly, and would continue to do so.

During July, Fuchs and Gold met yet again, "near an Art Museum" on the West Side according to Gold. Fuchs had important news. "We went for a long walk, almost entirely in Central Park and in the many winding roads and small paths leading through the park itself. This meeting took at least an hour and a half and was a very leisurely one." Fuchs told Gold he might be transferred, later in 1944 or early in 1945, "somewhere to the Southwest." Gold was sure later that he had heard Fuchs say Mexico; Fuchs was adamant that he had said *New* Mexico.

Fuchs revealed during the walk in Central Park, says Gold, "that his brother, Gerhard, was now in Switzerland and was convalescing as a result of having been only recently released from a German concentration camp." Gold gathered that Gerhard, like Fuchs, was a dedicated Communist. If

Fuchs imagined Gold to be his inferior, Gold considered Fuchs fragile and otherworldly and undertook to shelter him. "I also told Klaus that it would be perfectly all right, should his sister come to New York, for him to take an apartment together with her and the children. Actually, I had not mentioned the matter to John at all, but had taken it upon myself to tell Klaus that such a proceeding was O.K."

Then Klaus Fuchs disappeared. He was scheduled to meet with Gold in Brooklyn at the end of July, in front of the Bell Cinema, close to the Brooklyn Museum of Art. He did not make the meeting. It was standard procedure to schedule backup meetings in anticipation of missed connections. Fuchs also failed to appear at the backup meeting he and Gold had arranged at around 96th Street and Central Park West. Gold's maternal instincts kicked in: "On this second occasion I became very worried, particularly since the area is very close to a section of New York where 'muggings' often occur, and also the fact that Klaus was of slight build and might seem an inviting prey."

Gold met with John. They discussed the problem of Fuchs's disappearance for two hours. "Our principal trouble was to decide whether Klaus, for some reason, was unable to keep the meetings if he was still in New York, or whether he had actually left New York." Apparently they reached no conclusion. They met again in late August 1944, early on a Sunday morning, near Washington Square. John sent Gold to Fuchs's apartment to ask the physicist's whereabouts. Gold bought a book along the way, Thomas Mann's *Joseph the Provider,* wrote Fuchs's name and address in it and invented returning it to its "owner" as a pretext for his inquiry. At Fuchs's building, the building superintendent and his wife informed Gold that the physicist had left town. Gold met John again later that morning; they walked along Riverside Drive and "talked at great length." Stymied, John told Gold to " 'sit tight.' "

At a meeting in early September 1944, another long discussion, Gold finally thought to mention "that Fuchs had a sister who lived in Boston. Now it may be possible that John himself may have brought up the matter of Fuchs' sister. . . . In any event, John told me that he thought that there lay our best line of inquiry." By mid-September, John had turned up the name of Mrs. Robert Heineman. She lived, he told Harry, in Cambridge, Massachusetts.

On a Sunday in late September, Gold took the train to Boston and the T subway to Cambridge, found the Heinemans in the phone directory, walked out to their house and knocked on the door. A housekeeper answered; the family, she said, was still away on vacation and was not expected back until October. Gold returned to Philadelphia. When he next met John in New York the Soviet agent was "highly pleased" that they had at least located Fuchs's sister.

Sometime in October 1944, John dictated to Gold a message for Fuchs.

Gold printed the message "in engineering lettering" on a card and sealed the card into an envelope. The message consisted of a name—six years later, Gold remembered uncertainly that the first name may have begun with a "J" and that the last name might have been something like "Kaploun" —a Manhattan telephone number and "the information that Klaus was to call the phone number given, any time—on any morning between the hours of 8:00 and 8:30—and was to give the following message: Merely to say, 'I have arrived in Cambridge and will be here for _____ days.' " (Gold's revelation of a Manhattan phone contact adds another operative to the list of Soviet espionage agents active around the Manhattan Project. Based solely on Gold's partial recollection of the contact's name, a candidate for this contact might be Judith Coplon, a 1943 Barnard graduate whom Robert Lamphere later established to be involved in Soviet espionage. Coplon was living in New York at this time, working in the Justice Department Economic Warfare Section. She has not previously been identified in this context.)

Gold remembered carrying John's message to Cambridge to leave with Kristel Heineman in early November 1944. Fuchs's sister remembered Gold visiting her for the first time in late January or early February 1945. Neither Kristel, Fuchs nor Gold ever quite straightened out when their various Cambridge meetings occurred, but other records make it possible to establish some of them at least approximately.

Whenever it was that Gold visited her that winter, Kristel remembered looking out the window of her house and noticing a man whom she did not know walking down the street. It was just before noon. The man came to her door and rang the bell. She answered the door. The man asked her if she was Mrs. Heineman, the sister of Klaus Fuchs. She said she was and the man gave his name. She was never able to remember his name, but six years later, when she was shown a photograph of Harry Gold, she immediately and positively identified him as the man who rang her bell that day and returned twice more to her house in Cambridge.

Harry told Fuchs's sister that he was a chemist who had worked at one time with her brother. He was anxious to see Klaus, he said. The Heineman children came home for lunch then and Kristel invited Gold to join them. He mentioned that he was tired from a long train ride.

Kristel Heineman remembered telling Gold during lunch, in FBI paraphrase, "the approximate dates between which Klaus Fuchs would visit the Heineman home"—dates presumably in February 1945. Gold, to the contrary, remembers her mentioning Christmas:

Mrs. Heineman told me that Klaus had been transferred somewhere in the Southwest United States, but that she expected him here about Christmastime. I believe that she indicated that she had received several letters from him. She said that she thought that he would certainly be home about

Christmas, as he usually made a great event of bringing presents for the children.

If Kristel did not yet know that Fuchs would not visit Cambridge for Christmas, then she had not yet received a letter Fuchs wrote her from Post Office Box 1663, Santa Fe, New Mexico, on December 15:

Dear Kristel,

Many thanks for your letter. I am afraid I have been very busy during the last few weeks and I expect that will go on for a little time longer. But I do hope that I shall be able to take a holiday some time at the end of January. I have not even been able to do any Christmas shopping . . . I expect Marcia and Steve will be cross if my Christmas parcel does not arrive on time. But I trust you will be able to pacify them.

We have lots of snow around here and I am itching to get on skis. But before I do so I shall have to pacify my conscience as an uncle and get the parcel for your kids off.

With best wishes
Klaus

Placing Gold's visit in November or early December would also explain Yatzkov's urgency in dispatching him later, when word from Fuchs finally came, a month and a half after Christmas. But whenever Gold visited Cambridge, he accomplished his mission—he left the sealed envelope and went on his way.

Fuchs was indeed "very busy." The previous summer, on July 14, 1944, the German emigré physicist had met in Washington with James Chadwick, the Nobel laureate discoverer of the neutron and the head of the British Mission in the United States. Chadwick had informed Fuchs that his services had been requested at Los Alamos, the secret laboratory in northern New Mexico where the first atomic bombs were being designed, "provisionally until the end of December." Los Alamos was in turmoil and needed help.

The laboratory had been planning to build weapons that assembled critical masses of U235 or plutonium239 using a gun configuration: firing one subcritical piece of nuclear material up the barrel of a cannon to join it with a subcritical ring fitted to the muzzle. The worry with such an assembly mechanism was predetonation. Both uranium and plutonium fissioned spontaneously, as Georgi Flerov and K. A. Petrzhak had first demonstrated in the case of uranium. Secondary neutrons released by such random spontaneous fission might start a chain reaction prematurely within the barrel of the cannon, as the "bullet" approached the target ring, before the two pieces had time fully to assemble. If the mass of nuclear material thus predetonated, it would still explode, but it might do so inefficiently. Instead of exploding with a force equivalent to ten thousand tons or more of TNT, it might fizzle at the equivalent to no more than a few hundred pounds of TNT—no better than a conventional high-explosive bomb could do. The United States was spending some $2 billion to make three atomic bombs; a fizzle would be an unconscionable waste of money.

Pu239 was known to fission spontaneously at more than double the rate of U235. Another isotope of plutonium, Pu240, which turned up as a contaminant in Pu239, was even more unstable. Assembling a critical mass of Pu239 within the barrel of a cannon had appeared from the beginning to be problematic. The plutonium bullet would have to travel up the barrel several thousand feet per second faster than would the bullet in the uranium gun. Until April 1944, a plutonium gun assembly had looked barely attainable. But the experiments so far conducted at Los Alamos had used microgram quantities of plutonium transmuted laboriously in a cyclotron, which produced primarily Pu239. The first gram quantities of reactor-produced plutonium arrived at Los Alamos early in the spring of 1944 from Oak Ridge. A nuclear reactor generates far more neutrons than a cyclotron. That higher neutron flux had transmuted more of the uranium in the reactor to Pu240. The spontaneous fission rate of reactor-produced Pu239, with its greater admixture of Pu240, turned out to be five times greater than that of cyclotron-produced plutonium, unacceptably high for gun assembly. Even at the highest attainable muzzle velocities, a plutonium bullet would melt before it had time to mate with a target assembly.

By July 1944, when Fuchs talked to Chadwick, Los Alamos had decided that the plutonium gun would have to be scrapped. The uranium gun, Little Boy, a conservative and reliable but inefficient design, would require as much of the rare uranium isotope as could be separated through 1945. Unless Los Alamos worked out a way to assemble a critical mass of plutonium without predetonation, the Manhattan Project, which by then was approaching the US automobile industry in number of employees and capital investment, would be able to deliver only one atomic bomb.

An alternative to the gun system had been proposed soon after the lab had opened its doors in April 1943, though many had doubted that it could be made to work. It was called implosion. In its first incarnation it depended on the fact that whether or not a mass of fissionable material is critical is determined not only by its volume but also by its geometry. Six kilograms of plutonium cast as two solid hemispheres would begin chain-reacting as soon as they were brought into contact; but the same six kilograms of plutonium configured as a hollow shell, from which secondary neutrons would more easily escape, would be essentially inert. Pack high explosives (HE) around such a shell, figure out a way to detonate the HE from a number of different points simultaneously, thus collapse the shell inward into a solid ball, and critical assembly might be achieved so rapidly that spontaneous fission would not have time to spoil the chain reaction. Slammed with high explosives, the walls of the shell would have to move only a short distance inward, and the HE would accelerate them together far faster than a cannon could do.

No one had ever used explosives to assemble something before; their normal use was blowing things apart. The first experiments conducted at Los Alamos using two-dimensional arrangements—pinching steel pipes with collar rings of HE—had been disastrous. Navy Captain William "Deke" Parsons, who was in charge of explosives research, scoffed that implosion was like trying to "blow in a beer can without splattering the beer." From each point of detonation a convex detonation wave moved through the explosive; when the various waves spread into contact they interfered with each other in complex patterns like the interference waves that passing boats produce when their wakes collide. Instead of uniformly closing the steel pipes down to a solid pinch, the colliding shock waves liquified jets of hot metal and blew the pipes cockeyed.

Implosion phenomena were too complex for cut-and-try; the experimenters needed theory to guide them. Someone needed to go to work calculating the hydrodynamics—the complex, dynamic fluid motions—of implosion. Someone needed to work out the number and best placement of detonators around the outside of the HE sphere. Someone needed to calculate the ideal geometry of the plutonium shell, whether larger or smaller, whether thicker-walled or thin. The head of the Theoretical Division at Los Alamos, emigré physicist Hans Bethe, turned to Edward Teller, who was recognized then and later as one of the most imaginative, creative physicists alive. Teller took over direction of a small implosion group in January 1944 and made valuable contributions through the rest of the winter. But as winter turned to spring he began to neglect implosion calculations. He believed he had more important work to do, including early theoretical study of the possibility of using an atomic bomb to ignite a mass of deuterium, a weapon he called the Super that might explode with force equivalent not to thousands of tons of TNT but to millions of tons. "[Bethe] wanted me to work on calculational details at which I am not particularly good," Teller wrote later, "while I wanted to continue not only on the hydrogen bomb, but on other novel subjects."

Bethe knew that Rudolf Peierls was in New York working with Kellex. He requested that Peierls transfer to Los Alamos to help out on implosion. Peierls agreed provided that he be allowed to bring along two assistants: a young Englishman named Tony Skyrme and Klaus Fuchs. If the god of war had wanted to provide Igor Kurchatov with a clear channel directly into the heart of the most important and secret work then underway at Los Alamos, he could not have chosen a more providential channel than Klaus Fuchs.

Robert Oppenheimer, who had become the wartime director of Los Alamos, said much the same thing later, after Fuchs had been exposed. General Groves had complained that Los Alamos was not compartmentalized adequately for security. "If Fuchs had been infinitely compartmentalized," Oppenheimer countered, "what was inside his compartment would have done the damage."

Fuchs arrived at Los Alamos on August 14, 1944. "One of the most valuable men in my division," Hans Bethe would call him, ruefully. Nicholas Metropolis, a mathematician in the Theoretical Division whose office was next to Fuchs's, noticed the German's diligence. "Whenever I walked in—and I would walk in early, like eight o'clock—he was always there. And when I left at night at five o'clock, five thirty, he was still in his office working away. He worked long, long hours." In October, Oppenheimer led a colloquium that Fuchs attended on a new approach to implosion using three-dimensional "lenses" of high explosives. The radical new concept, proposed the previous summer by British physicist James Tuck, offered a possible way to overcome the interference between detonation waves that made such a mess of steel pipes. A detonator stuck in a piece of explosive started a wave that expanded outward through the HE equally in every direction, convexly, like a swelling dome; but it might be possible to design a complex arrangement of carefully fitted pieces of faster- and slower-burning explosives that would retard or accelerate the passage of the convex detonation wave so as to allow the sides of the dome time to catch up with and pass the peak—like turning a beanie or a yarmulke inside out. With the right combination of shapes and explosives, a detonation wave diverging outward from a point might be converted to a detonation wave converging inward on a point: an explosion might be converted to an implosion, eliminating detonation-wave interference and smoothly squeezing a subcritical ball of plutonium to supercriticality.

As he had when consulting with Kellex on gaseous diffusion, Fuchs at Los Alamos once again produced a series of significant papers, but these dealt with the crucial question of how to make plutonium efficiently explode. The titles of some of the papers Fuchs wrote in his two years at Los Alamos reveal the extent to which he had tunneled fortuitously to the very center of the plutonium problem:

> Formation of Jets in Plane Slabs
> Jet Formation in Cylindrical Implosion
> Efficiency for Very Slow Assembly
> Theory of Implosion, Part I
> Theory of Implosion, Part II
> Theory of Implosion, Part III
> Theory of Implosion, Part IV
> Theory of Implosion, Part V

Fuchs also worked on theoretical studies concerning a small but crucial component of an implosion bomb, a device Los Alamos called an "initiator." In September 1944, the physicist Robert Christy had proposed reducing the jetting problem by using as a bomb core not a shell of plutonium but a nearly solid subcritical ball (in the form of two fitted hemispheres). With a solid instead of a shell, nothing would be collapsing; the imploding detonation wave would simply squeeze the solid mass to criticality. It was a conservative, brute-force solution that would be much less efficient than a shell system and more dangerous as well—in its final incarnation it would be barely subcritical within a heavy natural-uranium tamper and would have to be safed with a removable cadmium wire—but it was a far simpler design.

Unfortunately, a solid core would necessitate adding in another complicated component. Implosion would reduce the core diameter by half, increasing the density of the solid metal by a factor of eight. In the few millionths of a second when the shock wave had squeezed the implosion assembly to maximum density, before the assembly began to rebound and disassemble, it needed a squirt of neutrons to start the chain reaction. The initiator was the first device used in atomic bombs to supply those neutrons, by knocking them out of a shell of beryllium foil with alpha particles from another shell of hot, highly alpha-radioactive polonium. It was a small nugget of exotic metals to be set exactly at the center of the bomb, nested in a cavity within the two hemispheres of plutonium. It was difficult to design because it had to remain inert, releasing no neutrons, until precisely the right moment and then unfailingly do its work. If it produced neutrons prematurely it might cause the bomb to predetonate. If it produced neutrons belatedly they would fly out uselessly through the rebounding wreckage. The initiator was nearly as difficult to design as the larger bomb around it, layers within layers, and its ingenuities were compressed within a gadget no bigger than a walnut. Fuchs would write three papers on initiator theory.

Fuchs attended seminars that winter on various alternatives to implosion. By February 11, 1945, when he left the mesa in northern New Mexico to visit his sister and her family in Massachusetts, he knew as much as anyone at Los Alamos about plutonium bomb design.

Sometime after Fuchs arrived in Cambridge, Kristel told him about Gold's November approach. Her brother "seemed surprised and somewhat annoyed," she remembered, ". . . but . . . he did not comment beyond saying, 'Oh, it's all right.'" She gave him the envelope Gold had left. He called the contact in Manhattan.*

---

* Fuchs denied making such a call without explaining how otherwise Gold would have known he was in Cambridge, but he was always careful during interrogations to deny contacts that had not yet been identified; he avoided identifying Gold until he surmised that Gold had confessed, and he only identified Sonia after he was in prison and she had left England for East Germany.

Before seven, one weekday morning, Yatzkov telephoned Gold just as Gold was getting ready to leave for work:

> With some difficulty he described to me the fact that he was in a gasoline station, near what I finally determined to be [the] Oxford Circle section of Philadelphia. John wanted to know if I would come down there and meet him. I did so. It was a very snowy morning, I recall it well, and John was wet. We got on the [street] car again and went down to the terminal in Frankford, where John told me that he had just the previous day received notification that Fuchs was now at Cambridge. . . . He then told me that I must, as soon as possible, go to Cambridge. I did so. I believe that I met John on a Tuesday or a Wednesday, and that I arrived in Cambridge on most likely a Friday.

In 1945 there was one Friday between Sunday, February 11, and Thursday, February 22, the day Fuchs left Cambridge to return to New Mexico; he and Gold most likely met on February 16.

"I went directly to the Heineman home," Gold remembered. "This was in the morning, and when I knocked I was admitted by, I believe, a servant girl. Klaus was there and welcomed me."

7

'Mass Production'

IF KLAUS FUCHS was the most productive spy delivering information on the Anglo-American atomic-bomb program to the Soviet Union from North America, he was by no means the only agent at work. Not many were ever exposed. Only a few of those who became known were brought to trial and convicted. But the collective record, limited and fragmentary though it is, corroborates Igor Gouzenko's characterization of Soviet espionage during and after the Second World War as "mass production," demonstrates its methodology and reveals patterns and practices that tend to support espionage revelations that many Americans understandably questioned in the poisoned atmosphere of the high Cold War years.

Two shocking quantitative measures of the extent of Soviet wartime atomic espionage emerge in contemporary and retrospective accounts. In a letter to Lavrenti Beria dated September 29, 1944, Igor Kurchatov refers to "new, very extensive [espionage] materials ... concerning the uranium problem" he has been reviewing—that is, materials that had been acquired after the large collection he had already reviewed—and notes parenthetically that these materials constitute "(about 3,000 pages of text)." And the Soviet physicist Yakov Petrovich Terletsky reports that when he joined the special department of the NKVD set up after the end of the war to deal with atomic espionage, he found "about 10,000 pages of ... reports in the safes ... for the most part American classified reports (there were also British materials). They outlined the content of the basic experiments on determining the parameters of nuclear reactions, reactors, and the description of various types of uranium reactors, the description of gaseous-diffusion installations, journal entries on the testing of the atomic bomb and so on.".

One early focus of Soviet espionage was the Radiation Laboratory of the University of California at Berkeley. In 1941, under the direction of the Nobel laureate American physicist Ernest Lawrence, the inventor of the cyclotron, physicists at the Radiation Laboratory began developing electro-magnetic isotope separation, a technology eventually enlarged to industrial scale at Oak Ridge that processed most of the U235 used in the Little Boy bomb. Robert Oppenheimer guided early work on atomic- and hydrogen-bomb theory from offices at Berkeley before he moved to Los Alamos in 1943 to direct actual bomb design. Oppenheimer's wife Kitty had been a member of the Communist Party during the 1930s; his brother Frank and Frank's wife Jackie were members from 1937 to 1941. Oppenheimer himself was "a fellow traveler," as he put it, until 1942, who contributed to Commu-nist causes.

Kitty Oppenheimer's first husband, Joe Dallet, had been a Communist Party official who had volunteered to fight in the Spanish Civil War. In 1937, Kitty had gone to Spain to meet Dallet. One of her husband's comrades-in-arms, Steve Nelson, a naturalized American born in Croatia who was a lieutenant colonel in the Abraham Lincoln Brigade, met her instead and broke the news that her husband had been killed during the siege of Madrid. Nelson had joined the American Communist Party in the late 1920s. He had trained at the Lenin Institute in Moscow in the early 1930s and was known there to be affiliated with the OGPU, the predecessor to the NKVD. He had worked for the Communist International in Shanghai during the same pe-riod when Sonia was active there for the GRU; Arthur Ewert, an agent high in the ranks of the Communist Party of Germany, was a significant connection between them. After the Spanish war, when he may have attended the Barce-lona Intelligence School with Morris Cohen, Nelson turned up in Berkeley, a friend of Kitty Oppenheimer "assigned," according to a congressional committee investigation, "as organizer for the [Communist] Party in the Bay area. . . . He was also given the underground assignment to gather informa-tion regarding the development of the atomic bomb."

Nelson made contact with several of the younger physicists working at Berkeley. Manhattan Project security officers observed him acquiring and passing information on electromagnetic isotope separation to the Soviets:

Late one night in March 1943, a scientist at the University of California, who identified himself as "Joe," went to the home of Steve Nelson. . . . Nelson was not present but arrived at about 1:30 on the morning of the following day. Upon his arrival at his home, Nelson greeted Joe and the latter told him that he had some information that he thought Nelson could use. Joe then furnished highly confidential information regarding the experiments conducted at the [Radiation Laboratory] of the University of California at Berkeley. . . .

Several days after Nelson had been contacted by Joe, Nelson contacted the Soviet consulate in San Francisco and arranged to meet Peter Ivanov, the Soviet vice consul, at some place where they could not be observed. Ivanov suggested that he and Nelson meet at the "usual place."

... The meeting [took] place in the middle of an open park on the St. Francis Hospital grounds in San Francisco. At this meeting, Nelson transferred an envelope or package to Ivanov. A few days after this meeting ... the third secretary of the Russian Embassy in Washington, a man by the name of Zubilin ... met Nelson in Nelson's home and at this meeting paid Nelson 10 bills of unknown denomination. ...

Nelson apparently explored the Oppenheimers' susceptibility to espionage. "Nelson later reported [to his Soviet contacts] that neither the physicist nor his wife were sympathetic to communism," the congressional committee found. If Nelson approached the Oppenheimers, neither of them ever reported the contact.

An approach to Oppenheimer through a different intermediary also failed, but Oppenheimer delayed reporting it, identified the intermediary only reluctantly and later changed his story, vacillations which eventually caused him great trouble. The intermediary was one of his Berkeley friends, a professor of French named Haakon Chevalier; Chevalier was acting on behalf of an Englishman named George Eltenton who was, Oppenheimer would testify, "a chemical engineer ... [who] had spent some time in the Soviet Union" and worked for Shell Development.

In his first version of the events, which he offered in August 1943 to Colonel Boris L. Pash, a Manhattan Project security officer, Oppenheimer connected the Eltenton/Chevalier approach to the Soviet Consulate in San Francisco:

A man whose name I never heard, who was attached to the Soviet consul, has indicated indirectly through intermediate people concerned with the project that he was in a position to transmit without any danger of a leak or a scandal or anything of that kind information which they might supply.

Oppenheimer identified Eltenton; "if you wanted to watch him," he told Pash, "it might be the appropriate thing to do." The physicist added that he did not know "the name of the man attached to the consulate. I think I may have been told and I may not have been told. ... He is and he may not be here now—these incidents occurred in the order of about five, six or seven months ago." Five to seven months before August 1943 would place the incidents around the time Igor Kurchatov, in Moscow, was reviewing isotope-separation technology and assigning research. Kurchatov asked the distinguished Soviet physicist Lev Artsimovich to explore electromagnetic isotope separation; it would have been logical to give Artsimovich any information available on the subject, and according to a Russian scientist, Artsimovich "was introduced to American [espionage] materials on electromagnetic isotope separation."

With the phrase "these incidents," Oppenheimer made evident in 1943 what he would later characterize as "a pure fabrication," "a piece of idiocy": that a military attaché at the Soviet Consulate, through intermediaries, had approached several people connected with the Manhattan Project who had subsequently moved to Los Alamos, and those people had in turn come to Oppenheimer for advice. "I might say the approaches were always made through other people who were troubled by them," Oppenheimer explained, "and [who] sometimes came and discussed them with me and that the approaches were quite indirect." Oppenheimer added: "I know of two or three cases, and I think two of the men are with me at Los Alamos. They are men who are closely associated with me. . . . They told me they were contacted for that purpose [i.e., for information]."

The rationale Oppenheimer's troubled colleagues reported to him, as the physicist described it to Pash, was the standard rationale that Soviet intelligence offered scientists:

> Let me give you the background. The background was, well, you know how difficult it is with relations between these two allies and there are a lot of people that don't feel very friendly towards Russia. So the information, a lot of our secret information, our radar and so on, doesn't get to them, and they are battling for their lives, and they would like to have an idea of what is going on, and this is just to make up in other words for the defects of our official communication. That is the form in which it was presented. Of course, the actual fact is that since it is not a communication that ought to be taking place, it is treasonable.

Oppenheimer himself believed that the world would be safer in the long run if the issues raised by the development of the atomic bomb could be discussed among the Allies, including the Soviet Union, before the end of the war—but he did not believe espionage was the proper channel for such a discussion:

> To put it quite frankly, I would feel friendly to the idea of the Commander in Chief . . . informing the Russians who [sic: that we?] are working on this problem. At least I can see there might be some arguments for doing that but I don't like the idea of having it moved out the back door.

Oppenheimer told Pash that the agent who tempted his colleagues had been careful to present his proposal not as espionage but as a facilitation of existing US policy—an allusion, probably, to Lend-Lease:

> But it was not presented in that method. It is a method of carrying out a policy which was more or less a policy of the Government. The form in which it came was that couldn't an interview be arranged with this man

Eltenton who had very good contact with a man from the Embassy attached to the consulate who is a very reliable guy and who had a lot of experience in microfilm or whatever.

Here were the usual mechanisms of Soviet espionage, paralleling those that Elizabeth Bentley and Harry Gold made notorious in later public testimony: a man from the embassy, a non-Soviet cut-out, an appeal to guilt and rationalization, microfilm. (Igor Gouzenko notes independently the "varied approaches made on Soviet instruction when atomic bomb information was demanded. Astonishingly enough it was shown there that when it comes to something really big, the money appeal isn't used. The appeal to 'higher feelings' such as the 'good of the world' proved most effective for Soviet Intelligence.")

After the war, Oppenheimer would claim that the story he told Pash, except for the name Eltenton, was "wholly false." His revised 1954 version of what happened at Berkeley disconnected him from "microfilm," from the Soviet consulate and from the wider knowledge of espionage approaches that he described to Boris Pash in 1943:

One day . . . in the winter of 1942–43, Haakon Chevalier came to our home. It was, I believe, for dinner, but possibly for a drink. When I went out into the pantry, Chevalier followed me or came with me to help me. He said, "I saw George Eltenton recently." Maybe he asked me if I remembered him. That Eltenton had told him that he had a method, he had means of getting technical information to Soviet scientists. He didn't describe the means. I thought I said "But that is treason," but I am not sure. I said anyway something. "This is a terrible thing to do." Chevalier said or expressed complete agreement. That was the end of it. It was a very brief conversation.

But the FBI interviewed Eltenton in 1946, and Eltenton confirmed a story closer to the original version that Oppenheimer had told Pash:

[Eltenton] admitted being approached by [Soviet military attaché] Peter Ivanov for the purpose of obtaining information as to what was going on "up on the hill [i.e., at the Berkeley Radiation Laboratory]." Eltenton admitted approaching Haakon Chevalier, who he knew was friendly with J. Robert Oppenheimer and requested Chevalier to approach Oppenheimer concerning the project. He advised that Chevalier agreed to the approach and then subsequently advised that there was no chance whatsoever of obtaining the information.

FBI agents interviewed Chevalier the same day in June 1946 that they questioned Eltenton. Chevalier offered a version of events different from

Eltenton's and identical to Oppenheimer's exculpatory 1954 version. Oppenheimer gave Chevalier's version for the first time to the FBI in September 1946; between June and September the two friends had met and had opportunity to concert their stories.

Eltenton may even have maneuvered to approach Oppenheimer directly before Chevalier came to call. So at least an investigator suspected, and seems to have had surveillance to corroborate:

Q. Had you met Eltenton on many other occasions?
A. Oh, yes. . . .
Q. Where?
A. I don't remember.
Q. A social occasion?
A. Yes.
Q. Can you recall any of them?
A. No.
Q. Do you recall who introduced you to him?
A. No.
Q. Did Eltenton come to your house on any other occasion?
A. I am quite sure not.
Q. Did he come to your house in 1942 on one occasion to discuss certain awards which the Soviet Government was going to make to certain scientists?
A. If so, it is news to me. I assume you know that this is true, but I certainly have no recollection of it. . . .
Q. Let me see if I can refresh your recollection, Doctor. Do you recall him coming to your house to discuss awards to be made to certain scientists by the Soviet Government and you suggesting the names of Bush, Morgan, and perhaps one of the Comptons?
A. There is nothing unreasonable in the suggestions.

Lavrenti Beria evidently put uncommon faith in the persuasive power of awards.

————————

Igor Gouzenko was posted to Canada from the USSR in June 1943. Officially he would be a civilian employee of the Soviet Embassy in Ottawa; in fact he was a cipher clerk on the staff of the military attaché, Colonel Nicolai Zabotin, the head of Soviet military intelligence (GRU) in Ottawa (his organization called the NKVD "the Neighbor"). Zabotin—"tall, handsome, personable," writes Gouzenko, someone whose "magnetic personality attracted contacts"—organized a phalanx of Canadian agents among politicians, bureaucrats and scientists working on explosives, electronics and atomic energy.

Israel Halperin, a mathematician who was Canadian-born of Russian parents, was attached to the Canadian Directorate of Artillery and reported to the GRU on weapons and explosives under the code name "Bacon." He carried Kristel Heineman's Cambridge address and Klaus Fuchs's British address in his address book and had supplied Fuchs with science journals when Fuchs had been interned in Canada in 1940.

Edward Wilfred Mazerall, a Canadian electrical engineer, worked on radar. "I did not like the idea of supplying information," he testified. Echoing Oppenheimer, he noted: "It was not put to me so much that I was supplying information to the Soviet Government, either. It was more that as scientists we were pooling information, and I actually asked if we could hope to find this reciprocal."

There were dozens of such conspirators tunneled into the Canadian political and defense establishment whose information Gouzenko coded for forwarding to Moscow, including a Russian-born member of the Canadian Parliament, Fred Rose; Elizabeth Bentley had serviced Rose's correspondence with Jacob Golos in New York a few years earlier through a mail drop. The most significant two among the twenty Canadian agents later identified were the physicists Alan Nunn May and Bruno Pontecorvo.

Nunn May, whom his friends described as "a charming, shy little man with a dry sense of humor" who wore old-fashioned glasses with round lenses, was another Cambridge product, a 1933 graduate who had been recruited by Donald Maclean. He had been a reader in physics at London University in May 1942 when he was asked to join the British atomic-energy program, which was code-named Tube Alloys Research. He had come to Canada from England in January 1943 as a member of a research team headed by John Cockcroft, a senior Cambridge physicist who would win a 1951 Nobel Prize. Joining an existing organization in Montreal, Cockcroft's team carried out research adjunct to the atomic-bomb development work going on in the United States; the Canadians were building a large heavy-water-moderated natural-uranium reactor at Chalk River, three hours north of Ottawa. "Before coming to Canada," a postwar Canadian investigation revealed, "[Nunn May] was an ardent but secret Communist and already known to the authorities at Moscow." Nunn May communicated with Zabotin under the cover name "Alek." He perceived his espionage idealistically, *à la russe*. "The whole affair was extremely painful to me," he would confess, "and I only embarked on it because I felt this was a contribution I could make to the safety of mankind. I certainly did not do it for gain." He was a member of two committees in Montreal which gave him access to secret reports.

In January 1944, Nunn May visited the Metallurgical Laboratory of the University of Chicago, the center of US nuclear-reactor research. He met

General Groves, who had authorized his visit. He returned in April. At that time, Groves reported after the war, "he worked on a minor experiment at the Argonne Laboratory, where the original graphite pile was, and is, located, and where a small-scale heavy water pile had also been constructed." Nunn May visited Chicago again in late August, Groves wrote, "conferring with officials of the Chicago Laboratory on the construction and operation of the Argonne pile and the proposed Montreal pile." On a third and last visit, Groves writes, for the entire month of October 1944, "he carried on extensive work in collaboration with our scientists in a highly secret and important new field." By then, Groves concluded, "May had spent more time and acquired more knowledge at the Argonne than any other British physicist." Groves barred further visits because he felt Nunn May, as a member of the British Mission, knew as much as he ought to know about "later developments."

The "highly secret" work in which Nunn May participated concerned making an atomic bomb using an isotope of uranium, U233, which is even rarer than U235 but which can be transmuted from thorium, element 90, a soft, silvery radioactive metal discovered in Sweden in 1829 and available for refining from monazite sand, of which there were major deposits in Brazil and North and South Carolina. If U233 proved to be bomb material, it could be bred from thorium in a nuclear reactor much as plutonium was being bred from U238, and like plutonium it could then be chemically separated from its parent matrix much more easily than U235 could be physically separated from U238.* Nunn May worked with the American experimental physicist Herbert Anderson in October 1944 trying to determine U233's cross sections for fission. The two physicists used foils of U233 for their cross-section measurements, foils that were extremely rare at the time because the U233 had to be transmuted laboriously in a cyclotron.

Groves thought Nunn May at Argonne had probably learned about the important phenomenon of reactor poisoning, discovered during the start-up of the first big production reactor at Hanford late in September 1944. There is Soviet evidence from the postwar period that the British physicist either did not know of reactor poisoning or did not communicate the information to Soviet intelligence. Other significant Nunn May contributions, however, were yet to come.

Bruno Pontecorvo, handsome as a movie star, was an Italian protégé of Enrico Fermi, one of Fermi's young, vigorous Rome group which had

---

* As it turned out, U233 was not good bomb material. Reactor transmutation of thorium breeds another rare uranium isotope, U232, along with the U233. U232 emits copious alpha particles, which knock unwanted stray neutrons from impurities in the material that encourage predetonation. The United States eventually tested a number of U233 bombs, however.

systematically worked its way through the periodic table in the mid-1930s bombarding the elements with neutrons to identify artificial radioactivities and had barely missed discovering nuclear fission. Pontecorvo, who was Jewish, had escaped France at the time of the German invasion and had found passage through Lisbon to New York. He joined the Anglo-Canadian research group in Montreal in 1943. He was an exceptional physicist, and made himself an expert on heavy-water reactors.

Donald Maclean arrived in New York on May 6, 1944. He was married now, to an American woman named Melinda; his wife was pregnant with their second child and traveled with him. "He is six foot tall," she had described him in a letter to her mother in 1940, when he was courting her in France, "blonde with beautiful blue eyes, altogether a beautiful man." But even then Maclean was drinking too much, partly in response to the stress of his double life; if he had to have a "drinking orgy," Melinda wrote him at that time in concern, "why don't you have it at home—so at least you will be able to get safely to bed?" Harry Gold and Klaus Fuchs also found release in periodic bouts of heavy drinking.

Maclean had served as third secretary at the British Embassy in Paris from September 1938 until the fall of France, in the midst of which he and Melinda had married; they had escaped to England on a tramp steamer. Back in London in wartime, Maclean was stuck in the Foreign Office General Department, bored with matters of shipping, supply and economic warfare, until he left for the United States. Throughout the war he continued his work of espionage. His control, Anatoli Gorsky, attaché and then second secretary at the Soviet Embassy in London, also controlled Anthony Blunt. Blunt had found his way into MI5, the British FBI. In 1940, Maclean met twice with Kim Philby, who had lost contact with his Soviet control. Maclean arranged a renewed connection. Philby, who had worked as a freelance correspondent in Spain during the Spanish Civil War, was beginning his remarkable career in British counterintelligence as a propaganda expert for the Special Operations Executive (SOE), the British counterpart to the US Office of Strategic Services (OSS), the predecessor to the Central Intelligence Agency. By the time Maclean left for the US, Blunt had become responsible for the security of the various governments in exile in London. Philby directed the Iberian section of the counterespionage branch of MI6, the British CIA.

Maclean shipped for America to work at the British Embassy in Washington as a member of the joint Anglo-American secretariat of the Combined Policy Committee (CPC). The CPC had been established at the 1943 Quebec Conference between Winston Churchill and Franklin Roosevelt to facilitate British, US and Canadian collaboration on the atomic bomb. One of its first results was the transfer to the United States of the group of British scientists that included Klaus Fuchs. Another result, which James Chadwick had recommended, would be the development of the Chalk River heavy-water reactor.

Under the CDC, a subordinate body known as the Combined Development Trust (CDT) had taken over work that General Groves had begun late in 1942 buying up rights to corner the world market in high-grade uranium and thorium ores. For Groves, ore was fundamental. Control the supply of high-grade ore, he believed, and other countries, the Soviet Union in particular, could not build atomic bombs. Groves's organization, code-named the Murray Hill Area, had reviewed some 67,000 volumes, more than half in foreign languages, reporting occurrences of uranium ores, had developed the first lightweight, portable Geiger counters for field investigation, had sent out geologists to explore ore fields in the US and abroad and had completed fifty-six geological reports covering more than fifty countries. Groves reported to Secretary of War Henry Stimson on behalf of the CDT in late November 1944 that the US and Britain would control more than 90 percent of the world supply of high-grade uranium ore if Belgium gave them exclusive rights to the output of its Shinkolobwe mine in the Belgian Congo. Before the end of the war, the Belgians agreed. The Soviet Union, the Murray Hill Area investigators had concluded, had only "medium-grade ore. [A] few hundred tons' production. Potential possibilities could be great."

Donald Maclean was in position to communicate such high-level policy information to the Soviet Union. By the time he transferred to Washington, the NKVD had assigned atomic-bomb espionage first priority; Maclean made contact with Anatoli Yatzkov, and would frequently travel to New York to deliver information. If Stalin needed evidence that the nations that called themselves his allies were colluding against him to deny him nuclear weapons while they built up an arsenal, Donald Maclean could supply it. Someone did; a discussion of "the question of the existence and reserves of uranium deposits" and who controlled them turned up in a general NKVD review of Anglo-American bomb development that went to Beria on February 28, 1945.

Nor was the Soviet Union the only country interested in knowing more about American work on the atomic bomb. The work had started in Britain, the British were US allies and had shared their secrets freely, but it was US policy to restrict and compartmentalize British access to American research and development. Thus, for example, General Groves refused to authorize revealing to scientists in Canada the process that Glenn Seaborg and his co-workers at the University of Chicago had developed for separating and purifying plutonium. "As a gesture in their direction," the official Manhattan Project history reports, straight-faced, "Groves agreed to permit a limited amount of irradiated uranium in the form of slugs from [Oak Ridge] to go to Montreal so that the group there could work out independently the methods of plutonium separation and purification." Similarly, Fuchs had not been told that a full-scale gaseous-diffusion plant was under construction in Tennessee.

But the British had decided, probably before their scientific team left England, that they would have to develop their own atomic bomb after the war. John Anderson, who directed British Tube Alloys Research, said as much to the scientists on his staff in January 1944. "We simply could not acquiesce in an American monopoly on this development," postwar British Foreign Minister Ernest Bevin would write. Churchill told Roosevelt of the British decision in February 1945, which raises the interesting question of the extent to which US political leaders tacitly endorsed the British project. When Rudolf Peierls moved to Los Alamos to direct the British group there, James Chadwick asked Peierls to keep him informed:

> I therefore wrote letters at regular intervals in which I summarized, to the best of my knowledge, what was going on. I was a little doubtful about the appropriateness of this, because no secret information was supposed to be sent out from the laboratory without special permission. . . .
>
> Then one day Richard Tolman, a distinguished elder statesman of physics who assisted Groves . . . asked to see me, as he had a message from Groves. When he started, "I understand you have been writing letters to Chadwick about the work of the laboratory," I felt that here my chickens were coming home to roost. But he continued, "General Groves finds that Chadwick is often better informed than he is, and wondered if he could have copies of your letters." He added that, if these letters referred also to purely domestic problems of the British group I could of course omit the relevant passages from the copies for Groves. This made it clear that the intention was not to censor my letters. I was relieved, and highly amused.

Ironically, Peierls was shocked to learn, after the war, of Fuchs's Soviet espionage. Peierls's charming story conceals a serious point: that Groves, who was not only rigorous about security but also a notorious Anglophobe, made an exception to his rules in the case of the British Mission at Los Alamos. He may have felt that limiting British and Canadian access to knowledge of how to separate U235 and plutonium made knowledge of bomb design academic. But not only Soviet agents spirited secret information out of Los Alamos during the Second World War.

What, if anything, the NKVD learned about the Manhattan Project from Morris Cohen's friend "Perseus" is more difficult to assess. According to Yatzkov/

Yakovlev, Perseus was posted to Los Alamos when it opened in April 1943 and Lona Cohen traveled to Albuquerque twice during the war to meet him. In the last months of her life, Lona Cohen confirmed to an American historian that she collected intelligence information from "a physicist" in Albuquerque at least once.

Harry Gold independently confirmed Yakovlev's link with Lona Cohen many years before Yatzkov/Yakovlev went public. "On at least two occasions," the FBI paraphrases Gold's 1950 testimony, "Yakovlev told him he would introduce Gold to a young woman, whose husband was in the United States Army, who would perform the function of doing leg work between Yakovlev and Gold. He recalled that she lived in upper Manhattan . . . and she may have been Russian-born, or of Russian descent, although he never met her."

But nothing in the documents released from Russian archives after the demise of the Soviet Union is identifiable as a Perseus contribution except, possibly, the compilation of 286 papers delivered in 1942 which Igor Kurchatov reviewed on July 3, 1943. All the revealed Los Alamos materials match known contacts between Klaus Fuchs and Harry Gold. If Perseus passed Lona Cohen the "secrets" of the atomic bomb, as Yatzkov claims, the information was redundant. On the other hand, Soviet foreign intelligence thrived on redundancy. Igor Gouzenko sent out the same questions to twenty or more addresses around the world when he worked as a cipher clerk in Moscow. Elizabeth Bentley sometimes suggested to Jacob Golos that one of her less fruitful and more fearful Washington contacts, "Bill," who passed her fragments of information about the activities of the War Production Board jotted down furtively on small scraps of paper, should be dropped from espionage work. " 'No,' [Golos] would say firmly. 'While the material he is producing is not outstanding, it does help to corroborate or supplement what we are getting through [other sources]. And, besides, there is still the possibility that we can push him into a really good position.' " For an institution as cautious and thorough as the NKVD, serving masters as paranoid as Beria and Stalin, redundancy provided independent evidence of the authenticity of the information its spies gathered. Fuchs and Nunn May passed many pages of documents, but not all ten thousand.

Jacob Golos was a harried man. He was not only responsible for the dozens of contacts Elizabeth Bentley serviced in Washington and for operating the travel agency that served as a front for his espionage activities. She understood that he also controlled other cells of spies. He was usually careful not to reveal his other contacts to her, since doing so would cross-link different lines of his espionage network if she were ever exposed. But early in the war Golos had used Bentley as a courier for another operation he directed, and significantly, she first reported the contact in 1945, volunteering the information to the FBI long before any of her or Golos's sources had been made public:

> Another group of whose existence I became aware sometime in the early summer of 1942 was composed of several engineers who, when I first learned of them, were located in New York City. I recall that on one occasion while I was driving through the lower East Side of the City of New York with Golos to keep a dinner engagement, he stopped the car and told me he had to meet someone. I remained in the car and saw Golos meet an individual on the street corner. I managed to get only a fleeting glimpse of this individual and I recall that he was tall, thin, and wore horn-rimmed eyeglasses. Golos told me that this person was one of a group of engineers and that he had given this person my residence telephone number so that he would be able to reach Golos whenever he desired. He did not elaborate on the activities of this person and his associates nor did he ever identify any of them except that this one man to whom he gave my telephone number was referred to as 'Julius.' However, I do not believe this was his true name. I received two or three telephone calls from Julius telling me he wanted to see Golos and relayed the message to Golos. . . . Approximately six months prior to the death of Golos [in November 1943], he told me that he was turning over Julius and that group to some other Russian whom he did not identify.

From her conversations with "Julius" and with Golos, Bentley learned that the tall, studious engineer lived in a housing development in lower Manhattan, Knickerbocker Village. She remembered his calls, spread across the next year, because "they always came after midnight, in the wee small hours. . . . I got waked out of bed. . . . This particular party always started his conversation by saying 'This is Julius.' " Julius would turn out to be the man's real name, Julius Rosenberg. In 1948, when Bentley went public with her story, Rosenberg told one of his espionage contacts, Morton Sobell, that he knew Elizabeth Bentley, had spoken to her by phone, but that everything was all right because she did not know who he was. He confirmed to his brother-in-law David Greenglass in 1950, in Greenglass's words, "that . . . he knew Jacob Golos, this man Golos, and probably Bentley knew him."

Julius Rosenberg was born May 12, 1918, in New York City, one of five children of parents who had emigrated to the United States from Poland. Harry and Sofie Rosenberg hoped their son might become a rabbi, and Julius showed promise, but he discovered politics in high school and chose to major in electrical engineering when he went on to college in 1935. At City College of New York he joined the Steinmetz Club, the campus branch of the Young Communist League, participating with a group of young engineering students that included several who would later be active in Soviet espionage. At a New Year's Eve benefit for the International Seaman's Union during Julius's undergraduate days he met a dedicated, determined young woman, Ethel Greenglass; the two soon fell in love. Ethel, born in 1915, had grown up in poverty in an unheated tenement apartment on the Lower East Side. She had skipped several grades to graduate from high school at fifteen; at nineteen she had organized a strike of some 150 women at the shipping company where she worked—the women finally blocked the company's trucks by lying down in the street. When the shipping company subsequently fired her, Ethel sought and won redress from the National Labor Relations Board and found a better job. Her brother Samuel would testify that she and Rosenberg became "violent Communists" in those Depression years who "maintained that nothing is more important than the Communist cause." They worked to convert Ethel's younger brother David, then a teenager, whom Ethel had already begun proselytizing. At first David disliked his sister's boyfriend and resisted the couple's politics. According to Samuel Greenglass, the gift of a chemistry set won David over. "Samuel Greenglass said that he became so concerned about the Communist influence of Julius and Ethel over David Greenglass," the FBI reports, "that he offered to pay the transportation to Russia . . . if they would agree to stay there. He said that they declined this offer, saying that they desired to remain in the United States."

Rosenberg graduated from CCNY in 1939, a watershed year for him; he and Ethel married on June 18, and he joined the Communist Party on December 12. (Not even the Rosenbergs' sons, who have long protested their parents' innocence of espionage, still dispute the fact of Ethel Rosenberg's CP membership, but the date she officially joined the Party has never been established.) The party cell of which the Rosenbergs became members, Branch 16B of the Industrial Division, included other engineers from Julius's CCNY group, among them Joel Barr and Alfred Sarant, who later defected to the Soviet Union.

After college, Rosenberg went to work for Williams Aeronautical Research in New York. He took a tool design course at Brooklyn Polytechnic and studied aeronautical dynamics and aviation engine design at the Guggenheim Aeronautical School at New York University. In the summer of 1940, moving into position for espionage, he became a civilian junior inspection engineer for the US Army Signal Corps. To do so he had to deny his Communist Party membership. Elizabeth Bentley's 1945 FBI testimony independently corroborates that Rosenberg was working as a Soviet espionage agent by 1942. Julius and Ethel were still active in Branch 16B at that time—in fact, Julius was chairman of the cell.

One of Rosenberg's classmates, Max Elitcher, remembered asking him in 1948 how he had started spying:

He told me that he had a long time ago decided that this was what he wanted to do and he made it a point to get close to people, people in the Communist Party . . . and he kept getting close from one person to another, until he was able to approach someone, Russian . . . who would listen to his proposition.

Rosenberg's proposition was evidently to supply information himself and to recruit engineers from his circle of classmates and acquaintances for espionage as well. He moonlighted his espionage at first—hence the late-night calls to Elizabeth Bentley—but in the longer run he hoped to operate full-time through a front. "I've got powerful friends," he told David Greenglass in 1943, "and we'll go into business after the war. They'll use us as a screen." Greenglass understood that his brother-in-law's friends were "Russians." He dated the beginning of Julius's efforts to "condition" him for possible espionage from that 1943 conversation, which took place, he re-called in 1979, in Manhattan at the Capitol Theater on Broadway. In 1943, Greenglass had thought Julius meant that they would work together after the war and he had been "not so sure" what the work would be. "I suspected espionage," he said in 1979. "I suspected going into business as the back-ground for espionage."

When Harry Gold had told Sam Semenov that Abe Brothman wanted the Soviets to set him up in business legitimately, Sam had called the notion "damned fool nonsense." But if legitimate financing was ludicrous to an agency which was organized, after all, to steal, front operations were not. Jacob Golos's travel agency was one such front. Igor Gouzenko reports a front drugstore in Montreal where the GRU processed espionage film. The expectation that Julius Rosenberg shared with David Greenglass in 1943 was reasonable. It also baited Greenglass with the tantalizing possibility that if he cooperated, he might become the business partner of an older brother-in-law whom he respected and admired.

By the time of his discussion with Rosenberg at the Capitol Theater, Greenglass had been drafted into the Army. He was inducted in April. He had just turned twenty-one—a loud, garrulous young man with a hearty appetite, born on a Lower East Side kitchen table, a machinist like his elderly, Russian-born father, brighter than average, brash, loyal and improvi-dent. The previous November, when he realized that he would be drafted, Greenglass had married his childhood sweetheart, Ruth Printz, a small, pretty nineteen-year-old. Both David and Ruth were members of the Young Communist League, though neither of them ever joined the Communist Party. Ruth was a new convert. From basic training in Aberdeen, Maryland, at the end of April, Private David Greenglass rallied his bride to the cause: "Although I'd love to have you in my arms," he wrote her, "I am content without so long as there is a vital battle to be fought with a cruel, ruthless foe. Victory shall be ours and the future is socialism's." Ruth responded on May 2, after her first May Day, with similar zeal:

> Well darling here it is Sunday and I went to the rally. Well sweetheart all I can say is that I am sorry I missed so many other May Days when I had the opportunity to march side by side with you. The spirit of the people was magnificent. . . . Perhaps the voice of 75,000 working men and women that were brought together today, perhaps their voices demanding an early invasion of Europe [i.e., the second front that the Soviet Union was urging on its Allies] will be heard and then my dear we will be together to build —under socialism—our future.

When David shipped out to Fort Ord, California, to work in a machine shop repairing tanks, the Greenglasses continued their ardent political correspondence. By then, late 1943, Julius and Ethel Rosenberg had quietly dropped out of the Party, but neither David nor Ruth understood them to have withdrawn in disaffection. In a January 1944 letter, Ruth regretted missing Ethel at a rally at Madison Square Garden where Earl Browder, the chairman of the American Communist Party, announced the party's possible dissolution when the war was over because, wrote Ruth, "the people won't be ready to accept socialism and all its reforms." The news made David feel "terribly let down"; he asked Ruth to send him a copy of Browder's speech and to "find out from Ethel what she and Julie think about it. Ask her to get the literature [for me]. Darling, I love you and no matter what happens in America politically. In the end it will be Europe and a large part of Asia that will turn Socialist and the American end of the world will of necessity follow in the same course. So, dear, we still look forward to a Socialist America and we shall have that world in our time."

Around June 1944, Julius Rosenberg traveled to Washington, DC, and called his old CCNY classmate Max Elitcher, who was working for the Navy Bureau of Ordnance on gun fire-control systems. Elitcher invited Rosenberg over. In the course of the evening, Elitcher later testified, Rosenberg asked Elitcher's wife Helene to leave the room and pressed the standard scientific recruiting line on the tall, stoop-shouldered engineer:

> Rosenberg told Elitcher what the Soviet Union was doing in the war effort and stated that some war information was being denied the Soviet Union. Rosenberg pointed out, however, that some people were providing military information to assist the Soviet Union, and that [Elitcher's friend Morton] Sobell was helping in this way. Rosenberg asked Elitcher if he would turn over information of that type to him in order to aid the Soviet Union.

The information would be passed along for evaluation, Rosenberg explained, "taken to New York in containers that would protect it and would be processed and returned before it was missed." According to Elitcher,

Rosenberg's June 1944 contact was the first of some nine attempts to recruit him for espionage. After Rosenberg returned to New York, a coded cable reporting the contact went out from the New York NKVD *rezidency* to the Soviet Union; a copy passed to the Army security agency, which filed it along with thousands of other such undeciphered—and, at the time, indecipherable—messages.

David Greenglass was transferred to Jackson, Mississippi, in the spring of 1944 to work as a machinist at the Mississippi Ordnance Plant. The work gave him time to read, he wrote Ruth on June 29:

> Darling, I have been reading a lot of books on the Soviet Union. Dear, I can see how far-sighted and intelligent those leaders are. They are really geniuses every one of them. . . . Having found out all the truth about the Soviets, both good and bad, I have come to a stronger and more resolute faith and belief in the principles of Socialism and Communism. I believe that every time the Soviet Government used force they did so with pain in their hearts and the belief that what they were doing was to produce good for the greatest number. . . . More power to the Soviet Union and a fruitful and abundant life for their peoples.

Early in July, the Army cut orders to transfer six men from the Mississippi Ordnance Plant to Oak Ridge for assignment to the Manhattan Engineer District. Greenglass's name was not on the list. One of the six men was absent without leave, however, and on July 14 the ordnance plant requested permission to substitute Greenglass for the soldier gone AWOL. Special orders for Greenglass came through on July 24. "I had been conditioned [to consider passing information to the Soviet Union] a long time before," Greenglass recalled in 1979. "Then when I got to Oak Ridge, I said, 'Gee.'"

Oak Ridge was a secret installation, not even marked on public maps. In an isolated region of parallel valleys in the hills of eastern Tennessee, the MED was building a vast gaseous-diffusion plant and a series of electromagnetic isotope-separation units to enrich uranium for atomic bombs. Yet Julius Rosenberg had heard of the installation and thought he knew its purpose. "Julie was in the house," Ruth wrote David on July 31, "and he told me what you must be working on. Sweets, I can't discuss with you (and certainly no one else either) but when I see you I'll tell you what I think it is and you needn't commit yourself."

But Greenglass spent less than two weeks at Oak Ridge. The isotope-separation facilities did not need machinists. Los Alamos did. By August 4, Greenglass was on his way to Santa Fe. In Kansas City he paused to mail Ruth a cautionary letter:

> Dear, I have been very reticent in my writing about what I am doing or going to do because it is a classified top secrecy project and as such I can't

say anything. . . . Darling, in this type of work at my place of residence there is censorship of mail going out and [censorship of] all off-the-post calls. So dear, you know why I didn't want you to say anything on the telephone. That is why I write C now instead of comrade.

The Greenglasses had signed their letters "Your sweetheart, wife and comrade" and "Your husband, lover and comrade," and David had proselytized his buddies. Now that he was traveling to secret work he understood that he needed to keep his political commitments to himself.

David Greenglass arrived at Los Alamos on August 5, 1944, nine days before Klaus Fuchs. "I don't think I . . . ever [saw] him," Greenglass would testify. But the two men shared a common activity: both had been transferred to the Hill (as its occupants called Los Alamos) to help develop implosion. Greenglass joined the Second Provisional Special Engineering Detachment —the SEDs, the technically skilled enlisted men were called—and was assigned to Group E-5 under explosives expert George Kistiakowsky. At first he worked on high-speed cameras and did not realize that the ultimate goal of the project was developing the atomic bomb. "About a month or two after I was assigned there," he recalled after the war, "I heard it among the employees." By October he was machining high-explosive lenses in Group X-1 under Walter Koski, which did flash photographic studies of imploding cylindrical shells. "The group also weighed the advantages and disadvantages of various explosives and explosive arrangements," notes a technical history of Los Alamos. The theoretician who analyzed Koski's photographs was Klaus Fuchs.

Once Greenglass knew what he was working on, he tried to alert Julius Rosenberg, apparently by telegram. He followed up his telegram with a letter to Ruth on November 4:

I am worried about whether you understand what my telegram is about? I really shouldn't because I know that you are intelligent and will understand. I was happy to hear that you spent a pleasant day with the Rosenbergs. My darling, I most certainly will be glad to be part of the community project that Julius and his friends have in mind. Count me in dear or should I say it has my vote. If it has yours, count us in.

"Community project" was "that business with the Capitol Theater," Greenglass clarified in 1979, "that time I suspected espionage." "Friends," as before, were "the Russians."

The Greenglasses missed each other. Their first wedding anniversary was November 29; they decided to rendezvous in Albuquerque to celebrate it.

Before Ruth left, she had dinner with the Rosenbergs. "I got invited to Eth's house for supper," she wrote David on November 15, 1944, confirming the occasion, "so I went home with them. . . . I had a very lovely evening at Eth's as you can imagine. . . . We spoke about several hundred things." Among those several hundred things, Ruth testified later, she and the Rosenbergs discussed espionage and the atomic bomb:

> Julius Rosenberg told me that I might have noticed that he and his wife . . . in recent months had not been attending any Communist Party meetings or any functions that had what he described to be a "Red" tinge to them, and that Ethel . . . had not been buying the *Daily Worker* at her usual newsstand. . . . [He] said he always wanted to do more than to be just a member in the Communist Party and that, therefore, he had searched for two years to place himself in contact with a group which I believe he described as a "Russian underground." In this way . . . [he] felt that he could do the work that he was slated for. . . . He . . . wanted to do something directly to help Russia. . . .
>   Julius . . . then told me that my husband David was at that time working at the place where the atom bomb was being made. . . .

Ruth knew her husband's work was secret, but she had not known its purpose. "I asked [Julius Rosenberg] how he knew and he said he just knew, his friends told him. He knew about it and he wouldn't go into it any further." It excited him. "Then he said that it was the biggest thing yet, that it was top secret." It was more dangerous than any weapon ever used, he added. "He also told me that there were radiation effects from the bomb."

Having identified the quarry, Rosenberg next offered Ruth his standard rationalization for why two American citizens twenty and twenty-two years old should volunteer for criminal espionage:

> He felt it was information that should be shared, that all countries should have it, you know, to their mutual benefit and that Russia was not being given this information and that just on a basis of exchanging mutual scientific information he felt that he was going to do his part to obtain it for them and he asked if I would relay that to David and ask if he would participate.

Ruth Greenglass testified that she objected. "I didn't like the idea." At that point, in Ruth's recollection, Ethel Rosenberg spoke up in support of the project. "When I stated my reluctance, Ethel felt that this would be something that [David] would want to do, that I should mention it to [him], at least I could deliver the message. . . . She said she felt it would be something he would want to know. . . . She urged me to tell David about it, because she felt that he would be willing to do it." Whatever Ruth's reluctance, she agreed to carry the Rosenbergs' message. Julius Rosenberg sweetened the deal with cash. Before Ruth left for Albuquerque, he gave her "about $150 to help pay the expenses of my trip."

Travel was difficult in wartime and Ruth had trouble getting tickets. She took a chance on a seat opening up out of Chicago, left New York early, hung around the Santa Fe ticket window until a ticket agent took pity on her and made it to Albuquerque on Sunday, November 26, two days early. David joined her at the Franciscan Hotel on Tuesday evening on a three-day pass; they stayed together through the weekend. Besides renewing their marriage and celebrating their anniversary they did some shopping; Ruth noticed after David left that she had "accumulated plenty of junk to take back."

Ruth waited until late in the vacation to deliver the Rosenbergs' message. "We went for a walk out on Route 66," David would testify, "past the ... Albuquerque City limits, and not yet to the Rio Grande River, and my wife started the conversation." Ruth began by telling her husband that he was working on the atomic bomb. "I was very surprised," he recalled. "David asked me how I knew about that," Ruth said, "because he had never divulged any information, and I told him that Julius told me." She described her dinner with the Rosenbergs and their proposal. "She said that my brother-in-law explained that we are at war with Germany and Japan and they are the enemy and that Soviet Russia is fighting the enemy and is therefore entitled to the information." Ruth also told her husband, in his words, "that she didn't think it was a good idea ... and that she didn't want to tell me about it." "I felt that we had taken something into our hands that we were not equipped to handle," Ruth explained her misgivings, "[that] we were tampering with things that were beyond our knowledge and understanding. ..." She asked her husband what he thought about it. Reality was different from vague promises of going into business after the war, David remembered feeling; "you're jumping into cold water." "At first I was frightened and worried about it and I told ... my wife that I wouldn't do it." But he thought about it overnight, consulting "memories and voices in my mind," and loyalty won out over caution. "I felt it was the right thing to do ... according to my philosophy at the time," he would testify. "... I started to have doubts almost as soon as I said that I was going to give the information. ... [But] I had a kind of hero worship there and I did not want my hero to fail, and [by refusing to cooperate] I was doing the wrong thing by him. That is exactly why I did not stop the thing after I had the doubts." His hero, he said, was Julius Rosenberg. The next morning he told Ruth he was in.

"She asked me for specific things that Julius had asked her to find out from me," David remembered. "She asked me to tell her about the general layout of the Los Alamos atomic project, the buildings, number of people and stuff like that; also scientists that worked there, and that was the first information I gave her." Among other names, David remembered mentioning his superior, George Kistiakowsky, as well as Robert Oppenheimer and Niels Bohr.

Rosenberg had asked Ruth to determine the bomb laboratory's situation. Surprisingly, David took her to see it—to see, in her words, "how it was located, whether it was camouflaged, whether you could see it easily. And I remember it now, as I saw it while I was there: it was very high on a hill, the place had been a school for horseback riding—a girls' school [sic: Los Alamos had been a private boys' school before the Army requisitioned it]. It couldn't be seen or easily detected until you were almost upon it. And of course it was guarded; there was a guard checking everyone going in and out."

More train trouble delayed Ruth's travel home. On Monday night she got a coach seat to Chicago. The train broke down in Newton, Kansas, and was late into Kansas City. She was stuck in Chicago until Wednesday; she finally returned to New York on Thursday. A few days later, Julius Rosenberg stopped by her apartment—"alone," she said. "He was almost always alone." By then she had written down what David had told her and what she had seen of Los Alamos. She gave her brother-in-law her notes; he told her he would discuss the information further with David when the young machinist came home on furlough.

David Greenglass returned to Los Alamos from his second honeymoon alert to learn more about the novel technology he was helping develop, but he quickly realized that he lacked a frame of reference. "I didn't exactly know what I was looking for," he testified; "I didn't have a conception of how the bomb was made. . . ." He began paying attention, listening, questioning the men with whom he worked. "The scientists would come into the shop, and the man who was in charge would assign a man to work with him. Three of us would stand around and talk . . . and after something was decided upon, the machinist who was given the job would do the job. . . . That way, of course, I did get to learn a lot about what was going on." He knew something by then about high-explosive lenses, having machined lens molds—forms in which to cast HE—for imploding-cylinder experiments in Walter Koski's group. It was a beginning, something to sustain his hero, something to carry back to New York.

T/5 David Greenglass, Army Serial Number 32882473, left Los Alamos on furlough on December 30, 1944, and arrived in New York on New Year's Day. The Greenglasses had no telephone; Julius Rosenberg turned up at their apartment soon after David got home. "We were trying to enjoy our furlough," Ruth recalled impatiently, "and . . . he came to our house for the purpose of discussing [the atomic bomb] with David. We were a little peeved with him because we felt that he was interrupting. . . ." David remembered a more productive morning*:

---

* In a series of testimonies at various times, in 1950 and after, which I compile here into one coherent statement; for sources, cf. Notes.

Rosenberg described to me generally how the atom bomb functions. . . . He said, Now I will explain and you [will] understand what we are looking for; you tell us what has gone on in the making of the bomb, give us materials, methods of use, experiments necessary. . . . He didn't tell me . . . who gave him the information. [I asked him.] . . . He ignored [my question]. . . . He said there was fissionable material at one end of a tube and at the other end of the tube there was a sliding member that was also of fissionable material and when they brought these two together under great pressure . . . a nuclear reaction would take place. That is the type of bomb that he described.

Rosenberg had described Georgi Flerov's "cannon" design, a uranium gun like the gun that Los Alamos was developing that would be nicknamed Little Boy. Greenglass had not worked on uranium-gun development and knew nothing of gun design; he had been working on HE lens development for the implosion bomb. As of early January 1945, NKVD *rezidents* had apparently not yet been made aware of the problem of plutonium predetonation or of implosion.

Rosenberg asked Greenglass what he was doing at Los Alamos. Greenglass told him he was working on high-explosive lenses. "He told me to write up anything that I knew about the atomic bomb," David testified, "write it up at night . . . and he would be back the following morning to pick it up." Rosenberg also asked for a list of Los Alamos scientists and of possible espionage recruits.

That night Greenglass wrote out his lists and drew "a number of sketches showing various types of lens molds." The only sketch he reproduced that was subsequently made public was what he called "the flat type lens mold," which was used at Los Alamos to mold two-dimensional HE assemblies for experiments imploding cylinders. The mold was shaped something like a four-leaf clover. "It has four curves on it," Greenglass would testify, ". . . it is hollow at the center and it was used to pour HE into it. . . . The HE took on the shape of the mold and the mold was removed and you had a high-explosive lens." The two-dimensional HE lens (which had other components besides the molded explosive Greenglass sketched) fit around a length of pipe like an Elizabethan collar with detonators at the apex of each of the four clover leaves; when the detonators were fired, the HE shaped an inward-moving detonation wave that pinched the pipe shut. It was a long way from imploding cylinders to three-dimensional lensed implosion systems, but in fact the two-dimensional experiments proved crucial to the design of the small device at the center of the implosion system—the initiator—that produced a burst of neutrons at the right time to start the chain reaction.

The next morning, Rosenberg came to pick up the lists and sketches that Greenglass had prepared and invited David and Ruth to dinner.

The Rosenbergs rented a modest one-bedroom eleventh-floor apartment, G-11, at 10 Monroe Street in Knickerbocker Village. When the Greenglasses arrived for dinner, they found another guest on hand, a woman named Ann Sidorovich. The Sidorovichs were friends of the Rosenbergs—Mike, Ann's husband, was an engineer—who lived in the New York suburb of Chappaqua. Ruth had seen Ann at the Rosenbergs' apartment several times before, but David had never met her. That evening before dinner, Ruth recalled, "she was there for a while and then she left and we remained. After she had gone, Julius said she was going to come to New Mexico to get the information from David. He said it would be either Ann or someone else, and I asked how [David] would know anyone else if she didn't show up. . . . At that point we were in the kitchen and [Julius] cut this Jello boxtop and he said one-half would be an identification [for] whoever came and he gave me the other half. . . . [Ethel] was standing behind him in the kitchen. . . . She saw it and heard it. . . . I slipped [the boxtop half] into my wallet."

Ruth kept the boxtop half because she was moving to New Mexico. At about the time of her November visit, Los Alamos had authorized enlisted men to quarter their families nearby. After dinner, David testified, "the Rosenbergs told my wife that she wouldn't have to worry about money because it would be taken care of. . . . She would be able to get out there and live out there, if she wasn't able to work, and money would be forthcoming."

David and Julius discussed high-explosive lenses. Julius was keen to know more about how they worked and so was his Soviet control. "[Julius] said that he would like [me] to meet somebody who would talk to me more about lenses." David was willing. Rosenberg briefed David on protocols, Ruth remembered. "I recall him telling [David] that he wanted him not to be obvious or take anything [such as] sketches or blueprints or material but that he should relay whatever he knew from information he had been working on and saw around him."

That evening, or at some other time during David's January furlough, Julius filled in the Greenglasses on some of his own activities:

Rosenberg told me that the Russians had a very small and a very poor electronics industry, that is, of course, another name for the radar industry, and that it was of the utmost importance that information of an electronics nature be obtained and gotten to him. Things like electronic valves (vacuum tubes) capacitators, transformers, and various other electronic and radio components were some of the things he was interested in. Rosenberg also told me that he gave all of the tube manuals he could get his hands on to Russia, some of which were classified Top Secret.

Elizabeth Bentley notes the curious Soviet penchant for gifts and awards. "For some strange reason," she writes, "it was a tradition in the NKVD that at Christmas everyone who worked for them—no matter in what capacity—

received a gift." She was another of those who received an Order of the Red Star. Her new control after Jacob Golos's death, Anatoli Gromov (as Gorsky now called himself), who had followed Donald Maclean to America, told her the Order "entitles you to many special privileges; . . . you could even ride on the street cars free." The Rosenbergs also received gifts, David Greenglass testified, and Julius had received a citation:

> [Julius] stated that he had gotten a watch as a reward. . . . He [showed me the watch.] His wife received also a watch, a woman's watch, and I don't believe it was at the same time. . . . [It was] later, at a later date. . . . I believe they [also] told me they received a console table from the Russians. . . . [Julius] said he received a citation. . . . He said it had certain privileges with it in case he ever went to Russia.

So Julius Rosenberg, like Harry Gold and Elizabeth Bentley, was assured of free trolley rides in Moscow.

A few days later, Greenglass remembered, Julius "asked to see me one night. I had a previous appointment of a social nature to see some personal friends and cut the appointment short in order to meet my brother-in-law." Greenglass borrowed his father-in-law's car, a 1935 Oldsmobile, and around eleven-thirty at night, "drove to the vicinity of about First Avenue somewhere above East 42nd Street but below East 59th Street," up the block from a brightly lit saloon. "I parked the car at the curb. . . . Julius Rosenberg walked over to the car and told me to wait. Then he walked away and came back with a man and introduced him to me by a first name which I do not recall. Then the man got into the car and I drove around."

He drove "all over that area," Greenglass testified. The man—"a Russian" —"just told me to keep driving and he asked questions about lenses. . . . He wanted to know . . . the formula of the curve on the lens; he wanted to know the HE used, and means of detonation; and I drove around . . . and being very busy with my driving, I didn't pay too much attention to what he was saying, but the things he wanted to know, I had no direct knowledge of and I couldn't give a positive answer." Greenglass nevertheless concluded that the man was technically trained and that, in FBI paraphrase, "the high-explosive lens approach to the problem of constructing an atomic bomb was an entirely new one to him." Greenglass's information on implosion, however limited, was the first news the Soviets had of the radical new approach.

Greenglass returned the Russian to their starting point. Rosenberg was waiting. " 'Go home now,' " Greenglass testified Rosenberg told him. " 'I will stay with him.' He was going to have something to eat with him." The Russian got out and the two conspirators went off together. Greenglass drove home and told his wife about his unusual encounter.

The identity of this mysterious Russian has never been established. He was almost certainly not Yatzkov/Yakovlev, since Greenglass noticed that he spoke almost accentless English, while Yatzkov had begun learning English only three months before he came to the United States. Sam Semenov spoke excellent English, having attended MIT, but he had left for Vladivostok through Kalama, Washington, on September 30, 1944. Yatzkov is nevertheless the likeliest person to have sought the information, whomever he sent to collect it, since he was evidently managing atomic-bomb espionage out of New York City at the time.

David Greenglass returned to Los Alamos on January 20, 1945, prepared to observe and to memorize. With Julius Rosenberg's explanation of how an atomic bomb worked, he testified, "I knew what to look for." Now Los Alamos sheltered at least two active Soviet spies, both of them positioned fortuitously at the very heart of the project.

8

Explosions

WALKING from the Moscow subway station to Laboratory No. 2 for the first time, one morning in 1944, the Soviet physicist Anatoli Alexandrov lost his way and stopped to ask a gang of neighborhood children for directions. "It's over the fence where they're making the atomic bomb," one of the children told him. Work proceeded slowly at the secret laboratory, paced by the exigencies of the war and the limited support that the Soviet bomb program had managed to win from Molotov. "These talented scientists and engineers," comments chemical industry commissar Mikhail Pervukhin, "started theoretical work aimed at determining the critical masses of U235 and plutonium despite having on hand not a single milligram of either substance." Igor Kurchatov had begun designing a first small graphite–natural uranium reactor in July 1943, but the Soviet Union lacked industrial sources of metallic uranium and high-purity graphite and would not produce sufficient supplies of either material until after the defeat of Germany. When physicist Boris G. Dubovsky joined the lab in 1944 the staff was still, he recalled, "very small—only several dozen people. There was enough nuclear 'virgin land' for all of us to plow. Work on the main problem—the nuclear reactor—had already begun. We were supposed to confirm the theoretical concept of the possibility of a chain reaction. The same reactor was meant to produce the first weighable quantities of the new nuclear fuel which is now known as . . . plutonium. . . ."

Other research toward a bomb was ongoing at Laboratory No. 2 and elsewhere in the USSR. Espionage may have been a source of ideas and information, but ultimately every experiment would have to be replicated and every number checked. "It looks as though we're going to live in Kharkov again," Eddie Sinelnikov wrote her sister in England on February 15, 1944:

As I wired you today, Kira has been appointed Director of the old Institute. . . . I'm not *very* enthusiastic about Kira being Director—with his health in such a state I'm not sure that it won't be too much of a strain. Things are difficult and everybody is "nervous" to put it politely. On the other hand I'm tired of traveling and it seems a terrible shame that an Institute like ours should just dissolve into thin air. . . . Kira will have to do a lot of traveling between Moscow, Kharkov, and Kiev, but when the war is over I hope things will be easier. . . . Kira is at present in Kharkov for ten days, and we are staying in Moscow with [Sinelnikov's sister] Marina [Kurchatov]. Jillikin's aunts utterly ruin her. She has had so many presents since we arrived here that her head is quite turned. I hope we shall be able to get to our old home in April so that it won't be too late to begin gardening.

Continuing a tradition he had begun at Cambridge, Peter Kapitza instituted seminars—Kapitza Wednesdays, they were called, something like an American journal club—to keep Soviet physicists up to date on unclassified aspects of the work. The experimental physicist Veniamin Aronovich Zukerman describes his debut on a Kapitza Wednesday in March 1944 on the same program with Yuli Khariton; both men's reports related to bomb research:

> The first report was given by Yu. B. Khariton. It was on mechanisms of explosive reactions. The second report—on flash [X-ray] radiography of explosions—was mine. Kapitza chaired. That was my first meeting with Peter Leonidovich Kapitza. I was struck by his engineer's grasp of subject matter and by his high voice. I remember he pronounced the Russian word *kondensàtor* like its English equivalent, condenser. The seminar room was crowded with well-known physicists—A. F. Ioffe, L. D. Landau, I. E. Tamm, N. N. Semenov, Ya. B. Zeldovich. . . . My report generated a lot of interest. Many present knew that this particular work had been nominated for a . . . Stalin Prize.

That year, Zukerman's group took up "intensely studying extremely sensitive explosive primers, such as lead azide and fulminate of mercury," dangerous objects which Zukerman often carried illegally in his pocket, "in a special container with shock-mounts," by streetcar from the institute that manufactured them to his laboratory. Zukerman's eyesight was deteriorating from retinitis pigmentosa, and one evening when he was transporting lead azide primers and his streetcar was late his fellow passengers had to help him find his way. When his colleague Lev Altshuler heard about Zukerman's adventures he commented, "For a few hours there, you were just a roaming torpedo, weren't you." ("During the last year of the war," Zukerman explains, "the seas and oceans were full of torpedoes that had missed their targets; they were christened 'roaming torpedoes.' There were many incidents where military and merchant vessels were blown up by such torpedoes.")

The big Leningrad cyclotron was rebuilt and operating by the time Boris Dubovsky arrived at Laboratory No. 2 in August 1944, and using it, in October, Boris Kurchatov produced the first micrograms of plutonium transmuted outside the United States. "Just look at this date, please," Dubovsky appeals. "The end of 1944. The war has just moved from our territory. Half of the country lay in ruins. The fascist beast is still alive and thousands of people are dying on battlefields and in concentration camps." Soviet scientists did better than overburdened wartime industry. "At that time," says State Defense Committee science deputy Sergei Kaftanov, "we practically possessed no raw materials. . . . The country's existing uranium mines had been flooded and abandoned. . . . We had to restore them and we had to look for new uranium deposits." As late as May 1944, V. I. Vernadski complained in a letter to the government Committee on Geological Affairs that he had "not received from you, in spite of your promise, news of the results of the pumping-out of Tiuia-Muiun. Money was allocated in sufficient quantity, there is ore, why the delay? This ought to have been done long ago." The Soviet reactor would need roughly fifty tons of purified uranium, Kurchatov told Mikhail Pervukhin. The first bags of uranium ore came out of the central Asian mines on the backs of donkeys. The State Institute of Rare Metals purified a first small piece of metallic uranium only in November 1944, and graphite production had not yet begun at Moscow Electrode.

With the Anglo-American invasion of Normandy on June 6, 1944, Stalin finally had his Second Front; the Allies, Soviet and Western, now pushed from opposite directions toward Berlin. "The roads are all cluttered up with the traces of a German retreat," Konstantin Simonov wrote back from the advancing Soviet front:

> . . . I am amazed day after day by the quantity of machines . . . abandoned by the Germans. Here are the notorious Tigers and Panthers, burnt and whole, and tanks of older types, and self-propelled guns, and huge armored carriers, and small carriers with one driving wheel looking like motorcycles, and huge, snub-nosed Renault trucks stolen from France, and numberless Mercedes and Opel staff cars, wireless units, field kitchens, antiaircraft installations, disinfection-chamber vans—briefly everything that the Germans had thought up and utilized in their past impetuous advances. And all that is now smashed, burned, or simply abandoned, stuck in the mud of these roads.

After the early disasters, the Soviet advance seemed almost miraculous: the Leningrad blockade broken in January 1944, the breakthrough to Romania in February and March, Odessa liberated in April, the Crimea completely cleared in May, Finland finished in June, the western Ukraine liberated in July all the way to Warsaw, Romania surrendered in August, Estonia and Latvia cleared in September, Hungary, eastern Czechoslovakia and northern Norway entered in October. American Lend-Lease was feeding several million Soviet civilians and half the Red Army. Stalin would acknowledge that about two-thirds of his major industries were being rebuilt with US equipment or technical assistance. But the blood that was spilled on the way west to Berlin was Russian blood. In 1943, Franklin Roosevelt's adviser and aide Harry Hopkins had noted that the Soviet Union "is the decisive factor in the war . . . [and] without question . . . will dominate Europe on the defeat of the Nazis. . . ." Certainly Stalin meant to do so. He also understood, he told Milovan Djilas one evening in March 1944, that the West would resist him:

> Stalin then invited us to supper, but in the hallway we stopped before a map of the world on which the Soviet Union was colored in red, which made it conspicuous and bigger than it would otherwise seem. Stalin waved his hand over the Soviet Union and, referring to the British and the Americans, he exclaimed, "They will never accept the idea that so great a space should be red, never, never!"

The prospect of an eventual end to the terrible war stirred old enmities. Averell Harriman, for one—since October 1943 the US ambassador in Moscow—took the Soviet determination to collect its spoils and secure its dominance as a threat. "What frightens me [about Soviet policy toward Poland and Eastern Europe]," he wrote Secretary of State Cordell Hull on September 20, 1944, "is that when a country begins to extend its influence by strong-arm methods beyond its borders under the guise of security it is difficult to see how a line can be drawn. If the policy is accepted that the Soviet Union has a right to penetrate her immediate neighbors, . . . penetration of the next immediate neighbors becomes at a certain time equally logical." Harriman's analysis was an early version of the domino theory that would shape American thinking about the Soviet Union for most of the rest of the twentieth century. It was hardly logical from a military point of view, since control and supply both attenuate with distance. Nor could it take into account what Harriman was not yet aware of, the coming US monopoly on the atomic bomb. But Harriman had seen ravaged Europe and knew Britain was nearly bankrupt; he had smelled the excitement in Moscow at the prospect of territorial gains and bounteous reparations; and he understood that the supply lines would be even longer from the United States.

Winston Churchill was more pragmatic or more cynical. Meeting with Stalin in Moscow in October 1944, he proposed that the two leaders "settle about our affairs in the Balkans. . . . Don't let us get at cross purposes in small ways. So far as Britain and Russia are concerned, how would it do for you to have ninety per cent dominance in Romania, for us to have ninety per cent of the say in Greece, and go fifty-fifty about Yugoslavia?" Churchill wrote out the percentages, adding "Hungary . . . 50–50%" and offering Stalin 75 percent dominance in Bulgaria, and pushed the paper across the table. "There was a slight pause. Then [Stalin] took his blue pencil and made a large tick upon it, and passed it back to us. It was all settled in no more time than it takes to set down." It was hardly settled at all, if only because Stalin expected to dominate the nations on Churchill's list, with the possible exception of Greece, not fifty or seventy-five or ninety but a full one hundred percent.

In February 1945, Soviet agents in North America delivered a rich harvest of atomic espionage to Moscow Center. Alan Nunn May weighed in first. Colonel Nicolai Zabotin, the GRU officer in Ottawa, had assigned a young lieutenant on his staff to control Nunn May after orders came from Moscow sometime late in 1944 to reactivate the British scientist, who had not been approached since he left England. The young officer, whose name was Angelov, had simply gone to Nunn May's apartment on Swail Avenue in Montreal, knocked on the door and identified himself. Renewed contact disturbed Nunn May, who seems to have imagined he could withdraw his services unilaterally; he told Angelov that his old connection had been severed and that he was under observation by Canadian security. Angelov thought Nunn May "a man who seemed to be trapped," but he was not impressed; he had a job to do. "I told him quite bluntly that I didn't believe him and that Moscow had an assignment for him," the officer bragged afterward to Igor Gouzenko. "If he refused the assignment it would be his worry, not mine. He seemed to shrink up before my eyes. Finally, he asked me what I wanted. I told him Moscow wanted a report on atomic bomb research in Canada and the United States." Nunn May asked for a week to prepare the report. They met a second time a week later at Nunn May's house.

Igor Gouzenko saw the document that Nunn May prepared when Zabotin passed it to him for ciphering. He described it in 1948:

The report obtained from Dr. May was extensive and comprehensive. It came in two sections. . . .

One part, covering the technical processes being followed in the bomb's construction, was ten single-spaced typed [pages]. . . .

[The second part] was a general description of the atomic project's organization in Canada and the United States. It explained the structure of the whole Manhattan Project and the War Department officials and scientists in charge. . . .

Zabotin was particularly delighted over Dr. May's naming of the highly hush-hush plants and the nature of the work being done at Oak Ridge,

Tennessee, at the University of Chicago, at Los Alamos, New Mexico, and at Hanford, Washington.

Gouzenko advised Zabotin that the technical part of the document, with its new and unfamiliar terminology, would be difficult to cipher and decipher without "costly errors." Zabotin decided to send it by diplomatic pouch. Gouzenko proceeded to cipher the general description, which was transmitted to Moscow by cable. The GRU shared it with the NKGB—the NKVD foreign intelligence division—and Vsevolod Nikolayevich Merkulov, the NKGB head, incorporated it into a summary of the Anglo-American program that went to NKVD commissar Lavrenti Beria on February 28, 1945. Besides the details of organization and personnel that delighted Zabotin, Merkulov's summary mentioned "two methods under development for activating the bomb: (1) the ballistic method and (2) the method of implosion"—another reference, independent of David Greenglass, to the radical new technology Los Alamos was inventing for assembling a critical mass with high explosives. The NKGB summary also included a discussion of sources of uranium ores and of American efforts to gain "unlimited control over mining of uranium ores in the Belgian Congo." The likeliest source of this information was Donald Maclean.

Igor Kurchatov reviewed espionage material on March 16, 1945, that appears to have included the first part of Nunn May's report. "The material is of great interest," Kurchatov wrote with excitement: "Along with methods and schemes which we have developed independently it discusses possibilities which we have not yet considered." One possibility concerned making a bomb with a nuclear core diluted with hydrogen—with uranium or plutonium hydride, that is. Because the hydrogen would slow secondary neutrons, increasing the number of fissions and therefore reducing the amount of uranium or plutonium needed (by a factor of twenty, the espionage document estimated), Edward Teller had championed such a scheme at Los Alamos. Further examination had made clear to the Americans what Kurchatov immediately deduced, that a hydride core, with its slower reaction rates, would blow itself apart before the reaction could chain through enough generations for an efficient explosion. Work on a hydride gun essentially ended at Los Alamos in August 1944, but someone like Nunn May, collecting information far from the source, might not have known that. Kurchatov was eager to know if this odd bomb design had been studied only through calculations or experimentally—if experimentally, then "that would mean that the atomic bomb has already been realized [by the Anglo-Americans] and that U235 has already been extracted in large quantities." He suggested "obtaining several grams of highly-enriched uranium from the American laboratories mentioned in the espionage material" he was reviewing. By "obtaining," of course, he meant stealing.

The more significant possibility discussed in the materials Kurchatov re-

viewed on March 16, 1945, concerned implosion. Kurchatov gave no indication that he had heard of implosion before reviewing the documents in
hand even though implosion is mentioned briefly in Merkulov's February
28 summary. The Soviet physicist was impressed:

> The "implosion" method uses tremendous pressures and velocities created
> by explosion. It is said in the [espionage] material that this method makes
> it possible to increase the relative velocities of particles up to 10,000 meters
> per second, providing that symmetry is achieved, and hence, this method
> is preferable to the gun method.
>
> Now, it is difficult to assess whether this conclusion is correct or not, but
> without doubt "implosion" is of great interest, is correct in principle and
> should be subjected to serious theoretical and experimental study.

If the information that plutonium bred in a natural-uranium reactor could
be a shortcut to the bomb was the first Anglo-American breakthrough that
the Soviet espionage network delivered to Soviet scientists, the information
that implosion was superior to gun assembly was the second. But whether
this information came from Alan Nunn May or from some other source, as
yet unknown, the declassified Soviet record does not reveal. It almost certainly did not come from Klaus Fuchs, who arrived at Los Alamos after the
hydride gun was abandoned, and who knew, by the time he visited his sister
in Cambridge in February 1945, what the documents Kurchatov reviewed
on March 16 apparently failed to report: that implosion was not only desirable for plutonium assembly but also necessary, because all Pu239 bred in
a reactor, whether American or Soviet, would be contaminated with Pu240,
and a gun bomb loaded with such material would detonate prematurely.

"I went up to Cambridge and saw Klaus there," Harry Gold remembered of
his February 16, 1945, meeting with Klaus Fuchs. It was winter in Massachusetts and there was heavy snow on the ground. Gold stopped along the way
to buy a book for Kristel Heineman—a piece of froth titled *Mrs. Palmer's
Honey*—and candy for the Heineman children. Gold had bragged of his
own children on one of his earlier visits, imaginary children made up as a
cover but elaborated into a fantasy of the family life that the lonely bachelor
chemist never knew. Essie and David, Gold would confide to Abe Brothman's secretary, those were the children's names—twins, a boy and a girl,
and his wife was a former Gimbel's model. Long afterward, defenders of
the Rosenbergs would cite Gold's family fantasies as evidence that he had
concocted his tales of espionage, but not even Harry Gold could have invented Harry Gold.

"Mrs. Heineman stated that she brought the chemist into her living room," the FBI paraphrases, "where Fuchs was then sitting." Kristel excused herself, Gold testified, "saying 'I have to pick up the children from the school.' Klaus asked me to go upstairs with him to his room, which as I recall was the front one looking out on the street, and we sat there for possibly fifteen or twenty minutes."* Fuchs briefed Gold on his move to Los Alamos and described the place. They had made tremendous progress, Fuchs said. Gold remembered that "he . . . made mention of a lens, which was being worked on as a part of the atom bomb." Fuchs told Harry "that he was getting along very well [at Los Alamos], but that he was strictly limited in regard to being able to leave. . . . He said that it had only been with the greatest difficulty and due to the fact that he had gotten a bit ahead of schedule on his work, as regards the rest of the group, that he had been able to wangle time off to come to Cambridge." Gold proposed meeting again in Boston along the Charles River, a prearrangement with Fuchs that Yatzkov had mentioned to Gold when they had met a few days previously in Philadelphia. "[Klaus] told me that such would be impossible; that he was certain that it would be a very long time, possibly even a year, before he could again leave Los Alamos, and that the next meeting would have to take place in Santa Fe." Fuchs mentioned April. Gold told him "that I could not possibly get to Santa Fe in April." They settled on early June.

To identify a location for the June meeting, Fuchs gave Gold a map: "a yellow folder," the FBI describes it. "Outside of this folded circular are printed the words 'Santa Fe The Capital City Different in the Land of Enchantment.' Both sides of this circular contain maps. One side contains a Chamber of Commerce map of the City of Santa Fe, New Mexico, which was compiled April 1940. This side of the folder shows a complete layout of the Santa Fe streets, public buildings, churches, hotels, restaurants, and auto courts. On the reverse side of this pamphlet is a map of the area surrounding Santa Fe." Fuchs pointed out the Castillo Street Bridge (over the Santa Fe River) and proposed to meet there at four in the afternoon on the first Saturday in June.

At that point, Fuchs passed Gold, in Gold's words, "a quite considerable packet of information." The contents of the packet, Fuchs would confess, covered everything he knew up to that time about bomb design:

Fuchs wrote a report . . . summarizing the whole problem of making an atomic bomb as he then saw it. This report included a statement on the special difficulties that would have to be overcome in making a plutonium bomb. He reported the high spontaneous fission rate of plutonium[240]

---

* Fuchs later claimed that he only passed documents to Gold in Boston, but he was obviously lying to protect his sister, who independently confirmed Gold's version of events.

and the deduction that a plutonium bomb would have to be detonated by using the implosion method rather than the relatively simple gun method. . . . He also reported that the critical mass for plutonium was less than that for U-235 and that about five to fifteen kilograms would be necessary for a bomb. At this time the issue was not clear as to whether uniform compression of the core could be better obtained with a high-explosive lens system, or with multipoint detonation over the surface of a uniform sphere of high explosives. He reported the current ideas as to the need for an initiator, though these, at the time, were very vague, and it was thought that a constant neutron source might be sufficient. Finally . . . he referred only to the hollow plutonium core for the atomic bomb as he did not then know anything about the possibility of a solid core.

Fuchs also reported the outer dimensions of the high-explosive lens system (which were effectively the outer dimensions of the bomb), the timing sequence for implosion and the plans for building and producing bombs at Los Alamos to the extent he knew them. From memory, he incorporated into his report portions of his two most recent Los Alamos technical studies: *Jet formation in cylindrical implosion with 16 detonation points* and *Formation of jets in plane slabs.* These studies were based on Walter Koski's work, in which David Greenglass was participating; the drawing Greenglass gave Julius Rosenberg in January depicted a mold for a cylindrical implosion lens with four-point detonation. Fuchs's and Greenglass's common references would have served Yatzkov as independent confirmation of the authenticity of the information his Los Alamos spies were passing.

"Mrs. Heineman had returned" by then, Gold says, "and one of the children peered curiously into the room. Mrs. Heineman called the child back. . . ." With the information in hand that he had come for, Gold was ready to leave, but he had one more duty to perform.

It was standard NKVD practice to try to buy even the organization's most high-minded spies. Yatzkov, probably nervous about the long hiatus between contacts with Fuchs, had given Gold the munificent sum of 1,500 1945 dollars—about $30,000 in 1995, more than poor Harry ever got at one time—to pass to Fuchs, with the caution "that I must proceed very delicately . . . so as not to offend him and that under no circumstance must I insist upon or make an issue of this matter." Harry also had an NKVD "Christmas present" for Fuchs, in the tradition that bemused Elizabeth Bentley, "a wallet of the very thin dress or opera type." Fuchs accepted the wallet, "but looked somewhat bewildered, and when I made some very tentative inquiries concerning whether he needed any money either for himself or possibly for his sister, the reply was so cold and final that I went no further with the matter. It was quite obvious that by even mentioning this, I had offended the man." "Fuchs held the envelope containing the 1,500 dollars as if it were an unclean thing," Gold remembered at another time, "and flatly refused to accept it." Five years later, Fuchs was still insulted. "He turned down this offer," he told the FBI, "and stated he would not do such a thing."

Gold backed off: "I left shortly thereafter and returned to New York."

Gold passed Fuchs's report to Yatzkov/Yakovlev and told him about the high-explosive lens that Fuchs had mentioned. At their next regular meeting in March, the Soviet *rezident* was hungry for more. "[He] told me to try to remember anything else that Fuchs had mentioned during our Cambridge meeting about the lens. Yakovlev was very agitated and asked me to scour my memory clean so as to elicit any possible scrap of information about this lens."

Yatzkov/Yakovlev was following the right trail. The day when Fuchs had returned to work at Los Alamos, February 28, 1945, the leaders of the Manhattan Project—including Groves, Office of Scientific Research and Development section head and Harvard president James Bryant Conant, Hans Bethe, George Kistiakowsky and Richard Tolman—had met in Robert Oppenheimer's office and decided tentatively to develop the lensed, solid-core Christy implosion design as a combat weapon. Exploding-wire electric detonators —physicist Luis Alvarez's new invention, far more reliable than lead azide or fulminate of mercury—would fire the complex arrangement of HE lenses. The "Christy gadget" would need a modulated initiator, a device still being engendered that drew on Walter Koski's studies of jet formation (as interpreted by Klaus Fuchs) for its design; the group agreed to review its decision May 1, by which time it hoped a reliable initiator would be in hand. "Now we have our bomb," Oppenheimer had concluded. The uranium gun design had been completed and tested that month as well.

At about the time that Harry Gold and Anatoli Yatzkov were meeting in New York, a plan General Groves had set in motion in Germany made life harder for Igor Kurchatov. Groves had sent a scientific intelligence mission to Europe to follow immediately behind the advancing western front and determine once and for all if the Germans had been working on the bomb. In Strasbourg, Groves's Alsos Mission had found documents identifying a metal-refining plant in Oranienburg, about fifteen miles north of Berlin in what would be the Soviet zone of postwar Germany, as the source of cubes and plates of uranium metal intended for a German nuclear reactor. The Red Army was then advancing from the east dismantling factories *en passant* and shipping them back to the USSR. "Since there was not even the remotest possibility that Alsos could seize the [Oranienburg] works," Groves writes in his memoirs, "I recommended to [Army Chief of Staff] General Marshall that the plant be destroyed by air attack." The ostensible purpose of the attack was to prevent Nazi Germany from completing an atomic bomb, but

Groves knew with some certainty by then that the Germans had not even begun work on nuclear weapons; evidently his purpose was to deny the facility to the Soviets. Groves sent an officer to London to confer with Carl "Tooey" Spaatz, the USAAF general, who commanded the strategic air forces in Europe at that time. "We did not have any target maps," one of Spaatz's intelligence officers, Lewis F. Powell, Jr., later an associate justice of the US Supreme Court, recalls. "I did obtain a city map of Oranienburg by a hectic flight to London at night and going to the British War Office there." The mission was laid on for the afternoon of March 15, 1945. "In a period of about thirty minutes," Groves concludes, "612 Flying Fortresses of the Eighth Air Force dropped 1,506 tons of high explosives and 178 tons of incendiary bombs on the target. Post-strike analysis indicated that all parts of the plant that were above ground had been completely destroyed." Groves was nothing if not thorough; if the Soviets desired uranium, he wanted them to start from scratch.

Ironically, Stalin at that time still anticipated that the USSR and its allies might come to accommodation postwar. In February 1945, while he was meeting with Winston Churchill and a mortally ill Franklin Roosevelt at Yalta, in the Crimea, to further that purpose, his generals had offered him the opportunity of crashing through to Berlin in a matter of days, shortening the war by months. To their fury, Stalin had overruled them, telling them that such an uncoordinated advance would be rash and dangerous. He knew that the Western leaders, Churchill in particular, feared the Red Army might overrun Europe, and held his armies back so as not to alarm them. "It was . . . a hard decision for Stalin to take," writes Alexander Werth. ". . . In the end, it cost the Russians hundreds of thousands of lives. Between February and April, the Germans had time to build powerful fortifications between the Oder and Berlin, and the final Russian victory was incomparably more costly to them than it would have been three months earlier."

The mood in the Soviet Union in those final months of the European war was a giddy mixture of triumph and tragedy. "Russia was a devastated, almost a ruined, country," Werth observes, "with a formidable task of economic reconstruction ahead of her. But on the other hand, she was sitting on top of the world, having won the greatest war in her history. . . . Among many of those who now dreamed of . . . a happy Russia there also existed the idea that the survival of the Big Three alliance after the war would, somehow, tend to liberalize the Soviet regime." Ilya Ehrenburg, writing for the regime, to the contrary expressed hardline menace:

When the Red Army inflicted a heavy defeat on the Germans in Belorussia last summer some American observers explained the Russian victory by the weakness of the Germans. . . . I hope that the Americans, with the in-quisitiveness peculiar to them, will study our country. It is time to drop the kind of talk that says the Russians are winning only because the Russian soldier has always been brave ... [or] that the Russians can fight only on their own soil. ... The sooner Americans learn that we are a strong and completely modern country, that our victories are not accidental gains but the fruit of striving and of toil, the better will it be for us and for America and for the world.

So there were hints at Yalta that the Soviet Union would look kindly on a loan for postwar reconstruction—the figure Molotov had proposed in January in an *aide mémoir* to Harriman was $6 billion—but no offer of a quid pro quo in Poland, which the Soviets were moving to dominate. Roosevelt understood how limited were his Eastern European options. "The Russians had the power in Eastern Europe," Assistant Secretary of State Dean Acheson quoted the President as telling a group of senators in January, "and there was little he could do to change this. Economic aid, he argued, did not 'constitute a bargaining weapon of any strength,' because the only instrument available was Lend-Lease and to cut it back would hurt the United States as much as it hurt the Russians. He also feared that an attempt to use economic pressure for political ends might jeopardize military cooperation at a time when it was 'obviously impossible' to break with the Russians." The US believed it needed the USSR to achieve victory against the Japanese, who still fielded an army of 700,000 men in Manchuria; Roosevelt's forbearance at Yalta, which would be criticized later as a sellout, followed in part from American efforts to hold the military alliance together long enough to finish the Pacific war.

Julius Rosenberg lost his job as a civilian inspector for the Signal Corps in February 1945. He feared at first that the government had discovered his espionage work; when he learned he had been fired because of his Communist Party affiliation he fought back, arguing that "I am not now, and never have been a Communist member. I know nothing about Communist branches, divisions, clubs or transfer. ... Either the case is based on a case of mistaken identity or a complete falsehood." The Signal Corps did not reinstate him—Army intelligence had collected photostats of his Communist Party membership card and other identifying documents—but the Emerson Radio Corporation almost immediately hired him to work as an engineer on some of the same military projects that he had inspected previously for the Signal Corps.

Before Ruth Greenglass left for Albuquerque in mid-February, Rosenberg dropped by her apartment with arrangements for an espionage contact. At dinner with the Rosenbergs during David's January furlough, the conspirators had discussed Ruth traveling to Denver to rendezvous with Ann Sidorovich. As an alternative, David would recall, they planned to meet "in front of a Safeway store on Central Avenue in Albuquerque." Rosenberg instructed Ruth to show up for the Safeway rendezvous during the last week in April and the first week in May.

Housing was hard to find in Albuquerque in wartime and for a while Ruth lived in hotels. "I think I stayed at the El Fidel . . . [for] five days," she said after the war. "Then I stayed in every hotel [in Albuquerque] until I found a place to live." David Greenglass worked with a fellow Special Engineering Detachment enlisted man, another New Yorker, William Spindel, whose wife had moved to Albuquerque; Sara Spindel took Ruth in. Eventually, on March 19, Ruth rented a place of her own at 209 North High Street, a second-floor front apartment, and David began driving down on Saturday nights for the one day of rest that the accelerating pace of work at Los Alamos allowed. David was promoted from Tec/5 to Tec/4—from private to corporal—on April 1. A week later, the Albuquerque branch of the federal Office of Price Administration hired Ruth as a clerk-stenographer.

Ruth befriended an older neighbor, Rosalea Terrell, who found the young New Yorker "a very nice considerate person" and "liked her very much." Boisterous David was another matter. He "had not been very well liked at the apartment house," Terrell would report, "because he was rather loud or noisy, slammed doors going in and out of the house and his apartment and made a lot of noise going up or down the stairs no matter what time of day or night it was." Terrell was curious to learn from Ruth that the Greenglasses had "packages of kosher food" shipped to them from New York. West of the Hudson River was new territory for Ruth; she told Terrell that "she had lived in big apartment houses all of her life, had never seen vegetables or farm produce being grown, and several times mentioned that she would like to quit office work and get some sort of a job on a farm while living in that area. . . ."

A week after she began work at the OPA, Ruth had a miscarriage—"on the couch in my wife's apartment," William Spindel recalls. Ruth wrote Ethel Rosenberg that she would not be able to keep her appointment outside the Safeway store. According to Ruth, Ethel replied "that she was sympathetic about my illness and that a member of the family would come out to visit me the last weeks in May, the third and fourth Saturdays." Ruth kept both appointments, the second time with David, but no one showed up.

In Moscow, Igor Kurchatov was reviewing the information that Klaus Fuchs had passed to Harry Gold on February 16. On April 7, 1945, the leader of the Soviet bomb program reported his preliminary conclusions. "Very valuable material," Kurchatov began. "The data on spontaneous fission of heavy nuclei are of exceptional importance." He was surprised at the high probability for spontaneous fission in Pu240; it was "very important to receive additional information on these matters."

The espionage material included a table of U235 and Pu239 fission cross sections for fast neutrons at various energies. "This table," Kurchatov noted, "makes it possible to define reliable figures for the critical mass of the atomic bomb" and confirmed that "the formula given for the critical radius may be correct within 2 percent, as the text indicates."

Kurchatov was puzzled at the accuracy of the cross-section measurements, since it implied that the US had access to large amounts of U235 and plutonium—which suggests that he was not yet aware that the Manhattan Project was now producing uranium and plutonium in kilogram quantities.

The larger part of the document concerned implosion, "about which," Kurchatov wrote, "we have learned only recently and work on which we have only begun." Yatzkov's eagerness to learn more from David Greenglass and then from Harry Gold about HE lenses presumably emanated from Moscow Center. "But already," Kurchatov added, "the advantages of this method over the gun method are clear."

Kurchatov briefly summarized the basics of implosion that the espionage document discussed. "All this is very valuable," he went on, "but most essential are the indications of the conditions necessary to achieve a symmetric explosion. The material describes the interesting phenomena of irregularities in the detonation wave"—these were the troublesome jets which Fuchs had studied that formed where detonation waves collided and intersected—"and describes how these irregularities may be avoided by the proper distribution of detonators and by using interlayers of explosives with different actions"—"interlayers" meaning explosive lenses. "This part of the material also deals with important questions of techniques of experimenting with explosives and the optics of explosive phenomena."

"Since research on implosion has not advanced much here," Kurchatov concluded, "it is not possible yet to formulate questions [to guide espionage]. This can be done after serious analysis of the material." Kurchatov suggested that a portion of the top secret text—"from page 6 to the end except for page 22"—should be shown to "Professor Khariton." To a limited extent, then, Yuli Khariton was aware from at least spring 1945 that espionage was supplying significant input into the Soviet program.

Kurchatov's desire in mid-March for "several grams of highly-enriched uranium" was partly satisfied a month later when Alan Nunn May passed Lieutenant Angelov, in Nunn May's words, "a slightly enriched sample [of U235] in a small glass tube [consisting] of about a milligram of oxide." Unlike Fuchs, Nunn May was willing to accept compensation; Angelov gave him two bottles of whiskey—a scarce luxury in wartime—and two hundred dollars.

The Red Army offensive against Berlin began in mid-April. Marshal Georgi

Konstantinovich Zhukov directed the battle as he had directed the battles of Moscow and Stalingrad, and described it at a press conference immediately afterward:

> I attacked along the *whole* front, and at night. . . . [The Germans] had ex-
> pected night attacks, but not a *general* attack at night. After the artillery
> barrage, our tanks went into action. We had used 22,000 guns and mortars
> along the Oder, and 4,000 tanks were now thrown in. We also used 4,000
> to 5,000 planes. During the first day alone there were 15,000 sorties.
>
> The great offensive was launched at 4 a.m. on April 16, and we devised
> some novel features: to help the tanks find their way, we used searchlights,
> 200 of them. These powerful searchlights not only helped the tanks, but
> also blinded the enemy, who could not aim properly at our tanks.
>
> Very soon we broke through. . . .

American and Soviet troops joined hands at Torgau, one hundred kilometers due south of Potsdam, a few days before Adolf Hitler's suicide on April 30. The Nazi dictator's personal staff burned his body in the garden of the *Führerbunker* and buried the remains in a shallow grave. Berlin fell on May 2. The Soviets had suffered 300,000 casualties in the final battle of the war. Three hundred thousand German soldiers surrendered in the course of the battle; another 150,000 were killed.

The day Berlin fell, a team of Soviet industrial managers and physicists flew in to Templehof airfield to explore German atomic-bomb research. Lieutenant General Avrami Pavlovich Zavenyagin, deputy director of the NKVD and the developer of the vast Magnitogorsk Steel Combine, led a group that included Lev Artsimovich, Isaak Kikoin and Yuli Khariton. The team established its headquarters in Berlin-Grünau. "A remnant of [German] scientists remained in Berlin and willingly talked to us," Khariton recalls. "From these discussions it was clear to us that German progress along these lines had been slight. Kikoin and I told Zavenyagin what we'd gathered, and told him it would be prudent to find out whether the Germans had accumulated any stockpiles of uranium. . . . It was entirely likely that uranium supplies in Belgium had been seized and taken out by the Germans. Zavenyagin approved of our idea and put an automobile at our disposal."

In fact, a mixed British-American strike force led by Lieutenant Colonel John Lansdale, Jr., who was Groves's liaison officer with the British, had moved into what would soon be Soviet-occupied eastern Germany on April 17 to strip a Stassfurt factory of what Lansdale believed to be all the remaining Belgian Congo ore in Germany, 1,100 tons stored above ground in broken barrels. "The plant was a mess," Lansdale reported to Groves, "both from our bombings and from looting by the French workmen. . . . By the evening of 19th April we had a large crew busily engaged in repacking the material and that night the movement of the material to [the railhead] started." Groves sent US Army Chief of Staff George Marshall a memorandum confirming the recovery on April 23. The Manhattan Project commander noted that in 1940 the German Army had confiscated "about 1200 tons of uranium ore" in Belgium, described Lansdale's operation, and concluded that "the capture of this material, which was the bulk of uranium supplies available in Europe, would seem to remove definitely any possibility of the Germans making use of an atomic bomb in this war."

Now, two weeks later, the Soviet team was scouring the same ground. "Through our discussions with the German scientists," says Khariton, "we discovered that there was a certain building in Berlin . . . where a card catalogue was kept with records of everything the Germans had plundered in the countries they'd occupied." Such was Nazi greed that the card catalogue filled the six-story building. The catalogue staff refused to cooperate with the Soviet expedition. "After prolonged and excruciating attempts to get our bearings," Khariton continues, "we managed to determine that there was in fact uranium oxide, but we couldn't come up with its location." Then other, more cooperative Germans directed them to affiliated card catalogues in other cities. They went from city to city; eventually they found the uranium-oxide reference in a warehouse card catalogue. "But it turned out that some military personnel must have shipped it as a pigment—uranium oxide is, after all, bright yellow in color." Finally they learned that a quantity of oxide had been sent to a tannery west of Berlin. The Soviet commander of that district told the physicists that the tannery was on American-occupied territory, but Khariton claims the tannery "turned out to be on our territory, right on the border with the American occupation zone":

The tannery was in the control of an antifascist group. It was made up of workshops and warehouses, some of which were crammed with sheepskins, raw material awaiting production. In one of these last warehouses we came across a fair number of small wooden barrels. There was a scrap of cardboard on one of them with an inscription, U^3O^8. We sighed with relief. We informed Zavenyagin of our excursion, and arrangements were made for shipping the uranium oxide to the Soviet Union. The net quantity was in the vicinity of 130 tons.

Between the American team rushing to remove uranium ore from Soviet-occupied territory and the Soviet team rushing to remove the remaining ore from "right on the border" of American-occupied territory, the two operations accounted for all the Belgian Congo stock in Germany and about half the existing world supply. The American requisition became U235 for Little Boy; the Russian requisition, Kurchatov later told Khariton, "hastened the startup of the first [Soviet] industrial reactor for obtaining plutonium by about a year." Such parallels seem enigmatic, but in fact the two bomb programs ran in parallel because the raw materials, the processing and the technology depended upon universal physical fundamentals that both sides could determine independently. At that basic level, there never was any "secret" of how to make an atomic bomb. Knowledge derived from espionage could only speed up the process, not determine it, and in fact every nation that has attempted to build an atomic weapon in the half-century since the discovery of nuclear fission has succeeded on the first try.

Back in Berlin, Soviet troops dismantled the laboratories of the Kaiser Wilhelm Institute for Physics, next door to the Institute for Chemistry where nuclear fission was discovered, and shipped the equipment to Moscow. In Vienna on May 5, a Soviet colonel wrote out a receipt for four hundred kilograms of uranium metal and a quantity of heavy water confiscated from the Institut für Radiumforschung. All the first-rank German scientists involved in atomic research—Nobel laureate theoretician Werner Heisenberg and Nobel laureate radiochemist Otto Hahn, among others—had fled into southwestern Germany in the closing days of the war to avoid being captured by the Soviets. Zavenyagin's team drafted a number of lesser German scientists to work in the Soviet Union, however, and others volunteered. They joined what Alexander Solzhenitsyn would call the First Circle of the Soviet gulag—scientific research centers staffed with political prisoners, in this case laboratories for developing uranium processing and isotope separation technologies at Sinop, near Sukhumi on the Black Sea, and at nearby Agudzeri (both laboratories in Beria country, where security staff personally loyal to the Georgia-born NKVD chief could keep an eye on them). One of the Germans, Nikolaus Riehl, who called his Soviet experience "ten years in a gilded cage," was seized along with his complete Auer Gesellschaft laboratory; with his capture, the Soviets acquired crucial knowledge of how to purify uranium metal.

The war in Europe ended in a schoolroom in Rheims early on the morning of May 7, 1945, when Colonel General Alfried Jodl signed the act of military surrender. "The world now sees the shining face of victory," Ilya Ehrenberg had written proudly a few days before, "but let the world remember how this victory was born: in Russian blood, on Russian soil. . . ." May 9, Alexander Werth records, "was an unforgettable day in Moscow:

> The spontaneous joy of the two or three million people who thronged the Red Square that evening—and the Moscow River embankments, and Gorki Street, all the way up to the Belorussian Station—was of a quality and a depth I had never yet seen in Moscow before. They danced and sang in the streets; every soldier and officer was hugged and kissed; outside the US Embassy the crowds shouted "Hurray for Roosevelt!" (even though he had died a month before); they were so happy they did not even have to get

drunk, and under the tolerant gaze of the militia, young men even urinated against the walls of the Moskva Hotel, flooding the wide pavement. Nothing like *this* had ever happened in Moscow before. For once, Moscow had thrown all reserve and restraint to the winds. The fireworks display that evening was the most spectacular I have ever seen.

But Americans at least hardly knew the Russian tragedy, and would not long remember it, though there was still an afterglow of popular goodwill for brave Ivan and steadfast Uncle Joe. In Washington the mood was already darkening; the new administration of Harry Truman was concerned immediately with Soviet determination to impose a puppet government on Poland, and there was mounting opposition in the US Congress to contributing further to Soviet support even though the US military believed it needed the Red Army's help in Manchuria to finish the war with Japan. It had always been intended to end Lend-Lease once the war was over, but on May 11, through a combination of miscommunication and zealous overreaction, Lend-Lease officials abruptly cut off ship loadings to the USSR and even called back ships at sea. Though the order was modified within days, the Soviets were outraged. Stalin told Harry Hopkins later that month that he thought the high-handed cutoff had been "unfortunate and even brutal."

With the end of the war in Europe, the men who directed the Soviet bomb program sought to improve the program's priority and accelerate the pace of the work. In the autumn of 1944, after he reviewed the three thousand pages of new espionage material the NKVD had collected, Kurchatov had written Lavrenti Beria complaining of the "completely unsatisfactory" Soviet program. "The situation with raw materials and questions of [isotope] separation is particularly bad," he told Beria. He was critical of Molotov's management. "The research at Laboratory No. 2 lacks an adequate material-technical base. Research at many organizations that are cooperating with us is not developing as it should because of the lack of unified leadership." He asked Beria "to give instructions for the work to be organized in a way that corresponds to [its] possibilities and significance." In May 1945, Pervukhin and Kurchatov carried their complaint directly to Stalin, writing that Molotov had not given the program the support it deserved.

Neither Beria nor Stalin chose to respond. As the historian David Holloway points out, Beria distrusted the atomic-bomb information his *rezidents* were collecting and distrusted the Soviet scientists as well. "From the very beginning," Yatzkov would recall, "he suspected disinformation in these materials and thought that our adversaries were trying to drag us into tremendous expenditures of resources and effort on work which led nowhere. . . . Beria was suspicious about the espionage information even when the work in the Soviet Union had achieved large scale. [An NKVD official] recalls that once, when he was reporting to Beria on the latest [atomic] intelligence, Beria threatened him: 'If this is disinformation I'll throw you all into the cellar.' "

Though the collection of atomic espionage documents filed at the Lubyanka approached ten thousand pages, Stalin, Beria and Molotov evidently did not yet believe in the atomic bomb. Untested, it was still an abstraction to them; where espionage was concerned, they valued only the tried and true.

9

'Provide the Bomb'

HARRY GOLD MET "JOHN"—the name by which he knew Anatoli Yatzkov/Yakovlev—in Volk's Bar at Third Avenue and 42nd Street in Manhattan late on the Saturday afternoon of May 26, 1945, "so that he might verify that I was going to see Fuchs in Santa Fe." Yakovlev was concerned to confirm the trip because Gold had been having trouble getting time off from work; the chemist had finally arranged to take part of his vacation early. They also needed to schedule their contacts after the Santa Fe trip, Gold testified, "one meeting at which I would transfer information which I was supposed to receive from Fuchs; then there would be a second meeting some time later, at which I would give Yakovlev a detailed report as well as a verbal account of exactly what would have transpired at this meeting with Fuchs." The two men had a drink standing at the bar. Gold suggested they take a walk, but walking exposed them to surveillance and Yakovlev had a lot of business to transact. He steered them instead to the back of the bar, to "a circular place with some tables in it, fairly secluded. . . . We sat down there and the waiter brought us a drink."

The two conspirators talked for most of an hour. Yakovlev told Gold he wanted him to go on to Albuquerque after he saw Fuchs and make a second rendezvous. Gold immediately protested such a flagrant violation of espionage protocol. "I told Yakovlev that it was highly inadvisable to endanger the very important trip to see Dr. Fuchs with this additional task." It rankled the Soviet professional to be lectured by an amateur. He told Gold "that the matter was very vital and that I had to do it. He said that a woman was supposed to go in place of me"—a reference, presumably, to Ann Sidorovich; that very day Ruth and David Greenglass were waiting for her in vain outside an Albuquerque Safeway store—"but that she was unable to make the trip." Then Yakovlev erupted. "I have been guiding you idiots through every step," he berated Gold. "You don't realize how important this mission to Albuquerque is." He bluntly ordered Gold to go to Albuquerque. "And that was all," Gold testified. "I agreed to go."

The mission was sufficiently important that Yatzkov/Yakovlev gave Gold his instructions typed on a sheet of paper, only the second time in fourteen years of espionage work that Gold remembered being briefed other than orally (the first time having been the summer of 1944, when Yatzkov had given him the typewritten pages of garbled questions about the US atomic-bomb program at which Fuchs had taken offense). Gold would variously recall the information on the paper. The name "Greenglass" was typed on it, he testified. "Then a number [on] 'High Street' . . . and then underneath that was 'Albuquerque, New Mexico.' The last thing that was on the paper was 'Recognition signal. I come from Julius.' " In other testimony, Gold had recalled the recognition name as "Frank Kessler" or "Frank Martin," aliases he had used previously, and as "Ben from Brooklyn." Rosenberg supporters would make much of Gold's inconsistency, but his testimony followed the events by five years, long enough to have forgotten what would have been a minor detail at the time. Nor was it normal Soviet practice to compromise security by using a spy's real name—witness Elizabeth Bentley's assumption that Julius was not the real name of the engineer who lived in Knickerbocker Place who woke her with his post-midnight calls.

Gold received verbal instructions from Yatzkov as well:

> John told me that there existed in Albuquerque a man who was employed in the atomic energy project. I assumed that he meant a civilian. He told me that after I had seen Dr. Fuchs, that I should return to Albuquerque, and that on that Saturday night I should visit this man and pick up certain information which he had prepared. I further was instructed that should this man not be in Albuquerque, that his wife would be there and would have information for me. In addition, I was given the sum of $500, and was told that should either the man or his wife evince any need for the money, that I should give it to them.

Gold also recalled Yatzkov giving him a recognition device corresponding to the tennis ball and gloves that he and Fuchs had carried to their first meeting: "I was to tender a piece of cardboard cut in an irregular manner; this piece of cardboard was to be matched by a second piece which the person whom I met would have." The "piece of cardboard" was half of Julius Rosenberg's Jello boxtop; Ruth Greenglass had carried the matching half to Albuquerque in her wallet.

Gold departed Philadelphia at the end of May for Chicago. In Chicago he managed to arrange an upper berth to Albuquerque. Yatzkov had instructed him to follow a more circuitous route through Arizona and Texas, but Gold recalled being "extremely short of money" and having "to watch what I had very carefully"—he had about four hundred dollars left from expense money Yatzkov had given him in February or March—and being short on time as well. Fuchs had recommended that Gold get off at Lamy, New Mexico, the usual stop for Santa Fe–bound passengers (since the New Mexican capital lacks a railroad terminus), saving the backtrack bus ride from Albuquerque sixty miles further south. But Gold had surmised that "the only people going to Santa Fe [from Lamy] would be those connected with the atomic energy project and they might wonder who this stranger was in their midst." As these and other decisions make clear, Gold was not easily swayed by other people's opinions—a characteristic he demonstrated most radically by his commitment to espionage.

The bus pulled into Santa Fe at about two-thirty Saturday afternoon, June 2, 1945. Fuchs and Gold had arranged to meet on the Castillo Bridge at four. "I had considerable time to spare," Gold remembered, ". . . and to avoid drawing attention to myself, I went as any ordinary tourist would, to the rather large historical museum located in Santa Fe." At the museum he asked for a map and got one identical to the yellow brochure Fuchs had given him in Cambridge. With the map, at the appointed time, the thirty-four-year-old industrial chemist found his way to the bridge. Fuchs drove up late in the battered old two-door gray Buick he had bought second-hand —you had to hold the gearshift to keep the transmission from jumping out of gear going downhill, its next owner, physicist Anthony French, remembers. "Klaus arrived . . . possibly two or three minutes late," Gold says, "during which two or three minutes I became extremely uneasy, as the area around the Castillo Street Bridge was extremely sparsely settled." Both men independently remembered what followed; in Fuchs's words, "I . . . picked up Raymond and we drove across the river bridge, turned into a lane which ended at a gate in an isolated place, and there we continued our meeting." They talked for about half an hour, thirty minutes of fateful significance for the Soviet atomic-bomb program.

"Klaus told me that he was getting along very well with his work in Los Alamos," Gold remembered, "and told me that he did not, however, believe —and that was a reiteration of his statement which he had made several times before, once in Cambridge and at least once or twice in New York— that the atomic energy project would be completed in sufficient time for use in the war against the Japanese." Everyone was working hard, the physicist reported, almost night and day; "he himself put in an average of from eighteen to twenty hours a day."

Getting down to business, Fuchs says he told Gold "the names of the types of explosives to be used in the bomb [information important to the design of high-explosive lenses]; the fact that the Trinity test explosion was to be made, with the approximate site indicated, soon in July, 1945, and that this test was expected to establish that the atom bomb would produce an explosion vastly greater than TNT and the comparative estimated force of this explosion was indicated in detail with relation to TNT." Fuchs put the expected Trinity yield at about ten kilotons. The explosives, Fuchs said, were "Baratol" and "Composition B"; he knew little about them, and did not understand, he said later, what their use meant in terms of high-explosive technology. He was aware at the time that a uranium gun bomb was under development, but it was outside his area of expertise and he apparently did not mention it to Gold. The two men discussed meeting again. Fuchs wanted to meet in August—"due to . . . some important development," Gold recalled without remembering the July test that would determine the effectiveness of implosion—but Gold "demurred, and we finally set [the meeting] for the 19th of September 1945."

Following his usual cautious practice, Gold accepted from Fuchs last of all what he called "a considerable packet of information." Fuchs emphasized the importance of the packet, and probably added measurably to Gold's anxiety, by telling the chemist "that among the data he had given me was a sketch of the atomic bomb itself." Fuchs later described the contents of that considerable packet in detail:

> I delivered . . . confidential and classified written information in a paper or document, which I had personally written in longhand. Included in this written paper were the following items . . . : a description of the plutonium bomb, which had been designed and was soon planned to be tested at Alamogordo; a sketch of the bomb and its components with important dimensions indicated; the type of core; a description of the initiator; details as to the tamper; the principle of the IBM calculations; and the method of calculating efficiency.

"He reported that the bomb would have a solid plutonium core," a physicist who interrogated Fuchs in 1950 specifies, "and described the initiator which, he said, would contain about fifty curies of polonium. Full details were given of the tamper, the aluminum shell, and of the high explosive lens system." Fuchs's sketch, which he later reproduced for the FBI, depicted a cross section of the Fat Man implosion design, a *matrioshka* of nested concentric shells. It revealed the relationship among the device's various parts. Significantly, it gave the thickness of each of the shells and reported the crucial information that an aluminum shell had been interposed between the explosive layers and the uranium tamper to dampen the hydrodynamic instability that would otherwise have developed when the light explosive mixed with the heavy metal, a phenomenon known for its English delineator Geoffrey Taylor as Taylor instability. Thus burdened with historic knowledge, a courier as cosmically mischievous as Mercury, Gold said his goodbyes and immediately got out of the physicist's car and walked away.

Gold walked to the Santa Fe bus station and took the next bus to Albuquerque, arriving there around eight or eight-thirty Saturday night. "I went to the place whose address had been given to me by Yakovlev," he recalled. The house at 209 High Street had a large screened porch; on the porch, Gold testified, "I was met by a tall, elderly, white-haired and somewhat stooped man," probably P. M. Sherer, the father of the Greenglasses' landlady. "The old gentleman . . . told me the Greenglasses had gone out for the evening; and, on my further inquiry, said that he thought they would be in the following morning."

"So," Gold continues, "Saturday night in a town in wartime. Just try to get a room in a hotel without a reservation (at one dignified old place called, I think, the Franciscan, they actually laughed at me). I had not, of course, expected to stay over and was, it may be believed, most anxious to get away as soon as possible from the area of Santa Fe and Albuquerque." In the course of his hunt for a place to stay that night, Gold gave his name at the Albuquerque Hilton:

Finally, about [midnight] . . . the Hilton advised me that there was such a long waiting list ahead of me that they were certain no room would be available that night. I thereupon wandered through Albuquerque and finally, upon asking a policeman, he directed me to a private home near the main street . . . which had been temporarily converted into a rooming house. The only space that these people had, and I with difficulty talked these people into letting me stay there, was in the hallway on the second floor . . . where a makeshift screen was put up around a very rickety cot. I spent the night there. . . .

The chemist slept badly, if at all: he had secrets to protect. "Now, with servicemen on the loose," he remembered long afterward, "police sirens kept screaming all night. And every time one did, I was jarred instinctively reacting with the thought that they might be coming for me—because of that fat package from Fuchs in my possession. It was [a] traumatic experience. . . ."

Sunday morning, June 3, 1945, Gold moved uneasily to his rendezvous. "I clearly remember that on leaving the rooming house that morning I was anxious to get to High Street before the Greenglasses might go out again. So, likely I checked my bags at the [Santa Fe railroad] station, as it was right in the direction I was going. . . ." He knocked at the Greenglass apartment. David Greenglass opened the door. "We had just completed eating breakfast," the young soldier would testify, "and there was a man standing in the hallway who asked if I were Mr. Greenglass, and I said 'Yes.' He stepped through the door and he said, 'Julius sent me,' and I said, 'Oh.' And I walked to my wife's purse, took out the wallet and took out the matched part of the

Jello box. . . ." Gold offered the part Yatzkov/Yakovlev had given him; the two parts matched. David introduced Ruth.

"The whole setup smelled wrong to me," Gold would recall. Not only was he "jeopardizing an already accomplished mission with Fuchs" by meeting with David Greenglass, but "the man was a G.I. Yakovlev had made no mention of this. As Greenglass opened the door I saw that he had on a pajama top and Army trousers; and on the wall to the right there was hanging a (non-com's) coat with stripes." Gold never explained why Greenglass's military status troubled him; he may have worried that military personnel were watched more closely than civilians.

David testified that he offered Gold something to eat. Ruth, to the contrary, asked later by the FBI if she offered Gold a cup of coffee, snapped, "I didn't like the situation well enough to be friendly." David liked the situation, but he was unprepared. "He just wanted to know if I had any information, and I said, 'I have some but I will have to write it up. If you come back in the afternoon I will give it to you.'" The garrulous New Yorker tried to start a conversation. "I started to tell him [a] story about one of the people I [was going to] put into the report"—a buddy at Los Alamos who he imagined might be "good material for recruiting into espionage work." Appalled, Gold "cut him very short indeed. I told him that such procedure was extremely hazardous, foolhardy, that under no circumstances should he ever try to proposition anyone on his own into trying to get information for the Soviet Union." Years later, Gold still shuddered to remember his dismay at the young soldier's brashness: "Greenglass was not only young, but at once impressed me as being frighteningly naive, particularly in his eager volunteering of the idea of approaching other people at Los Alamos as potential sources of data. I was horrified at his total inexperience in espionage, especially considering what we were after." David was chagrined. "He agreed with me," Gold says. "He did not seem angry or taken aback by the rebuke. He said, yes, I was right, that just previous[ly] a man whom he knew at [Los Alamos] had been broken to the ranks and had been sent elsewhere. . . ."

For the rest of Gold's brief twenty-minute visit, the conspirators confined themselves to small talk, some of it significant. "Mrs. Greenglass told me that she had seen and spoken with a Julius in New York, just prior to her coming to Albuquerque in April 1945. . . . Greenglass told me . . . that he expected to be furloughed and would take the opportunity to go home to New York. He told me I could get in touch with him about Christmas time by calling Julius." Evidently both the Greenglasses assumed Gold knew David's brother-in-law. Associating themselves with Julius Rosenberg further cross-linked and compromised the two separate lines of Yatzkov's operation. He had only himself to blame for breaking with protocol and sending Gold to Albuquerque in the first place.

After Gold left the Greenglasses, he stopped at the railroad station to ask about reservations eastward. He thought later that he may also have stopped for breakfast. (Though Greenglass remembered Gold responding to his invitation to eat by saying he had already eaten, the chemist was probably offering an excuse to cover his discomfort with a situation that "smelled wrong.") Then Gold went on to the Hilton to take a room for the day, standard operating procedure to stay out of sight that Gold appreciated. "With all that material from Fuchs on me, wandering an entire day around a relatively small town such as Albuquerque was a risk to be avoided," he explains; he was tired and "under a strain from the whole mission." There was as well an airlines office nearby, Gold was "most anxious to get away from New Mexico and I had to have some sort of address at which I could be called should space be available." He camped out in the hotel lobby, "waiting for people to check out. . . . There was crowding around the registration desk . . . , confusion and jostling." At 12:36 he registered and went to his room.

In the meantime, in his small apartment back at High Street, David Greenglass prepared to write down what he knew:

I got out some eight-by-ten ruled white line paper, and I drew some sketches of a [high-explosive] lens . . . and how they are set up in an experiment, and I gave . . . a description of this experiment. . . . I gave sketches relating to the experiment[al] setup: one showing . . . the face of the flat-type lens. . . . I showed the way [a high-explosive lens] would look with this high explosive in it with the detonators on, and I showed the steel tube in the middle which would be exploded by this lens. . . . I showed . . . a schematic view of the lens . . . set up in an experiment.

To clarify his sketches, which depicted configurations and experiments concerning cylindrical—two-dimensional—implosion, David keyed them to a detailed description that he wrote out by hand. These, a discussion of "the growth of the project" and "a pretty substantial list of names of both possible recruits and of scientists who worked there," went into a large letter-size envelope. He was not well-informed. He thought that Hans Bethe, a staunch patriot, was a possible espionage recruit, and he thought that Harold Urey, the Columbia chemist who had guided gaseous-diffusion research and development, was head of the Manhattan Project.

Harry Gold returned to High Street midafternoon. David gave him the envelope and briefed him verbally as well. "David and this man discussed how the atom bomb was detonated," Ruth Greenglass remembered, "and . . . this man told David that he was a chemical engineer. I also recall that David and [Gold] discussed lenses and high-speed cameras."

Gold said later that Greenglass asked to be paid. Both the Greenglasses remember Gold to the contrary offering them money unasked that they accepted with shame. Given Gold's lonely solicitude for family life—his own family, the family he invented as a cover, Kristel Heineman and her children—it seems likely that he offered the money directly in return for Greenglass's information. "Gold told me that I was living in a rather poor place," Greenglass described the exchange, "and said I could probably use some money. I answered that I could use some money. Gold then gave me an envelope containing $500 in currency." Gold remembered that Greenglass looked disappointed. The young soldier put the envelope into the pocket of his army blouse. "[Gold] said, 'Okay?' 'Yes,' I said, 'it will be enough.' . . . He said something to the effect that he would be back. I said okay. . . . I remember saying that my wife had just had a miscarriage and cost me a lot of money for doctor bills and medicines, etc. He was very sympathetic about that and about the place we lived in. . . . I said something about, 'I guess I need it.' "

Now that he was holding espionage documents, Gold wanted to leave immediately, following his standard protocol. David discouraged him. "I said, 'Wait, and we will go down with you,' and he waited a little while." There was small talk as the Greenglasses got ready. Gold remembered them telling him "that they had regularly had food packages containing delicatessen items sent to them from New York. . . . I particularly recall the mention of . . . salami and pumpernickel bread. . . ."

"We went down," David Greenglass testified, "and we went around by a back road and we dropped him in front of the USO. We went into the USO and he went on his way. As soon as he had gone down the street my wife and myself looked around and we came out again and back to the apartment and counted the money."

"The taking of the money made David and me feel worse," Ruth Greenglass confessed. "I was under the impression at first that Julius said it was for scientific purposes we were sharing the information, but when my husband got the $500, I realized it was just C.O.D.; he gave the information and he got paid." Five years later, David was still rationalizing the transaction. "I furnished [Gold] with information concerning the Los Alamos project," he insisted, "although I did not do it for the promise of money. . . . I felt it was gross negligence on the part of the United States not to give Russia the information about the atom bomb because she was an ally."

Gold headed for the railroad station "to see if Pullman space had been verified . . . [or] maybe it was getting near train time." A long Roman Catholic religious parade blocked his way. "So I leaned on a low stone wall watching it, till I finally could get across [the street]."

En route to Chicago, somewhere in Kansas, Gold surveyed his treasures, with what solitary rapture he never divulged. "On the train . . . I examined the material which Greenglass had given me. I just examined it very quickly. . . . I put it into an envelope, into a manila envelope, one of the kind with a brass clasp, and in another manila envelope I put the papers which Dr. Fuchs had given me. I labeled the two envelopes. On the one from Fuchs I wrote 'Doctor.' On the one from Greenglass I wrote 'Other.' . . ." For thirty-six hours, at least, Harry Gold was another roaming torpedo, the only person on earth in private possession of the plans for the world's first atomic bomb.

Monday, June 4, 1945, Ruth Greenglass opened a savings account in her own and her husband's name at the Albuquerque National Trust and Savings Bank with an initial deposit of four hundred dollars in cash.

In Chicago that Monday morning, Gold caught a flight to Washington, as near as he could get to New York, "to save time . . . since the wait for a train would involve a stay in Chicago until late evening." From Washington in the afternoon he continued by train to New York. He was rushing to rendezvous with Yatzkov/Yakovlev to pass on the incriminating documents:

> I met Yakovlev along Metropolitan Avenue in Brooklyn . . . where Metropolitan Avenue runs into Queens. It was a very lonely place, particularly at that time of night. . . . It was about 10 o'clock. . . . This meeting had been arranged at Volk's cafe [in May]. . . . [It] lasted about a minute, that was all. . . . We met and Yakovlev wanted to know if I had seen the both of them, "The doctor and the man." I said that I had. Yakovlev wanted to know had I got information from both of them and I said that I had. Then I gave Yakovlev the two manila envelopes.

Two weeks later, Gold met with Yatzkov/Yakovlev again, "at the end of the Flushing elevated line in Flushing," to report on his trip to New Mexico. "The time was in the middle of the evening. . . . Yakovlev told me that the information which I had given him some two weeks previous[ly] had been sent immediately to the Soviet Union. He said that the information which I had received from Greenglass was extremely excellent and very valuable. Then Yakovlev listened while I recounted the details of my two meetings, the one with Fuchs in Santa Fe, the one with Greenglass in Albuquerque." They talked for two and a half hours.

In Moscow on July 2, an NKVD officer briefed Igor Kurchatov on the progress of the Manhattan Project. The undated notes on that briefing contain details of implosion bomb design that correspond to those Fuchs confessed passing to Harry Gold on June 2. They also contain information on current supplies of fissionable materials that Fuchs was in a position to know. Yatzkov asserted late in life that the source of this summer 1945 briefing information was Perseus and that the information was drawn from material that Lona Cohen had successfully spirited out of Albuquerque, but nothing in the document itself was outside Klaus Fuchs's provenance at Los Alamos. The document justifies reproduction in its entirety; it marks the first transmission to the Soviet Union of details of atomic-bomb design:

TOP SECRET

BOMB TYPE "HE" (HIGH EXPLOSIVE)

The first test explosion of an atomic bomb is anticipated in July of this year.

Design of the bomb. The active material of the bomb is element 94 without the use of uranium-235. The so-called initiator—a beryllium-polonium source of alpha particles—is situated in the center of a 5-kg plutonium ball which is surrounded by 500 pounds of "tube alloy,"* which serves as a "tamper." All this is put into a shell made of aluminum, 11 cm thick. This aluminum shell, in its turn, is surrounded with a layer of explosive "penthalite" or "composition C" (according to other information "Composition B") 46 cm thick. The case of the bomb housing the explosive has an internal diameter of 140 cm. The total weight of the bomb, including the penthalite, the case, etc., is around 3 tons. The anticipated yield of the bomb is equal to 5,000 tons of TNT (efficiency factor—5-6%). Number of "fissions" $75 \cdot 10^{24}$.

STOCKS OF ACTIVE MATERIALS.

a) Uranium-235. By April of this year the amount of uranium-235 was 25 kg. Its production now constitutes 7.5 kg per month.
b) Plutonium (element 94). There are 6.5 kg of plutonium on hand at Compound Y [i.e., Los Alamos]. Its production is organized. The plans for production are overfulfilled.

The explosion is anticipated on approximately July 10 this year.

------

* *Tube alloy—code name for uranium (commercial radium [sic] tubealloy)* (*handwritten: Not clear if tube alloy means 235 or natural uranium.*) (*handwritten: This material compiled for the oral orientation of Academician Kurchatov.*)

The briefing officer's information was accurate so far as it went, but it was less than complete, as Kurchatov would have realized. The officer gave only a general idea of the initiator and made no reference to explosive lenses or to detonators and their placement. But Kurchatov learned vital information, most crucially that there was enough plutonium on hand at Los Alamos to make at least one bomb, that the United States believed it knew how to do so and that enough plutonium was in the pipeline to use up five precious kilograms on a test (the Trinity test device actually used a little more than six).

The Anglo-American Combined Policy Committee met formally and secretly in Washington on Independence Day to carry out a significant provision of the 1943 Quebec Agreement: the British officially gave their approval that day for the use of atomic bombs against Japan, as the agreement provided they must before the United States could act. Donald Maclean was positioned to pass along the information to the Soviets.

Sometime that summer, Yatzkov learned that Abe Brothman was under suspicion of having engaged in espionage. At his regular monthly meeting with Harry Gold in early July, perhaps anticipating problems in maintaining contact if Brothman was questioned and confessed, Yatzkov had Gold prepare a recognition signal "whereby," says Gold, "some Soviet agent other than himself could get in touch with me." Like the Rosenberg/Greenglass Jello boxtop, the recognition signal was a piece of ephemera—in this case a memorandum sheet from a laboratory supply house that Gold happened to have in his pocket—on which Gold wrote a street address and which he then divided with his Soviet control. Yatzkov outlined a procedure to follow to make contact using the torn memorandum sheet; Gold would be alerted by two tickets to a New York theatrical or sporting event mailed to him in an otherwise empty envelope.

A test model of the plutonium implosion device on which Klaus Fuchs, David Greenglass and many others at Los Alamos had been working, and on which Igor Kurchatov had been briefed two weeks previously, exploded in its corrugated iron cab on a hundred-foot steel tower at Trinity Site, in the desert north of Alamogordo, New Mexico, at 5:29:45 A.M., July 16, 1945, just before dawn. I. I. Rabi, the tough-minded Columbia Nobel laureate physicist who visited Los Alamos from time to time as a consultant, was one of many on hand to observe the explosion:

> Suddenly, there was an enormous flash of light, the brightest light I have ever seen or that I think anyone has ever seen. It blasted; it pounced; it bored its way right through you. It was a vision which was seen with more than the eye. It was seen to last forever. You would wish it would stop; altogether it lasted about two seconds. Finally it was over, diminishing, and we looked toward the place where the bomb had been; there was an enormous ball of fire which grew and grew and it rolled as it grew; it went up into the air, in yellow flashes and into scarlet and green. It looked menacing. It seemed to come toward one.
>
> A new thing had just been born; a new control; a new understanding of man, which man had acquired over nature.

Fuchs was there to see the new thing he had caused to proliferate, the new control, but no one put a penny in his slot, so he left no record of how the unique experience affected him.

Forewarned with information from his spies, Stalin played dumb at the Potsdam Conference convened outside Berlin when Harry Truman came around the green baize table to inform him of the bomb on the afternoon of July 24. According to Jimmy Byrnes, Truman's new Secretary of State, the President was afraid that if Stalin understood the full power of the new weapon, understood that it might bring a swift end to the Pacific War, the Soviet dictator might expedite his declaration of war against the Japanese and gain a share of spoils he had hardly earned. Truman had confided his expectations to his private diary as soon as he heard of the successful test at Trinity: "Believe Japs will fold up before Russia comes in. I am sure they will when Manhattan appears over their homeland." So Truman intended to reveal no more of the bomb at Potsdam than was necessary to protect himself from a Soviet charge of perfidy. "I casually mentioned to Stalin that we had a new weapon of unusual destructive force," the President recalled in his memoirs. "The Russian Premier showed no special interest. All he said was that he was glad to hear it and hoped we would make 'good use of it against the Japanese.'"

"Stalin . . . pretended he saw nothing special in what Truman had imparted to him," Marshal Zhukov reports. "Both Churchill and many other Anglo-American authors subsequently assumed that Stalin had really failed to fathom the significance of what he had heard. In actual fact, on returning to his quarters after this meeting, Stalin, in my presence, told Molotov about his conversation with Truman. 'They're raising the price,' said Molotov. Stalin gave a laugh. 'Let them. We'll have to have a talk with Kurchatov and get him to speed things up.'" Since Kurchatov and Pervukhin had written to Stalin two months previously to complain of Molotov's unenthusiastic management, the American news must have made the old Bolshevik uneasy.

Molotov himself claimed to remember no such conflict, although his recollection sounds exculpatory. "Truman didn't say 'an atomic bomb,'" he contended, "but we got the point at once. We realized they couldn't yet unleash a war, that they had only one or two atomic bombs. . . . But even if they had had some bombs left, [so few bombs] could not have played a significant role." The more important point in this recollection is that in late July 1945, the Soviet leadership knew approximately how many atomic bombs the US had in its arsenal.

Soviet intelligence continued its work. A telegram from Moscow on July 28 asked Colonel Zabotin in Ottawa to "try to get from [Alan Nunn May] before [his] departure [to return to England] detailed information on the progress of the work on uranium." The atomic bombing of Hiroshima on August 6—seventy thousand dead from one bomb delivered from one bomber, less than a kilogram of fissioning matter destroying a large city by blast and fire—made that progress brutally clear. Nunn May obliged immediately, reporting that the Trinity test had been conducted in New Mexico, that the bomb dropped on Japan was made of U235, reporting the daily output of U235 and plutonium from Oak Ridge and Hanford. Zabotin noted that the British scientist "handed over to us a platinum [foil] with 162 micrograms of uranium 233 in the form of oxide in a thin lamina." The foil had been sent to Nunn May at Montreal for research, legally, following up the British physicist's work with Herbert Anderson at Argonne in October 1944. "Herb said he noticed later that about half the U233 was missing," recalls Anderson's colleague, physicist Alvin M. Weinberg. "He always wondered where it went."

Igor Gouzenko was on hand to record the excitement in the Soviet Embassy when what he calls Nunn May's "uranium samples" came in:

> I was working late in the cipher room the night Angelov brought them from Montreal. Zabotin placed the samples on his desk and excitedly called Lieutenant-Colonel Motinov to see the latest "prize catch."
>
> There was some discussion on how the samples should be sent safely to Moscow. The diplomatic pouch wasn't regarded as safe enough. Then it was decided to send the samples with Motinov who was due to return to Moscow shortly for reassignment to Washington. Motinov, of course, was delighted because bringing back uranium samples would more or less assure him a good reception.
>
> Zabotin was in high fettle. I heard him say excitedly: "Now that the Americans have invented it, we must steal it!"

Klaus Fuchs already had.

If Stalin knew as much about the bomb as Harry Truman, the Soviet dictator seems nevertheless not to have grasped its full import until word arrived of the destruction of Hiroshima. "I didn't see my father until August," Svetlana Alliluyeva reports, "when he got back from the Potsdam Conference. The day I was out at his *dacha* he had the usual visitors. They told him that the Americans had dropped the first atom bomb over Japan. Everyone was busy with that, and my father paid hardly any attention to me." She had borne him a grandson whom he had not yet seen and had given her son his name, Josef, but he was too preoccupied, or too indifferent to her, to respond. According to NKVD staff physicist Yakov Terletsky, who probably heard the story in the corridors of the Lubyanka, "after the explosion of the atomic bomb in Hiroshima, Stalin had a tremendous blow-up for the first time since the war began, losing his temper, banging his fists on the table and stamping his feet." Terletsky thought Stalin "had something to be angry about. After all, the dream of extending the socialist revolution throughout Europe had collapsed, the dream that had seemed so close to being realized after Germany's capitulation. Hiroshima seemed to highlight the 'negligence' of our atomic scientists headed by Kurchatov." It would have been entirely consistent with Stalin's character to blame the difficulties Kurchatov's underfunded and low-priority bomb project had encountered on Kurchatov himself. According to Anatoli Alexandrov, who worked with Kurchatov and probably heard the story from him, at some time during this immediate postwar period "Stalin summoned Kurchatov and accused him of not demanding enough for maximum acceleration of the work. Kurchatov answered, 'So much is destroyed, so many people perished. The country is on starvation rations and everything is in shortage.' Stalin said irritably, 'If the baby doesn't cry, the mother doesn't know what he needs. Ask for anything you need. There will be no refusals.'" August 7, Stalin met promptly with Lavrenti Beria and appointed Beria head of the bomb program. Once again, lacking faith in the patriotism of his scientists, Stalin would rely on his whip.

The Soviet press delayed announcing the Hiroshima bombing until the morning of August 8 and underplayed the story, *Pravda* publishing only an excerpt from Truman's statement at the bottom of the foreign page. But the event did not go unnoticed. "On my way to the bakery," Andrei Sakharov remembered, ". . . I stopped to glance at a newspaper and discovered President Truman's announcement. . . . I was so stunned that my legs practically gave way. There could be no doubt that my fate and the fate of many others, perhaps of the entire world, had changed overnight. Something new and awesome had entered our lives, a product of the greatest of the sciences, of the discipline I revered."

Just as Truman had feared, Stalin moved up his intervention in the Far East from the mid-August launch he had promised at Potsdam to August 8. "I declared war on Japan," Molotov bragged. "I called the Japanese ambassador to the Kremlin and handed him the note." Molotov received the press that night to pass along the text of the Soviet declaration of war without a word about the atomic bomb. A member of the Polish Provisional Government in Moscow at the time, Stanislaw Mikolajczyk, asked Molotov at supper if the bomb would affect the international situation. "This is American propaganda," Molotov snapped. "From a military point of view it has no important meaning whatsoever." The Soviet people knew better, Alexander Werth reports:

> Yet the bomb was the one thing everybody in Russia had talked about that whole day. . . . Although the Russian press played down the Hiroshima bomb, and did not even mention the Nagasaki bomb until much later, the significance of Hiroshima was not lost on the Russian people. The news had an acutely depressing effect on everybody. It was clearly realized that this was a New Fact in the world's power politics, that the bomb constituted a threat to Russia, and some Russian pessimists I talked to that day dismally remarked that Russia's desperately hard victory over Germany was now "as good as wasted."

There had been a great victory parade in Red Square on June 24, hundreds of Nazi flags captured in the march westward to Berlin flung down on the steps of Lenin's tomb at Stalin's feet in a driving rainstorm, and a celebration that night at the Kremlin when Stalin entertained several thousand officers and soldiers of his victorious army. But Ilya Ehrenburg described a harsher reality that month, a nation in ruins:

France recently commemorated by a day of mourning the anniversary of the destruction of Oradour-sur-Glan. In Czechoslovakia President Benes drove out to the ashen ruins of Lidice. I think about our own Oradours and Lidices: how many are there? If you proceed west from Moscow to Minsk, or south to Poltava, or north to Leningrad, you will see everywhere ruins, ashes, graves, and after removing your cap you will not put it back on again. And everywhere the surviving inhabitants will tell how men swung from gallows, how mothers attempted to save babes-in-arms from the executioners, how houses with live people in them were burned to the ground.

One-tenth of the Soviet population—some twenty million human beings —had died in the war; millions more were invalids. The NKVD under Lavrenti Beria had murdered at least another ten million Soviet citizens, a slaughter more extensive than that of the Holocaust. "In the age groups that had borne arms," writes Werth, "there were at the end of the war only 31 million men left, as against 52 million women." The Germans had destroyed 1,700 towns, 70,000 villages, 84,000 schools, 40,000 hospitals, 42,000 public libraries. Twenty-five million people were left homeless. Coal production compared to 1941 was down 33 percent; oil down 46 percent; electricity down 33 percent; pig iron down 54 percent; steel down 48 percent; coke down 46 percent; machine-tool production down 35 percent. Thirty-one thousand industrial enterprises had been destroyed; overall, Soviet industry had been razed to one-half its prewar level. "Ninety-eight thousand collective farms and 1,800 state farms were destroyed or looted," Molotov reported in 1947; "7 million horses, 17 million head of cattle, 20 million pigs, 27 million sheep and goats had vanished." Meat production was down 40 percent; dairy production down 55 percent. The Red Army was the strongest force in Europe, but the Soviet people were exhausted and nearly starving.

And now the battered nation would have to gear up to build the atomic bomb. At the Yalta Conference in February 1945, Molotov had whispered that the Soviet Union would look with favor upon a US loan of $6 billion for postwar reconstruction. Building the industry necessary to manufacture atomic bombs had cost the United States more than $2 billion. That much would have to be subtracted from the crippled Soviet economy to win a similar capability for the USSR in the years after the war. Ten days after Hiroshima, the Supreme Soviet ordered the State Planning Commission and the Council of People's Commissars to begin work on a new Five-Year Plan. In mid-August, Stalin called together People's Commissar of Munitions Boris Vannikov and his deputies. Kurchatov walked in and they knew why they were summoned. "A single demand of you, comrades," Stalin told them, "provide us with atomic weapons in the shortest possible time. You know that Hiroshima has shaken the whole world. The equilibrium has been destroyed. Provide the bomb—it will remove a great danger from us."

10

A Pretty Good Description

IN AUGUST 1945, the most destructive war in history ground to an end, having claimed 55 million human lives. The Japanese armies in Manchuria quickly collapsed before the Soviet advance that began at midnight on August 8.* A United States B-29 atomic-bombed Nagasaki on August 9. For the next several days, Japanese military factions maneuvered unsuccessfully to prevent a humiliating surrender. Emperor Hirohito, in an unprecedented broadcast to his people on August 15, announced that surrender, which Japanese officials signed in Tokyo Bay aboard the United States battleship *Missouri* on September 2 with Curtis LeMay among the officials on hand to watch. Two great powers emerged from the ruins. The United States and the Soviet Union were both young nations forged in revolution, both ethnically diverse, organized on abstract principles rather than historically evolved, both vast in extent and rich in resources. They contested no territory, which led Enrico Fermi to ask dryly once when someone insisted that the two countries would go to war one day, "Where will they fight?" But their physical similarities did not obscure an intractable difference between them. They were opposite experiments in the large organization of people and natural wealth, the one through liberty and competition, the other through terror and centralized control, an open society and a closed—a crystal of quartz and a crystal of onyx, Robert Oppenheimer once contrasted them —and each was convinced that the other side's intentions were malevolent.

* In the brief Soviet-Japanese conflict, eighty thousand Japanese combatants died and 594,000 were taken prisoner. Official Soviet casualties totaled eight thousand dead and twenty thousand wounded (the real figures were probably higher). Soviet forces found the Japanese only lightly armed, primarily with rifles. Werth (1964), p. 1040. More than the US atomic bombings, the Soviet declaration of war influenced the Japanese decision to accept unconditional surrender; the Japanese leadership understood that without a neutral Soviet Union it no longer had an influential intermediary through which to negotiate surrender conditions.

Astride the ruins of Europe and Asia they were positioned peacefully to organize the world, but they could not agree on how the world should be organized.

Their mutual victories thus became mutual warnings. The Red Army— thousands of tanks, artillery pieces and mobile rocket launchers and ten million foot soldiers—had smashed west across Eastern Europe and Germany; it might as inexorably roll on to the Atlantic in a matter of days and confine all of Europe behind what Winston Churchill had already, before the end of the war, called an "iron curtain." The much less numerous US forces in Europe would certainly be overwhelmed by such an advance, but America knew how to build atomic bombs, and the Soviet Union understood that its recent ally had developed that capability clandestinely and had not hesitated to use the cruel new weapons of mass destruction against an enemy which had no such weapons of its own.

Yet neither side seems to have wanted—or expected—war, at least not in the short run. Each demobilized rapidly, the Soviet Union reducing the strength of its armies by the end of 1946 from 11.5 million to fewer than three million, the US by mid-1947 from more than twelve million to fewer than 1.6 million. The long run was more problematic. Since the Russian Revolution, the wealthy elite of the United States had feared that the red tide of Communism would flood across the world if it was not resolutely stanched. The Soviet victory over Germany, Stalin's evident determination to dominate Eastern Europe, his reluctance to quit northern Iran, all reinforced Western fears. " 'Give [the Germans] twelve or fifteen years,' " the Soviet dictator had predicted to a delegation of Yugoslavs over dinner in Milovan Djilas's presence in the final winter of the war, " 'and they'll be on their feet again.' " "[Stalin] got up," Djilas writes, "hitched up his pants as though he was about to wrestle or to box, and cried out almost in a transport, 'The war shall soon be over. We shall recover in fifteen or twenty years, and then we'll have another go at it.' " But Stalin was nothing if not cautious—"he regarded as sure only whatever he held in his fist," Djilas adds, "and everyone beyond the control of his police was a potential enemy"—and the Americans had the bomb.

"A lot of urgent long sittings were held," Igor Golovin reports of those first months after the war. "At one of the first sittings Stalin asked how much time would be necessary to create the bomb. [Isaak] Kikoin answered: 'five years.' The first priority in the State was given to the solution of the atomic problem. . . ." "Until 1945," Yuli Khariton and Yuri Smirnov note, "this program was carried out by only a few researchers who had scarce resources. The project gained real momentum only after the first American atomic explosions. It was precisely at that time that the Soviet atomic industry and technology could be developed on a broad footing, with large installations and combines."

The State Defense Committee formally enacted Stalin's decision on August

20, 1945, naming a Special Committee on the Atomic Bomb headed by Lavrenti Beria charged with coordinating all work on nuclear energy. The committee's membership included rising Politburo star Georgi Malenkov; Boris Vannikov; Avrami Zavenyagin, the Red Army general and senior NKVD officer who had led the roundup of uranium ore and scientists in defeated Germany; Mikhail Pervukhin, the chemical industry commissar; Peter Kapitza; and Igor Kurchatov. "Stalin's word decided the fate of the project in general," says Anatoli Alexandrov. "One gesture of Beria was sufficient to make any of us to disappear. But Kurchatov was on the very top of the pyramid. It was our great luck then that he combined competence, accountability and power."

Beria led the atomic-bomb project more actively than Alexandrov's comment implies, Khariton has insisted:

> Once the project passed into Beria's hands, the situation changed completely. Beria understood the necessary scope and dynamics of the research. This man, who was the personification of evil in modern Russian history, also possessed great energy and capacity for work. The scientists who met him could not fail to recognize his intelligence, his will power, and his purposefulness. They found him a first-class administrator who could carry a job through to completion. It may be paradoxical, but Beria —who often displayed great brutishness—could also be courteous, tactful and simple when circumstances demanded it.

The US War Department with Truman's approval released a report on the Manhattan Project on August 12 that supplied the Soviet Union with information on atomic-bomb development nearly equivalent to all the information it had acquired laboriously during the war through espionage. The *General Account of the Development of Methods of Using Atomic Energy for Military Purposes,* written by Princeton physicist Henry D. Smyth and quickly nicknamed the Smyth Report, confirmed the validity of that espionage. It discussed the problems the Manhattan Project had encountered in separating uranium isotopes, building reactors, breeding plutonium and designing the bomb and identified the most effective solutions. General Groves had ordered the report written to draw a line of declassified information beyond which Manhattan Project scientists, whom he thought irresponsible, could not trespass, but the result belied his intention. When Alan Nunn May eventually confessed to espionage, he sought to minimize the damage he had done by comparing his indiscretions to the Smyth Report: "I also gave . . . a written report on atomic research as known to me. This information was mostly of a character which has since been published. . . ." Smyth's dry, semitechnical study did not mention implosion, but the Soviets already knew about that technology in detail.

On September 5, in Canada, disaster struck Soviet foreign intelligence. Igor Gouzenko, the code clerk who had transferred to Ottawa in June 1943, had barely avoided being sent back to the Soviet Union in September 1944. Since then he had been preparing to defect. In the late summer of 1945, he was twenty-six and married, with a pregnant wife, Svetlana ("Anna"), and a young son, and Canadians who knew him then and later noticed that he was wide-eyed at the freedom and prosperity of the West. When he had first come to Canada, at a stop on the way to Ottawa, he and a fellow clerk had impulsively bought a crate of oranges. Gouzenko had only once before in his life tasted an orange; on the train he gorged on them—oranges spilling in the aisle, an orgy of oranges. "You take your wife to the movie at night," he told an acquaintance after his defection. "I don't do that. Anna and I go to the IGA [grocery] and just look at the things in the store, just to see all these things and to know we could take this can and this bag and buy these things."

Once he had decided to defect, Gouzenko, a small man whom one of his neighbors thought "a very quiet and well-behaved gentleman," began surreptitiously tagging cables in his cipher files that revealed the activities of GRU espionage agents in Canada, including physicist Alan Nunn May, Member of Parliament Fred Rose, National Research Council scientists Durnford Smith, Edward Wilfred Mazerall and Israel Halperin and a dozen others. Gouzenko had tagged an accumulation of some 109 documents by the time it appeared that he was about to be transferred back to Moscow, including the cables that reported Nunn May's transfer of samples of U235 and U233 to Gouzenko's superiors. The young cipher clerk returned to the embassy after an evening out with the boys on the warm Wednesday night of September 5, gained access on a pretext, stuffed his loose shirt full of the cables he had tagged and nervously walked out the front door.

Gouzenko was naive enough to imagine that the Soviet Union's recent allies were eager to know of its espionage activities against them and would welcome him with open arms. From the Soviet Embassy he took a streetcar downtown to the *Ottawa Journal,* intending to see the editor. When a woman on the elevator recognized him—she asked him if there was news breaking at the embassy—he panicked, rode the elevator back to the ground floor and bolted. Eventually he caught a streetcar home. His wife, who had agreed to defect with him, calmed him down and advised him to try again. Time was of the essence. "You still have several hours before the Embassy learns what has happened," she told him.

Back to the *Ottawa Journal* offices Gouzenko went, his shirt still stuffed with the incriminating cables. The City Room was crowded. An office boy told Gouzenko that the editor was gone for the night and led him to an older man wearing a green eyeshade. Gouzenko walked up to the man—the night city editor, Chester Frowde—and signaled that he wanted to speak to him in private. Frowde led Gouzenko, "short, with a tubby build, and . . . white as a sheet," into the newspaper morgue, where Gouzenko's first words were, "It's war. It's war. It's Russia." But he chose not to tell Frowde his story. "He just stood there apparently paralysed with fright," Frowde recalled. Gouzenko, for his part, "could see from the man's expression that he thought I was crazy." Frowde walked him to the elevator. Gouzenko claims Frowde brushed him off. Frowde says Gouzenko refused even to give his name.

Where next? Gouzenko asked himself out in the street. The Justice Building, he decided, to see the Minister of Justice. A young Royal Canadian Mounted Police officer guarding the door told him to come back in the morning and turned him away. Frightened, he went home. His wife tucked the documents into her purse and hid it under her pillow and then the two of them lay awake the rest of the night worrying.

In the morning, Gouzenko took his visibly pregnant wife and his young son with him back to the Justice Building. The receptionist sent the Gouzenkos to see a clerk, to whom Igor explained that he could only speak to the Minister of Justice himself. The clerk called ahead and then led the little family to the Parliament Building, where Gouzenko explained his mission to another clerk. The second clerk sent the message along. The Gouzenkos waited for two hours. "They were all panicking," Svetlana Gouzenko concludes. "They were all just in a panic. Didn't know what to do." Finally the Minister of Justice sent out word that they should go back to the Soviet Embassy and return the documents. The Gouzenkos assumed that Soviet agents within the government must have made so stupid and deadly a decision. In fact, it came directly from the Prime Minister of Canada, Mackenzie King, who seems to have been terrified that he might stir up trouble with the Soviet Union.

Svetlana proposed they try the *Ottawa Journal* once more. This time a reporter at least interviewed them, a woman named Elizabeth Fraser. "[Gouzenko] was utterly agitated," Fraser recalled, "almost incoherent. He blurted out on our first encounter: 'It's dess [death] if you can't help us' and then proceeded to try to convince me that his situation was indeed as dangerous as he felt it to be. He said he had evidence with him of terrible Soviet spying activities against the western countries and that he wanted to save Canada from their perfidy, all of which, given the political climate of the time, sounded to me utterly fantastic." Fraser went to a senior editor for advice. "I am terribly sorry," Gouzenko reports she told him when she came back. "Your story just doesn't seem to register here. Nobody wants to say anything but nice things about Stalin these days." Svetlana asked Fraser what they should do. Fraser suggested they talk to the Crown Attorney about filing for naturalization. "That should prevent the Reds from taking you back," she theorized.

Desperately they trudged once again to the Justice Building. "The day was getting hot," says Gouzenko. ". . . Anna was obviously growing weary." They were told the applications clerk had gone to lunch. They went to lunch themselves and then took their son home and left him in the care of a neighbor. "Back we went to the Crown Attorney's office." Only after they had wasted more time filling out naturalization papers did they learn that the process would take months.

So it went for the rest of the day. A woman in the Crown Attorney's office, Fernande Coulson, tried to help them, even to the extent of calling a reporter she knew and having Gouzenko translate extracts from the documents that referred to the atomic bomb. The reporter demurred. "It's too big for us to handle," he told the Gouzenkos. ". . . It's a matter for the police or the government." But neither the police nor the government expressed interest in the first important atomic espionage breakthrough since the beginning of Anglo-American atomic-bomb development. By the end of the day, Fernande Coulson had managed to convince an RCMP inspector to see the Gouzenkos—the next morning. Exhausted, knowing that by now Igor must have been missed at the embassy, the Gouzenkos went home. Coulson watched them out the window as they stumbled down the street and boarded a streetcar. "I said to myself: 'That man may not be alive tomorrow.'"

Gouzenko sent his wife and child to the next building to hide while he reconnoitered his apartment. There were two men sitting on a bench in the park across the street, watching his windows. He thought they were probably NKVD and he took his wife and son around the back way. No sooner were they settled when someone began pounding on the door of the apartment and calling Gouzenko's name. Gouzenko recognized the man's voice—it was Zabotin's chauffeur. The Gouzenkos froze. Eventually the chauffeur went away. The men in the park were still watching. Gouzenko remembered that his neighbor next along the rear balcony, Harold Main, was a corporal in the Royal Canadian Air Force. He found the Mains on their balcony escaping the heat and asked if they would take care of his son if something happened to him. When the good sergeant learned what was wrong— Gouzenko told him the NKVD was making an attempt on their lives—he offered to take in Svetlana and the boy and summon the police. "He was a military man," Svetlana Gouzenko comments, "and to tell him that one man can kill another was not new to him." Fetched by Harold Main, the constables promised to keep the building under surveillance. In the meantime, Mrs. Main having objected to sheltering the Gouzenkos, another neighbor across the hall took them in.

At about ten o'clock that night, an NKVD officer and three embassy men —"three or four of them with a Russian-movie sleaze look to them," remembers a neighbor who saw them out his window—pounded on Gouzenko's apartment door and then shouldered it open. After the Soviet raiding party had started searching the apartment, the constables stepped in with their guns drawn and demanded to know what was going on. The NKVD officer, invoking diplomatic immunity and insisting that the apartment was Soviet property, ordered the constables to leave the apartment. Instead they called in an inspector and the Soviets slunk away. The next morning the Mounties took the Gouzenkos into protective custody and questioned Gouzenko for five hours. The men watching from the park turned out to have been Mounties. "You weren't quite as neglected as you thought," one of them told the Soviet cipher clerk.

Gouzenko had heard another, later knock on his apartment door the previous night but had not revealed himself hiding in his neighbor's apartment across the hall. The later visitor was Sir William Stephenson, the director of British intelligence in the Western Hemisphere whose code name was Intrepid, who wanted to hear Gouzenko's story. Luckily for Gouzenko, Stephenson happened to be visiting Ottawa from his offices in New York. He had urged Mackenzie King's deputy Norman Robertson to take Gouzenko when King had decided to send the Soviet defector back to the Soviet Embassy with his documents. After visiting Gouzenko's building, Stephenson had gone to Robertson's home in the middle of the night and had convinced him to place the Gouzenkos in protective custody.

The Canadian government spirited the Gouzenkos away to a safehouse and began Igor Gouzenko's lengthy debriefing. "It was much worse than what we would have believed," King confided to his diary. Gouzenko's documents disclosed "an espionage system on a large scale." Robertson told King "he felt that what we had discovered might affect the . . . Council of Foreign Ministers [then meeting in London]; that if publicity were given to this it might necessarily lead to a break in diplomatic relations between Canada and Russia and . . . in regard to other nations as well, the U.S. and the U.K. All this might occasion a complete break-up of the relations that we have been counting on to make the peace. There was no saying to what terrible lengths this whole thing might go." The Canadians told the British about Gouzenko and called in the FBI.

Late in September, Mackenzie King flew to Washington to brief President Truman, whom he met in the White House Oval Office with Dean Acheson:

> Narrated the incidents regarding [Gouzenko]. . . . Told them of the extent of espionage in Canada. What we had learned about espionage in the United States. Mentioned particularly . . . information regarding the atomic bomb. . . . Spoke of the [Vice-]Consul at New York who apparently had charge of the espionage business in the United States. . . . What was thought to have gone from Chicago [i.e., the samples of uranium]. Also the statement that an assistant secretary of the Secretary of State's Department* was supposed to be implicated. . . .

---

* Gouzenko had reported the "assistant secretary" without knowing who he was, but the cipher clerk's account started the investigation that later pointed to Alger Hiss, special assistant to US Secretary of State Edward Stettinius at Yalta.—RR

Confronted with these shocking revelations, writes King, Truman asked that "above all nothing should be done which might result in premature action in any direction." For the time being the several security agencies involved would keep their secrets and set out their snares. Gouzenko's information in any case hardly shone glory on their guardianship of state secrets.

With the end of the war, David Greenglass received an early furlough. He and Ruth visited New York together in September and stayed in the cold-water flat in his parents' building where David had lived before he was married. The morning after the couple arrived, David would testify, Julius Rosenberg dropped by:

> He came up to the apartment and he got me out of bed and we went into another room so my wife could dress.... He said to me that he wanted to know what I had for him.... I told him "I think I have a pretty good description of the atom bomb."... He said he would like to have it imme- diately, as soon as I possibly could get it written up he would like to get it. ... During this conversation he gave me $200 and he told me to come over to his house.... He then left and I was there alone with my wife.... My wife didn't want to give the rest of the information to Julius, but I overruled her on that.... I said that "I have gone this far and I will do the rest of it, too."... We went down—it was late in the morning—we had a combina- tion breakfast and lunch, and I came back up again and I wrote out all the information and drew up some sketches and descriptive material.... I would say about twelve pages or so.

"I did not want [David] to give the information to Julius," Ruth Greenglass confirmed. "The bomb had already been dropped on Hiroshima and I real- ized exactly what it was and I didn't feel that the information should be passed on. However, David said that he was going to give it to him again."

The description of the implosion bomb which David Greenglass passed to Rosenberg that afternoon was garbled, but it contained useful informa- tion. He collected it in the course of his work and by being observant. He described the "lens molds"—the high-explosive lenses—as "pentagonal" in shape, incorrectly reported that there were thirty-six lenses (there were thirty-two) and indicated that the detonators were fired by capacitators. Most valuably, he described a key feature of the small initiator at the center of the assembly that supplied a burst of neutrons to start the chain reaction: "cone-shaped holes . . . , the apex of each cone being toward the periphery of the beryllium." In the initiator, at the appropriate moment, implosion mixed polonium210 with beryllium. Po210 is a powerful source of alpha particles, which easily dislodge neutrons from beryllium atoms. The cone-shaped holes in the initiator served to break the barrier of nickel plating between the polonium and the beryllium and improve their timely mixing. Their design took advantage of the Munroe effect, the principle on which the shaped charge is based that is used in such devices as armor-piercing rockets like the famous bazooka of the Second World War: a cone focuses a shock wave moving through such a configuration to an intensely penetrating high-speed jet. Applying the Munroe effect to initiator design was a direct outgrowth of the experiments with two-dimensional implosion and the studies of jets in which both Greenglass and Klaus Fuchs had participated. Initiator design, significantly, was one of the most difficult aspects of implosion development and effectively paced the plutonium implosion project. Greenglass had learned about the Munroe effect and shaped charges in conventional explosives during his training at Aberdeen. The cone configuration was an advanced design, different from the initiator used in the Trinity and Nagasaki Fat Man units. Its Los Alamos inventors patented it; the patent went jointly to experimental physicist Rubby Sherr and Klaus Fuchs.

But Greenglass recalled including even more valuable information in the handwritten pages he passed to his brother-in-law in September 1945. He described and supplied a rough sketch of an experiment, he said later, "which was concerned with the reduction of the amount of plutonium to be used in the atomic bomb." It was also concerned with increasing the efficiency of the explosion. "This experiment . . . consisted of one sphere of uranium inside of a larger sphere of uranium with a large air gap between the two spheres and stilts to hold the inner sphere apart from the outer sphere. I informed Rosenberg that the air gap was used to increase the speed with which the outer sphere is imploded. I told him this would result in a greater explosion with the use of less plutonium. . . . I made up portions of this experiment as one of my duties at Los Alamos."

The experiment Greenglass described to Rosenberg concerned two important improvements in implosion design: levitation and the composite core. Hans Bethe explains:

The solid core clearly was very hard to compress. Originally we had wanted a hollow shell, but we didn't trust the symmetry [of the implosion with that arrangement]. Then we decided, yes, we could after all have a hollow shell, and if we had a hollow shell, it was useful to have the very center solid. So that the hollow shell was pasted on the [tamper], so to speak. And the core [at the center] had to be levitated. The question was, could you make thin enough wires [levitating the core within the shell] which were strong enough, so that they could survive transportation by plane and [being dropped as a bomb]. They had to be strong enough, and yet small enough —thin enough—so that they wouldn't disturb the spherical symmetry. Because the spherical symmetry is all-important. Only [if the symmetry of the implosion is maintained] do you get the increase in density, which you bank on. In addition to that, with the hollow construction, you can put a little more material in. You can put in more than a critical mass. Which again increases the yield. And then already in the last month of [the war], we invented the composite [core], plutonium in the center and uranium outside, which [results in] a great increase in the yield. Plutonium was much more expensive than separated uranium. Three times maybe. Therefore, we would get a much better arsenal by having the mixtures.

Levitation gave the imploding shell time to acquire momentum before it hit the core. Nuclear-weapons designer Theodore B. Taylor explained the principle to the writer John McPhee once without naming it: "The way to get more energy into the middle was to hit the core harder. When you hammer a nail, what do you do? Do you put the hammer on the nail and push?" The solid Fat Man core had been pushed; levitation hammered. And because it increased efficiency, levitation also made it possible to design bombs of smaller diameters than Fat Man, lighter weapons more easily transportable by plane. The composite core, as Bethe points out, used less plutonium and resulted in increased yield as well, both important advantages. (The core Greenglass described, with a uranium shell and a uranium center, was a substitute, probably of plentiful and non-chain-reacting natural uranium, used in implosion experiments.)

Greenglass asked Rosenberg why Harry Gold had contacted him in Albuquerque instead of Ann Sidorovich. "She couldn't make it," Rosenberg told him unhelpfully.

Max Elitcher remembered Julius Rosenberg phoning him from Washington's Union Station that September as well and coming over for a talk. Rosenberg told Elitcher he was still in business even though the war was over; the Soviet Union still needed information on military technology, he said.

Harry Gold had met with Anatoli Yatzkov in Brooklyn in mid-August to prepare for the planned trip out to New Mexico in late September to rendezvous again with Fuchs. Since Yatzkov had told him that the information David Greenglass passed in June was valuable, Gold suggested meeting again with the young Army machinist as well. This time it was Yatzkov's turn to invoke protocol; Gold says the Soviet agent "told me that it would be inadvisable to endanger the trip to see Fuchs by complicating it with a visit to the Greenglasses in Albuquerque."

Gold went out to New Mexico in mid-September. The trip had been difficult to organize; he had trouble getting time off from work and he was short of money. From the Palmer House in Chicago, where he spent the night of September 16, he called his old friend Tom Black in Newark and asked Black to wire fifty dollars to him care of the Albuquerque Hilton. Black was able to raise only twenty dollars but loyally sent it on.

With Curtis LeMay in the air in his B-29 somewhere eastward of Hokkaido,

Gold arrived in Albuquerque on Wednesday, September 19, 1945, and signed in at the Hilton. He was scheduled to meet Fuchs that day on the outskirts of Santa Fe, at a time Gold remembered as "very late in the afternoon, about six o'clock." Once again he caught a bus up from Albuquerque. Fuchs arrived at the rendezvous in his dilapidated Buick uncharacteristically late—"fully twenty or twenty-five minutes tardy," says Gold, who always felt exposed when he was waiting for a contact—and they drove off into the Santa Fe hills. Fuchs told the portly chemist apologetically that he'd had difficulty getting away. As Gold remembered it, the physicist and his "friends with whom he worked at Los Alamos" were having a party "that very evening . . . to celebrate the successful use of atomic energy in the form of a weapon"; Fuchs was hauling a supply of liquor. The occasion must have been the formal British Mission party held at Los Alamos not that evening but three days later on Saturday, September 22, "in celebration of the Birth of the Atomic Era," as the formal invitations proclaimed. The party featured a footman announcing guests, steak-and-kidney pie served on paper plates, several hundred paper cartons of trifle, a full-scale pantomime, dancing and many toasts—enough work to keep the British staff on the Hill busy throughout the week making preparations. Fuchs evidently contrived to get away to his rendezvous with Gold by volunteering to pick up the liquor. Gold could not have learned about the British Mission party from a source outside Los Alamos. That he approximately recalled Fuchs's late-September circumstances five years later is significant confirmation of his veracity.

Fuchs had only just finished writing the report he was delivering to Gold. "En route . . . for this planned meeting . . . ," he would confess, "I stopped somewhere on the way in the desert, drove off the highway to a solitary place, and wrote a part of the . . . paper . . . which I planned to deliver. . . ."

Driving, Fuchs had much to say. He had attended the Trinity test, he told Gold, and learned later that the flash of the explosion had been visible all the way up at Los Alamos, two hundred miles northwest, despite overcast skies and rain. "He himself was rather awestricken by what had occurred," Gold paraphrases Fuchs. ". . . Frankly, he had not been too certain that the project might not have been abandoned before it was completed, and . . . certainly he had grievously underestimated the industrial potential of the United States. . . . He was also greatly concerned by the terrible destruction which the weapon had wrought." Fuchs was not the only one at Los Alamos disturbed by the death tolls at Hiroshima and Nagasaki. His great concern was evidently not sufficient to lead him to reconsider proliferating the destructive new technology to the Soviet Union.

At some point Fuchs pulled off the road and parked—"a fair distance away" from town, says Gold, "because below us I could barely see the lights of Santa Fe in the distance." The physicist continued his recitation of marvels. "He told me that whereas, before, the townspeople in Santa Fe had regarded them, the people of Los Alamos, as a sort of 'boondoggling' outfit engaged in work which they could not comprehend, that now they were hailed on all sides as conquering heroes. . . ." A Los Alamos security officer, Fuchs said, told him casually one day that Army intelligence realized there were "hundreds" of Soviet agents in the US and England, but the British and the Americans together had "only one" agent in the Soviet Union. Fuchs laughed at the discrepancy, Gold remembered.

From derisive, Fuchs turned somber. He had been barred from some sections of the project, he reported; "the relationship between the British Mission and the United States, which once had been extremely cordial and free, had now become somewhat strained, and . . . there was no longer the free exchange of information between the two groups." He expected to be returning to England before the end of the year or early in 1946, "where he would again resume work on atomic energy, exclusively for [the British]." British intelligence had notified him that they were trying to contact his father and might bring Emil Fuchs to England. Fuchs was "very much concerned" about his father's "welfare and health," but he was also worried that his father would talk too much: "Klaus told me that as far as he knew the British had no inkling about his past as it related to his Communist activities, and he was anxious that this continue so." Gold told him "to proceed as he thought best"; possibly, Gold consoled him, "he was greatly overestimating the extent to which the old man would talk and also the extent to which the British might be interested in Klaus' past." As it turned out, Gold was right on both counts.

Yatzkov, with his high-level informants in Washington, had heard that Fuchs would soon be returning to England and had prepared Gold with a London contact protocol, which Gold now passed along to Fuchs. Fuchs mentioned that he might stop off to visit his sister in Cambridge again around Christmastime "and that the best way of ascertaining his whereabouts was to make an inquiry shortly before that time." Fuchs drove Gold back to Santa Fe. "The last event that transpired before Klaus dropped me off . . . was that [he] gave me the packet of information relating to atomic energy." Fuchs drove away; Gold headed for the bus station. "After a period of anxious waiting, about an hour and a half, I finally obtained a bus going back to Albuquerque."

American Airlines woke Gold at the Albuquerque Hilton at two-thirty Thursday morning to confirm a seat as far as Kansas City. From Missouri Gold took the day coach to Chicago, caught a train late that night to New York and from New York, still carrying Fuchs's packet, commuted home to Philadelphia. He returned to New York the next day, September 22, to meet Yatzkov and transfer Fuchs's packet, but Yatzkov, alerted by now to the Canadian espionage debacle, failed to appear. At a backup meeting in Queens early in October, Yatzkov finally took the incriminating documents off Gold's hands. He met regularly with Gold throughout the rest of the year, but by December Gold noticed that he was "very touchy and very apprehensive." Yatzkov told him "they had to be extremely careful." Gold got the impression Yatzkov "had the wind up."

Fuchs's report added additional details to the full description of the Trinity plutonium implosion design that he had passed Gold in June. He noted that the production rate of U235 was up to a hundred kilograms per month and of plutonium to twenty kilograms per month and gave the critical mass of each material so that the Soviets could calculate roughly how many bombs the US was capable of stockpiling. He communicated important information about plutonium phases—different crystalline states, each with unique properties.

Plutonium is a bizarre metal. Determining its metallurgical properties had given Los Alamos metallurgists much trouble. "Plutonium is so unusual," its discoverer, Glenn Seaborg, once told a reporter, "as to approach the unbelievable. Under some conditions, plutonium can be nearly as hard and brittle as glass; under others, as soft and plastic as lead. It will burn and crumble quickly to powder when heated in air, or slowly disintegrate when kept at room temperature. It undergoes no less than five phase transitions between room temperature and its melting point. Strangely enough, in two of its phases, plutonium actually *contracts* as it is being heated. . . . It is unique among all of the chemical elements." Phase differences made large differences in density and thus in the volume a given weight of plutonium occupied and in critical mass. "During our first trials of shaping methods," writes Los Alamos chief metallurgist Cyril Stanley Smith, ". . . a beautifully flat sheet of [plutonium] would curl up like a saucer as it transformed, and cylinders would develop strongly concave ends." Fuchs's revelations made it possible for Soviet scientists to approach the difficult business of plutonium fabrication well-informed. In particular, he reported importantly that the delta phase—the densest phase that is still malleable—could be stabilized at room temperature by alloying Pu with the rare metallic element gallium.

Fuchs reported the results of the Trinity test, described his work on initiators and explained how the Fat Man design had been preassembled with one HE lens left out and a removable plug drilled through the tamper to form a passageway to the center of the assembly through which the core could be inserted. He reported that the uranium-separation filters ("barriers") developed for the Oak Ridge gaseous-diffusion plant were made of sintered nickel, a material that had only been identified after a long and difficult search. On this occasion and later, he passed along information about composite core design, emphasizing the economic advantage to the United States of drawing on both isotope separation and plutonium production for bomb materials. He knew about levitation as well and probably reported it. Once again David Greenglass's ad hoc information would usefully corroborate Fuchs's scientifically accurate account; in this instance Greenglass had even reported the new developments first.

The head of Soviet foreign intelligence, Vsevolod Merkulov, sent Lavrenti Beria a detailed plan of the Fat Man plutonium implosion bomb on October 18, 1945. The top secret seven-page document began with a summary:

GENERAL DESCRIPTION OF THE ATOMIC BOMB

The atomic bomb is a pear-shaped projectile with maximal diameter of 127 cm and 325 cm long, fins included. The total weight is around 45,000 kg. The bomb consists of the following parts:

    a. Initiator
    b. Active material
    c. Tamper
    d. Aluminum layer
    e. Explosive
    f. 32 explosive lenses
    g. Detonating device
    h. Duralumin shell
    i. Armor steel shell
    j. Fins

All the above-listed parts of the bomb, except for the fins, detonation device and external steel shell are hollow balls, inserted into each other. For instance, the ball of active material itself is inserted into the tamper (moderator), which is also a hollow ball. The ball of the tamper in its turn is inserted into another hollow ball, made of aluminum, which, in its turn, is surrounded by a spherical layer of explosives.

The layer of explosives, which also contains the lenses, is surrounded by the duralumin shell, to which the detonation device is attached, and which is covered by the outer shell of the bomb, made of armored steel.

There followed a systematic discussion of each of the important parts of the bomb, beginning with the initiator. David Greenglass had described an initiator with shaped-charge cones machined into its beryllium shell to facilitate shock-wave mixing of its beryllium and polonium; the document Beria received in October 1945 described another initiator design, "a hollow beryllium ball with wedge-like grooves on the internal surface of the ball" with "the axes of all grooves . . . parallel to each other." This design, the document noted, was called "Urchin." Physicist Rubby Sherr at Los Alamos had nicknamed the grooved initiator the Urchin. Because of its grooves, its other nickname was the "screwball." The Trinity implosion device used an

Urchin initiator. So did the Fat Man exploded over Nagasaki; so would all US atomic bombs for half a decade to come. The October 18 document gave the Urchin's precise measurements and described its operation in detail— its two parts, its gold and nickel plating, the way its various layers and parts interacted to generate neutrons to start the chain reaction. The collapsing grooves produced a "Munroe jet," the document noted, referring to the Munroe effect that David Greenglass was later to cite as the principal mechanism by which initiators mix beryllium with polonium.

Of the "active material" the document noted:

2. ACTIVE MATERIAL

The active material of the atomic bomb consists of the element plutonium of delta phase with specific weight of 15.8 [sic: the modern value is 15.7]. It is manufactured in the form of a hollow ball consisting of two halves, which like the outer ball of the initiator are pressed in an atmosphere of nickel-carbonyl. The external diameter of the ball is 80–90 mm. The weight of the active material, initiator included, is 7.3–10.0 kg [the Fat Man core without initiator weighed 6.2 kg]. A gasket of corrugated gold 0.1 mm thick is located between the halves of the sphere, which prevents penetration to the initiator of high-speed jets moving along the junction planes. These jets might otherwise prematurely activate the initiator.

There is an opening 25 mm in diameter for the purposes of inserting the initiator into the center of the active material, where it is fixed on a special bracket [the initiator was levitated]. After the initiator is inserted, the opening is closed with a plug also made of plutonium.

The gold foil was a significant detail. Metallurgist Cyril Stanley Smith had developed it to resolve a jetting problem. The foil served to true the surfaces of the two plutonium hemispheres after plating and improve the fit. In retirement, Smith would sometimes exhibit a spare foil to visitors—a small circular sheet of pure gold with a large hole in the middle which he kept at home in a plain white cardboard jewelry box. After his death in 1992, the National Museum of American History acquired Smith's spare and placed it on permanent exhibit.

The Merkulov document next discussed the tamper—the heavy shell of natural uranium that surrounded the core and served as both a neutron reflector and an inertial restraint on explosive disassembly. Besides the tamper's composition and dimensions, and the fact that it also was plugged for core insertion, the espionage report included a significant detail that is invariably omitted in US official descriptions of the Fat Man design: "The external surface of the tamper is covered with a layer of boron which absorbs thermal neutrons emitted by the radioactive materials of the system that are capable of causing predetonation." The muddled description of the implosion bomb that David Greenglass had passed to Julius Rosenberg in September had included this neutron-absorption system, although Greenglass had misidentified it as barium, not boron. At his trial in 1950 he called the boron layer a "barium plastic sphere"; in one of his confessions the previous July he had correctly positioned the boron layer, which he called "a plastic shield," "between the plutonium [sic: uranium tamper] and the high explosives."

The Merkulov document described the Fat Man high-explosive configuration in detail, but omitted crucial information about the precise curve of the HE lens assemblies:

5. THE LAYER OF EXPLOSIVES AND LENSES

The layer of explosives consists of 32 blocks of special form. It follows the layer of aluminum. The internal surface of the blocks, facing the center, is spherical and has a diameter equal to the external diameter of the aluminum layer. There are special slots in the external surface of the blocks which provide for the insertion of 20 lenses of hexagonal and 12 lenses of pentagonal shape. A felt lining $1/16$ of an inch thick is located between the surfaces, perpendicular to the axis of the sphere. Voids between radial surfaces are filled with blotting paper. The air gap between layers of explosives and lenses shouldn't be greater than $1/32$ inch, since a larger gap could lead to an increase, or, alternatively, a decrease of detonation depending on the orientation of these gaps. Each lens consists of two types of explosives—one fast-burning, another slow-burning. The lenses are cast in special casings made of cellulose acetate. The lenses are installed so that the fast-detonating part contacts the layer of explosive. Total weight of explosives is around 2 tons.

One detonator is attached to each lens, which for higher reliability is provided with two electric fuses. There are 64 wires in total, divided into 4 quadrants, 16 wires in each. Two wires lead to each lens, but from separate quadrants.

The document concluded with a brief description of the bomb's duralumin shell and a careful discussion of its assembly, noting in particular that "since plutonium and the radioactive substances of the initiator generate heat and warm themselves to temperatures higher by 90 degrees centigrade than the temperature of the environment, they are transported to the bomb assembly site in special containers fitted with cooling systems."

This historic document was evidently specially prepared for Lavrenti Beria, who had broad experience in industrial management but no scientific training; it was not a verbatim transmission of the more detailed information Klaus Fuchs had passed in June and late September 1945.

Beria was vulnerable. He had been assigned a vital project which had been given the highest priority of the state, but he lacked the knowledge necessary to judge its progress. He was at the mercy of scientists, intellectuals, people he viscerally distrusted. "With all of Beria's apparent power," writes a Russian historian, "he understood nothing about physics and he remained silent when the subject came around to uranium, plutonium, the separation of isotopes, 'items'. . . . And the success of the work . . . also meant the destiny of the leader's adviser himself, who bore personal responsibility for the creation of nuclear weapons under Stalin." "At first all the problems were solved through Kurchatov," says Yuli Khariton. "[Eventually] [Beria] was forced to pay attention to us."

So Beria sought ways to decrease his vulnerability. He sent security officers to Japan to film the destruction at Nagasaki. He began developing a stable of "backup" scientists—with fewer Jews among them—whom he might call upon to replace the Kurchatov team if it proved to be treacherous.

Already on October 3, Peter Kapitza had written Stalin boldly complaining of Beria's leadership of the atomic-bomb project. "Is the position of a citizen in the country to be determined only by his political weight?" Kapitza had asked rhetorically, adding, "It is time . . . for comrades of Comrade Beria's type to begin to learn respect for scientists." Beria first tried to smooth over the disagreement. After Stalin called him about Kapitza's letter, he telephoned Kapitza and invited the physicist over for a talk. When Kapitza refused to visit Beria at the Lubyanka, Beria sent Kapitza an elegant gift, a Tula shotgun. Kapitza was not assuaged, however, and the argument developed through the rest of the year. It concerned an issue on which Harry Gold could have advised Kapitza from his experience supplying industrial espionage: whether, as Beria was proposing, the first Soviet atomic bomb would be a copy of the Fat Man design, or whether, as Kapitza had proposed, the Soviets should proceed to develop a more sophisticated design of their own. "Peter Leonidovich's point of view," comments Anatoli Alexandrov, "was that if we followed the same road the Americans followed then we would never be ahead of them. It was necessary to find our own way." Alexandrov explains: "People like Beria could see only the bomb itself. He had no idea of the fundamental and multi-faceted character of the research. For example, Beria forbade development of a nuclear reactor for ships. He wanted the bomb first; everything else later."

Kapitza was not risking his life opposing Beria merely for a matter of prestige. He had confronted the same issue before, in 1935, when he had been detained in Moscow and denied permission to work abroad. Then he had written to his wife, who was still in England:

All the efforts [of the state] are now directed to the accumulation of the material basis on which the socialist society will be built. This accumulation is going on at a terrific pace, at such a pace that nobody could predict. But it is going so smoothly because its base is imitation; the country spends almost nothing on the creation of new technical forms. Research is all directed to the solution of different secrets and the mastering of different processes of general character which are very well known and mastered in Western Europe. For this work one does not require any special depth of thought or qualification, but the results are very spectacular.... How long this phase will continue I cannot say, but it is clear that the position of pure science, if not completely nil, is not far from it....

I am certain [that] when we shall enter in our socialist development into the period of original thought, then all will completely change.... Then the inventive mind and creation will have freedom in front of it; originality of mind will then be valued more highly than organizing gifts, as is the present position.

For Kapitza, that is, the shift that he believed must come from what he called "coarse imitati[on]" to original science would be a movement from dogmatism to intellectual freedom. The bomb must have seemed to him a vital opportunity to demonstrate to the Soviet leadership what a significant contribution science could make. But between the physicist and his dream stood Lavrenti Beria.

"[Comrade Beria], it is true, has the conductor's baton in his hands," Kapitza wrote Stalin again dangerously in November, pursuing his argument. "That's fine, but all the same a scientist should play first violin. For the violin sets the tone for the whole orchestra. Comrade Beria's basic weakness is that the conductor ought not only to wave the baton, but also to understand the score. In this respect Beria is weak."

"I told him straight out," Kapitza added: " 'You don't understand physics. Let us scientists judge these matters.' And to that he retorted that I knew nothing about people."

After Kapitza's assault, Beria asked Stalin if he could arrest him, which would certainly have been the physicist's death warrant. Stalin had begun to be wary of Beria's power. "I will remove him for you," the Soviet dictator responded, "but don't you touch him." In December, Stalin allowed Kapitza to resign from the Special Committee. Through an intrigue arranged by Beria to which Stalin presumably acquiesced, Kapitza in August 1946 was stripped of his scientific positions, including his directorship of the Institute for Physical Problems, and placed under house arrest, where he languished for the next eight years.

Beria wanted a bomb that would be guaranteed to work, even if less efficiently than his scientists would prefer. He knew the American implosion design would work. It had already been tested, twice. Once it had exploded on a tower in the New Mexican desert and turned night into day. The second time it had destroyed Nagasaki.

The Council of Foreign Ministers met in London in September 1945 to continue the work of postwar settlement that the Allies had begun at Yalta and Potsdam. Secretary of State Jimmy Byrnes represented the United States; British Foreign Minister Ernest Bevin, Great Britain; Molotov, the Soviet Union. Byrnes meant to rely on the US atomic-bomb monopoly to lever concessions from the Soviets. To his surprise, Molotov was unmoved. Did Byrnes have "an atomic bomb in his side pocket"? Molotov asked the South Carolinian when Byrnes tried to push him. "You don't know southerners," Byrnes attempted to joke, "we carry our artillery in our pocket. If you don't cut out all this stalling and let us get down to work, I'm going to pull an atomic bomb out of my hip pocket and let you have it." That night at a cocktail party Molotov let the West have it. "At one point of the occasion," a US security officer reported afterward, "Mr. Molotov was taking great delight in teasing Mr. Bevin, first on one thing and then on another. During the course of this badinage Mr. Molotov stepped out of the room for a minute and then suddenly reappeared with the statement, 'You know we have the atomic bomb.'" Whereupon the Soviet ambassador to Great Britain led Molotov from the room. Byrnes took Molotov's remark for evidence at best of interest in acquiring atomic weapons. Beria, Fuchs, Harry Gold, David Greenglass or Julius Rosenberg would have recognized that Molotov was not bluffing. The Soviets did not literally have an atomic bomb. But they had the plans; they knew how to make one.

PART TWO

———

New Weapons
Added to
the Arsenals

If atomic bombs are to be added as new weapons to the arsenals of a warring world, or to the arsenals of nations preparing for war, then the time will come when mankind will curse the names of Los Alamos and of Hiroshima.

J. ROBERT OPPENHEIMER

11

Transitions

JUST WHEN THE SOVIET UNION began a crash program to build an atomic bomb, the American program "essentially came to a grinding halt," Los Alamos experimental physicist Raemer Schreiber remembers. Schreiber, a handsome, confident man with warm blue eyes who grew up on an Oregon farm, had been one of the crew of scientists assigned to Tinian to assemble the first atomic bombs. Los Alamos "was stopped by the time I got back," he says, "which was early in September [1945]. People were tidying up jobs. A few of the research projects were being finished up. We were about fifty percent staffed by the Special Engineer Detachment [enlisted men] and Navy officers and other military people. And, of course, all they wanted was out. A lot of the civilian staff were just as eager to go out and take their newfound knowledge and go back and start the programs at their universities. So there really wasn't much useful work going on.... It was a very severe transition period."

If the atomic bomb had shocked the Japanese, it had also shocked America. Materializing from secrecy to such conquering effect, it seemed a mysterious and almost supernatural force. It was a new fact dropped into the world—"a new understanding of man, which man had acquired over nature," as I. I. Rabi called the first explosion at Trinity—and no one at first knew quite what to do with it. The discovery of how to release nuclear energy was a technological revolution, most of all a revolution in war; like all revolutions, its meaning would not necessarily accord with hopes or theories or prophecies, but would reveal itself over time as individuals and governments maneuvered to exploit its energies and adapt it to their goals.

The scientists who worked on the bomb also materialized from secrecy and found it necessary to explain themselves. "It kind of felt like you were caught out in the street without any clothes on," Schreiber recalls. "I mean, we were so accustomed to having this all so hush-hush, to have it all out in public took a little getting used to.... That was also the time one realized that it would be impossible to simply say, those are fine gadgets, they ended the war and now let's just lock everything up and forget about it. Really one had to live with this situation from here on." But even that seemingly obvious conclusion was debatable. "We all felt that, like the soldiers, we had done our duty," Hans Bethe writes, "and ... deserved to return to the type of work that we had chosen as our life's career, the pursuit of pure science and teaching.... Moreover, it was not obvious ... that there was any need for a large effort on atomic weapons in peacetime." Ernest Lawrence, James Chadwick, Niels Bohr, Enrico Fermi, University of Chicago Metallurgical Laboratory director Arthur Compton and Robert Oppenheimer had attended a dinner with General Groves at Los Alamos before the end of the war where postwar developments had been discussed. Groves had worried about maintaining US military strength in peacetime. People had talked about developing nuclear power. "And Fermi," Oppenheimer writes, "said, thoughtfully: 'I think it would be nice if we could find a cure for the common cold.'" Bethe went back to Cornell University; Fermi accepted a professorship at the University of Chicago, where he had worked during the war. Both men continued to serve Los Alamos as consultants.

Richard Feynman, who had driven his roommate Klaus Fuchs's old Buick down to Albuquerque the previous June, in the midst of the final effort to finish the bombs, to keep vigil with his young wife Arlene while she died of tuberculosis, found himself lost between worlds. Before he left Los Alamos he had thought about what the bomb meant and had made some notes. He had calculated that Little Boys in mass production would cost about as much as B-29s. "No monopoly," he had written. "No defense." And: "No security until we have control on a world level.... Other peoples are not being hindered in the development of the bomb by any secrets we are keeping. ... Soon they will be able to do to Columbus, Ohio, and *hundreds* of cities like it what we did to Hiroshima. And we scientists are clever—too clever —are you not satisfied? Is four square miles in one bomb not enough? Men are still thinking. Just tell us how big you want it!" The twenty-six-year-old widower may have seen too much of death. He sat in a bar in Manhattan one afternoon in the months after the war looking out the window at all the people going by and shaking his head, thinking how sad it was that they didn't realize they had only a few years to live. On Bethe's recommendation, Cornell snapped him up, but creative work eluded him in that time of grieving until he took to heart the advice that the Hungarian mathematician John von Neumann had tendered him at Los Alamos during the war. "We used to go for walks on Sunday," Feynman recalls. "We'd walk in the canyons.... It was a great pleasure. And von Neumann gave me an interesting idea: that you don't have to be responsible for the world that you're in." He had not been able to fix Arlene; why should he presume he could fix the world?

Robert Oppenheimer had directed the work at Los Alamos to spectacular success, but after the atomic bombings of Hiroshima and Nagasaki he seems to have fallen into a period of doubt and even of guilt. Tall, rail-thin, chain-smoking, "an extraordinary man," as a colleague would describe him in 1947, "who worked very hard and always seemed to be on the verge of a nervous breakdown," Oppenheimer faltered for a time between personal burden and visionary advocacy. "You will believe that this undertaking has not been without its misgivings," he wrote an old friend two weeks after the end of the war; "they are heavy on us today, when the future, which has so many elements of high promise, is yet only a stone's throw from despair." He had supported and even promoted using the bombs. "We were concerned," he told an audience a year later, "we were rightly and somewhat desperately concerned, that these weapons . . . should be manifest to all men to see and understand, that they might know what future war would be. . . . It would not have been a better world if the unrealized possibility of these terrible weapons had been a secret shadow on our future." Edward Teller had carried a petition into Oppenheimer's Los Alamos office in July 1945 opposing use, and had come away bearing just that message, advising the petitioner, his fellow Hungarian Leo Szilard, that "our only hope is in getting the facts of our results before the people. This might help to convince everybody that the next war would be fatal. For this purpose actual combat-use might even be the best thing." The advice was pure Oppenheimer, and for the rest of his life Teller would resent having parroted it, even claiming in old age that doing so had been a ruse, "not very nice . . . but we had censorship in Los Alamos, and I felt sure that Oppenheimer would see the letter. . . . and I did not care to contradict Oppenheimer too strongly." In fact, Teller sent the letter to Oppenheimer for approval with an obsequious cover note.

The weekend after the Nagasaki bombing, Oppenheimer met at Los Alamos with the three other members of the scientific panel—Lawrence, Arthur Compton and Fermi—that advised the Interim Committee that Henry Stimson, the Secretary of War, had assembled to consider the postwar disposition of the atomic enterprise. Lawrence found Oppenheimer weary, guilty and depressed, wondering if the dead at Hiroshima and Nagasaki were not luckier than the survivors, whose exposure to the bombs would have life-time effects. The four men worked to prepare a letter that went to Stimson on August 17; it warned that the new weapon that seemed so absolute in monopoly would eventually, when it got around, pose a threat not only to an enemy but also to the United States. "We are convinced," the scientists wrote, "that weapons quantitatively and qualitatively far more effective than now available will result from further work on these problems. . . . Nevertheless we have grave doubts that this further development can contribute essentially or permanently to the prevention of war. We believe that the safety of this nation . . . cannot lie wholly or even primarily in its scientific or

technical prowess. It can be based only on making future wars impossible."
If the letter was not clear enough on the urgency of a political solution,
Oppenheimer personally carried it to Washington at the end of the month
and elaborated on its message to whoever would listen, as he reported to
Lawrence:

> I . . . had an opportunity . . . to explain in more detail than was appropriate
> in a letter what our common feelings were in this all important thing. I
> emphasized of course that all of us would earnestly do whatever was really
> in the national interest, no matter how desperate and disagreeable; but that
> we felt reluctant to promise that much real good could come of continuing
> the atomic bomb work just like poison gases after the last war. . . . In the
> end this will have to be based on a national policy which is intelligible in
> its broad outlines to the men who are doing the work. . . . I do not come
> away from [i.e., I still feel] a profound grief, and a profound perplexity
> about the course we should be following.

Before he left Washington, Oppenheimer heard from Jimmy Byrnes that "in
the present critical international situation there was no alternative to pushing the program full steam ahead."

Oppenheimer disagreed. During September and October he communicated his disagreement forcefully to officials at the highest levels of the US
government. He was certainly secure in his conviction, but he was also
inevitably gauging the political influence of his authority as the magus who
had guided the invention of the miraculous bombs. From obscurity before
the war and invisibility during the war years he would soon appear in *Time*
magazine, celebrated as "the smartest of the lot."

On September 24, the forty-one-year-old physicist met with Acting Secretary of State Dean Acheson and Stimson aide George L. Harrison. "Dr. Oppenheimer philosophized at great length about the work of the scientists,"
Harrison dictated afterward for the record, "their objectives, their prejudices
and their hopes. There is distinct opposition on their part to doing any more
work on any bomb. . . . He says that much of the restiveness in his laboratory
is not so much due to the delay in [atomic energy] legislation as to a feeling
of uncertainty as to whether they are going to be asked to continue per-
fecting the bomb against the dictates of their hearts and spirits. This is true
particularly in terms of a better one, but the feeling persists even as to
continuing the manufacture of the present one. Mr. Acheson seemed much
interested in this." A tough-minded patriot, Acheson took the threat of a
revolt of the scientists in stride. Harrison caught on and started filling behind
him the hole Oppenheimer was opening; after he shepherded the physicist
to a meeting with Robert Patterson, the Undersecretary of War, he advised
Patterson to talk to Groves "and obtain his views which are quite different
from those of the scientists."

A month later, Oppenheimer got a chance to present his case to the President himself. Intricate urbanite and blustery farmer, Harvard elitist and Midwestern autodidact, they repelled each other snappishly. "In the winter of 1945–46," Oppenheimer told an interviewer two decades later, "hysteria centered on our hypercryptic power and the hope of retaining it. I saw President Truman and he told me he wanted help in getting domestic legislation through. 'The first thing is to define the national problem,' he said, 'then the international.' I said, 'Perhaps it would be best first to define the international problem.' "

"I feel we have blood on our hands," Oppenheimer remembered adding, and Truman replying, "Never mind. It'll all come out in the wash." Closer to their meeting, Truman still felt indignant at Oppenheimer's presumption, writing Acheson in 1946 that Oppenheimer was a " 'cry baby' scientist . . . [who] came to my office . . . and spent most of his time [w]ringing his hands and telling me they had blood on them because of the discovery of atomic energy." It was Truman, after all, who had decided to drop the bomb and had blood on his hands.

Truman's indignation disguised his own great uneasiness about bombs that destroyed entire cities. In the last days of the war, Los Alamos had cast a third plutonium core for shipment out to Tinian, where a Fat Man high-explosive assembly was ready to receive it. Truman decided not to authorize its use and told his Cabinet why; Secretary of Commerce Henry Wallace noted the President's reason in his diary: "Truman said he had given orders to stop the atomic bombing. He said the thought of wiping out another 100,000 people was too horrible. He didn't like the idea of killing, as he said, 'all those kids.' "

In October, not long before his confrontation with Oppenheimer, already impatient with Soviet intransigence in Eastern Europe, Truman had complained to his budget director, Harold D. Smith, "There are some people in the world who do not seem to understand anything except the number of divisions you have." Smith had rejoined in the Jimmy Byrnes mode, "Mr. President, you have an atomic bomb up your sleeve." And Truman had concluded somberly, "Yes, but I am not sure it can ever be used."

Oppenheimer decided to leave Los Alamos and return to teaching and research. He had offers from at least Columbia, the Institute for Advanced Study, Berkeley, Caltech and Harvard; he chose Caltech, but he soon found himself traveling to Washington almost weekly as the government discovered his talent for advice.

On his last day as director, October 16, at an outdoor ceremony that nearly everyone on the mesa attended, he received a certificate of appreciation for the laboratory from Groves on behalf of the Secretary of War, expressed pride in the work the laboratory had done, and then shadowed the bright day with strong words about the potential consequences. "Today that pride must be tempered with a profound concern," he told the men and women who had teased the prepotent mechanisms into existence. "If atomic bombs are to be added as new weapons to the arsenals of a warring world, or to the arsenals of nations preparing for war, then the time will come when mankind will curse the names of Los Alamos and of Hiroshima. The peoples of this world must unite, or they will perish." By their works, he said, they were committed, "committed to a world united, before this common peril, in law, and in humanity." That the peril of atomic war was common to all the nations of the world was an idea Niels Bohr had brought to Los Alamos; the charismatic Oppenheimer was fast becoming Bohr's spokesman.

Was there, in fact, opposition at Los Alamos to working on the bomb, as Oppenheimer had warned Acheson and Patterson? Most of the civilian staff at the laboratory signed a public statement early in September warning of the danger of an atomic arms race and urging efforts at international control, but that is not the same thing as opposition. Norris Bradbury, the vigorous, Berkeley-trained Navy physicist whom Oppenheimer nominated in September to take over the laboratory's direction, offered the only report of opposition the record contains, and it sounds like Oppenheimer's. "There was one school of thought," Bradbury said, "which held that Los Alamos should become a monument, a ghost laboratory, and that all work on the military use of atomic energy should cease." Anyone who might have been opposed was free to go, of course, and presumably, like Oppenheimer, did so. Those who stayed remember primarily confusion and insecurity. There was "continual uncertainty about the future," says John Manley, who had helped Oppenheimer organize and run the place. ". . . It was a miserable time." Manley recalls that "Oppenheimer thought I should leave at the end of the war. I didn't take his advice." Whatever he was advising privately, in formal council Oppenheimer stressed the need for continuity, as two British observers reported: "Oppenheimer made it clear that any large exodus would be a limitation on the future freedom of action of the Project and should be avoided."

Edward Teller remembers Oppenheimer encouraging him to leave. The first team was going, Oppenheimer needled his fractious colleague. Teller did not want to see Los Alamos close up shop. Bethe proposed that he consider taking over the Theoretical Division and Teller poured out his frustrations. "In this conversation," Bethe remembers, "for the first time in my recollection, he expressed himself as terribly pessimistic about relations with Russia. He was terribly anti-Communist, terribly anti-Russian. . . . Teller said we had to continue research on nuclear weapons. . . . It was really wrong of us all to want to leave. The war was not over and Russia was just as dangerous an enemy as Germany had been."

Without question, Edward Teller was consistently and vocally anti-Communist throughout his long life. Reading Arthur Koestler's *Darkness at Noon* at Los Alamos soon after the laboratory opened its doors in the spring of

1943 finished the process of determining him in that conviction. But he also had a personal stake in seeing Los Alamos continue its work postwar: he had passionately championed the development of a thermonuclear explosive— a "superbomb" based on hydrogen fusion that might be a hundred or a thousand times more destructive than the atomic bombs had been—since Enrico Fermi first suggested the idea to him at Columbia University one afternoon in September 1941. And about the superbomb in particular, the scientific leaders of the Manhattan Project were clear. Although the scientific panel, in its report to Stimson on August 17, found "quite favorable technical prospects of the realization of the superbomb," Oppenheimer told George Harrison the next day that (in Harrison's paraphrase) "the scientists prefer not to do that . . . unless ordered or directed to do so by the Government on the grounds of national policy." In a long report finished on September 28 proposing research and development in the field of atomic energy, the scientific panel recommended "that no such effort [comparable to the Manhattan Project] should be invested in [the thermonuclear] problem at the present time, but that the existence of the possibility should not be forgotten, and that interest in the fundamental questions involved should be maintained." Which, translated, meant research into the basic physics of thermonuclear fusion but no development. Arthur Compton put the matter even more plainly in a letter to Henry Wallace summarizing the panel's findings. "We feel that this development should *not* be undertaken," Compton wrote, "primarily because we should prefer defeat in war to a victory obtained at the expense of the enormous human disaster that would be caused by its determined use." The Nobel laureate physicist suggested reassessing the question in ten years—that is, in 1955. "Perhaps there may be then, an international government adequate to make its development under world auspices safe or perhaps unnecessary for further consideration." Even Groves, according to Oppenheimer, thought his mandate did not reach so far. "General Groves told me very briefly that he had been told by Byrnes . . . that, with things as they were, the work at Los Alamos ought to continue, but this did not apply to the Super."

Fermi knew that Teller, who had led thermonuclear research at Los Alamos during the war, disagreed with the scientific panel's findings, and encouraged the Hungarian physicist to write him a letter of record summarizing his position. Teller did so on October 31. In 1944, Teller had briefed James Bryant Conant on the superbomb. Conant, who was supervising atomic-bomb development, reported following that briefing that a hydrogen bomb was "probably at least as distant now as was the fission bomb when . . . I first heard of the enterprise." That estimate—between four and five years—was already optimistic compared to the estimate generally accepted at Los Alamos, partly because the thermonuclear looked like a hard case, partly because fission bombs would have to be better understood and considerably improved before they could be made efficient enough—hot enough—to serve as thermonuclear detonators. Now, in late 1945, framing his dissent, Teller altered his estimate, formulating for the first time in a report many of the arguments for pursuing technological security that he would elaborate through the decades to come.

"When could the first super bomb be tried out?" the Hungarian physicist asked rhetorically. He answered with two numbers, the second an early example of what has come to be called threat inflation:

> It is my belief that five years is a conservative estimate of this time. This assumes that the development will be pursued with some vigor. The job, however, may be much easier than expected and may take no more than two years. In considering future dangers it is important not to disregard this eventuality.

How soon could another country produce such a bomb? Faster than the United States, apparently, despite his adopted country's lead: "The time needed . . . may not be much longer than the time needed by them to produce an atomic bomb." What about moral objections? They were meaningless before the onrush of technology:

> There is among my scientific colleagues some hesitancy as to the advisability of this development on the grounds that it might make the international problems even more difficult than they are now. My opinion is that this is a fallacy. If the development is possible, it is out of our powers to prevent it.

Teller thought defensive measures such as the dispersal of cities might prove effective against atomic bombs but "very much less so against super-bombs." He could not yet offer detailed plans for the peaceful use of thermonuclear explosives. "But I consider it a certainty that the superbomb will allow us to extend our power over natural phenomena far beyond anything we can at present imagine."

By the time he wrote his letter to Fermi, Teller had already talked to Norris Bradbury about the future of the lab. "I said we either should make a great effort to build a hydrogen bomb in the shortest possible time or develop new models of fission explosives and speed progress by at least a dozen weapons tests a year." Bradbury knew he would have his hands full simply keeping Los Alamos alive; at the moment it was foundering in legal limbo. The authority of the US Army had officially terminated with the war, and Congress was then in the midst of debating what legal entity might assume it. The new director told Teller neither of his programs was realistic.

Teller decided to leave. "I was not willing to work without backing," he writes. He had an offer to go with Fermi to the University of Chicago. Oppenheimer encouraged him, telling him, "You are doing the right thing." At a farewell party Oppenheimer added, "We have done a wonderful job here, and it will be many years before anyone can improve on our work in any way." The insensitivity of Oppenheimer's remark rankled Teller, as its ambiguity confused him; he quoted it repeatedly in the years to come, always to demonstrate its lack of foresight. It might have meant: the Russians will not soon build a bomb. Or it might have meant: the Oppenheimer team had accomplished in fission weapons what a Teller team could not soon improve in fusion. Teller would read it both ways. "It was obvious and clear to me," he concluded, "that Oppenheimer did not want to support further weapons work in any way." Teller left for Chicago in February 1946.

How many years it would be before anyone matched the work of the Manhattan Project was crucial to the question of what role Los Alamos should play now that the war was over. Though good intelligence was lacking on where the Soviet project stood, no one in authority felt much urgency. The public statement that the Los Alamos civilian staff had signed in September had argued:

> The development of the atomic bomb has involved no new fundamental principles or concepts; it consisted entirely in the application and extension of information which was known throughout the world before intensive work started. Furthermore, deposits of basic materials for atomic bombs have been found, even before the war, in many parts of the world and new deposits will undoubtedly be discovered. It is therefore highly probable that with sufficient effort other countries, who may in fact be well underway at this moment, could develop an atomic bomb within a few years.

Henry Stimson had commented similarly, in an August 29 memorandum to Truman on atomic arms control, that US possession of the atomic bomb would "almost certainly stimulate feverish activity on the part of the Soviet[s] toward the development of this bomb, and there is evidence to indicate that such activity may have already commenced." Citing scientific authority, Stimson told Truman "that it is as certain as any future pronouncement can be that the method of manufacture of these bombs as now known by the United States, cannot be preserved as a secret from other nations beyond a relatively short time."

The Joint Chiefs of Staff Joint Intelligence Staff, in a report on "Soviet

Capabilities," quantified that "short time" in November 1945. The intelligence staff found that Soviet control of Eastern Europe would "probably remain high during the next several years," that "the Soviet economy will probably remain incapable of alone supporting a major war during the next five years" and that the Soviet Union was therefore "likely to avoid the risk of such a war during that period." But, concluded the intelligence staff, "the Soviets are believed to be capable of developing atomic weapons within five to ten years, and will make every effort to do so as soon as possible." This conclusion, so much more conservative than that of the young Los Alamos scientists, followed from a pessimistic assessment of Soviet industrial capability. "The evidence in Soviet industrial history," the military intelligence staff believed, "does not warrant the assumptions[:] that the USSR can accomplish the research, planning and designing stages with modern technical efficiency; that they can execute a huge construction program without appreciable delays; or that they will be able promptly to eliminate the bugs in initial production which impede full-scale manufacture."

Of even these conservative conclusions Groves was not convinced. The Manhattan Project commander asked his assistant, Brigadier General T. F. Farrell, to review the question of Soviet capability with some of the industrialists and engineers who had developed the uranium isotope-separation installations for the Manhattan Project at Oak Ridge, Tennessee. Farrell reported back on October 12, 1945, that "starting now in an 'all-out' effort, [the Soviet Union] could successfully make an atomic bomb in a relatively few years." The likeliest Soviet approach, Dobey Keith of the Kellex Corporation thought, was plutonium production via an operation like Hanford. Another Kellex man thought the Soviets would develop uranium isotope separation by gaseous diffusion, which a third Kellex man thought they could achieve within three and a half years. But Groves held out for a much longer interval of US atomic monopoly. Whenever asked, year by year after the war, he always said "Twenty years."

Groves based his argument partly on an extremely conservative assessment of world resources of high-grade uranium ore and the Combined Development Trust's success at cornering the existing world reserves of that ore. In April 1944, Alvin Weinberg remembers, Los Alamos physicist Philip Morrison had reported "that not more than 20,000 tons of uranium was in sight in the whole world." It was generally assumed within the Manhattan Project, says Weinberg, that "*separated* Pu^{239} and U^{235} was always going to be rare and expensive"—one reason why nuclear power generation, even for military applications such as driving submarines, appeared to be a distant prospect in 1945. Groves forwarded a slightly more generous estimate of world resources to Ernest Lawrence in Berkeley a few days after the end of the war—20,000 tons of ore of 10 percent or better uranium content in sight, 50,000 tons presumptive; 30,000 tons of 0.1 to 10 percent ore in sight, 100,000 presumptive; and 400,000 tons of 0.05 to 0.07 percent ore presumptive, with lower-percentage ore essentially unlimited*—but noted sourly, "I have no confidence in [these figures]."

Groves understood that ore supplies were plentiful at lower percentages for any country which chose to invest in refining them. Far more significant from his point of view was what he believed to be the primitive state of Soviet industrial technology. The Soviet Union, he wrote dismissively in the *Saturday Evening Post* in 1948, "simply does not have enough precision industry, technical skill or scientific numerical strength to come even close to duplicating the magnificent achievement of the American industrialists, skilled labor, engineers and scientists who made the Manhattan Project a success. Industrially, Russia is, primarily, a heavy-industry nation; she uses axle grease where we use fine lubricating oils. It is an oxcart-versus-automobile situation."

Soviet mistrustfulness would get in the way as well, Groves thought. If the US had shipped "the complete blueprints of the Manhattan Project to Russia on V-J Day, they would waste a couple of years searching suspiciously for a gimmick in the plans, which, they would be confident, some American had fiendishly inserted to assure Russia the privilege of blowing herself off the map." He was more right than he knew, but his timetable was off. A fair portion of the complete blueprints had been shipped well before victory in Japan, and the couple of years that Stalin's and Beria's suspicions had cost the Soviets had already expired.

Nor was Soviet technological and industrial ability taken seriously elsewhere within the US government in the late 1940s. Herbert York, who worked on electromagnetic isotope separation at Oak Ridge as a graduate physics student during the war, remembers a joke popular in Washington in those days. The Russians couldn't deliver an atomic bomb in a suitcase, the joke went, because they didn't know how to make a suitcase.

Norris Bradbury's vigorous advocacy saved Los Alamos. "He was a very complex man," Raemer Schreiber describes him. "In outward appearance, he was not particularly impressive. Medium build, rather on the skinny side if he kept his tendency toward a pot belly under control, rather craggy features, hair short, grayish-blond and sparse, clothes casual and tending toward bagginess. Even in casual conversation he was not very exciting; he was not good at small talk. But when he was on laboratory business, the words came out pell-mell and his brain generated them faster than he could

---

* Uranium at low percentages is ubiquitous in the earth's crust, which is why so many houses have radon in their basements—radon, a gaseous product of the radioactive decay of uranium, seeps from stone foundations.

articulate. . . . He could also sit back patiently and let people argue at great length. . . . He was no orator, but spoke with quiet self-confidence."

Bradbury encouraged people to stay. He told them they were needed. "The use of nuclear energy may be so catastrophic for the world that we should know every extent of its pathology," he said. "How bad *can* this bomb . . . be? . . . One studies cancer—one does not expect or want to contract it—but the whole impact of cancer on the race is such that we must know its unhappy extent. So is it with nuclear energy released in this form. It can be a terrible thing; we cannot hide our head in the sand; we must know how terrible it is." He told them the nation needed the bomb. "The project cannot neglect the stockpiling or the development of atomic *weapons* in this interim period. Strongly as we suspect that these weapons will never be used; much as we dislike the implications contained in this procedure, we have an obligation to the nation never to permit it to be in the position of saying it has something which it has not got. The world now knows we have a weapon. How many or how good it does not know. To weaken the nation's bargaining power in the next few months during the administration's attempt to bring about international cooperation would be suicidal."

As a technician, Bradbury was offended by the crudeness of the weapons they had designed. "We had only scratched the surface of atomic bombs," he would recall. "We had, to put it bluntly, lousy bombs. We had a set of bombs which were totally wrongly matched to the production empire." They would go to work "engineering . . . a new weapon whose aims should be . . . increased reliability, ease of assembly, safety, and performance; in short, a better weapon. . . . Possibly in six months, possibly in a year—maybe in a few years, weaponeering will stop, but our present lead is our chief weapon in procuring a peace—we must not lose it until that peace and that cooperation is established." In the meantime they would "stockpile the current [Fat Man] up to a number of 15," but they would "develop internal modifications, possibly in the method of fusing, almost certainly in the method of detonating." They would also "develop a levitated model."

Consistent with the recommendations of the Interim Committee scientific panel, but not with the moral qualms Compton had expressed, Bradbury proposed "that the fundamental experiments leading to the answer to the question 'Is or is not a Super feasible?' be undertaken. These experiments are of interest in themselves in many cases; but even more, we cannot avoid the responsibility of knowing the facts, no matter how terrifying. The word 'feasible' is a weasel word—it covers everything from laboratory experiments up to the possibility of actual building, for only by building something do you actually finally determine *feasibility*. This does not mean we will build a Super. It couldn't happen in our time in any event. But someday, someone must know the answer: Is it feasible?"

So Los Alamos had a new leader, and a program, and a staff of several thousand younger men and women, about half its wartime complement. They would get busy again, working on improvements. Bradbury set everyone to work writing down what he or she had learned during the war. The resulting multi-volume series of technical reports became a significant historical record of the development of the first atomic bombs. By October 1945, Los Alamos had procured hardware (but not uranium, plutonium or initiators) sufficient for sixty bombs and had begun developing an improved implosion design with a levitated composite core. Maybe the lab would survive.

The "urgent long sittings" in Moscow of September and October gave way to action. "It was necessary to employ more and more people and to choose the staff for work on the nuclear enterprises," Mikhail Pervukhin recalled. "There was no nuclear industry as such and there were no trained personnel for it. But we had chemists, metallurgists and other specialists. We needed engineers and workers for the nuclear enterprise. We explained to people that we needed them for a new field that was very important to the state. Not everyone understood immediately. It was difficult to negotiate with ministers. 'You are taking our people from us,' they would say, 'while we have our own tasks to do, our own state plans. We will not give away our people!' In such cases the Central Committee apparatus was very helpful. It was their job to explain everything in the right way and to attract the necessary people."

On the scientific front as well, Igor Golovin notes, "every institute capable of helping solve the atomic problem was called upon to mobilize its scientific resources and contribute under an integrated scientific plan. New institutes were brought into being to develop research that had not existed before the war (for example, uranium and plutonium metallurgy)."

The task they faced was daunting. "Yesterday I met the physicists and the radiochemists from the Radium Institute," Boris Vannikov, who was responsible under Beria for the industrial part of the enterprise, told one of his deputies in September. "For the present we are still speaking different languages. Or more precisely, they are speaking while I blink. . . . We engineers are used to touching everything with our hands and seeing everything with our eyes, and in extreme cases a microscope will help. But here it is powerless. It makes no difference, you won't see an atom, and even less will you see what is hidden inside it. And on the basis of this invisible and intangible thing, we have to build factories and organize industrial production." They pushed ahead. Averell Harriman in Moscow later that autumn reported a Soviet contact with a Westinghouse engineer that offered "possible indications that the USSR is studying equipment for the manufacture of the A-bomb."

In September, Soviet troops had occupied Japanese mining sites in North

Korea and had begun a preliminary survey of the ores found there, which included useful sources of uranium and thorium. On November 19, a reliable source in Czechoslovakia cabled the US State Department that "the Czechoslovakian Government has been officially requested to furnish uranium ore to the Soviet government." The Soviet Union concluded a secret agreement with Czechoslovakia on November 23 granting it exclusive rights to all uranium mined within the country and began expanding mining around Jáchymov, the old site in the Ore Mountains where the ores were dug from which uranium was first isolated in 1789. Marie and Pierre Curie had extracted the first polonium and radium from Jáchymov ores (U238 decays to radium and polonium along the way to becoming lead). Czechoslovakia would get part of the radium recovered from the ore in return. Under Soviet supervision, sixty-four German (presumably Nazi) political prisoners first worked the Czech mines in 1946, increasing to nearly twelve thousand in 1953 (by that time all Czech). The Czech government organized some seventeen forced-labor camps at its mines over the years; Czech ore deposits met about 15 percent of Soviet uranium requirements through 1950. Soviet geologists also began extensive exploration throughout the USSR and found additional deposits in southwestern Siberia. By Groves's standards the Siberian ore was of low quality, but the Soviets sorted out the best pieces by hand during mining and concentrated the material to 1 percent or better locally before shipping it to the refinery. Domestic sources met about one-third of Soviet uranium needs through 1950.

Although the small F-1 reactor at Kurchatov's Laboratory No. 2 was still in the planning stage, a government commission in October inspected and approved a location east of the Urals for the Soviet Hanford, where the first big plutonium-production reactor and extraction facility would be built. The Chelyabinsk Tractor Plant in Chelyabinsk province, Kurchatov's ancestral home, had merged during the war with the evacuated Kharkov Diesel Works and parts of the Leningrad Kirov Plant to become a major tank production complex known popularly as Tankograd. To supply the complex and dozens of other armament works in the area, a huge new power station had gone up in 1942 from which electricity could be drawn. Chelyabinsk province, particularly around the small town of Kyshtym, was also a major gulag station, with some twelve labor camps in the area. In November 1945, site studies began for the plutonium-production complex, to be known as Chelyabinsk-40, some fifteen miles east of Kyshtym, in the area around Lake Kyzyltash in the upper drainage basin of the Techa River. The first buildings for what would become a city—Beria, it was gloomily named—also went up that month. There were four gulags in the immediate area of the site, three for men, one for women. Prisoners started cutting down forests by hand; army tanks fitted with bulldozer blades graded out roads.

In September 1945, Lavrenti Beria appointed Pavel Sudoplatov, an NKVD officer who had previously specialized in organizing assassinations (including Trotsky's) and guerrilla warfare, to head a new Department S ("S for Sudoplatov," Sudoplatov claims). Beria charged the new department with reviewing, translating and communicating to Soviet scientists the vast information collected on the Anglo-American atomic-bomb program that only Kurchatov and a few select assistants, including Khariton, had been allowed to see before. According to the Russian Foreign Intelligence Service (the successor agency to Soviet intelligence), Sudoplatov's department "had no direct contact with the agents' network" and Sudoplatov himself "had access to atomic problems during a relatively brief period of time, a mere twelve months [i.e., from September 1945 to September 1946]."

Sudoplatov evidently initiated almost immediately an ad hoc, bungled NKVD attempt to extract technical information from Niels Bohr, who had returned from the United States to his institute in Copenhagen immediately after the end of the war in Europe. The NKVD officer's recollection of the incident, recorded late in life, is garbled, and his co-authors' attempts to clarify it in the 1994 book *Special Tasks* garbled it further and not incidentally libeled Bohr. Bohr did inadvertently communicate information of use to the Soviet program; but the information he communicated came from the Smyth Report, a public source, and its value was a consequence of General Groves's attempts to suppress sensitive technical information from one edition of that report to the next.

The Smyth Report—the detailed report on the science behind the Manhattan Project—was released to the press in a lithoprint edition reproduced from a typed manuscript (the xerographic copier had not yet been invented) on August 11, 1945. Six copies of that first edition went to Tass, the Soviet news agency, in mid-August. Tass immediately passed this compendium of valuable information about the US bomb program to Soviet intelligence.

Princeton University Press then published a typeset, hardbound edition of the Smyth Report titled *Atomic Energy for Military Purposes* on September 1. Copies of the Princeton edition also went to Soviet intelligence, which was preparing a translation. Between the lithoprint and typeset editions, however, a significant change occurred. "General Groves," remembers physicist and security adviser Arnold Kramish, "was horrified to discover that the lithoprint version contained some items that he considered sensitive. They were deleted from the Princeton edition." The most important of these items was a single sentence:

In spite of a great deal of preliminary study of fission products, an unforeseen poisoning effect of this kind very nearly prevents operation of the Hanford piles, as we shall see later.

The sentence refers to the near-disaster the plutonium-production complex at Hanford, Washington, faced on September 27, 1944, when its first big production reactor, the B pile, started up successfully, ran for about twelve hours, mysteriously died, started up again spontaneously after a delay and twelve hours later began another decline. Princeton theoretician John Archibald Wheeler, on hand for the start-up, worked out the problem in an all-night marathon review of fission physics. Uranium (92) does not always break up into barium (56) and krypton (36) (56 + 36 = 92) when it fissions. It frequently breaks up into other fragment sizes instead—iodine (53) and yttrium (39), for example. Wheeler realized that the high neutron flux of the B pile, the first large reactor built anywhere in the world, was creating a fission product that was poisoning the chain reaction by soaking up needed neutrons. After working through the possibilities he decided on iodine[135], a radioactive isotope of iodine, and calculated that it would decay with a half-life of about six hours into a previously unknown daughter product, xenon[135], which had a nine-hour half-life. Wheeler estimated that Xe[135] had an appetite for pile neutrons that was a whopping 150 times as great as the most absorptive element previously known, cadmium, the metal of which the pile's control rods were made. The big production pile, it seemed, would start up normally and chain-react; fission would produce increasing quantities of iodine[135]; the iodine would decay to Xe[135]; as the Xe[135] built up, its atoms would absorb neutrons one for one; and slowly the pile would be poisoned until there were not enough free neutrons left circulating to sustain the chain reaction. The Xe[135] would then decay into a nonabsorptive daughter product; the flux of free neutrons would build until finally the pile had enough neutrons circulating to begin chain reacting again, at which time the cycle would repeat. "Xenon," Wheeler writes, "had thrust itself in as an unexpected and unwanted extra control rod."

The solution to the problem at Hanford was to increase the pile's reactivity by adding more uranium slugs until the sheer number of free neutrons available from fission overrode the poisoning effect. But that solution was only possible because the pile had been designed deliberately with a third again as many uranium channels drilled through its massive graphite block as calculations had indicated it needed. And such a generous margin of error had been possible in turn only because the United States had acquired ample supplies of uranium ore by 1944 and had mastered the production in quantity of highly purified graphite and of uranium metal. Had the B pile been designed with minimal tolerances, as someone might design a production reactor whose supplies of graphite and uranium were limited, it would have had to be completely rebuilt, delaying plutonium production by months or even years.

All this science, engineering and industry lay hidden behind that one fugitive sentence of Henry Smyth's report. And Groves's deletion of the sentence from the typeset Princeton edition highlighted its importance as surely as if the general had waved a red flag. Yuli Khariton and Yakov Zeldovich had warned presciently, in their March 7, 1940, paper on chain-reaction kinetics, that just such a problem might occur. ("As examples of such factors [affecting criticality] which need investigation we may note . . . the appearance of new nuclei capable of capturing neutrons in decay. . . .") Obviously fission-product poisoning of the Hanford reactor was a phenomenon Kurchatov needed to know more about as he began to design the first Soviet production reactor destined for Chelyabinsk-40. Which product caused the poisoning effect? At what stage of operation did it occur? How did the US overcome it? On all these vital questions, the Smyth Report was silent.

At some point during autumn 1945, someone within the Soviet atomic-bomb establishment noticed the discrepancy. Evidently he or she did so prior to Sudoplatov's Bohr caper. Department S was responsible for translating the Smyth Report; its technical editor, comparing the two American editions sentence by sentence, was probably the first to notice the discrepancy. In his 1994 memoir, Sudoplatov remembered the issue of reactor poisoning in garbled form. "A pivotal moment in the Soviet nuclear program occurred in November 1945," he writes. "The first Soviet nuclear reactor had been built [sic], but all attempts to put it into operation ended in failure [sic], and there had been an accident with plutonium [sic]. How to solve the problem?"

Shortly after the end of the war, Bohr had publicly expressed his hope that the scientific knowledge developed in the United States during the Manhattan Project years would be shared internationally as part of an agreement to forestall an atomic arms race. Superficially, Bohr's position on secrecy sounded like the argument that NKVD agents had used to prospect successfully for espionage in Britain and America during the war. To people of Beria's and Sudoplatov's mentality, Bohr's vision of an open world pioneered by scientific sharing seemed an appeal for collaboration. "We decided to turn to Bohr," Sudoplatov recalls. "We took a young worker from my Department S . . . a young theoretical physicist, and we sent him to Bohr. Denmark, at the time, had recently been liberated from the Germans by the Red Army [sic: Denmark was still under German occupation on V-E Day], and attitudes in general to Soviet Russians were especially warm." The young Soviet physicist, Yakov Petrovich Terletsky, had been drafted into service with the NKVD from outside the Soviet atomic-bomb program.

As Terletsky tells the story, Sudoplatov drafted him with Beria's approval to review translations of the voluminous intelligence materials the NKVD had collected and to brief Kurchatov's scientists. It was Terletsky who learned, on his first day on the job, October 11, 1945, that some ten thousand pages of espionage materials lay on hand at the Lubyanka. These "were photocopies of scientific reports typed on a typewriter," he writes. "At the top of every report was a standard stamp from the American state security agencies, warning that the report was secret." After only four days of reviewing this unfamiliar material—Terletsky's specialty was statistical physics, not nuclear physics—the young scientist was ordered to report to the bomb project technical council on what he had learned. He was warned not to reveal the source of his information; he was to say it came from a fictitious "Bureau No. 2," implying a parallel bomb program. Most of the scientists and managers who were members of the technical council knew at least informally of the extensive NKVD collections, however; few would have been fooled.

The Bohr caper followed a week after Terletsky's technical council report, when a messenger rousted him from sleep on a Saturday night and drove him to the Lubyanka for a meeting with Beria that never materialized. Sudoplatov turned up instead and asked Terletsky if he knew Niels Bohr. "What physicist didn't know Bohr!" Terletsky writes with literal-minded incredulity. "From further hints it became clear that a meeting was to take place with Niels Bohr. . . . I was sent home and warned that I should maintain readiness and not leave town, even on Sunday. But where would I have gone at that time anyway?" The following week, Sudoplatov briefed Terletsky in detail. He would meet with Bohr in Copenhagen and ask him questions about the American project. Sudoplatov imagined that Bohr "was inclined against the Americans, and it could be expected that he would help us." Kapitza would supply a letter of introduction. Terletsky met with Kapitza, who pointedly advised him not to ask Bohr many questions but to listen to what Bohr had to say.

Before he left for Copenhagen, Terletsky had his worn-out wartime clothes replaced with NKVD tailoring, "starting with underwear . . . at some top-secret tailor's shop" and met with Lavrenti Beria. To get to Beria's office, Terletsky remembers, he had to pass through a room "filled with armed officers who carefully looked us over" and then wait in an outer office "which reminded me of nothing so much as the dressing room adjoining a public bath," with Sudoplatov and other Department S representatives already at hand and a four-hundred-pound torture specialist and Beria confidant named Bogdan Kabul—"egg-shaped," Terletsky calls him. The gang filled the time joking about the hot Scandinavian girls Terletsky was likely to meet. Finally Beria received them:

When we entered, Beria got up from behind his desk, which was deep within an enormous room, and went up to a large conference table. . . . Then I was introduced to the People's Commissar. He was of average height, aging, with a skull that narrowed slightly toward the top, with severe features and no shadow of warmth or a smile. Beria did not give the impression I had expected from seeing his portraits before, of a young, energetic member of the intelligentsia wearing a pince-nez. Everyone sat down at the big conference table. In the middle of the table was a large white marble ash tray in the shape of a polar bear with little ruby eyes. That was the only object on the long table . . . [and] it was obvious that no one used it.

Beria questioned Terletsky about his background. He asked the young physicist and Sudoplatov how they envisioned their assignment. Terletsky says he "was completely unclear as to what I was supposed to ask Bohr." Sudoplatov professed to have no idea. Then it emerged that Terletsky's English was too poor for conversation with Bohr and that the NKVD officer who would be accompanying him, Lev Vasilevsky, Sudoplatov's deputy, only knew French. Beria assigned a translator. Terletsky got the impression that no one had thought his mission through. To cobble together a list of questions, he writes, Beria summoned Vannikov, Vannikov's deputy Avrami Zavenyagin and "the scientists." Less then an hour later Kurchatov, Khariton, Kikoin and Artsimovich arrived at the Lubyanka. When Khariton heard that a novice like Terletsky was being sent to question Bohr, he told Beria bluntly that it would be better to send Zeldovich. "He would worm all the fine points of the atomic problem out of Bohr," Terletsky remembers Khariton explaining. "Beria cut him off," Terletsky continues, "saying in his harsh Georgian accent, 'We don't know who would worm more out of whom.'" Beria had no intention of sending someone out of the country who had knowledge of the Soviet program and who might be kidnapped or might defect. He ordered the scientists to prepare a list of questions. They went off to do so. In the hallway, Khariton tried to talk Terletsky out of the mission. Terletsky understood all too well that he had no choice but to go.

Sudoplatov required Terletsky to memorize the questions the scientists had prepared and Beria had approved and charged the naive young operative not to deviate from the list. Since writing down Bohr's answers might make the Dane suspicious, Terletsky was expected to memorize the answers as well.

Off Terletsky, Vasilevsky and the translator went to Copenhagen, Terletsky marveling along the way not at the girls but that "the people on the streets were well-dressed, the stores filled with goods, and all sorts of food, including sweets and fruits . . . in abundance." A Communist member of the Danish parliament approached Bohr and asked him to meet with Terletsky, telling him the Soviet physicist carried a letter from Kapitza, Bohr's old colleague at Cambridge University, and wished to deliver it to Bohr in confidence. Bohr's response was blunt: he could not agree to secret arrangements of any kind; if a Soviet scientist wished to speak with him, the meeting would have to be open. Bohr immediately communicated the contact to British intelligence, which notified Groves.

Terletsky in Copenhagen followed up with a letter on November 13 asking for an open meeting. Bohr agreed to meet the next morning. As a precaution, he asked his twenty-four-year-old son Aage, also a physicist, to sit in with him, and stationed another son in an adjoining room with a pistol in case the Soviets had kidnapping in mind. Aage Bohr remembers the encounter:

> Terletsky brought with him a letter of introduction (dated October 22, 1945) from Kapitza. . . . Kapitza sent along recent scientific publications from his Institute. The conversation with Terletsky first dealt with Kapitza and other personal acquaintances among Russian physicists. Terletsky then raised some technical questions concerning atomic energy, to which my father answered that he was not acquainted with details and referred Terletsky to the report recently published by the US (Smyth Report). . . .

According to Terletsky, Niels Bohr talked at length—about Lev Landau, with whose political safety he apparently continued to be concerned, and about preventing nuclear war—and gave the Soviet delegation a tour of his institute. Terletsky remembered vividly, twenty-five years later, Bohr saying "that in his opinion, all countries should have the atomic bomb, and Russia first of all. Only by extending this powerful weapon to other countries could we guarantee that it would not be used in the future." Bohr never once expressed such a reckless idea anywhere in the West, nor, it seems, did he promote nuclear proliferation to Terletsky. In Terletsky's contemporary notes of his questions and Bohr's answers, he quotes the Danish physicist on just this issue. In response to the question if there were any defenses against the atomic bomb, Bohr told him that "only international cooperation, exchange of scientific discoveries, internationalization of the achievements of science can lead to the elimination of wars and thus to the elimination of the very necessity to use the atomic bomb. This is the only rightful method of defense. . . . All scientists believe that this greatest discovery must belong to all the nations and serve the unprecedented progress of humankind. . . . Atomic energy, once discovered, cannot remain the asset of one nation since a country which does not possess this secret can soon discover it independently." "Independently," of course, does not mean giving the bomb away. Terletsky heard what he wanted to hear.

He seems not to have understood the terms Bohr had set for the meeting; only after the group finished the tour did he realize that he would not be allowed to talk to Bohr privately, without Aage present, and hastened to start on his list of questions. His allotted hour ended before he was finished. He was frantic; "I already knew what happened when you didn't obey Beria's orders," he writes, and Beria's orders were to ask Bohr all the questions on the list. Bohr must have understood the reason for Terletsky's discomfiture; he agreed to meet again on November 16.

Terletsky completed his questioning that Friday—"the Niels Bohr Interrogation," Sudoplatov christened the black comedy sarcastically. Then, says Terletsky, "at his father's orders, Aage brought us a unique present, the report by Henry Smyth." Terletsky thought in mid-November that the Smyth Report "had only just been declassified and we were probably the first Soviet people to see it." Had Terletsky seen it in Moscow, he would have realized that the document answered most of the questions he had been assigned to ask Bohr. Ironically, Bohr had written Robert Oppenheimer on November 9 praising "the decision to publish the account of the pioneer work" but wishing that "further steps as regards release of information about purely scientific matters could soon be taken."

Buried deep in Terletsky's list was the crucial question about reactor poisoning, placed there, Terletsky's account implies, by one of the scientists who made up the list:

QUESTION 15:

Is there a process of slowing down the reactor due to the accumulation of waste from the fission of the light isotope of uranium?

On the way back to the Soviet Embassy from Bohr's institute, Terletsky conferred with the translator to reconstruct Bohr's answer:

ANSWER:

Pollution of the reactor with waste as the result of the fission of the light isotope of uranium takes place, but so far as I know, the Americans do not make special stops to clean the reactor. The reactor is cleaned when the [uranium] rods are removed for the extraction of plutonium.

Bohr was no reactor expert. The reactor was never "cleaned" to remove xenon; it was enlarged to override the poisoning effect. Kurchatov got less information than he needed. But at least Bohr (having seen the lithoprint version of the Smyth Report where the information appeared) had confirmed that poisoning took place and that the problem had a solution. Further espionage might reveal the poisonous isotope. Bohr, for his part, promptly reported the episode to Danish, British and US security. Terletsky went back to digging through the vast Lubyanka accumulation, concluding correctly that "Bohr told us nothing new beyond the Smyth Report...." (When he reported these disappointing results to Beria, the People's Commissar was disgusted; Terletsky says Beria interjected "crude curses addressed to Bohr and Americans.")

Andrei Sakharov remembered that *The British Ally* began serial publication of the Smyth Report" early that winter. "... I would snatch up each new issue of the *Ally* and scrutinize it minutely...." (Sakharov was a graduate student by then at FIAN, the Physics Institute of the Soviet Academy of Sciences, studying under Igor Tamm. He was not yet a member of Kurchatov's team, but he did not go unnoticed. "The institute's staff got younger by the autumn [of 1945]," his engineering colleague L. V. Pariskaya remembers, "the youths came back from the front, the women became livelier. We cleaned up as much as possible: the junk and old crates were thrown away, the age-old dirt was scrubbed off the parquet floor, the damned [blackout] curtains were pulled off, the windows were washed. The laboratories became light and spacious...." Pariskaya and Sakharov together washed the enormous window at the end of the institute's main corridor and Sakharov was proud of his work. "Now I've learned to wash windows," he told Pariskaya. "This may come in handy." Here's a man who will always be himself, Pariskaya decided.)

The Russian translation of *Atomic Energy for Military Purposes,* edited by G. N. Ivanov (whose real name was G. N. Kolchenko, a member of the Department S staff), was published in Moscow ostensibly by the State Railway Transportation Publishing House in an edition of thirty thousand copies on January 30, 1946. The text followed the Princeton edition but included the deleted sentence on reactor poisoning. Arnold Kramish, by then on the staff of the US Atomic Energy Commission, discovered the discrepancy in 1948, correctly concluding that "at least one Soviet technical man has screened the Smyth Report in great detail and it is very unlikely that some of the references which we have hoped 'maybe they won't notice' have not been noticed. With particular regard to ... fission product poisoning ... we must realize that that information most certainly has been compromised."

"With the discovery of fission," comments C. P. Snow, "... physicists became, almost overnight, the most important military resource a nation-state could call upon." Stalin still had suspicions about his physicists, but he now arranged for their comforts. Kurchatov met with Stalin, Beria and Molotov at the Kremlin late on the night of January 25, 1946, to discuss the bomb program. Kapitza was on someone's mind that night, probably Beria's; Kurchatov notes chillingly that "a question was asked about Ioffe, Alikhanov, Kapitza and [FIAN director Sergei] Vavilov, and about the utility of Kapitza's work. Misgivings were expressed: who were they working for, and to what were their activities directed—the good of the Motherland or not?" Stalin decisively rejected Kapitza's proposal of an independent Soviet program, which the physicist had argued would be cheaper than the US approach. "In the course of the conversation," write Khariton and Smirnov, "Stalin advised avoiding side issues or wasting time looking for inexpensive solutions to problems. He stressed that the work should be done 'on a broad front, on a

Russian scale,' and that he would give it full support. Stalin mentioned that our scientists are modest people and 'sometimes don't realize that they don't live well enough.' " Kurchatov noted at the meeting that "in respect to scientists, Stalin was concerned to . . . improve their everyday conditions and to reward them for major achievements—for example, for the solution to our problem. . . . He proposed that we report on the measures necessary to speed our work, to list everything that we need." Exiled geneticist Zhores Medvedev comments:

> Financial resources for science were increased sharply. The average salary for scientists was doubled or tripled, and in a country where food and consumer goods were still rationed, scientists found themselves in the highest privileged group.
>
> Almost half of the western part of the Soviet Union was in ruins, and the farmers of many destroyed villages lived in dugouts . . . on the sites of their war-burnt homes, but scientists suddenly became the privileged elite of the country, their living standards having been raised much higher than the pre-war level. The new institutes multiplied like cells in a culture, and almost all demobilized soldiers who had a secondary education . . . were absorbed by the enlarged network of higher technical schools and universities. The number of students, which was 817,000 just before the war, reached more than 1,500,000 in 1948–1949.

"Our state has suffered very much," Kurchatov paraphrased Stalin in his notes on their January 25 meeting, "yet it is surely possible to ensure that several thousand people can live very well, and several thousand people better than very well, with their own dachas, so that they can relax, and with their own cars." Stalin had a special gift for Igor Kurchatov. The Soviet dictator authorized building the project director a house on the grounds of Laboratory No. 2. An Academician architect designed an elegant eight-room, two-story Italianate structure with a classical pediment, large windows, parquet floors, marble fireplaces, fine wooden paneling and a sweeping central stairway. Building began early in 1946, with Italian craftsmen imported to finish the interior. The Kurchatovs occupied the house the following November. In contrast to the usual shoddy socialist construction, the opulent structure that Kurchatov's staff nicknamed ironically "the forester's cabin" was comparable in the quality of its workmanship to the Kremlin, which was built, of course, for the czars. From his "cabin," Kurchatov could stroll through a forest aromatic with birch and pine to the site he had chosen nearby for the F-1 reactor, which would be shielded below ground in a ten-meter pit dug under tenting, but later fitted with a brick laboratory building of its own, code-named "Assembly Workshops." Highly purified uranium metal and graphite were still in short supply in 1945; excavation for the F-1 reactor had not yet begun at the end of the year.

12

Peculiar Sovereignties

IN THE MONTHS immediately postwar, United States military and intelligence organizations wheeled their attention like heavy artillery around from Germany and Japan to the Soviet Union. Not only did real Soviet forces on the ground in Europe challenge, by their continued presence, the demobilizing Western defense; the Soviet Union was also the only theoretical enemy visible, as far ahead as it was sensible to look. In a first working estimate of the number of atomic bombs the US should stockpile, for example, confined to the years 1945–1955 when conventional bombers would still be the only available means of delivery, US Army Air Forces Major General Lauris Norstad pointed out that "during this period Russia and the United States will be the outstanding military powers," and for that reason the estimate used "the destruction of the Russian capability to wage war . . . as a basis upon which to predicate the United States atomic bomb requirements." To General Groves, who continued by default to direct the atomic weapons program, the Soviet Union had always been the ultimate adversary; from the beginning, Groves had guided the Manhattan Project in the direction not of making a few bombs to end a war but of developing a broad industrial capability to turn out atomic weapons in quantity after the war was won. Regardless of political views, responsible contingency planning required military leaders to consider from which direction war might come and what forces and strategy they would need to forestall it or to claim victory. This planning proceeded even as the United States government attempted to negotiate through the United Nations a program of international control of atomic energy. Such cross-wired confusion about the application of nuclear energy to war and international relations would trouble American atomic policy for years to come.

On August 8, 1945, between the Hiroshima and Nagasaki bombings, USAAF General Carl Spaatz, anticipating "plans for [a] post-war atomic-bomb program," stressed in a memorandum that "the atomic bomb is essentially an air weapon" and "there must be a plan for orderly transition from the present to a post-war basis which envisions our ability on short notice to deliver atomic bombs...." Spaatz proposed that the 509th Composite Group, which had been organized under Colonel Paul Tibbets to drop the first atomic bombs, "should remain intact as a nucleus for an expanded program." Spaatz was commander of the Pacific Air Force at the time; in 1946 he became commanding general of the air forces. The 509th, renamed the 509th Bomb Group, which operated the only aircraft equipped to carry atomic bombs, moved to Roswell Army Air Base in Roswell, New Mexico, soon after the war.

The US Joint Chiefs of Staff met secretly before the atomic bombings of Japan and approved a new policy of "striking the first blow"—surprise attack —in the event of an atomic war. The first-strike policy found embodiment subsequently in a planning document issued on September 20, 1945, which stressed that during a crisis, while diplomacy proceeded, the military should be "making all preparations to strike a first blow if necessary." Surprise attack went against previous US military policy, which had been formally defensive, as well as national tradition, but the change was not gratuitous. To the Joint Chiefs it seemed to follow logically from a realistic assessment of the destructiveness of nuclear weapons: whoever struck first with such powerful weapons was likely to carry the day. "Offense," the Joint Chiefs would assert two years later, "recognized in the past as the best means of defense, in atomic warfare will be the only general means of defense." In that spirit, by October 1945, the JCS Joint Intelligence Committee began drafting a plan for a first strike on the Soviet Union of twenty to thirty atomic bombs, a number based on a realistic assessment of currently available resources of ore and manufacture. The plan foresaw two scenarios that might require such a strike: in retaliation for Soviet aggression or, when the Soviet Union became capable of attacking the United States or of repelling a US attack, as preventive war.

Groves also mulled preventive war in those first heady months of nuclear monopoly. "If we were truly realistic instead of idealistic, as we appear to be," he wrote in a secret report, "Our Army of the Future," "we would not permit any foreign power with which we are not firmly allied, and in which we do not have absolute confidence, to make or possess atomic weapons. If such a country started to make atomic weapons we would destroy its capacity to make them before it has progressed far enough to threaten us." The Joint Intelligence Committee plan explored doing just that.

What the US military planned contingently and some military leaders vigorously advocated was not official policy. The US government never endorsed or authorized preventive war. Harry Truman evidently found the idea morally repugnant as well as politically suicidal. "Such a war is the

weapon of dictators," the President said publicly in 1950, "not of free democratic countries like the United States." But the extreme conviction that the
only sure way to protect America from what one Air Force general, Nathan
Twining, would call "the whims of a small group of proven barbarians" was
to destroy the industrial capacity of the USSR preemptively—to strike Soviet
cities by surprise with atomic bombs, that is, with the potential loss in a few
apocalyptic days of tens of millions of human lives—persisted within the
military, the USAAF in particular.

Norstad's more ambitious study of September 1945, which incorporated
the strategic chart of Russian and Manchurian urban areas that Groves had
seen in late August, found that the Soviet Union could be defeated at the
outset of a war if the United States destroyed sixty-six Soviet "cities of
strategic importance," neutralized a few air bases the Soviets might use
outside the USSR and isolated "the battlefield" by atomic-bombing such
tactical targets as the Dardanelles and the Kiel and Suez Canals. For these
purposes, and estimating that only 48 percent of the bombs would get
through and find their targets, Norstad concluded that the United States
would need to stockpile 466 atomic bombs of Nagasaki scale. The USAAF
general sent his study to Groves for comment. Groves dismissed this first
air effort impatiently. It underestimated the destructiveness of atomic
bombs, he told Norstad, and overestimated how destructive they would
need to be to disable a city. "My general conclusion would be that the
number of bombs indicated as required, is excessive."

The day before Norstad sent his study to Groves for review, September
14, he and USAAF Lieutenant General Hoyt Vandenberg had been appointed
to a board headed by Carl Spaatz charged to report on "the effect of the
atomic bomb on the size, organization, composition, and employment of
post-war Air Forces." A few of Norstad's findings made their way into the
report the Spaatz Board issued in October, but overall its conclusions were
cautious. It noted that the USAAF knew very little about atomic bombs
because of Manhattan Project secrecy, which the President had recently
extended postwar. The weapon was large, heavy, "enormously expensive
and definitely limited in availability." For these and other reasons, the Spaatz
Board recommended that the USAAF wait and see, concluding that "the
atomic bomb does not at this time warrant a material change in our present
conception of the . . . Air Force." The board proposed assigning a blood-
hound to follow the trail—a new Deputy Chief of Air Staff for Research and
Development—and recommended appointing Curtis LeMay, just back from
Hokkaido in his long-range B-29.

Production that autumn from Oak Ridge and Hanford confirmed the limits
the Spaatz Board had assessed. Oak Ridge separated 1.063 kilograms of U235
per day at a daily cost of $158,300. The Little Boy uranium gun used sixty-
four kilograms, which was two months' production (six Little Boys per year),

and with composite cores in the offing for the implosion bomb, Groves decided to stockpile the U235 rather than make it up into wasteful and obsolete guns. Hanford produced about four to six kilograms of plutonium per month, enough for about ten to twelve Fat Man bombs per year (with just over 6 kg of plutonium per core), but composite cores would need only 3.2 kg of plutonium each (plus 6.5 kg of U235). So the only bomb assemblies Los Alamos produced for the rest of the year and during 1946 were Fat Man designs, now called Mark IIIs, which could accommodate a solid Christy core or a new composite. The composite, however, could not be certified for military use until the design had been tested at full scale, and no such test was in the offing. Effectively, then, the US production of U235—by far the larger quantity of fissionable material—was long-term reserve with no short-term military application.

Before Curtis LeMay took up his new duties of research and development, he went on leave, back to his native Ohio, "spending [a] few weeks with my family," he said, "getting acquainted with them once more" in the midst of "the most beautiful Indian Summer I have ever seen." On his way to Washington in November he found time to speak to the Ohio Society of New York. Like most returning veterans, he was full of feeling and of resolve.

He could not describe, he told his fellow Ohio State alumni, "the difference between the bomb-blackened ruins and the desolation of our enemies' cities and the peaceful Ohio cities and landscape, untouched and unmarred by war. I can only say to you, 'If you love America, do everything you can to make sure that what happened to Germany and Japan will never happen to our country.'"

On leave, he had thought through the last four years of war, he said. America had not been prepared. "She escaped the ruin visited upon other nations because she was given time to prepare, and because of distance." But in the next war, LeMay warned, "distance will be academic and there will be no time for preparation." The next war would be launched in the air. It would be fought with fantastic new weapons. "December 7, 1941, will seem like a quiet day in the country in comparison with the first day of the next war." The next war would be a war of "rockets, radar, jet propulsion, television-guided missiles, speeds faster than sound and atomic power." They had not had enough bombers at the beginning of the last war, LeMay recounted in horror; "American unpreparedness . . . extended to the point where on September 1, 1939, the day Hitler smashed into Poland, United States strategic air power consisted of *nineteen poorly-equipped heavy bombers.*" Before the next war "the air force must be allowed to develop unhindered and unchained. There must be no ceiling, no boundaries, no limitations to our air power development."

Then LeMay came to the contradiction that he would chew over for years to come. First he offered one of his touchstone concepts: "No air attack, once it is launched, can be completely stopped." In late 1945, that meant to LeMay that the US would have to have an air force *"in being"* that could move immediately to retaliate if the country was attacked. But if no air attack can be completely stopped, then retaliation would not protect the country —it would only destroy the enemy's country in turn. That threat of retaliation, that preparation, might be sufficient to prevent attack in the first place. "If we are prepared it may never come. It is not immediately conceivable that any nation will dare to attack us if we are prepared." So in November 1945, Curtis LeMay was already thinking in terms of what came to be called deterrence. But it was only not *immediately* conceivable that an enemy might attack us if we were prepared. Evidently it was still conceivable if conditions changed. "International gangsters have twice made the mistake of leaving America until last. They will not make the same mistake a third time." And if they did not—if they were not deterred—what then?

Moving on to the Pentagon after his stop in New York, LeMay remembered feeling like a square peg in a round hole at first, out of his depth technically in research and development, "but it didn't take me long to become mighty interested." To glean the spoils of advanced technology from defeated Germany, he organized Project Paper Clip, which brought German scientists to the US as prisoners of war and put them to work continuing the research they had begun for the Third Reich. LeMay's chief catches were Walter Dornberger and Werner von Braun, who had developed the V-2 rocket, the first long-range ballistic missile. He built a wind-tunnel complex for advanced aircraft research in Tullahoma, Tennessee. He commissioned the consulting company that became the Rand Corporation—the first "think tank"—and assigned it the research that resulted in May 1946 in its first report, *Preliminary Design of an Experimental World-Circling Spaceship* (that is, a satellite). "The reason for having Rand do the study," comments Herbert York, "was to get a jump on the Navy, which also was studying satellites. LeMay was determined that it wasn't going to be a Navy program, it wasn't going to be a joint Navy–Air Force program, it was going to be an Air Force program." The report noted pointedly that "the development of a satellite will be directly applicable to the development of an intercontinental rocket missile." LeMay also lobbied the USAAF during his R&D tenure to develop the long-distance capability to detect an atomic explosion, but his proposal fell on deaf ears. Why bother to look for what everyone knew the Soviets couldn't soon make?

In the midst of these duties LeMay won another assignment. An aide to Secretary of the Navy James Forrestal, an investment banker with long-standing connections in the physics community named Lewis L. Strauss, had recommended shortly after the end of the war that the Navy "test the ability of ships of present design to withstand the forces generated by the atomic bomb" to scotch "loose talk to the effect that the fleet is obsolete in the face of this new weapon." Such talk, Strauss feared, would "militate against appropriations to preserve a postwar Navy." An alert young Democratic senator, Brien McMahon of Connecticut, made a similar proposal on August 25. The Navy negotiated with the USAAF. The two services agreed to a joint project to atomic-bomb a fleet of surplus Japanese, German and US ships moored at anchor. The Joint Chiefs appointed a subcommittee to plan the exercise. To head the subcommittee they chose Curtis LeMay. Operation Crossroads, to be carried out on Bikini atoll in the Marshall Islands, would be LeMay's second atomic-bombing command.

The destruction of Hiroshima and Nagasaki by single bombs shocked and frightened the world. Americans in particular felt vulnerable for the first time to devastating attack, a reaction in part to having been the first to inflict such attack on another people. A great debate started up in the United States in the months after the war about war and peace. There were calls to outlaw the bomb, calls for world government, visions of "one world or none." Even Edward Teller briefly embraced internationalism, writing in the new *Bulletin of the Atomic Scientists* that "nothing that we can plan as a defense for the next generation is likely to be satisfactory; that is, nothing but world-union."

After a time of confusion about appropriate legislation, the Truman administration backed civilian control of atomic energy domestically. Internationally, in an Agreed Declaration with Britain and Canada concluded on November 15, 1945, it supported action "to prevent the use of atomic energy for destructive purposes" and "to promote the . . . utilization of atomic energy for peaceful and humanitarian ends." Such action required a plan; Jimmy Byrnes as Secretary of State got the job of devising it. Byrnes appointed a protesting Dean Acheson chairman of a committee that included Groves, the two civilian wartime leaders of atomic-bomb development, Vannevar Bush and James Bryant Conant, and, from the War Department, Henry Stimson's former aide John J. McCloy (Stimson had retired). Acheson in turn appointed a five-man board of expert consultants. David Lilienthal, a lawyer who was head of the Tennessee Valley Authority, agreed to serve as chairman. The other four members were Monsanto Chemical vice-president and plutonium specialist Charles Thomas, General Electric engineering vice-president Harry Winne, New Jersey Bell president Chester Barnard and Robert Oppenheimer.

Oppenheimer came prepared. He had explored the complexities of international control not only with Niels Bohr at Los Alamos and with Conant but also with shrewd, tough I. I. Rabi. "Oppenheimer and I met frequently and discussed these questions thoroughly," Rabi told an interviewer many years later. "I remember one meeting with him, on Christmas Day of 1945, in my apartment [on Riverside Drive west from Columbia University in Manhattan]. From the window of my study we could watch blocks of ice floating past on the Hudson. We were then developing the ideas that became the basis of the Acheson-Lilienthal Report." The work of formulating a plan was a way beyond misgivings and despair. "Once [Oppenheimer] got interested in something," Rabi observes, "he went right in to become the leader of it." Gordon Arneson, the State Department's specialist on atomic matters, says Oppenheimer became "the chief teacher for the Acheson-Lilienthal group."

To Lilienthal, the group's work was "one of the most memorable intellectual and emotional experiences of my life." The men met first in Washington. Oppenheimer gave them a ten-day course in nuclear physics, properly taking control, as the only real expert, of defining the technical basis of the problem, but other than serving as their *savant* he kept his own council at first. They moved next to New York to talk to a group of scientists, including Luis Alvarez, who had explored for Groves a scheme of control by inspection alone. Discussion intensified. Ideas came from every side—these were men of diverse background and conviction—and they debated them night and day. When patience gave way to exasperation and someone proposed simply outlawing the bomb, which happened frequently, Lilienthal always waved a newspaper clipping about the Agreed Declaration to remind them that their government had already committed itself to international control. Back to Washington to study geology. They made progress. Then they got seriously stuck. Lilienthal proposed they tour Oak Ridge and Los Alamos. Whiskey on the train down to Knoxville and a hungover tour of the vast gaseous-diffusion plant, where supervisors prowled among the surrealistic piping on bicycles, warmed their friendship.

They flew to Los Alamos in Groves's private C-54, which Lilienthal in his diary called "a luxurious army transport plane." The President was trying to reach Lilienthal—to offer him a Cabinet post as Secretary of the Interior, the TVA chairman thought—but not even that provocation dulled him to the significance of the secret mesa, "with the high mountains forming a majestic backdrop," where they "went into casual little buildings, saw things only few men have seen, talked with soft-spoken, gentle, intelligent men about the things they had done. . . . Now I have a sense that this thing of atomic bombs is *real. . . .* " Herbert Marks, Acheson's personal representative to the board, who accompanied the men on their travels, caught the stenchy note of brimstone that accented the Faustian essence:

It wasn't a large place . . . and it wasn't a spectacular one. I looked around me and there were the same materials, colors, textures, and fabrics you might see in any warehouse. I saw the receptacles that contained the labor of God knows how many men, the cargoes of thousands of freight cars, the mental triumphs of gifted scientists born in a dozen countries. The receptacles were small, and I thought to myself, hell, I could walk out of here with one of them in my pocket. Not that I could have. Too many soldiers outside and inside the vault were watching us closely—tough troops who looked as though they kept their rifles cleaned. And supposing I had got away with one, what could I, an ordinary layman, have done with it? In a way, the same was true of so much of the whole Manhattan District. It bore no relation to the industrial or social life of the country; it was a separate state, with its own airplanes and its own factories and its thousands of secrets. It had a peculiar sovereignty, one that could bring about the end, peacefully or violently, of all other sovereignties.

What they came to was a radical proposal. Remarkably, it won their common agreement. When Bohr read it he wrote Oppenheimer of his "deep pleasure." In every word of it, he said, he found "just the spirit which I think offers the best hopes for the development in which we all put our whole faith."

The Acheson-Lilienthal Report, as the board's proposal came to be called, found that outlawing atomic bombs by inspection was unreliable: "Any system based on outlawing the purely military development of atomic energy and relying solely on inspection for enforcement would at the outset be surrounded by conditions which would destroy the system." To the contrary, "every stage in the activity, leading from raw materials to weapon, needs some sort of control." That control might work "if the element of rivalry between nations were removed." The way to do that was to assign "the intrinsically dangerous phases of the development of atomic energy to an international organization responsible to all peoples. . . ."

If, for example, only this Atomic Development Authority could legally own and develop uranium ore, "then . . . not the purpose of those who mine or possess uranium ore but the *mere fact of their mining or possessing it becomes illegal,* and national violation is an unambiguous danger signal of warlike purposes. The very opening of a mine by anyone other than the international agency is a 'red light' *without more;* it is not necessary to wait for evidence that the *product* of that mine is going to be misused."

Three kinds of activity were intrinsically dangerous: acquiring raw materials, producing fissionable uranium and plutonium and using them to make atomic weapons. Other kinds of activity were safe: using radioactive tracer materials in medicine and science; operating small research reactors; possibly operating power reactors as well if the fuel were properly "denatured" to make converting it to an explosive more difficult. Safe activities might be licensed to states, reducing the necessary scale of the Atomic Development Authority bureaucracy. Intrinsically dangerous activities it would have to control and operate exclusively.

There would be a significant advantage in such operation: the scientists and engineers on the Authority staff could be more than policemen. They would work at the cutting edge of a new and exciting field of technology.

The work would be intellectually satisfying. It would keep them keen. It would also keep them at least abreast of potential violators.

These were sufficiently radical ideas, particularly coming from hard-headed businessmen like Chester Barnard and Harry Winne. Winne even felt compelled to defend the group against a charge of radicalism, contributing to the report an assessment that "it may seem too radical, too advanced, too much beyond human experience. All these terms apply with peculiar fitness to the atomic bomb." In the final pages of the report the group introduced a truly radical idea. Radical for its time, it would still seem radical today. It concerned the crucial question of sanctions.

If a nation attempted to violate such an agreement as the group was proposing, how could it be punished? Would the authority maintain an army? Would it stockpile atomic bombs? No. Instead, the authority as it developed would spread its mines and factories around. Its Oak Ridges and Hanfords, its research laboratories and nuclear power reactors would need to be spread around anyway so that their benefits could be dispersed. Spreading such development around would be a way of transitioning from national to international control, and "a systematic plan" would probably have to be written into the authority's charter "governing the location of the operations and property of the Authority so that a strategic balance may be maintained among nations." Whereupon the system would become self-policing:

This will . . . be quite a different situation from the one that now prevails. At present with Hanford, Oak Ridge, and Los Alamos situated in the United States, other nations can find no security against atomic warfare except the security that resides in our own peaceful purposes or the attempt at security that is seen in developing secret atomic enterprises of their own. Other nations which . . . may fear us, can develop a greater sense of security only if the Atomic Development Authority locates similar dangerous operations within their borders. . . . [Then] a balance will have been established. It is not thought that the Atomic Development Authority could protect its plants by military force from the overwhelming power of the nation in which they are situated. Some . . . guard may be desirable. But at most, it could be little more than a token. The real protection will lie in the fact that if any nation seizes the plants or the stockpiles that are situated in its territory, other nations will have similar facilities and materials situated within their own borders so that the act of seizure need not place them at a disadvantage.

This remarkable idea—spreading the intrinsically dangerous mines and factories around—is indistinguishable from what has come to be called nuclear proliferation, except that the agent of proliferation in the board's design would have been an organ of the United Nations rather than individual states, and the technology that proliferated would have been infrastructure alone rather than infrastructure and stockpiled weapons. Though the report does not belabor the point, it notes more than once that true security is incompatible with secrecy. Its proposal for a radical system of self-policing makes starkly clear what Bohr's open world would be: a world where how to design atomic bombs would be public knowledge; a world, as it were, where the guns have all been laid out together in the open on a table but disassembled and arranged so as to be within everyone's equal reach. Bohr liked to ask of new ideas in physics whether they were crazy enough to be truly original. Here was the logic of openness extended into a practical proposal, and it looked odd indeed. Would it have worked? In a much more unstable and dangerous form, as nuclear proliferation, it did. In the form in which the Acheson-Lilienthal Report proposed it in March 1946, it never had a chance.

Around this time—early 1946—a remarkable conjunction of statements and events revealed and contributed to a darkening of relations between the US and the USSR. The first of these was a speech Stalin read before a crowd of four thousand Party members, government officials and Army officers at the Bolshoi Theater, across from the Kremlin in the center of Moscow, on the evening of February 9. The occasion for the speech was ostensibly an upcoming election; deputies to the Supreme Soviet were to be elected (on one-party ballots) for the first time since December 1937 and Stalin desired to express his appreciation for having been renominated. In fact the speech reclaimed the USSR for the Party. No "brothers and sisters" now, no "my friends," as Stalin had embraced the Soviet people in the wake of the German invasion on July 3, 1941. The dictator addressed himself now to "comrades"; he informed them that their victory meant, "in the first place, that our Soviet system has won." No gratitude for gallant allies; the war had been "an inevitable result...of modern monopoly capitalism." The war had demonstrated that the Soviet system was a "perfectly viable and stable form of organization...a form of organization superior to all others." The Red Army had proved itself first-class. The Communist Party had managed war production "with the utmost success." But now the country needed to be restored. A new Five-Year Plan would squeeze into half a decade the development that had been planned for the decade truncated by the war.

Echoing at the Bolshoi what he had recently told Igor Kurchatov, Stalin expressed "no doubt that, if we give our scientists proper help, they will be able in the near future not only to overtake but to surpass the achievements of science beyond the boundaries of our country." There would be "extensive construction of scientific research institutes." He would soon remove rationing and the country would produce more consumer goods. But there would be no relief from the grinding toil and poverty of the war years, as millions dreamed:

> Our Party intends to organize a powerful new upsurge of the national economy which would enable us, for instance, to raise the level of our industry threefold, as compared with the prewar level; only under such conditions can we regard our country as guaranteed against any eventualities. That will require perhaps three new Five Year Plans, perhaps more.

The country had lost 30 percent of its national wealth and 25 million people were still homeless, but "Stalin used the victory," writes biographer Dmitri Volkogonov, "consciously and resolutely to preserve the system."

In Washington a few days later, Navy Secretary James Forrestal asked Supreme Court Justice William O. Douglas if he had read Stalin's speech and what he thought of it. "The Declaration of World War III," Douglas said. Forrestal agreed.

The second defining statement during this period was a long telegram from George Kennan, the American chargé d'affaires in Moscow, to the State Department on February 22, 1946, analyzing the Soviet mentality. Kennan was bedridden at the embassy "with cold, fever, sinus, tooth trouble and . . . the aftereffects of the sulpha drugs administered for the relief of these other miseries" when he dictated his acerbic dispatch. His illness followed the unsuccessful meeting of the Council of Foreign Ministers—Molotov, Ernest Bevin of Britain and Jimmy Byrnes—in Moscow in December, in which he had participated. There were few places on earth less pleasant in winter than the Soviet capital; Kennan calls the time "these unhappy days," and associates his five-part telegram with "an eighteenth-century Protestant sermon." For a year and a half, he wrote thirty years later, still exasperated, "I had done little else but pluck people's sleeves, trying to make them understand the nature of the phenomenon with which we in the Moscow embassy were daily confronted and which our government and people had to learn to understand if they were to have any chance of coping successfully with the problems of the postwar world. So far as official Washington was concerned, it had been to all intents and purposes like talking to a stone." Kennan's mood of exasperation—with the Soviets and with Washington— colored his analysis. In his *Memoirs* he claims "horrified amusement" at rereading his long telegram and mocks its pretension—"much of it reads exactly like one of those primers put out by alarmed congressional committees or by the Daughters of the American Revolution, designed to arouse the citizenry to the dangers of the Communist conspiracy." He was evidently in dead earnest at the time.

"The USSR still lives in antagonistic 'capitalist encirclement,'" Kennan began his analysis ominously, "with which there can be no permanent peaceful coexistence." He ascribed to the Kremlin a "neurotic view of world affairs," which "at bottom" was the "traditional and instinctive Russian sense of insecurity." Russian rulers, he said, "have always feared foreign penetration, feared direct contact between [the] Western world and their own, feared what would happen if Russians learned [the] truth about [the] world without or if foreigners learned [the] truth about [the] world within." Marxism simply supplied "justification for their instinctive fear of [the] outside world, for the dictatorship without which they did not know how to rule, for the cruelties they did not dare not to inflict. . . ." The Soviet rulers were not necessarily insincere; "who, if anyone, in this great land actually receives accurate and unbiased information about [the] outside world" was an "unsolved mystery." An atmosphere of "Oriental secretiveness and conspiracy" pervaded the government, so that "possibilities for distorting or poisoning sources and currents of information are infinite."

In consequence, "we have here a political force committed fanatically to the belief that with [the] US there can be no permanent modus vivendi, that it is desirable and necessary that the internal harmony of our society be disrupted, our traditional way of life be destroyed, the international authority of our state be broken, if Soviet power is to be secure." How to cope with such a country "should be approached with [the] same thoroughness and care as [the] solution of [a] major strategic problem in war."

Kennan believed the problem could be solved, however. The Soviets were militarily weaker and their system was not necessarily stable. War was not the solution. Rather, for the indefinite future, the Communists would have to be contained:

> Soviet power, unlike that of Hitlerite Germany, is neither schematic nor adventuristic. It does not work by fixed plans. It does not take unnecessary risks. Impervious to the logic of reason, it is highly sensitive to the logic of force. For this reason it can easily withdraw—and usually does—when strong resistance is encountered at any point. Thus, if the adversary has sufficient force and makes clear his readiness to use it, he rarely has to do so.

In the years since this historic telegram sparked along the cables from Moscow to Washington, George Kennan has claimed more than once that he was writing less of military containment than of political. Washington did not read his sermon that way, and its language makes clear why. When the US naval attaché in Moscow alerted the Chief of Naval Operations to the document, for example, he noted that it revealed the "utter ruthlessness and complete unscrupulousness of [the] Soviet ruling clique." James Forrestal embraced it, writes Kennan, "had it reproduced and evidently made it required reading for hundreds, if not thousands, of higher officers in the armed services." Byrnes thought it a "splendid analysis." Louis Halle, a State Department official, remembered that "it came at a moment when the Department, having been separated by circumstances from the wartime policy towards Russia, was floundering about, looking for new intellectual moorings. Now, in this communication, it was offered a new and realistic conception to which it might attach itself. The reaction was immediate and positive. There was a universal feeling that 'this was it,' this was the appreciation of the situation that had been needed. Mr. Kennan's communication was reproduced for distribution to all the officers of the Department. . . . We may not doubt that it made its effect on the President." Truman read it at a time when he had concluded that "unless Russia is faced with an iron fist and strong language another war is in the making. . . . I do not think we should play compromise any longer. . . . I'm tired of babying the Soviets." Nothing much happened immediately as a result of Kennan's analysis except that he won recognition as a master strategist and saw his "official loneliness [come] . . . to an end," but in fact the long telegram had defined a new American policy toward the Soviet Union nearly as ideologically rigid as the policy Stalin had defined toward the West.

The third defining statement may have been the most important of all. Winston Churchill, now seventy-two, deposed as Prime Minister of England during the Potsdam Conference the previous July, traveled to Florida early in January 1946 to continue his recuperation from the hard years of war and the shock of losing election. He arrived with an invitation in his pocket to speak at a small Protestant men's college in Fulton, Missouri, a town of Southern ways in rich farm country north of the Missouri River in the center of Truman's home state. Truman's military aide, General Harry Vaughan, was a Westminster College graduate and had promoted the invitation with the President, who had endorsed it with a promise to travel to Fulton with Churchill and introduce him.

The former Prime Minister was worried about the hastening Soviet take-over of Eastern Europe. Americans seemed still to trust the Russians and he believed that trust dangerously misplaced. He decided to write a deliberately shocking speech, to give the West a wake-up call, as he had tried to wake up England to the Nazi threat in the decade between the world wars.

Perhaps to his surprise, he found that "the dire situation with which the insatiable appetites of Russia and of international Communism were confronting us was at last beginning to make a strong impression in American circles."* Byrnes and Truman both read Churchill's speech before he

---

* Churchill's visit coincided with a sea change in US attitudes. "When the war ended in the summer of 1945," notes historian John P. Rossi, "60 percent of the American people polled expressed confidence about cooperation between Russia and the Western Allies. By February 1946, on the eve of Churchill's speech, that figure had dropped to 35 percent." Rossi (1986), p. 117.

delivered it, although Truman afterward denied having done so because it advanced a harsher line than he was yet prepared to take. Byrnes in late February delivered a tough anti-Soviet speech at the Overseas Press Club. Truman's exasperation with "babying the Soviets" evidently warmed him to Churchill's views. "The President invited me to travel with him in his train the long night's journey to Fulton," Churchill recalled. "We had an enjoyable game of poker." Since Truman "seemed quite happy" about Churchill's "general line," the doughty Prime Minister "decided to go ahead."

Forty thousand people had descended on Fulton that warm Tuesday afternoon, March 5, 1946. Westminster is set on a hill westward several blocks from the Fulton town square. Churchill, robed in scarlet for an honorary degree, spoke on national radio in the college gymnasium; loudspeakers carried the orotund Churchillian rumble to audiences in other buildings and to the crowds outdoors.

This was Churchill's famous speech "The Sinews of Peace," when the former Prime Minister defined in a memorable phrase what had happened in Europe since the end of the war:

> From Stettin in the Baltic to Trieste in the Adriatic, an iron curtain has descended across the Continent. Behind that line lie all the capitals of the ancient states of central and eastern Europe. Warsaw, Berlin, Prague, Vienna, Budapest, Belgrade, Bucharest and Sofia, all these famous cities and the populations around them lie in the Soviet sphere and all are subject in one form or another, not only to Soviet influence but to a very high and increasing measure of control from Moscow.

Bluntly Churchill specified Moscow's measures of control: "Communist parties . . . raised to pre-eminence and power far beyond their numbers," "police governments . . . prevailing," "Turkey and Persia . . . both profoundly alarmed and disturbed," "a pro-Communist Germany" that would "cause new serious difficulties in the British and American zones." "Whatever conclusions may be drawn from these facts," Churchill summarized, ". . . this is certainly not the liberated Europe we fought to build up. Nor is it one which contains the essentials of permanent peace."

The British statesman nevertheless found it possible to "repulse the idea that a new war is inevitable; still more that it is imminent. . . . I do not believe that Soviet Russia desires war. What they do desire is the fruits of war and the indefinite expansion of their power and doctrines." The answer, he thought, must be Western strength and Anglo-American alliance. "From what I have seen of our Russian friends and allies during the war, I am convinced that there is nothing they admire so much as strength, and there is nothing for which they have less respect than military weakness." The "western democracies" would have to "stand together." If they did so, "no one is likely to molest them."

The condemnation of a former ally was shocking—to Stalin as well as to Americans. ("Mr. Churchill has now adopted the position of a warmonger," Stalin soon told *Pravda*. "Mr. Churchill and his friends bear a striking resemblance to Hitler and his friends. . . . Mr. Churchill also [has] . . . a racial theory, asserting that English-speaking nations are the only nations of full value, and must rule over the remaining nations of the world.") But most shocking of all to US public opinion, judging by the immediately subsequent reaction to the speech in the American media, was a paragraph now largely forgotten:

It would . . . be wrong and imprudent to intrust the secret knowledge or experience of the atomic bomb, which the United States, Great Britain and Canada now share, to the [United Nations], while it is still in its infancy. It would be criminal madness to cast it adrift in this still agitated and ununited world. No one in any country has slept less well in their beds because this knowledge and the method and the raw materials to apply it are at present largely retained in American hands. I do not believe we should all have slept so soundly had the positions been reversed and some Communist or neo-Fascist state monopolized, for the time being, these dread agencies. The fear of them alone might easily have been used to enforce totalitarian systems upon the free democratic world, with consequences appalling to the human imagination. God has willed that this shall not be, and we have at least a breathing space before this peril has to be encountered, and even then, if no effort is spared, we should still possess so formidable superiority as to impose effective deterrents upon its employment or threat of employment by others.

Which was nothing less than a call to an all-out atomic arms race.

Walter Bedell Smith, Dwight Eisenhower's aide during the Second World War, who had just been appointed the new US ambassador to the Soviet Union, stopped in to pay his respects to the British statesman in New York shortly after the Fulton speech. The street outside Churchill's hotel was noisy with picketers demonstrating solidarity with the Soviet Union. "I found him in the bathtub," Smith sets the scene, "but he called me in; and as he dressed himself I read the speech which he was to make that night [in Manhattan]:

The former Prime Minister obviously was disturbed by the picketing outside. It was his first experience, in America, of any sentiment other than friendship. But he stood firm in his belief in the correctness of his Fulton analysis, declaring, "Mark my words—in a year or two years, many of the very people who are now denouncing me will say, 'How right Churchill was.' "

Such was the clamorous valley of the shadow into which the members of the Acheson-Lilienthal board of consultants descended from their mountaintop.

Just as the board was finishing its seven weeks of intense deliberation, Truman moved to appoint a conservative to head the delegation that would present the US proposal to the United Nations. International control of atomic energy would mean a significant concession of national sovereignty; Truman concluded that only a certified conservative could carry such a measure through Congress. On Jimmy Byrnes's recommendation, he chose the multi-millionaire financier Bernard Baruch, whose legendary advice to Presidents and statesmen from a humble park bench and careful but expansive allotment of campaign funds had won him political authority. (Acheson, for his part, thought Baruch's "reputation was without foundation in fact and entirely self-propagated" and that Byrnes had "fallen victim to Mr. Baruch's spell.")

When Baruch, freshly appointed, read about the Acheson-Lilienthal Report in the newspapers—it had been leaked from Brien McMahon's new Joint Committee on Atomic Energy, in consequence of which the State Department released it officially on March 28—he was enraged. When he learned, a day or two later, that Acheson had said the report would be the basis for discussion at the United Nations, he was furious. "I [could not] see the purpose of my appointment," he writes, "if the United States plans on atomic control had already been decided upon. . . . I told [Acheson] plainly that he would then have to find another messenger boy, because Western Union didn't take anybody my age. I had never served as a messenger or mouthpiece before, and did not intend to start now." From the startled Acheson, Baruch swerved to Truman. Truman claims he dressed down "the only man to my knowledge who has built a reputation on a self-assumed unofficial status as 'adviser' ": "I had asked him to help his government in a capacity of my choosing. I had no intention of having him tell me what his job should be. I made that clear to him, in a very polite way." Baruch claims to the contrary that Truman "was most affable, and plainly anxious for me not to withdraw from the task. When the question arose about who was to draft the atomic proposals, he made this exact and characteristic reply: 'Hell, you are!' "

Baruch did so, and his first concern, predictably, was that the Acheson-Lilienthal Report "did not deal with the problem of enforcement—a problem which I considered crucial." For Baruch, who was then seventy-six years old and who had served as a technical adviser on the American delegation to the 1919 peace conference following the First World War, "swift and sure punishment" would determine whether an agreement on atomic energy was effective or only "another in the long line of history's empty declarations and gestures. . . . If I had learned anything out of my experiences in international affairs, it was that world peace is impossible without the force to sustain it." He was in charge and he would insist on "sanctions against those who violated the rules." And since the veto power that the permanent members of the United Nations Security Council enjoyed could block enforcement of such sanctions, he would further insist that the veto be suspended in matters of atomic control.

Acheson was appalled:

The 'swift and sure punishment' provision could be interpreted in Moscow only as an attempt to turn the United Nations into an alliance to support the United States threat of war against the USSR unless it ceased its efforts, for only the United States could conceivably administer 'swift and sure' punishment to the Soviet Union.... 'Swift and sure punishment' for violation of the treaty, if realistically considered, seemed uncomfortably close to war, or certainly to sanctions that under the United Nations treaty were subject to the veto of permanent members of the Security Council. Did it seem likely that they would forgo it here? The only practicable safeguard in case of violations would be clear notice and warning that they were occurring.

Baruch did not agree. His view prevailed. His Baruch Plan also proposed that the United States would give up its stock of atomic bombs only as the other nations of the world fell into line. The Soviet Union countered the Baruch Plan with a proposal for immediate and universal nuclear disarmament without inspection. The United Nations Atomic Energy Commission to which Baruch was the American delegate discussed the proposals until December, when it voted a plan largely modeled on the American plan, the Soviet Union and Poland abstaining. "There," writes Acheson, "the matter died." Nothing came of the effort except bad will.

Oppenheimer told an interviewer long afterward that the day Jimmy Byrnes appointed Bernard Baruch "was the day I gave up hope." Rather than speaking out publicly, as he might have done, he agreed to serve Baruch as a scientific adviser: "That was not the day for me to say so publicly. Baruch asked me to be the scientific member of the delegation, but I said I couldn't. Then Truman and Acheson told me it might not look right if I got out now, so I said I would be present at meetings." Baruch assesses Oppenheimer's contribution more fulsomely: "Once I got to work . . . I found scientists who were willing to help." Oppenheimer, writes Baruch, "rendered invaluable aid to me by serving on the Scientific Panel. . . . His is one of the most brilliant minds I have ever encountered."

Rabi, who was no man's fool, assessed Oppenheimer's conflicts realistically:

I found him excellent. We got along very well.... I enjoyed the things about him that some people disliked. It's true that you carried on a charade with him. He lived a charade, and you went along with it. It was fine—

matching wits and so on. Oppenheimer was great fun, and I took him for what he was. I understood his problem. . . . [His problem was] identity. . . . He reminded me very much of a boyhood friend about whom someone said that he couldn't make up his mind whether to be president of the B'nai B'rith or the Knights of Columbus. Perhaps he really wanted to be both, simultaneously. Oppenheimer wanted every experience. In that sense, he never focussed. My own feeling is that if he had studied the Talmud and Hebrew, rather than Sanskrit, he would have been a much greater physicist. I never ran into anyone who was brighter than he was. But to be more original and profound I think you have to be more focussed.

"I think [my brother] felt that he wanted to make a big difference," Frank Oppenheimer commented of this period in Robert Oppenheimer's life. "I argued with him quite a lot after the war. I felt that the kind of big difference would happen if one really taught people a lot about the dangers of the bomb, about the possibilities of cooperation. He said there wasn't time for this. He'd been in the Washington scene. He saw that everything was moving. He felt that he had to change things from within."

Both Rabi and Oppenheimer came to question whether the US proposal had been offered entirely in good faith. "Whether we really wanted to turn the bomb over [to the UN]," Rabi wondered late in life, "I don't know. Baruch didn't believe in it." Oppenheimer in a 1948 postmortem noted that even before the US took a position, "doubts [about Soviet and Anglo-American policy] pointed rather strongly to the need for discussion between the heads of state and their immediate advisors, in an attempt to re-open the issue of far-reaching cooperation. The later relegation of problems of atomic energy to discussions within the United Nations, where matters of the highest policy could only be touched upon with difficulty and clumsily, would appear to have prejudiced the chances of any genuine meeting of minds." Had there been "more reality to the plans," Oppenheimer concluded, ". . . we ourselves, and the governments of other countries as well, would have found many difficulties in reconciling particular national security, custom and advantage with an over-all international plan for insuring the security of the world's peoples." The physicist vividly remembered a conversation he had with Truman during this period; in biographer Nuel Pharr Davis's transcription, it went like this:

"When will the Russians be able to build the bomb?" asked Truman.
"I don't know," said Oppenheimer.
"I know."
"When?"
"Never."

At some level, for Harry Truman, US monopoly mooted the issue of international control.

———————

Yuli Khariton visited Veniamin Zukerman's radiography laboratory in late December 1945. Zukerman recalled the occasion vividly:

> Khariton arrived at the laboratory and asked us without preamble, "Have you read Smyth's book?"
> "Of course we have."
> "Then you understand what an enormous amount of work will have to be done before our country, too, has the secret of atomic weapons. I would like your laboratory, which has been working on radiography of explosions and detonations, to engage itself fully with the atomic problem. Don't give a thought to the formal side of the matter, all of that will be properly drawn up in due course. All I need from you is your consent." We requested two or three weeks' time to think the proposal over, although we did give him our tentative consent.

The following month, January 1946, Zukerman and Lev Altshuler received State Prizes for inventing flash radiography of explosive phenomena—reinventing, actually, since flash radiography had been developed secretly at Los Alamos during the war to observe implosion, the same purpose to which Khariton hoped to apply it. Khariton reappeared in February. "This time," says Zukerman, "he'd come for a working conference." In the course of the conference, he told them, "I can't rule out the possibility that certain experiments will be called for in the course of your development work that will be difficult to conduct in Moscow conditions. It could be that you'll have to relocate to different regions of the country for six months, a year. But it's still early to be talking about that."

It was not too early for Khariton to be doing something about that. While Stalin was complaining to *Pravda* in mid-March about Winston Churchill's Iron Curtain speech, Khariton remembers searching for a suitable location for a bomb laboratory:

> It was clear that creating the bomb would require colossal pressures to squeeze fissionable materials, and also that pressures of the required magnitudes could be created by large-scale explosions. Moscow was hardly a suitable place for carrying out work of this sort. And finding a suitable, unpopulated area not far from Moscow was not a simple matter. We spent a lot of time surveying the grounds of various munitions factories that had been active during the war. Finally, after long searching, on April 2, 1946, Pavel Mikhailovich Zernov (the future head of the Institute, once it was

organized) and I arrived at the small town of Sarov. Here there was a small factory that had produced shells for *katyushas* and other munitions during the war. On all sides were dense wooded territories. This made it possible for us to carry out the necessary explosions, since there was plenty of space well isolated from populated points.

Boris Vannikov had recommended the location, Khariton notes. "We immediately took a liking to the place. . . . Looking down from a high riverbank, Pavel Mikhailovich began planning the locations for our production buildings and the future town. I was struck by the ease with which he could do that. Most of the plans he indicated then were eventually realized."

Sarov, four hundred kilometers due east of Moscow, was the site of a famous monastery. The new Soviet state took over the monastery in the 1920s and used it to house war orphans, a mission celebrated in a popular Soviet motion picture. In the 1930s the monastery became a prison camp. After Khariton and Zernov selected it, a detachment of Beria's troops ringed it with a double fence of barbed wire and it disappeared from the map. The record is silent on the fate of its population of several thousand souls. At various times it would be known as the Volga Office, KB [i.e., Design Bureau]-11, Installation No. 558, Kremlev, Moscow Center 300, Arzamas-75 and Arzamas-16. To its scientific occupants it was always known simply as Sarov. Khariton and Zernov had found the Soviet Los Alamos.

13
Changing History

KLAUS FUCHS DID NOT RETURN to England in December 1945, as he had expected to do. With the decision to stage a test of the effect of atomic bombs on ships at Bikini and with so many American scientists leaving Los Alamos to return to teaching, Norris Bradbury asked the hardworking physicist to stay on into the spring to help with preparations. Harry Gold insisted later that he had no further contact with Fuchs after their September 19 meeting in Santa Fe, but Fuchs himself confessed in 1950 to "several further meetings with [a Russian agent] in Santa Fe in the autumn of 1945 and spring of 1946." If Gold was telling the truth—and he is unlikely to have lied at a time when he had already revealed espionage episodes of which he was deeply ashamed—then Fuchs must have been in contact with some other courier.

He never identified that contact. To the contrary, he maintained that "Raymond" was his sole cut-out during his years in the US. His denial is unconvincing. After his arrest in 1949, he denied recognizing Gold until Gold confessed. He steadfastly denied the telephone contact in New York that allowed Gold to reconnect with him in Cambridge after he moved to Los Alamos. He denied knowing "Sonia's" identity until after her escape to East Germany in 1950. Nor is it credible that Soviet intelligence would ignore Fuchs at the very time when the Soviet bomb program was expanding to full scale under Beria's new authority. If Beria was willing to risk revealing the existence of that program by pursuing Niels Bohr (and, in 1946, the German Nobel laureate physicist Werner Heisenberg*), why would his foreign intelligence *apparat* not further exploit the services of its best-placed and most cooperative spy? Some of the information on postwar developments within the American program that Fuchs claimed he passed after

---

* Heisenberg refused, saying, "The fox notices that many tracks lead into the cave of the bear, but that none come out." Quoted in Walker (1989), p. 184.

returning to England probably went to the Soviet Union from the United States between October 1945 and June 1946.

The likeliest courier in contact with Fuchs at Los Alamos after September 19, 1945, would have been Lona Cohen. Anatoli Yatzkov has claimed that Lona Cohen visited Albuquerque twice to contact "Perseus," who supposedly moved to Los Alamos during the war. Cohen herself admitted only to meeting "a physicist." Perseus may be mythical or a composite, and Cohen's adventures may well concern Klaus Fuchs. Yatzkov had good reason for disconnecting Harry Gold from Fuchs as well as from David Greenglass. Not only had Igor Gouzenko's defection seemed to put them all at great risk of discovery, but Gold, as Yatzkov had warned him during the summer, was dangerously linked to Abe Brothman, whom Yatzkov knew to be under active surveillance. If Yatzkov disconnected Gold from Fuchs after September 1945 because of Gouzenko's defection and the Brothman connection, the Soviet *rezident*'s instincts were shrewd; on May 13, 1946, ignoring Yatzkov's cautions, Gold blithely signed on to work with Abe Brothman in Brothman's small Long Island commercial laboratory.

Before then, Gold had ample occasion to notice Yatzkov/Yakovlev's discomfort. At a meeting between the two men late in 1945, says Gold, "Yakovlev ... told me that I should be very careful, much more careful than ever before.* He related to me an incident which had taken place [recently]. He said that a very important person who had upon him information on the atom bomb had come to New York ... and that he, Yakovlev, had tried to get in touch with that person over a period of time, a period of a few days, but that the man had been trailed by Intelligence men continually, so that Yakovlev had to give up the idea of getting in touch with this source of information." Who this "very important person" might have been has never been established, although Donald Maclean traveled to New York from time to time to deliver espionage materials and had participated in planning British Prime Minister Clement Attlee's visit to Washington in November to discuss postwar Anglo-American atomic policy. Yatzkov/Yakovlev, in the FBI's paraphrase of Gold's testimony, "was very touchy and very apprehensive ... during this meeting [with Gold]. ... [He] made a couple of other appointments with [Gold] at that time ... but ... did not keep any of these appointments."

In February 1946, the Royal Canadian Mounted Police arrested twenty-two Canadian citizens and residents suspected of espionage on the basis of Igor Gouzenko's documents and testimony. Among those arrested was Israel Halperin, in whose address book the Mounties had found Klaus Fuchs's intern-camp and Edinburgh addresses and Kristel Heineman's address in

---

* Gold does not explain why he and Yakovlev were meeting or what assignment he was supposed to be careful about.—RR

Massachusetts. Learning that Soviet GRU (military intelligence) had given Alan Nunn May a new contact in London, the RCMP allowed him to return to England undisturbed, where British authorities arrested him after several delays on March 4.

For reasons Gold never explained, and in violation of the unwritten regulations to which he was otherwise fastidiously devoted, he decided after missing his meeting with Yatzkov early in 1946 to look up Fuchs at Kristel Heineman's house in Cambridge, as Fuchs had told him he could do. He arrived on another snowy February morning. "On previous trips to Cambridge the only persons I had (variously) encountered there were Kristel Heineman . . . and her small children, a maid, a housekeeper and Klaus Fuchs himself. But this time, unluckily for me, the husband, Robert Heineman, was present, and along with him a friend, a Greek. . . ." The Greek, Konstantin Lafazanos, lived with the Heinemans; he was a professional graduate student, Robert's pal and (according to the FBI) Kristel's lover. The two men were home because of a power failure at Harvard. Gold stayed for lunch; Lafazanos recalled later that they talked about vitamins, about which Gold was knowledgeable. (In 1943, Gold and a friend had applied unsuccessfully to the Corn Exchange Bank in Philadelphia for a loan to start a vitamin assay laboratory, one of Gold's many unfulfilled dreams.) Gold told them he was a biochemist from Pittsburgh with a wife and two children. Though he never said so, he presumably learned during his visit to the Heinemans that Klaus Fuchs was still at Los Alamos.

Whether Fuchs passed information to Lona Cohen in New Mexico in the winter and spring of 1945–1946 as well as to another courier, as he confessed, "soon after he returned to the United Kingdom," much of what he passed added further detail to his previous communications about implosion, plutonium metallurgy, initiator design and bomb effects. He attended talks at Los Alamos on experimental data developed on levitated implosion and the composite core on March 11, 1946; on the possibility of thermonuclear reactions in water and air on March 12; on nuclear breeder and power reactors on March 21; on the processing of plutonium from nitrate to metal on April 1, all information of value to Soviet research. But information about the thermonuclear—Teller's superbomb—became available to Fuchs as well before he left Los Alamos in June 1946. Fuchs confessed to communicating some of what he learned to Harry Gold and to his postwar espionage contact in Britain. He is unlikely to have passed less than everything he knew.

The idea of a superbomb exploiting the fusion of light elements as well as the fission of heavy elements was a logical extension of basic ideas in nuclear physics known to physicists throughout the world. As Oppenheimer drummed into the Acheson-Lilienthal board of consultants (until Herbert Marks could recite it by heart), "Only in reactions of very light nuclei, and in reactions of the very heaviest, has there ever been, to the best of our knowledge, any large-scale release of atomic energy." In May 1941, University of Kyoto physicist Tokutaro Hagiwara, in a lecture on "Super-explosive U235," had commented that the fissionable uranium isotope "has a great possibility of becoming useful as the initiating matter for a quantity of hydrogen." Hagiwara was the first scientist on record to notice that an explosive fission chain reaction might generate enough energy to force hydrogen to fuse to helium, with the potential for producing a far larger nuclear explosion than fission could yield alone.

Ernest Rutherford and two of his younger colleagues at Cambridge, Marcus Oliphant and Paul Harteck, had discovered the hydrogen fusion reaction in 1934. In a paper titled *Transmutation effects observed with heavy hydrogen* they described bombarding hydrogen2—deuterium, in the form of concentrated heavy water—with deuterium-accelerated nuclei. (A hydrogen nucleus contains a single proton, making it the lightest of all elements. Deuterium is an isotope of hydrogen with a neutron in its nucleus as well and is therefore twice as heavy.) Acceleration gave the deuterium nuclei of the 1934 experiment enough energy to overcome the positive electrical repulsion between the nuclei of probe and target. The result, to the experimenters' surprise, was "an enormous effect," specifically "the union of two [deuterium nuclei] to form a new nucleus of...helium...." Driven into proximity by the energy of acceleration, which is essentially a form of heat, the deuterium nuclei had fused together to form the next-lightest element in the periodic table, helium, with two protons and one neutron in its nucleus. Neutrons, heat and intense gamma radiation came out of the reaction as well as the new nucleus adjusted its energy level and stabilized.

"This was another of that long catalogue of scientific papers which came before their time," writes historian David Irving. "In retrospect, [this] paper can be seen to have been of little less moment that Hahn and Strassmann's 1939 paper on the fission of the uranium nucleus." Because the fusion reaction depended on heating the nuclei until their thermal motion overcame their electrical repulsion, the reaction came to be called "thermonuclear fusion." It could be created a few nuclei at a time in particle accelerators such as Cambridge's Cockcroft-Walton generator or Berkeley's cyclotron. In 1938, Hans Bethe identified a sequence of thermonuclear reactions proceeding from hydrogen to carbon as the source of the energy that lit the sun and stars. But until a fission chain reaction became feasible, no one had imagined that a large-scale thermonuclear fusion reaction could be kindled on earth. ("In the center of an exploding fission bomb," notes theoretical physicist Herbert York, "temperatures substantially exceeding 100,000,000 degrees are produced, and so at least one of the conditions necessary for igniting a thermonuclear reaction under the control of man seemed to be within reach.")

If Hagiwara was the first, his insight fell on fallow ground—Japan in wartime lacked the resources even to develop an atomic bomb, much less to explore a thermonuclear. But the same idea occurred to Enrico Fermi at Columbia University and he passed it along to young Edward Teller in September 1941. Fermi wondered if an atomic bomb might serve to heat a mass of deuterium sufficiently to kindle a full-scale thermonuclear reaction. If so, then cheap deuterium distilled from seawater could be added to a critical mass of expensive U235 or plutonium. Each gram of deuterium converted to helium should release energy equivalent to about 150 tons of TNT, 100 million times as much as a gram of ordinary chemical explosive and eight times as much as a gram of U235; theoretically, twelve kilograms of liquid deuterium ignited by one atomic bomb would explode with a force equivalent to one million tons of TNT—one megaton; a cubic meter of liquid deuterium would yield ten megatons. Teller made the realization of Fermi's idea the focus of his life.

At a secret seminar on atomic-bomb development that Robert Oppenheimer chaired and Teller, Bethe, Robert Serber and other theoretical physicists attended at Berkeley in the summer of 1942, the possibility of a thermonuclear bomb was discussed at length. During that discussion one of the participants, a young theoretician from Indiana University named Emil Konopinski, suggested mixing another isotope of hydrogen, radioactive hydrogen3, tritium (one proton, two neutrons), into the thermonuclear fuel. Tritium (T) was much rarer than deuterium (D) but because of its nuclear characteristics ought to kindle thermonuclear reactions at a far lower ignition temperature, 40 million degrees rather than 400 million. The cross section for fusion (a measure of probability) of $D+T$ turned out to be one hundred times greater than the cross section for fusion of $D+D$. Teller began systematic theoretical studies of the thermonuclear at Los Alamos in autumn 1943, devoting his full time to the project with Oppenheimer's approval in the last year of the war. Among others in his group, Teller signed on a Polish mathematician named Stanislaw Ulam, whom John von Neumann had recommended, to help with the work; on Ulam's first day on the job, late in 1943, Teller asked him to study the exchange of energy between free electrons and radiation in a hot gas, one process that might cool a fusion reaction sufficiently to prevent it from propagating. Oppenheimer arranged for a small quantity of tritium to be bred in an Oak Ridge reactor for cross-section measurements and other research. Manhattan Project heavy-water production was intended for thermonuclear studies as well as for reactor research.

Since kindling a thermonuclear explosion required setting off an atomic bomb, its progress could not be studied in the laboratory, as fission had been studied with tabletop near-critical assemblies and at full criticality with diluted cores in the notorious Dragon critical-mass experiments. Instead of real experiments, Teller and his group had to depend on mathematical calculations of unprecedented complexity which nevertheless greatly oversimplified the phenomena they modeled. To establish the initial conditions for the thermonuclear explosion, Los Alamos needed to understand the fission explosion that preceded it in great detail: the behavior of the immense flux of neutrons which the fission explosion produced (the neutronics), of the immense flux of heat released (the thermodynamics) and of the fluid flow of particles and radiation released in the explosion (the hydrodynamics). These fission calculations had been started by Richard Feynman and others during the war as part of the work of implosion research, using IBM punch-card machines to automate the thousands of necessary repetitions. The calculations were repetitive because they followed the histories of dozens or hundreds or thousands of individual particles through cross-sectional slices of time as the explosion bloomed—like catching the successive positions of a hall full of dancers with the quick pulses of a strobe.

Understanding the fission explosion was only the first step in thermonuclear explosion calculations, however, and by the end of the war even that step had been advanced only tentatively and crudely, for a small sample of particles through relatively thick slices of time. Thermonuclear calculations added significantly higher levels of complexity, Stanislaw Ulam writes:

> All the questions of behavior of the material as it heated and expanded—the changing time rate of the reaction; the hydrodynamics of the motion of the material; and the interaction with the radiation field, which "energywise" would be of perhaps equal importance to that of the thermal content of the expanding mass—had to be formulated and calculated. . . .
>
> To realize . . . the magnitude of the problems involved, one should remember that, even only mathematically, the problem of the start and explosion of a mass of deuterium combined a considerable number of separate problems. Each of these was of great difficulty in itself, and they were all strongly interconnected. The "chemistry" of the reaction, i.e., the production, by fusion, of new elements not originally present and the appearance of tritium, [the helium isotopes] He_3, He_4, and other nuclei, together with the increasing density of neutrons of varied and variable energies in this "gas" influences directly the changing rate of the reaction; and it does so also by changing the values of the density and temperature. Simultaneously, the radiation field is increasingly present and influences, in its turn, the motion of the material.

"The work of the 'Super' group on the visualization and on quantitative following of these processes," Ulam concludes, "constituted a veritable monument to the imagination and skill of theoreticians. . . ."

It had been possible to develop a crude mathematical model of implosion during the war using desktop mechanical calculating machines and IBM punch-card sorters. The Super calculations that Teller's group needed to do exceeded the capabilities of such machines. Here John von Neumann, one of the great mathematicians of the twentieth century, intervened creatively. Von Neumann, a prodigy who could recite whole chapters verbatim of books that he had read only once and a lightning mental calculator, had taken up theoretical physics as a sideline before the Second World War and had made himself an expert on shock and detonation waves. "The story used to be told about him at Princeton," writes a colleague, Herman Goldstine, "that while he was indeed a demi-god he had made a detailed study of humans and could imitate them perfectly. Actually he had great social presence, a very warm, human personality, and a wonderful sense of humor." It was von Neumann who had calculated the complex shape of the high-explosive lenses in the Fat Man bomb.

In 1944, Goldstine and a small group of engineers at the University of Pennsylvania's Moore School of Engineering had been building a new type of calculating machine with government funding that used vacuum tubes rather than gears to run calculations. They called it the ENIAC, an acronym that summarized its functions as an electronic numerical integrator and computer. "Sometime in the summer," Goldstine remembers, ". . . I was waiting for a train to Philadelphia on the railroad platform in Aberdeen [Maryland, the location of the US Army's Aberdeen Proving Ground] when along came von Neumann." The Penn mathematician had never met his legendary colleague. "It was therefore with considerable temerity that I approached this world-famous figure, introduced myself and started talking. . . . The conversation soon turned to my work. When it became clear to von Neumann that I was concerned with the development of an electronic computer capable of 333 multiplications per second, the whole atmosphere of our conversation changed from one of relaxed good humor to one more like the oral examination for the doctor's degree in mathematics. Soon thereafter the two of us went to Philadelphia so that von Neumann could see the ENIAC."

It was just what von Neumann and Los Alamos had been looking for. The Hungarian-born mathematician embraced the machine and the concept of the machine, and soon abstracted from its crude vacuum-tube technology a logical system for manipulating and processing information, mathematical or otherwise. Goldstine believes von Neumann's 101-page draft report, written that final winter and spring of the war, was "the most important document ever written on computing and computers." The ENIAC as the Moore School group had designed it had to be prepared for each new problem by physically rearranging its circuit wires, plugging and unplugging what looked like old-fashioned telephone switchboards. In his draft report, von

Neumann formulated for the first time the idea of a stored operating program—and defined in the process the basic organization of the digital computer: "The logical control of the device, that is, the proper sequencing of its operations, can be most efficiently carried out by a central control organ. If the device is to be ... *all purpose,* then a distinction must be made between the specific instructions given for ... a particular problem, and the general control organs which see to it that these instructions—no matter what they are—are carried out."

The first problem assigned to the first working electronic digital computer in the world was the hydrogen bomb. Los Alamos mathematician Nicholas Metropolis (writing in the third person) recalled participating in the breakthrough:

> In early 1945, as the construction of the ENIAC was nearing completion, von Neumann raised the question with [physicist Stanley] Frankel and Metropolis of using it to perform the very complex calculations involved in hydrogen bomb design. The response was immediate and enthusiastic. Arrangements were made by von Neumann on the basis that the "Los Alamos problem" would provide a much more severe challenge to the ENIAC on its shakedown trial. . . .

The ENIAC ran a first rough version of the thermonuclear calculations for six weeks in December 1945 and January 1946. Los Alamos prepared a half million punched cards of data, enough to keep a hundred people busy for a year at mechanical desktop machines.

The outcome appeared promising, writes Ulam:

> It seemed at that time that the feasibility of the thermonuclear bomb was established, according to the opinion of the author [i.e., Ulam]. Even though the work was of necessity incomplete, and had to omit certain physical effects, the results of the calculations had great importance in leaving open the hopes for a successful solution to the problem and the eventual construction of an H-bomb. One could hardly exaggerate the psychological importance of this work and the influence of these results on Teller himself and on people in the Los Alamos laboratory in general. . . . I well remember the spirit of exploration and of belief in the possibility of getting trustworthy answers in the future. This [was] partly because of the existence of computing machines which could perform much more detailed analysis and modeling of physical problems.

Los Alamos published a *Super Handbook* on October 5, 1945, collecting together technical data and computations, and three days later issued a further technical review of the Super program that recommended investigating producing tritium in quantity. In December 1945, Teller filed a disclosure of invention for a related device, a "boosted" atomic bomb that would use fusion neutrons generated from a small quantity of deuterium and tritium gas confined in the core of an implosion system to accelerate the fission chain reaction and increase the explosive yield.

To review wartime work on the Super and to propose a course of further studies, Los Alamos scheduled a secret three-day conference for April 18–20, 1946. Just before the conference began, the laboratory issued its first major technical report on the thermonuclear, *Prima facie proof of the feasibility of the Super*. ("Prima facie"—"at first sight"—means upon examination of the existing evidence, without further investigation.) The fifty-nine-page report, the work of Teller and six of his Los Alamos colleagues, asserted that "present knowledge of the physics of the Super is sufficient to indicate with reasonable certainty that an operable Super model can be made" and that "a large-scale theoretical and experimental program for the development of a thermonuclear bomb is justified" and proposed undertaking the production of tritium "concomitant with this program. . . ."

Along with Teller, Konopinski, Philip Morrison, von Neumann, Canadian theoretician J. Carson Mark, Metropolis, Robert Serber, Ulam and twenty-three other scientists, Klaus Fuchs attended the April Super Conference. On the first morning, in Norris Bradbury's office, the group heard Edward Teller review the prima facie arguments of the April 15 technical report and then describe his proposed design. It came to be called the "Super" and the "classical Super" to distinguish it from the booster and other, later designs. Its configuration has never been made public in detail, but Carson Mark, who took over direction of the theoretical division at Los Alamos after the war, outlined it in an interview:

> The classical Super was the idea that deuterium could be set burning if you got it hot enough and [that] perhaps . . . a fission bomb might provide the sort of temperature level that you would need. So we have long pipe full of liquid deuterium and we have a fission bomb which we set off at one end of it with the idea that we will heat that end sufficiently that a burning wave will get started and proceed along the pipe. The burning wave being a deuterium reaction. Now there is the classical Super. There are some answers that have to be filled in. Amongst them, if you heat one end of a deuterium pipe like that, will a burning wave in fact run along it that will detonate like a stick of high explosives? That's a central question. Of course you would have to ask, how hot do I have to make it before that will happen, but even if I make it very hot, [you would have to ask] *will* that happen?

An important difference between a fission bomb and a thermonuclear bomb was that except in its fission trigger, the thermonuclear would require no critical mass. As a fission bomb exploded, it disassembled its critical mass, at which point fissioning stopped. This disassembly process set a natural limit to the size of fission explosions of about one megaton. A thermonuclear explosion, however, if it could be made to ignite and sustain thermonuclear burning, would proceed like a nuclear version of a chemical explosion, continuing to burn so long as it had access to thermonuclear fuel. The stars—thermonuclear furnaces thousands and millions of times as large as the earth—made it obvious that there were no inherent physical limits to the size of thermonuclear explosions. The report of the conference just then getting underway would emphasize that the Super's "scale is limited—if it proceeds at all—only by the amount of deuterium fuel provided. Thermonuclear explosions can be foreseen which are not to be compared with the effects of the fission bomb, so much as to natural events like the eruption of [the] Krakatoa [volcano, in the Sunda Strait between Java and Sumatra in 1883]. . . . Values like [the energy released in] the San Francisco earthquake may be easily attained." During the war, Serber remembers, "on Edward Teller's blackboard at Los Alamos I once saw a list of weapons— ideas for weapons—with their abilities and properties displayed. For the last one on the list, the largest, the method of delivery was listed as 'Backyard.' Since that particular design would probably kill everyone on earth, there was no use carting it elsewhere."

At the afternoon session on the first day of the Super Conference, a member of Teller's group discussed a tamper of beryllium oxide that was a feature of the fission trigger of the Super design (beryllium is fertile with loose neutrons that can be mobilized to enhance a fission chain reaction; most modern nuclear weapons have beryllium tampers). Cylindrical implosion was also discussed, probably as a way of incorporating the more efficient implosion mechanism into the pipelike configuration of the Super. According to Los Alamos records, "Dr. von Neumann suggested the ignition of a 'Super' bomb through the employment of an implosion process. . . ." Fuchs would claim—"laughingly," an FBI interrogator notes—that it was in fact *his* idea to use implosion to ignite the Super. Apparently he and von Neumann developed the idea jointly; they filed for a patent together on May 28, 1946.

The next day, April 19, mathematicians Nicholas Metropolis and Anthony Turkevich reviewed their ENIAC calculations of Super hydrodynamics. The terms of the calculations had been simplified from three dimensions to two and excluded significant physical processes that tended to cool the reaction by bleeding off energy. Under those idealized conditions, Metropolis and Turkevich reported, a Super fueled with deuterium would produce a substantial energy release. (Bethe and Teller had debated just such problems at the 1942 Berkeley summer conference, Serber remembers: "Edward first thought [the hydrogen bomb] was a cinch. Bethe, playing his usual role, knocked [Teller's idea] to pieces. Edward had figured the energy that would be released, how hot it would heat the gas, and so forth. Everything looked fine until Bethe pointed out that you would get radiation [from such intense heating]; you had to be in equilibrium with the black-body radiation, which goes up with the fourth power with temperature, drains the heat right off, and cools everything down. You start feeding [the fire] and bingo, everything goes into electromagnetic radiation. Edward hadn't allowed for that. Bethe thought of a mechanism that really drained the energy off fast—we called it the inverse Compton effect—that knocked Edward's calculations into a cocked hat, and they never actually recovered." The inverse Compton effect was one of the significant physical processes excluded from the optimistic calculations that Metropolis and Turkevich reported.)

That afternoon and the following morning, another participant discussed the compression properties of deuterium and deuterium-tritium mixtures. Teller wrote later that the idea of compressing the hydrogen fuels with radiation was discussed at the Super Conference. (A fission bomb produces radiation—light, that is, primarily soft X rays—in such copious quantities that it is capable of significantly compressing matter.) It was not obvious how such a mechanism might work, Bradbury would recall. "I can remember very vividly in 1946 exploring in great detail this original idea with members of my technical staff. At the time we saw absolutely no way to make it into a usable system."

Teller proposed an experimental program that second afternoon and discussed studying the various fusion reactions a Super might kindle—"tritium plus tritium, helium plus deuterium, hydrogen plus deuterium, and the like. . . ." As he had been in 1942 despite the wide safety factors that Bethe's cooling effects established, Teller was still concerned about the possibility of a fusion explosion igniting the atmosphere of the earth, which is predominantly nitrogen, in a thermonuclear Armageddon; he thought the "nitrogen plus nitrogen reaction" should certainly be studied.

At a final meeting on the morning of April 20, Teller opened the floor to a discussion of peaceful applications of thermonuclear fusion. Four years later, Fuchs could not even remember what the meeting had been about.

Teller and his thermonuclear group drafted a conference report that he circulated among the conferees in May 1946. It observed that deuterium was a more effective explosive than U235 or plutonium, yielding energy "several times that from the same weight of fissionable material," while its unit cost, estimated at twenty cents per gram, was comparable not with U235 and plutonium, which cost several hundred dollars per gram, but with ordinary uranium. The report noted that hand calculations done in parallel with the ENIAC problem "indicated that . . . the system would ignite." But whether, once ignited, thermonuclear burning would continue and propagate, the calculations had not "conclusively demonstrated."

1. Only days after the end of the Second World War, Los Alamos physicist Robert Serber, who designed the Hiroshima bomb, studied bomb effects in Hiroshima and Nagasaki. In Tokyo he posed with a Japanese fire pump.

2. Physicists Luis Alvarez and Harold Agnew flew the Hiroshima mission. "This terrible weapon . . . may prevent further wars," Alvarez wrote home prophetically.

3. The Air Force disagreed. Generals Curtis LeMay, Emmett "Rosie" O'Donnell, Jr., and Barney Giles flew three B-29s nonstop from Japan to Chicago to demonstrate intercontinental strategic air power. (Second from right: US Army Air Forces commanding general Henry H. "Hap" Arnold.) The war was over; the troops were coming home.

4. Lavrenti Beria, secret police head and gulag master, with Soviet dictator Josef Stalin and Stalin's daughter, Svetlana. After the war, Stalin charged Beria to develop a Soviet atomic arsenal.

5. Young Soviet physicist Georgi Flerov pushed for atomic-bomb development early in the war, but the beleaguered nation had no resources to spare.

6. Nuclear physicist Igor Kurchatov as a student at Baku, 1924. Kurchatov assumed scientific leadership of the Soviet bomb program in 1943.

7. In a late-1942 report to Kurchatov, Flerov proposed assembling two hemispheres of U235 by firing one into the other. Espionage from America would deliver a better idea.

8-9. British diplomats Guy Burgess and Donald Maclean passed high-level Anglo-American secrets throughout the war and into the early Cold War. They defected to the Soviet Union in 1951.

8   9   10   11

10-11. American spies Morris and Lona Cohen reported the US atomic-bomb program to the Soviets as early as 1941.

12. Washington overruled Lend-Lease dispatcher George Racey Jordan when he challenged massive Soviet espionage shipments. Red Army officer Anatoli Kotikov arranged a promotion.

13. German Communist "Sonia" built her own shortwave radio to pass secrets from Britain. Her most valuable source was trusted physicist Klaus Fuchs.

14. The Soviet Union fought for its life against the devastating 1941 German invasion. As many as thirty million Soviet citizens died in the war—ten million at the hands of Beria's brutal NKVD.

15

16

15. Spy Klaus Fuchs moved to Los Alamos in 1944 and mastered bomb design.

16. NKVD New York *rezident* Anatoli Yatzkov directed a major spy network.

17. American chemist Harry Gold traveled to New Mexico in 1945 to collect bomb data from Fuchs and David Greenglass. Julius Rosenberg had recruited Greenglass.

17

18

19

18-19. David and Ruth Greenglass, 1949.

20-21. Julius and Ethel Rosenberg, 1950.

20

21

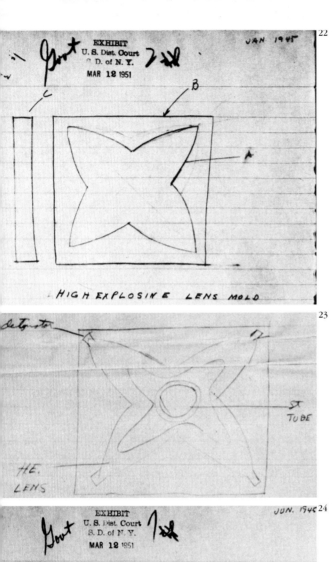

EXHIBIT
U. S. Dist. Court
D. of N. Y.
MAR 12 1951

JAN 1945

B

C

A

HIGH EXPLOSIVE LENS MOLD

detonator

ST
TUBE

H.E.
LENS

EXHIBIT
U. S. Dist. Court
S. D. of N. Y.
MAR 12 1951

JUN. 1945

C
B

D

E

F

22–24. David Greenglass gave Gold drawings of a high-explosive lens mold, a lens arrangement and a two-dimensional implosion experiment at their meeting in summer 1945. In 1950, he duplicated the sketches for the FBI.

25. Implosion—squeezing a nuclear core to supercriticality with high explosives—was the crucial secret necessary to make an atomic bomb with plutonium. Both Fuchs and Greenglass independently passed the secret to the Soviets.

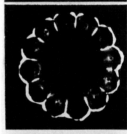

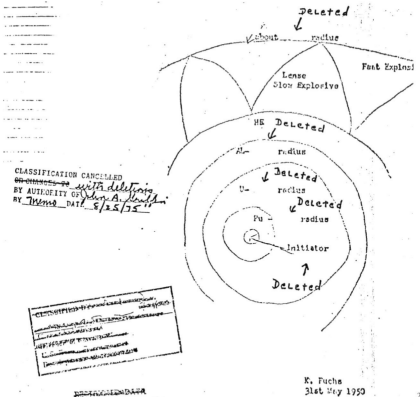

26-27. Meeting Harry Gold in Santa Fe in June 1945, Fuchs gave the Soviets a precise drawing and measurements of the Fat Man bomb, here partly disassembled to show blocks of high explosive and heavy uranium tamper. Fuchs sketched his drawing for the FBI in 1950, which deleted classified measurements.

28. "Compiled for the oral orientation of Academician Kurchatov": actual Soviet intelligence briefing document describing the Fat Man bomb, dated July 2, 1945.

28

29

29. After a successful July test of the implosion design, a Fat Man was readied in August on Tinian Island in the Marianas for Nagasaki.

30. Winston Churchill, Harry S. Truman and Josef Stalin at Potsdam in late July. Truman told Stalin about a "new weapon of unusual destructive force." Stalin already knew.

30

31-33. The first mushroom cloud over Japan marked a transition to what Niels Bohr would call "a completely new situation that cannot be resolved by war." The *Enola Gay* returned to the cheers of men who no longer had to fear being killed in an invasion. But there was sickening devastation at Hiroshima—a first look at atomic war.

34

THE BRITISH MISSION

INVITES YOU TO A PARTY IN CELEBRATION OF

THE BIRTH OF THE ATOMIC ERA

FULLER LODGE

SATURDAY, 22ND SEPTEMBER, 1945

DANCING, ENTERTAINMENT,
PRECEDED BY SUPPER AT 8 P. M.

Mr & Mrs C. Critchfield

R.S.V.P. TO MRS. W. F. MOON
ROOM A-211 (EXTENSION 250)

34. Fuchs arranged to pick up liquor for the British Mission party at Los Alamos as an excuse to meet Gold a second time in Santa Fe in September 1945.

35. Beria sent a Keystone Kops delegation to Copenhagen in autumn 1945 to pry secrets from Niels Bohr. Bohr shared only information the US had already published.

36. The bomb Fuchs gave away: the Fat Man test device.

37. The Soviet bomb program went
into high gear after the war under
Igor Kurchatov's vigorous leadership. His colleagues called him "the
Beard."

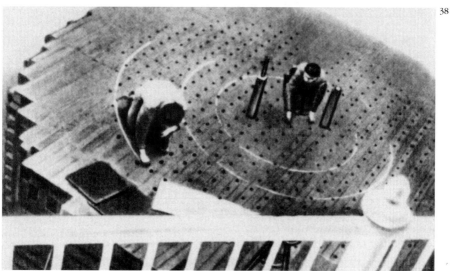

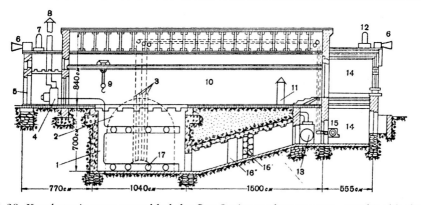

38–39. Kurchatov's team assembled the first Soviet nuclear reactor—graphite blocks
set with uranium slugs—partly underground in a special building in Moscow (code-
named "Assembly Workshops") in early winter 1946. Its stolen design matched the
US 305 reactor built at Hanford, Washington, in 1944.

40-41. Scientific Director Yuli Khariton established the Soviet Los Alamos at Sarov (Arzamas-16), east of Moscow. A famous old monastery supplied a base of buildings.

41

40

43

44

45

42

42-43. Under Khariton, experimental physicists Veniamin Zukerman and Lev Altshuler led teams at Sarov studying implosion.

44-45. Fuchs also passed information on early US hydrogen-bomb research. Soviet physicist Yakov Zeldovich (above) began work on H-bomb design in 1946; young Andrei Sakharov joined the work in 1948.

46

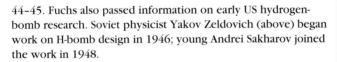

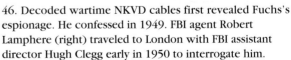

46. Decoded wartime NKVD cables first revealed Fuchs's espionage. He confessed in 1949. FBI agent Robert Lamphere (right) traveled to London with FBI assistant director Hugh Clegg early in 1950 to interrogate him.

47. Opposite page: Joe 1, the first Soviet atomic bomb, tested in Kazakhstan on August 29, 1949, was a carbon copy of the US Fat Man. Khariton had a better design on the drawing boards, but Beria took no chances.

48. Andrei Sakharov (here with Igor Kurchatov) invented a limited-yield H-bomb in 1948, a "layer-cake" nest of concentric shells of uranium and hydrogen. After the Joe 1 test in 1949, Beria pursued it.

49

49. A mock-up of Joe 4, the Soviet layer-cake thermonuclear. The demon aboard for the ride is a doorstop from the Kurchatov house in Moscow.

50. Joe 4 (left), forty-kiloton Joe 2 and Joe 1 models in the bomb museum at Sarov, 1993. Standing between the models (left to right): physicists Yuri Smirnov, Yuli Khariton and Victor Adamsky. Portraits of Soviet nuclear-weapons-program heroes line the wall.

50

51. Joe 4, August 12, 1953. The single-stage thermonuclear device, which contained tritium as well as U235, U238 and lithium deuteride, yielded four hundred kilotons, 15 to 20 percent from fusion. By then the US had a bigger atomic bomb and megaton hydrogen bombs. Despite the fears of H-bomb enthusiasts, the US was never "behind."

52. The Soviet Union tested a staged, megaton-range hydrogen bomb November 22, 1955, three years after the first such US test.

53. Soviet SS-4 missiles with three-megaton warheads menaced the US from Cuba during the Cuban missile crisis. They could have destroyed Washington.

The report found in conclusion:

It is likely that a super-bomb can be constructed and will work.

Definite proof of this can hardly ever be expected and a final decision can be made only by a test of the completely assembled super-bomb....

The detailed design submitted to the conference was judged on the whole workable. In a few points doubts have arisen concerning certain components of this design.... In each case, it was seen that should the doubts prove well-founded, simple modifications of the design will render the model feasible.

In a final paragraph, the report noted that undertaking the "Super Bomb Project" would use up a fair portion of the national nuclear-weapons budget for some years to come. It followed, the report argued, "that further decision in a matter so filled with the most serious implications as is this one can properly be taken only as part of the highest national policy." It did recommend that tritium production should be organized, starting at a gram per day, to be increased as development required.

Serber remembers complaining that the draft he saw of the Super Conference report was far too optimistic. He thinks others may have complained as well. (Ulam comments that "the promising features of the plan were noticed and to some extent confirmed, but there remained great questions about initiation of the process and, once initiated, about its successful continuation.") Serber recommended changes that he assumed would be incorporated into the final document. When the report was issued as quoted here on June 12, though it claimed to represent "essentially the unanimous opinions of those attending the conference," Teller had left it virtually unchanged from the first optimistic draft.

The Super Conference report did not initiate a national policy discussion in the United States. At the time the report was issued, weapons as destructive as earthquakes and volcanoes were not on the national agenda. No military requirement for such a weapon would be established for years to come. Bernard Baruch was putting the finishing touches on his sanction-heavy plan for the international control of atomic energy, which he would present to the United Nations on June 14 (beginning melodramatically, "We are here to make a choice between the quick and the dead"). Los Alamos was busy with the Bikini tests soon to commence, needed to improve what Bradbury called its "lousy" fission bombs—among other reasons, to develop one hot enough to serve as a thermonuclear blasting cap—and was still struggling to survive. The report gathered together the work that had been accomplished on a thermonuclear up to that point; for the next several years, Carson Mark's theoreticians would devote about half their time to thermonuclear calculations while the rest of the laboratory worked on improving atomic bombs.

Sometime in 1946, however, Soviet physicists Isai I. Gurevich, Yakov Zeldovich, Isaak Pomeranchuk and Yuli Khariton prepared a special report for the Soviet government titled *Utilization of the nuclear energy of the light elements.* Gurevich implied long afterward that the report was offered unsolicited: "I think that at that time they simply waved us away. Stalin and Beria laid extreme emphasis on the creation of an atomic bomb. Moreover at that time we had not yet commissioned an experimental reactor, and here the scientific 'wise guys' are pestering us with new projects, while it is still not known whether they could be realized. . . ." The timing of the Gurevich et al. report argues against spontaneous generation. So does a Kurchatov espionage gloss dated December 31, 1946, in which Kurchatov mentions reviewing "an American work on the super-bomb" which he believed to be "probably true and of great interest for our work in this country." Igor Golovin says Fuchs gave the Soviet Union Teller's Super concept. Khariton confirms that "Zeldovich looked at the plans, which were Teller's work."

In their brief presentation, Gurevich and his colleagues explained why light elements might serve for nuclear explosives and reviewed some of the problems that would have to be solved to make such explosives work. They did not propose using tritium, probably because the design they had in mind did not require that exotic material. Tritium, which is radioactive and has a half-life of only 12.5 years, does not exist in nature; like plutonium, it must be bred in a nuclear reactor, a machine the Soviet scientists had yet to build.

Gurevich and his colleagues outlined a design for a thermonuclear explosive different from the Super, a design which was a logical extension of the implosion system which Fuchs had passed in 1945: "In order to improve the conditions of ignition it appears possible to use uranium charges of increased sizes and of a special shape (cumulation [i.e., concentric shells]) and to introduce into the deuterium heavy elements near the initiator which might be capable of receiving the radiation pulse." They also noted that "the greatest possible density of deuterium is desirable, and this should be realized by using it under high pressure" and proposed employing "massive enveloping shells [to] delay . . . the dispersal [of the material]." Teller invented a similar spherical, layered design in mid-1946 which he called the "Alarm Clock" (because it might "wake up the world" to the possibility of a new generation of nuclear explosives). A committee of Los Alamos senior scientists, including Bradbury, Teller and John Manley, judged in 1950 that Fuchs "was probably familiar with [the] 'booster' idea and may have obtained vague information on the very early proposals of the 'alarm clock' from another member of the British Mission."

Soviet physicist Yuri A. Romanov remembers that "research on the [thermonuclear] problem was begun" as a consequence of the 1946 Gurevich report, suggesting that Gurevich's claim that Stalin and Beria "waved us away" is at least ingenuous. "A small group of associates . . . under Zeldo-vich's direction was soon formed at the Chemical Physics Institute of the USSR Academy of Sciences. . . ." The United States had begun work on a thermonuclear weapon in 1942. Sometime in 1946, the Soviet Union joined in the quest for that unholy grail. Romanov notes that "the status of their research [at that point] was about identical." It should have been, since Klaus Fuchs had shared American progress with Soviet intelligence.

Lavrenti Beria had his hands full with the Soviet bomb program. He was flying blind and it made him snappish. His deputy Avrami Zavenyagin re-membered being lashed one day by Stalin's whip:

> We had received a telegram from Czechoslovakia, reporting that the pro-gram for uranium extraction . . . would need many funds. Some obviously incorrect, astronomical figures were quoted. Reading the telegram, Beria became indignant and began to swear. I couldn't stand it and I replied, Enough [theatrics], we are executing a decision of the government by which we are charged to make an agreement with the Czechs on a program for extraction. These stupid jacked-up figures named here are incorrect, we will correct them. Again swearing: "Oh, we have a hero." I reply, No hero, and no fool, you've no business duping people. "Get lost." I had to get lost. Then Beria softened up and tried to smooth things over.

Beria soon had another reason for apprehension. Stalin moved to diffuse his power. "One day Stalin suddenly asked Beria why all his generals and security staff seemed to be Georgians," writes historian Robert Conquest. "Beria answered that they were devoted and loyal. Stalin said angrily that not only Georgians but also Russians could be loyal." In March 1946, the aging Soviet dictator made Beria a full member of the Politburo and deputy chairman of the Council of Ministers. He raised the NKVD and NKGB from commissariats to ministries—they became the Ministry of Internal Affairs (MVD) and the Ministry of State Security (MGB)—but he removed Beria from direct authority over them and replaced Beria's men with Russians. Stalin's mistrust made Beria's successful delivery of an atomic bomb at the earliest possible date that much more imperative.

Living conditions worsened in the Soviet Union in 1946. A drought began in Moldavia, east of Romania on the Black Sea, at the end of March and spread north through the Ukraine. By the middle of May, according to a Soviet government report, "the drought . . . embrac[ed] a very considerable part of the whole of the European part of the USSR." The country had not seen such widespread and prolonged drought since 1891. Famine followed. Children's bellies bloated with protein deficiency and there was widespread starvation. The United Nations Relief and Rehabilitation Administration (UNRRA), headed by the former mayor of New York, Fiorello La Guardia, distributed food aid in the Ukraine that year. When Stalin went south on vacation that summer for the first time since 1937, his daughter writes, "the housekeeper . . . told me . . . how upset he was when he saw that people were still living in dugouts and that everything was still in ruins." His answer was to retreat even further into the Kremlin.

The US and Britain refused to support an extension of the UNRRA past 1946, preferring direct charity to countries the former Allies favored. La Guardia was shocked. "Does the government of the United States," he countered, "intend to adopt a policy which will make innocent men and women suffer because of the political situation which makes their government unacceptable to the United States?" It did, another sign that the Cold War was polarizing the world.

In June 1946, the Soviet Council of Ministers established a new industrial entity, the First Chief Directorate, headed by Boris Vannikov, to manage the atomic-bomb program. Vannikov reported directly to Beria; Mikhail Pervukhin and Igor Kurchatov became his deputies, as they had been under the Technical-Scientific Council established the autumn before. Vannikov had been Beria's prisoner in the dread Lubyanka cellar in October 1941, one of a number of senior military officers arrested in the purges. "Merciless beatings were administered," Beria would admit in 1953; "it was a real meat grinder." The NKVD executed many of the officers when it evacuated Moscow in the general exodus of October 15–16. Vannikov, one of the two most senior officers in the cellar, was spared. Stalin apparently realized that he needed the officers who were left to run his armies and ordered the interrogations ended. Vannikov was rehabilitated to serve as commissar of the munitions industry during the war.

"Plants were going up not by the day but by the hour," remembers Pervukhin of that early postwar period when the bomb program got underway. "Enterprises for the extraction of uranium were created. Production of slugs of metallic uranium was organized—these slugs were needed for the nuclear reactors." In 1946, the CIA estimated five years later, the Soviet Union invested 270 million rubles in atomic-bomb development.

Spies no less than other veterans had to resettle their lives now that the war was behind them. David Greenglass, despite Julius Rosenberg's encouragement, rejected a Los Alamos solicitation to work on the atomic tests at Bikini because he did not want to be separated from his wife. He was honorably discharged from the army on February 29, 1946. He and Ruth moved back to Manhattan; in April he began working with Rosenberg in the business his brother-in-law had promised to set up. When Greenglass applied for telephone service that spring, he listed his occupation as a machinist with the G & R Engineering Company of 300 East 2nd Street—"G & R" presumably standing for Greenglass and Rosenberg. Greenglass knew enough about his brother-in-law's espionage activities to feel comfortable shaking him down from time to time. "I got money from Julius whenever," he would confess. ". . . Julius had money, I went to Julius, [I said] 'Look, I need money' and he would give me money." In the next three years Greenglass collected "about a thousand dollars all told." According to Greenglass, Rosenberg encouraged him to think about going to college, perhaps to train as a scientist who might work with the scientists he had met at Los Alamos. "He wanted me to go to school full time . . . and be obligated to him. My wife and I had discussed that a number of times and we agreed to stall." Ruth Greenglass claimed later that she wanted to do more than stall: "I told my husband in 1946 that I wanted to go to the FBI with the story [of their involvement in espionage]. However, there had been nothing happening, everything was very peaceful, and we thought perhaps it would die down and the thing would never come to light, so we did nothing about it."

Klaus Fuchs's last act before he left Los Alamos was to review every paper in the Los Alamos document archives on thermonuclear weapons design. Thus fortified, he departed Los Alamos on June 14, 1946, traveling first to Washington, then to Cambridge to visit Kristel, then to Cornell with his sister to talk physics with Hans Bethe. (Fuchs took his sister along to Ithaca, the FBI paraphrased him later, "to give her a little outing"; they flew part of the way "in order that she might have this [new] experience.") He traveled on to Montreal alone and left Montreal for England in a British bomber on June 28. The British flew rather than shipped Fuchs home because they wanted him promptly at Harwell. A former Royal Air Force Base located south of Oxford on the Berkshire Downs, Harwell was being reactivated as a research center; the British were preparing secretly to build their own atomic bomb, and what Fuchs knew was valuable to them. He thus became a vector for nuclear proliferation to England as well as the Soviet Union. "He is the only physicist I know who truly changed history," Hans Bethe comments. Fuchs gave his first scientific paper at Harwell (on fast reactors, which were essentially slow bombs and were valuable for critical-mass and chain-reaction studies) in August 1946.

In 1947, Moscow Center connected Fuchs with a London-based Soviet agent, Alexander Semonovich Feklisov, who linked up with the physicist one evening at the Nags Head pub in North London. "At his own initiative," Feklisov wrote in retirement, "Fuchs brought important materials on the technology of plutonium production [to that first meeting] which he had failed to acquire in the United States." Fuchs's espionage continued at Harwell throughout the decade. "In 1947–1949," writes Feklisov, "I met with Fuchs once every three or four months. All the meetings were carefully prepared; Moscow approved each plan." The questions Feklisov asked Fuchs from quarter to quarter enabled the physicist to estimate the progress of the Soviet bomb program. At one meeting Fuchs turned the tables and queried Feklisov, "Is it true that your 'baby' will be born soon?" The nonplussed Soviet agent denied knowing. "I do see that my Soviet colleagues are proceeding," Feklisov says Fuchs explained. "No one among the American and British scientists expects the Soviet Union to build its gadget for years to come." Feklisov thought Fuchs sounded joyous.

Abe Brothman operated his commercial chemistry laboratory as a sideline; he hired Harry Gold in May 1946 to supervise the work of the laboratory staff. He was trying to develop commercial processes that he could sell. He and Gold approached the Soviet purchasing agency Amtorg that spring with a legitimate offer to design a synthetic vitamin plant. The two chemists had developed what Gold calls "a scheme of synthesis which would not conflict with any existing patents"; they gave Amtorg estimates on what the plant would cost. "Nothing ever came of it," Gold says. He should have known nothing would.

Privately, Gold was elaborating his espionage cover story into a fantasy of family. To Brothman's secretary and mistress, Miriam Moskowitz, "he had frequently spoken . . . of his beautiful wife and twin children, Essie and David. He claimed his wife was a tall, redheaded girl who had formerly been a model for Gimbel's. . . . On one occasion he even pulled out his wallet in order to show Miss Moskowitz a picture of the twins, but then replaced it almost immediately after opening it stating that he must have left the picture at home." Moskowitz discovered Gold's deception when she asked him how many dependents she should list on his income-tax withholding form. Forgetting his fantasy family, Gold told her he had no dependents. Why wouldn't he list his wife and children? Moskowitz persisted. "Gold thereupon became 'hysterical,' " the FBI quotes the secretary, "and insisted that she forget his family." To cover his mistake, Gold embellished his deception further with Brothman: "In 1946, when [Gold] first came to work for [him], Gold told Brothman that his wife had left him and that he was so despondent over it that he wanted to commit suicide. Brothman said that he offered his services to effect a reconciliation and even offered to borrow money and give it to Gold if this would help in any way. Gold, however, rejected both of these offers."

A detailed report of the Canadian investigation by a Royal Commission, published in Ottawa that June, reproduced many of the documents Gouzenko had pirated and caused an international scandal. The encroaching Cold War felt apocalyptic to serious men and women on both sides of the Iron Curtain. Just as Richard Feynman in a Manhattan bar had been unable to shake his sense of doom, so also Lev Altshuler, working now with Veniamin Zukerman at Laboratory No. 2, thought his country felt "defenseless and alarmed":

I remember one day in the summer of 1946 when I was strolling around Moscow with an acquaintance who had commanded an artillery division during the war. It was a clear, sunny day. Looking at the passersby, my companion wiped his face with his hand and said unexpectedly: "I look at these Muscovites and before my eyes they turn into shadows of people who have gone up in smoke in the fire of an atomic explosion."

Sometime that spring, Yuli Khariton and his scientific team prepared a one-tenth scale model of the Fat Man design—a pear-shaped *matrioshka* of metal shells about fourteen inches in diameter—and delivered it to Beria for Stalin to see. A statement of the bomb's technical requirements for the Council of Ministers followed on July 25, 1946. One of Beria's aides invented a name for the device, RDS-1. The scientists joked ironically that RDS was an acronym for "Russian Made," but in fact RDS stood for *Reaktivnyi dvigatel Stalina,* "Stalin's Rocket Engine," as deliberately meaningless a name as "Fat Man" or "tuballoy."

Beria sent two observers to the US tests at Bikini that July—a Radium Institute physicist and an MGB geologist, the latter traveling as a *Pravda* correspondent. The deputy chairman wanted someone nearby who had seen an atomic explosion. The Soviet observers saw two, one an air drop, the second a spectacular underwater detonation that sent a column of radio-active water and steam thicker than a fleet of battleships high into the air and capped it with a polluted mushroom cloud. Nearly a year after the end of the war, the Bikini bombs used two of only three cores in the US stockpile. Nine more Christy cores would go into production at Los Alamos in mid-August 1946 at the rate of two per week, after which the lab planned to switch to producing some forty Mark IV levitated composite cores. All would fit the handmade Fat Man HE assemblies that operated on forty-eight-hour lead-acid batteries, that had to be stored disassembled, that required two days to assemble, for which there was no permanent assembly team available and no regular supply of short-lived polonium for initiators. If Lev Altshuler had known the disarray of the US stockpile, he and his friend might have felt less alarmed.

Bikini, Lewis Strauss's bright idea and Curtis LeMay's albatross, supplied a name for a scandalous new French bathing suit. Otherwise it was a technical disaster. The air drop on July 1 fell a quarter mile from its aiming point, the battleship *Nevada.* General Joseph W. Stilwell, "Vinegar Joe," a hard-bitten hero of the war in China, saw the battleship still afloat from his observer plane and swore, "The damned Air Corps has missed the target again." Radiation monitor David Bradley heard his Navy co-pilot growl, "Well, it looks to me like the atom bomb is just about like the Army Air Force[s]— highly overrated." The snafu followed in part from interservice rivalry, a taste of things to come. The USAAF, resentful of Navy incursion into the atomic weapons field, had not supported 509th training with significant priority; of six B-29s assigned to Bikini training, no more than two were usually available at a time because of mechanical troubles. But part of the problem was simply that for reasons of safety, shipboard observers, including the media and foreign dignitaries, had been positioned too far away. The MGB geologist, Simon Alexandrov, dismissed the distant spectacle contemptuously: "Not so much." The second, underwater shot twenty-four days later was more impressive, especially since it polluted the target ships and the entire area with radiation. By then, most of the observers had gone home. The USAAF blamed Los Alamos for the first misdropped bomb, but the morning of the drop Paul Tibbets, the wartime 509th commander, had checked the drop crew's bombsight calibration with his Hiroshima and Nagasaki bombardiers and found "obvious miscalculation." He had informed the Bikini aircraft commander, who "listened a bit impatiently, thanked me for the advice and then told me politely that they were satisfied with their own results." Tibbets's people ran the miscalculated numbers and predicted the miss—1,600 feet short of the *Nevada*—within two hundred feet.

Bikini also fouled the message of peaceful intentions that Baruch was proclaiming at the United Nations. *Pravda* called the tests "common blackmail," sneering that "if the atomic bomb at Bikini did not explode anything wonderful," it did "explode something more important than a couple of out-of-date warships; it fundamentally undermined the belief in the seriousness of American talk about atomic disarmament." Former Secretary of the Interior Harold L. Ickes asked publicly if "the Bikini experiment" was "diplomacy by intimidation." Navy Secretary James Forrestal, who attended the tests, countered lamely that they were "not a gesture of war, aggression or threat."

One benefit at least that the tests provided was a successful experiment in long-range detection, carried out by the USAAF and Standard Oil's California Research Laboratories. An analyst reported to Groves after the experiment that it was "possible by monitoring the air currents at various points around the world to determine if an atomic bomb has been detonated in the air. By detailed analysis of wind conditions, it may also be possible to determine the direction to the blast and, by additional judicious reasoning, approximately when and where it was detonated. . . . It is . . . reasonable to expect positive results at ranges of 2000 miles or less."

However much the missed target may have chagrined Curtis LeMay—and he makes no mention whatsoever of his Bikini responsibilities in his memoirs—the two explosions were the first he had seen. He was impressed. So was the blue-ribbon evaluation board, including Vinegar Joe Stilwell, that the Joint Chiefs had sent out to observe. A year later, the board issued a report that LeMay took pains to summarize with emphasis to Carl Spaatz:

(1) Atomic bombs *in numbers conceded to be available in the foreseeable future* can nullify any nation's military effort and demolish its social and economic structures.

(2) In conjunction with other mass destruction weapons *it is possible to depopulate vast areas of the earth's surface, leaving only vestigial remnants of man's material works.*

(3) The atomic bomb emphasizes the requirement for the most effective means of delivery. *In being there must be the most effective atomic bomb striking force possible.*

"Experience, experience," LeMay writes of what he was learning at this time. "Again I affirm it: no substitute for experience."

A third test had been planned in the Bikini series, using the third available Mark III core. Groves proposed calling it off and the Joint Chiefs agreed. More than long telegrams or abandoned food relief, Groves's reason for canceling the third test defines the changing political conditions one year after the end of a war when the US and the USSR had been allies and had defeated every power that might threaten them except each other:

I wish to call to the attention of the Joint Chiefs of Staff that even a single atomic bomb can be an extremely important factor in any military emergency. It is imperative that nothing interfere with our concentration of effort on the atomic weapons stockpile which constitutes such an important element in our present national defense.

14
F-1

In July 1946, while Los Alamos was busy at Bikini, Igor Kurchatov, physicist Igor Semenovich Panasyuk and a dedicated crew of scientists and workmen began assembling the first nuclear reactor outside of North America at Laboratory No. 2.

Now Kurchatov's Moscow research center was a thriving enterprise. "The brand-new building housing [the laboratory] was flooded with sunshine and glowed with a whiteness of fresh construction unseen during years of war," recalls Igor Golovin. The three-story building had a cyclotron on the first floor, busy laboratories and offices, enthusiastic staff. Veniamin Zukerman remembered distinctly the first scientific seminars he attended there:

> Usually about twenty people would meet in an empty room; each would bring some sort of chair. The chairs would make up a motley, unmatched lot, but somehow there would always be an old-fashioned upholstered armchair with carved elbow-rests, matching legs, and a high back waiting for Igor Vasilievich [Kurchatov]. The seat and back were upholstered with bright-green plush.
>
> He would arrive for the seminar after a sleepless night, fresh from a shower, his hair still wet. He would listen almost without interrupting, without intruding, although the discussion, as a rule, would be on subjects of interest to him.
>
> When he chaired conferences, they were lively and provocative. He made a point of drawing out a clearly formulated opinion from everyone present. He would survey everyone in turn: "Your opinion? Yours?" If he was pleased with the answer, his encouragement would take the form of his inimitable "Quite right, quite right," in which he came down especially hard on the prolonged, rolling "r." ... It seems that even now I can hear his powerful voice.

He was very fond of witty sayings and turns of phrase, and often came up with them himself. During one session we were discussing a technical project that was going to require the participation of industry and considerable financial expense. "Let's ring up the adminiboys right now," said Kurchatov, dialing the phone. I bent to my neighbor at the table and asked quietly, "Who are the adminiboys?" "It's his own little abbreviation, the administrative boys, that's how he dubs everyone starting with the deputy minister."

Kurchatov's beard was regal now, dark and thick and squared off like a Pharaoh's; the men and women who worked for him called him the Beard. Anatoli Alexandrov gave him a giant straight razor as a joke and regularly pursued him for a date when he would shave off his unique embellishment. "What kind of Beard could I be without a beard?" Kurchatov would counter. He was a heavy smoker who wore old-fashioned spectacles for reading and wrote with a steel-nibbed pen. His wife Marina played the small grand piano in the living room of their elegant house or they listened to the recordings of Russian masters and Mozart that they collected, filling the house with music. "But administrative duties distracted him," says Golovin sympathetically. The truth was, Kurchatov worked eighteen-hour days, seven days a week, starting at eleven A.M. and going until early morning as all the Soviet leadership did to accommodate Stalin's vampire office hours. He "was taking enormous personal risks," Alexandrov notes, "by deciding on the construction of large plants without possessing finished technological solutions or products. The majority of the experiments were carried out with microgram quantities of materials, not even test-tube quantities. The results of these microgram-quantity experiments were scaled up directly to industrial scale, to special equipment which had never been used before and which resembled nothing that already existed."

Leslie Groves had taken similar risks, as a result of which the special equipment necessary to process uranium and plutonium for atomic bombs did in fact exist in the US. Because Kurchatov had at least partial access, through espionage, to information about that equipment, his task was made somewhat less burdensome, if no less dangerous (the MGB might help with information, but Beria was still looking over his shoulder). An important case in point is the first Soviet reactor itself.

Kurchatov and young Panasyuk had prepared the initial proposal for the reactor they called F-1 ("Physics-1") back in July 1943. "We see the reactor to be designed as a sort of rationally assembled installation," they wrote, "composed of a uranium-graphite grid, graphite, control rods activated from a distance, with experimental channels and wells, to be installed in a pit dug in the ground for reasons of radiation safety." For three years, the challenges of producing sufficiently pure graphite and metallic uranium had delayed realizing this proposal.

"We needed graphite a thousand times purer than was available then," Alexandrov comments. "There were not even methods of measurement for purity of that degree. These methods were developed then; we developed them." Alexandrov mentions the senior Soviet radiochemist V. G. Khlopin in particular as a leader in developing the technology to produce metallic uranium from ore. "The Germans had solved that problem," notes Nikolai Ivanov, one of those who built the F-1 reactor, who eventually became its chief engineer. "Because we didn't know the process we had to wait until the war was over to acquire it." The Auer Gesellschaft's Nikolaus Riehl, captured with his laboratory in Germany at the end of the war, supplied the Soviet scientists with crucial information on German uranium-processing technology. According to Riehl, the Soviets acquired full information on the American technology through espionage* at some point after the war and switched over from the less efficient German approach sometime after November 1946. The uranium for F-1 was probably purified by the German process, since the US process yielded extremely high purity, while the batches of F-1 material were irregular.

Calcium of high purity was a crucial ingredient in uranium metallurgy. The Soviets lacked facilities for producing pure calcium; they looked to the Soviet zone of occupied Germany for support. In the autumn of 1945 they began operating a small calcium plant at Bitterfeld; calcium from that pilot operation presumably purified the first supply of refined uranium that arrived at Laboratory No. 2 in January 1946. In April they began processing calcium at full scale at an I. G. Farben plant in Bitterfeld. "The specifications of that date," notes a 1951 CIA review, "show little real awareness of the required final purity of uranium for use in graphite piles. . . ."

If the Soviet Union was forced to turn elsewhere for its industrial chemistry, its leaders had only themselves to blame. The Leningrad physicist Sergei Frish comments:

> Before the war the Soviet Union had put its main efforts into heavy industry. We lagged behind in precision instrument manufacture. But there were still other negative circumstances [as well]. . . . Due to mistakes committed by ignorant administration, several important branches of physics and chemistry had been abandoned. . . . The director of our Institute of Optics, Chekhmataev, forbade the study of rare earths on the grounds that they occur only rarely in nature and thus could present no practical interest. Unfortunately, Chekhmataev, in his administrative zeal, was not alone. It turned out that, as if nature had deliberately planned it that way, creating an atomic industry called for profound knowledge of the physical and chemical properties of precisely the rare earths.† Luckily for us . . . Khlopin

---

* The American source of this espionage information has never been identified.

† The chemistry of uranium and plutonium resembles that of the rare earths—the elements in the periodic table from cerium (58) through lutetium (71).—RR

at the Academy of Sciences Radium Institute, despite all the prohibitions, continued working on rare earths. When a speedy solution of certain complex chemical problems of these elements was called for [i.e., uranium metallurgy], it was none other than Khlopin who rendered invaluable service. Had he . . . not shown the stubbornness with which a scientist will pursue his favorite "irrelevant" research, it would undoubtedly have taken longer to create our atomic bomb.

On the other hand, Frish observes, "the ability to carry out slated assignments quickly and as ordered, to throw our resources at the 'bottlenecks' and to close gaps as they arise"—abilities Frish attributes to the Soviet system of centralized planning—served them in good stead in their atomic enterprise. Nor did skeptical Americans like Groves take into account "the experience we had acquired during the war in the mounting of contemporary armaments production and the advances we had made in our precision-instruments industry as a result of captured German technology."

The implication of Kurchatov and Panasyuk's 1943 proposal was that they were describing an original invention, but one month earlier, in June 1943, a small group of physicists at the University of Chicago Metallurgical Laboratory—the Met Lab—had begun designing a graphite–natural uranium reactor that was intended for use at Hanford testing the purity of the fuel elements destined for the big production reactors being built there. The Hanford 305 test reactor (so called because it was destined for building number 305 at Hanford) was the fourth nuclear reactor built in the United States, completed at Hanford in March 1944 and operated beginning in April. The Soviet F-1 reactor as it was ultimately built shared many of the 305 reactor's significant physical parameters. Physicist Arnold Kramish, then an analyst for the Rand Corporation, noticed the parallels when the Soviet Union declassified details of the F-1 design in 1955; in a Rand study he compared them in a table:

	Hanford 305	F-1
Power	10 watts	10 watts
Diameter	18–20 feet	19 feet
Lattice spacing	8.5 inches	8 inches
Loading	27 tons uranium	45 tons uranium
Rod diameter	1.448 inches	1.2 to 1.6 inches

The two reactors differed slightly in their control-rod arrangements. The Soviet F-1 design had three control rods—two for emergency scrams, one for adjusting reactivity—that descended vertically into the reactor. The American 305 had three horizontal control rods, Kramish notes, but three vertical holes as well. "A single boron-steel safety rod was suspended above the reactor. There were two other vertical holes, to accommodate a smaller metal safety rod, which could shoot small boron-steel pellets downward in an emergency." The decision to forgo horizontal control rods in the F-1 design appears to have followed from the difference in siting; the 305 was built above ground, but F-1 was assembled in a pit with only about a meter of clearance around the reactor—not enough room to maneuver control rods horizontally. Both reactors were air-cooled. Both were designed to test the purity of graphite and uranium for the industrial-scale production reactors they supported. To that purpose, both reactors were built with channels bored through them into which samples of materials could be introduced. Both used internal and external boron ion chambers to measure activity from which their power was computed. Both used an unusual measurement technique involving computed reactivity to determine the purity of the materials being tested.

"The first Soviet reactor," Kramish concludes, "was practically a carbon copy of the American 305 reactor built at Hanford. . . . The similarity of construction is interesting. Is it coincidental, or were details on the 305 reactor obtained through espionage?" In the opinion of reactor experts, coincidence in so many particulars is not even remotely possible. Pioneer reactor physicist Alvin Weinberg points out that lattice spacing, at least, is determined by physical constants that Soviet scientists could have calculated as well as American—as indeed they did for F-1, since it was Kurchatov's practice to require his staff to work out the science and engineering independently of his espionage information in order to confirm its authenticity and to build a base of knowledge for further Soviet development. But reactor technology was still in its infancy in 1946 and there was no common body of technical literature available to consult; coincident identity among so many major design variables is unlikely. "To come up with the same design blind?" questions physicist Charles Till, a reactor inventor and designer and in 1995 the associate director of Argonne National Laboratory. "I wouldn't put any money on it, and I'm a betting man."

If the F-1 reactor design was based on espionage, who was the spy? Alan Nunn May could have had access to 305 details, although he arrived on the scene several months after Kurchatov and Panasyuk began planning F-1 (not a fatal exception: plans for the reactor could have changed to reflect new information, as plans for the bomb did when Kurchatov learned through espionage of the virtues of plutonium). Or someone else at the Met Lab could have passed the plans—several other Met Lab scientists were observed during the war in contact with known Soviet agents, though they escaped prosecution. Whoever passed the information made a careful selection among possible designs, bypassing CP-1 and the Argonne heavy-water reactor for a system that was practical given Soviet uranium and graphite supplies and that served the further purpose of testing materials for the big production reactor being developed for Chelyabinsk-40. Uranium loading was a significant difference between the two designs; the 60 percent larger Soviet loading is evidence of the difficulties the USSR had purifying uranium.

No one connected with the Soviet atomic-bomb program has ever acknowledged that espionage was the basis for the F-1 design.

Beria's hand is evident in the functional similarity between the 305 test reactor and F-1—the fact that both machines were intended for materials testing. If the F-1 design was based on the 305, Kramish comments, then "rather than attempt an independent and improved materials testing reactor, the Soviets chose to copy a sure thing. This suggests at this stage of the Soviet program both a sense of urgency and a lack of technical self-confidence." V. S. Fursov, a physicist who worked on F-1, admits that "there was some reason to consider the proposed construction . . . as something other than a completely guaranteed undertaking." Even if Kurchatov had been confident of the purity of the materials available to him and of the accuracy of the physical measurements his team had accomplished with limited resources, copying a sure thing was standard operating procedure in Soviet industrial espionage, as Harry Gold knew all too well.

Igor Kurchatov may have based F-1 on the 305 test reactor, but he and his colleagues still had to repeat the measurements, experiments and tests that the Met Lab reactor physicists had carried out. The Soviet scientists had to do so not only to check for disinformation. They also needed to learn the craft of reactor design and construction, which simply reproducing a copy could not teach them. Their resources were different as well—less pure and less dense graphite, less pure uranium metal—and dictated differences in design. "We paid serious attention to the choice of materials and the design of every detail of the reactor," Pervukhin insists. ". . . We had to deliver; we couldn't afford to take the time to improve the design later." In that serious attention lay the difference between Lavrenti Beria's uninformed approach to the "atomic problem" and the scientific approach Kurchatov followed. Beria wanted copies, as if nuclear reactors and atomic bombs were no more complicated than a jeep or a B-29; Kurchatov insisted on accumulating a working body of knowledge to guard against present failure while building a future. It was Kapitza's point, but Kapitza had been expendable. Kurchatov was not.

"On a plot of grass outside Kurchatov's office window," Golovin remembers of the time when construction began on F-1, "two army tents were erected." Soldiers began digging a pit under the canvas for the reactor while "I. S. Panasyuk began to stack graphite columns to measure the absorption and moderation of neutrons in graphite." The pit was supposed to be dug ten meters deep (thirty-three feet), but at seven meters the workmen encountered groundwater and had to stop short; the reactor would project a little way out of the ground.

For a uranium metal and graphite reactor to operate at all, the uranium cross section for slow-neutron absorption had to be no more than 4–5×10^{-27} square centimeters. "The tests with the first batches of graphite," recalls Panasyuk, "gave values of 50–500×10^{-27} cm^2." The chemists who analyzed the various batches of graphite reported kinds and levels of impurities that failed to account for the unacceptably high cross sections that Panasyuk found. "Kurchatov advised us not to become desperate but to wait until the results of tests for all the batches of graphite [being processed] became available." Kurchatov knew from espionage what he could not directly tell Panasyuk—that sufficiently purified graphite would work and therefore the chemical analyses must be wrong. With specific exceptions, cleared in advance, Kurchatov was required to keep the members of his team in the dark about the espionage information he received. "He was not authorized to tell them," confirms Anatoli Yatzkov, "so he didn't." Russian astrophysicist Roald Sagdeev, one of the younger generation of Soviet physicists, heard the stories:

> Kurchatov used American materials for the dual purpose of double-checking the scientific results obtained by members of his team, and for evaluating the probability that the stolen secrets might contain purposely planted disinformation. Inside the Russian nuclear establishment legends were told of how his subordinates—the theoretical physicists—would report to Kurchatov with freshly calculated formulas. According to their accounts, Kurchatov would look carefully at their work, then silently open the safe with the precious stolen American secrets to compare the results. "No, it is not right," he would say. "You have to work more and come again."

And as for the theoretical physicists, so also for the chemists who analyzed the purity of F-1 graphite. "Data of this sort [i.e., unacceptably high cross sections] . . . made the physicists depressed," writes Panasyuk, "but they didn't stop measuring new batches of graphite from various sources of raw materials that were processed under various technological conditions. Finally, measurements that had just been finished of a recent shipment were put on Kurchatov's desk. Victory! For the first time a cross section of 8.6–0.4×10^{-27} cm^2 was obtained!" Kurchatov then personally checked a few samples from the earlier "bad" batches against the pure new graphite and proved that the chemists were wrong. "The method for measuring impurities was improved," concludes Panasyuk of this discouraging episode, "and it appeared that the 'bad' batches were slightly contaminated by boron and rare-earth elements."

The quality soon stabilized; the new graphite that began arriving was not only purer but also denser. By August there was enough graphite on hand —about five hundred tons—to assemble the reactor. To make sure the material was adequate, Kurchatov had the entire mass stacked together in the pit under the tent into a vast black cube twenty feet on a side to measure its average absorption cross section. "Measurements and calculations," notes Panasyuk, "determined that cross section to be 4×10^{-27} cm^2." Loaded with sufficient uranium, the reactor should work. Teams began drilling single blind holes in the graphite blocks for the uranium slugs, a total of thirty thousand holes by the time the work was done. Each uranium-loaded block would be surrounded on four sides by a set of plain graphite blocks to make a grid spacing of about eight inches—room for fission neutrons from one slug to bounce off carbon atoms in the graphite and slow down enough to resist absorption by U238 nuclei until they encountered a fresh U235 nucleus in the next slug along.

In the meantime, the Assembly Workshops went up around the reactor pit, a handsome brick building 130 feet long, fifty feet wide and two stories high. The pit floor and walls were lined with poured concrete, the sandy soil itself serving as the reactor's primary shielding. Construction workers dug a passageway from the sub-basement-level floor of the pit up past a baffle of special walls made of lead blocks and hollow bricks filled with a mixture of boron and paraffin to an underground control room. A laboratory on the first floor, into which the dome of the reactor would protrude, went unshielded. Radioactive gases would be removed through a fan system and vented into the Moscow air.

As a first step, workmen suspended three cadmium control rods above the center of the pit. Steel wires from the control rods ran over reels hung in the attic framing above the main laboratory and down into the underground control room, which was fitted with both electrically and manually operated winches. A bright brass submarine periscope in the control room would allow the F-1 operators to observe the notches on the control rod above that would indicate its position within the reactor block. The two emergency rods were unnotched; they would be positioned either entirely out or fully within the reactor and could be scrammed ("emergency dropped," the Soviet scientists called the operation) with the winches to quench the chain reaction. Workmen also installed red lights and sirens in the pit connected to radiation monitors. Electricians wired in backup electrical supplies to support the winches and the control panel.

Just as Enrico Fermi had done building the first man-made nuclear reactor in Chicago in the fall and early winter of 1942, Kurchatov proceeded toward a full-scale assembly by directing the construction of a series of smaller, subcritical assemblies. These would not contain enough uranium and graphite to achieve a self-sustaining chain reaction, but they would carry measurements and calculations incrementally forward in that direction and enable the scientists to learn what to expect from the novel process as they went along. Panasyuk calls these smaller assemblies "model" assemblies (Fermi's people called theirs "exponential" since an exponent entered into the calculation of their approach to critical mass). It was standard laboratory procedure in experimental physics in those days before computer simulation to build such functioning models to accumulate data that could be extrapolated to predict the operation of a full-scale machine.

"Kurchatov," writes Golovin, "proposed to attain the critical dimensions by . . . increasing the diameter of the sphere each time and using all the available uranium prepared up to then." Supplies and approximations would thus keep pace across the weeks. The reactor team laid down a flooring of graphite blocks a meter deep and began building the first model assembly in the center of the flooring. The core layers got the uranium and graphite of the highest purity, with less pure materials reserved for the periphery. When they dismantled the model assemblies, they stored the graphite blocks and uranium slugs on the floor of the main hall upstairs, moving the dirty, greasy, heavy materials by hand onto and off of a belt lift.

Kurchatov's team completed the first of the four model assemblies on August 1, 1946, using 1.4 tons of uranium and 32 tons of graphite. "Layer by layer they would put in graphite and uranium," team member Boris Dubovsky recalls, "conducting measurements at the same time and processing the results. The work, as a norm, went on all around the clock." Because of the backwardness of the Soviet radio industry,* the measurements group used electronics salvaged from German aircraft shot down during the war.

The third model assembly "alarmed everyone," says Golovin. It demonstrated hardly any increase in neutron multiplication over model number two despite its larger loading of uranium; they feared some "essential miscalculation." Kurchatov ordered additional measurements, which showed "that the third batch of uranium was considerably less pure than the rest." "This discovery," writes Panasyuk, "called for urgent reorganization of physical quality control for all batches of uranium produced by our industry." The reorganization was effective; Golovin says the fourth model assembly, completed in early November, "reassured everyone that success was imminent."

Kurchatov's team began assembling the full-scale reactor on November 10, 1946. By now the work was routine. The assembly materialized a layer at a time, swelling outward in a roughly spherical configuration that began to crowd the pit, looming overhead as it reached ground level, the black graphite soaking up the light, graphite dust adding to the gloom. They did not feel gloomy. They worked with increasing excitement.

---

* "It is known, for example, that the USSR electronics industry during the [Second World War] was incapable of standardizing such items as vacuum tubes to the extent that spares and replacements could be used interchangeably. It is doubtful that they could have overcome all such difficulties since the war." JCS 1952, 1, 21.xii.48, appendix, p. 11, in Ross and Rosenberg (1989), n.p.

There was barely enough uranium. Before they were through they had used up all the uranium metal available in the USSR at that time—forty-five tons. It was not enough. They had ninety kilograms of uranium oxide and 218 kilograms of powdered metallic uranium on hand that they had used for physical measurements in the years before Vannikov organized Soviet production of uranium metal. They pressed the impure materials into briquettes, loaded the briquettes into graphite blocks and laid the blocks around the periphery of the lattice. Three tons of graphite added near the end, a cockade of sorts, was American Lend-Lease graphite shipped during the war for searchlight electrodes.

Fortunately, the reactor approached criticality well before their rough initial calculations had predicted it would. That consequence, like the oxide loading and much else in the Soviet program, repeated the earlier American experience and defines the boundaries of what remote and fugitive espionage could pirate. "The density of neutrons in the effective center of . . . the reactor approximately doubled between layer 53 and layer 58," notes Panasyuk; "because of this change, it was clear already at layer 58 that the . . . forecast dimensions (76 layers) were highly exaggerated."

"It became obvious," Golovin reports, "that there was going to be a chain reaction. The final layers of uranium were stacked behind extra shielding in case of an unforeseen runaway reaction." They finished layer 61 on December 24, 1946, in the evening. On the graph of neutron intensity that Kurchatov had been maintaining, it was obvious that layer 62 would cross the threshold of criticality. Kurchatov sent the workers home for a rest. People began trickling back to the pit in the middle of the night. Layer 62 was laid with all three control rods inserted into the reactor by two in the afternoon on Christmas Day. "Kurchatov was in another building at the time," Panasyuk remembers. "We telephoned him and notified him that the reactor was ready."

Kurchatov arrived in the underground control room to direct the start-up. He was sufficiently concerned about a possible accident that he cleared the building even of guards and had the area cordoned, authorizing only his crew of four immediate assistants, which included Boris Dubovsky and Igor Panasyuk, to remain. Fermi had trusted his calculations more; there had been a crowd on hand when CP-1 had started up in Chicago four years previously, even though Fermi's reactor had pioneered the technology.

Kurchatov and Panasyuk sat down at the F-1 control panel late in the afternoon, Dubovsky remembers:

All radiation-measuring equipment was switched on. We checked the functioning of the control and protection systems. The emergency rods were pulled completely out of the reactor but not locked; from this position they could be dropped into the reactor in one second. Then Kurchatov personally began winching out the control rod. We were all very anxious, of course. . . . Everyone was silent. The only sounds were the clicks of neutron counts from the loudspeakers and Kurchatov's brisk orders.

Kurchatov stopped the control rod with 280 centimeters (9 feet) still inserted into the reactor. The clicks of neutron counts and the flashes of red light from the gamma-ray counters multiplied. "Everybody got excited," reports Panasyuk. Kurchatov watched the light on the galvanometer connected to the main neutron counter. "The beam did not move even after ten minutes had elapsed," Panasyuk continues. "The frequency of the clicks and flashes increased and then steadied." Kurchatov checked the graph of neutron intensity Dubovsky was plotting, declared that they had not yet achieved a self-sustaining chain reaction and immediately shut the reactor down.

They took a ten-minute break. When they were all back in their places, Kurchatov winched the control rod 10 centimeters (4 inches) farther than before. ("Ever since his Leningrad days," reminisces Kurchatov's colleague V. A. Davidenko, "taking measurements, Igor Vasilievich would always start a stopwatch or flip a switch to his favorite countdown: 'Ready? *Dzik!*' and on '*dzik!*' he would activate.") This time the neutron count took an hour to level off. Kurchatov dropped the two emergency rods into the reactor and then lifted the control rod another 10 centimeters farther out.

Another ten-minute break. Panasyuk:

Kurchatov quickly removed the two emergency rods from the reactor. As the seconds passed, the graph showed an almost linear growth of reactor power. For the first time the sound turned into a roar. The indicator lamps no longer blinked but burned with a reddish-yellow light. Everyone watched Kurchatov with excitement while he studied the graph. After awhile he declared that although the effective multiplication ratio had reached 1, that did not yet prove that the reactor was functional. It was necessary to repeat everything from the beginning.

One final run-up. Kurchatov dropped the emergency rods and withdrew the control rod only 5 centimeters more. Panasyuk does not say they took a break. Out came the emergency rods:

In thirty minutes all the sound indicators were roaring, the light indicators glowing and the galvanometer beam from the boron trifluoride counter [inside the reactor] deflecting not in an even way, as in the previous series, but in an accelerating progression. The tension became extreme when the second boron trifluoride counter, which was located within the underground control room, began producing more frequent clicks than its background of two or three per minute—an increase which meant that neutrons from the reactor had penetrated the thick layers of earth and cement and reached the room. . . .

Kurchatov pressed the button that dropped the emergency rods.

Dzik. "Well, we have reached it," Kurchatov announced laconically. It was six o'clock in the evening, December 25, 1946. The first Soviet nuclear reactor was operational. For years to come, Kurchatov and his team would run it night and day.

On December 26, they measured its neutron doubling time at 134 seconds. Within a few days, says Panasyuk, "this level of supercriticality no longer suited either the creators of the reactor or its increasing number of users." More uranium had become available by then; Kurchatov decided to enlarge the uranium loading, adding several more layers to reduce the doubling time to twelve seconds.

F-1 was their first concrete achievement and they were proud of it. They used it to test the purity of the uranium and graphite being prepared for Chelyabinsk-40: they pushed samples of the test material through a channel into the center of the reactor and measured the extent to which the test material affected F-1 reactivity. They used the little reactor to determine the optimal grid spacing for the production reactor and to test its control and monitoring equipment. Most crucially, operating from a backup control room a kilometer away from the reactor building on weekends and over holidays, they ran the reactor remotely at one-thousand-kilowatt power levels to breed plutonium.

A few days after the start-up, Beria came to see what his scientists had wrought. They had shut the reactor down so that he could watch it come alive. With Beria in the underground control room, write Golovin and Smirnov, "Igor Vasilievich put the winch in motion, lifting the control rod. The clicks increased and gradually transformed into a continuous roar. The light beam of the galvanometer ran off the scale. The people on hand shouted 'It's going!' meaning the chain reaction." Beria had seen massed assemblies of Stalin's Organs fire racks of screaming Katyushas; here were only clicking counters and flashing lights. "And that's all?" he challenged Kurchatov. "And nothing else?" He wanted to see this famous machine for himself and asked to approach the reactor. "No," Kurchatov told him. "You can't go down there now. It's dangerous to your health." That made Beria more suspicious: where was the danger? "Beria began suspecting that Kurchatov was swindling him," Golovin and Smirnov report.

Beria was habitually suspicious, of course; his rise to power in the criminal enterprise that was the Soviet Communist Party and his duties along the way demanded that he be. Anatoli Yatzkov says Beria questioned "the idea of nuclear weapons even when nuclear weapons work in the Soviet Union had achieved large scale." His doubt may have arisen from his ignorance of science and from his intellectual and parvenu insecurities. If he suspected the scientists were swindling him, Kapitza's assault on his leadership can only have deepened his suspicions. But Beria even doubted the veracity of the information that his own espionage apparatus had collected, Yatzkov notes: "L. R. Kvasnikov says that once, when he was reporting to Beria on intelligence information, Beria threatened him, saying, 'If it's disinformation I'll throw you all into the cellar.' When the head of intelligence, P. M. Fitin, hinted that it was high time to reward agents for their accomplishments, Beria berated him that it was not yet clear if they deserved reward or punishment."

Beria seems to have been incredulous that the United States would allow its most strategic military technology to be stolen so easily, that spies so perfectly positioned as Fuchs had been would simply walk in off the street and volunteer, that privileged citizens would willingly subvert their nation's commanding military advantage in the service not of fear or cold cash but of insubstantial ideals. The bribes and the medals that his subordinates kept throwing at their surprised, amused or indignant charges imply a need to mark these volunteers with some feculence Beria could recognize. (In England after the war, Fuchs accepted a gift of one hundred pounds from his Soviet contact because "security precautions had been tightened up after the exposures in Canada [as one of his interrogators paraphrases him] and he felt that in accepting this money he was more or less assuring his contact of his loyalty.") And if scientists in England and the United States were such naïfs, how could he trust his own?

Kurchatov and Alexandrov had their work cut out for them convincing Beria that the plutonium-production reactor should be different from the reactors the United States had built at Hanford. The Hanford production reactors were horizontal cylinders of graphite bored with several thousand horizontal channels into which aluminum-clad fuel elements could be loaded. When the fuel elements had been irradiated long enough to transmute some of their U238 into plutonium, they could be pushed through the reactor out the other end for plutonium separation. Military security was crucial to the Soviet project, however—Chelyabinsk was potentially within range of American B-29s—and in pursuit of that security the Chelyabinsk reactor would be sheltered underground, in a huge pit, like F-1. The pit, as well as the scientists' first experience with F-1, argued for a vertical design. "Every day," Pervukhin recalls, "there were heated arguments in the reactor section concerning the choice of design for the industrial reactor. After lengthy discussions, the vertical design was chosen. But the project for a horizontal graphite reactor was prepared as well, and only after considering all the arguments for and against did everyone agree on the advantages of the vertical design." It is significant that those advantages primarily concerned security, a subject Beria understood.

In the meantime, the scientists were learning what it meant to work for

Lavrenti Beria. "The successful startup [of F-1] brought confidence to all who worked on the problem," says Pervukhin, "and showed them they were on the right track. That was very important, because among the people mobilized for the work in this field, not all were confident of the positive results of our efforts." Golovin and Smirnov characterize the scientists' predicament more bluntly. "Everyone understood that if the bomb didn't explode, the whole team would be in trouble."

———

While the Soviets started up their first reactor, the United States was shutting reactors down. The three production reactors at Hanford had sickened with what came to be called "Wigner's disease" (after the Hungarian-born theoretical physicist Eugene Wigner, who designed the reactors and predicted the effect): graphite bombarded intensely with neutrons stored the acquired energy by rearranging its crystal lattices, which caused it to swell and occlude the reactor fuel-element channels. The US produced plutonium in the Hanford reactors, of course, but a more immediately critical concern was the polonium it also bred in the reactors for bomb initiators. Polonium210 loses half its radioactivity every 138.3 days; if the Hanford piles broke down and polonium production ceased, the atomic bombs in the nation's small stockpile would become unreliable within a year. DuPont, which operated Hanford until September 1946, when General Electric took over, cut one reactor back to 80 percent of rated power and shut down and unloaded a second; standbying one reactor would stop its Wigner's disease infection from progressing and offer an emergency capability for polonium production if the other two reactors failed.

The cutback had an immediate effect on bomb production. Los Alamos had planned to begin manufacturing forty Mark IV levitated composite cores in August 1946 at the rate of two per week. Associate Director Darol Froman had not checked with Hanford in framing that plan. When he did so, he discovered "that this rate of production could be maintained only for about 13 weeks. Thereafter we would be limited by Hanford production to less than one Mark IV unit per week. . . . I would suggest that the schedule be revised to a production of one Mark IV and ½ Mark III (or equivalent Mark IV) cores per week." They would still accumulate their forty cores, Froman believed, but it would take eight months.

Kurchatov learned of Hanford's problems with Wigner's disease; in the December 31, 1946, espionage gloss where he reported reviewing "an American work on the superbomb" he also reported reviewing material concerning "certain particularities of the operation of the nuclear piles at Hanford"; like the superbomb report, he found the Hanford information to be "probably true and of great interest for our work in this country." Evidently someone at Hanford was on the Soviet payroll.

Core production was not the only problem Los Alamos faced as the dust settled at Bikini. Bradbury reminded Groves in August 1946 that "the presence of a stockpile of all weapon components does not insure a state of readiness. The absence at the present time of an organization capable of assembling an atomic weapon on short notice must be regarded as the most serious stockpile deficiency. In a state of declared national emergency, practically every component now short could be produced at its maximum rate in a time shorter than an assembly crew could be collected and trained from personnel now scattered." The military was interested in building a small arsenal of Little Boy uranium guns for tactical use destroying bunker-like submarine pens, but the scattering of personnel had also expunged Los Alamos's working knowledge of Little Boy manufacture: "All personnel involved in this activity have left the Project," Bradbury cautioned Groves, "and the 'know-how' for this weapon will have to be completely relearned." ("The uranium gun was a very, very rugged beast," remembers Los Alamos engineer Jacob Wechsler. "That thing with some velocity on it could go through huge amounts of concrete. All you needed to do was put good enough detents on it so that it didn't try to inertially assemble. You could use it for other things but it was really designed to be a sub pen penetrator carried by torpedo bomber, to come in there and send that thing sailing right on in. And give off, what was it, 15 kilotons?") Los Alamos also had problems with initiator production.

The availability of housing no less than fissionable materials paced the revitalization of the lab. During the war, the Army had thrown up temporary construction; to attract and hold staff, Los Alamos needed more than flimsy fourplex apartments with tin showers and coal furnaces. Bradbury was building a new technical area and a new residential section of prefabricated housing up on the beautiful but inaccessible mesa, but "authorization of new hires," writes Raemer Schreiber, "was frequently delayed pending a housing vacancy." Housing set the pace of Los Alamos technical programs for years to come, a mundane fact of life that congressional and military nuclear-stockpile enthusiasts found it convenient to ignore.

The lab had to refocus its program from its narrow wartime goals, Schreiber explains. "What was really new was the stockpiling of bombs and the recognition that hardware produced by the laboratory would have to survive long-term storage and be 'GI-proof' when handled under field conditions. Whereas the Little Boy and the Fat Man had been meticulously built from selected parts by the people who helped create them, stockpiled weapons had to work reliably under much less ideal circumstances. Thus the laboratory found itself entering into a relatively unexplored field of interchangeability of parts, quality control, environmental testing, preparation of specifications, production of manuals and training of personnel." Schreiber helped train groups of young Army, Navy and AAF officers who billeted at Los Alamos for six months at a time attending lectures on nuclear physics, observing weapons R&D and learning to assemble atomic bombs in the Ice House—a real ice house from Los Alamos Ranch School days when the school cut ice from Ashley Pond. "It was a substantial stone building," Schreiber recalls, "so it was equipped with a vault door and used for our nuclear assembly laboratory."

Burdened with these challenges, Los Alamos had only limited time for Super research, Bradbury noted in November 1946: "Since the program for such a weapon would involve a laboratory fully as extensive as Los Alamos at the peak of its [wartime] activity, and would require as well large developments in other portions of the Manhattan District, the interest of the laboratory [has been] restricted to determining the feasibility of the weapon and to research and theoretical calculations bearing towards this end." Calculations on Teller's Alarm Clock design turned out to be "not very promising" according to Hans Bethe, and were set aside at the year's end.

After long and often acrimonious debate, mounted in part by a remarkable lobby of former Manhattan Project scientists led by Leo Szilard, Congress passed an Atomic Energy Act that Truman signed on August 1, 1946. The act was based on a bill introduced by Brien McMahon, a florid, fleshy pol of Irish descent who believed in good tailoring and good staffing. Attorney James R. Newman had helped write McMahon's bill; irreverent physicist Edward Condon, who would direct the National Bureau of Standards, had been an adviser. "I remember a famous occasion," Condon told an interviewer once, "—this was just before the [1945] Christmas holidays, and McMahon was going to take his gorgeous beautiful blonde wife down to Bermuda for a vacation and lie on the sand and get sun. And I of course had . . . very little knowledge of Washington, except high school civics which is a very deceptive subject. So I was surprised at James Newman, because he says, 'Now, listen, God damn it, Brien, we're going to carry on a national publicity campaign about this bill and how wonderful it is, and so for Christ's sake be sure you read it while you're down at Bermuda. Because when you get back to Washington, people will expect you to know what's in it. Because you wrote it, see?' So McMahon said he would, but I don't think he did, because he never was very familiar with the specifics. . . . He gradually got familiar with it." On his return, McMahon told the Senate that the atomic bomb was the greatest event since the birth of Christ.

The act Truman signed nationalized all aspects of atomic-energy development, much as the Acheson-Lilienthal Report would have internationalized such development; Newman and co-author Byron Miller called the new authority frankly "an island of socialism in the midst of a free enterprise economy." Most uniquely, the act provided civilian control of atomic energy through a board of five appointed commissioners rather than military control. Senator Arthur Vandenberg had preserved military influence with an amendment establishing a Military Liaison Committee to the new Atomic

Energy Commission; Congress kept a connection through a requirement that the new commission inform it in a timely manner of its deliberations. A new congressional Joint Committee on Atomic Energy, drawing members from both the House and the Senate, acquired unique responsibilities for overseeing the AEC's activities. McMahon, a Democrat, became the JCAE minority leader in a Republican Congress; conservative Republican Bourke Hickenlooper of Iowa served as the committee's first chairman. ("He's a nice fellow," Vandenberg characterized Hickenlooper, "a very nice fellow. He thinks he and he alone stands between the security of the nation and disaster.") The act authorized a General Advisory Committee of scientists appointed by the President to provide the AEC commissioners with technical guidance.

Truman nominated David Lilienthal, the TVA chairman, to head the AEC; Lewis Strauss, the wealthy investment banker and reserve Navy admiral who had proposed the Bikini series, would be a commissioner, as would physicist and former Los Alamos Gadget Division director Robert Bacher. The General Advisory Committee would include James Bryant Conant, Enrico Fermi, I. I. Rabi, metallurgist Cyril Smith, plutonium discoverer Glenn Seaborg and Robert Oppenheimer. At the opening of the first GAC meeting, on January 3, 1947, before the former Los Alamos director even arrived, Conant nominated Oppenheimer for the chairmanship; when Oppenheimer made it through a snowstorm to join the group the next day, he discovered he had won unanimous election.

Truman at that time was transitioning from internationalist to cold-warrior. "Reds, phonies and . . . parlor pinks can see no wrong in Russia's four and one half million armed forces," he complained to his diary petulantly in September 1946, "in Russia's loot of Poland, Austria, Hungary, Rumania, Manchuria. . . . But when we help our friends in China who fought on our side it is terrible. When Russia loots the industrial plant of those same friends it is all right. When Russia occupies Persia for oil that is heavenly." He wrote former Vice-President John Nance Garner that month that there was "too much loose talk about the Russian situation. We are not going to have any shooting trouble with them but they are tough bargainers and always ask for the whole earth, expecting maybe to get an acre." He fired Secretary of Commerce Henry Wallace for publicly advocating conciliation with the Soviet Union at a time when Secretary of State Jimmy Byrnes was negotiating in Moscow. He reviewed a remarkable report that his young special counsel, Clark Clifford, and Clifford's assistant George Elsey had assembled in consultation with Byrnes, the Joint Chiefs, Dean Acheson, James Forrestal, the Attorney General, George Kennan, Central Intelligence Group director Sidney Souers and others that put teeth in the proposals Kennan had advanced in his long telegram.

The Clifford-Elsey report detailed the sorry record of Soviet intransigence in Eastern Europe and elsewhere in the world, documenting and extending Churchill's Iron Curtain indictment; it also, as Clifford wrote the President in a cover letter, brought together and focused opinion at the highest levels of the administration, "the simultaneous definition by so many government officials of the problem with which we are confronted...." The hundred-page report argued that Soviet foreign policy was "a direct threat to American security" that was "designed to prepare the Soviet Union for war with the leading capitalistic nations of the world." The Soviets, it asserted, were developing "atomic weapons, guided missiles, materials for biological warfare, a strategic air force [and] submarines of great cruising range...." The Red Army was building up "large reserves" and mechanizing. Not diplomacy but military power should be the US's "main deterrent.... It must be made apparent to the Soviet Government that our strength will be sufficient to repel any attack and sufficient to defeat the USSR decisively if a war should start. The prospect of defeat is the only sure means of deterring the Soviet Union." To defeat its new enemy, "the United States must be prepared to wage atomic and biological warfare.... A war with the USSR would be 'total' in a more horrible sense than any previous war...." The US could not afford to lose its technological edge: "Any discussion on the limitation of armaments should be pursued slowly and carefully with the knowledge constantly in mind that proposals on outlawing atomic warfare and long-range offensive weapons would greatly limit United States strength, while only moderately affecting the Soviet Union." Truman found the Clifford-Elsey report so incendiary that he confiscated the twenty extant copies Clifford had prepared but not yet distributed and locked them up permanently in the White House safe. "I read your report with care last night," he told Clifford. "It is very valuable to me—but if it leaked it would blow the roof off the White House, it would blow the roof off the Kremlin." Truman was ready to blow roofs, but he judged that the American people were not.

His military felt its vulnerable disarray. The US had fewer than two Army divisions and twelve air groups in Europe. From twelve million men at the end of the war, the armed forces had declined to 1.5 million; military spending had plunged from $90 billion to $11 billion per year as the Truman administration moved to reduce the deficit accumulated during the war. "The same thing happened here as everywhere else," a disgusted Curtis LeMay would write a friend from Europe the following year. "Everyone dropped their tools and went home when the whistle blew. The property is in terrible shape and we do not have enough people left in the theater to properly take care of it." USAAF General Lauris Norstad, briefing the President on the situation in Europe at the end of October 1946, told him, "Simple arithmetic dictates hasty withdrawal in the event of an emergency. It is our estimate that it would require good fortune as well as good management to retain as much as a small bridgehead on the continent of Europe if the Russians should decide to strike. . . . At this time, [the USSR] appears not only the *most* probable, but in fact the *only* probable source of trouble in the foreseeable future." Fewer than half of the forty-six 509th Silverplate B-29s—so-called because they were the only US aircraft modified to carry atomic bombs—were still operational, and even those were airworthy only half the time. George Marshall, who replaced Byrnes as Secretary of State in January 1947, told a Pentagon audience some years later, "I remember, when I was Secretary of State, I was being pressed constantly, particularly when in Moscow, by radio message after radio message, to give the Russians hell. . . . When I got back I was getting the same appeal in relation to the Far East and China. At that time, my facilities for giving them hell—and I am a soldier and know something about the ability to give hell—was one and a third divisions over the entire United States. That is quite a proposition when you deal with somebody with over 260 and you have one and a third."

Truman was sensible to the disparity, but apparently believed the atomic bomb evened it out—two bombs, after all, had ended a war. When Clark Clifford had presented the President with a new design for the presidential seal a few weeks after the Japanese surrender, Truman had suggested adding lightning emanating from the arrowheads in the eagle's left claw as a "symbolic reference to the tremendous importance of the atomic bomb." (Clifford convinced Truman that the addition would ruin the design.) "This," comments Clifford, "was the first time I heard the President talk about the atomic bomb, and I thought it told more than he publicly acknowledged about the impact the bomb had had on his thinking." In October 1946, Truman's press secretary, Eben Ayers, told the President that radio commentator Drew Pearson had claimed the US had moved atomic bombs to Europe. "The President said what Pearson had said was a 'lie.' He said there were no bombs, either with or without detonators, in [Europe]. In fact, he said, none had gone out of the United States except those used in the Bikini tests and those dropped on Japan. He said he did not believe there were over a half dozen in the United States, although, he added, that was enough to win a war."

Groves may have been the source of Truman's conviction. According to Bradbury, "Groves had a mystic number [of weapons he believed should be available in the stockpile], which when we got to it was all he gave a damn about. When we got there I think he was also leaving . . . and at that time I think he thought the [United States] was in control [of the world]." Groves left when the AEC took over; the official stockpile at that time stood at seven Mark III bombs.

The newly appointed AEC commissioners flew out to Los Alamos in January 1947 to lay hands on the weapons their organization had inherited from the Manhattan Engineer District. "I was very deeply shocked," commissioner Robert Bacher told an interviewer in 1953. ". . . I actually went into the vaults

... and selected at random cartons and various containers to be opened. These I then inspected myself.... Judging by the consternation which appeared on some of the faces around there, I concluded that this must have been about the first detailed physical inventory that had been made.... With weapons, the situation was very bad. We did not have anything like as many weapons as I thought we had."

Bradbury protests:

I knew that number every hour on the hour!... Every time that number would quote-unquote change, and believe me for technical reasons that number would sometimes go down, I'd get a call from the good general [Groves] saying, "What the hell happened now?"

... There were technical reasons. I don't mean that we lost anything, but little details like corrosion or something like that would go wrong and the number of available devices would change....

We were up with the production of material at that time, but there were a lot of things that ... we hadn't solved, particularly in respect to plutonium. ... I mean, sure, you could make a device that would stay pretty and beautiful for a week, two weeks, three weeks, a month. Then it began to look like it had the measles. That didn't really affect its performance, but it sure affected a lot of other things. We had to solve a lot of purely practical, and also very important, problems in corrosion metallurgy, materials stability, all kinds of nasty things which in wartime we never had to solve.... [But] when you ask if I knew what the number was, I sure as hell did.

The scene at Los Alamos sickened David Lilienthal. "Probably one of the saddest days of my life," he remembered, "was to walk down in that chicken-wire enclosure; they weren't even protected, what [bombs] there were.... I was shocked when I found out.... Actually we had one [bomb] that was probably operable when I first went off to Los Alamos; one that had a good chance of being operable."

Dutifully, on April 3, 1947, Lilienthal and the other commissioners carried the news to Truman. "We walked into the President's office at a few moments after 5:00 P.M.," Lilienthal recorded in his journal within hours of the meeting. "I told him we came to report what we had found after three months, and that the quickest way would be to ask him to read a brief document." Everyone at that time, says Lilienthal, "the President included, [assumed] that America had a supply of atomic bombs. In fact, Winston Churchill was declaiming that it was our atomic 'stockpile' that restrained the Soviet Union from moving in on an otherwise defenseless Europe. What we of the new AEC had just discovered ... was that this defense did not exist. There was no stockpile. There was not a single operable atomic bomb in the 'vault' at ... Los Alamos.... Nor could there be one for many months to come." "This news was top secret," Lilienthal adds, "the biggest secret of that time, so secret that I did not commit it to paper." The AEC chairman whispered the number to the President: zero. The lightning struck Truman then: the eagle's claw was empty of arrows. "He turned to me," says Lilienthal, "a grim, gray look on his face, the lines from his nose to his mouth visibly deepened. What [did] we propose to do about it? He realized the difficulties."

"We had lots of capsules—nuclear cores—I guarantee you," Jacob Wechsler explains. "But we didn't have weapons, we had lots of pieces. The idea was, if there was a threat, you would start putting them together. The fusing systems weren't there, the initiators needed to be changed, the detonators had to be stored in desiccated boxes and you put them in when you needed them and then put them on again. And it went on and on this way. We didn't have any weapons, we had piles of pieces. That's what Lilienthal was going in and saying—'We don't have weapons.'" But the President of the United States only knew what his Atomic Energy Commission told him.

15

Modus Vivendi

LEV ALTSHULER FIRST WENT OUT to Sarov in December 1946. He traveled by train with a longtime colleague of Yuli Khariton, a woman named Tatyana Vasilievna Zakharova. "Our future place of work . . . was several dozen kilometers from the train station," Altshuler remembers. "We made this leg of the journey by bus, garbed in thoughtfully provided sheepskin coats. Small villages which recalled Russia from before [the] time [of Peter the Great] flashed by the windows. . . . At the designated place we saw monastery churches and farmsteads . . . and the inescapable companions of the era— 'zones' (prison camps). . . ." The contrast reminded Altshuler of a poem by Mikhail Lermontov and he recited a line: "The land of slaves, the land of lords." Tatyana Vasilievna accused him: "You do not love Russia." "I did not know what to say," Altshuler writes. "The question, 'What does it mean to love Russia?' is the same as the question in the Gospels, 'What is truth?'— there are no answers. Or, in any case, there is more than one answer." So Altshuler's first tour of the installation where he would make his life for the next two decades introduced him to its essential dissonance. As Los Alamos had been for scientists working in the United States on the atomic bomb during the Second World War, Sarov would be a place of comradeship, of exciting and original work, of patriotic fulfillment. But it was also "one of many islands of the 'White Archipelago,'" as Altshuler calls it, Beria's chain of installations for making the bomb, and beyond the domes of the old monastery "the harsh reality was the columns of convicts going through the settlement on their way to work in the morning and on their way to the zone at night."

And in truth the scientists who began streaming out to Sarov in the spring of 1947 were prisoners themselves, even if their cage was gilded. "It was like this," remembers physicist Victor Adamsky, who joined the staff a few years later: "you come and you don't even communicate. If you are sick and you need a doctor then they will send you somewhere." Altshuler elaborates:

Leaving the Site for personal, and even for business reasons, was very difficult to arrange. In a gloomy moment one of the local poets wrote a ballad beginning with these words:

A plane flies from Moscow to Sarov.
Whoever comes here will never go back.

The regime of secrecy had an oppressive effect as well. It was not merely a regime, but a way of life, which defined people's behavior, thoughts and spiritual condition. I often dreamt the same dream, from which I would awaken in a cold sweat. I dreamt that I was in Moscow, walking down the street carrying top secret and extremely top secret documents in my briefcase. I was killed because I could not explain why I had them.

Altshuler had worked during the war with his friend Veniamin Zukerman; at Kazan, where so many scientific institutes had been evacuated, they had met Yakov Zeldovich and Yuli Khariton. Zukerman, also a specialist in radiography, had started the war developing a rifle-mounted launcher for batch-manufactured glass Molotov cocktails, a weapon of desperation against German tanks. At a mortar factory near Moscow in the summer of 1942, the factory manager had shown Zukerman a strange new captured German munition with a metal-lined conical cavity shaped into its explosive charge —"an event," says Zukerman, ". . . that would shape my own destiny and the destiny of our laboratory for many years to come." The cavity reduced the volume of explosives in the shell, yet the shell pierced armor three or four times thicker than a conventional shell of the same caliber could penetrate. Zukerman was fascinated. He realized that night, lying awake during an air raid, that he could use flash X-ray radiography to film the explosion of the strange German shell in progress. Back in Moscow he looked up Yuli Khariton, who directed him to officials who sponsored his subsequent work in microsecond radiography for the study of explosion and detonation phenomena. The German shell turned out to be a shaped charge; Zukerman renewed his acquaintance with the Munroe effect. In 1946, Khariton recruited Zukerman and Altshuler to work under Zeldovich on implosion. Before they were through, Altshuler had been nicknamed "Levka the Dynamite Man."

They moved out to Sarov with their families in May 1947, Zukerman recalls:

Everything we saw came as a surprise to us: the dense woods, the fine hundred-year-old pines, the monastery with its cathedrals and white bell tower on the high river bank, and—in sharp contrast—the gray ranks of strictly-guarded prisoners marching back and forth across the village day and night. The local folklore abounded in stories of innumerable crowds of pilgrims, miraculous cures performed by Saint Serafim of Sarov, a personal visit to the monastery by the Tzar. . . . There was a small factory here, with tracts of dense forest on all sides. The place was well isolated. This made it possible for us to carry out the necessary explosions.

Finnish cottages [i.e., temporary wooden housing] very soon went up in the wooded area. Our family settled in one of them.

A pre-revolutionary three-story red-brick building served as an administrative center and cafeteria. The small factory yielded a smithy, a foundry and a supply of tools. Zukerman and Altshuler set up laboratories in the factory complex. Laborers from the prison zone—*zeks*—built reinforced-concrete bunkers and explosive-confinement barrels in the woods for explosives research. The physicists conducted their first experimental explosion in one of the barrels that May. Pavel Zernov was the chief at Sarov, Yuli Khariton scientific director—Robert Oppenheimer's counterpart. Khariton appointed Zukerman director of the explosion diagnostics division, responsible as well for detonator development. Altshuler got a separate division with his own staff to study implosion hydrodynamics. "We have to learn five times, ten times more than we need to know today," Khariton counseled them. "That's the only way we can lay a decent scientific foundation for later work." It was Kurchatov's and Kapitza's perspective as well, if not Beria's. But Beria was a realist; he would do whatever it took to get the job done.

"I came there to Sarov," Altshuler characterizes the support Khariton arranged for them, "I was given five or six assistants, and that enabled us to conduct experiments."

I had my own workshop. If I suddenly had an idea, I would make a sketch and take it to the mechanics and they would produce the device. We would go to the forest, to a small test site where there were bunkers—the experimentalists and their equipment within and the tested charges outside. We would explore the results ourselves and we would also show them to Khariton. It was very quick. We would work fourteen hours per day, all of us. There were slogans there addressed to the prisoners. *Work For Your Early Release* was one of them. We joked that it was meant for us.

We would meet often in family settings. We had a slogan then too: *To Out-Khariton Oppenheimer.* We felt very insecure, all of us, that only the United States possessed nuclear weapons. The United States found a way to destroy two Japanese cities. We were very nervous about that—nervous that you could use similar bombs against us. It was a feeling of constant concern.

"No one had any decent record players or tape recorders," Zukerman reminisces. He had his family piano shipped out from Moscow, the first musical instrument at Sarov. "I would sit down at the piano and play foxtrots, tangos, waltzes.... Sometimes while I was playing ... we would look out the windows and see couples dancing on the asphalt lane outside." Another Sarov physicist, Boris I. Smagin, recalls that "there was no respect for rank in our small town; Academicians as equals participated in the life of our young and fun-seeking neighborhood." Like many of his Los Alamos counterparts, Smagin remembers those times as "the best days of our lives." They were not the best days for Altshuler, who mentions the *zeks* and the double security perimeter of barbed wire and says he found Sarov "depressing." Someone altered a verse of Pushkin to capture the surrealism of Igor Kurchatov's far-flung scientific enterprise:

> *The Bearded One is rich and famous,*
> *His installations can't be numbered;*
> *Scientists wander there in herds—*
> *Free they may be, but ever guarded.**

They were plagued with shortages. Even two years after the end of the war, coming north from Yugoslavia in the winter snow, Milovan Djilas found "limitless desolation and poverty" in the workers' paradise: "The Ukraine and Russia, buried in snow up to the eaves, still bore the marks of the devastation and horrors of war—burned-down stations, barracks, and the sight of women, on the subsistence of hot water ... and a piece of rye bread, wrapped in shawls, clearing tracks.... We stopped in Kiev only briefly, to be switched to the train for Moscow.... Soon we were on our way into a night white with snow and dark with sorrow."

But Khariton had trained at Cambridge in the string-and-sealing-wax tradition of British physics; the Sarov experimenters made do. Electricity at the forest test site came from an American Lend-Lease five-hundred-kilowatt generator. Exposure to salt water on the sea route to Vladivostok had damaged its windings; Zukerman and a young protégé, Arkadi Admovich Brish, personally repaired it. Brish, twenty-nine in 1946, "blond, with gray eyes nearly the color of steel" in Zukerman's recollection, later the director of a leading institute, was a man so vigorous that the Sarovians cooked up a unit of productive activity—the Brish—in his honor and measured their own contributions comparatively in milli- and micro-Brishes.

---

* The original, from *Poltava*, went:
> Kochubei is rich and famous,
> His fields vast and boundless;
> Herds of his horses
> Graze there, free and unguarded.

"Physicists in the Soviet Union were still snipping flat rubber gaskets for their vacuum instruments out of auto-tire inner tubes, of all things," Zukerman notes. "Getting vacuum hosing was a problem. For a small section of Leibold red vacuum tubing people would offer up anything—from rare liquors to a precision galvanometer. Ramsay's oil, as a rule, would be 'home-brewed' as required out of rubber, wax, and vaseline." Zukerman's first rotating-disk camera for photographing explosions at Sarov depended on a motor cannibalized from a home vacuum cleaner bought in a Moscow second-hand store. To compensate, Zernov roamed the compound cutting through red tape. When Zukerman needed 150 kilos of castor oil, Zernov delivered a two-hundred-kilo barrel in two days from Bulgaria. When Zukerman worked out a way to destroy cheap mirrors rather than expensive camera lenses photographing explosions, the Sarov director commandeered a handsome specimen that had just been hung in the installation barbershop. Zukerman soon replaced the mirror from a shipment he had ordered from Moscow, but he was persona non grata at the barbershop for months and had to shave himself with an old safety razor—blind, as by then he was almost totally.

"The summer of 1947 was a hot one," Zukerman concludes, "both literally and figuratively. Our scientific divisions quickly grew and came up to full strength."

The summer of 1947 was a hot one for spies as well. Elizabeth Bentley had stunned the FBI two years previously by walking in off the street and confessing her extensive activities as a Soviet courier. Among many others, she had identified Abe Brothman. In the spring of 1947, she began testifying secretly before a federal grand jury. When the grand jury investigated Abe Brothman, he implicated Harry Gold.[*]

Before that, on December 26, 1946, when Gold had been out of touch with Yatzkov for ten months, the chemist took a call in Brothman's office. An attempt to contact Gold by mailing him tickets to a boxing event earlier that month had failed because the ticket envelope had been misaddressed and had arrived too late. But Gold had given Yatzkov the phone number for emergency use if they lost contact, and now he found Yatzkov at the other end of the line. The Soviet agent asked if Gold had been all right. Gold would testify that the question meant: Was he under surveillance? Gold answered that he had been fine. Yatzkov told Gold to meet him that night in

---

[*] Probably in response to Bentley's revelations, one of Julius Rosenberg's associates, the engineer Joel Barr, abruptly left the United States for Europe in 1947. In 1949, Barr defected to the Soviet Union, where as Iozef Veniaminovich Berg he became chief engineer of the Leningrad Institute of Semiconductors. cf Kuchment (1985).

the men's room of the Earl Theater, a movie house in the Bronx. Gold agreed to the rendezvous. It occurred to him later that Yatzkov/Yakovlev must not have known that he was calling Abe Brothman's office; Gold had not identified the location of the phone.

Dutifully Gold trudged out to the Earl Theater. The FBI paraphrases his account of the meeting:

> He was ... approached by an unknown Russian, who had a torn slip of paper, which Gold had ... given to Yakovlev with the understanding that this paper would be used for identification in the event Yakovlev could not make a meeting with Gold. ... The unknown Russian came out of the men's room and walked directly up to him and showed him a portion of this [paper], saying in broken English, "You Harry, you have material from the doctor [i.e., Fuchs]." Gold answered no, and he was then told by the unknown Russian to go to the Third Avenue Bar [on 42nd Street and Third Avenue in Manhattan] to meet Yakovlev. ...

Gold never explained why the Soviet courier—he was in fact Yatzkov/Yakovlev's boss Pavel Ivanovich Fedosimov—expected him to have a package from Klaus Fuchs, just as he never explained why he had gone to Cambridge to look up Fuchs after Yatzkov failed to appear at an earlier Earl Theater rendezvous the previous February. Fedosimov and Yatzkov probably ran Harry all the way out to the Bronx to separate any handoff of material from the Yatzkov meeting; Igor Gouzenko had blown Yatzkov's cover and the FBI was watching him. The Soviets probably also wanted Gold to meet Fedosimov so that the two men would be able to recognize each other in future contacts.

Yatzkov met Gold at the bar, apologized for the ten-month interruption in their work "but said it was unavoidable, that he had to lie low during that period," told Harry "that he was very glad that I was now working in New York" because "it would put much less of a strain upon me as regards to meeting my Soviet contact." Yatzkov paid Gold several hundred dollars for past expenses and told him to begin planning for a mission to Paris in March to meet a physicist. He gave Gold "a sheet of onionskin paper," listing the hotel where Gold was supposed to stay and presumably identifying the physicist Gold was supposed to meet. They would meet in the Paris Metro, the physicist would pass Gold some documents and he would carry them to England and pass them along.

Sitting in the bar, sharing drinks, the Soviet agent and the American courier began inventing strategies Gold could use to get off work to travel to Paris. Maybe he could write a number of French chemists expressing interest in their research; if he got a response, he could use it as an excuse to make the trip. Gold innocently mentioned a timing problem and all hell broke loose:

I told Yakovlev that once the pressure of work at Abe Brothman and Associates had eased up a bit—and then Yakovlev almost went through the roof
of the saloon. He said, "You fool!" He said, "You spoiled eleven years of
work." He told me that I didn't realize what I had done, and he told me
that I should have remembered that some time in the summer of '45 he
had told me that Brothman was under suspicion by the United States
government authorities of having engaged in espionage. . . . Yakovlev threw
down on the table where we were sitting, the bar, an amount of money
which was about two or three times the actual cost of the drinks which we
had had, and he dashed out of that place. I walked along with him for
awhile, and he kept mumbling that I had created terrible damage and that
he didn't know whether it could be repaired or not. Yakovlev then told me
that he would not see me in the United States again, and he left me.

Yatzkov/Yakovlev left the United States the next day on the S.S. *America*
bound for Cherbourg. Harry Gold seems to have believed his unwitting
revelation forced Yatzkov's departure. In fact, the Soviet vice-consul was
scheduled to leave to take up new duties at the Soviet Embassy in Paris and
was traveling with his family. Meeting with Gold was part of closing out his
work in North America. Gold's linkage with Abe Brothman was still an
evident disaster.

Yet nothing came of it, at least not in the short run. When two FBI agents
questioned Brothman for the federal grand jury on May 29, 1947, after
Elizabeth Bentley's damaging revelations, Brothman concocted a story of
legitimate dealings through the woman he knew as Helen but implicated
Gold as a party to his Soviet contacts after "Helen" bowed out. After the
agents had left Brothman at his offices in Long Island City, Gold happened
to drop by from the Brothman laboratory in Elmhurst where he worked. He
found his boss nearly hysterical:

Brothman was in a state of great excitement; he immediately [came] forward to meet me. The first thing Brothman said was, "The FBI were here
—they know everything—they know all about us—they know you were a
courier—they have a photograph of you and me together in a restaurant!
Look, we don't have much time. Look, Harry, you've got to get this straight.
You have got to tell the same story I told of how we met."

The story Brothman had told required Gold to have known Bentley's
deceased lover Jacob Golos—Gold had never met him—and depicted both
men as innocent experts who had merely evaluated chemical processes for
Golos and solicited legitimate business, in the course of which they made
blueprints available. Ever accommodating, Gold memorized Brothman's description of Golos and agreed to make up a story to explain how he knew
the Soviet agent. "[In] about the middle of this limited conversation," Gold

notes wryly, "Brothman said to me, 'Someone has ratted—it must be that bitch Helen!' " Brothman warned Gold that FBI agents were on their way to the Elmhurst laboratory to talk to him. "I promised them I would not talk to you," Brothman cautioned, "so don't let on that we've talked about this. You've got to cover me up and tell them the same story that I told you." Gold did not want to leave until he had thought the problem through; he had just realized "the full import of it. . . ." Brothman insisted he get going. On the subway, Gold quickly invented a plausible story: that an acquaintance had introduced him to Golos at a meeting of the American Chemical Society in Philadelphia in 1940. The acquaintance was a man Gold knew would not talk; he was dead and buried.

At the Elmhurst laboratory, the FBI team questioned Gold from five in the afternoon until nine at night. Brothman and his secretary, Miriam Moskowitz, picked up the shaken chemist at nine-thirty and the three conspirators went to dinner at a Chinese restaurant in Queens. "Harry," Brothman asked Gold guiltily over dinner, "you don't blame me for having brought your name into this, do you?" Brothman rationalized that he expected the FBI to identify Gold sooner or later and that it was better to bring him in at the beginning. Back at the laboratory, where Harry still had work to do, Moskowitz went for coffee while Gold and Brothman finished conflating their stories. Brothman was evidently worried about the extent of Gold's involvement. He knew Gold had been awarded a Red Star—more than the Soviets had done for him. "I [have] got to know all about you," he told the Philadelphia chemist. "What can they find out that I don't know?" Considering the extent of Gold's courier work, his response was either disingenuous or self-absorbed: "I then told Brothman that in reality I had never been married, and further, that my brother was still alive and had not been killed in the Pacific, and that I lived with my family in Philadelphia. . . . Brothman made many recriminations for my having told these falsehoods, but he said that he did not think these points would be serious." Brothman tried once more, expressing "concern . . . that I might have had other dealings in my association with the Soviets with which he was not familiar." Gold said not a word.

FBI agents visited Gold's home over the Memorial Day weekend at his invitation. He had hurried to Philadelphia ahead of them and had destroyed several incriminating documents, including the sheet of onionskin Yatzkov had given him at their last meeting. He told the agents that he had found no blueprints and they seemed to accept the story.

Gold and Brothman met several more times before they testified individually before the grand jury. Brothman nursed a grudge against Gold that grew with each meeting. He blamed Gold for involving him with Tom Black's activities, challenged Gold to tell him "everything" and worried that he might be "caught short while testifying." When Brothman's summons arrived, he threatened to "tell the grand jury the true story of his work for the

Soviet Union. . . ." Gold, Moskowitz and Brothman's attorney succeeded in dissuading the volatile chemist. After Brothman testified, he bragged over dinner that he had "neither cringed, flinched, or begged." The grand jury, he warned Harry, had been "stuffed to the gills with stories of spying."

Then it was Gold's turn. He was nervous, perhaps frightened. He and Brothman drove around aimlessly for most of the night before his grand jury appearance on July 31, 1947. Gold wanted to review Brothman's testimony. "Abe kept brushing me off and went into a great dissertation on political theory and the declining state of capitalism." They stopped to eat watermelon "and other time-killing incidents." It was four in the morning by the time they arrived at Brothman's parking garage—Harry was staying with Brothman while Abe's wife summered in Peekskill. An attendant took the car and the two men wandered the neighborhood on foot. Gold came close to telling his boss and nominal business partner about his other courier work, mentioning "trips by railroad and plane" that had created records the FBI might locate. He "did not come out and say that these trips were in connection with my Soviet espionage activity, [but] it was certainly understood. . . ." Brothman told him not to worry and coached him on how to play to the grand jury. Gold got two hours' sleep. He looked in on Brothman before he left to testify. Brothman's parting words were, "Look, Harry, you don't hold it against me for having brought your name into this, do you?" After Gold appeared before the grand jury, he told Brothman "that I thought that I had succeeded in putting across . . . the fact that I was a blunderer [rather than a spy]. . . ." Far from blundering, Gold's skillful evasions had thrown off the grand jury and Brothman as well.

None of Bentley's charges stuck. Lacking corroborating evidence, the grand jury returned a no bill. Gold now had a file at the FBI that identified him as a possible Soviet espionage courier and included his physical description, his background and profession and his affiliation with Brothman. But there were no charges pending against him on the American side. Since Yatzkov had fled him as if he were a leper he had heard no more from the Soviets. As of September 1947, Harry Gold was a free man, to the extent that he could ever be. A loss marked that unshackling. On September 27, his mother died. Everyone attended the funeral: Tom Black, Abe Brothman, Miriam Moskowitz, his friends from Pennsylvania Sugar. They met no red-headed Gimbel's-model wife or twin children, only Harry, overweight and thirty-six years old, his forbearing father and his brother Yosef, the brother who he told Moskowitz had died a paratrooper in the Pacific War. Celia Gold had dominated her husband and her sons. Her death and the suspension of his espionage work would open Harry's life to the possibility of change. What he called his "repressed longing for a family" would finally emerge to expression. So would guilt. "I had the leisure to reflect at length," he writes, "and to evaluate the damage I had done, the full implications involved in this spying, and inevitably, [I came] to the horrible and sickening realization that it had all been . . . a tragic and irremediable mistake."

Donald Maclean, for whom espionage was lavatory-attendant duty, approached the very Augean stables of atomic policy information in February 1947 when he was appointed British co-secretary of the Combined Policy Committee in Washington. By then he had already had access to Tube Alloys files at the British Embassy for more than a year. He also had a pass to the Pentagon and would soon receive a coveted unescorted pass to the offices of the US Atomic Energy Commission.

British and American ambitions had just come into open conflict. The wartime Quebec Agreement that Winston Churchill had arranged with Franklin Roosevelt in August 1943, still in force, provided that their two countries would never use the bomb against each other but also "that we will not use it against third parties without each other's consent"—effectively giving the British a veto over US sovereignty in the matter of atomic war (dropping the bombs on Hiroshima and Nagasaki had not been only a US decision; the CPC with its British and Canadian representatives had met formally in furtherance of the Quebec Agreement and approved the use of atomic bombs against Japan). In September 1944, Churchill and Roosevelt had enlarged their previous agreement to include "full collaboration . . . in developing Tube Alloys for military and commercial purposes . . . after the defeat of Japan. . . ." Truman, Clement Attlee and Mackenzie King had reaffirmed "full and effective cooperation" in November 1945. During the war, all production of Belgian Congo uranium ore, the richest source in the world, had flowed directly to the United States, although the British had paid for half. Attlee had properly demanded dividing the annual production now that the war was won; Truman had agreed even though half the Congo supply plus supplies from sources in the US and Canada were not even sufficient to support the plutonium production reactors at Hanford, one problem the new AEC commissioners mentioned in the shocking report they asked Truman to read on April 3. The British for their part had no immediate use for the ore. They were stockpiling it against the time when their own secret bomb program could put it to use. Congress had passed the Atomic Energy Act in ignorance of these secret agreements; one provision of the act forbade sharing information on the design and manufacture of nuclear weapons with foreign governments. Agreements were therefore in conflict with US law and would have to be reconciled. And Donald Maclean was positioned to observe and to participate in these revealing negotiations.

The sorry state of the US atomic arsenal was one secret at least that Donald Maclean evidently did not learn. Halfway through 1947, feeling the pressure of US Cold War initiatives, the Soviet Union began rebuilding its military forces. From three million men at arms at the beginning of the year, Soviet forces began a gradual increase to more than five million. "Immense Russian superiority in conventional arms," Alexander Werth interprets the expansion, "was the only effective antidote [the Soviet Union] had against an atomic attack on Soviet territory." It would not have been much of an attack.*

Lilienthal had briefed Truman on the stockpile disarray just weeks after the President had proposed (on March 12, 1947) the Truman Doctrine that offered "to support free peoples who are resisting attempted subjugation by armed minorities or by outside pressures." The doctrine, which amounted to a declaration of Cold War, had evolved into a major US policy position from a British appeal to the US to take over assisting Greece and Turkey, something the British, victorious but nearly bankrupt, could no longer afford to do. Secretary of State George Marshall had been meeting with the Council of Foreign Ministers in Moscow at the time of Truman's declaration. What to do about Germany—divided into Soviet, US, British and French zones of occupation and still nearly moribund—had been once again the central issue on the agenda. Charles Bohlen served at Marshall's translator:

> During our visit . . . Stalin took a relaxed attitude toward the failure of the conference to achieve any results. Doodling the inevitable wolf's heads with a red pencil, he asked what difference it made if there was no agreement. "We may agree the next time, or if not then, the time after that." To him, there was no urgency about settling the German question. We should be detached and even relaxed about the subject. . . .
>
> Stalin's seeming indifference to what was happening in Germany made a deep impression on Marshall. He came to the conclusion that Stalin, looking over Europe, saw that the best way to advance Soviet interests was to let matters drift. Economic conditions were bad. Europe was recovering slowly from the war. Little had been done to rebuild damaged highways, railroads, and canals. Business alliances severed by years of hostilities were still shattered. Unemployment was widespread. Millions of people were on short rations. There was a danger of epidemics. This was the kind of crisis that Communism thrived on. All the way back to Washington, Marshall talked of the importance of finding some initiative to prevent the complete breakdown of Western Europe.

In the midst of these transitions, with terrible timing, the British unwisely chose to make an issue of the wartime agreements. At the end of January

---

* As chairman of the General Advisory Committee to the AEC, Robert Oppenheimer knew how few bombs the US had stockpiled. The fact that Stalin fielded two million men to oppose what he believed to be a plentiful US stockpile is clear evidence, if any is required, that Oppenheimer was not a Soviet spy, as J. Edgar Hoover believed and as retired KGB general Pavel Sudoplatov has alleged.

1947, Roger Makins, deputy chairman of the CPC, raised the question with Dean Acheson of exchanging information about reactor design. Acheson suggested in turn dropping British consent to use the atomic bomb. Makins returned to England to report. Acheson briefed Lilienthal and Marshall on the exchange, emphasizing that "some action is urgently needed."

Congress had to be informed. "The somewhat incredible truth," Acheson would note in his memoirs, "was that very few knew about [the Quebec Agreement] . . . and those who did thought of it as a temporary wartime agreement. . . ." Lilienthal and the other AEC commissioners met with the Joint Committee on Atomic Energy for the first time on May 5 and taught the committee members the facts of life:

> There was some alarm expressed that England is getting half of the Belgian [Congo] uranium output and some surprise at learning that Great Britain and Canada actually had had men participating . . . in the development of the bomb itself, during the war. Senator [Tom] Connally [of Texas] said that then you mean that England knows how to make the bomb. The answer is certainly "Yes." I explained that we had been concerned that the Joint Committee . . . had not previously been informed of our agreements with England and recommended that they learn of these basic agreements directly from the Secretary of State and the Under Secretary. This apparently they will do.

A little later in his journal Lilienthal adds: "This was in accordance with a prior understanding with Acheson. They were considerably shocked to learn that we have so little in this country in the way of raw materials."

Acheson briefed the JCAE on May 12. "When he read the terms of the agreement," recalls State Department atomic energy specialist Gordon Arneson, "the hearing room erupted in indignation and anger. Several members walked out at the very thought that we'd have to ask anybody's permission to use these weapons." Two Republican senators in particular—Arthur Vandenberg and Bourke Hickenlooper—were outraged at the British veto as well as the ore arrangements. Vandenberg, a former newspaper editor, had swung from isolationism to vital support of bipartisan foreign policy during the Second World War and had been crucial to enlarging US alliances in the early years of the Cold War. He believed, as he had written a friend in February, "that I am best serving my country when I provide the tightest possible 'public control' of atomic energy in the United States." Hickenlooper, in contrast, was a Red-baiting ultraconservative. In the months to come, both men pushed to improve the US position.

England's financial situation gave them leverage. Back from the Moscow conference, Marshall quietly raised the issue of European aid in a commencement speech at Harvard University on June 5, 1947 (Robert Oppenheimer sat behind him, a fellow candidate for an honorary degree). "The

United States should do whatever it is able to do," the Secretary of State proposed, "to assist in the return of normal economic health in the world, without which there can be no political stability and no assured peace. Our policy is directed not against any country or doctrine but against hunger, poverty, desperation and chaos.... The initiative, I think, must come from Europe. The role of this country should consist of friendly aid in the drafting of a European program and of later support of such a program so far as it may be practical for us to do so." Prodded by British Foreign Secretary Ernest Bevin, the Europeans quickly responded. The program that resulted would be called the Marshall Plan.

Marshall's offer came none too soon for the British. "It is very secret indeed," Acheson told Lilienthal privately on June 28, "but the British are almost out of dollars. The way things are going, unless something can be done ... in a few months they will have exhausted about all but $500 million of the British loan of 7½ billion." Lilienthal immediately took the point. "I asked if this situation might make it worth their ... while to consider changing the allocation of [uranium] ore so we would be buying and paying for the ore? I pointed out that we could use it because we had the plants, but they can't, perhaps for some time." Acheson was skeptical. "No, he didn't think there was a chance of that. We had offered sometime back ... to have us buy it all and hold it in this country subject to later allocation, but they would have nothing of it."

Lilienthal never says he talked to Vandenberg and Hickenlooper about forcing a *quid pro quo.* The AEC chairman and the two senators may have arrived at the same conclusion independently, but by Lilienthal's own report, the approaching British bankruptcy was "very secret indeed." In any case, by August 1947, Hickenlooper was writing to Marshall bluntly that "the present agreement [with the British], in view of all the circumstances, is intolerable." He wanted it revised. If the British refused, the Iowa senator threatened the Secretary of State, "I shall oppose, as vigorously as I can, and publicly if necessary, any further aid or assistance to Britain." When Lilienthal had breakfast early in September with Bernard Baruch, the two men talked about "some important matters concerning raw materials—this was at [Baruch's] insistence—and its relation to the Marshall Plan." Lilienthal goes on to endorse explicitly the idea of a *quid pro quo:* "This made sense to me and fits into plans we have under way." And when Lilienthal and his fellow commissioners discussed ore supply estimates at AEC meetings later in September, they agreed that the US needed the entire free world supply and reviewed a policy proposal that would require both the British and the Canadians to give up not only future allocations but also their existing stocks of ore. Since only the United States was in a position to process the ore into atomic bombs, the proposal reasoned, such a change of policy would benefit mutual security.

It would look bad to blackmail a friendly sovereign state, Marshall argued in a meeting at the Pentagon in mid-September with the new Secretary of Defense, James Forrestal. Fearing a damaging outcry, the State Department wanted to decouple aid and ore. Forrestal agreed. (The National Security Act, which Truman signed on July 26, 1947, after months of brutal infighting between the Army and the Navy, replaced the War Department with a Department of Defense and created a separate Air Force, a Central Intelligence Agency and a National Security Council. As Navy secretary, Forrestal had stubbornly opposed the new arrangement and had succeeded in weakening the authority of the Secretary of Defense. So that the combative Forrestal would not subvert the work of the new office, Truman gave him the job.)

Hickenlooper pursued the problem as vigorously as he had promised. He bawled out Lilienthal for two hours in mid-October. "Got me so disgusted and mad during the first bitter scolding that I almost lost control and told him off," Lilienthal reprised the argument in his journal, noting however that "there was some point in some of his complaint. . . ." Twice that month, Hickenlooper's office nudged the Joint Chiefs of Staff to make up their minds, two years after the war, about how many atomic bombs the military needed. The Joint Chiefs decided that the United States could use no more than 150 "Nagasaki type" bombs, basing that number on a Pentagon study that envisioned "attacks on approximately 100 different urban locations" in the USSR. "The efficient utilization of atomic bombs," the JCS reasoned, "will dictate the use of one bomb only in any one attack on an objective area. Therefore, the maximum which would be dispatched in any one attack under present conditions is unlikely to exceed 100." In October 1947, that is, the US military believed officially that 150 atomic bombs with a total yield of three million tons of TNT equivalent—3 megatons—would be sufficient to defend the United States and defeat the USSR, its most likely enemy. In the years to come, one hydrogen bomb would develop four times that total yield and the number of such weapons in the US arsenal would multiply to four and then five figures, the extent of the threat always increasing in official estimates to match the rate of manufacture. For the time being, however, the US had the opposite problem.

As yet ignorant of the secret US debate, the British continued to push for more information on bomb and reactor technology. The three CPC nations convened a three-day conference in Washington in mid-November "to determine," explains a British diplomat, "which wartime secrets could now be declassified." The point was to declassify information by common agreement rather than piecemeal and inconsistently. Both Donald Maclean and Klaus Fuchs attended the declassification conference. Presumably neither man knew that the other was a spy. Among other subjects, the conferees discussed atomic weapons. One participant, Berkeley physicist Robert L. Thornton, told the FBI in 1950 "that Fuchs had on occasion exasperated some of the panel members because of his conservative attitude with regard

to the advisability of declassifying certain atomic energy type information.
... Dr. Thornton concluded that if Fuchs was attempting to determine what
the panel members considered as holding forth the greatest promise for
future military development in the field of atomic energy, he could not have
been more 'damnably clever.' "

A congressional debate on the Marshall Plan was approaching. If the
Departments of State and Defense had decoupled aid and ore, Vandenberg
and Hickenlooper had not (nor, except cosmetically, had the AEC). In a
meeting at the Pentagon on November 16, 1947, of Undersecretary Robert
Lovett for the State Department, Forrestal for the Defense Department, Vandenberg and Hickenlooper, the two senators reaffirmed their insistence on
a change in Anglo-American arrangements. Vandenberg called the wartime
agreements "astounding" and "unthinkable." "He said that failure to revamp
the agreements," his son paraphrases his contemporary diary, "would have
a disastrous effect on Congressional consideration of the Marshall Plan. ...
Both Senators ... said that a satisfactory conclusion must be reached before
final action on the Marshall Plan program." A few days later, State Department staff members George Kennan and Edmund Gullion tried unsuccessfully to move the AEC commissioners toward a broader cooperation—
"something that smacks of an alliance," Lilienthal recorded distastefully,
adding, "I shall do everything I can to discourage and prevent this; it is not
the way to approach this matter." (In the midst of these discussions, William
Donovan, whose Office of Strategic Services was about to lose its authority
over US intelligence to the new CIA, sent "a young fellow" to see Lilienthal
bearing "a weird story, brought in from eastern Czechoslovakia (the Car-
pathians) by a Jewish rabbinical student. ..." The story "purport[ed] to concern Russian efforts in atomic energy." Lilienthal was unimpressed that the
Soviets might be mining ore in Czechoslovakia. "Turned him over to [an
aide]," he noted in his journal.)

Vandenberg was restive. In late November, he said publicly (in Lilienthal's
paraphrase) "that before the Marshall Plan was passed the United States
would insist on Belgium's uranium," a statement a Communist newspaper
in Belgium immediately headlined. Lilienthal finally had a proposal for the
British to show the JCAE, however—"it has been months and months in
coming"—and on November 26 he and Lovett met with Vandenberg and
Hickenlooper to preview it. "To my great surprise," writes Lilienthal, "it was
received almost without question, in every essential." It ought to have been:
it proposed abrogating the wartime agreements, diverting all Belgian Congo
ore production to the United States and acquiring the entire British stockpile.

Lilienthal and Lovett presented the proposal to the full Joint Committee
on December 5, 1947. The problem from State Department's point of view
was to find some face-saving interest other than Marshall Plan aid to seem

to exchange for British capitulation. The exchange that the AEC proposed was technical information on nuclear energy development. The JCAE "accepted the idea that [the proposal] should not be explicitly tied to economic relief to Britain—though, of course, the two cannot be wholly separated in reality," Lilienthal records. Vandenberg told them bluntly, according to Lilienthal's notes, "I don't recognize that [British] veto [on the use of the bomb]. . . . It can be a source of desperate embarrassment. Horrified and have been since I learned of it. Vitally essential that we wipe that off beyond any dispute." Lovett would handle the negotiations; Vandenberg gave him a deadline of December 17, twelve days away.

Lovett cabled the American ambassador in London immediately after the meeting ended, and his cable explicitly linked aid and ore: "Vandenberg and some others strongly feel that further aid to Britain . . . should be conditioned on Britain's meeting our terms with respect to allocation of atomic raw materials." The Marshall Plan could be at risk, he warned. The British were infuriated, but they had their backs to the wall.

Donald Maclean participated in the next two weeks of hard negotiations, conducted through the Combined Policy Committee. Information exchange as the negotiators agreed to define it boiled down to subjects peripheral to bomb development: health and safety, research radioisotopes, fundamental physics. The Americans agreed to share information on designing natural-uranium power reactors, which was the information Roger Makins had sought almost a year previously when he opened the can of worms. It was harder to agree on ore allocations. The Joint Chiefs weighed in on December 17 to bolster the US position with a convenient finding that for their 150 atomic bombs, "preemption of the world's supply of usable ores and stockpiling within the United States is an urgent and at this time paramount consideration for national security." At the last minute, the British delegates had to return to London to report. Vandenberg and Hickenlooper waited impatiently. The delegates returned with reluctant agreement: the British veto on using the atomic bomb would be surrendered, Britain would ship two-thirds of its ore stockpile to the US and the US would get all Belgian Congo ore for at least the next two years. The negotiators named the agreement a *modus vivendi*. (Gordon Arneson explains: "A *modus vivendi* says, 'Mr. A sits here and says I would propose to act in this way.' B sits there, and he says, 'Yes, I intend to work the same way.'")

Both sides initialed the *modus vivendi* at what Lilienthal calls an "undramatic and unimpressive" ceremony on January 7, 1948. The AEC commissioners had met that very morning to review an agreement vital to their work that they had not yet seen; Lewis Strauss, always fractious, had resisted allowing even the marginal technical cooperation with the British that the *modus vivendi* described until limits were written into the AEC minutes. "I never felt more ashamed of the internal workings of an organization," Lilienthal complained. At the ceremony, Maclean pulled Gullion aside to comment acidly, in Arneson's recollection, " 'You know, this is hardly an agreement among people who agree; it's an agreement among people who disagree.' Ed replied, 'Yes, that's about right.' " Gullion chose not to take offense. Since the United Nations charter required all international agreements to be registered (which would alert the Soviet Union), and the US Atomic Energy Act forbade sharing atomic-energy information with foreign governments (Strauss's objection), the hard-won *modus vivendi* was doubly illegal. A British diplomat says it "had to be kept under wraps." In fact it was kept secret even from Congress outside the JCAE.

By acceding to the *modus vivendi,* the British effectively delayed their program to build an atomic bomb by several years. The negotiations, says Dean Acheson, "left the British with a sense of having been ungenerously, if not unfairly, treated." The information Maclean gleaned on ore supplies and U235 and plutonium production rates might have been valuable to the Soviet Union, but the United States had not even tabled that information honestly during the negotiations. The AEC reported to the FBI several years later that "the estimates of raw materials supply that were used in [the] Combined Policy Committee calculations of 1947 were much under the actual supplies received in that period." By 1947, with fewer than 1.5 million men under arms, the United States had made the atomic bomb its first line of defense (however thin the line) as it began to mount a broad, worldwide challenge to what it perceived to be Soviet expansionism; under such conditions, as it inevitably does for sovereign states, the end of national security justified almost any means.

16

Sailing Near the Wind

JOHN VON NEUMANN once told Françoise Ulam that he had never met anyone with as much self-confidence as her husband—"adding," writes Stanislaw Ulam puckishly, "that perhaps it was somewhat justified." Ulam, a handsome, aristocratic Polish mathematician of first rank, had emigrated to the US in the 1930s with von Neumann's encouragement, had been a Junior Fellow at Harvard, had taught at Harvard and the University of Wisconsin (where he and Françoise married) and had found his way to Los Alamos late in 1943. Theoretical physics was a new field to him then; on his first visit to the Los Alamos Tech Area, when he had encountered von Neumann discussing theory with dark, intense Edward Teller, the "tremendously long formulae on the blackboard" had scared him. "Seeing all these complications of analysis, I was dumbfounded, fearing I would never be able to contribute anything." But the equations stayed on the board from day to day, which meant to Ulam that the pace of invention was relatively slow, and he soon regained confidence. "I found out that the main ability to have was a visual, and also an almost tactile, way to imagine the physical situations, rather than a merely logical picture of the problems."

He was soon engaged with Teller, attempting to derive more rigorously a formula Teller had roughed in for the inverse Compton effect that plagued Teller's dreams of a thermonuclear—the effect that cooled the reaction by spilling out radiation and seemed to prevent it from propagating. Ulam appraised the volatile Hungarian perspicaciously:

When I first met Teller, he appeared youthful, always intense, visibly ambitious, and harboring a smoldering passion for achievement in physics. He was a warm person and clearly desired friendship with other physicists. Possessing a very critical mind, he also showed quickness, sense, and great determination and persistence. However, I think he also showed less feel-

ing for true simplicity in the more fundamental levels of theoretical physics. To exaggerate a bit, I would say his talents were more in the direction
of engineering, construction, and the surveying of existing methods. But
undoubtedly he also had great ingenuity.

George Gamow, the Russian emigré theoretician who had offered Teller the
position at George Washington University that brought him to the US before
the war, told Ulam later that Teller had been a different person then—
"helpful, willing, and able to work on other people's ideas without insisting
on everything having to be his own. According to Gamow, something
changed in [Teller] after he joined the Los Alamos Project."

With the war over and Teller and so many others going, the Ulams decided to leave as well. They had both lost family in the Holocaust—Françoise's mother at Auschwitz—and they had both become American citizens,
so there was no question of returning to Europe. Ulam accepted an associate
professorship at the University of Southern California in Los Angeles, a place
the couple found strange—"I used to say that any two points in Los Angeles
were at least an hour's drive apart," Ulam writes. They settled in without
quite settling down. Then Ulam was struck with violent illness, a "fantastic
headache" that was "the most severe pain I had ever endured." When Françoise finally roused a doctor and got her husband to the hospital, he was
vomiting bile. "The surgeon performed a trepanation not knowing exactly
where or what to look for. He did not find a tumor, but did find an acute
state of inflammation of the brain. He told Françoise that my brain was
bright pink instead of the usual gray. These were the early days of penicillin,
which they applied liberally."

Ulam lapsed into a coma. His wife, his doctors and his friends worried
about brain damage. When he awoke a few days later, Ulam worried about
it even more. "One morning the surgeon asked me what 13 plus 8 were.
The fact that he asked such a question embarrassed me so much that I just
shook my head. Then he asked what the square root of twenty was, and I
replied: about 4.4. He kept silent, then I asked, 'Isn't it?' I remember Dr.
Rainey laughing, visibly relieved, and saying, 'I don't know.' "

Recovery was slow. Ulam spent weeks in the hospital. Nicholas Metropolis,
who had set up the ENIAC problem, traveled all the way from Los Alamos to
visit him and let slip that "the security people . . . had been worried that in
my unconscious or semi-conscious states I might have revealed some atomic
secrets." The peripatetic mathematician Paul Erdös turned up just as Ulam,
still shaky, was leaving the hospital. On the way home they talked mathematics, whereupon Erdös pronounced Ulam "just like before." When Ulam beat
Erdös at chess, he decided he had escaped damage.

On leave from the university, resting at home during his extended recovery, Ulam amused himself playing solitaire. Sensitivity to patterns was part

of his gift. He realized that he could estimate how a game would turn out if he laid down a few trial cards and then noted what proportion of his tries were successful, rather than attempting to work out all the possible combinations in his head. "It occurred to me then," he remembers, "that this could be equally true of all processes involving branching of events." Fission with its exponential spread of reactions was a branching process; so would the propagation of thermonuclear burning be. "At each stage of the [fission] process, there are many possibilities determining the fate of the neutron. It can scatter at one angle, change its velocity, be absorbed, or produce more neutrons by a fission of the target nucleus, and so on." Instead of trying to derive the expected outcomes of these processes with complex mathematics, Ulam saw, it should be possible to follow a few thousand individual *sample* particles, selecting a range for each particle's fate at each step of the way by throwing in a random number, and take the outcomes as an approximate answer—a useful estimate. This iterative process was something a computer could do.

Ulam was happy to be invited to the Super Conference in April 1946. Associating Los Angeles with his illness, he was even happier to be invited to return to work at Los Alamos. When he told von Neumann about his solitaire discovery, the Hungarian mathematician was immediately interested in what he called a "statistical approach" that was "very well suited to a digital treatment." The two friends developed the mathematics together and named the procedure the Monte Carlo method (after the famous gaming casino in Monaco) for the element of chance it incorporated.

Two years after the end of the war, thermonuclear research at Los Alamos was still almost entirely theoretical. The model system on which calculations were based continued to be Teller's Super—a pipe of liquid deuterium with an atomic bomb screwed to one end. The medium that would transfer energy from the atomic bomb to the deuterium would be the copious flood of neutrons that the fission explosion produced. Implosion appeared to be an unsuitable geometry for the Super's fission component, however. Implosion generated neutrons at the center of an imploding mass of hydrogenous material—the chemical explosives that supplied the initial squeeze that assembled the critical mass—and the material interacted with the outward flow of neutrons, soaked up energy and otherwise got in the way. "The radiation just would not get out," Hans Bethe comments, "because there's all that high explosive around. First the uranium [tamper] and then the high explosive." Nor was the Mark III implosion bomb hot enough to do the job. Instead of an implosion system they decided they would need a uranium gun—a system that assembled essentially bare critical masses—and a big, powerful uranium gun at that, one that would yield not the nominal 13.5 kilotons of the Hiroshima device but several hundred kilotons. No uranium gun had yet been tested under experimental conditions, on a tower with instrumentation; Little Boy, the Hiroshima bomb, was the only such device so far exploded.

It had also been clear for some time that deuterium alone required too high an ignition temperature to propagate on its own, that some amount of rare and expensive tritium would have to be incorporated into the design, not necessarily mixed with the deuterium throughout the pipe but at least mixed with the deuterium at the end nearer the fission bomb. "The main question," comments John Manley, who became associate director of Los Alamos in 1947 as well as secretary to the AEC's General Advisory Committee, ". . . was whether you could under any circumstances get a propagating reaction in a straight deuterium mixture. And the general conclusion I knew: that you couldn't. Then the question [was] how much would you gain from t[ritium] addition . . . and any other tricks. . . ."

Teller, doing basic physics with Enrico Fermi at the University of Chicago, kept in touch, meeting travelers passing through Chicago and spending summers on the mesa as a consultant. "One might be going to Washington," Carson Mark recalls, "and stop off in Chicago and Edward was always interested in what people were doing and what progress might have been made. He could always be counted on for suggestions. . . . He followed things really quite closely, spent very generous amounts of time here and was of course very helpful. He was also personally very interested. Not only in the work on thermonuclear problems, possible thermonuclear systems, but also on what was going on in the fission bomb studies. . . . He really expressed an interest in everything. But the thing on which he was most inclined to try to work was the Super problem."

After his 1947 summer consultancy, Teller assessed the H-bomb effort at Los Alamos in a technical report, "On the Development of Thermonuclear Bombs," a marker of the lab's progress and of Teller's own contemporary judgment of where the work stood. The report found that the Super was "probably feasible" but judged that "its complex construction gives little hope that it can actually be made to work in the next three to four years." Mentioning what Carson Mark calls "some adverse effects not previously taken into account," it doubled the April 1946 Super Conference estimate of the amount of tritium the Super would require.

Teller's memorandum also reviewed problems with his 1946 Alarm Clock system—the system that incorporated layers of fusion materials into the concentric shells of an enlarged implosion device. Whether or not such a device was feasible, Teller concluded, depended on how much the light fusion materials would mix with the heavy fission materials in the course of the implosion. Physicists refer to an element's atomic number generically as Z. Hydrogen has a Z of 1, helium 2, and so on to uranium at $Z = 92$ and plutonium at $Z = 94$. Elements of higher Z radiate more rapidly at higher temperature than elements of lower Z. So mixing of high-Z and low-Z materials within an Alarm Clock system, while such mixing would compress the fusion materials and enhance their reaction, would also increase the mixture's cooling by radiation. Hans Bethe thought Teller's Alarm Clock conclusions in this 1947 report were "most pessimistic." On the other hand, Teller noted in the report that the Alarm Clock was a possibility "that may be open to our competitors as well as ourselves," although he thought a Soviet Alarm Clock was "not very probable."

For the Super or the Alarm Clock or both, Teller proposed exploring the production and use of a gray salt-like compound of a lithium isotope and deuterium, lithium[6] deuteride, as a fuel alternative to liquid deuterium. Lithium, a soft, silvery-white metal, atomic number 3, was already in use in the American bomb program in the form of lithium fluoride slugs, which were irradiated in the Hanford reactors to produce tritium. Theoretically, lithium in a bomb would pick up neutrons from $D + D$ reactions or from fission and make tritium *in situ;* the T would then react with D, releasing energy and making more neutrons, which would repeat the cycle. The advantage of using lithium in a thermonuclear device would be at least twofold: it would generate tritium at hand, reducing or eliminating the need for incorporating expensive reactor-bred tritium into the design; and it was a solid at room temperature and did not require maintaining within a bomb at several hundred degrees below zero (with all the elaborate bottling and insulating that would entail) as liquid deuterium did. But lithium deuteride had the serious disadvantage that its Z was three times that of deuterium; it radiated at nine times the rate of the hydrogen isotopes and therefore appeared to be far more difficult to ignite. Assuming the ignition problem could be overcome, Teller thought that hundreds of kilograms of Li^6D might need to be produced.

Teller concluded significantly that both the Super and the Alarm Clock designs needed to be explored further before it would be possible to choose between them. That exploration would require calculation on electronic digital computers that were only then being developed. With such development in mind, Teller proposed a deliberate delay: "I think that the decision whether considerable effort is to be put on the development of the [Alarm Clock] or the Super should be postponed for approximately two years; namely, until such time as these [proposed] experiments, tests and calculations have been carried out." Carson Mark explains why calculations were so important for a device that could only be tested at full scale: "The very proof of feasibility required the fully detailed calculation of [the Super's] behavior during an explosion. Without this, no conclusive experiment was possible short of a successful stab in the dark, since a failure would not necessarily establish unfeasibility, but possibly only that the system chosen was unsuitable, or that the required ignition conditions had not been met." Testing a system that sputtered and fizzled would not teach them how to design one that caught fire.

In the meantime, they needed a detailed calculation of the full course of a *fission* explosion from beginning to end, something they had not had time or sufficient calculating power to do during the war and something that was important to understanding better not only how to design fission bombs but also how to design an Alarm Clock and the fission component of the Super. Preparing for that work fell to Robert Richtmyer, whom Carson Mark had just then—in autumn 1947—replaced as leader of the theoretical division, a position Richtmyer had filled since shortly after the end of the war. Richtmyer, Stanislaw Ulam remembers, "was tall, slim, intense, very friendly, and obviously a man of great general intelligence" who was also interested in music and cryptography. ("Richtmyer took a great interest in the Super," Teller notes, "and he and I developed the Alarm Clock.") No computer had yet been built that could run the fission problem calculations; what the former division leader would do, for the next two years, was plan what Mark calls "a fully detailed machine calculation of the course of a fission explosion" against the time when a successor to the ENIAC became available. In this regard also, Los Alamos was still a long way from knowing how to build a hydrogen bomb.

In the meantime, atomic-bomb production had begun to pick up. Although the official stockpile number for 1947 is thirteen Mark series plutonium bombs, by the end of the year there were in fact fifty Mark series cores at hand of which nine were all-plutonium Christy (solid) cores, thirty-six were composite Christy cores and five were levitated composite cores (a design which had still not been tested). Eleven more Mark series cores were in the process of being certified. The stockpile contained fifty Class A initiators with another twenty-two being certified and thirteen Class B older initiators in which the polonium had partly decayed. There were sufficient nonnuclear components on hand to make 104 complete Fat Man (FM) assemblies; fifty-four more were being certified. The stockpile was short of aluminum pusher shells for FM assemblies, with only sixty-three on hand. Ten Little Boys (LB) were being certified and six LB initiators; the program to manufacture these penetrator weapons, which could also be used as free-fall bombs, was ahead of schedule. In an emergency, then, the United States at the end of 1947 had available an arsenal of at least fifty-six atomic bombs (fifty FM and six LB with initiators), each one sufficient, in the judgment of the Joint Chiefs of Staff, to destroy a city. But available bomb-assembly teams would need up to thirty days to put together even twenty bombs, and delivery would be problematic. The US Air Force (as the former Army Air Forces were now called) had thirty-five Silverplate B-29s operational at the end of 1947 to deliver atomic bombs with thirty flight crews available to man these planes, but only twenty crews were fully trained in handling atomic weapons.

With Robert Oppenheimer's election in January 1947 as chairman of the AEC's General Advisory Committee, the charismatic physicist came into his own as an influential government adviser. "In the early days [of the AEC]," he testified later, "we [on the GAC] knew more collectively about the past of the atomic energy undertaking and its present state, technically and to some extent even organizationally . . . than the Commission did. . . . It was very natural for us not merely to respond to questions that the Commission put, but to suggest to the Commission programs that it ought to undertake." As of early 1947, though negotiations were still ongoing at the UN, prospects were fading for agreement on international control. "The problem that we faced then," Oppenheimer noted, "was to devise a program which would regain some of the wartime impetus and vigor, and above all to make available the existing know-how, the existing plant, the existing scientific talent . . . in the form of actual military strength. It was not so available as of the first of January 1947." Without debate—"I suppose with some melancholy"—they concluded immediately "that the principal job of the Commission was to provide atomic weapons and good atomic weapons and many atomic weapons." The AEC should look into atomic power, the GAC advised, as well as military uses of atomic energy such as submarine propulsion, and it should stimulate basic science. But its primary job was to make bombs and better bombs and more bombs. The GAC set to work to facilitate those goals, meeting monthly throughout the year, pushing initiator manufacture, production reactor development and improvement and bomb tests.

Oppenheimer was still trying to teach at Caltech, but a new opportunity had presented itself. AEC commissioner Lewis Strauss was a member of the board of trustees of the Institute for Advanced Study in Princeton, a place Oppenheimer would come to characterize as "an intellectual hotel," and the institute needed a new director. Its faculty, an independent centenary of geniuses that included John von Neumann, Albert Einstein and mathematicians Oswald Veblen and Kurt Gödel, thrashed out a list of five recommendations. Oddly, Strauss's name appeared as a fifth choice on the list (though he was a man of considerable intelligence, Strauss was entirely self-educated and a specialist only in making money; he had left school at sixteen to sell shoes). Oppenheimer was the IAS faculty's first choice, a choice in which the trustees soon concurred. Strauss offered Oppenheimer the directorship late in 1946 when the financier visited the Berkeley Radiation Laboratory as part of the AEC review of its Manhattan Project inheritance. Oppenheimer dithered through spring 1947—"I regard it as a very open question whether the Institute is an important place, and whether my coming will be of benefit," he commented loftily at the time—but he and his wife Kitty heard news of his acceptance over their car radio while they were driving across the Bay Bridge one night in April and let that leak of accident decide them.

Life magazine welcomed them to Princeton with a feature story headlined "The New Director" that babbled tidbit non sequiturs:

The new director has a sharp, selective mind, and his friends sometimes feel that he wins arguments too quickly. He and his family live in an 18-room, white colonial house near Fuld Hall, and Oppenheimer stops work at about 6:30 every evening to go home and play with his children, Peter, 6, and Katherine, 3. On Sundays he and his wife, who was a biologist, take the children out to hunt four-leaf clovers. Mrs. Oppenheimer, whose thinking is also direct, keeps her children from cluttering the house with four-leaf clovers by making them eat all they find right on the spot.

An accompanying full-page photograph showed a youthful Oppenheimer in a three-piece suit tenting his hands beside an amused Einstein comfortable in a sweatshirt; the great relativist, *Life* captioned, was telling Oppenheimer "about his newest attempts to explain matter in terms of space."

The inconsistencies in Oppenheimer's security record followed him into his new responsibilities. As part of the transfer of authority from the Army to the new civilian agency, the AEC laboriously reviewed personnel security files. Oppenheimer's was thick. When the FBI had interviewed him on September 5, 1946, about George Eltenton and Haakon Chevalier, the former Los Alamos director had not only aligned himself with Chevalier's version of the disputed 1943 espionage contacts in contradiction to Eltenton. He had also denied that Kitty Oppenheimer's Lincoln Brigade friend, Steve Nelson, had ever approached him for information. The FBI, however, had a wiretap of Nelson telling another suspect physicist, Joseph Weinberg, "that he had previously approached Oppenheimer for the purpose of securing information concerning the project at the Radiation Laboratory but . . . that Oppenheimer had refused him the information." Nelson's wiretap might have helped exonerate Oppenheimer from suspicions of disloyalty, but his prevarications raised further doubts. It took the AEC until late August 1947 to vote Oppenheimer a high-level Q clearance.

Lewis Strauss never reported when he first began to despise Robert Oppenheimer. Strauss was both intellectually insecure and thin-skinned, two weaknesses that would have made him especially vulnerable to the physicist's notorious arrogance. Finding his name at the bottom of the IAS faculty list of recommendations with Oppenheimer's at the top was a good start on enmity. Disagreeing politically added to their mutual ill-will; Strauss was a conservative Republican preoccupied with keeping atomic secrets, Oppenheimer a liberal who championed openness. Oppenheimer's friend Joseph Alsop, the influential journalist and columnist, would write of Strauss a few years later that he was a "natty, energetic, ambitious, and intelligent man" who was "all pliability" with his "chiefs" but who "likes no argument" from "equals and subordinates. . . . One of his fellow commissioners has said of him, 'If you disagree with Lewis about anything, he assumes you're just a fool at first. But if you go on disagreeing with him, he concludes you must be a traitor.' " With such a man as Strauss, Alsop concludes, "Oppenheimer was fated from the first to get on badly."

More elementally than political differences, Strauss seems to have been repelled by what he characterized as Oppenheimer's immorality. When Edward Teller, some years later, wanted to write that Oppenheimer had been "magnificent," Strauss rebuked him waspishly: "Is a man magnificent who is what JRO was by his own admissions in respect to his veracity and personal morals? (Did Ernest Lawrence ever tell you what he did in the Tolman household?) Some other word maybe, Edward, but not 'magnificent.' " What Oppenheimer had done in the Pasadena household of Caltech senior physicist Richard Tolman, before the war, had been to sleep with Kitty when she was still married to the English physician Stewart Harrison; that Strauss was still indignant two decades later at such gossip is a measure of the financier's psychological rigidity. "[Strauss] was a contrary person," Herbert York confirms, "and very obstinate."

Oppenheimer's nonchalant disregard of Judaism—the assimilationism that I. I. Rabi caricatured as Oppenheimer's inability to make up his mind "whether to be president of the B'nai B'rith or the Knights of Columbus"—threatened Strauss as well. In Washington in that era when anti-Semitism was routine within the political and social establishment, Strauss successfully presented himself as a secular administrator with a special competence in science, but he was also a prominent Jewish layman, president for a decade of Temple Emanu-El, the largest Jewish congregation in New York. "Strauss spoke with God, you know," commented forthright Italian emigré physicist Emilio Segrè, Fermi's close friend and fellow Nobel laureate, adding, perhaps redundantly, "I didn't like him."

Strauss voted to approve Oppenheimer's Q clearance—in his memoirs he blames Groves for advising the commissioners to do so—but the two men were never friends. An early disagreement was foreign distribution of radioisotopes bred in US reactors for medical and industrial research. The GAC favored such distribution, believing it would build goodwill and foster the progress of science; Strauss disapproved. When the AEC commissioners met in August 1947 at the Bohemian Grove, north of San Francisco, the minutes report that Strauss found himself "the only member who was opposed to such action at this time. . . . He . . . expressed doubt that the proposed restrictions to be placed on the shipment of radioisotopes abroad could effectively prevent their use for unfriendly purposes or their divergence to unfriendly hands which might endanger our national security." For good measure, Strauss announced his opinion that the publication of the Smyth Report had been "a serious breach of security." "The debate within the Commission was so bitter," Groves gloated in a contemporary memo, "that Admiral Strauss almost resigned."

Small quantities of radioisotopes for scientific research were never even remotely a threat to national security, but Strauss bullied the question obsessively, carrying tales to James Forrestal and leaking them to the *New York Times*. Finally, David Lilienthal, fearing "a bad rift in the Commission . . . after all my months of working to create solidarity," confronted him:

> He . . . is very sensitive to moods and was obviously worked up. . . . As I told him my story he said he still felt we are wrong about the foreign isotope matter. . . . He said . . . he thought the best thing for him to do was to resign. I said that was absurd; but that it was essential that he realize how dangerous and fatal to everything we are doing it was for him to oppose Commission action *outside* the Commission. . . . He was agitated, said I was a saint, etc. I tried to turn it back to the issue of what is permissible and what is not. . . . He said, "No, I'm through with it. I will forget it, or try to, though I am still not convinced I'm not right." He said something more about how terrible he felt. I said, "Don't criticize yourself that way, you just didn't realize what you were doing." He turned and grinned in what seemed a very genuine way and said, "No, I'm old enough; I knew exactly what I was doing." Which I believe is about the size of it.

If Strauss could be obstinate, he could also be perspicacious. In 1939, advised by Leo Szilard, he had been one of the first to realize the implications of the discovery of nuclear fission. Almost his first act as a newly confirmed AEC commissioner, in April 1947, had been to ask what the US was doing about the long-range detection of foreign atomic tests. "It is to be presumed," he had lobbied his fellow commissioners, "that any other country going into a large-scale manufacture of atomic weapons would be under the necessity of conducting at least one test to 'prove' the weapon. If there is no [US] monitoring system in effect, it is incumbent upon us to bring up the desirability of such an immediate step and, in default of action, to initiate it ourselves, at once." The commissioners asked Strauss to look into the problem; he assigned his former Navy aide, William T. Golden, a businessman of independent wealth, to serve as AEC representative to a Long-Range Detection Committee organized at the AEC's instigation under the Air Force. Thereafter the USAF cooperated with the AEC to develop a capability to detect nuclear explosions anywhere in the world.

Curtis LeMay, whose August 1946 recommendation of long-range detection had preceded Lewis Strauss's by eight months, transferred abroad in October 1947 as commander of the United States Air Forces in Europe (USAFE), picking up a third star along the way to become a lieutenant general. He found the Germans "still in a state of utter shock" more than two years after their defeat, "like the walking dead," and USAFE forces in not much better condition. "At a cursory glance it looked like USAFE would be stupid to get mixed up in anything bigger than a cat-fight at a pet show. We had one Fighter group, and some transports, and some radar people, and that was about the story. I had to shake things up right quick, and I kept working day and night to shake them."

LeMay was particularly concerned with the vulnerability of American supply lines, through Bremerhaven well to the northeast toward the Soviet zone of Germany, "away out in front of our troops. . . . All the Russians had to do was to whiz forward and they could cut our supply lines before they even made contact with our troops." To sidestep the complex and time-consuming politics of a Germany divided into four zones, each under the authority of a different country, LeMay resorted to extralegal arrangements. Just as he had negotiated privately with Mao Zedong in China, he now negotiated privately with the French and Belgian Air Force Chiefs of Staff:

> I told them that I wished to have some fields well in the rear of our troops, back in Belgium and France, all stocked up. Ammunition, gas, food, bombs, mechanical equipment: every type and condition of supplies which we might need. . . .
>
> What the hell was I doing now? I was breaking other nations' laws into bits.
>
> You couldn't *have* any foreign troops stationed in France in peacetime. Nothing was more illegal to the French mind or the French code. Same thing for Belgium. . . .

To disguise the transfer of matériel, LeMay put the supplies on trains and shuttled them around Western Europe. "We zigzagged our trains from hell to breakfast," parked them on sidings, reworked their bills of lading. Once the origin of the trains was sufficiently obscured, LeMay sent in troops in civilian clothes to unload them at the bases the French and Belgian military agreed to share. "What this amounted to, in effect, was that we had our own private little NATO buzzing along, there in West Germany and France and Belgium, before the North Atlantic Treaty Organization ever existed." Seared with the memory of the United States's lack of readiness at the beginning of the Second World War, neglect he had paid for in his squadrons with the lives of too many young men, the USAFE commander was determined to subvert the renewed complacency of victory.

James Forrestal took a longer view when he analyzed the military situation in early December 1947. Defending the billions the US was preparing to spend on the Marshall Plan in a letter to the chairman of the Senate Armed Services Committee, he wrote:

We are keeping our military expenditures below the levels which our military leaders... estimate as the minimum [necessary to]... ensure national security. By doing so we are able to increase our expenditures to assist in European recovery. In other words, we are taking a calculated risk. ...

During those years [of calculated risk]—of which the exact number is indeterminate—we will continue to have certain military advantages which go far toward covering the risk. There are really four outstanding military facts in the world at this time. They are:

(1) The predominance of Russian land power in Europe and Asia.
(2) The predominance of American sea power.
(3) Our exclusive possession of the atomic bomb.
(4) American productive capacity.

As long as we can outproduce the world, can control the sea and can strike inland with the atomic bomb, we can assume certain risks in an effort to restore world trade, to restore the balance of power—military power—and to eliminate some of the conditions which breed war.

This incisive analysis was not disingenuous, but Forrestal also had reason to believe that Stalin had no desire to start a war. Averell Harriman—former ambassador to the Soviet Union and now Secretary of Commerce—had testified to the President's Air Policy Commission in September that he was "convinced that [the Soviets] will not take any steps which they feel would bring them into a major conflict in the foreseeable future." In October, the new Secretary of Defense had asked Harriman's successor in Moscow, Walter Bedell Smith, where the Soviets stood and Bedell Smith had told him, "Stalin said, we do not want war but the Americans want it even less than we do, and that makes our position stronger."

Milovan Djilas, riding with Stalin in the Soviet leader's car in Moscow in dark December, noticed "his... hunched back and the bony gray nape of his neck with its wrinkled skin above the stiff marshal's collar." To the Yugoslavian diplomat it was "incomprehensible" how much Stalin had aged since he had seen him last, two years before. At dinner, "Stalin spoke up about the atom bomb: 'That is a powerful thing, pow-er-ful!' His expression was full of admiration, so that one was given to understand that he would not rest until he, too, had the 'powerful thing.' " That night and again a few nights later, Stalin predicted that Germany would not be reassembled from its pieces. "The West," Djilas heard him say, "will make Western Germany their own, and we shall turn Eastern Germany into our own state."

The Soviets were biding their time, recovering, turning inward, working toward the bomb. In autumn 1947, the Soviet press began to attack Byrnes, Forrestal and even Truman. "In the American expansionist circles there is a new kind of religion," Molotov railed in an important November speech.

"There is no faith in internal strength, but there is a fanatical faith in one thing—the secret of the atom bomb—even though, for a long time now, no such secrets have existed." On December 15, to soften the impact of monetary reform, the Soviet government abolished rationing. But a cruel new repression began in the same season, a "vigilance campaign." "There is still a good deal of toadying to the West in our country," Molotov rebuked the Soviet people in his November speech, "and far too much slavish admiration for capitalist culture. We have no use for this. . . . Even in the old Bolshevik Party the enemy had his spies and agents—Trotskyites, Rightists, and so on, and in the present international situation Soviet citizens must be particularly vigilant." The campaign "soon developed into a new kind of spy-mania," writes Alexander Werth.

The vigilance campaign was cruelly utilitarian. A Canadian woman, Suzanne Rosenberg, a gulag victim whose parents had moved to the USSR before the war to support the Communist experiment and who had been caught up in the Great Terror, saw through the new xenophobia:

> [The MVD committed] no worse felonies than the rearrest, on the same charge, of people who had completed their sentences and been released, as well as the non-release of prisoners after they had served their full terms. Those arrested during the bloody years of 1937 and 1938 were either shot or sentenced to ten years; longer terms, especially the ubiquitous twenty-five years, were instituted during and after the war. This meant that large numbers of political prisoners would have to be released in 1947, and the slave labor force so badly needed for the post-war industrial economy greatly depleted.
>
> An ingenious solution was to rearrest the released prisoners and, ten years later, start a new wave of terror—the 1947–1953 mass arrests—to prevent the shutting down of the camps and to provide the new building projects with penal labor. Politically, too, it was most expedient to keep the innocent victims of 1937 and 1938 in continued isolation, lest they besmirch the good name of the Party. It was killing two birds with one stone.

Beria needed camp dust to build his Manhattan Project.

Igor Kurchatov's brother Boris had succeeded by then in extracting the first few micrograms of Soviet plutonium from uranium slugs irradiated in the F-1 reactor; a team at the Radium Institute under V. G. Khlopin had begun enlarging the extraction technology to industrial scale. With the coming of freezing weather that autumn, Kurchatov and Vannikov traveled out to the Chelyabinsk-40 building site. "A large city had already sprung up there," writes Golovin, "populated by thousands of workmen, technicians, and engineers of various categories. The place where the pile was under construction was over ten kilometers from town, and Vannikov, who had recently suffered a heart attack, decided that daily trips over this distance would be hard on him. He moved into a railroad car right next to the construction site. Kurchatov stayed with him and bore the discomforts of the icy winter without complaint." Chelyabinsk-40 became Kurchatov's base. Ironically, before the Revolution the local Kyshtym mining and metallurgical industry had been managed by a young American engineer named Herbert Hoover, Lewis Strauss's earliest mentor in government service in the years after the First World War.

Not only directors and generals shipped out for Chelyabinsk-40. A conscript soldier remembers being assigned there to guard the *zeks* who were digging the pit for the first *objekt*—the graphite production reactor, the A reactor, nicknamed Anotchka ("Little Anna"). "The town . . . had a triple belt of barbed wire," the soldier writes: "the external fence with guard towers, then the zone with the settlement 'Techa' (after the river Techa) where the scientists lived, then the central 'zone' with the work force—soldiers, prisoners and released prisoners—and finally, the *objekt* itself, also surrounded by barbed wire. They carried out the construction in parts. Some people worked on the initial stage, then others went on with it and yet others finished. That way, no one knew what was being built."

Graphite for the production reactor was manufactured on the site, diverting most of the Soviet supply that year; US intelligence discovered "a serious shortage of graphite electrodes" in the USSR in 1947 and early 1948. (The US had continued selling graphite to the Soviet Union after Lend-Lease ended—about 5,500 tons in 1946. "With the development of the cold war," the CIA noted in 1951, "exports were restricted to 1,500 tons in 1947 and stopped altogether in 1948 after the delivery of 700 tons.") The graphite core of the A reactor would be about 30 × 30 feet in height and width, drilled with 1,168 vertical channels for aluminum-clad natural-uranium metal slugs, which would be dropped in at the top, irradiated and then gravity-discharged out the bottom into a spent-fuel pool. The core would be set below ground in a pit sixty feet deep. *Zeks* dug the pit by hand, shoring it with timber, until they got to bedrock; the rock, says the soldier, "was cut by explosives and we loaded the fragments onto trucks which carried them away." When the pit was finished they lined it with water tanks, then poured a concrete well with walls ten feet thick to encase the graphite block. Over the reactor went a substantial building in the Soviet neoclassical style with stone facing on the first floor and two-story columnar facings above.

They began assembling the reactor block in March 1948. Kurchatov delivered a speech. "You and I are founding an industry not for one year," he told the staff, "not for two [but] . . . for centuries." He hoped that a city would grow on the site with "kindergartens, fine shops, a theater" and that their children would succeed them at the work. "And if in that time not one uranium bomb explodes over the heads of people, you and I can be happy!

And our town can then become a monument to peace. Isn't that worth living for!"

"When reactor assembly started," writes Mikhail Pervukhin, "B. L. Vannikov, I. V. Kurchatov and E. P. Slavsky [the director of Chelyabinsk-40, a metallurgist] were always on hand. [MVD general] Avrami Zavenyagin would sometimes visit the site, as would A. N. Komarovsky, who was in overall charge of construction.... I also visited several times. We inspected the assembly work very carefully, especially in the part of the reactor that would later become radioactive. We entered the reactor through a special manhole to check the quality of the work, the welding in particular. Since Vannikov, Zavenyagin and Komarovsky wore general's uniforms, the construction workers called the entry we used 'the General's manhole.' " The workers were welding pipes to carry water through the reactor to cool it, a change from the F-1 system made necessary by Anotchka's higher power level.

At Sarov, two hundred kilometers nearer Moscow, Zukerman's and Altshuler's groups worked regularly with high explosives at their site in the woods, studying implosion. "During this very early stage," Zukerman writes, "not a lot of attention was paid to safety procedures. A charge [of HE] contained in an ordinary string shopping bag would be hung in front of the armored bunker. A few preliminary radiographs would be taken, which served to confirm that the charge was oriented in line with the X-ray beam. Then Maria Alekseevna Manakova would come out of the bunker and bang a hammer against a scrap of rail hung from a tree branch.... These signals meant that an explosion was imminent and everyone on the field was to take cover. Sirens, telephones and similar 'miracles' of signaling and communications technology appeared only later." Once, the high-voltage X-ray equipment induced a current in a cable connected to an electrodetonator while two members of the staff were still out on the field. "Suddenly there was a powerful explosion. Everyone in the bunker instantly understood: the charge Anya and Boris had been working on had blown up. Hearts stopped. A few seconds passed that seemed an eternity; then, in the doorway to the bunker, first Anya appeared, looking agitated, and behind her followed Boris, imperturbable as always. 'Nothing to worry about,' he said, 'that was our charge going off because of your induction. We had already stepped away from the barrel.' " When an experiment worked, they tallied it for the Soviet Union; when one failed they scored it "in favor of Harry Truman."

For compression and initiator experiments, Altshuler devised a quick, cheap substitute for the precisely fitted two-piece lenses that would shape the implosive wave in the finished bomb: spheres of HE glued to a solid HE shell. This "charge for three-dimensional implosion of very simplified design," as Altshuler calls it, made it possible "to realize up to twenty experiments a month." They assembled the rough-and-ready charges in the field, heating their pot of glue over an open fire. On one occasion, writes Zukerman, "they were readying an experiment involving a large explosive charge, over one hundred kilograms. Suddenly the charge caught fire. In such cases, the burn can trigger a detonation, with all its consequences. [The group leader] stayed calm and collected. He led his brigade to the bunker and phoned the dispatcher to order everyone to keep away from the area. This time, nature was kind: there was never an explosion and the charge burned down without incident." Accidents were acts of sabotage in Berialand. The scientists attributed the fire to spontaneous combustion—a passing bird had shat on the charge, they claimed, and the splash of liquid had functioned as a lens to focus the sunlight. It was a story only technological illiterates would swallow, and the bosses did.

In the winter and spring of 1948, the Soviet Union and the Western allies gave up any remaining pretense of continuing their wartime collaboration. When a Council of Foreign Ministers meeting broke down in London on December 15, 1947, in disagreement over the future of Germany, Ernest Bevin proposed to George Marshall "the formation of some form of union, formal or informal in character, in Western Europe backed by the United States and the Dominions." Bevin started the process in January by pursuing an alliance with the Benelux countries—Belgium, the Netherlands and Luxembourg—and France.

The Soviets, on their side, confident that war was not an immediate threat ("America may pull on our leg," Georgi Malenkov told a group of Italian Communists, "but war is out of the question now"), moved to consolidate their control over Eastern Europe. The Marshall Plan was already a success, Marshall had reported to the US Cabinet in November—"the advance of Communism has been stemmed and the Russians have been compelled to make a reevaluation of their position." Stalin resented the plan bitterly. Czechoslovakia in particular had attempted to take advantage of Marshall Plan aid. At a meeting with the Czech Prime Minister, who complained that his country needed foreign exchange, Stalin laughed in his face: "We know you have enough." Still laughing, he turned to Molotov: "They thought they could lay their hands on some dollars, and they didn't want to miss the chance." On February 25, 1948, Soviet forces occupied Prague. Bedell Smith cabled Marshall from Moscow on March 1: "Full information on and explanation to . . . Congress of significance [of] recent Soviet moves in Czechoslovakia and Finland may result in speeding consideration and adoption [of] universal military training and building programs for Army, Navy, and particularly Air Force."

Forrestal wanted a renewal of the draft and increased funding for the armed services. The US aviation industry was in nearly fatal trouble that winter as well, approaching bankruptcy unless defense contracts could be authorized immediately and funding accelerated. Marshall needed approval of $5.3 billion in Marshall Plan aid. The country required a good war scare to rally a recalcitrant Republican Congress. The Soviet takeover of Czechoslovakia helped, but not enough; in March, Marshall, Forrestal, the State Department's Robert Lovett and others colluded to add fuel to the fire. At a luncheon on March 4 that included Cabinet officials and senators, Forrestal records in his diary, "Marshall talked over the war situation.... Everyone present agreed that the public needed information and guidance on the deterioration of our relations with Russia."

Besides speeches, the war scare group called on General Lucius Clay, the intense, imperious military governor of the American zone of occupied Germany. Clay had predicted in November 1947 that the Soviets might decide to move against Berlin. The former capital of Nazi Germany, divided into four zones like the German nation, was embedded behind the Iron Curtain a hundred miles deep into the Soviet zone, beyond the range of more than token US military protection. The day after the war situation luncheon, Clay sent an Eyes-Only cable to Washington that Forrestal parlayed into a major event:

For many months, based on logical analysis, I have felt and held that war was unlikely for at least ten years. Within the last few weeks, I have felt a subtle change in Soviet attitude which I cannot define but which now gives me a feeling that it may come with dramatic suddenness. I cannot support this change in my own thinking with any data or outward evidence in relationships other than to describe it as a feeling of a new tenseness in every Soviet individual with whom we have official relations. I am unable to submit any official report in the absence of supporting data but my feeling is real. You may advise the chief of staff of this for whatever it may be worth if you feel it advisable.

The same day he sent this will-o'-the-wisp message, which made headlines across America when Forrestal reported it in congressional testimony on March 8, Clay wrote a senator that American personnel were "as secure here [in Berlin] as they would be at home...." Clay told a biographer long afterward that he sent his war-scare cable because the Army's director of intelligence had come to see him in Berlin in late February and told him "that the Army was having trouble getting the draft reinstated, and they needed a strong message from me that they could use in congressional testimony." In the upshot, Marshall got his aid and the aviation industry was rescued. With the Truman administration worried about inflation, the military fared less well.

In the meantime, on March 17, Britain, France and the Benelux countries committed themselves to alliance in the Treaty of Brussels. Then it was the

United States's turn to begin negotiating a military alliance with Britain and Canada, writes the British diplomat Robert Cecil, who attended the first secret meetings:

> Bevin judged that the time had come to propose to the State Department that the US government, having already decided to invest in Western Europe's recovery through the Marshall Plan, should agree to undertake the defence of its investment, if this should prove necessary. It was essential, however, to proceed with caution, since Republicans dominated Congress and Congress had never sanctioned a military alliance in time of peace. On 22nd March, with no fanfare and no accompanying staff, Sir Gladwyn Jebb (Lord Gladwyn) came out from London and Lester Pearson from Ottawa. The State Department, fearing that their joint arrival on their doorstep with delegations drawn from their Embassies might provoke press enquiries, decided to hold the initial meetings at the Pentagon, which in those days was less haunted by the press corps. Further to discourage premature leakage, it was stipulated at the opening meeting that no notes should be taken and that we should not disperse for lunch, which was eaten at the long table where discussion took place.

The meetings continued through April 1. The US had rejected inviting the French because the State Department feared they might compromise security, but Cecil's senior partner at the negotiations was Donald Maclean. Maclean certainly communicated the substance of the discussions to the MGB. The Polish newspaper *Zycie Warszavy* published an article on a North Atlantic alliance on April 4 that described the secret Anglo-American plans; a British Foreign Office internal memorandum noted at the time that the article "did sail pretty near the wind." The secret discussions included an agreement among the parties to aid each other militarily in the event of an armed attack; Canada tabled a proposal that "among others, Western Germany and Western Austria might join" the alliance. The possibility that the West might rearm Germany may well have precipitated the Soviet decision to move against Berlin. The Russian scholar Sergei Goncharov, reviewing Soviet archives, reports that "during the prolonged Berlin crisis . . . Stalin was riveted by the possible inclusion of West Germany in the developing American alliance structure. That move would greatly enhance the West's potential for encirclement, and he sought to block it at all costs." Cecil notes that "at that date negotiations with the USSR for a peace treaty with Austria were well advanced; for reasons that have never become clear the Russians went into reverse; seven years elapsed before they were finally convinced that we envisaged a neutral, independent Austria on the model already under discussion in 1948."

Berlin had been unstable even before the secret Pentagon meetings; Marshal Vassily Sokolovsky, the Soviet military governor, Clay's counterpart, had walked out of the Allied Control Council in Berlin on March 20, whereupon the Soviets began to manipulate Allied rail access. Even this first Soviet sally raised the question of atomic war in the minds of US military leaders. Major General Kenneth Nichols, an Army engineer who had been Groves's deputy during the war, had succeeded Groves as commander of the Armed Forces Special Weapons Project (AFSWP) on January 1 (Groves retired from the Army on February 29). On March 31, over lunch, Nichols briefed Forrestal, the Joint Chiefs (which then included Dwight Eisenhower) and the armed-service secretaries:

> Clay had reported further restrictions on our land transportation, and he threatened to confront the Russians if they stopped and boarded any train, to shoot if necessary. I was at the meeting to supply information about whether, if the crisis grew worse, we were in a position to deliver any atomic weapons. We were not. I told them that the only assembly teams, military and civilian, were at Eniwetok for the *Sandstone* test and that the military teams were not yet qualified to assemble atomic weapons. I was told in very definite terms by Eisenhower to accelerate training and improve the situation at once. . . . Action was initiated to perfect plans for transfer of atomic weapons to the military in case of emergency and to expedite training and equipping the military assembly teams.

Sandstone was the designation of the first test series of new atomic weapons that Los Alamos was then preparing at Eniwetok atoll in the Marshall Islands. *Sandstone* X-Ray, a composite-core, levitated implosion device, would be exploded on April 15, 1948, with a yield of thirty-seven kilotons. Two other tests would follow: Yoke, on May 1, another composite, levitated core that yielded 49 KT, the largest yield yet coaxed from any atomic weapon, almost four times that of the Hiroshima bomb; and Zebra, on May 15, a levitated all-U235 core that yielded 18 KT. An uninvited Soviet warship would be on hand to watch the proceedings from twenty miles off, as well as at least one submarine. The *Sandstone* tests demonstrated that small amounts of fissionable material could develop large yields. A levitated composite core typically used less than half as much plutonium as a solid Christy core and ten times less U235 than a Little Boy gun. "The most immediate military effect of *Sandstone*," the independent scholar Chuck Hansen notes in his authoritative history of US nuclear weapons development, "was to make possible within the near future a 63 percent increase in the total number of bombs in the stockpile and a 75 percent increase in the total yield of these bombs. . . . *Sandstone* also proved conclusively that implosion of U235 was far more efficient than assembling it in a gun-type weapon and demonstrated that current implosion theory was sound. . . . The results of

Sandstone were characterized [in contemporary reports] as 'radical' and representing 'substantial' improvement in the military position of the US." Carson Mark would note that the tests "marked the end of the day of the atomic device as a piece of complicated laboratory apparatus rather than a weapon.... The number of units which could be made from the existing stock of plutonium and uranium could [now] be increased appreciably merely by refabricating the fissile parts of the weapons by hand." While the Soviet Union challenged the United States over access to Berlin, however, the crews Los Alamos had trained to assemble atomic weapons were off in the Pacific six thousand miles from Albuquerque, where the 509th would pick up the US's meager store of bombs were there someone on hand to assemble them.

Without atomic backup, the Joint Chiefs reined Clay in; his modest forces in Germany avoided confrontation by taking to the air. "During one period of eleven days in early April," LeMay recalls, "when the Soviets demanded the right to search and investigate all military shipments by rail, we flew small quantities of food and other critical supplies into Berlin; something like three hundred tons." Clay discontinued the little airlift in mid-April when the Soviets eased back. The Marshall Plan had become US law on April 3.

From February to May 1948, the Joint Chiefs had been charting and revising a series of emergency war plans. They approved BROILER in March, modified it to FROLIC and finally approved HALFMOON in early May. HALF-MOON included an Air Force atomic annex, HARROW, which envisioned dropping fifty atomic bombs (the entire stockpile that spring) on twenty Soviet cities, causing "immediate paralysis of at least 50 percent of Soviet industry." Such paralysis would not be sufficient to stop the Red Army, HALFMOON foresaw; the plan expected that Soviet forces would overrun Western Europe at the outset of any conflict. Truman's chief of staff, Admiral William D. Leahy, briefed the President on HALFMOON on May 5. Truman was not happy to be saddled with a war plan that counted on atomic weapons. He told Leahy that such weapons might be outlawed before war came and that the American people in any case would not tolerate using atomic bombs for "aggressive purposes" and he ordered an alternative plan developed that depended on conventional forces alone.

Any immediate possibility of outlawing atomic weapons was foreclosed on May 17 when the United Nations Atomic Energy Commission, still carrying on the negotiations that Bernard Baruch had begun for the US the previous summer, announced that it had reached an impasse and recommended suspending its work. By then the US and the Soviet Union had danced one more diplomatic dance. On May 4, Bedell Smith in Moscow delivered a note to Molotov intended to reassure the Soviet leadership after the March war scare. The note protested that "the United States has no hostile or aggressive designs whatever with respect to the Soviet Union" and declared in a key sentence that "as far as the United States is concerned, the door is always wide open for full discussion and the composing of our differences." Bedell Smith believed that the US note was merely "a statement for the record," but the Soviets eagerly rejoined. Molotov handed Bedell Smith a response on May 9 that found the Soviet government "in agreement with the proposal to proceed . . . toward a discussion and settlement of the difference existing between us." In the next several days the Soviets released an edited text of the US note—moving it beyond denial—while Moscow radio announced that the Soviet government had accepted the US proposal. The Soviet move caught the Truman administration by surprise; the President and the Secretary of State immediately backpedaled. The *New York Times* reported of a Marshall press conference on May 12 that he "threw more cold water . . . on the Soviet proposal for a United States–Russian 'peace' conference"; of Truman, the press reported on May 13 "that the recent exchange of views with Russia has not increased his hopes for peace."

Common opinion at the upper level of the Truman administration attributed the Soviet response to propaganda. But in any case the United States no longer had any desire for a peace conference. The time for negotiation had passed; the US was preparing to establish a separate West German government, a decision approved in a meeting on May 24 that Marshall, Forrestal and Omar Bradley, among others, attended at the State Department. Marshall believed Germany was a crucial barrier to the spread of Soviet power across Western Europe, a position he had made clear as early as February, when he cabled the US ambassador to Britain that the US was determined "not to permit reestablishment of German economic and political unity under conditions which are likely to bring about effective domination of all Germany by [the] Soviets. It would regard such an eventuality as the greatest threat to [the] security of all Western nations, including [the] U.S."

In mid-June 1948, the Senate endorsed associating the United States with "regional and other collective arrangements," preparing the way for public negotiations toward a North Atlantic Treaty Organization to follow on from the talks of late March that the participants imagined to have been secret.

The US, Britain and France were preparing in June to reform the currency in the German zones they occupied—to stop inflation, choke off the black market, improve the banking system and promote economic recovery. "The old currency was so valueless," noted a contemporary State Department memorandum, "that cigarettes had in practice replaced it in many areas." If the new currency became legal tender in the Western sectors of Berlin as well, it would wreak havoc with the money supply in the Soviet zone. Lucius Clay had been attempting to negotiate some common currency for Berlin. He understood his instructions to preclude accepting Soviet currency in common unless the Soviets allowed the Western allies to participate in its control. "Since Sokolovsky offered no such participation," Clay writes, "I knew his proposal was unacceptable to our government. Therefore I replied immediately to Sokolovsky [on June 23, 1948] that we could not accept his proposal and that I would join with my colleagues in placing West marks in circulation in the western sectors of Berlin." The Soviets installed new currency measures that same day, twenty-four hours before the West mark was due to become legal currency.

The day of Western currency reform dawned to crisis, Clay reports:

When the order of the Soviet Military Administration to close all rail traffic from the western zones went into effect at 6:00 A.M. on the morning of June 24, 1948, the three western sectors of Berlin, with a civilian population of about 2,500,000 people, became dependent on reserve stocks and airlift replacements. It was one of the most ruthless efforts in modern times to use mass starvation for political coercion. . . .

I called General LeMay on the telephone . . . and asked him to drop all other uses of our transport aircraft so that his entire fleet of C-47s could be placed on the Berlin run.

The first direct confrontation of the Cold War between the United States and the Soviet Union had begun.

17

Getting Down to Business

WHEN THE SOVIET UNION blockaded Berlin on June 24, 1948, the United States Air Force in Europe had at hand 102 C-47 cargo planes, each of less than three tons capacity, and two C-54s of ten tons capacity. The British Royal Air Force in Germany had a few C-47s as well. Berlin before the blockade had imported 15,500 tons of food and fuel daily to feed and warm its more than two million people. It needed a minimum of four thousand tons per day. Lucius Clay expected at the outset of his airlift to be able to supply a maximum of seven hundred tons per day. "I didn't ask Washington [for permission to initiate an airlift]," Clay remembered. "I acted first. I began the airlift with what I had, because I had to first prove to Washington that it was possible."

For a few days, the airlift flew ad hoc. Then Clay called Curtis LeMay to a meeting in Berlin. LeMay flew a B-17 from Wiesbaden into Tempelhof early on the afternoon of June 27, 1948, reports the daily diary that his aide maintained, "and proceeded immediately to Gen. Clay's house where a consultation was held concerning the situation in Berlin and the feasibility of supplying all of the Western Sectors (i.e., American, British and French) with all necessities." Clay and LeMay decided to ask for more aircraft: a C-54 group, a P-51 fighter group in case the cargo planes had to be defended from Soviet fighters and several squadrons and groups of B-29s to be moved to England in case LeMay had to bomb Soviet zone airfields. "These decisions," the LeMay diary continues, "were based on Gen. Clay's policy of remaining in Berlin at all costs, using force if necessary . . . to support his plan, which has the complete backing of the State Department." Truman approved the June 27 decisions the next day. "The President . . . [said that] we were going to stay, period," James Forrestal reports Truman's endorsement.

June 29, a Tuesday, LeMay flew a C-47 up to Tempelhof and back to Wiesbaden to inspect the airlift operation. In Berlin, Clay asked him about airlifting coal for electrical generation and, in the winter, for heating. "It was decided that the only means of moving sufficient quantities of coal would be by B-29," LeMay's diary notes bizarrely, "and since Tempelhof... could not accommodate this type aircraft, the cargoes would have to be dropped from low altitude." LeMay was a bomber man, not a transport officer. Fortunately for the people of Berlin, he was not reduced to blitzing their city with coal.

"No one in authority at the time expected that the Airlift would last very long," observes USAF Major General William H. Tunner, who would soon play a leading role in the operation. "It was President Truman's opinion that the Airlift would serve only to stretch out the stockpile of rations in Berlin and thus gain time for negotiations." "The Russians are convinced that they hold all the cards," a British official reported from Moscow early in July, "and will be able to manoeuvre us into a position where ... we have no choice but to withdraw from Berlin. Our current air effort may have disconcerted them. But I doubt whether they believe we can keep it up indefinitely and on a sufficient scale."

Ten-ton capacity C-54s began arriving from the US on July 1, 1948. The airlift's limiting factor was the number of landings Templehof could handle (the British had taken over Berlin/Gatow for their parallel airlift), so the greater the carrying capacity of the aircraft, the more tons of supplies could be flown. Another request for C-54s went out on July 7 after a conference among the Western zone military governors and service commanders. (Official negotiations toward a North Atlantic Treaty Organization had begun in Washington the previous day, following upon the secret negotiations of the previous spring.) By now, the American press had swarmed to Germany to cover the unprecedented confrontation and the airlift had a name. It was formally Operation Vittles, but informally people called it LeMay's Coal and Feed Company. "Nobody regarded the enterprise very soberly at first," LeMay comments dourly.

Truman ordered Clay back to Washington in mid-July to report. The Soviets had replied to a salvo of diplomatic notes, writes Walter Bedell Smith, "that Berlin was in the center of the Soviet zone and was part of that zone" and that "the Soviet high command had been compelled to take urgent measures to protect the interest of the German population." Truman concluded "that the blockading of Berlin by the Russians was a major political and propaganda move. . . . [They] were obviously determined to force us out of Berlin."

Clay told the National Security Council, which in those days included the key military and diplomatic secretaries as well as the Joint Chiefs of Staff, that abandoning Berlin, in Truman's words, "would have a disastrous effect upon our plans for Western Germany." Given the planes, Clay said, they could supply Berlin indefinitely; he asked for 160 C-54s, more than half the

USAF's entire existing transport capacity. Truman remembers that USAF Chief of Staff Hoyt Vandenberg resisted Clay's request, arguing that "an emergency would find us more exposed than we might be able to afford." Truman says he disagreed and ordered the planes sent. "Truman realized that the Berlin crisis was a political war," Clay praised the President in a late interview, "not a physical military war. I am not being critical of the Joint Chiefs of Staff, because I think they visualized it as a military operation."

Sixty B-29s had already made headlines by then, moving from Florida and Kansas to East Anglia. They were bombers of the new US Strategic Air Command and the government made a point of revealing that they were atomic-capable and hinted that they carried atomic bombs—"bringing nuclear weapons," a newspaper man would write, "for the first time directly into the system of diplomacy and violence by which the affairs of people were henceforth to be regulated." The implied nuclear threat, the first of the Cold War, was a bluff; none of the planes were atomic-capable Silverplates, nor were their crews trained in bomb assembly, nor did they carry atomic bombs. (Silverplate B-29s never left North America in those years except for one squadron that trained out of Japan during the *Sandstone* tests; the US had no intention of losing its specially equipped aircraft, which carried secret radar-jamming systems, where the Soviets in particular might be able to salvage them, as they had salvaged and copied three B-29s lost over Siberia during the war.) Forrestal used the occasion of the B-29 transfer to ask Truman to review "the question of custody of atomic weapons"—whether the AEC should physically hold the weapons, as it did, or whether they should be transferred to the military. ("[Weapon] storage bases [are] built by the Corps of Engineers," AFSWP commanding general Kenneth Nichols explained custody arrangements to the National War College that year; ". . . when completed they are turned over to the Atomic Energy Commission. They, with our assistance, place the bombs in storage. They have the keys to the igloos, and we have the guard around the igloos. . . . So the question is, Just who is the boss of the storage base?") Forrestal writes that Truman responded that "he wanted to go into this matter very carefully and he proposed to keep, in his own hands, the decision as to the use of the bomb, and did not propose 'to have some dashing lieutenant colonel decide when would be the proper time to drop one.'"

Truman went into the matter on July 21 in a meeting crowded with AEC commissioners and defense officials. David Lilienthal believed it was "one of the most important meetings I have ever attended." He thought the President "looked worn and grim . . . and we got right down to business." Legally, Truman could transfer atomic weapons to the military whenever he judged such transfer necessary, but in those days the weapons lacked locking mechanisms; whoever possessed them—Truman's "dashing lieutenant colonel" —could detonate them. Lilienthal argued for keeping the weapons in civilian hands. Stuart Symington, the Secretary of the Air Force, a tall, handsome loose cannon from Missouri, offered up a string of foolish rebuttals. "Our fellas . . . think they ought to have the bomb," was one of Symington's lines, Lilienthal reports. "They feel they might get them when they need them and they might not work." Have they ever failed to work? Truman responded sharply. Symington "left that one," Lilienthal says, and went on to cite "one fellow" he had spoken to at Los Alamos who thought the law prevented the military from having the bomb, "I forgot his name . . . I don't believe he thought we ought to use it anyway." Truman took up that question with remarkable candor, revealing the sense of Solomonic burden that agonized him:

> I don't think we ought to use this thing unless we absolutely have to. It is a terrible thing to order the use of something like that ["Here he looked down at his desk, rather reflectively," Lilienthal interjects] that is so terribly destructive, destructive beyond anything we have ever had. You have got to understand that this isn't a military weapon. . . . It is used to wipe out women and children and unarmed people, and not for military uses. So we have got to treat this differently from rifles and cannon and ordinary things like that. . . . You have got to understand that I have got to think about the effect of such a thing on international relations. This is no time to be juggling an atom bomb around.

Forrestal disagreed with the President's decision. He believed atomic war with the Soviet Union was inevitable and wanted the military fully prepared. A week after the custody debate, the increasingly grim Secretary of Defense ordered the JCS to restore HALFMOON planning rather than pursue the conventional war plan alternative that Truman had demanded. As authority for this illegal action, Forrestal cited his own; he told the Joint Chiefs he would take full responsibility. He pointedly asked Truman at a presidential briefing in September if he was prepared to use the atomic bomb if Berlin came to war. "The President said that he prayed that he would never have to make such a decision, but that if it became necessary, no one need have a misgiving but what he would do so." Forrestal found support for his belligerence over dinner the evening after the briefing at the house of the publisher of the *Washington Post,* Philip Graham, when the gathering of newspaper editors and publishers from throughout the United States expressed "unanimous agreement that in the event of war the American people would not only have no question as to the propriety of the use of the atomic bomb, but would in fact expect it to be used." The thought of using the atomic bomb again depressed Truman, however; in a private memorandum he wrote after the Secretary of Defense questioned him, he lamented: "Forrestal, [Omar] Bradley, [Hoyt] Vandenberg, Symington brief me on bases, bombs, Moscow, Leningrad, etc. I have a terrible feeling afterwards that we are very close to war. I hope not. Discuss situation with Marshall at lunch. Berlin is a mess." The President, Lilienthal observed in his journal, "is blue now, mighty blue." In the new calculus of the atomic age, enthusiasm for using atomic weapons varied inversely with responsibility for doing so.

Truman might have been less blue had he known Stalin's conviction that blockading Berlin was a low-risk strategy. "I believe that Stalin . . . embarked on that affair in the certain knowledge that the conflict would not lead to nuclear war," Andrei Gromyko, at that time the Soviet Deputy Foreign Minister, said later. "He reckoned that the American administration was not run by frivolous people who would start a nuclear war over such a situation." In the first direct confrontation of the Cold War, both leaders were improvising strategies for challenging each other's commitments without escalating to full-scale conflict.

US Army Lieutenant General Albert Wedemeyer, who had commanded the China theater during the Second World War and who was now the Army's director of plans and operations, inspected LeMay's Coal and Feed Company in the first month of the operation and found it wanting. Wedemeyer had special competence for reviewing the airlift: the AAF had supplied his army in China by airlift over the Hump in the last years of the war. The Hump airlift was not LeMay's improvised early B-29 operation out of China but a major air transport operation over the Himalayas that William Tunner had commanded. From Germany, Wedemeyer sent an Eyes-Only message to Vandenberg arguing that an airlift could break the blockade or sustain Berlin during extended negotiations, but that Tunner should run it because he had run one before and knew how. Clay and LeMay resisted Wedemeyer's recommendation. The Army general met personally with Vandenberg and prevailed.

Tunner, a steady, solid man to whom Hap Arnold had offered the presidency of a civilian freight service after the war, flew to Germany in late July 1948 to take over the airlift. He found what he called "a real cowboy operation." "Pilots were flying twice as many hours per week as they should," he writes. ". . . Everything was temporary. . . . Confusion everywhere. Planes had been scraped up from all over Europe. . . . My chief of operations . . . was going to have plenty of headaches. Back on the Hump, we had thirteen bases in India feeding planes into six bases in China. . . . But here in Berlin all planes had to land at two airfields."

With Tunner in command, after a few false starts, the Berlin Airlift got underway in earnest. The transport expert established three unvarying rules that steadied the schedule and maximized the delivery of goods: crews would stay with their planes on the ramp at Tempelhof or Gatow while the planes were being unloaded; all missions would follow instrument flight rules ("You can fly by instruments in clear weather," Tunner writes, "but you sure can't fly by visual rules in the North German fog"); and any pilot who missed a landing would not go around for a second attempt but would return directly to his home base without unloading. "All planes under my command," Tunner summarizes, "would fly a never-changing flight pattern by instrument rules at all times, good weather or bad, night or day." The airlift expert hoped to make his deliveries predictable to the point of boredom—every three minutes around the clock regardless of the weather.

Truman authorized an increase to two hundred C-54s early in September. By September 25, Clay was talking to his British counterpart about delivering eight thousand tons of supplies per day, though the airlift never achieved that capacity, building up slowly across the months ahead to a five-thousand-ton average. An aide to Admiral William Leahy, Truman's chief of staff, toured Operation Vittles in late September and reported his impressions directly to the President. "The airlift is the greatest feat of its kind in the history of air transport," the aide wrote enthusiastically. ". . . The efficiency with which the operations are being conducted now, and the plans that are being made for future operations during bad weather months are outstanding." The aide understood that the airlift would meet Clay's minimum requirement of 4,500 tons per day even "during the winter months when flying conditions will be at their worst." He judged that "newspaper reports of Russian interference in our air corridors have been exaggerated." These two points together argued for the airlift's eventual success.

If the prospect of direct conflict had receded, LeMay nevertheless had prepared a private war plan. "In his mind," the aide reported, "he envisages that the Western States would decide upon a position such as the Rhine Valley, to which the Allied forces could make an orderly withdrawal in the event of hostilities, and behind which we will have established in advance the necessary air bases, dumps, depots, motor pools, etc. to meet a Russian advance." LeMay estimated that the combined British, French and American forces in Europe were almost equal to Soviet forces, a much more optimistic estimate of the European situation than the JCS view. The USAFE commander then revealed his little-NATO arrangements to the aide, who passed them on to the President: "He has . . . secured one base in France and one base in Belgium to which he could fall back, if necessary, and support the ground forces," although the bases only held a "10-day level of supply."

The aide had picked up incontrovertible evidence that the Soviets were not preparing to go to war over Berlin:

Because the Russians have assembled a formidable fighting force in Germany, they will require a tremendous logistical effort in order to launch any large-scale and sustained offensive. Lines of communication to the eastward are essential to its success. I was told at the G-2 [intelligence] briefing that the Russians have dismantled hundreds of miles of railroads in Germany and sent the rails and ties back to Russia. There remains, at the present time, so I was told, only a single track railroad running eastward out of the Berlin area and upon which the Russians must largely depend for their logistical support. This same railroad line changes from a standard gauge, going eastward, to a Russian wide gauge in Poland, which further complicates the problem of moving supplies and equipment forward.

"Neither Stalin nor Molotov believed that the airlift could supply Berlin," writes Walter Bedell Smith. "They must have felt sure that cold and hunger, and the depressingly short, gloomy days of the Berlin winter would destroy the morale of the Berlin population and create such a completely unmanageable situation that the Western Allies would have to capitulate and evacuate the city." Tunner thought the Luftwaffe's failed airlift to the German armies trapped in the Stalingrad cauldron in 1943 had prejudiced the Soviet leaders. "The Russians had never had an airlift themselves," he observes, "and they didn't take ours seriously until it was too late." The airlift commander also believed the Soviets underestimated the significance of instrument flying—of navigating with compass and attitude indicator without ground reference, a skill American military aviators had developed in the 1930s, well before radio or radar guidance systems came along. "The Russians were good pilots, capable of all kinds of stunts, and they flew in the lousiest weather conceivable—but always beneath the clouds, never on instruments. I am convinced that the Russian unfamiliarity with instrument flying led them to take our airlift too lightly. . . . They did not think we could do it."

James Arthur Hill, a line pilot during the Berlin Airlift who served later as a USAF Vice-Chief of Staff, remembers training in the C-54 at Great Falls, Montana, Racey Jordan's old turf. Hill flew from Rhein/Main to Tempelhof:

Two hours up, two hours back, reload, two up, two back. I never once got out of the airplane. Not one time. I never set foot on the tarmac at Tempelhof in all those months. At Rhein/Main I'd go to an airplane with ten tons of coal in burlap sacks with a Hungarian loading party, displaced people. I'd fly to Tempelhof, up the corridor, land, leave two engines running on the right hand side. A chute would be put down and people would come aboard and start manhandling those burlap sacks onto the chute, onto a flatbed. In about twelve to fifteen minutes, the door would close and I'd fire up the other two engines while I was moving out to takeoff position. Ground time was often less than twenty minutes.

Hill always carried coal. "Some wings carried flour," he says. "Some carried chocolate or mixed loads of sugar and flour. Staples. Staples or energy." There was a seven-story apartment building in the final approach at Tempelhof and landings could be hairy:

I went for months and never saw the ground after departure, until I broke out over the apartment building in Berlin. We were landing on about four thousand feet of pierced steel planking, so it was a very steep approach. The building was well marked with strobe lights, the 1948 version of strobe lights. You could see fairly well at about a quarter of a mile, but many, many times, we would come over that apartment building and never see the lights. We would get down to a hundred feet and I never was waved off. You had to haul the power off and jam your nose down, round it out very sharply in order to stick it on the first third of the runway. And then jump on the brakes to get stopped. We had a few cases of overruns, people running into the fence at the far end. No sweat. I was twenty-six years old. I was bulletproof.

Tunner's bulletproof pilots delivered. An anonymous American fired a burst of gallows doggerel at the Gatow air controllers one night that caught the spirit of the operation:

> *Here comes a Yankee with a blackened soul,*
> *Headin' for Gatow with a load of coal!*

By October 1948, the people of Berlin collected rations of heat, light and food under a continual drone of Allied aircraft.

Igor Kurchatov's team finished assembling the A production reactor at Chelyabinsk-40 at the end of May 1948. After a week of instrumentation testing, Kurchatov initiated a dry criticality run on June 7 (since the cooling water was a neutron absorber and would lower the system's reactivity, the reactor could be nudged to low-power criticality more simply without it). Sometime after midnight, Kurchatov had the system running at ten kilowatts and shut it down. After additional uranium loading, the reactor achieved full criticality on June 10. "We were all triumphant," Mikhail Pervukhin recalls, "and we congratulated Kurchatov and his colleagues."

The A reactor reached its designed power of 100,000 kilowatts on June 22. "At the beginning, our reactors were not powerful," Georgi Flerov recalled late in life, ". . . and there was only one. We were afraid to go larger. This one could produce about 100 grams of plutonium in twenty-four hours." That would be one solid Christy core—about 6.2 kilograms—every sixty days, but in fact the first Soviet bomb core was not ready until the following spring. Espionage had missed a critical physical process that soon shut the Chelyabinsk reactor down—not Wigner's disease, which Kurchatov knew about from Beria's collections, but the swelling of uranium metal slugs in high-flux reactors. Uranium metal swells under intense neutron bombardment because some products of fission—argon and other gases—accumulate within the structural space of the metal and deform it. Twisted, rippled slugs in the Chelyabinsk A reactor stuck in the discharge tubes and blocked them. Beria came running, alleging sabotage. "Kurchatov was able to parry the blow," reports Golovin, "[by convincing] the necessary people of our approach to the unknown area of natural phenomena, high-power neutron fields, where various surprises could be expected." They stopped the reactor, drilled out the channels, dissolved the entire loading of uranium slugs, extracted the accumulated plutonium, studied the swelling, redesigned and replaced the slug channels throughout the reactor and manufactured a new loading of uranium. The disaster delayed the operation until the end of the year. The big remote-controlled chemical plant needed to extract the plutonium was still under construction nearby in any case and would not be finished until December.

Earlier in 1948, Kurchatov, Yuli Khariton, Yakov Zeldovich and Khariton's deputy Kirill Shchelkin had met formally and decided that they would use the US Fat Man design that Klaus Fuchs had supplied them for their first bomb, RDS-1, whereupon they temporarily stopped work on a parallel independent design that would be physically smaller and would use less plutonium. "Given the tension between the Soviet Union and the United States at the time," write Khariton and Yuri Smirnov, "and the scientists' need to achieve a successful first test, any other decision would have been unacceptable and simply frivolous." They had delayed a final decision, Smirnov reports, until they "had conducted the research and experiment necessary to confirm that the information provided by intelligence was true and not disinformation. The decision was adopted for political reasons rather than technical." Vannikov concurred in it. "Beria, no doubt, knew about it," Smirnov reports, "but it remains unclear if Stalin was aware of it." Smirnov and Khariton are honorable men, but Beria's record of disagreement with Peter Kapitza and his long-standing rule that his agents should steal only conservative, tested technology dispute the physicists' independence of decision. If their decision was political, its politics were domestic more than international. Beria would not have tolerated an original design; his neck was also on the block.

Nor had Klaus Fuchs's report of the April 1946 Super Conference fallen on fallow ground. During 1947, a group under Zeldovich at the Institute of Chemical Physics had explored Edward Teller's Super design. Now the Soviet government authorized thermonuclear weapons research at the Physics Institute of the Soviet Academy of Sciences (FIAN) under senior Soviet theoretical physicist Igor Tamm. Tamm immediately recruited young Andrei Sakharov:

Toward the end of June 1948, Tamm, in a rather furtive manner, asked me ... to remain behind after his Friday in-house seminar. As soon as we were alone, Tamm shut his office door and announced his startling news: by decision of the Council of Ministers and the Party Central Committee, a special research group had been created at FIAN. . . . Our task would be to investigate the possibility of building a hydrogen bomb and, specifically, to verify and refine the calculations produced by Yakov Zeldovich's group at the Institute of Chemical Physics. (I gave it no thought at the time, but I now believe that the design developed by the Zeldovich group for a hydrogen bomb was directly inspired by information acquired through espionage. . . .)

Besides Sakharov and another Tamm protégé, Semyon Belenky, a specialist on gas dynamics, the team included Vitaly Ginzburg—"extremely talented," says Sakharov, "and one of Tamm's favorite students"—and young researcher Yuri Romanov. "A few days later," Sakharov writes, "after recovering from shock, Belenky remarked lugubriously that: 'Our job is to kiss Zeldovich's ass!' "

They moved to rooms on FIAN's newly built third floor. "Guards sat by our doors," one of their two calculators, L. V. Pariskaya, remembers. "We were given new German-made Mercedes [calculating] machines. They were good and convenient but rather noisy. Sakharov stated immediately that he would work only with me and requested others to give me no assignments."

"During the first months," Yuri Romanov recalls, "we began to familiarize ourselves with the new field of technical physics; we studied the published literature, went to the Institute of Chemical Physics to meet Zeldovich and his colleagues, became familiar with their work, and studied the problems confronting us on the drawing board. In this way we laid the foundations of a new science." Romanov worked under Sakharov. "At twenty-seven this simple, modest, childlike man already enjoyed authority in scientific circles. He distinguished himself through the clarity and correctness of his thought, and the conciseness of expression of his ideas. He dedicated himself with energy to the new problems of national defense. . . ."

Sakharov worked "feverishly," Pariskaya says:

It seemed to me often that he was deadly tired: either he worked at night or did not sleep well. Once he came late. I came to him with the work. He looked at me with eyes so empty that I just asked: "What's the matter?" He was silent. Suddenly he clutched his head with both hands and whispered: "But you don't understand! This is horrible, horrible! What am I doing?" —he added very softly: "You know, I have internal hysterics. I can't do anything . . ."

It was then that I told him: "Go right home and go to bed. Go!" He thought for awhile, agreed and left. He came back the next day and said to me triumphantly: "You know, I slept for thirteen hours in a row . . ."

"Despite summer's distractions," Sakharov writes, "we worked with a fierce intensity. Our world was bizarre and fantastic, a striking contrast to everyday city and family life, and to normal scientific pursuits." They were convinced that their work was "*essential*. . . . We were possessed by a true war psychology."

By the end of the summer, Sakharov had discovered an alternative, more promising approach. "I radically changed the direction of our research by proposing an alternative design for a thermonuclear charge that differed from the one pursued by Yakov Zeldovich's group in both the explosion's physical processes and the basic source of the energy released." In his memoirs, Sakharov calls this alternative design the "First Idea." The quiet physicist who did his thinking looking out the window had independently reinvented Teller's Alarm Clock.

Sakharov called his First Idea a "layer cake"—"alternating layers of light elements," writes Romanov, "(deuterium, tritium and their chemical compounds) and heavy elements (U238)." Sakharov was essentially proposing to enlarge the natural-uranium tamper of a Fat Man implosion system to incorporate a layer of light elements.* The fissioning of the system's plutonium core would heat the tamper materials to thermonuclear temperatures. Under such extreme conditions, matter is almost completely ionized—bare nuclei stripped of their electrons, that is—and such ionization would equalize pressures between the layers of heavy and light elements. "This means," writes Lev Altshuler, "that the light substance should be very much compressed, which is the main condition for a fusion reaction." In Soviet weapon-design circles, Altshuler adds, "this particular phenomenon came to be called 'sakharization' "—"sugarization" (*sakhar* is "sugar" in Russian) —in free translation, "caramelizing." High-energy neutrons released in fusion would then immediately fission the U238 tamper nuclei mixed with the fusing hydrogen nuclei, greatly increasing the yield in a system that might be no bigger than a Fat Man system. This chain of energy-releasing processes, notes Altshuler, "later became common for all future design variants: fission-fusion-fission." (U238 can be fissioned with high-energy neutrons. It does not chain-react, but neither does its fissioning require a critical mass. It serves as a fuel in U235 fission-hydrogen fusion-U238 fission systems much as deuterium does, thus overcoming to some extent the basic limitation of a pure fission device—the disassembly of the critical mass as it heats up and expands that stops the chain reaction and limits the yield.)

---

* The 1946 Gurevich, Zeldovich, Pomeranchuk and Khariton paper *Utilization of the nuclear energy of the light elements* discussed using "uranium charges of increased sizes and of a special shape (cumulation) and [introducing] into the deuterium heavy elements near the initiator which might be capable of receiving the radiation pulse." Although this description sounds like a hypothetical spherical hydrogen bomb, it lacks the crucial layering that Sakharov proposed.

Carson Mark comments:

The great virtue of Sakharov's First Idea was that feasibility didn't have to be established. It wasn't open to argument as to whether the process would work or not. If you take one look at the layer cake and think of what needs to go on, you don't have to doubt that something will happen. You say, If I heat this up, something is bound to happen. You might question, Will it happen to an exciting extent or only to a poor extent? That you have to work to get a feeling for. But you don't have to establish the feasibility of the process, whereas if you look at the results of the Super Conference in April of '46 or look at what Zeldovich was doing, the first thing to ask yourself is, My God, will it work that way or not? And that's what Sakharov said very quickly, that his First Idea had the lovely feature of feasibility.

Igor Tamm embraced the new design as soon as he heard of it, Sakharov remembers; "he'd been skeptical from the start about the earlier approach." Zeldovich "saw the merit of my proposal" as soon as Sakharov found a way around Zeldovich's mistrustful assistant to tell him. "[Zeldovich and I] discussed both our projects at length and agreed that Tamm's group would concentrate on the new proposal, while his team would continue work on the earlier design, at the same time providing any help we might need, since there were still many gaps in our knowledge." That early, Sakharov believed, Zeldovich decided to request Sakharov's transfer to Sarov, although the talented young physicist would continue to work at FIAN for another year and a half.

What Sakharov calls a Second Idea added to the attractiveness of his layer-cake design: Vitaly Ginzburg suggested using lithium deuteride (LiD) instead of deuterium and tritium in the fusion layer. Ginzburg, writes Romanov, "affectionately named the . . . LiD 'Liddy'. . . . I. V. Kurchatov . . . efficiently organized its production." Sakharov got a raise for his breakthrough and a meeting with one of Beria's generals, who complimented him and urged him to join the Party. The physicist had the presence and the prescience to tell the general that he was unable to do so "because a number of its past actions seemed wrong to me and I feared that I might have additional misgivings at some future time."

Sakharov was lonely in his new work, a FIAN colleague, Matvei S. Rabinovich, remembers.

When in a confiding mood, he sometimes said, "You know, you are the only person I can have a word with." Once, he told me, "This sort of thing happens: I'm often asked to the Kremlin to a meeting. It goes on usually until four in the morning; then they all go to their cars, but I haven't got a car, and nobody knows that I haven't got a car, and I don't tell anyone. It means that I've somehow or other to get from the Kremlin to Oktyabrskoye

Polye, and that's at least 12 kilometers and perhaps 15." If he couldn't get a taxi, he had to walk it.

In Vannikov's office with Tamm, early in 1949, Sakharov received an offer from the leadership he could not refuse. Vannikov proposed to transfer Sakharov permanently to Sarov to work with Yuli Khariton. Tamm resisted —he wanted to save Sakharov for pure science and not limit him to weapons research. "The direct Kremlin line rang. Vannikov answered and then tensed up. 'Yes, they're here with me now,' he said. 'What are they doing? Talking, arguing.' There was a pause. 'Yes, I understand.' Another pause. 'Yes sir, I'll tell them.' Vannikov hung up and said: 'I have just been talking with Lavrenti Pavlovich [Beria]. He is *asking* you to accept our request.'"

"There was nothing left to say," Sakharov concludes.

A tall, smart, methodical twenty-nine-year-old FBI agent from the Coeur d'Alene mining district of Idaho, Robert Lamphere, took up cryptanalysis work at Bureau headquarters in Washington late in 1947. The wartime cables from the Soviet Consulate in New York to Moscow Center that the Army Security Agency had copied during the war still awaited decoding. Working on them at ASA was a brilliant cryptanalyst and linguist named Meredith Gardner, whom Lamphere soon befriended.

Gardner had made a little progress, a few words here and there. The cables had been coded using one-time pads, a system which was usually unbreakable, but Gardner had a copy of a partly burned NKVD codebook. The Finns had recovered it from a battlefield in 1944 and had sold some 1,500 pages to the OSS. Shocked that the US might be spying on its wartime ally, Secretary of State Edward Stettinius had insisted that the OSS return the cipher material to the Soviet government. The agency did so, but not before it had clandestinely copied the codebook. The NKVD had assumed the OSS was shrewder than the Secretary of State and immediately in May 1945 had changed its codes. The codebook was therefore a window into NKVD cable traffic that opened in 1944 and closed in 1945. Gardner and Lamphere had no way yet of knowing that those were crucial years for Soviet atomic espionage.

Early in 1948, Gardner asked Lamphere if he could supply him with the plain text of some of the cable traffic. Lamphere forwarded the request doubtfully to the New York field office. But New York had pulled a bag job on the Soviets in 1944, burglarizing their offices; back came a stack of documents. "This, then," Lamphere exults, "was the beginning of an important new phase in our breakthrough, for in a short while Meredith began to give me some completely deciphered messages. . . ."

Among the messages Gardner deciphered were exact copies of telegrams from Winston Churchill to Harry Truman and a report "that someone (designated by a code name) had approached a man named Max Elitcher and had requested that Elitcher provide information to him on his current work at the Navy's Department of Ordnance." The Elitcher contact was dated June 1944 (which, as Elitcher would eventually independently corroborate, was when Julius Rosenberg had traveled to Washington and pitched espionage to him with the argument that the Soviet Union was being denied technical information vital to its war effort). Lamphere had Elitcher checked out and found that he still worked for the Navy's ordnance organization, now called the Bureau of Ordnance, in 1948. Further checking uncovered a connection between Elitcher and a fellow Navy employee named Morton Sobell, both of whom had been suspected at one time of Communist connections. "Background checks revealed that Elitcher had attended the City College of New York from 1934 to 1938, and had graduated with a degree in electrical engineering." (So had Rosenberg, although Lamphere was not yet aware of him.) Sobell had been one of Elitcher's classmates and a roommate in their bachelor days.

Another message fragment concerned two possible espionage contacts or agents. One was Joel Barr, Julius Rosenberg's fellow Communist cell member who had moved to Europe in 1947 at the time of the Elizabeth Bentley grand jury investigation. Lamphere went looking for Barr and discovered that the electronics specialist had been a project engineer at Sperry Gyroscope in 1946 but was now living in Finland, playing the piano to support himself. "He'd been in the same CCNY undergraduate electrical engineering department as Sobell and Elitcher," Lamphere also learned, "and at the same time, graduating in 1938."

The other possible contact was a woman. Based on the fragmentary information Gardner had decoded, Lamphere in June 1948 was able to conclude that either Barr or the woman might have "acted as an intermediary between [a] person or persons who were working on wartime nuclear fission research and for MGB agents (1944)." What Lamphere knew about the woman went into a profile he circulated on June 4:

> Christian name, ETHEL, used her husband's last name; had been married for five years (at this time) [i.e., 1944]; 29 years of age; member of the Communist Party, USA, possibly joining in 1938; probably knew about her husband's work with the Soviets.

Barr was a bachelor, but Lamphere checked out his girlfriends. None fit the profile. "We came to a dead end on the investigation into 'Christian name, ETHEL,' in 1948," the FBI agent recalls.

Max Elitcher, whose marriage was failing and who may have become aware that he was under FBI surveillance, had decided to leave the Bureau of Ordnance that summer and find civilian work. He visited Sobell, who told him not to give up his BuOrd job until he talked to Julius Rosenberg. He met Rosenberg in New York—"on the street," the FBI paraphrases his testimony—and "Rosenberg told Elitcher that it was too bad Elitcher had decided to leave because he, Rosenberg, needed someone to work at the Bureau of Ordnance for espionage purposes. Sobell was present at this meeting." Elitcher and Rosenberg went on to dinner at Manny Wolfe's Restaurant, "where they continued to talk about Elitcher's desire to leave his job." That was when Elitcher had asked Rosenberg how he had become a Soviet agent.

Driving from Washington to New York with his wife in July to stay with the Sobells and look for a house, Elitcher had noticed that he was being followed. ("On the trip from Manhattan to the Sobells' home," the FBI agents following the Elitchers subsequently reported, "it was confirmed without a doubt that the Elitchers were 'tail conscious' and, therefore, the surveillance was discontinued.") Sobell was furious that the Elitchers might have led the FBI to his house. That evening, worried about a raid, he took Elitcher with him to deliver a can of 35 millimeter film to Rosenberg in Knickerbocker Village. Elizabeth Bentley had recently testified publicly and sensationally before the House Un-American Activities Committee; on the way back to Queens, Elitcher asked Sobell if Rosenberg had known Bentley. Sobell, Elitcher recalled, said Rosenberg "once talked to Elizabeth Bentley on the phone but he was pretty sure she didn't know who he was and therefore everything was all right." Elitcher moved to Queens in October, to a house adjoining Sobell's, and joined Sobell working at the Reeves Instrument Company.

Harry Gold was in love. He was no longer working for Abe Brothman. Sometime in 1947, Brothman had stopped paying him. He had tried unsuccessfully to borrow five hundred dollars from his former boss at Pennsylvania Sugar, telling the man that Brothman was nearly bankrupt—not the best collateral. On June 5, 1948, he gave up on Brothman and quit. Brothman was still worried about the grand jury testimony that the two men had concocted, Gold recalled. "On the occasion when I finally left A. Brothman and Associates . . . Abe told me that he wanted to go over my story one more time, but I told him there was no point in it because I was well acquainted with the story. One of Abe's final remarks was, 'Remember when the Rover Boys come around, you'll want to tell the same story you did before.' [Brothman] . . . appeared to imply a threat. . . ."

Word that Gold was looking for work eventually reached Beatrice Schied, a woman he had known and tried several times to date at Pennsylvania Sugar during the war. In 1948, Schied, a lab technician, was employed at the Philadelphia General Hospital Heart Station, a laboratory devoted to cardiology research. The laboratory had received a grant from the US Public Health

Service to add a biochemist to its staff. Schied recommended Gold. He got the job in August and started work the next month. That was when he encountered the love of his life:

> I fell in love with Mary Lanning when I first met her in Dr. Samuel Bellet's laboratory at P.G.H., on Wednesday, September 10, 1948. It really happened so simply: just like that; I knew that here was the girl I had been searching for all my life—as banal as this sounds. And, as we started to go out together and I got to know her well, this feeling only increased—and the wish to make her my wife became an overpowering drive in my life. Her unassuming manner, forthright honesty and complete lack of artificiality, and her snub nose—completely captivated me. I could go on for hours.

Yosef, Gold's brother, probably listened to Harry going on for hours; he told his supervisor at the Naval Aviation Supply Depot, where he now worked, "that his brother, Harry, had a serious romantic interest in a Gentile girl who lived in Germantown." Gold and Mary Lanning dated regularly from September onward. She evidently accepted him as a serious suitor, but she sensed, Harry knew, that he was holding something back:

> Even in the very beginning a warning bell sounded: suppose that the Grand Jury Investigation in 1947 is really not the end of all inquiry into my life, and who knew better than I on what a precarious house of cards my whole life rested? And from the very first I realized, and Mary often remarked on it, that I never could be completely relaxed and at ease in her presence. But she never suspected the real cause.

He could no more have told her about his years of espionage work than he could have confessed to murder, though he seems to have edged toward disclosure at least once, Mary Lanning recalled for the FBI:

> She . . . said that at one time during the period of her acquaintanceship with Gold he had mentioned a visit to New Mexico. She said that she recalled his mention of having been in the city of Santa Fe. The exact dates of those visits were not known to her, but she believed that it was during the period he was employed by Pennsylvania Sugar Company, as he indicated that Pennsylvania Sugar had had some interest in a Coca Cola bottling plant in that area.

So Harry threw himself into heart research, receiving regular promotions, and uneasily courted the snub-nosed girl of his dreams.

In the same season—on September 1, 1948—Donald Maclean departed New York with his family for reassignment in England. Before he left, the

Atomic Energy Commission gave him a farewell luncheon at the elegant old Hays Adams Hotel.

————————

The crisis in Berlin had spurred the US Air Force to review its readiness to go to war. The Strategic Air Command in particular—the United States's only deterrent in 1948—was notably below standard. Hoyt Vandenberg, the senator's nephew and since April 1948 the USAF Chief of Staff, asked the distinguished aviator Charles Lindbergh to study the Air Force's atomic squadrons and recommend their improvement. Lindbergh flew with SAC aircrews through the first summer of the Berlin airlift. He found them ill-trained and overworked. Even the 509th contrived to inflate its training record. During July, it made 386 visual bomb drops from under 25,000 feet with an average circular error of 353 feet but only forty-four drops from above 25,000 feet with an average circular error of more than one mile. It made four visual drops for each radar drop. Since the SAC atomic mission would be to bomb the Soviet Union by radar at night from above 25,000 feet, the 509th training program corresponded to shooting fish in a barrel. Lindbergh's indictment was blunt. "The personnel for atomic squadrons were not carefully enough selected," he found, "the average pilot's proficiency is unsatisfactory, teamwork is not properly developed and maintenance of aircraft and equipment is inadequate. In general, personnel are not sufficiently experienced in their mission."

Lauris Norstad retraced Lindbergh's SAC investigation and confirmed his findings. Three years after the war, with the Soviets actively probing US intentions, the only atomic striking force was still not combat-ready. Norstad insisted that Vandenberg appoint a new commander. "Vandenberg asked whom he would recommend," writes a military historian, "and Norstad responded with a question: Who would you want in command of SAC if war broke out tomorrow? The chief of staff quickly replied: LeMay."

Perhaps restless at USAFE with Tunner running the airlift, LeMay signed on enthusiastically on October 19, 1948. He arrived at SAC headquarters at Andrews Air Force Base, outside Washington, with a roar. "The first morning," one of his staff officers recalled, ". . . General LeMay said, 'As the first order of business, I want to review the war plan.'" There was no war plan, LeMay storms:

> Then I asked about the status of training: "Let me see your bombing scores." "Oh," was the response, "we are bombing right on the button." They produced the bombing scores, and they were so good I didn't believe them. The same was true of the radar bombing scores. Then, looking a little further, I found out that SAC wasn't bombing from combat altitudes, but from 12,000 to 15,000 feet. I looked at the radar picture, and the planes

weren't at altitude. There had been trouble with the radars working at altitude. Instead, the crews bombed down where the radars would work. Instead of bombing a realistic target, they were bombing a reflector on a raft out in the ocean. . . . It was completely unrealistic.

"We didn't have one crew," LeMay adds in his memoirs, *not one crew* in the entire command who could do a professional job."

"The day that was bloody was the first day or two that General LeMay was there," the staff officer, Jack Catton, continues. "[The General was] so disappointed and frustrated with what [he] found in the Strategic Air Command when [he] came back to command it, that it got right bloody. General LeMay assembled the staff and advised who was going to stay and who was going to go. He restaffed himself right then. So it was bloody, but it was necessary and appropriate, and we really got a head of steam going." Catton survived the purge.

"Everybody thought they were doing fine," the new SAC commander understood. "The first thing to do was convince them otherwise." After moving the command from Andrews to Offutt Air Force Base, near Omaha, Nebraska (a basing decision concluded before he arrived), LeMay ordered a maximum-effort mission against Wright Field at Dayton, Ohio—"a realistic combat mission, at combat altitudes, for every airplane in SAC that we could get in the air." Since Air Force intelligence could supply only vintage prewar aerial photographs of Soviet cities, LeMay gave his crews 1938 photographs of Dayton. He instructed them to bomb by radar from thirty thousand feet and to aim for industrial and military targets, not radar reflectors.

"Oh, I'll admit the weather was bad," he recalled in retirement of the January 1949 mission. "There were a lot of thunderstorms in the area; that certainly was a factor. But on top of this, our crews were not accustomed to flying at altitude. Neither were the airplanes, far as that goes. Most of the pressurization wouldn't work, and the oxygen wouldn't work. Nobody seemed to know what life was like upstairs." Not many crews even found Dayton. For those who did, bombing scores ran from one to two miles off target, distances at which even Nagasaki-yield atomic bombs would do only marginal damage.

LeMay called the results of the Dayton exercise "just about the darkest night in American military aviation history. Not one airplane finished that mission as briefed. *Not one.*" Offutt AFB was a disaster as well. "There wasn't much to Offutt except a big bomber plant and a cockeyed runway ending in a steep bank—just about as silly a runway as you could imagine." The wartime base of "flimsy barracks [and] tar-paper shacks" lacked family housing, which discouraged the reenlistments SAC needed to build experience in its crews. "My goal," LeMay said, "was to build a force so professional, so strong, so powerful that we would not have to fight. In other words, we had to build a deterrent force. And it had to be good." The National Security Council had adopted deterrence formally as United States policy in November 1948, concluding that the US must "develop a level of military readiness which can be maintained as long as necessary as a deterrent to Soviet aggression." Later that year, however, Truman impounded $822 million in supplemental Air Force funds that Congress had voted and limited overall military spending to $14.4 billion, only 60 percent of Forrestal's original request. The sturdy, phlegmatic new SAC commander, who chewed a cigar stub to disguise the Bell's palsy he'd caught flying high and cold that drooped one side of his lower lip, who "remembered the horrible experience that we all had . . . of going to war with nothing," had a load of work to do.

The knot of the Berlin crisis finally loosened over the 1948 Christmas holidays. The winter had been severe, fog in particular limiting deliveries, and stockpiles in Berlin were running low. "It looked like curtains," Army Undersecretary William Draper remembered. "If that fog had stayed another three weeks we probably would have had to run up the white flag. We probably couldn't have gone on. You can't have people starving and keep on with the occupation. But the weather lifted about the fifth of January . . . and immediately we restored the situation. The Russians knew they were licked right away. . . ." At the same time, an economic counterblockade had pinched the Soviet zone of Germany severely, reducing needed imports in 1948 by 45 percent.

Stalin sent a signal of capitulation at the end of January 1949. Kingsbury Smith, an American journalist, had telegraphed the Soviet leader a series of questions. Stalin responded to one which asked if he was prepared to lift the Berlin blockade if the US, Britain and France would postpone establishing a West German state until the Council of Foreign Ministers could renew its meetings. Stalin answered that his government was prepared to do so on those terms if the counterblockade was also lifted. Charles Bohlen, one of the State Department's Soviet experts, noticed that Stalin had not mentioned the currency problem, which had stymied previous diplomatic efforts.

Truman had won the November election against all predictions and had just been inaugurated for his first full term as President; he had appointed Dean Acheson his new Secretary of State, following upon George Marshall's retirement because of illness. Acheson brought Stalin's signal to Truman's attention; the American leaders responded with a signal of their own at an Acheson press conference on February 2. Secret negotiations began soon thereafter to end the crisis, and rail traffic into Berlin resumed in May.

Why did the Berlin blockade not come to war? At the outset of the conflict, Winston Churchill, brooding out of office, had propounded a much more belligerent course to the US ambassador to Britain, Lewis Douglas. "When and if the Soviets develop the atomic bomb," Douglas had reported Churchill's views, "war will become a certainty. . . . He believes that now is the time, promptly, to tell the Soviet[s] that if they do not retire from Berlin and abandon Eastern Germany, withdrawing to the Polish frontier, we will raze their cities. It is further his view that we cannot appease, conciliate or provoke the Soviet[s]; that the only vocabulary they understand is force; and that if, therefore, we took this position, they would yield." LeMay had favored military action at the beginning as well:

> [Army] General [Arthur G.] Trudeau, who commanded the constabulary, and I concocted a plan where he would run a small military force up the autobahn and open Berlin by force. I would have a communications van, and when he started up, I would have the B-29s based in England in the air over Germany with the fighters that I had also moved up closer. If General Trudeau made the decision that he was at war, instead of just pushing through token resistance, then I would let the air force go and hit the Russian airfields. The Russians were all lined wingtip to wingtip on their airfields. We presented this plan to General Clay . . . and he sent it to Washington, but the answer was "No."

Instead of military confrontation, with caution and restraint, the two suspicious adversaries had limited themselves to an extended, nonviolent exploration of their mutual positions and commitments, improvising communications as they went along. Though both were revolutionary systems with messianic pretensions, they had found it expedient to cooperate first of all during the Second World War. Through that four years of cooperation, the Soviet leadership had learned just how immense was the productive capacity the United States could deploy in war—capacity sufficient to sustain the USSR with Lend-Lease while fighting a two-ocean war, capacity sufficient also to absorb the immense cost (as the Soviets were now learning at first hand) of developing a capability to manufacture atomic bombs. If the US had removed all but token forces from Europe since the victory, it had continued to enlarge its atomic capability, as its tests at Bikini and Eniwetok in 1946 and 1948 confirmed. And however much Soviet leaders publicly belittled atomic weapons, Stalin evidently judged them sufficiently valuable to invest a major portion of his limited resources in acquiring them as rapidly as possible.

Faced with Western determination not only to establish a separate West German government but also (as Donald Maclean had probably reported) to rearm that traditional Russian foe, Stalin in blockading Berlin chose a significant but peripheral challenge to US authority. He calculated that the United States would not consider access to Berlin a cause sufficient to justify going to war. He was right, although he failed to anticipate the effectiveness of an airlift to thwart his blockade.

What restrained the US leadership from issuing a Churchillian ultimatum or probing the blockade with Clay and LeMay? Operationally, lack of readiness—atomic or conventional—made the US cautious and limited its response. But Truman's reluctance to reinforce a precedent he himself had introduced—of pursuing military goals with weapons of mass destruction —should not be underestimated as an influence, possibly decisive, on US restraint. "This isn't just another weapon," the President told Lilienthal again in February 1949, "not just another bomb. People make a mistake when they talk about it that way. . . . Dave, we will never use it again if we can possibly help it." That early in the Cold War, the President as yet had no experience with Soviet attitudes toward atomic weapons and still assumed the worst, adding, "But I know the Russians would use it on us if they had it." Truman's conviction that the Soviets would use atomic weapons on the West if they had them boded ill for the approaching time when there would be two atomic powers in the world. But it probably also gave him an additional reason to resist using them to resolve the confrontation over Berlin. Which suggests that a degree of mutual deterrence had already been installed between the United States and the Soviet Union even before the Soviet Union finishing building its first atomic bomb.

18

'This Buck Rogers Universe'

SINCE CURTIS LEMAY had found no war plan on that bloody day in October 1948 when he arrived to take over the Strategic Air Command, he set out to prepare one. No one in the world knew more about strategic bombing than he did; only he had actually commanded and carried out a successful full-scale strategic bombing campaign against an enemy nation. The Air Force had analyzed the firebombing and atomic-bombing of Japan carefully in the years since the end of the war; LeMay had sifted the bombing results thoroughly for revelation. "The fact that Japan," he told an audience in 1946, "while still in possession of a formidable and intact land army, surrendered without having her homeland invaded by enemy land forces, represents a unique and significant event in military history." LeMay was convinced that the heavy bomber, devastating "Japan's cities, factories, transportation and shipping," had been a principal factor in determining the Japanese surrender.

Time and space had favored the United States in the last war, LeMay believed. "We had space between us and our enemies which could not be spanned by the then-existing weapons. . . . This in conjunction with our allies who fought the holding battle gave us 'time'—time to build a fighting machine to defeat our enemies." Since the war, however, the US had begun development of long-range bombers capable of carrying ten thousand pounds of bombs ten thousand miles. Our "potential enemies" could do likewise, LeMay warned. "Super rockets" were coming as well. These developments meant that "our space factor has disappeared." So had our time factor. "If there is another war, we will be first, instead of last to be attacked, and the war will start with bombs and missiles falling on the United States."

LeMay judged that new technology and new political conditions were moving the US into the same circumstances of exposure that had made

Japan vulnerable during the Second World War. "Since the problems are similar," LeMay proposed, "Let us place ourselves in the position of the Japanese. What could they have done to prevent atomic bombs being delivered on Japanese targets?" LeMay saw "three possible solutions," all military (he did not consider negotiation or surrender). One solution would have been to build fighter and antiaircraft defenses sufficient to shoot down the bombers. But "throughout the war, not one of our attacks was ever turned back by enemy action." Defense, then, was "the most inefficient method, and the one least likely to succeed."

A second solution, LeMay thought, would have been to destroy the B-29s at their bases in the Marianas. But it would have been "virtually impossible to destroy every single airplane, and even if that happened, others could be flown in and staged through the battered fields." With atomic bombs the effort would be futile, since "one airplane does the work of hundreds."

LeMay concluded that only his third solution might have saved the Japanese: "Destroy our factories and laboratories that were producing the bombs." To do so, Japan would have needed a long-range bombing force, which it had not built. "So the Japanese found themselves without an answer and went down in defeat before modern methods, even though [they] still had intact over seven million men under arms, most of which were never committed to combat. Let us hope that we never find ourselves in a similar position."

Now, taking over SAC, the new commander was assuming responsibility for making sure that his country did not find itself without an answer to the challenge of atomic attack by long-range bomber. "Preparation" had been the answer he had offered the Ohio Society in New York shortly after the end of the war—deterrence. But then he had skirted a mortal question: what if the enemy was not deterred?

LeMay was prepared now to face that question. He had remembered that the decisive attacks on Japan, 91 percent of the total bomb tonnage dropped, had been concentrated into the last five months of the war. He associated that successful blitzkrieg with his old Iron Ass axiom, "Hit it right the first time and we won't have to go back." Atomic bombs made hitting the target right the first time far more probable than ordinary high explosives had allowed. Given a "war aim of complete subjugation of the enemy," the Air Force war plans division had recently concluded, "it would be feasible to risk an all-out atomic attack at the beginning of a war in an effort to stun the enemy into submission." The distinguished board of military and civilian experts that had evaluated the Bikini test results, headed by MIT president Karl Compton, had pushed the blitzkrieg concept a dangerous step farther, arguing that since "offensive measures will be the only generally effective means of defense . . . the United States must be prepared to employ them before a potential enemy can inflict significant damage upon us"—arguing, that is, for first-strike preventive war.

So when LeMay took his ideas for a SAC war plan to USAF Chief of Staff Hoyt Vandenberg in November 1948, he proposed that "the primary mission of SAC should be to establish a force in being capable of dropping 80% of the stockpile in one mission." By then he was confident, he told Vandenberg, that "the next war will be primarily a strategic air war and the atomic attack should be laid down in a matter of hours." Vandenberg agreed; in the first SAC Emergency War Plan that LeMay delivered in March 1949, his November proposal became a goal to increase SAC capability "to such an extent that it would be possible to deliver the entire stockpile of atomic bombs, if made available, in a single massive attack." Fitted to the most recent JCS war plan, LeMay's plan for SAC meant destroying seventy Soviet cities within thirty days with 133 atomic bombs, causing at least 2.7 million civilian deaths and another four million casualties. (This scale of destruction corresponds notably to that of the firebombing of Japan, which resulted in the burning out of sixty-three Japanese cities and the killing of 2.5 million civilians. Such a relatively modest atomic-war plan, limited not by strategic restraint but simply by the exigencies of the atomic stockpile, reinforced the protective delusion that atomic war would differ from conventional war primarily in efficiency. But the bombing of Japan had been a maximum effort, while the atomic campaign could and would increase in destructive scale as the stockpile grew.) The Air Force high command signed on to LeMay's plan at a conference at the Air University in December, allotting SAC top budget priority.

American air-power strategists had a name for such an attack as LeMay was proposing: "killing a nation." SAC had a long way to go before it could kill a nation, as the Dayton debacle would soon demonstrate. In December, LeMay told General Roger Ramey, who commanded one of SAC's two air forces, that the 509th atomic bomber group was "no damn good." "Since the 509th had fallen so desperately low in efficiency," LeMay's aide paraphrases, "he considered it not operational and directed that a major turnover in the entire personnel be made and other drastic action be taken to get this group into operational efficiency." At the same meeting, LeMay ordered Ramey to modify tankers for aerial refueling "as fast as possible but to keep quiet about this as he did not desire any publicity on this whatsoever." He also wanted the number of airmen in radar school doubled to compensate for lost reenlistments and the intelligence section improved.

"My determination was to put everyone in SAC into this frame of mind," LeMay writes: *"We are at war now.* So that, if actually we did go to war the very next morning or even that night, we would stumble through no period in which preliminary motions would be wasted. We had to be ready to go *then."*

We took the 509th Group, the original atomic outfit. I said: "Okay, we will start with that one." We cleaned out the supply warehouses and stocked the things that the unit needed. We equipped all the planes with the things they were supposed to have on them. Some of the airplanes didn't even have guns on them; since it was peacetime, they supposedly didn't need them, and didn't have them. We put all the things on the airplanes they were supposed to have, and then started cleaning out the people who didn't belong there, and getting people in who did.

LeMay had told Vandenberg that he intended to establish a mobile operational force by January 1, 1949, that would include two atomic medium groups* and one atomic heavy group, cannibalizing the rest of SAC of the best pilots and crews to do the job. "This will barely give us the capability of meeting [our] mission," he reported. By June, he intended to double this primary mobile force. The rest of SAC he would build one group at a time as resources became available. As of January 1, SAC had ninety special crews and 124 aircraft modified to carry atomic bombs; its overall force included thirty-five of the huge new six-engine B-36s that could carry 86,000 pounds of bombs and fly above forty thousand feet, thirty-five B-50s—improved atomic-capable B-29s—and 486 B-29s. The official US atomic stockpile now numbered fifty-six Mark III bombs; new Mark IVs would begin entering the stockpile early in 1949. The Mark IV could be stored in final assembled form; Kenneth Nichols of the Armed Forces Special Weapons Project called it "the first engineered atomic weapon."

At the beginning of 1949, and for at least a decade to come, SAC had an ace in the hole, the same ace that LeMay had counted on in 1945 when he planned the firebombing of Japan: like Japan before it, the Soviet Union was defenseless against strategic bombing. Vandenberg had reported Soviet vulnerabilities to Forrestal in December 1948. They were appalling:

> Soviet antiaircraft artillery consists mainly of 88 mm heavy guns and 37 mm automatic weapons. The maximum effective ceiling of the 85 mm gun is 25,000 feet. . . .
> Jamming operations by Allied aircraft will be conducted against gun-laying radar and will greatly reduce its effectiveness. . . .
> There is no evidence that the Soviets possess any aircraft specifically designed for night or all-weather fighter operations. . . .
> Of the known types of fighter aircraft in existence or under development [in the USSR] none are suitable for such operations against B-29 type aircraft. . . .
> An adequate perimeter [fighter and antiaircraft] defense [of the USSR] is clearly out of the question. . . .

---

* A group consisted of three to four squadrons of about fifteen aircraft each: forty to sixty operational aircraft in all, plus spares.

Even if [Soviet] fighters are scrambled before identification is made, no fighter passes are possible before [a US] bomber releases its bombs. . . .

The Soviets do not have the capability to interfere with the effective use of our airborne radar bombing equipment. . . .

It is not believed that the Soviets have the capability of making United Kingdom bases untenable before D + 45 to 60 days at the earliest. . . . The strategic air offensive would delay considerably or deny completely this capability.

Vandenberg reported an estimated loss of US aircraft in initial atomic attacks against the Soviet Union of 25 percent. LeMay revealed the real estimate over lunch at the National War College a year later when two Navy officers ragged him about the effectiveness of air defense. "General LeMay . . . stated," his aide records, "that in his estimation, if the 'bell were to ring now,' certain targets could be penetrated and attacked with very little loss and that the overall losses would not exceed ten percent." SAC only used 25 percent, LeMay observed, "for a logistical planning figure."

However formidable on the ground, from the air the Soviets were naked unto their enemies. Building SAC would exploit that mortal vulnerability. The way to build SAC was to fly. LeMay reports sending a B-36 in December 1948 "over 8,000 miles in about 35 hours from Fort Worth, Texas, to Honolulu and back, carrying a useful load of simulated bombs which were dropped . . . in the ocean off Honolulu." In February 1949, the *Lucky Lady II,* a SAC B-50, flew nonstop around the world with aerial refueling in ninety-six hours. "Flying," writes LeMay, "going through each vital motion except for the physical act of releasing live bombs from the shackles—we attacked every good-sized city in the United States. People were down there in their beds, and they didn't know what was going on upstairs." SAC intelligence concluded that Baltimore "more closely resembled European and Soviet cities than any other urban area in America," a military historian reports. "Reconnaissance aircraft then overflew the city from every angle, photographing hundreds of [radar] scope presentations." The photographs taught crew members how their targets would look on radar. San Francisco was another favorite target; SAC once faux-bombed it more than six hundred times in one month. As if hardening itself to fatality, the Strategic Air Command prepared to kill the Soviet nation by practicing on its own.

———

At Chelyabinsk-40, the conscript soldier was assigned to construction now, labor that seemed to him not much better than the *zeks*'s. The men lived in long, dark one-story barracks that prisoners had vacated, with three levels of wooden berths in four rows. "In the morning," the soldier remembers, "after a light breakfast of oat or millet porridge and tea, we were formed up into teams and marched through the forest to work. Our job was to dig trenches for pipelines and cables. The norm per person was established at 2.5 cubic meters of hard and frozen soil to be cut with sledgehammer and chisel. We were organized in teams so that one member could hold the chisel and the other swing the sledgehammer. Only after fulfilling our norms could we get a meal. After such hard work we were subjected to humiliating drills, goose-stepping or crawling on our bellies in the snow. People were exhausted and unhappy."

The soldier's regiment comprised some seven hundred men; there were four such regiments in the central zone at Chelyabinsk-40 as well as four prison camps (including one for political prisoners and one for women). They were helping build B installation, the vast remote-controlled chemical plant that would dissolve the irradiated uranium discharged from the A reactor and chemically separate the kopek's-weight of plutonium contained in each ton. "We worked in three shifts," the soldier says. "There was no shortage of manpower." After blasting into the bedrock, the *zeks* loaded the shatter onto trucks which entered the deep, canyon-like pit down a spiral access road. The pit was completed early in the winter of 1948 and lined with timber forms for concrete work when disaster struck:

We were wakened by an alarm signal and ran to the pit. It was all on fire, vast flames shooting out. We were ordered to shovel gravel into the pit to contain the fire, but it was so hot we were unable to approach the edge. So we returned to our barracks. The next day, Lavrenti Pavlovich Beria himself flew in with his staff—to investigate the circumstances of this sabotage, so we were told. Of course, nobody ever told us the results of this investigation.

Damage from the B installation fire was quickly repaired; by December 1948 the plant was built and operating. Soviet health officials had established radiation standards for A and B installation workers, but Russian nuclear experts report that "during the start-up period the radiation conditions at both facilities were very hard." It was impossible to work at the plutonium separation plant without dangerous radiation exposure. Sixty-six percent of B installation workers received subclinical but excessive doses in the first year of operation of up to 100 rem; an unlucky 7 percent received above 100 rem where clinical signs such as vomiting and blood changes begin to appear. (For comparison, the average lifetime dose of workers in the US and British nuclear-weapons industry has been estimated at from 3 to 11 rem.) "The first cases of radiation sickness [showed] up as early as the beginning of 1949," the Russian experts write. Workers as well as managers recognized the danger, the experts add, but claim "they realized that the country needed nuclear arms desperately and often put their safety at risk." One manager at least supports the experts' claim of patriotic sacrifice. "We used to wake up every morning and cup an ear toward the west," Boris Brokhovich, one of the first engineers on site at Chelyabinsk-40 and later its director, told an American physicist in retirement. "We expected to hear your B-29s coming over the Urals. We couldn't believe you would allow us to build this place and make a bomb." But safety in Berialand was a secondary consideration at best; B installation, by design, discharged its intensely radio-active fission wastes directly into the Techa River. By 1951, radioactivity from Chelyabinsk-40 had been measured in river water discharging into the Arctic Ocean more than a thousand miles north.

New Year's Eve. "We were strictly forbidden any access to the 'Techa' settlement where the scientists lived," the conscript soldier reports. "In our zone there was only one shop, selling odds and ends, including Troynoi ["Triple"] brand eau-de-Cologne. We bought some and saved it to [drink to] celebrate the New Year, sitting on our berths and wondering why we had been sent to such a place. Up on the third level of the wooden berths, three Georgians were also drinking Troynoi and singing sad Georgian songs. Such 'festivals' were very rare, life was mostly routine."

But life was far worse for the zeks, if testimony from a gulag contemporary with the Chelyabinsk-40 camp applies. Vassily Erchov, an agronomist and decorated Red Army colonel, described to an international commission in Brussels the conditions he had observed in 1947 at a Soviet factory gulag:

[Looking] out the [factory] window . . . I saw some observation posts with guards on them. A bit to the left, I saw some people, but I couldn't tell what kind of people they were. They were not human beings, but heaps of rags. . . . They were wearing some sort of torn padded shirts and skirts made of khaki-colored shreds. Some of them were wearing old clogs on their feet; others had old ankle-boots tied with string. The women had shaved, dirty heads. The director [of the factory] said one could not go near these people because they stank so. They were rotting away. . . .

The women slept fifty to a barrack, on litters, without mattresses or blankets; they had only the rags they worked in during the day to cover themselves with and put under their heads. The walls were red, as though covered with blood. The reason was that, though they worked during the day, they spent their nights crushing bedbugs. It was impossible to talk with these women, and whenever the manager tried to start a conversation with them, he got nowhere, as these people were so degraded that they had almost lost the use of their tongues. They answered all questions with oaths. Why had they become what they were? They said frankly that they had lost their entire past and had nothing more to hope from the future.

The first plutonium nitrate solution from B installation went to a tempo-rary purification facility, "Shop No. 9," on February 27, 1949. By late spring,

Anatoli Alexandrov, who was responsible for plutonium separation, was nickel-plating two hemispheres of plutonium—the first Soviet bomb core —when Pervukhin arrived with a platoon of generals:

> They asked what I was doing. I explained, and then they asked a strange question: "Why do you think it is plutonium?" I said that I knew the whole technical process for obtaining it and was therefore sure that it was pluto-nium and could not be anything else! "But why are you sure that some piece of iron hasn't been substituted for it?" I held up a piece to the alpha counter and it began to crackle at once. "Look," I said, "it's alpha-active." "But perhaps it has just been rubbed with plutonium on the outside and that is why it crackles," said someone. I grew angry, took that piece and held it out to them: "Feel it, it's hot!" One of them said that it did not take long to heat a piece of iron. Then I responded that he could sit and look till morning and check whether the plutonium remained hot. But I would go to bed. This apparently convinced them, and they went away.

From Chelyabinsk-40, Kurchatov had the bomb core carried to Sarov for criticality tests. Andrei Sakharov heard a whispered conversation about it between Boris Vannikov and a senior manager when he visited the research station for the first time in late June. ("Is it here?" "Yes." "Where?" "In the storehouse.") "Zeldovich later told me," writes Sakharov, "that when he saw those ordinary-looking pieces of metal, he couldn't help feeling that a multitude of human lives had been compressed into each gram: he had in mind not only the prisoners who worked in the uranium mines and at the nuclear installations, but also the potential victims of atomic war."

Yuli Khariton recalled one criticality test that was inadvertent but convinc-ing. To test a core for criticality, physicists build a shell around it with material that reflects neutrons, measuring the increasing neutron multiplica-tion from fission as they go. Any light-element material will do for a reflector —cubes of beryllium, blocks of paraffin, even body fat. "Vannikov appeared at one of the final tests," says Khariton. "He came closer and began to read the gauges. He was a large man, very fat. He went back and forth and read the gauges. . . . So during this episode we understood: the bomb would definitely work."

The first Soviet bomb core never traveled to Moscow for Stalin to touch, as Soviet-era myths purport, but the project leaders were called there that spring to report. "The specialists were invited to Stalin's office one by one, and Stalin attentively listened to each," Khariton and Smirnov write. "The first report was delivered by Kurchatov, followed by Khariton and the others. This was Khariton's only meeting with Stalin. Stalin asked Khariton: 'Couldn't two less powerful bombs be made from the plutonium that is available, so that one bomb could remain in reserve?' Khariton, who knew that only the precise amount of plutonium required for the American-designed weapon was available . . . responded negatively." Another source reports Stalin complaining, "We may bully the Americans while having nothing in reserve in the warehouse. What if they press on with their atomic bombs, and we have nothing to contain them?"

Stalin's concern that the first Soviet test of an atomic bomb might challenge the United States while leaving him empty-handed prompted a decision to delay a test until a second bomb core could be processed and made ready: at a hundred grams per day, the Chelyabinsk A reactor could breed enough plutonium in sixty days; allowing for processing, that would result in a second core in time for a test in late August. Kurchatov moved out to the test site, on the windy steppes in Kazakhstan about sixty miles northwest of the town of Semipalatinsk, in May 1949, the same month his scientists organized a drama theater at Sarov, as if they needed any more drama in their lives.

James Forrestal had been a boxer at Princeton; his flattened, twice-broken nose emphasized his wiry Irish aggressiveness. After Princeton he had made a fortune on Wall Street, married a Ziegfeld Follies girl turned *Vogue* editor, moved to Washington as a wartime presidential assistant, won appointment as Undersecretary and Secretary of the Navy and then as the first Secretary of Defense. With each shift upward in his fortunes, he cut himself off further from intimacy and friendship until finally he could be found working in his Pentagon offices even on Christmas Day. He had disputed US support for the new Jewish state of Israel, backed centralizing intelligence and championed a more robust defense. He felt the burden of the world on his shoulders, as his promulgation of George Kennan's long telegram and his determination to rescue the atomic war plan from Truman's disapproval revealed.

In the winter of 1948–1949, Forrestal succumbed to mental illness. The syndicated columnist Drew Pearson had begun attacking him viciously and personally with covert support from the big, loud American Legion commander and Truman fund-raiser, Louis Johnson, who meant to succeed him at the Department of Defense. Truman suspected him of supporting Republican Thomas Dewey in the tough presidential campaign Truman had just won. Ordinarily alert and decisive, Forrestal sank into depression. "Jim calls me ten times a day," Truman complained to a naval aide in January, "to ask me to make decisions that are completely within his competence, and it's getting more burdensome all the time." By the end of the month, Forrestal was becoming delusional, claiming that "Jewish or Zionist agents" were following him and that the FBI had "tapped my wires." He told his friend William O. Douglas, the Supreme Court justice, "Bill, something awful is about to happen to me."

Truman asked for Forrestal's resignation, which came on March 28. Forrestal's friend Ferdinand Eberstadt found him in his darkened and shuttered house that afternoon whispering of Communist, Zionist and White House conspiracy, floridly paranoid. Eberstadt bustled him off to Florida for a rest, but when vacationing Robert Lovett met his plane, joking about golfing, Forrestal told the Undersecretary of State, "Bob, they're after me."

Forrestal's friends called in William Menninger of Kansas's famed Menninger Clinic, who diagnosed severe depression—"of the type," a Navy doctor subsequently explained, "seen in operational fatigue during the war." Forrestal was worn out with stress, including the stress of trying to hold himself together during the months of his cumulative breakdown. The Menninger Clinic had successfully treated hundreds of cases of combat fatigue during the war, but Forrestal's wife, Eberstadt, Menninger and the Navy doctor, Captain George Raines, decided to send the former Secretary of Defense to the US Naval Hospital at Bethesda, Maryland, where his mental illness would be deniable. Under Raines's care at Bethesda, early on the Sunday morning of May 22, 1949, after copying out half of Sophocles' desolate poem "The Chorus from Ajax" as a valediction (" 'Woe, woe!' will be the cry...."), James Forrestal tied one end of his bathrobe sash to the radiator of the diet kitchen across the hall from his sixteenth-floor room, tied the other end around his neck, removed the screen from the window above the radiator and jumped. He hung by the sash long enough to claw the framework below the window. Then the sash gave way at the radiator and he fell to his death on a third-floor roof below.

Forrestal's suicide may have been idiosyncratic, but there was more than enough fatality left over from the war and threat thickening from postwar conflict to make Washington somber in the late 1940s and push its mood toward paranoia. Before the war, the MIT physicist Jerrold Zacharias would observe a few years later, "a lot of people did not regard [Communism] as the threat that it turned out to be. Russia was small, it was experimental, it was backward.... I do not think any people who were backing it then knew that it would capture half the globe...." Communism did not seem small and experimental in Washington in the decade after the war. It seemed an enlarging menace that backwardness only made more brutal. Twelve Western nations signed a document creating the North Atlantic Treaty Organization in early April 1949 to join together in defense against that menace, and two days after the signing, Truman said publicly for the first time that he would use the atomic bomb again if he had to. The Nationalist Chinese retreated to Formosa on May 8, leaving the vast Chinese mainland to a Communist revolutionary army that many believed the Soviets controlled. (Truman, David Lilienthal discovered at a meeting around that time, was more philosophical. "Well," the President told the AEC chairman, "nothing can be done about China until things kind of settle down.... The dragon is going to turn over and after that perhaps some advances can be made out of it.") Dean Acheson remembered that his "own first-hand attempt to work out something in regard to Germany in May of 1949 added me to the list of those whose experience convinced them that so long as it appeared in Russian eyes that there were soft spots, those soft spots would be probed."

Edward Teller reacted to the changing American perception of Soviet threat. After leaving Los Alamos, Teller had settled into a satisfying life at the University of Chicago. His wife Mici had borne him a second child, a daughter, in the summer of 1946 and he devoted more time to his family. He was contributing again to basic science, work deeper and more fulfilling than weapons research. "The years after Los Alamos," his friend and colleague Eugene Wigner believed, "and until the renewal of his preoccupation with national security, were perhaps Teller's most fruitful years scientifically." Teller taught, co-authored thirteen scientific papers, regularly visited Los Alamos to consult and wrote articles for the *Bulletin of the Atomic Scientists.* He praised the Acheson-Lilienthal Report in the *Bulletin* as "a bold and dangerous solution . . . ingenious, daring and basically sound." He described with compassionate horror the terrible devastation of Hiroshima: "One is struck by the picture of fires raging unopposed, wounds remaining unattended, sick men killing themselves with the exertions of helping their fellows." It was even possible to imagine, he wrote, "that the effects of an atomic war will endanger the survival of man." He thought in December 1947, in the wake of Soviet rejection of the Baruch Plan, that "agreement with the Russians still seems possible"; the Danes, he noted waggishly, were once similarly imperialistic and ambitious. "We must now work for world law and world government. . . . Even if Russia should not join immediately, a successful, powerful, and patient world government may secure their cooperation in the long run. . . . We [scientists] have two clear-cut duties: to work on atomic energy and to work for world government which alone can give us freedom and peace."

But hardly anyone was listening. The Cold War was picking up momentum. Oppenheimer had found his way into the high councils of government; he was internationally famous, a household name. Fermi kidded Teller about the implications of his origins: "Edward-a how come-a the Hungarians have not-a invented anything?" A plaintive footnote in a 1948 Teller *Bulletin* review of the AEC's first year's work indicates his isolation from power at that time: "Due to the limited experience of the author the account is necessarily incomplete."

Teller would not easily wrench himself away from his good life in Chicago. As late as July 1948 he could still write in the *Bulletin* that "world government is our only hope for survival. . . . I believe that we should cease to be infatuated with the menace of this fabulous monster, Russia. Our present necessary task of opposing Russia should not cause us to forget that in the long run we cannot win by working against something. We must work for something. We must work for World Government." But a crucial reason for Washington's and Teller's sense of security in the immediate postwar years was America's sole possession of the atomic bomb, and the physicist had begun to suspect that the US monopoly was eroding. In a memorandum to Norris Bradbury in September 1948, Teller conjectured that the Soviets were "likely to find production of either bomb material (Pu-239 and U-235) quite difficult," but feared they might instead "concentrate on radiological warfare" using "radioactive poisons" bred in heavy-water reactors. "If the probability of such a plan is admitted," he concluded, "one . . . may feel less certain about our continued superiority in atomic warfare."

That summer, Bradbury had decided that Los Alamos needed Teller's help. "Norris was rather diffident in his approach to the scientists who had left," Ulam recalls. "He felt that they should recognize by themselves how important for the country and the world it was for them to come back. As a result, although he wanted to, he did not like to ask people like . . . Teller to visit. It was actually left to me, with his consent, to write such invitations. . . . Thus, in a way I was instrumental in bringing Teller back to Los Alamos." Oppenheimer would testify that he endorsed Teller's return to the lab, testimony Teller corroborated: "Oppenheimer had talked to me and encouraged me to go back to Los Alamos and help in the work there." Teller wrote Bradbury at the end of the 1948 summer that he was "giving most serious consideration to this possibility. . . . The main reason that attracts me is the great importance of the work on the atomic bomb. I fully realize the menacing international situation and I believe that the United States must develop its military strength to the utmost if we are not to succumb to the danger of communism. This is the main reason why I consider to interrupt my scientific work in Chicago in spite of the fact that I cannot hope to work as happily and with as much immediate satisfaction in a field of applied science. . . ."

Teller was disturbed by the Soviet coup in Czechoslovakia, by the Berlin blockade and by the impending Communist victory in China. A more personal tribulation was the fate of Hungary. As a child in the years immediately after the First World War, he had lived through the first Communist revolution in his native country, and it had scarred him. "Russia was traditionally the enemy," Teller's Hungarian colleague John von Neumann explained. ". . . I think you will find, generally speaking, among Hungarians an emotional fear and dislike of Russia." In the wake of the Second World War, the Central European nation had briefly experienced democratic government as a republic under the protection of an Allied Control Commission. But the Red Army had remained in occupation and by 1948 the Communist Party had maneuvered itself into power. A one-slate election on May 15, 1949, finished the job. Teller's father, mother, sister and nephew had survived the destruction of Hungarian Jewry and still lived in Budapest. Now they were cut off from him. Intending to spend a year on leave of absence from Chicago, Teller returned to rejoin the Los Alamos staff in July.

Another alarmed and influential participant that winter and spring of 1949 was a twenty-eight-year-old Yale College and Yale Law graduate and former bomber pilot named William Liscum Borden—a small man with a square jaw, blond, with blue eyes. Bright, ardent and utopian, Borden had been an isolationist who had converted to interventionism shortly before Pearl Harbor. He had enlisted in the US Army immediately after graduation in 1942, volunteering to fly bombers, and saw three years' service flying out of England with the Eighth Air Force. He had lost a college roommate and a relative to the war. Of the "men who died," he wrote angrily in the months after victory, "many of them would be alive today had a little more honest realism been displayed before Pearl Harbor." The honest realism Borden had in mind was "to think realistically about the worst that could befall as well as the best." He had seen a V-2 rocket "streaming red sparks and whizzing past us" on its way to London one night in 1944 when he was returning in his B-24 from a mission to Holland. Hiroshima had a further "galvanic effect," he said later, and he had "decided instantly that this was the most important thing in the world." Between his military discharge and his entry into law school, Borden began writing a book that would "think straight about the strategic implications of the new weapons." He called it, urgently, *There Will Be No Time*.

The title of the second chapter of Borden's book summarized its essential argument: "The Certainty of War Amidst Anarchy." The anarchy the young strategist had in mind was the international anarchy of contending nation-states. "War," Borden wrote, "is the inescapable by-product of a system based on separate sovereignties." Others might argue that "war has become so horrible . . . that no people will turn aggressor for fear of retaliation; or, to phrase the contention in more sophisticated language, possession of atomic bombs by both potential belligerents will act as a mutual deterrent." Borden did not agree. To the contrary, he argued:

> We are witness . . . to a momentous race between World War III on the one hand and a voluntary world federation on the other. Unless a federation intervenes in time, war is certain and inevitable. The two great rivals in the post-Hiroshima world are Soviet Russia and the United States. If these two remain dominant—and unless they can unite into a single sovereignty— war between them is as inescapable as the physical law that oil and water do not mix.

And if not Soviet Russia, Borden went on, then China or India later, or even Germany or Japan once they recovered. "The essential point is that an armed peace cannot persist indefinitely, that either war or voluntary federation must resolve the truce."

What follows if you believe that atomic war is inevitable? It followed, Borden thought, that the United States had to become as strong as possible.

He believed that "giving away atomic information is a form of unilateral disarmament" and he excoriated the "group of American liberals [which he believed had] argued that other nations should be given our atom-bomb secret, no strings attached." There was safety only in superior arms. Like LeMay, Borden understood that productive capacity had been the United States's most important resource in wartime; now, Borden wrote, "as wartime weapons, cities and industries are obsolete. Their mission must be accomplished before the fighting begins, or not at all." War in the future would begin with a "rocket Pearl Harbor." But paradoxically, military preparation might deflect the blow: "If the United States is strong, no cities will be damaged. The initial targets will all be fortresses on land, warships at sea, and our island outposts." The enemy, that is, attacking with "many hundreds of intercontinental rockets, each carrying an atomic warhead," would first strike US retaliatory systems.

Beyond this point, Borden's imagination failed him. He thought preventive war proposals "quixotic" for the simple reason that "the American people will never strike first," but he could imagine no more decisive resolution to his scenario than the very deterrence he had dismissed at the outset. American strength would at least postpone Armageddon, Borden prayed, and "who knows what hopes the passage of time may bring to realization?" The Soviet Union might become "sufficiently industrialized to create consumer goods on a vast scale," permitting "a more liberal regime." Or, when it acquired atomic bombs of its own, it "may be so impressed by their destructive force as to reevaluate plans for stringent international control."

Borden published *There Will Be No Time* to modest sales in 1946. After graduating from law school the following year, he returned to Washington, where his family lived, to work for the Justice Department. From there, it seems, pursuing his utopian visions, he and two law school classmates wrote Brien McMahon an alarming letter. They named it "the Inflammatory Document," as if they were christening a new bomber; it proposed that the United States should take advantage of its atomic monopoly while that monopoly still existed and simply give Stalin a nuclear ultimatum: "Let Stalin decide— atomic peace or atomic war." McMahon was Borden's parents' neighbor; the putative author of the Atomic Energy Act took Borden to lunch, rejected the young lawyer's Churchillian brinkmanship but tendered him a job as legislative assistant. Borden joined McMahon's Senate staff in August 1948. Congress went Democratic in the November elections that year and McMahon replaced Bourke Hickenlooper as chairman of the Joint Committee on Atomic Energy. In January 1949, the former bomber pilot who believed that only massive atomic strength could delay war with the Soviet Union became executive director of the Congressional committee that oversaw the development and production of the United States's atomic arsenal.

Darker than these movements to arms was Washington's gathering mood of suspicion. Hickenlooper attacked the Atomic Energy Commission unmercifully in spring 1949, even demanding David Lilienthal's resignation; the attacks led to JCAE hearings beginning in late May into what Hickenlooper called the AEC's "incredible mismanagement." At one of those hearings, in early June, Lewis Strauss repeated his complaints about distributing radioisotopes abroad for research. A week later, at another session with Strauss in attendance, Robert Oppenheimer countered with his patented brand of arch ridicule. "No man can force me to say you cannot use these isotopes for atomic energy," he testified. "You can use a shovel for atomic energy. In fact you do. You can use a bottle of beer for atomic energy. In fact you do. But to get some perspective, the fact is that during the war and after the war these materials have played no significant part and in my knowledge no part at all." Lilienthal would remember Strauss's reaction: "There was a look of hatred there that you don't see very often in a man's face." Hauled before the House Un-American Activities Committee that June, Frank Oppenheimer testified (in FBI paraphrase) "that he joined the Communist Party in 1937 in Pasadena, California, under the name Frank Fulsom. He stated that he dropped his Party membership in 1940 or 1941." Frank's wife Jackie testified similarly. Exposing Robert Oppenheimer's brother may have been a way of chastening the GAC chairman; more likely it meant HUAC judged him too powerful to challenge. Robert himself had appeared before HUAC the previous week and the committee had let him off with easy questions. Young HUAC Congressman Richard Nixon had even praised him for his work, saying they were all "tremendously impressed . . . and . . . mighty happy we have him in . . . our program." The University of Minnesota had hired Frank Oppenheimer as an assistant professor of physics two years previously after he had taken a loyalty oath denying that he had ever been a Communist. An hour after Frank testified, the university asked him to resign.

HUAC had been investigating the Berkeley Radiation Laboratory, the institution Ernest O. Lawrence had founded and still led, and Frank Oppenheimer had seen his troubles coming. Before the hearings got around to him, he had applied for reappointment to the laboratory where he had spent part of the war working hard for Lawrence on electromagnetic isotope separation, Lawrence's extravagant contribution to the Manhattan Project. "Lawrence thought the world of him and had every reason to," Hans Bethe recalled. "Frank had every reason to look to him for help, but as soon as Frank had any difficulty, Lawrence forbade him even to come to the lab and visit." Lawrence, a man of narrow and unquestioning patriotism, was probably stung by the exposure of his laboratory as a wartime hotbed of Soviet espionage. "[Frank] did come to Berkeley," Robert Oppenheimer told Lawrence's biographer, "and [physicist] Ed McMillan and [his wife] Elsie invited him to dinner and Ernest asked them not to have him. That was sometime in the spring. We were out there during the summer [of 1949] and when we ran into Ernest at one of these infinite parties that someone else was giving, I said something about it. I don't think Ernest minded that, but as is often the case, my wife said something sharper and I think maybe he minded that." I. I. Rabi thought Lawrence had never forgiven Oppenheimer in any case for leaving Berkeley for Caltech and then the Institute for Advanced Study after the war. Lawrence's abandonment of Frank shredded away all but a vestige of the old friendship between Lawrence and Robert Oppenheimer. "I think there was probably warmth between us at all times," Oppenheimer said later, "but there was bitterness which became very acute in '49 and which was never resolved. . . ."

A chill wind had picked up from the distant storm front of Soviet progress. In January, the CIA had reported to Hickenlooper "fragmentary information . . . that [the Soviets] are attempting a plutonium bomb." The intelligence agency knew that the Soviet program had not begun "until late 1945" and that "sufficient uranium was available for the operation of [only] one production pile." Based on those facts, it concluded that "mid-1950 is the earliest possible date for the Soviets to complete their first atomic bomb, and . . . mid-1953 is the most probable date for completion." Depending on which of these dates proved out, the CIA estimated that the Soviets would have a stockpile by 1955 of either fifty or twenty atomic bombs. But the wind out of the East made people nervous. The chemist Wendell Latimer, a fervent anti-Communist who had been at Berkeley since 1919 and who had directed the study of plutonium chemistry there in the early days of the Manhattan Project, remembered feeling as early as 1947 "that it was only a question of time [until] the Russians got the A-bomb. . . . It seemed to me obvious that they would get the A-bomb. It also seemed to me obvious that the logical thing for them to do was to shoot immediately for the super weapon. . . . As time passed, I got more and more anxious . . . that we were not prepared to meet . . . a crash program of the Russians. . . . They knew that they were behind us on the A-bomb, and if they could cut across and beat us to the H-bomb or the super weapons, they must do it." Worries like Latimer's became general among the more militant members of the atomic energy community, Lewis Strauss's aide William Golden recalls:

As the year 1949 came along and as people felt that the Russians were getting nearer to a weapon . . . there was more thought given to the efforts on the Super. . . . There was a clear scale of interest with Lewis Strauss being on the high side of thinking [that] greater emphasis should be put on it because of the feeling that it would be dreadful if the Russians got it first. . . . It was not that we needed a stronger weapon, but [rather a concern for] what would happen if it could be made. . . . The feeling . . . was that if it could be made, the United States should have it first. Ernest Lawrence was another who thought so. I'm sure some of the military people did. These feelings, these subsurface rumblings, were [evident] in the spring of 1949. ... Nothing had catalyzed it and there were strong feelings against it. ... But the feeling was strong.

William Borden sensed the subsurface rumblings. In a conference later in the summer with the AEC's director of military applications, General James McCormack, Jr., the young JCAE executive director "asked whether or not sufficient emphasis is being placed upon the so-called super bomb and experiments in that connection." McCormack reassured Borden that work involving "refinement of the existing implosion weapon ... is a necessary condition precedent to achievement of the fantastic temperatures necessary to create a thermonuclear reaction. ... In short, he gave the impression that we were traveling along the road toward a thermonuclear reaction as rapidly as is possible. By the same token, the Soviets will hardly succeed in by-passing extensive work on uranium bombs and moving directly to super bombs." So the AEC reassured its nervous constituents, but those in particular who were not nuclear physicists continued to doubt. Since the United States did not yet know how to make a thermonuclear weapon, they judged, it could not be certain that the Soviets would not find a shortcut.

Robert Oppenheimer, on the other hand, continued to believe that only diplomacy could avert eventual catastrophe. When the Oak Ridge statistician Cuthbert Daniel wrote him in the spring of 1949 proposing a new course, he responded with a profession of faith:

On the general advantage of unilateral and inspiring example, I cannot be in greater agreement. ... That the greatest hope lies in what the United States does and can make intelligible, I deeply agree; and that this is a moral, as well as a practical problem, I agree also.

In the midst of his technical duties helping assure the development of the American atomic arsenal, Oppenheimer had not abandoned Niels Bohr's conviction that sacrifice would be necessary before negotiation could begin.

A US response to these many menacing uncertainties emerged, as it often does in government, from the canonical precincts of military security. The *Sandstone* tests of 1948—of levitated composite core technology in particular—had established that a much larger atomic arsenal might be forthcoming within the near future, no less than a 63 percent increase in the total number of bombs in the stockpile and a 75 percent increase in their total yield. The AEC accordingly reported to the Joint Chiefs of Staff in October 1948 that four hundred atomic bombs would be available by the beginning of 1951. Because the AEC, not the military, paid for them, atomic weapons looked like a bargain to the JCS at a time when the Truman administration had limited and capped the defense budget. Late in 1948, the Joint Chiefs decided to work around their budget limitations by increasing the military requirement for atomic bombs.

Finessing the defense-budget ceiling does not appear among the reasons the JCS gave for increasing the military requirement, but their other reasons were persuasive. They had learned that atomic bombs could probably be used economically against "relatively small targets." They had learned in the NATO negotiations "the full extent of the military prostration of our Western European allies." They had concluded with LeMay that "it would be unsound to rely on accelerating production in time of war," that to the contrary "it was in the national interest to possess on D-Day the fissionable material necessary . . . to carry out our strategic plans." They judged to be threatening "the failure of our proposals in the United Nations on the control of atomic energy, coupled with the fact that the time was approaching when the United States would lose its monopoly position in the atomic weapon field." They settled on 1956 as a target date, since the AEC would have to build new facilities, and probably asked for an arsenal enlarged by one full order of magnitude.* They needed the expansion, they argued, not simply to bomb cities, as of yore, but to allow the conduct of:

> the air offensive against Soviet industrial potential; a counter offensive against anticipated Soviet atomic forces during the period under consideration; limited atomic operations against offensive Soviet Armed Forces and their lines of communication, as a direct contribution to the defense of the signatory nations of the North Atlantic Pact;

and needed as well

> a small general reserve; and an equally small post-hostilities stockpile for the purpose of guaranteeing the peace by serving as a deterrent to aggression.

The US Army, it seemed, was joining what John von Neumann liked to call "this Buck Rogers universe"; for the first time the Joint Chiefs had proposed a requirement for tactical as well as strategic atomic weapons. They did not, however, ask for hydrogen bombs.

Their proposal landed on David Lilienthal's desk by way of the Military Liaison Committee on May 26, 1949. The MLC informed Lilienthal that the AEC would need more production facilities to meet the new national military requirements. How much more only became clear as the AEC commis-

---

* Specifically how many weapons the JCS in 1949 asked to have added to the arsenal was still a military secret four decades later, but by 1956 the US stockpile contained 3,620 atomic bombs.

sioners met around the disruptive schedule of Hickenlooper hearings: at least a new gaseous-diffusion isotope separation plant that would cost some $300 million, half again as much as the entire 1948 AEC budget. Lilienthal resented what he took to be an attempt by the military to preempt civilian authority over the atomic stockpile. He wrote the MLC on June 28 that an expansion of such magnitude would require the approval of the President.

The *modus vivendi* that the United States had negotiated with Britain and Canada the previous year was due to expire at the end of 1949. Since the US still needed British cooperation to assure itself a supply of uranium ore, Truman called a meeting at Blair House in mid-July to massage the senators and congressmen whose support he would need to broaden Anglo-American atomic relations. Into a small room hung with a huge painting of Franklin Roosevelt crowded Truman ("looking like a tired owl," Lilienthal thought), Acheson, Louis Johnson (the new Secretary of Defense, large and bald), Eisenhower (who had just given up thirty-five years of chain smoking and was still twitching), and a phalanx of legislators including Arthur Vandenberg, McMahon, Hickenlooper and young Henry "Scoop" Jackson. Most of the discussion rehashed the earlier arguments that had resulted in the *modus vivendi*. But Truman revealed a significant change in his attitude toward atomic policy that evening. "I am of the opinion we'll never obtain international control [of atomic energy]," he told the leaders. "Since we can't obtain international control we must be strongest in atomic weapons." Truman appointed a Special Committee of the National Security Council on July 26, 1949, consisting of the Secretary of State, the Secretary of Defense and the chairman of the AEC to study and evaluate the new JCS atomic weapon requirements. All three had attended the Blair House meeting, however, and heard Truman's opinion, which thus effectively foreordained the outcome: the United States would respond to the failure of international diplomacy by building more atomic bombs.

No one had briefed the press on the purpose of the Blair House meeting. Reporters filled the vacuum with a fantastic story: that the meeting had been called to consider "the detonation of three atomic bombs in Russia." Apropos this revelation, Henry Jackson told one of Borden's aides in late July that he did not take "any stock in the possibility of the Russians actually exploding the bomb." Jackson had good reason not to do so. On July 1, 1949, the CIA had issued its top secret annual report on the status of the Soviet atomic energy project. The intelligence agency repeated its previous estimates— that the Soviet Union "might be expected to produce an atomic bomb [by] mid-1950 [with a] most probable date [of] mid-1953"—but noted new evidence of Soviet work on gaseous diffusion which suggested "that their first atomic bomb cannot be completed before mid-1951." With a test of that first atomic bomb only weeks away, Lavrenti Beria would have been amused to know how badly the CIA was misinformed.

19

First Lightning

BILLOWING COAL SMOKE, a train left Sarov in the late summer of 1949 carrying soldiers, scientists and the carefully husbanded components of the first Soviet atomic bomb. The first American bomb, of which RDS-1 was a precise copy, had gone under canvas by open truck from Los Alamos to the desert of southern New Mexico in the summer of 1945, but Russia required rails and trains—an old, rugged technology of iron and steam—to quarter its immensity; from Sarov to Kazakhstan the bomb train drove down cleared and guarded tracks two thousand miles. "We traveled quickly," Veniamin Zukerman remembers, "stopping only at major train junctions to change engines and to check the rolling stock. We were struck by how completely deserted the platforms were." Yakov Zeldovich and some of his young assistants jumped out at one stop for a stretch and a game of pickup volleyball; a grumbling MVD colonel chased them back aboard. "But then we were at our destination," Zukerman concludes. "The engine slowly drew the rolling stock into the zone between two barbed-wire barricades. We drove off . . . for an inspection of our country's first atomic proving ground."

The town of Semipalatinsk was sited on high, short-grass steppe land on the Irtysh River in northeastern Kazakhstan. Fifty miles northwest along the river, Mikhail Pervukhin had caused a small settlement to be built, designated Semipalatinsk-21, with laboratories nearby. A road led from the settlement thirty miles south to the test site, in a valley between two small hills. "Along the way there were neither houses nor trees," writes an eyewitness. "Around was the stony, sandy steppe, covered in feather-grass and wormwood. Even birds here were fairly rare. A small flock of black starlings and sometimes a hawk in the sky. Already in the morning the intense heat could be felt. In the middle of the day and later there lay over the roads a haze, and mirages of mysterious mountains and lakes." The American test had been code-named Trinity; the Soviets called their test First Lightning.

As Oppenheimer's men had done on the Jornada del Muerto, the Soviets built a hundred-foot tower on which to explode their bomb, but beside the tower they constructed a hall of reinforced concrete with rails and a traveling crane where the device could be assembled indoors. "A freight elevator could lift the car carrying the bomb to the height of the tower," Zukerman notes. "Personnel could also climb up to the test platform by external staircases." At Trinity site the desert had been ornamented only with instruments, but the Soviets wanted to study the potential destruction. Besides instruments in the barrens among the mirages, writes historian David Holloway, "one-story wooden buildings and four-story brick houses were constructed near the tower, as well as bridges, tunnels, water towers and other structures. Railway locomotives and carriages, tanks, and artillery pieces were distributed about the surrounding area.... Animals were placed in open pens and in covered houses near the tower, so that the effects of initial nuclear radiation could be observed." Trucks brought the bomb components down from Semipalatinsk-21 in the middle of August.

In the next weeks, Kurchatov conducted two rehearsals. Then Lavrenti Beria arrived with Avrami Zavenyagin, leading a state commission to observe the test. Beria brought with him the two men whom he had sent to witness the US Bikini tests in 1946, the only Soviets who had ever seen the explosion of an atomic bomb. They would report whether the spectacle Beria's scientists were preparing for him was authentic.

Assembly began in the hall beside the tower on August 28 with Beria, Kurchatov, Zavenyagin, Arzamas-16 director Pavel Zernov, Georgi Flerov and Yuli Khariton observing. The assembly crew put together the lower half of the lensed high-explosive shell first—a five-foot brown belly of waxen explosive blocks like a broken geode set into an aluminum case mounted on a wheeled trolley. Into that dark manger of high explosives went the aluminum pusher shell, the heavy purple-black uranium tamper, then the first of the two shining nickel-plated plutonium core hemispheres. Khariton personally checked the nickel-plated initiator for neutron activity before he clicked it into place in the cavity at the center of the core. The second plutonium hemisphere closed the core assembly. Over that, lowered with the traveling crane, went the other half of the uranium tamper and the aluminum pusher shell. It was night before the upper layers of solid and lensed HE blocks had been fitted into place. At two A.M. on the morning of August 29, 1949, the workers finally wheeled the bomb on its trolley out the doors to the dimly lit freight elevator below the tower black in the overcast darkness. The bomb was scheduled to ride alone to the top of the tower. Beria expressed surprise and Zernov quickly stepped onto the freight elevator—"holding onto a crosspiece," says Golovin, ". . . moving up in an imposing posture." Four men followed, including Flerov; two of them inserted the sixty-four detonators one by one through openings in the aluminum case into the outer HE blocks and connected them with cables to the big capacitor bank that would fire them. Flerov and a colleague checked the counters that would monitor the neutron background within the bomb. Flerov was the last to descend from the tower. By then, Beria had gone off to a cabin near the command bunker, five miles from ground zero, for a few hours sleep.

"It drizzled all night," Zukerman reports. As at Trinity, weather forced an anxious postponement; the shot had been scheduled for six A.M. but was moved forward to seven. They had luck with the weather. "Towards morning," Zukerman continues, "there was a slight clearing. Although the sky was still overcast, there would be no visibility problems." Kurchatov, Khariton, Pervukhin, Flerov, Zavenyagin and other scientists and managers gathered at the command bunker. Beria and his entourage arrived shortly after the automatic countdown began at T minus thirty minutes and the bunker filled up with generals. Before Beria came out to Kazakhstan, Kurchatov had ordered the earthen berm between the command bunker and the shot tower banked higher, which had blocked the view of the tower; now he opened the glass-paneled bunker door on the opposite side so that they would be able to see the light from the bomb reflected from the distant hills. The shock wave, slower than the light, would hit the bunker after a thirty-second delay, time to close the door again. As he always did, Kurchatov droned as he paced: "Well, well, well. Well, well, well." At T minus ten minutes, Beria spat a last curse into the mutter. "Nothing will come of it, Igor," he sneered. Kurchatov reddened at the slur. "I don't think so," he countered. "We'll certainly get it."

An eyewitness at an observation post seven miles to the north had no berm blocking his line of sight. "In front of us," he reports, "through the gaps in the low-lying clouds could be seen the toy tower and assembly shop, lit up by the sun. In spite of the multilayered cloud and wind, there was no dust. . . . Waves of fluttering feather-grass rolled away from us across the field. 'Minus five' minutes, 'minus three' minutes, 'one,' 'thirty seconds,' 'ten,' 'two,' 'zero.'" In the command bunker, Kurchatov turned abruptly to the open door. Light flooded the steppes. "It worked," Kurchatov said simply. "What remarkable words," writes Zukerman—" 'It worked. It worked!' " If it had not worked, one of them told a German chronicler later, they would all have been shot.

Flerov closed the door just before the shock wave hit the bunker with a roar and shattered the glass. "Right then Beria rushed to hug me," Khariton recalls. "I could barely manage to tear myself away from him. The only thing I felt during those moments was relief." The eyewitness at the north observation post saw the explosion:

The white fireball engulfed the tower and the shop and, expanding rapidly, changing color, it rushed upwards. The blast wave at the base, sweeping in its path structures, stone houses, machines, rolled like a billow from the center, mixing up stones, logs of wood, pieces of metal and dust into one chaotic mass. The fireball, rising and revolving, turned orange, red. Then dark streaks appeared. Streams of dust, fragments of brick and board were drawn in after it, as into a funnel. Overtaking the firestorm, the shock wave, hitting the upper layers of the atmosphere, passed through several levels of inversion, and there, as in a cloud chamber, the condensation of water vapor began. . . . A strong wind weakened the sound, and it reached us like the roar of an avalanche.

"The shock wave shook the command bunker," Pervukin recalls. ". . . We rushed out and saw the fire cloud and the column of dust and soil rising from the ground following the cloud and forming an enormous mushroom. The explosion was successful and we congratulated each other, kissing and hugging." Beria hugged Kurchatov and shouted with the rest of them, but suspicion surged back to drown his exultation. He rushed into the bunker and ordered a call put through to one of his Bikini observers at the north post. "Did it look like the American one?" Golovin has the gulag master shouting into the phone. "How much? Haven't we slipped up? Doesn't Kurchatov humbug us? Quite the same? Good! Good! So may I report [to] Stalin that the experiment was a success? Good! Good!" Eagerly, Beria ordered the general on duty to connect him through to Stalin. It was two hours earlier in Moscow; Stalin's secretary, A. N. Poskrebyshev, warned Beria that the dictator was still asleep. "It's urgent, wake him up," Beria demanded, flying high. "People present at the conversation heard the angry voice of Stalin" then, Golovin and Smirnov report:

"What do you want? Why are you calling?"
"Everything went right," Beria said.
"I know already," replied Stalin, and he hung up the phone.
Beria went wild and attacked the general on duty. "Who has told him? You are letting me down! Even here you spy on me! I'll grind you to dust!"

Within ten minutes, observers rolled into ground zero in lead-lined tanks to take soil samples. Kurchatov and others followed in four-wheel-drive Gazik vans, one of them recalls:

There were no traces left of the central tower. The surrounding columns and towers were damaged and listing to the side, the walls of the nearby buildings had collapsed, and the roofs had either been torn off or had caved in. There remained a distorted reminder, as in a nightmare, of the technical orderliness of the structures—everything had been mowed down, uprooted, and set afire.
Sparkling bluish black in front of us was the surface not of a crater but

of a plate left by the explosion. One was almost not aware of the depression because it was very broad and gently sloping. And the surface was covered with slag, melted smooth and sparkling, that was formed from the soil scorched by the fire of the explosion. At the very center the melted surface was unbroken and as one moved away from the center one could see uneven and broken areas, and finally there were individual fritter-shaped pieces either formed on the spot or spilled out of the center. . . . At a good speed the Gazik cut into the crust of slag crunching under the wheels . . . ; two of us jumped out on the right and left, cut off hefty pieces, stuck them into a sack and the Gazik took us back.

At Trinity the desert sand had melted to a green glaze that came to be called trinitite; at First Lightning the slag was blue-black. The mushroom cloud blew away to the south, the north post observer reports, "losing its outlines and turning into the formless torn heap of clouds one might see after a gigantic fire."

The First Lightning bomb had yielded twenty kilotons, comparable to its American counterparts at Trinity and Nagasaki. Kurchatov sent a handwritten report to Moscow the same day. The Soviet Council of Ministers soon secretly decreed medals and honors, cash prizes, automobiles and dachas for Kurchatov and his scientists, free education for their children at state expense, free transportation anywhere in the Soviet Union for them and for their families. The physicists' bomb had not been a swindle after all, and now Stalin had a bomb on his hip and the West could no longer blackmail him. But he authorized no public announcement while his warehouse was still nearly bare.

Lewis Strauss had continued to push for a US capability to detect long-range atomic explosions after his initial proposal in April 1947. As a result, the Central Intelligence Group (CIG), the predecessor to the CIA, had formed a committee to study the problem. The long-range detection committee established at its first meeting, in May, that no such capability existed and concluded that detection was possible three ways—by listening for the sound of the explosion, by monitoring the earthquake-like seismic wave that an explosion would cause and by collecting and measuring the gaseous or particulate radioactivity. CIG director Admiral Roscoe Hillenkoetter issued a report at the end of June 1947 recommending setting up a monitoring system but estimating that doing so would take two years. Two years was too long, David Lilienthal rejoined for the Atomic Energy Commission: "We cannot regard a two-year period as acceptable or realistic."

Strauss had turned then to James Forrestal for support (when he told the Secretary of Defense that the US had no monitoring, Forrestal had re-

sponded incredulously, "Hell! we must be doing it!"). By September, Forrestal had passed the matter on to Hoyt Vandenberg and the US Air Force,
which had the necessary planes and airfields, in particular three squadrons
of weather-modified B-29s, the most significant of them the Alaska-based
375th Weather Reconnaissance Squadron that regularly patrolled from
Alaska to Japan and over the Arctic, windward of the USSR. What the Air
Force lacked was the scientific expertise. For that its new office of atomic
testing—AFOAT-1—turned to the AEC, which surprised it by warning that
long-range detection was not an established technology but an undeveloped
field. In December 1947, an advisory committee of which Robert Oppenheimer was chairman warned of "grave doubts that the techniques and the
instruments for detection . . . are available either potentially or actually." The
Air Force, one of its generals testified later, "was frantic." It had accepted a
grave responsibility only to discover that there was no good way to carry it
out.

By accident, an engineer for a private radiological laboratory, Tracerlab,
had learned of the Air Force's difficulty. The small, two-year-old lab had
gone after the long-range detection business eagerly, a Tracerlab scientist
recalled. At a first meeting with the military, Tracerlab's sales manager "was
busy counting up all the scalers, cutie pies, survey meters and so forth that
they would need" when the scientist had asked, "Who was going to man all
this stuff? They said they had nobody to run it. So I said I'd collect the staff.
. . . It was our equipment . . . and who is more competent to keep it running
than our people? This was the sales pitch." The lab won a contract.

It soon discovered that Los Alamos, where the AEC's expertise was concentrated, had been dragging its feet because Oppenheimer and others, as
the Tracerlab scientist explains, "had come to the conclusion that there was
no way you could detect radioactive isotopes at long distances from an
explosion. It was a waste of time to even try since they were going to be
atomized." Oppenheimer had decided before the Hiroshima bombing that
an air-burst weapon which did not pick up debris from the ground would
dissociate its components to atoms—essentially to gas—which would
quickly mix with the air to undetectable dilution. That was why the United
States had repudiated Japanese claims of dangerous fallout at Hiroshima and
Nagasaki: because both bombs had been air bursts.

In March 1948, Oppenheimer told the AFOAT-1 scientists face-to-face,
with Edward Teller on hand to back him up, that seismic detection might
work but that sonic and particularly radiological detection were nonsense.
One of the Tracerlab men returned to his hotel room after the meeting and
spent the evening working out calculations to show that the atoms of irradiated bomb material should stick together and agglomerate to particles of
detectable size. He shared his calculations with Teller and others the next
day (Oppenheimer was unavailable) and won agreement that agglomeration

to detectable size was at least possible. Experiment would test the calculations; preparations for *Sandstone* were then underway.

AFOAT-1 had the sense to mount the experiment even though the *Sandstone* tests were all tower shots and therefore dirty—their fireballs would reach the ground and scoop up debris. Sniffer planes equipped with fuselage ducts that scooped air through paper filters flew below the jet stream thousands of miles from Eniwetok and collected fallout. Even the scientists at Tracerlab were surprised by what they found—"not only particulate matter [i.e., dirt particles] but [also] matter in terms of shiny metallic-looking spheres that were just beautiful." The microscopic metallic spheres were radioactive agglomerations of gasified atoms from the *Sandstone* bombs. By July 1948, the Air Force was able to tell the AEC that it "was confident of being able to detect by radiological means an atomic air burst." By January 1949, when the Air Force briefed Strauss on its program, it was running regular sniffer flights on its own, and Strauss could report to Forrestal that "the door is no longer left unguarded."

Oppenheimer was still unhappy, however. He speculated that the Soviets might conduct their tests underground to avoid releasing telltale radioactivity and wanted AFOAT-1 to develop a seismic detection system before it officially deployed its network. Strauss added Oppenheimer's resistance to relying on radiological detection alone to his list of suspicions, concluding that the mercurial physicist was trying to thwart long-range detection. It was not the first time Oppenheimer, a man who jumped to finish other people's sentences, had outsmarted himself. When Luis Alvarez had carried the news of nuclear fission to Oppenheimer at Berkeley in January 1939, the theoretician's first response had been to tell Alvarez, "That's impossible" and to give Alvarez "a lot of theoretical reasons why fission couldn't really happen. When I invited him over to look at the oscilloscope later, when we saw the big pulses [of ionization from fission], I would say that in less than fifteen minutes Robert had decided that this was indeed a real effect and . . . that you could make bombs and generate power. . . . It was amazing to see how rapidly his mind worked, and he came to the right conclusions." But first, as later in the matter of agglomeration and long-range detection, Oppenheimer got it wrong.

To enlarge its reach, the Air Force had decided after *Sandstone* to let the British in on the secret detection program, which required working around the Atomic Energy Act. The US Navy also began developing a ground-based system that included collecting rainwater off the roof of the Naval Research Laboratory in Washington, DC. By the time the Air Force began flying sniffer planes out of Alaska for the AFOAT-1 program in April 1949, the British were in business over the North Atlantic and the Navy was filling rain barrels and scanning for gamma radiation in Alaska, the Philippines and Hawaii as well as in Washington. Between April and August, the WB-29s of the 375th

Weather Reconnaissance Squadron trapped radioactivity on their filters greater than the natural background a total of 111 times. The suspect filters went to Tracerlab in Berkeley, which dissolved them, chemically separated a selection of fission products such as radioactive isotopes of barium, cerium, molybdenum, zirconium and lead, carefully measured the rates of radioactive decay of the isotopes and counted back to establish when each isotope had been created—its radioactive birthday. Only if all the birthdays were identical could the isotopes have been created in an atomic bomb. The 111 samples all turned out to be natural in origin, released from the earth (where the spontaneous fission of uranium creates fission products with randomly different birthdays) by earthquakes and volcanoes. But they gave Tracerlab the opportunity to perfect its techniques.

A WB-29 flying east of the Kamchatka Peninsula picked up radioactivity on September 3, 1949. The Kamchatka filter paper showed activity 300 percent greater than the level Tracerlab had established as an alert; measurements quickly confirmed that the radioactivity was fission-derived. Off went the paper to Berkeley. "More evidence came in over that weekend," writes GAC secretary and Los Alamos associate director John Manley. "During the next week the radioactive air mass was followed across the [US]." A physicist at Tracerlab remembers measuring radioactivities "night and day for a period of two weeks. I didn't sleep more than four hours a day. Our little group was working around the clock." The AEC alerted the British on September 9 that the air mass was approaching England. Routine sniffer flights out of Gibraltar and Northern Ireland before that notice had found no radioactivity, so the British sent a Halifax bomber north from Scotland on the evening of September 10 and followed up the next morning with several Mosquitos that sniffed along the Norwegian coast. These explorations quickly confirmed radioactivity. In the meantime, the Navy had flocculated rainwater from the roof of its Washington laboratory and further confirmed the Air Force findings. "The 'cloud' which drifted over the Pacific and the US," a JCAE staff member would note of the Navy's operation, "split up over midwestern Canada, the southern part of the cloud drifted down over Washington and hung there for two or three days, during which the rain brought down the material. The northern part of the 'cloud' traveled on out over the Atlantic where it was detected in Scotland by the English."

"By September 14," Manley continues, "95 percent of the experts analyzing the data were convinced [that the samples represented bomb debris] and had dubbed the event 'Joe One' after Stalin."* Los Alamos estimated an explosion no more than thirty days prior to September 13; the Navy and the British came within a week on each side of the actual date. Tracerlab, with a cleaner lab and more practice teasing information out of microscopic sam-

---

* Arnold Kramish coined RDS-1's American name.

ples, estimated the Soviet explosion to have taken place at 0000 Greenwich Mean Time on August 29—six A.M. in Semipalatinsk, only an hour off the actual event. The commercial laboratory also established that the bomb used a plutonium core and a natural uranium tamper.

Secretary of Defense Louis Johnson refused to believe the expert ninety-five percent, however. "I recall going with [William] Webster [chairman of the Military Liaison Committee] to see Secretary Johnson," Kenneth Nichols said many years later. "Bill was telling him that we had intelligence that the Russians had exploded an atomic bomb. The Secretary sort of pooh-poohed the idea. He said, 'I don't believe too much in intelligence.' I remember I spoke out of turn, I said, Well, Mr. Secretary, you better get ready to believe this one. It's not of the creep type [of intelligence], it's firm." Johnson's doubt settled on the possibility that a Soviet reactor might have exploded and he refused to authorize a public announcement. Truman knew of the technical intelligence by now but was not prepared to overrule Johnson. The AEC responded by assembling an expert committee under Vannevar Bush, the wartime US science czar, that included Robert Oppenheimer; William Penney, the English physicist who was directing the British bomb program; physicist and former AEC commissioner Robert Bacher; and Hoyt Vandenberg. The committee concluded on September 19 that the Soviets had exploded an atomic bomb, which confirmed the AEC in its conviction that the event ought to be made public as soon as possible. The AEC commissioners decided the same day that Lilienthal should carry the AEC case past Louis Johnson to Truman and win the President's agreement.

Exhausted by his summer of battles with Hickenlooper, Lilienthal had retreated to Martha's Vineyard to rest (and had made up his mind to resign). The AEC sent General James McCormack, Jr., its Director of Military Application, off to New England in a C-47 to fetch Lilienthal. Returning from dinner in Edgartown in a Vineyard fog at eleven that Monday night, Lilienthal encountered McCormack hatless in the middle of the road, pretending to thumb a ride, "as if I frequently found him on a windswept moor, in the dead of night, on an island, outside a goat field. . . . No questions; said he had lighted a candle in our house. Had he parachuted; what was this?" Inside, "General Jim" gave Lilienthal the news, by the light of a kerosene lamp.

The AEC chairman flew to Washington the next morning, September 20, to find Bacher "deeply worried," Oppenheimer "frantic, drawn," but also characteristically "positive [of the evidence for a Soviet test], unequivocal." Lilienthal slipped through the White House back entrance a little before 3:45 P.M. and encountered an Oval Office scene out of a genre painting: "The President was reading a copy of the *Congressional Record,* as quiet and composed a scene as imaginable; bright sunlight in the garden outside, the most unbusy of airs. Started talking about it. . . ." Truman gave Lilienthal a long list of political reasons why he was not yet ready to announce a Soviet bomb, but his basic argument seemed to be that he doubted the intelligence. "German scientists in Russia did it," Lilienthal paraphrases him at one point, "probably something like that." At another: "Can't be sure, anyway. I stepped into that: [it] is sure, substantial. . . . Really?—sharp look, question. . . ." Gordon Arneson, the State Department AEC liaison, remembered getting a call from the National Security Council's Sidney Souers, who was close to Truman, when the news had first reached the White House:

> Admiral Souers . . . called me urgently on the telephone and said: "Come over right away. I have some 'hot' intelligence." It was a beautifully bright fall day in Washington, the kind that makes you glad you were alive whatever the day's news. But as I walked over to Old State, I had a premonition —"This was it." . . . Admiral Souers expressed the idea that the radioactive cloud might have come from an accident, a blow-up of a nuclear reactor. I told him I hoped he might be right but it seemed unlikely.

Souers may have supplied Louis Johnson's reactor story or the other way around, but either way, Truman was skeptical of the Soviet achievement; he told a senator later that he could not believe "those asiatics" could build so complicated a weapon as an atomic bomb.

Lilienthal argued for candor. The President took the decision under advisement. There was nothing more the AEC chairman could do. Back at his office, Lilienthal found Oppenheimer "feeling badly." "We mustn't muff this," he quotes the physicist in his diary; "[it was a] chance to end the miasma of secrecy—holding a secret when there is no secret." "Weren't you surprised?" Lilienthal asked Oppenheimer. "Yes, yes." "A good deal?" "Yes, a good deal," Oppenheimer explained. "Always hoped, half-thought our troubles would have"—but Lilienthal does not finish Oppenheimer's halfthought. Did the former Los Alamos director imagine that Soviet scientists could not duplicate so soon, or at all, what he and his colleagues had accomplished? Lilienthal for his part droned back to Martha's Vineyard that evening in a B-25 to tell his wife about his extraordinary twenty-four hours, ending his diary entry in the style of Samuel Pepys, another government official who kept a personal record of the dissonances that reverberate between public and private life:

> 10:30 P.M. Long visit with Helen before the fireplace, from the light of limbs from the old dead apple trees, the wind blowing like mad, the Wuthering Heights touch again as it goes through the loose house; to bed.

Because Los Alamos was involved with fallout analysis, the word went round the laboratory before any public announcement. It caught up with

Stanislaw Ulam returning from a trip to Washington. Ulam and others—Nicholas Metropolis, Teller, John von Neumann when he was in town—played small-stakes poker once a week, Ulam explains, "a bath of refreshing foolishness from the very serious and important business . . . of Los Alamos." Landing at the airstrip that ran off the eastern end of the mesa, Ulam remembered being "met by several people, Nick Metropolis and others, who told me two things. One, the Russians have exploded an atomic bomb. Two, [a poker buddy] had won 150 dollars in poker. So I believed only the poker story." No one had expected a Soviet bomb so soon, Ulam points out. "Teller and others thought it would come sooner or later, but not that quickly. . . . It was quite a successful shot and it really shocked people."

Teller had enjoyed a leisurely late-summer visit to England. "He saw a lot of [Klaus] Fuchs at that time," the FBI paraphrases him. ". . . Fuchs had met him at the American embassy in London shortly after Teller's arrival in England and . . . later they had had official contact at Harwell. One evening during Teller's stay at Harwell, he had spent several hours with Fuchs at Fuchs' flat." In September, the Hungarian-born physicist had sat long at dinner at Caius College, Cambridge, with James Chadwick, the notoriously taciturn discoverer of the neutron and the wartime director of the British Mission in the United States. Responding to a question from Mrs. Chadwick, Teller made the mistake of disparaging General Groves, whereupon Chadwick spoke for a solid hour on Groves's conscientiousness and reliability, concluding, "I hope you will remember what I have said tonight." Teller sailed home, learned at the Pentagon what Chadwick already knew and took Chadwick's oration for a warning that the scientists who had directed the Manhattan Project lacked the drive and conviction to frame a proper response to the challenge of the Soviet bomb.

Before Truman would tell the world, Herbert York reports, he made Lilienthal and the members of the detection committee personally sign a statement "to the effect [that] they really believed the Russians had done it." After briefing Congressional leaders and the Joint Chiefs, the President announced on the morning of September 23, 1949, "that within recent weeks an atomic explosion occurred in the USSR." Though its arsenal by then numbered at least a hundred atomic bombs, the United States no longer enjoyed an atomic monopoly.

———

Before Truman's announcement, US intelligence and security agencies had already begun to explore the possibility that espionage had contributed to the timely Soviet success. At an FBI meeting in early September, Robert Lamphere and other agents reviewed the three major wartime Soviet probes on which they had files: the effort through Steve Nelson to recruit scientists around Robert Oppenheimer at the University of California Radiation Laboratory, which the FBI believed it had countered; a recruitment of minor value at Chicago; and the hemorrhage out of Alan Nunn May in Canada. "Then, in mid-September," Lamphere writes—"still before the President's announcement—I found a startling bit of information in a newly deciphered 1944 KGB message." The information Lamphere found in the New York cable was a partial summary of a theoretical paper on gaseous diffusion. The decode immediately established that a spy had been at work within the United States program as well as the Canadian. More analysis of the intercept revealed that the spy had operated out of the British Mission in New York. By the time of Truman's announcement, the AEC had identified Fuchs as the author of the paper. The identification did not confirm that Fuchs was a spy, and for a few days Lamphere carried Rudolf Peierls on his suspect list as well. Then he reviewed another intercept that mentioned a British agent whose sister was attending an American university. Fuchs's sister Kristel had attended Swarthmore. Lamphere discovered other clues in Fuchs's file that the Bureau had previously overlooked: a captured Gestapo document identifying Fuchs as a German Communist and the names and addresses of the physicist and his sister in the address book of Israel Halperin, one of the GRU agents Igor Gouzenko had exposed. "And so I became convinced that Klaus Fuchs was the prime suspect," Lamphere concludes. On September 22, the FBI agent opened a case file, alerted New York to begin an investigation and wrote British intelligence, discussing not only Fuchs but also, because of a clue in one of the intercepted Fuchs cables, Abe Brothman.

Two days later, a Soviet agent named Filipp Sarytchev traveled to Philadelphia and knocked on Harry Gold's door. It was Saturday night and Gold was asleep on the living room couch, his father sleeping upstairs, his brother Yosef out for the evening. In July, Gold had received a code signal by mail that had called him to a rendezvous at a Manhattan bar, but he and his unknown contact had missed connections. Now a stranger greeted him in an accent so thick Harry started to shut the door. "Remember John and the Doctor in New York," the stranger rushed to say. Gold let him in. Sarytchev sat down and immediately asked Harry if he had any material. Gold, surprised, told him it had been many years. The Russian—Gold was sure the man was Russian—berated him for missing the meeting in July, then briefly went over Harry's 1947 grand jury testimony. They arranged to meet again on October 6.

Harry had recent, anguishing evidence that his years of espionage were a curse. He had taken two weeks' vacation from Philadelphia General at the beginning of August determined to convince Mary Lanning to marry him. He probably knew at that time that he was being promoted to chief research chemist at the Heart Station; the promotion became official on August 16. He and Mary had taken a walk along Wissahickon Creek in Fairmount Park and he had proposed. Mary had not given him an answer, but she had told him that for the first time he seemed "completely natural; at this time she came very close indeed to accepting me." Harry pressed his case and they decided to go off together:

> But on our next meeting several days later, during a trip to the Poconos, I "froze" completely—yes, I froze as badly as a tyro on a high scaffold. And Mary complained she did not believe that I really loved her and cited my "lack of ardor" as proof. But it was not lack of ardor, it was fear of exposure —and not fear for myself, but a horror at the thought that the revelation might come after we had been happily married for, say, three or four years, with children and a home of our own.

The snub-nosed girl of his dreams turned down Harry's second proposal then and they broke off their relationship; he spent the rest of his vacation at home.

Gold took the train to New York on the night of October 6 and met Sarytchev at nine o'clock outside a movie theater in Queens. It was raining hard. The Soviet agent was dressed for the weather, but Gold had brought no raincoat. Indifferently, Sarytchev ignored Gold's drenching; they walked in the rain and talked for three hours. The Soviet agent grilled Harry minutely about the 1947 grand jury, concluding by asking him what he believed the jury knew about him. "Gold advised," the FBI paraphrases, ". . . that he believed the grand jury thought that he was, at the most, a well-meaning dupe or possibly implicated to some small degree. At this remark . . . [Sarytchev] shook his head and smiled, indicating to Gold that he was wrong in his opinion of the grand jury's naiveness." Then Sarytchev advised Gold to begin planning on the possibility of having to leave the country. "The Russian . . . briefly pointed out that it could easily be handled by Gold leaving the United States, going to Mexico first and then eventually to one of the countries in Europe, which Gold construed to mean one of the Iron Curtain countries and not Soviet Russia." There would be plenty of money, Sarytchev promised him, "should an emergency arise." Gold "was horrified and practically speechless." Sarytchev, whom Gold concluded must have been a trained interrogator, left him soaked but not cold; he discovered when the man walked away that he was sweating.

The two men met once more at the Bronx Park IRT station on the night of October 23, 1949. Sarytchev told Gold to "lay low," which Gold took to be an order, not a request. They arranged a bimonthly schedule of regular meetings for the months ahead and an emergency system of contact. "Such meetings were desired on the part of the Soviets," the FBI reports Gold explaining, ". . . in order that they might be able to observe that he was still at large and had not been arrested."

In an effort to limit British intelligence access to the cable intercepts,

Lamphere had passed them a sensational decode late in 1948. "Several fragments of deciphered KGB messages," Lamphere writes, "indicated that someone in the British embassy in Washington in 1944–45 had been providing the KGB with high-level cable traffic between the United States and Great Britain." The agent's code name was "Homer." (Homer, not yet identified, was Donald Maclean.) Immediately thereafter, in February, the Soviets had broken off contact with Fuchs, a disconnection he later claimed to have initiated after a crisis of conscience. Evidently there was a Soviet mole within British intelligence, someone other than Kim Philby, since Philby did not become privy to British cryptanalysis information until after his appointment in August as liaison at the British Embassy in Washington between MI5, the FBI and the CIA. The Soviet July contact with Harry Gold probably followed from the earlier British breakthrough, but the late-September and October contacts evidently connected to Lamphere's unmasking of Fuchs and concomitant implication of Abe Brothman. Fuchs was probably also warned. In mid-October, he dropped by the office of Harwell security officer Henry Arnold to report that his father was moving into the Soviet zone of Germany that month to take up a professorship in theology at the University of Leipzig; he wondered if he ought to resign. Instead of giving Fuchs an answer, the security officer asked him what he would do if Soviet agents tried to contact him. Fuchs told Arnold he did not know. He spent the next several months thinking about leaving Harwell, but ego held him back—he imagined, as he later confessed, that his departure would "deal a grave blow to Harwell," as if his espionage had not already done enough.

Evidently British intelligence had not yet reported the FBI discovery to the Ministry of Supply, which administered the British atomic-bomb program; at the end of October the Bureau heard from what its records call "the British Authorities" that "they felt bound to advise the appropriate authorities in England that the continued employment of Fuchs in the Atomic Research Station at Harwell, England, represented a grave risk to security and that Fuchs should be consequently removed." They were concerned not to "jeopardize the Bureau's original informant"—that is, the fact of the cryptanalysis breakthrough. The FBI advised them "that they should feel free to take any action with respect to an interview with Fuchs that they might desire" but that "there was the necessity of protecting at all costs, the original informant." Hoping to allay his suspicions while they pondered what to do, the British gave Fuchs a promotion and a salary increase that autumn and one of the few and much-coveted detached houses at Harwell.

News of Truman's Joe 1 announcement on September 23 reached the General Advisory Committee that day in the midst of its first meeting since the AFOAT-1 detection. "I happened to be sitting next to Oppie on the day this information came in," metallurgist and GAC member Cyril Stanley Smith reminisced long afterward, "on his left in the room where we were meeting. He was called out and he asked me, because I was next to him, just to act as chairman for a few minutes while he was away. A few moments later, the [release] came in and I had the job of reading it to the committee. I very much remember my embarrassment, my general tension, almost my inability to read it clearly because of the emotional impact." Smith, who had prepared the plutonium core for the Trinity shot, knew what terrible news it was. So did I. I. Rabi, another GAC member in attendance that day, who worried immediately, the GAC minutes report, "that the Russian achievement brought the prospect of war much closer and therefore prompted the question as to what courses of action should be taken." Oppenheimer had returned to the room by then; he "mentioned some of the matters which had been discussed in this connection, but expressed the opinion that it is too early to attempt to reach conclusions until there is a better developed public reaction to the announcement."

That was also what Oppenheimer had told Teller when Teller had called him after hearing the news at the Pentagon upon his return from England: "Keep your shirt on." "My mind did not immediately turn in the direction of working on the thermonuclear bomb," Teller would testify. "I had by that time quite thoroughly accepted the idea that with the reduced personnel [at Los Alamos] it was much too difficult an undertaking."

In Italy on a long summer leave, William Golden had a different response:

> I was in Florence.... I was able to read enough Italian in the headlines of the paper to see that President Truman announced that the Russians had exploded a nuclear device, and I stayed up all night, I was so struck with that, I stayed up ... writing a letter to Lewis Strauss....
>
> I set forth some observations.... I thought they were probably superfluous.... But one thing I wrote him was, "We, the USA, should intensify our efforts toward development of superweapons. This is much more important than an increase in production rate of existing weapons. A quantum jump in intensity is called for as a matter of urgency comparable in every way to the wartime Manhattan Project.... This I regard as urgent and of supreme importance. I can conceive of, though I hope unwarrantedly, one or more of your fellow Commissioners wishing to go slow, awaiting some international-control arrangement with Russia. You must strive for concurrent development work with wartime urgency." ... I stayed up till three A.M. and I delivered [the letter] to the Consul-General in Florence in the morning and said, "Will you please send this in the [diplomatic] pouch?"

Strauss liked the phrase "quantum jump" and the idea of a new Manhattan Project. He began to compose a memorandum.

The Joint Committee on Atomic Energy started a series of urgent meetings on September 23 that extended through the following week. William Borden and his staff cobbled together a list of twenty-three "possible methods to increase or augment the production of atomic weapons" that included increasing the staff at Los Alamos, bringing DuPont back into the program and launching what the memorandum termed an "all-out" effort to build a hydrogen bomb. The JCAE took testimony on the possibility of pursuing a thermonuclear on September 29. (By then, Tass had belatedly announced that the Soviet Union had "the atomic weapon at its disposal" but asserted that bombs had been on hand since 1947.) Carroll Wilson, the AEC's general manager, an MIT-trained science administrator, pointed out that Los Alamos was working toward a 1951 test of an implosion bomb boosted with deuterium and tritium, "a step toward a possible thermonuclear bomb" that would take "all of the energy and efforts that can be expended on it between now and the test. . . ." The test series had just been named *Greenhouse* and a major building program had begun at Eniwetok, including construction of a large plant for liquefying deuterium for thermonuclear tests.

General McCormack confirmed the AEC general manager's timetable:

I think that is true. We reviewed about a year ago . . . the program leading toward a . . . thermonuclear weapon and as best as anyone could see then, the first and necessary step would be . . . the booster . . . and that would take two or three years to do. The thermonuclear weapon itself, according to our best scientific advice, is a really major endeavor and can certainly spread over a period of a number of years. . . . We have to achieve temperatures greater than we have achieved in any atomic explosion thus far even to trigger the thing if it can be triggered. This is a fact that we are trying to determine in the next test—can it be triggered? Because if it can, then there is a huge development program ahead of us.

Brien McMahon asked McCormack how destructive a true thermonuclear weapon would be. McCormack told him, "If all of the theory turned out, you can have it any size up to the sun or thereabouts that you wanted. I think one talks in terms of the super weapon as being one million tons or more of the TNT equivalent." It would be "a huge thing," the general observed. "Delivery by railroad train or perhaps boat seems to be in order." One Super design Los Alamos had reviewed early in 1949, notes historian Chuck Hansen, included "a fission trigger that in itself weighed 30,000 pounds; the overall length of the bomb was estimated at approximately 30 feet, with a diameter in excess of 162 feet. Under these circumstances, the scientists at Los Alamos preferred not even to estimate the gross weight. . . ."

AEC commissioner Sumner Pike, a tough-minded self-made millionaire from Maine ("Sardines?" he would be asked of his business interests at a hearing a few years later; "Sardines," he would answer), cautioned the JCAE that tritium production would be a serious problem. "It requires a great deal more reactivity than we have in any pile we have built or under contemplation unless we look forward to a considerable reduction in plutonium production." To made tritium in a reactor, the reactor has to be souped up to generate more neutrons. In a graphite reactor that means replacing a portion of the natural-uranium slugs with slugs of U235. Since U235 fissions with neutron capture rather than transmuting to neptunium and then plutonium, using more U235 reduces plutonium production. Breeding tritium in existing AEC reactors, Pike would explain at another time, meant that "the cost of producing tritium in terms of plutonium that might otherwise be produced looked fantastically high—80 to 100 times, probably, gram for gram." Unless it built a new generation of reactors, that is, the US would have to forgo eighty to one hundred kilograms of plutonium—enough for thirty to forty composite-core atomic bombs—for every kilogram of tritium it produced. Pike thought the time might have come to start building at least the first of that new generation of reactors. "If we look forward to a success of the booster in 1951, it is not too early to [begin]. . . . This would be doing quite a little finessing, and a month or so ago [i.e., before the Soviet test], I would have questioned even the sense of bringing it up. I don't question it now and I think that it should be put on the table."

McMahon, coached by Borden, found this collective recitation of brutalisms undaunting. Nor did it matter to the JCAE, as Pike would also point out that autumn, "that we had no knowledge that the military needed such a weapon." The JCAE did not even know at that time how many atomic bombs the US held in its stockpile. Borden had already concluded on the committee's behalf that the survival of the nation depended on building many more such bombs while simultaneously pursuing another of his utopian fantasies, the ultimate weapons system of a thermonuclear bomb delivered by a nuclear-powered bomber.

David Lilienthal wanted nothing to do with such technological fanaticism. The chairman of the US Atomic Energy Commission found nuclear weapons obscene. After Norris Bradbury and Los Alamos Associate Director Darol Froman had briefed him on the *Sandstone* tests in June 1948 he had denounced the "keen enthusiasm" of the Los Alamos men in his diary. "I don't object at all," he wrote, ". . . to expressions of satisfaction that the job . . . is being pushed and done well; but that there should not be even a single 'token' expression of profound concern and regret that we are engaged in developing weapons directed against the indiscriminate destruction of defenseless men, women and children . . . this bothered me." When Lilienthal returned from Martha's Vineyard and met with his fellow commissioners on the morning of October 5, he took up not the thermonuclear question but the proposal that the Joint Chiefs had instigated early in 1949 for a major

AEC production expansion. Strauss by then had drafted his thermonuclear memorandum and had tried it out on Pike and new commissioner Gordon Dean, a Seattle native and Nuremberg trial staff assistant who had been Brien McMahon's law partner. Since Lilienthal did not bring up the thermonuclear question, Strauss had no chance to present his memorandum and had to distribute copies to his colleagues after the meeting. Strauss felt, he informed them, that a "larger stockpile of weapons than the Russians" was not a sufficient response to the Soviet test because "our relative lead is likely to decrease." Instead, echoing Golden, Strauss proposed:

> It seems to me that the time has now come for a quantum jump in our planning (to borrow a metaphor from our scientist friends)—that is to say, that we should now make an intensive effort to get ahead with the super. By intensive effort, I am thinking of a commitment in talent and money comparable, if necessary, to that which produced the first atomic weapon. That is the way to stay ahead.

Strauss concluded that the commission should "immediately consult the General Advisory Committee to ascertain their views as to how we can proceed with expedition."

Never one to trust to open debate alone what he could more reliably transact by subterfuge, Strauss passed from the morning AEC meeting directly to lunch with his fellow Rear Admiral Sidney Souers. Half a decade later, Souers remembered the occasion well:

> Strauss came to me shortly after the Russians had exploded their atomic bomb. . . . I didn't know anything about this super-bomb, this H-bomb, until Strauss came into my office then. Strauss looked upon me as a fellow security man and asked me whether the President had ever had any information on the super-bomb, and if he had, had he made a decision with regard to going ahead with it. I said, "Lewis, as far as I know, he has had none, none from me anyway. Can we build one?" [Strauss] said, "Yes," and I said, "Then why in the world don't we build it?" He said, "Well, I don't think [the President] has been informed because Lilienthal is opposed to it." I said, "You should certainly see that it gets to the President so that he can get the facts and make a decision." . . . The next morning at our meeting I told [the President] about this conversation. . . . I . . . asked him if he had had any information on it. He said, "No, but you tell Strauss to go to it and fast." . . . I did call Strauss. I told him I had thought about it overnight, and that . . . he should expedite it. With that he got busy. . . .

Until October 6, 1949, the President of the United States had never heard of the hydrogen bomb.

20

'Gung-ho for the Super'

WENDELL LATIMER, the Berkeley chemistry dean, remembered that he "really got concerned" about the United States's military position after the Soviet Union tested Joe 1. On October 5, 1949, the same day Lewis Strauss in Washington proposed a "quantum jump" in nuclear firepower to his fellow commissioners on the AEC, Latimer discussed his concern with Luis Alvarez, Ernest Lawrence's protégé at the Berkeley Radiation Laboratory. Alvarez, a prolific and successful inventor as well as a physicist of Nobel Prize caliber, decided to start a diary that Wednesday, as he had done in the early days of Second World War radar research, for history and to establish patent priorities should inventions turn up along the way. "Latimer and I independently thought that the Russians could be working hard on the super," Alvarez opened his diary, "and might get there ahead of us. The only thing to do seems to be to get there first—but hope that will turn out to be impossible." Fundamental physical principles might make a thermonuclear impossible, that is, but if such a weapon could be built, Latimer and Alvarez believed the US should build it before the Soviet Union.

The next day, Latimer recalls, he "got hold of Ernest Lawrence and I said, 'Listen, we have to do something about it.' . . . I saw Ernest Lawrence in the Faculty Club on the campus." Apparently Latimer advised Lawrence to explore the question with Alvarez; the Nobel laureate physicist dropped by Alvarez's office at the Radiation Laboratory that same afternoon. "Talked with E.O.L. about the project," Alvarez noted in his diary afterward, "and he took it very seriously—in fact he had just come from a session with Latimer. We called up Teller at Los Alamos to find out how the theory had progressed in the last four years." Teller was excited to hear that two such influential colleagues were interested in the thermonuclear, but since the project was secret, he could say little on the phone. Lawrence happened to be traveling east that weekend for a meeting in Washington on Sunday of a panel on radiological warfare—"a subject," writes Alvarez, "which was very close to . . . Lawrence's heart." The Radiation Laboratory director asked Alvarez to join him and proposed they detour through Los Alamos to talk to Teller. The two men left San Francisco that evening and landed in Albuquerque at three A.M. The next morning they flew up to Los Alamos.

At the weapons laboratory that Friday, Lawrence and Alvarez talked with Teller, Associate Director John Manley, Stanislaw Ulam and visiting Russian emigré theoretical physicist George Gamow. "They give [the] project [a] good chance if there is plenty of tritium available," Alvarez noted in his diary. "There must be a lot of machine calculations done to check the hydrodynamics, and Princeton and L.A. are getting their machines ready." The machines Alvarez and Lawrence heard about were more capacious, all-electronic successors to the ENIAC computer, one being built at John von Neumann's direction at the Institute for Advanced Study in Princeton, a copy under construction at Los Alamos. Theoretical Division leader J. Carson Mark had lured Nicholas Metropolis from the University of Chicago in January 1949 to direct the computer project. As a joke, in what Metropolis calls "a conscious effort to *end* the naming of computers [that] . . . had just the opposite effect," the Chicago mathematician had named the Los Alamos computer the MANIAC. (Gamow decided that the pseudo-acronym stood for "*M*etropolis *a*nd *N*eumann *I*nvent *A*wful *C*ontraption.") Los Alamos needed the new computer, observes Carson Mark, because "we really couldn't make any headway with what is called the main Super problem in a finite time with the kind of computing power . . . [previously] available. . . ." The main Super problem was an elaborate calculation, which Ulam had prepared with Gamow, von Neumann and Los Alamos theoretical physicist G. Foster Evans, to determine if Teller's Super design would propagate thermonuclear burning. Without such a calculation, no conclusive test was possible; if a test failed, they would not know whether it had failed because a thermonuclear was not feasible or merely because the particular design was wrong or had malfunctioned. That was why Teller himself, in 1947, had proposed waiting to develop the Super further until sufficient computing power was at hand to run the problem. It would not be at hand until the MANIAC was finished, a year or more away.

Teller was nevertheless sanguine about the prospects for his Super design. "I was going on the assumption that [it] . . . would work," he says, based on the optimistic conclusions of the April 1946 Super Conference. Although much evidence had accumulated since then that Teller's Super design was problematic ("Every time he reported," GAC member Lee DuBridge comments, "we thought he'd taken a step backwards"), Teller's endorsement was good enough for Lawrence, always an optimist himself in the business of building new machines. "Ernest told me, under those conditions, you just have to work on it." But it was not yet clear what Lawrence and Alvarez could do to help. To pursue that question, Teller accompanied the two Berkeley physicists on their flight back to Albuquerque, where they continued talking in Lawrence's hotel room until bedtime, Lawrence frugally using the occasion to wash out his shirt. ("Now you will have to do a lot of traveling," Teller recalls Lawrence instructing him, "and you won't be able to do it unless you learn to wash your shirt!") "We agreed that a conference should be called at [Los Alamos] next month," notes Alvarez's diary, "to see what should be done. L.A. had been talking about one for early next year. We can't wait too long." During the evening, Teller mentioned tritium production and the problem of trading plutonium for tritium in the AEC's existing graphite reactors. "Ernest and I . . . decided we would get going at once," Alvarez writes in his memoirs, "to start a project to build production-scale heavy-water reactors for the AEC." Teller returned to Los Alamos that night; Lawrence and Alvarez flew on to Washington before dawn with a cause they could fight for.

By Monday, October 10, the two men had talked to AEC staff including Kenneth Pitzer, a priggish Berkeley chemist who was AEC director of research, and General McCormack, had discussed their plans at length with Robert LeBaron, the chairman of the Military Liaison Committee, and had attended Lawrence's radiological warfare panel. Lawrence, Alvarez explains, "had made serious proposals in the Defense Department that warfare could be waged effectively by the use of radioactive products." The radwar panel decided on Sunday that its program would require what Alvarez called in his diary a "gram of neutrons," a requirement he thought "ties in well with our program" for tritium production, which also needed neutrons. "Heavy water piles," Alvarez testified, "would provide, we hoped, considerably more than a gram of neutrons. Therefore, we would have available either tritium or [if the thermonuclear proved unfeasible] radioactive warfare agents." In the midst of his Sunday radwar discussions, Lawrence learned that his wife had just given birth to their sixth child.

Monday, Lawrence and Alvarez had lunch with two members of the JCAE, Brien McMahon and California Republican Congressman Carl Hinshaw, with William Borden on hand taking notes:

The two scientists expressed keen and even grave concern that Russia is giving top priority to the development of the thermonuclear super-bomb. They pointed out that the Russian expert, Kapitza, is one of the world's foremost authorities on the problems involving the light elements. This fact, along with the logic that Russia might experience great difficulty in competing with us in the production of "conventional" atomic bombs, means that she has every incentive to concentrate on being first to acquire the super-bomb.* Drs. Lawrence and Alvarez even went so far as to say that they fear Russia may be ahead of us in this competition. They declared that for the first time in their experience they are actually fearful of America's losing a war, unless immediate steps are taken. . . .

The Berkeley scientists proposed convincing Canada to convert its heavy-water reactor at Chalk River to tritium production while they built a production-scale heavy-water reactor in California. Lawrence criticized the existing Super program as well, grandly building a feasible thermonuclear in the heady congressional air: "Dr. Lawrence said that not nearly enough is being done on the super-bomb at present; that the contemplated Booster test in 1951 is only a mincing step; and that given an all-out effort to produce tritium on the necessary scale, an actual super-bomb test—rather than a mere preliminary experiment—might be held." For good measure, Lawrence warned "that the British have a committee actively engaged in work on the super-bomb project, and that [they] . . . too, may possibly be ahead of us."

The congressmen were impressed. "They were very happy to see some action in the field of thermonuclear weapons," Alvarez writes; "they both told us they thought we were doing the right thing." McMahon had already started to campaign for the Super. On October 1, a Saturday and the day of the founding of the People's Republic of China (a further Communist threat that many believed to be directed out of Moscow in monolithic conspiracy), the influential chairman of the JCAE had called a meeting at his house to discuss the Soviet bomb and the possible US response. Hinshaw would be making the rounds of the laboratories that month as part of a subcommittee of the JCAE to assess AEC technical resources. The two congressmen offered Lawrence and Alvarez their support; Hinshaw told them he would see them in Berkeley in ten days' time.

David Lilienthal gave the two physicists a colder reception that afternoon. The AEC chairman was waiting dispiritedly to hear Truman's decision on the proposal of the Special Committee the President had appointed that the US invest $319 million in expanding production of atomic bombs to meet the Joint Chiefs' increased military requirement. "We keep saying, 'We have no other course,' " an exhausted Lilienthal noted in his diary; "what we

---

* Kapitza, as it happened, was under house arrest at this time. The difficulty everyone had in mind in 1949 with producing atomic bombs, Soviet or US, was the imagined scarcity of uranium ore. Lawrence and Alvarez would soon propose a system for transmuting thorium to U233 in a particle accelerator of heroic dimensions to help the US resolve this difficulty; when the AEC finally realized that raising the price the government was willing to pay for ore would stimulate prospecting, US proven reserves increased dramatically. The Soviets had already identified plentiful reserves in East Germany as well as Soviet Asia.—RR

should say is 'We are not bright enough to see any other course.' " The day, he went on, "has been filled, too, with talk about supers, single weapons capable of desolating a vast area." At that point, Lawrence and Alvarez arrived. Alvarez remembered that Lilienthal was unwilling even to consider what they had to say: "I must confess that I was somewhat shocked about his behavior. He did not even seem to want to talk about the program. He turned his chair around and looked out the window and indicated that he did not want to even discuss the matter. He did not like the idea of thermonuclear weapons, and we could hardly get into conversation with him on the subject." Lilienthal, for his part, recoiled from what he took to be fanatic enthusiasm: "Ernest Lawrence and Luis Alvarez in here drooling over the [superbomb]. Is this all we have to offer?" After the two physicists left his office, Lilienthal heard that Truman had decided not to request a supplemental appropriation. The AEC chairman was relieved and impressed: "This despite the unanimous recommendation of [the] Joint Chiefs of Staff, etc., etc. He makes his *own* decisions." (In fact Truman had approved the expansion and wanted it started; he simply did not want to ask Congress for more money that year.)

The Berkeley men saw Strauss and the other AEC commissioners after they left Lilienthal; Strauss learned then not only of their enthusiasm for the Super but also of Teller's. Their next stop was New York, whence they hoped to book seats to Ottawa to visit Chalk River. Seats were unavailable, so they took the opportunity to run up to Columbia University to talk to Rabi. In his contemporary diary, Alvarez wrote that they found Rabi "very happy at our plans. He is worried too." In testimony five years later, Alvarez added that Rabi had told them, "It is certainly good to see the first team back in.... You fellows have been playing with your cyclotron and nuclei for four years and it is certainly time you got back to work...." Rabi remembered a more provisional exchange. "Following announcement of the Russian explosion," he testified, "... I felt that somehow or other some answer must be made in some form ... to regain [our] lead.... There were two directions in which one could look: either the realization of the super or an intensification of the effort on fission weapons ... to get a large variety and very great military flexibility." But Lawrence and Alvarez had already made up their minds, Rabi felt:

They were extremely optimistic. They are both very optimistic gentlemen. ... They had been to Los Alamos and talked to Dr. Teller, who gave them a very optimistic estimate about the thing.... So they were all keyed up to go bang into it....

I generally find myself when I talk with these two gentlemen in a very uncomfortable position. I like to be an enthusiast. I love it. But those fellows are so enthusiastic that I have to be a conservative. So it always puts me in an odd position [where I have to] say, "Now, now, there, there," and that sort of thing. So I was not in agreement in the sense that I felt they were, as usual . . . overly optimistic.

Alvarez flew home to Berkeley that night. Lawrence returned to Washington. He had another idea, perhaps evolved from learning about the success of the Joint Chiefs' move to increase the military requirement for atomic bombs. He looked up Kenneth Nichols of the Armed Forces Special Weapons Project and encouraged him to approach the JCS about establishing a military requirement for hydrogen bombs as well. The Joint Chiefs were scheduled to meet with the JCAE on October 14. McMahon also approached Nichols about the JCS position. Nichols decided he should brief the Joint Chiefs before they testified. He went to see Lauris Norstad, who took him to see General Vandenberg, who was scheduled to be the JCS spokesman at the JCAE hearing. Like Truman, it turned out, the USAF Chief of Staff had never heard of the hydrogen bomb. "I think I spent about two hours with Vandenberg," Nichols recalled, "and I explained as well as I could from the information I had what the thermonuclear weapon would be. Vandenberg wanted to know if it could be delivered and I said, well, it might take a B-36 that you might have to drone in [unmanned], but I was pretty certain it wouldn't be bigger."

The JCS were meeting that day, October 13, in preparation for the forthcoming JCAE hearing. Vandenberg sent Nichols with Norstad to talk to the other Chiefs:

He said he couldn't go, tell General Bradley that he was for the development of the H-bomb. . . . I briefed the Joint Chiefs. . . . General [Omar] Bradley [Army Chief of Staff] said, "Nichols, why haven't you been around here before, advocating it?" I said, "Well, General, the reason I haven't been around is that before the Russian explosion, I was under the impression that it would be very difficult to organize a real effort at Los Alamos." There just weren't enough scientists to work on it. I told him Edward Teller was interested, but to organize an effort before the Russian explosion would have been well nigh impossible. But I said I've been following it all these years, and . . . now I think enough scientists will be willing to work on it. And after a few more questions, General Bradley finally announced his opinion . . . that it would be intolerable for the Russians to have it first. . . . This was his first reaction, that if it can be done, that it would be intolerable to have us sit on our butts, not do anything, if the Russians should get it first.

The next day, according to a summary Nichols wrote, a handsome, confident Hoyt Vandenberg told the congressional committee authoritatively:

One of the things which the military is pre-eminently concerned with as a result of the early acquisition of the bomb by Russia is its great desire that the Commission reemphasize and even accentuate the development work on the so-called super-bomb. General Vandenberg discussed this subject briefly and stated that it was the military point of view that the super-bomb should be pushed to completion as soon as possible, and that the general staff had so recommended. In fact, his words were, "We have built a fire under the proper parties"—which immediately brought forth Senator Hickenlooper's comment, "Who are the right parties?" General Vandenberg replied that it was being handled through the Military Liaison Committee. He further stated that having the super weapon would place the United States in the superior position that it had enjoyed up to the end of September 1949 by having exclusive possession of the [atomic] weapon.

John Manley had been new to the ways of Washington at this time of Super debate. What he learned from the experience, he would comment sardonically many years later, was that, "quite contrary to the way I thought things were ... you don't do staff work and then make a decision. You make a decision and then do the staff work."

On the weekend when Lawrence and Alvarez had traveled to Washington, Robert Oppenheimer had gone up from Princeton to Cambridge, Massachusetts, to attend a meeting of the Harvard University Board of Overseers, of which he was a member. He stayed at Harvard President James Bryant Conant's house, where they had what Oppenheimer in a subsequent letter called a "long and difficult discussion having, alas, nothing to do with Harvard." Almost certainly they discussed the Soviet test and what to do about it. "When we last spoke," Oppenheimer would write Conant two weeks later, having not spoken to Conant in person in the interim, "you thought perhaps the reactor program offered the most decisive example of the need for policy clarification. I was inclined to think that the Super might also be relevant." At the outset of the Super debate, that is, Robert Oppenheimer considered the development of a hydrogen bomb one possible response to the Soviet test.

Conant's response—probably his contribution to the "long and difficult discussion" in Cambridge on October 9—had been sharp. Oppenheimer was in some doubt about the matter, the IAS director confirmed later, and had not made up his mind. "Conant told me he was strongly opposed to it. ... He told me what his views were before mine were clearly formulated." If it ever came before the General Advisory Committee, Conant told Oppenheimer, "he would certainly oppose it as folly."

The influential Harvard administrator based his opposition, as he testified later, "on a combination of political and strategic and highly technical considerations." Conant thought US reliance on atomic and potentially thermonuclear weapons "was sort of a Maginot Line psychology being pushed on us." The better answer, he believed, "was to do a job and revamp our whole defense establishment, put in something like universal military service [and] get Europe strong on the ground, so that [the US atomic deterrent] would not be canceled out [by the Soviet development of atomic weapons]." Conant, that is, believed that US reliance on a nuclear arsenal for cheap defense gave the nation a false sense of security, as the Maginot Line of fortified defenses had given France a false sense of security prior to the Second World War, and that pushing for a thermonuclear without a corresponding build-up in conventional forces would simply compound the mistake.

More deeply even than his strategic concerns, Conant believed that building the thermonuclear would be morally wrong. He had accepted the necessity of poison gas in the First World War and atomic bombs in the Second because, as he wrote in 1943, "the battlefield is no place to question the doctrine that the end justifies the means." When the war was over, however, he continued, "let us insist, and insist with all our power, that this same doctrine must be repudiated." War, he had repeated in the *Atlantic Monthly* as recently as January 1949, "is always totally different morally from peace." If someone attacks you, Conant meant, you were justified in doing whatever you had to do to defend yourself. Such a distinction might have made sense in an age of traditional war. Curtis LeMay and William Borden would have disputed its relevance to the new atomic age. In their view, the destructiveness of atomic weapons made it improbable that a nation could add to its arsenal after war began; any new weapons, however immoral, would have to be developed and stockpiled in peacetime or not at all.

Whether Conant discussed the morality of thermonuclear weapons with Oppenheimer at Cambridge is unclear. The GAC chairman learned of Conant's moral position soon enough as he began organizing a special meeting of his committee that Lilienthal had asked him to call in response to Strauss's "quantum jump" proposal. In the meantime he spoke with John von Neumann, whose experience of war, Communist revolution and fascist counter-revolution as a young man in Hungary left him in no difficulty about morality or strategy. "I remember von Neumann saying at this time," Oppenheimer testified: " 'I believe there is no such thing as saturation. I don't think any weapon can be too large. I have always been a believer in this.' He was in favor of going ahead with it."

Also in response to Strauss's proposal, the AEC asked Norris Bradbury to appear before it in Washington on October 19 to discuss the Los Alamos program. Bradbury called a major meeting of laboratory staff on October 13 to prepare for his appearance. Teller and Manley both wrote open letters for that Thursday meeting. Manley argued that the laboratory had "tacitly assumed that [a Soviet test] . . . would not occur before 1952, a date beyond the expected 1951 fruition of current programmatic work [i.e., the *Greenhouse* tests of a boosted fission weapon and of thermonuclear ignition scheduled at Eniwetok in 1951]. At the very least, therefore, the laboratory should consider that it has lost some three years of time" on thermonuclear work. Nor could it assume that the Soviets would progress no faster than the US had been and was progressing. It followed, Manley concluded, that "we should no longer assume any time scale for their developments but rather choose our action so as to strengthen our position as rapidly as possible." Manley stopped short of endorsing an all-out program to build the Super, however.

Teller took up that agenda in his open letter, outlining "why it is essential for us to develop a Super bomb at the earliest possible time or else be able to say with reasonable confidence that the Super is not feasible." Referring as Manley had to the question of the Soviet rate of progress, he went further and criticized the laboratory for its conservatism: "If the Russians continue to make actual progress faster and if we lose the atomic armament race, it will make little difference whether the reason has been the particular brilliance of Russian scientists or the exaggerated caution and thoroughness of our own group." Going further still, Teller argued emotionally that a thermonuclear breakthrough was indispensable to survival regardless of any other consideration: "If the Russians demonstrate a Super before we possess one, our situation will be hopeless." Why a nation with a burgeoning arsenal that now approached 169 atomic bombs would find itself in a hopeless situation if its enemy tested a prototype of a larger weapon, Teller did not specify. The Hungarian-born physicist proposed an "all-out" effort to build a Super, provided that the laboratory could "marshal the necessary support from Washington for a really vigorous program."

After meeting with the AEC in Washington, Bradbury and Manley stopped in at the Institute for Advanced Study on October 20 to brief Oppenheimer. The GAC chairman knew the lab had been working on thermonuclear theory and basic physical measurements since his own tenure as director. He would have heard about the calculation problems. He probed the progress of Teller's Super design to see if four years of work had improved it. The lab's political position, as Bradbury explained it several years later, was that "we did not wish to enter into the debate as to whether or not this course was wise or moral or politically sound. We regarded ours [to be] the technical responsibility to know as much as it was possible to know and as rapidly as it was possible to know it about what was broadly called the H-bomb."

That afternoon, or Friday morning, Oppenheimer wrote Conant to prepare him for the forthcoming GAC meeting scheduled for October 29–30 and for a meeting booked with Truman immediately afterward (which never took place). Playing the good son to powerful older men to cultivate their approval, Oppenheimer routinely saluted Niels Bohr in correspondence as "Uncle Nick" (derived from Bohr's Manhattan Project cover name, Nicholas Baker); now the forty-five-year-old physicist addressed the cool, formal fifty-six-year-old Harvard president as "Uncle Jim." He reminded Conant of their earlier discussion, then subtly moved to distance himself from his previous inclination that the Super might be relevant:

> On the technical side, as far as I can tell, the Super is not very different from what it was when we first spoke of it more than seven years ago; a weapon of unknown design, cost, deliverability and military value. But a very great change has taken place in the climate of opinion. On the one hand, two experienced promoters have been at work, i.e., Ernest Lawrence and Edward Teller. The project has long been dear to Teller's heart; and Ernest has convinced himself that we must learn from Operation Joe that the Russians will soon do the Super, and that we had better beat them to it.

Worse, Oppenheimer went on, "the joint Congressional committee, having tried to find something tangible to chew on ever since September 23, has at last found its answer: We must have a Super, and we must have it fast." The Joint Chiefs had signed on. Even "the climate of opinion among the component physicists" was showing "signs of shifting"—shifting presumably in favor of pushing the thermonuclear. Oppenheimer then moved fully into congruence with Conant's position as far as strategy was concerned, though he did not yet know or was not yet prepared to commit to Conant's moral position as well:

> What concerns me is really not the technical problem. I am not sure the miserable thing will work, nor that it can be gotten to a target except by ox cart. It seems likely to me even further to worsen the unbalance of our present war plans. What does worry me is that this thing appears to have caught the imagination, both of the Congressional and of military people, as the answer to the problem posed by the Russian advance. It would be folly to oppose the exploration of this weapon. We have always known it had to be done; and it does have to be done, though it appears to be singularly proof against any form of experimental approach. But that we become committed to it as the way to save the country and the peace appears to me full of dangers.

Oppenheimer had lunch Friday with LeBaron and McCormack and explored the military's position further. In the afternoon, Hans Bethe and Teller arrived together to discuss whether or not Bethe would return to Los Alamos to work on the Super.

Since meeting with Lawrence and Alvarez, Teller had gone on the road, though he probably did not wash out his own shirts, pilgrimaging to Cornell to try to convince Bethe to return to Los Alamos. "Bethe," he recalls unambiguously, "said he would come." Bethe remembered to the contrary having "the greatest misgivings when Teller first approached me...." Teller tempted Bethe with Super improvements, "some...technical ideas which seemed to make technically more feasible one phase of the thermonuclear program. I was quite impressed by his ideas. On the other hand, it seemed to me that it was a very terrible undertaking to develop a still bigger bomb, and I was entirely undecided and had long discussions with my wife." Rose Bethe, like her husband a refugee from Nazi Germany, had challenged him similarly in 1942 when they had first heard of Teller's thermonuclear vision. "What should I do?" Bethe continues. "I was deeply troubled what I should do. It seemed to me that the development of thermonuclear weapons would not solve any of the difficulties that we found ourselves in, and yet I was not quite sure whether I should refuse." Teller testified that Oppenheimer called at that point and invited the two physicists "to come and discuss this matter with him in Princeton." Unless Oppenheimer could read minds, the more likely circumstance would have been that Bethe called Oppenheimer.

In Princeton, Bethe says, they found Oppenheimer "equally undecided and equally troubled in his mind about what should be done. I did not get from him the advice that I was hoping to get. That is, I did not get from him advice...to decide me either way.... He mentioned that...Dr. Conant was opposed to the development of the hydrogen bomb, and he mentioned some of the reasons which Dr. Conant had given." Teller remembered Oppenheimer quoting Conant to the effect that the US would build a thermonuclear "over my dead body," but he conceded that Oppenheimer himself "did not argue against any crash program. We did talk for quite a while. ...At least one important trend in this discussion...was that Oppenheimer argued that some phases of exaggerated secrecy in connection with the A-bomb were perhaps not to the best interest of the country, and that if he undertook the thermonuclear development, this should be done right from the first and should be done more openly." Bethe, says Teller, "reacted to that quite violently" because he thought thermonuclear work should be done secretly. (Notice also that Oppenheimer, in Teller's recollection, was considering *leading* the thermonuclear effort at that point.) Teller, who had assumed Oppenheimer would oppose his thermonuclear campaign, was surprised to find the former Los Alamos director undecided. Before going to Princeton, Teller testified, "I am pretty sure that I expressed to Bethe the worry, we are going to talk with Oppenheimer now, and after that you will not come. When we left the office, Bethe turned to me and smiled and he said, 'You see, you can be quite satisfied. I am still coming.'"

Teller went one way then, Bethe another. Leo Szilard, the maverick Hungarian emigré theoretical physicist who had first conceived the idea of using a physical chain reaction to release the energy of the atomic nucleus, had organized an Emergency Committee of Atomic Scientists, with Albert Einstein as chairman, which was meeting at Princeton University that weekend. Bethe, a founding member of the ECAS, went over from the Institute for Advanced Study to the university to attend the meeting. "I remember that I walked into the meeting room and Leo Szilard greeted me by saying, 'Ah, here comes Hans Bethe from Los Alamos.' I protested that I was not at Los Alamos and didn't know if I wanted to go back there." Bethe's good friend Victor Weisskopf, an Austrian emigré theoretician, attended the meeting. Walking together on the university grounds that autumn weekend, the two men imagined together the horrors of a thermonuclear war. "Weisskopf vividly described to me a war with hydrogen bombs—what it would mean to destroy a whole city like New York with one bomb, and how hydrogen bombs would change the military balance by making the attack still more powerful and the defense still less powerful." "We both had to agree," Bethe added at another time, "that after such a war, even if we were to win it, the world would not be . . . like the world we want to preserve. We would lose the things we were fighting for. This was a very long conversation and a very difficult one for both of us." Bethe and Weisskopf continued their difficult conversation on the drive up to New York with another physicist, the sharp-tongued Bohemian emigré George Placzek. "In this conversation," Bethe remembers, "essentially the same things were confirmed once more." The issue engaged the three men so intensely that Bethe missed his plane back to Ithaca and Weisskopf and Placzek inadvertently traded coats. Bethe's conversations with his two friends and with his wife decided him against working on the thermonuclear. "A few days later," he says, "I told Teller over the phone that I would not join the project. He was disappointed. I felt relieved."

In Chicago on Monday, on his way southwest, Teller met Enrico Fermi at the airport. The Italian emigré Nobel laureate was exhausted after a long flight from Europe. Teller appealed to him to return to Los Alamos to help build the Super. Teller remembered in 1994 that Fermi "refused out of hand." But Alvarez heard from Teller by phone that day in 1949 and noted in his diary that Fermi had "no reaction." Teller also told Alvarez that he "felt Oppie was lukewarm to our project and Conant was definitely opposed."

From AEC commissioner Sumner Pike, Oppenheimer received a letter proposing questions that the General Advisory Committee might address at the meeting scheduled for the coming weekend in Washington. Broadly, asked Pike, was the commission doing things that might be slowed or stopped, and were there other things that ought to be done in addition or instead, "to serve the paramount objective of the common defense and security?" Pike asked specifically about civil defense, further production expansion, radiological warfare and "any new aspects of the international control of atomic energy which may be perceived as a result of the new situation," but he was most concerned about the Super, which he noted "would conflict" with fission weapons "in terms of demand for neutrons." (The commissioners had heard, a few days earlier, that by 1951, given present production rates, there would be only enough tritium available in the US program for the one D + T booster shot scheduled for *Greenhouse*— no more than a few grams.) Pike wondered first of all if the US would use a Super if it had one. He wondered what its military worth would be—"would it be worth two, five, fifty existing weapons?" He wondered what it would cost and how to relate its value to the value of improving existing fission weapons. Some of these questions appear oddly out of place in an agenda intended for a body of technical advisers who were none of them military experts—except that, as Glenn Seaborg comments, "The Atomic Energy Commission didn't tell us what to do. We told them what to do." Oppenheimer took Pike's agenda under advisement.

Teller had rushed back to Los Alamos because Carl Hinshaw and two other members of the JCAE subcommittee would arrive there on Thursday for a briefing. Bradbury led the briefing, describing a possible Super design to the congressmen that would weigh twenty thousand pounds and yield at least one megaton—one million tons of TNT equivalent, more than seventy times as powerful as the bomb that destroyed Hiroshima. Bradbury thought such a weapon could be ready by 1958, nine years hence, although a test might be possible by 1952. Evidently Bradbury was assuming, optimistically, that MANIAC calculations of the main Super problem would determine that a Super was feasible in the first place.

After Robert Serber returned from Japan at the end of the war, he had left Los Alamos and moved to Berkeley to become director of theoretical physics at the Radiation Laboratory. In that position, in 1949, he had witnessed Ernest Lawrence's agitation after the announcement of the Soviet test. "I warned him about Edward's Super," Serber recalls, "that it wasn't a practical idea at the moment. I told him if you want to really find out, you should talk to Bethe, but he never did. He was all gung-ho for the Super and he immediately—action was always the thing with Ernest—immediately asked what he could do and the thing to do was to build these reactors to make tritium." Now Lawrence delegated Serber, Oppenheimer's old friend, to make the case in Washington to the GAC. "Why was I representing Ernest and Luis [Alvarez]? The reason Ernest asked me was he thought I would get a more sympathetic hearing from Oppenheimer than Luis would. Lawrence was a dictator and if he asked you to do something you either did it or you got fired. But that wasn't the principal reason. It didn't seem like a bad idea to build some more reactors. To make neutrons which could be used to make tritium or to make plutonium or whatever. It seemed to me like a plausible thing to do at that point." But Alvarez also traveled east at the same time to lobby the GAC and the AEC commissioners informally, stopping in Chicago to discuss reactor design with Argonne Laboratory director and reactor pioneer Walter Zinn.

Serber arrived in Princeton on Thursday, October 27, to talk to Oppenheimer and to spend the night before the GAC meeting, which was scheduled to begin informally the following afternoon. By Thursday, Oppenheimer may have heard from Conant in response to his October 21 letter. Serber remembers Oppenheimer telling him about Conant's position and possibly showing him a letter. Conant, it seemed, had moved to a moral position against the Super in addition to raising technical and strategic questions. "I was astonished," Serber recalls. "Coming from Berkeley, I had no idea that people were thinking anything except pushing weapons development as fast as they could. Conant was the prime mover in that opposition." Serber's sense is that Oppenheimer was reporting to him what Conant intended to propose to the GAC, not that Oppenheimer himself had signed on to Conant's position.

Friday, October 28, Oppenheimer and Serber rode down to Washington together by train. George Kennan was scheduled to open the informal Friday afternoon session. Bell Telephone Laboratories president Oliver Buckley, Caltech president Lee DuBridge, Fermi, Rabi and Cyril Smith were able to attend. Manley served as secretary. Glenn Seaborg was in Sweden. In early September, Swedish Nobel laureate physicist Manne Siegbahn had invited the co-discoverer of plutonium to lecture during October in the country of his ancestors. "Well," Seaborg says with a grin, "I knew what they meant. They wanted to look me over for the Nobel. I wasn't about to miss that, so I went. But I wrote Oppenheimer a letter." In his letter, dated October 14, Seaborg deliberately softened his endorsement of an accelerated Super program. "Although I deplore the prospects of our country putting a tremendous effort into this," he wrote, "I must confess that I have been unable to come to the conclusion that we should not. . . . I would have to hear some good arguments before I could take on sufficient courage to recommend not going toward such a program." Cyril Smith remembered vividly that Oppenheimer showed Seaborg's letter to the GAC members before the meeting opened Saturday morning.

Kennan talked, says Smith, "of the general state of industry in the USSR and the attitude of the Russian government vis-à-vis America." The director of the State Department's Policy Planning Staff left Smith at least with the impression that it might be possible to negotiate a halt to the arms race with the Soviet Union. Kennan spoke for forty-five minutes; for the next hour the committee members discussed the issues. Next they heard Bethe, Smith recalls, "on the general state of thermonuclear research and on the probability that a working bomb could be built." Bethe stressed the many technical problems that remained to be solved; Smith remembers being "freshly and strongly influenced by what he . . . said." Bethe also found occasion in the course of the afternoon to talk to some of the committee members about the dark vision Weisskopf had conjured of thermonuclear war and the sense both men shared that such a war would destroy what it was supposed to preserve.

Serber rounded out the afternoon with a presentation on Lawrence's plan to build a heavy-water reactor under Radiation Laboratory sponsorship, carefully disassociating himself from the all-out Super crusade. "Fermi said to me, 'Why Berkeley?'" he remembers. "'It's the one place that has no experience at all building reactors.' I said I think the point is that Ernest is trying to stress what he considers the importance of building more reactors, to the point where he's even willing to build them himself, and if there's a better way of doing it, Ernest will be the first one to applaud it."

Conant was on hand when the GAC convened formally on Saturday morning, October 29, 1949. So was respected senior GAC member Hartley Rowe, one of the pioneering engineers who had built the Panama Canal. Downstairs in the lobby of the AEC building, Alvarez stationed himself by the door and watched people come and go. Oppenheimer formally—"very solemnly," says Rabi—opened the meeting by posing the question whether to recommend a major program to build the Super—"not whether we should make a thermonuclear weapon," Rabi interprets, "but whether there should be a crash program." Next, DuBridge testified, Oppenheimer "asked the members of the committee if they would in turn around the table express their views on the question." Oppenheimer was careful not to lead the committee members, but in DuBridge's recollection they had already made up their minds in any case:

> Dr. Oppenheimer did not express his point of view... until after all the rest of the members of the committee had expressed themselves. It was clear, however, as the individual members did express their opinions as we went around the table, that while there were differing points of view, different reasons, different methods of thinking, different methods of approach to the problem, that each member came essentially to the same conclusion, namely, there were better things the United States could do at that time than to embark upon this super program.... I suppose each person took five to ten minutes or thereabouts to express his views. After they were all on the table, the chairman said he also shared the views of the committee. We then discussed the question of how to state our views and our recommendations most effectively to the Commission. It was on this subject of how our general conclusions could be most effectively and clearly stated that a very substantial discussion went forward for the next day or two.

Not everyone would recall having made up his mind before the meeting began. Rabi had not; "there were some people," he would testify, "and I myself was of that opinion for a time, who thought that the concentration on

the crash program ... was the answer to the Russian [atomic] ... weapon."
Oppenheimer himself had not made up his mind before the meeting, he
testified under oath in 1954; his views changed "in the course of our discussion." They changed in the direction of Conant's views, an interviewer paraphrased Oppenheimer in 1957, as "a result of Conant's intervention. Conant
said he just wouldn't have this, and pointed out that a firm stand could be
expected to meet with the approval of various groups, [including] churches."
Hartley Rowe, on the other hand, felt strongly from the outset that the Super
was wrong:

> It was a pretty soul-searching time, and I had rather definite views. . . . I
> may be an idealist, but I can't see [how] ... any people can go from one
> engine of destruction to another, each of them a thousand times greater in
> potential destruction, and still retain any normal perspective in regard to
> their relationships with other countries and also in relationship with peace.
> . . . If a commensurate effort had been made to come to some understand-
> ing with the nations of the world, we might have avoided the development.
> . . . I think I arrived at my conclusions even before the discussion came
> before the committee.

Four of the five AEC commissioners joined the GAC then—Lilienthal,
Strauss, Gordon Dean and physicist Henry Smyth, the author of the Smyth
Report and Robert Bacher's recent replacement on the commission. Conant
was grim with righteousness, Lilienthal saw. "A dramatic setting: Oppenheimer at the end of the table. Conant looking almost translucent, so gray." For
an hour, the minutes report, the committee discussed "the super-bomb
program" with the commissioners. "Campbell's," Lilienthal coded the Super
in his diary ("standing for 'soup,' that is, 'super,'" he explains). "Subject,"
he introduces his notes on that weekend: "what now, centering around
'Campbell's.'" At eleven A.M. the Joint Chiefs arrived trailing LeBaron, Norstad and lesser aides in their wake; Alvarez had watched their progress
through the lobby, "the famous military men whom I recognized from their
pictures. . . ." In Lilienthal's diary summary the Chiefs told a curious tale.
They "no longer favor[ed] 'outlawing' the bomb ... because without it ...
there is nothing we can do to prevent or deter Russia taking over Europe. . . ." Some of their strategists believed "that war [with the Soviet Union]
is inevitable. . . . [The Joint Chiefs themselves], however, do not go so far;
they believe it's 'likely' in a relatively short time, four to five years. . . . [In
their view] 'Further negotiation with the Russians is useless.'" But Omar
Bradley was hard-pressed to explain what military value a thermonuclear
weapon might have, Manley writes:

> The reaction of the highest ranking military person [who attended the
> meeting], General Bradley, was most interesting. Instead of being infatu-

ated with the possibility of a bomb 1,000 times as powerful as our first A-bombs, he thought such a weapon would be useless against most military targets and that its value would be mostly "psychological."

Lilienthal also reports Bradley's estimate that "[the] chief value of such a weapon [would be] 'psychological,'" but adds significantly that "'Campbell's' made the[ir] eyes light up." What about simply increasing production of atomic bombs? Lilienthal asked the military leaders. "No answer; Norstad came to me later to say, 'We've got no answer; we are studying it once each six months.'"

By the time the Chiefs paraded out it was twelve-thirty. People went off in small groups to lunch. Serber, small and wiry, joined tall, ruddy Alvarez standing outside the AEC building and then Oppenheimer came along and swept them both up. He knew an intimate restaurant nearby. "That was the first time I heard Robert's views on the building of the hydrogen bomb," Alvarez writes. "He told me he didn't think the United States should build it. The main reason he gave was that if we built a hydrogen bomb then the Russians would build a hydrogen bomb, whereas if we didn't build a hydrogen bomb then the Russians wouldn't build a hydrogen bomb. I thought this point of view odd and incomprehensible. I told Robert that he might find his argument reassuring but that I doubted if he would find many Americans who would accept it." Oppenheimer would testify that he was reporting to Alvarez what the sense of the meeting was becoming, not expressing his own position. "I said [that] quite strongly negative things on moral grounds were being said." But he would add that these views "were getting to be" his views "in the course of our discussion." Alvarez concluded that the reactor program he and Lawrence had proposed "was dead"; after lunch, without waiting to hear the outcome of the GAC meeting, he left to return to California.

The full discussion unfolded in the afternoon. Lilienthal sketched it impressionistically in his diary that evening:

Conant flatly against [the Super] on moral grounds. Hartley Rowe, with him: "We built one Frankenstein." Obviously Oppenheimer inclined that way. Buckley sees no diff[erence] in [the] moral question x [compared to] y times x, but Conant disagreed—there are grades of morality. Rabi completely on the other side. Fermi, his careful enunciation, dark eyes, thinks one must explore it and do it and that doesn't foreclose the question: should it be made use of? Rabi says decision to go ahead will be made; only question is who will be willing to join in it. I deny there is anything inevitable in political decisions. Lewis [Strauss] says the decision won't be made by popular vote, but in Wash[ington]. Conant replies: but whether it will stick depends on how the country views the moral question.

Conant wanted the subject opened up to public debate, Lilienthal continues. "I said [the] President certainly could announce it if he wished to.... Cyril Smith strong for the Conant point, as was I. Lewis quite dubious, evidently." Conant, perhaps a man with a guilty conscience, felt a wave of déjà vu. "Conant says, 'This whole discussion makes me feel I was seeing the same film, and a punk one, for the second time.'"

Smyth and Gordon Dean both judged that the scientists were reacting viscerally. It had been one thing for the members of the General Advisory Committee to invent and deliver a weapon of unprecedented power in wartime, when they believed their most implacable enemy, Nazi Germany, with whom they were actually at war, might be racing to invent and deliver such a weapon first; it would be quite another to invent and deliver a weapon of no definable military use, a weapon of mass destruction, into a world not at war. But even setting aside their moral qualms, they could see no sense in it, as their conclusions made clear.

They came to those conclusions by drafting statements, late Saturday evening, and revising and concerting those statements on Sunday. Oppenheimer and Manley drafted the two-part main report. Conant and DuBridge drafted what came to be called (because Buckley, Oppenheimer, Rowe and Cyril Smith also signed it) the "majority annex." Rabi and Fermi drafted a "minority annex."

The main report embodied their technical recommendations. They recommended exploiting lower-grade ores and building more reactors and isotope-separation plants. They thought the AEC should work harder to make available tactical atomic weapons. They endorsed the production of neutrons, "a gram per day," not by the Berkeley Radiation Laboratory but at the Argonne National Laboratory, which had experience building reactors. They advised that this flood of neutrons should be used to make U233, to produce radiological warfare agents, to test reactor components, to convert U238 to plutonium, to make polonium for initiators, to produce tritium for boosting fission bombs and, last of all on their list, "for super bombs." Even within this descending series they recommended that "the super program itself should not be undertaken." Nor did they recommend building reactors uniquely to make tritium for the Super.

Part II of the main report concerned "super bombs." It stated specifically what they were discussing: "the question of whether to pursue with high priority the development of the super bomb." It stated what they recommended: "No member of the Committee was willing to endorse this proposal." Their reasons, it said, "stem in large part from the technical nature of the super and of the work necessary to establish it as a weapon." What that technical nature was in October 1949, and what work the most distinguished and knowledgeable body of scientific advisers available to the US government believed remained to be done "if the super bomb is to become a reality," Part II then specified, beginning with a description of the bare idea:

> The basic principle of design of the super bomb is the ignition of the thermonuclear DD [i.e., D + D] reaction by the use of a fission bomb, and of high temperatures, pressure, and neutron densities which accompany it. In overwhelming probability, tritium is required as an intermediary, more easily ignited than the deuterium itself and, in turn, capable of igniting the deuterium.

Thus Edward Teller's "Super," "classical Super," "runaway Super," the only design under consideration. To make the Super a reality, the GAC report continues, would require manufacturing tritium "in amounts perhaps of several hundred grams per unit." (No one really knew how much tritium the Super would need. "Edward promised people in Washington [at this time] that they'd get by with a certain amount," Carson Mark observes. "He had no particular basis for the amount that he mentioned except that it didn't appall them. He chose an amount that had that property. It didn't necessarily have the property of starting the reaction.") The Super also needed further theoretical studies—the MANIAC calculations—as well as design engineering and tests. The report warned that only a test could determine if a given model would or would not work and that many tests might be required. As a first estimate, the committee proposed that "an imaginative and concerted attack on the problem has a better than even chance of producing the weapon within five years"—that is, by 1954. ("When you are talking about something as vague as this particular thing," Rabi interpreted that estimate later, "you say a fifty-fifty chance in five years. . . . It was a field where we really did not know what we were talking about, except on the basis of general experience. We didn't even know whether this thing contradicted the laws of physics. . . . [Teller's Super] could have been altogether impossible.")

The report then emphasized the single most distinctive characteristic of a thermonuclear as opposed to a fission weapon: that if it could be built—if a runaway thermonuclear reaction could be initiated in deuterium—it would have essentially unlimited explosive potential. "This is because one can continue to add deuterium . . . to make larger and larger explosions. . . ." This characteristic distinguished it sharply from even such horrific weapons as atomic bombs. In deducing what followed from the superbomb's unique destructive potential, Oppenheimer and Manley staked out principled ground free of Conant's contentious moralizing:

> It is clear that the use of this weapon would bring about the destruction of innumerable human lives; it is not a weapon which can be used exclusively for the destruction of material installations of military or semi-military

purposes. Its use therefore carries much further than the atomic bomb itself the policy of exterminating civilian populations.

If the GAC had stopped there, it would have discharged its statutory and its ethical responsibilities admirably. But feeling had been running high that weekend and the eight men had stayed up late Saturday night drafting their annexes. The majority annex—Conant and DuBridge's language that Buckley, Oppenheimer, Rowe and Cyril Smith also signed—foresaw "possible global effects of the radioactivity" and feared that "a super bomb might become a weapon of genocide." It answered Omar Bradley's "psychological" justification by arguing that "reasonable people the world over would realize that the existence of a weapon of this type . . . represents a threat to the future of the human race," which would make "the psychological effect of the weapon in our hands . . . adverse to our interest." The majority annex urged that "a super bomb should never be produced. . . . To the argument that the Russians may succeed in developing this weapon, we would reply that our undertaking it will not prove a deterrent to them." Should they use such a weapon against the US, we had a "large stock of atomic bombs" with which to reply. "In determining not to proceed to develop the super bomb," the majority annex concluded, "we see a unique opportunity of providing by example some limitations on the totality of war and thus of limiting the fear and arousing the hope of mankind."

Fermi and Rabi, who had favored working on the Super at the outset of the weekend discussion, opposed the project now, but thought the President should announce the US renunciation and "invite the nations of the world to join us in a solemn pledge" not to build thermonuclear weapons. The two physicists judged it "highly probable" that a thermonuclear test could be detected "by available physical means." Rabi clarified later what they meant. "Fermi and I said that we should use this as an excuse to call a world conference for the nations to agree, for the time being, not to do further research on [thermonuclear weapons]. . . . [We] felt that if the conference should be a failure and we couldn't get agreement to stop this research and had to go ahead, we could then do so in good conscience. . . . Some of the others, notably Conant, felt that no matter what happened it shouldn't be made. It would just louse up the world."

But Fermi and Rabi also condemned their friend Edward Teller's Super in the strongest language that appears anywhere in the nine pages of the GAC report:

Necessarily such a weapon goes far beyond any military objective and enters the range of very great natural catastrophes. By its very nature it cannot be confined to a military objective but becomes a weapon which in practical effect is almost one of genocide.

It is clear that the use of such a weapon cannot be justified on any ethical ground which gives a human being a certain individuality and dignity even if he happens to be a resident of an enemy country. It is evident to us that this would be the view of peoples in other countries. Its use would put the United States in a bad moral position relative to the peoples of the world.

Any postwar situation resulting from such a weapon would leave unresolvable enmities for generations. A desirable peace cannot come from such an inhuman application of force. The postwar problems would dwarf the problems which confront us at present. . . .

The fact that no limits exist to the destructiveness of this weapon makes its very existence and the knowledge of its construction a danger to humanity as a whole. It is necessarily an evil thing considered in any light.

One way or another, most of the men on the General Advisory Committee had worked for international control of atomic energy. On the evidence of their October report, they had not left that intention so completely that they now viewed the Soviet Union as an obdurate, remorseless enemy. But those who urged racing to build the Super—Teller, Strauss, McMahon, Borden— did view the Soviet Union that bleakly. To Lilienthal, the men at Berkeley might appear to be "a group of scientists who can only be described as drooling with the prospect and 'bloodthirsty.'" In fact, their enthusiasm masked a profound fear. They were afraid of an enemy which was evidently ruthless, which had hidden itself and its intentions behind minefields and barbed wire, which fielded large and powerful armies and which had just successfully tested an atomic bomb. So were McMahon and Borden afraid; so were the Joint Chiefs, men whose responsibility it was to protect the nation. When the GAC argued that building the Super might unleash unlimited destruction, then, it unwittingly enlarged the scope of its opponents' fears and encouraged them to pursue the project with even greater urgency, because they immediately translated the weapon's destructive potential into a threat and imagined the consequences if the enemy should acquire it first. An arms race is a hall of mirrors.*

The GAC might also have weighed the impact of the strong language it used in its reports on those who disagreed with its conclusions. "Genocide" and "evil" are fighting words; "genocide" was even a new word then, coined only in 1944 and still fresh in its application to the Nazi destruction of the European Jews. Teller had lost members of his own family in the late sweep of the Holocaust through Hungary; Strauss was a prominent Jewish lay

---

* Teller eventually realized that there is a limit to the destructiveness of even thermonuclear explosions. At somewhere around a hundred megatons, he estimates, "it would simply lift a chunk of atmosphere—ten miles in diameter, something of that kind—lift it into space. Then you make it a thousand times bigger still. You know what would happen? You lift the same chunk into space with thirty times the velocity." Edward Teller interview, Los Alamos, x.93.

leader. It was not clear to either man how the GAC distinguished between an atomic bomb and a thermonuclear bomb in regard to mass destruction. Oppenheimer would be asked that question a few years later in the context of the October GAC report, and his answer, while profoundly ironic, hardly does justice to the destructiveness of the atomic weapons devised under his direction at Los Alamos during the Second World War:

Q. In fact, Doctor, you testified, did you not, that you assisted in selecting the target for the drop of the bomb on Japan?

A. Right.

Q. You knew, did you not, that the dropping of that atomic bomb on the target you had selected, [would] kill or injure thousands of civilians, is that correct?

A. Not as many as turned out.

Q. How many were killed or injured?

A. Seventy thousand.

Q. Did you have moral scruples about that?

A. Terrible ones. . . .

Q. Would you have supported the dropping of a thermonuclear bomb on Hiroshima?

A. It would make no sense at all.

Q. Why?

A. The target is too small.

The confusion and fear, the passionate intensity of proponents and opponents both, flooded the prudent advice contained in the GAC's main report and drowned it out. That advice concerned US security, not moral issues. Oppenheimer summarized it best, as he usually did, five years later:

The notion that the thermonuclear arms race was something that was in the interests of this country to avoid if it could was very clear to us in 1949. We may have been wrong. We thought it was something to avoid even if we could jump the gun by a couple of years, or even if we could outproduce the enemy, because we were infinitely more vulnerable [because more of the US population lives in large cities than does the Soviet population] and infinitely less likely to initiate the use of these weapons, and because the world in which great destruction has been done in all civilized parts of the world is a harder world for America to live with than it is for the Communists to live with. . . .

We thought [a US decision not to build the thermonuclear] . . . would make it less likely that the Russians would attempt [it] and less likely that they would succeed in the undertaking.

When Brien McMahon read the GAC report in the presence of the AEC commissioners on Monday evening, Manley writes, "there [was] a rather violent discussion." Lilienthal found the discussion with the Joint Committee chairman "pretty discouraging. What [McMahon] is talking about is the inevitability of war with the Russians"—Borden's conclusion in his book—"and what he says adds up to one thing: blow them off the face of the earth, quick, before they do the same to us—and we haven't much time." *There will be no time.*

McMahon went off to begin a campaign of letters and personal appeals to Truman to convince the President to authorize a crash program to build the Super. Lilienthal alerted Dean Acheson to the debate on Tuesday, November 1. "He was somber enough when I began; after a few questions he was graver still. . . . What a depressing world it is, said Dean, looking quite gray."

Teller began moving around the country in what Manley calls "a frenetic campaign to obtain converts." Teller wrote Alvarez that he felt like Sisyphus and needed some encouragement. They both blamed Oppenheimer; as far as Alvarez was concerned, the outcome of the GAC meeting meant that "Robert's views prevailed." When Teller saw the GAC reports, Manley recalls, the volatile physicist became "morose and almost silent (*very* unusual). . . . Edward offered to bet me that unless we went ahead with his Super . . . he, Teller, would be a Russian prisoner of war in the United States within five years!"

Midweek, the AEC commissioners split on their recommendation to the President. Strauss and Gordon Dean favored an accelerated Super program; Lilienthal, Pike and Smyth opposed Super development. Truman received the GAC reports along with the AEC recommendations. He blustered to Lilienthal, Manley notes in a contemporary diary, "that he was not going to be blitzed into this thing by the military establishment." That left Lilienthal believing Truman might oppose the Super program; "he came back feeling quite happy." But Strauss met with McMahon, then with Louis Johnson; Johnson met with Truman; whereupon Truman reappointed (on November 19) the Special Committee—Acheson, Johnson and Lilienthal—he had appointed earlier to consider expanding weapons production. At the same time, Truman cut off all public and most private debate by restricting discussion of the question to the Special Committee and its staff. "The habit of obsessive secrecy may be as significant here," writes former National Security Adviser McGeorge Bundy, "as any conscious intent to restrict the range of analysis and advice. . . . When the government decided to conduct [the staff] work through in-house, departmentally-based officials, it was in effect turning away from the exploration of unfamiliar suggestions."

Johnson professed his uncomplicated creed in a later public speech. "There is but one nation in the world tonight that would start a war that would engulf the world and bring the United States into war," he said.

"... We want a military establishment sufficient to deter that aggressor and sufficient to kick the hell out of her if she doesn't stay deterred." Acheson went through at least the motions of deliberating, Gordon Arneson recalls. "The Dean was a liberal, but on foreign policy he was very tough. He was also a good lawyer. He wanted to know all sides to the argument. He'd make up his mind quickly. He sought the advice and got the advice of several people not on this committee. ... He talked with Van[nevar] Bush, he talked with Lilienthal, for whom he had the greatest admiration—the two of them worked together on the Acheson-Lilienthal Report, an unprecedented proposal—and he talked to Oppenheimer." Bush, who had put scientists to work on military applications for Franklin Roosevelt during the Second World War, was tough and blunt; he told Kenneth Nichols at this time, when Nichols went to him worrying about the AEC's majority vote against the Super, "Nichols, be patient. You don't have to worry about this. ... The Commission is basically wrong, and it'll fall of its own weight." Nichols thought Bush had talked to someone by then and knew "the score." Bush had talked to Acheson. Oppenheimer, for his part, was unable to clarify his argument sufficiently to convince the Secretary of State of its virtue. "[Acheson] was deeply troubled," Arneson writes. "Finally, he said, 'You know, I listened as carefully as I knew how, but I don't understand what "Oppie" was trying to say. How can you persuade a paranoid adversary to disarm "by example"?' " Niels Bohr had offered the same argument to Roosevelt and Winston Churchill in 1944 when he appealed to them to talk to the Soviets about the atomic bomb—that it would be easier to negotiate a moratorium on a weapon no one had yet made—but Acheson no more took the point than Roosevelt and Churchill had before him.

Arneson thought domestic politics strongly influenced Acheson's recommendations. "His sense of realism prompted him to conclude that even if the Soviet Union refrained from undertaking a thermonuclear program as the result of our refraining—a non-existent prospect—the Administration would run into a Congressional buzz saw and the proposal would be stillborn." But Arneson remembers Acheson parodying George Kennan's opposition to the Super with a contemptuous comedy sketch. "George was in favor of not proceeding. ... [Acheson] said to George, 'If that is your view of the matter, I suggest you put on a monk's robe, put a tin cup in your hand, and go on the street corner and announce the end of the world is nigh.' Dean had made up his mind. ... Looking at the international situation, looking at the Cold War which was very much upon us then, he saw no hope in getting the Russians to agree on the time of day." That the end of the world might be nigh if the Soviet Union beat the US to a thermonuclear was Teller's position and Borden's.

The GAC met again in early December 1949 and augmented and reaffirmed its October conclusions. Several members submitted further statements arguing against a Super. When the Special Committee met for the first time, on December 22, Acheson saw such a gulf between Johnson and Lilienthal that he set no date for another meeting. Omar Bradley at the Special Committee meeting had repeated his argument that a Super would have "psychological value." Lilienthal demurred. "I say: I don't know what that means." To find out, the AEC chairman went to see Bradley privately. The Army Chief of Staff lamented the country's great vulnerability. "All we have right now," Lilienthal paraphrases Bradley, parenthetically filling in the blanks he had left in his diary for the sake of security, "but all, is (our A-bomb stockpile). Without that we are helpless to aid our friends and must, if they are overrun, try to hold our foes off from home base; never again able to normandize (invade Europe)." And then a sign that the US military had begun to understand that atomic weapons had changed war: "I asked: Is (A-bomb stockpile) with or without 'Campbells' really a deterrent—will it be five or ten years hence? He admitted [that] (bombs) [were] of declining value, but still [atomic] war would leave both sides beat up so bad [there was] little value in using them."

But the JCS response to the GAC that went to Louis Johnson on January 13, 1950, said nothing about mutual assured destruction. It said that the Joint Chiefs considered it "necessary to have within the arsenal of the United States a weapon of the greatest capability, in this case the super bomb. Such a weapon would improve our defense in its broadest sense, as a potential offensive weapon, a possible deterrent to war, a potential retaliatory weapon, as well as a defensive weapon against enemy forces." It argued that a superbomb "might be a decisive factor [in war] if properly used" and emphasized dryly that the JCS preferred "that such a possibility be at the will and control of the United States rather than of an enemy." Curtis LeMay had justified the firebombing of Japanese cities on the grounds that Japanese war industry was dispersed to workers' homes; the JCS now borrowed that prevarication to defend urban targeting, asserting that "They do not intend to destroy large cities *per se;* rather, only to attack such targets as are necessary in war in order to impose the national objectives of the United States upon the enemy." Beyond rebuttal, the JCS asserted their authority over that of civilian science advisers to judge the military effect of renouncing development of a potentially decisive new weapon:

The Joint Chiefs of Staff believe that the United States would be in an intolerable position if a possible enemy possessed the bomb and the United States did not. . . . It would be foolhardy altruism for the United States voluntarily to weaken its capability by such a renunciation. Public renunciation by the United States of super bomb development might be interpreted as the first step in unilateral renunciation of the use of all atomic weapons, a course which would inevitably be followed by major international realignments to the disadvantage of the United States. Thus, the peace of the world generally and, specifically, the security of the entire Western Hemisphere would be jeopardized.

Bypassing the Special Committee, the Secretary of Defense sent this memorandum directly to Truman.

And that was that. Truman told Souers on January 19 that the JCS memorandum "made a lot of sense and that he was inclined to think that was what we should do." When the Special Committee came in on January 31 to recommend to him that the nation proceed with the Super, Lilienthal was prepared to argue at length that the policy was not wise. Truman cut short the discussion. "What the hell are we waiting for?" he remembered telling them. "Let's get on with it." The President announced to the world the same day that he was directing "the Atomic Energy Commission to continue its work on all forms of atomic weapons, including the so-called hydrogen or super-bomb."

Certainly Truman learned from the months of debate within the government, but he may have learned only that a decision was urgent politically: that the military and a vocal, organized segment of Congress would fight him if he decided not to build the Super. Authoritative contemporary testimony affirms that his decision to build the hydrogen bomb was never in doubt in the first place, that the painful debates of autumn 1949 that left such a gulf of bitter division among American scientists were little more than a White House public-relations ploy. "The White House," Sidney Souers confirmed in 1954 when the events were still fresh in memory, "felt it was necessary to show the country that the President used an orderly process in arriving at his decisions, not snap judgments, which he has been accused of." Even so, Souers went on, "I am sure [the President's] mind was made up at the very beginning."

Eben Ayers, Truman's assistant press secretary, confirms Souers's impression in his contemporary diary:

February 4 [1950]. . . . The president said there actually was no decision to make on the H-bomb. He said this really was a question that was settled in making up the budget for the atomic energy commission last fall when $300 million was allotted. He said he had discussed that last September with David Lilienthal . . . Acheson . . . and Johnson. He went on to say that we had to do it—make the [H-]bomb—though no one wants to use it. But, he said, we have got to have it if only for bargaining purposes with the Russians.

Truman had consoled Lilienthal in similar terms back in early November, when the two men had discussed who to appoint to succeed Lilienthal as

AEC chairman. "We don't want a military-minded civilian," Lilienthal quotes Truman saying; "he must be someone who sees the necessary military setting, how it fits in, but he must be someone who doesn't regard that as our objective—and we're going to use this for peace and never use it for war—I've always said this, and you'll see. It'll be like poison gas (never used again)." Truman thus began what became a US presidential tradition of maintaining and enlarging a threatening nuclear arsenal he had no intention of using except for political leverage in international negotiations.

But that was not the worst result of the January 1950 decision. "I never forgave Truman," Rabi would identify the greater danger. ". . . He simply did not understand what it was about. . . . For him to have alerted the world that we were going to make a hydrogen bomb at a time when we didn't even know how to make one was one of the worst things he could have done."

The frightened men, Edward Teller first among them, who had advised the Truman administration that they knew how to build a hydrogen bomb that would save the country from disaster had started the clock on a new, ultimate arms race. Now they would have to deliver.

PART THREE

Scorpions in a Bottle

We may anticipate a state of affairs in which two Great Powers will each be in a position to put an end to the civilization and life of the other, though not without risking its own. We may be likened to two scorpions in a bottle, each capable of killing the other, but only at the risk of his own life.

ROBERT OPPENHEIMER

21

Fresh Horrors

"On January 27, [1950]," writes Gordon Arneson, "Sir Derek Hoyar-Millar, the Counselor of the British Embassy who served as liaison with us on atomic energy matters, asked urgently to see the Undersecretary [of State], Robert Murphy. This was clearly no run-of-the-mill matter, for if it had been he would have come to see me, entering my office with his customary greeting: 'Any fresh horrors today?' This time he had his own to tell. Ashen of face, his usual casual aplomb quite collapsed, he told us Klaus Fuchs had that day admitted to espionage on behalf of the Soviet Union. . . ."

The cryptanalysis information that Robert Lamphere had passed to British intelligence in September 1949 had been acquired through burglary and was therefore legally tainted. Klaus Fuchs, MI5 had concluded, would have to be coaxed into confessing, which was why it had taken until late January to expose the wary physicist. The delicate assignment went to William Skardon, an MI5 officer who specialized in handling traitors, "sort of a British *Columbo* character," Lamphere recalls him, "complete with disheveled appearance and an intellect that was sometimes hidden until the moment came to use it to point to incongruities in a suspect's story."

Henry Arnold, the Harwell security officer to whom Fuchs had spoken about his father's move into the Soviet zone of Germany, introduced Skardon to Fuchs on December 21. Skardon brought up Fuchs's father and listened to the physicist's explanation of his family history for more than an hour before confronting him directly. "Were you not in touch with a Soviet official or a Soviet representative while you were in New York?" the MI5 officer asked. "And did you not pass on information to that person about your work?" Fuchs was startled. "I don't think so," he answered ambiguously. Skardon told him there was "precise information which shows that you have been guilty of espionage on behalf of the Soviet Union." Fuchs again demurred: "I don't think so." Skardon pointed out the ambiguity of

Fuchs's reply and Fuchs finally mustered a denial. "I don't understand. Perhaps you will tell me what the evidence is. I have not done any such thing." The two men sparred for the rest of the day and through two more meetings in early January before Fuchs finally asked to see Skardon and confessed on January 24, 1950.

At least Lewis Strauss may have been aware of the Fuchs investigation during the late-1949 Super debate; the FBI had notified the AEC in early November. The General Advisory Committee learned of it after Sir Derek Hoyar-Millar offered Arneson his "fresh horror" on January 27. The Special Committee heard that Fuchs was about to be arrested during its final deliberations before reporting to Truman on January 31. The President himself did not learn of Fuchs's espionage until February 1, when J. Edgar Hoover telephoned the information to Sidney Souers. (Truman's response, writes David Lilienthal, was "tie on your hat!") Strauss, who had just resigned from the AEC effective April 15 (Lilienthal was leaving at the same time), hurried to flatter the President that "the recent word from the FBI . . . only fortifies the wisdom of your decision. The individual in question had worked on the super-bomb at Los Alamos." To Hoover, Strauss added (in Hoover's paraphrase): "It will make a good many men who are in the same profession as Fuchs very careful of what they say publicly." Lilienthal on February 2, the day he heard, felt no such *schadenfreude*:

> The roof fell in today, you might say. After the gay and happy time, the going-away party given me by the employees last night, this was quite a contrast. The news, I learned at about seven when I left the office, will be out when the man is arraigned in London tomorrow at about twelve our time. . . .
>
> It took place during the war project; but he had been here since, which drags us in. It is a world catastrophe, and a sad day for the human race.

That night, Lilienthal added the following day, "sleep wasn't too easy: the vision of 'the top blowing off things,' antagonism increased between US and Britain, witch-hunts, anti-scientist orgies. . . ." The Anglo-American *modus vivendi* would not be renewed.

When Hans Bethe heard about Fuchs, he phoned Ralph Carlisle Smith, the security officer at Los Alamos, to ask after the 1946 Fuchs–von Neumann thermonuclear patent:

" 'Is it all there?' Bethe asked.

" 'All,' said Smith.

" 'Oh,' replied Bethe. . . ." The patent application included a description of Teller's Super. Teller called Smith to ask the same question. He got the same answer but had a more venomous response. "I don't believe it," Smith reports him saying. "If it's all there, it's because *you* put it there."

Fuchs's arraignment on February 3 made headlines throughout the world. Six days later, in Wheeling, West Virginia, Wisconsin Senator Joseph McCarthy began the witch-hunt Lilienthal had feared with a speech in which he claimed to have a list of 205 Communists who worked in the State Department. McCarthy was following up and capitalizing on the perjury conviction of former State Department officer Alger Hiss on January 21 (for having denied passing secret documents to Whittaker Chambers) as well as the Fuchs case. McCarthyism drew its poisons in part from a pervasive American conviction that the US had been made vulnerable to Soviet atomic bombs because spies in its midst had conspired to give the "secret" away.

With Fuchs's confession, writes Lamphere, "all hell broke loose at Bureau headquarters." Fuchs had admitted he passed information to an American cut-out whom he knew only as "Raymond." Hoover immediately ordered an all-out, first-priority nationwide search for Fuchs's American counterpart, designated "Unknown subject"—"Unsub" in Hooverspeak.

Julius Rosenberg evidently knew that Fuchs was under suspicion even before the Harwell physicist confessed. David Greenglass had gone to see Rosenberg during the 1949 Christmas holidays, to talk about their failing machine-shop business, when Rosenberg shocked him with the news that he would "have to begin thinking about going away to Paris." Greenglass, Rosenberg warned, was "hot." Greenglass's first thought was, "I'll never be able to read Li'l Abner again."* He realized abruptly, he recalled many years later, how much he loved America. The two men withdrew to a nearby diner for coffee and Rosenberg outlined an escape route—New York to Paris, where Greenglass would contact someone. When Greenglass asked why he had to go into exile, Rosenberg told him that "something is happening which will cause you to leave the United States." Greenglass told Rosenberg he thought a New York departure was a bad idea that would expose him to the FBI. "Julius said that more important people than I had left by this route," Greenglass recalled. "When I asked who they were, Julius said, 'Joel Barr, for one.'" At a party among friends that New Year's Eve, Greenglass felt uneasy. His brother-in-law's warning meant he might never ring in the New Year with his friends again.

Greenglass had another scare the last week in January: an FBI man called him and asked to see him. "He came to my house," the young machinist testified; "he sat down at my table; I offered him a cup of coffee and we spoke. He did not say to me that he suspected me of espionage or anything else. He just spoke to me about whether I had known anybody at Los Alamos. . . . I didn't tell him [about my espionage activity], but I was pretty well on the verge to tell him." The visit must have unnerved Greenglass so soon after Rosenberg's warning, even though it concerned a lesser crime than

---

* The popular Al Capp comic strip.

espionage. Some of the Special Engineering Detachment (SED) men at Los Alamos, Greenglass included, had taken home natural-uranium dummy initiator spheres as souvenirs; the FBI was investigating the thefts.

Fuchs's arrest unnerved Harry Gold. He was scheduled to make a rendezvous on the first Sunday in February, February 5. He forgot which corner of the Jackson Heights intersection had been designated for the meeting and resorted to moving from corner to corner so that he would not be missed. His contact was supposed to be smoking a cigar. "Before leaving," he said later, "I noticed a man walk past me with a cigar in his mouth. As he walked past me he turned around and looked at me. He then kept on walking. . . . I placed no significance on this at the time." Later, from FBI photographs, Gold identified the man with the cigar as Julius Rosenberg. The walk-by was as close as anyone approached Gold that Sunday; it was the last time he tried to contact the Soviets, or they him.

Monday, still "completely panicked," Gold sought out his old friend Tom Black, the man who had recruited him for espionage fifteen years before. "It took me a full half-hour of walking through the dark side streets of downtown Philadelphia before I got up enough courage to tell him," Gold writes. Black, says Gold, "was dumbfounded and horror-stricken" at the news. "As nearly as I can recall," Black reported, "[Harry's] exact words were, 'The FBI is looking for Fuchs's American contact and I am that man.'" Gold remembered with gratitude that Black "did not express any concern at being himself implicated or involved because of his known friendship for me; his principal concern seemed to be for my welfare." Black: "He said that if he should be caught, he was going to take an overdose of sleeping pills. . . . I tried to persuade him from committing suicide." Gold tells the rest of the tale in FBI paraphrase:

> Black counseled Gold that if Gold [was] picked up and questioned he should deny everything because it would be one person's word against another's. Gold told Black that his principal concern was his family and were he arrested, Gold requested Black to visit Gold's family and cheer them up. . . . Gold and Black agreed that all future meetings between the two would be at the Franklin Institute in Phila[delphia] because no suspicion could be attached to meeting there.

Rosenberg came around to see Greenglass at about the same time, Greenglass recalled:

> We walked along the park [for] about forty-five minutes. . . . [He] said to me I would have to get out of the country with my family. I gave him the impression I was willing to go; only thing is, I didn't have the money to pay off my debts. He wasn't interested in that, wanted me to go and forget

about my debts; I said I can't do that; these people are not wealthy people, whatever I take from them is blood money. . . . At that time he wanted me to go to Czechoslovakia, a good job was waiting for me. . . . The reason he wanted me to leave the US is because . . . Gold was Fuchs's contact man. . . . He did not mention Gold by name—he said the same man, [he said] you remember that man out in Albuquerque. . . . He said that . . . this man knew me and that when Fuchs was taken . . . he would tell about Gold and he would lead them to me. . . . He wanted me to go with my whole family: pouf, disappear! . . . I figured I might [really disappear], so I better not go.

Greenglass suggested that Rosenberg contact "Dave"—Gold—and warn him to lie low; Rosenberg told his brother-in-law that Ethel had made the same suggestion.

Greenglass chose not to tell his wife about this conversation. It seemed unreal to him and Ruth was six months pregnant. Then accident foreclosed escape. An open gas heater warmed the Greenglasses' small Lower East Side apartment. Early on the Tuesday morning of February 14, Ruth approached too close to the grating. Her nightgown caught fire and blazed up around her. David rushed to help, batting out the flames with his hands. Ruth was admitted to Gouverneur Hospital suffering extensive first-, second- and third-degree burns. She was critical for two days; the burns required grafting and she spent almost a month in the hospital. She had miscarried in Albuquerque in 1945; this time, despite her injuries, she did not lose her baby. David sustained second-degree burns to his right hand. While Ruth was hospitalized, he worked the night shift and took care of their young son by day. The Greenglasses were in no shape to go anywhere.

A week after Ruth Greenglass's accident, a newly deciphered 1944 Soviet cable gave Robert Lamphere "reason to believe that someone in a lower-level position at Los Alamos, who had had furlough plans in late 1944 and early 1945, was a KGB agent." Lamphere requested the FBI Albuquerque field office to investigate.

Even before Harry Truman decided to announce work on a hydrogen bomb, Edward Teller, Stanislaw Ulam and George Gamow had organized an informal committee at Los Alamos to move the project along. Gamow, a tall, blond Russian, was a brilliant original and a wild man. To commemorate the informal committee he had drawn a witty cartoon of its three leaders miming their Super ideas while a winged, pipe-smoking Stalin flew into frame clutching a bomb labeled "Made in USSR" in his talons and Robert Oppenheimer, robed as a saint with a radiant halo, hovered on a cloud dangling an olive branch. Gamow depicted Ulam spitting wine into a spittoon, Teller wearing an Indian necklace hung with a womb symbol like the Greek letter omega ("vombb," Gamow pronounced it in his heavy Russian accent), Gamow himself holding up a cat and squeezing its tail. Ulam's spitting may have symbolized neutron transport into the main mass of deuterium; Gamow's tail squeeze referred to an invention of his called "the cat's tail"— possibly cylindrical implosion of the fission trigger; Teller's womb was his classical Super. "Both Gamow and I showed a lot of independence of thought in our meetings," Ulam reports, "and Teller did not like this very much." One of Ulam's independent thoughts, which he expressed at a committee meeting on January 21, was that the Super would probably require much more tritium than Teller had recently estimated and that ignition prospects looked "miserable." "Not too surprisingly," Ulam concludes, "the original 'super' directing committee soon ceased to exist." Ulam recalls that Gamow "was quite put out by this. I did not care, but I wrote him, prophetically it seems, that great troubles would follow because of Edward's obstinacy, his single-mindedness and his overwhelming ambition."

Gamow and Ulam may have blamed Teller unfairly for the demise of their committee; in the wake of the President's decision, Los Alamos necessarily moved to organize a formal Super committee in February 1950. "The idea . . . was to get everything that Teller could give, which was a lot," Technical Associate Director Darol Froman recalled, "but not to let him run the thing. Because he would sure push it too hard and too fast and get all kinds of people outside the Laboratory up on their ears. . . . People like the AEC." Teller got to be chairman of the twenty-five-man so-called Family Committee, with direct responsibility for the lab's thermonuclear programs, but he had to report to Froman, a blunt, no-nonsense administrator who had managed the *Sandstone* tests.

Teller immediately set to work recruiting, with mixed results. He thanked Oppenheimer in a February letter "for the help you are giving and are going to give us in this connection"—with recruiting, that is; Oppenheimer was not planning to work on the Super. Bethe was no longer interested. "I still believe that it is morally wrong and unwise for our national security to develop this weapon," the Cornell physicist wrote Norris Bradbury on February 14. "For this reason, if and when I come to Los Alamos in the future I will completely refrain from any discussions related to the super-bomb. . . . In case of war I would obviously reconsider my position." Emilio Segrè at Berkeley recalled "several conversations with Teller, whom I had known well since my time at the Physics Institute in Rome. I soon realized, however, that he was dominated by irresistible passions much stronger than even his powerful rational intellect." Segrè turned Teller down.

Most of the recruits for the Family Committee came from established Los Alamos staff, including Carson Mark and Weapons Division Leader Marshall Holloway, but Teller was able to enlist Princeton theorist John Archibald Wheeler; Indiana University's Emil Konopinski, who had first suggested using tritium back in 1942, Charles Critchfield, who had worked on solar thermonuclear reactions with Bethe before the war; and talented young Stanford theoretical physicist Marshall Rosenbluth. Rosenbluth decided to return to Los Alamos, where he had been an SED during the war, because of Fuchs. "I thought that what Fuchs knew would make a qualitative difference to the Russians," Rosenbluth recalls. "He would have told them that we were working on it—and he did. So I assumed that they would be working on it and I thought Stalin was just a terrible son of a bitch. If he ever got the bomb before we did it could be very dangerous. That was basically my motivation."

Despite these valuable additions to the Los Alamos theoretical staff, Teller complained bitterly to William Borden and anyone else who would listen that "the Laboratory had had some luck in getting the youngest and brightest of the physicists but the senior scientists who feel that such weapons are morally reprehensible have great influence over the younger men." At the beginning of March, Teller published the equivalent of a want ad in the *Bulletin of the Atomic Scientists,* a brief appeal titled "Back to the Laboratories." "Our scientific community has been out on a honeymoon with mesons,"* he scolded. "The holiday is over. Hydrogen bombs will not produce themselves."

"If things were going happily," says Carson Mark, "[Teller] was an exciting person to have in the group. . . . [But] there were just so many people and there were so many problems and Edward, who could throw out intriguing new ideas, was always doing so and each new one was . . . the important one today. It was very difficult to continue to make progress . . . if he was deciding, with as great a frequency as he did, that one should drop that and do this." Manley recalled "a tendency really to put up with Edward, but also to ignore him, in the Laboratory. I can't remember any real fights about it. . . . He didn't get along with his ideas of what the Lab should do, and yet he was tolerated. I don't believe anybody ever told him to 'get the hell out of here' for his views." Mark agrees: "It was a matter of trying to balance Edward."

Teller outlined further research on the Super in a seventy-two-page paper, *On the development of thermonuclear bombs,* published at Los Alamos in mid-February 1950. The report was evidently an updated version of Teller's similar report of September 1947. The 1950 version offered the Super design presented to the April 1946 Super Conference as a basis for thermonuclear development, but emphasized the Alarm Clock as an alternative. Only weeks after the President had endorsed publicly his scheme to answer the Soviet atomic bomb with a thermonuclear, Teller wrote pessimistically: ". . . It may be stated that the Super is probably feasible. Its complex con-

---

* The meson, a nuclear particle postulated theoretically before the Second World War, was discovered shortly after.

struction gives us little hope that it can be actually made to work in the next 3 or 4 years. It requires, furthermore, considerable amounts of tritium."

Teller evidently warmed to the Alarm Clock as an alternative for the same reason the layered design had appealed to Andrei Sakharov: because it was clearly feasible on basic physical principles. It could also be made with lithium deuteride and little or no tritium. (On the other hand, Bethe remembers that "nobody could see a way to compress the Alarm Clock sufficiently to get the desired high densities.") In the Soviet Union at that time, writes Victor Adamsky, "the idea [of a layer-cake thermonuclear] was mature enough to find its impending implementation. . . ." Sakharov and Igor Tamm would move to Sarov in March 1950 to pursue it, starting a new second theory department alongside Zeldovich's. A kiloton-range layer cake might form a system no larger than a Fat Man. With plutonium production modest and U235 just becoming available, the Soviet program was willing to settle for a kiloton-range thermonuclear that would stretch out supplies of fissile metals; 80 percent or more of an Alarm Clock's yield would derive from thermonuclear neutrons fissioning ordinary tamper U238.

But Los Alamos, which already knew how to make fission weapons with yields in the hundreds of kilotons, was focused on a megaton-range thermonuclear, and adding layers to a spherical Alarm Clock to push the yield into the megaton range resulted in a mechanism that would be prohibitively large and heavy. "Delivery of such an object by aircraft," Teller wrote in his February report, "is likely to remain impossible for quite some time to come." Rather than scale down his megaton ambitions, Teller proposed alternative methods of transportation. "We shall see, however," he went on, "that delivery by boat or submarine is capable of producing disastrous effects." To produce disastrous effects on so vast and land-locked a country as the Soviet Union with a weapon that had to be delivered by boat or submarine, Teller was willing to consider devising what he called "a one-billion-ton Alarm Clock," meaning a device with a one-thousand-megaton yield. Punching a hole in the atmosphere would come into play with such a device, however; "it seems to me likely that it will be difficult to destroy an area greater than approximately one thousand square miles by shock." Flash burns would be "very serious" at one hundred miles and "an area of 30 thousand square miles* would be affected"—but only if the weapon that no aircraft could deliver were somehow detonated more than a mile above the ground. Otherwise, the horizon would limit the damage. So a big shipborne Alarm Clock might destroy Leningrad, but unless it could be sneaked several hundred miles up the Volga River, it would not spread its blast and fire to the nerve center of the Soviet state. It would be a notably dirty bomb, however, and might reach Moscow with lethal radiation, "a strip 40 miles

---

* An area 173 miles on a side—about the size of Pennsylvania.—RR

broad and 400 miles long." For comparison, Teller considered "a bomb of this type dropped [sic] near Washington, D.C. Let us assume that the winds are blowing north along the Alleghenies, a condition quite frequently encountered. Then Washington, Philadelphia, New York and Boston could all be close to the path of the radioactive cloud and even the farthest point, Boston, would be within reach of the danger." Killing the Soviet Union with radioactive fallout from Alarm Clocks delivered by (remote-controlled?) Navy ships was not likely to appeal to the US Air Force when it could kill the nation with blast, fire and radioactivity from atomic bombs delivered by strategic bomber.

Teller looked forward to the *Greenhouse* tests scheduled for spring 1951 to provide information valuable to developing a thermonuclear, particularly the effect of the high-energy, fourteen million-electron-volt (MeV) neutrons produced in $D + T$ thermonuclear reactions on natural uranium, U235, plutonium and U233. But he estimated that it would be one and a half to two years before they could determine whether an Alarm Clock was feasible, and at least that long before the MANIAC under construction at Los Alamos became available. In 1947, Teller had proposed delaying for two years choosing between developing his classical Super and an Alarm Clock; the February, 1950, report repeated the same proposal:

> I think that the decision whether considerable effort is to be put on the development of the Alarm Clock or the Super should be postponed for approximately 2 years; namely, until such time as these experiments, tests and calculations have been carried out.

At a briefing for AEC, MLC and Laboratory managers a week later, Norris Bradbury noted other obstacles that the Super program imposed. (Truman had ordered that the "scale and rate of effort" of thermonuclear development be determined jointly by the AEC and the Defense Department; the Los Alamos meeting was an informal preliminary to that determination.) Even the booster looked doubtful at that point, Bradbury said, primarily because the modifications to the implosion system necessary to insert the mixture of deuterium and tritium gas into the bomb's core caused a loss of fission yield that the increased yield from fusion barely overcame. (The gases tended to diffuse into the U235 shell that enclosed them, a problem Los Alamos eventually overcame by sequestering them within a copper shell liner.) The lab would go ahead and use up the few grams of available tritium for a booster test at *Greenhouse* only because the booster might be "the only model which will be ready for thermonuclear experiments" by then.

More serious from Bradbury's point of view was the cost in time and fissionable material that the Super program imposed. They would have to withhold four to five hundred kilograms of fissionable material from the stockpile for experiments at Los Alamos and would use several hundred kilograms for experiments at Eniwetok. They would forgo creation of much more fissionable material when they began making tritium at Hanford. They were "abandoning research and leadership in important areas [of weapons development] except for that which is immediately applicable to a thermonuclear weapon." Overall, Bradbury thought, it would take "fantastic good luck for a period of time on the order of, say—three years—before a device can be produced which might ignite [deleted; probably "a cubic meter of"] D_2." Teller, after ritually warning of "grave danger that we have lost or are losing the atomic armaments race,"* described several experimental models of thermonuclear devices and "emphasized his agreement with Dr. Bradbury's observations as to the unpredictability of time and development in this field." Informing the world that you were planning to build a superbomb was one thing; evidently, building it was quite another.

While Los Alamos scraped its way to the bare realities, alarmist prognostications further terrorized Washington. Brigadier General Herbert B. Loper, a member of the MLC, disturbed by the Fuchs revelations, sat himself down and proceeded to estimate what he called "a measure of the outside bracket of Russian capabilities" on the assumption that the Soviets had begun "a nuclear energy development project" by 1943 at the latest that had benefited from espionage. To get to his "outside bracket," Loper assumed (despite good CIA information to the contrary from returning German scientists and engineers who had been Soviet prisoners of war) that the Soviets had begun exploring and mining in 1943 and had begun building isotope-separation and reactor facilities in 1945. It might follow, he thought, that the Soviets had established "the theoretical basis for developing the thermonuclear weapon" by 1945 and had tested an atomic *and* a thermonuclear weapon before the Joe 1 test that the US detected in September 1949. If so, the brigadier general concluded sensationally, then "the USSR stockpile and current production capacity [might be] equal or actually superior to our own, both as to yields and numbers," and "the thermonuclear weapon may be in actual production."

Loper's speculations, to which Kenneth Nichols lent his credibility in concurrence and which Loper sent along to MLC chairman Robert LeBaron, would not have stood up to even superficial scrutiny at Los Alamos, but no one at the weapons laboratory seems to have vetted the document. ("If [the Soviets] had been able to make any [H-bomb] advances on the basis of information given them by Dr. Fuchs," Robert Oppenheimer quipped at a

---

* The Soviet atomic bomb stockpile at the *end* of 1950, ten months after this meeting, consisted of five RDS-1 plutonium implosion bombs; the US stockpile at the end of 1950 totaled 298 engineered, levitated composite weapons. Neither country had thermonuclear weapons at this time.

meeting of the State-Defense Policy Group on February 27, "they were marvelous indeed.") LeBaron added further to the Loper memorandum's authority when he passed it to the Secretary of Defense by pointing out that even though it contained estimates that were much higher than the latest CIA estimate (which projected a Soviet arsenal of a hundred twenty-kiloton atomic bombs by "sometime during 1953"), "there are areas in Russia that are not covered by our agents. . . ." Louis Johnson alerted Truman to Loper's ideas and asked the Joint Chiefs for their response. They recommended, in Johnson's words, "an all-out program of hydrogen bomb development if we are not to be placed in a potentially disastrous position. . . ." ("Los Alamos has always been in a crash program," Norris Bradbury would grumble. "The word crash means merely everybody works just as hard as he can. This we have been doing since 1943.") Truman referred this latest imbroglio to the NSC Special Committee he had come to rely upon for military atomic energy decisions, with Henry Smyth now representing the AEC. The committee warned the JCS that they would have to accept reduced atomic-bomb production if they wanted a crash program for the Super and they agreed. On March 9, the Special Committee advised the President that Los Alamos was already going "all-out," but recommended he approve the principle that "the thermonuclear weapons program is regarded as a matter of the highest urgency" and a production goal of ten thermonuclear weapons per year using a total of one kilogram of tritium even though "the total estimated cost of these weapons, spread over a period of two or three years, will be on the order of 30 or 40 fission bombs." Truman did so the next day. Around this time James Conant wrote Bernard Baruch, "When I am in Washington, it seems as though I were in a lunatic asylum, but I am never sure who is the attendant and who the inmate. Nor am I even sure whether I am a visitor or a potential patient. However, I am trying to keep my sanity. . . ."

Klaus Fuchs came on for trial at the Old Bailey at 10:30 A.M. on March 1, 1950. He was not allowed a jury. His judge was Lord Goddard, the Lord Chief Justice of England, a rugged conservative in scarlet robes who believed in retribution. His prosecutor, Sir Hartley Shawcross, was the Attorney General and had been the chief British prosecutor at the Nuremberg Trials. The Duchess of Kent and the Mayor of London came to watch, as did two men from the American Embassy and representatives of some eighty newspapers and news services, including Tass. J. Edgar Hoover had hoped to send an FBI observer, but the British kept the Americans at arm's length to limit as much as possible public awareness of the appalling security breach; Fuchs would be charged with passing secrets in Birmingham, New York, Boston and Berkshire but not in Santa Fe. Until minutes before the trial began, the stoic physicist assumed the maximum penalty for his crimes was death. His barrister, who proposed to plead him guilty, corrected him sharply: since the Soviet Union had not been an enemy at the time of Fuchs's espionage, treason was excluded and the maximum penalty was fourteen years.

Shawcross went through Fuchs's career and fulminated at length on the evils of Communism. Fuchs's counsel, Derek Curtis-Bennett, protested that Fuchs was a human being with a divided mind and relied on Fuchs's confession to claim that the physicist "would only tell [the Soviets] things he found out himself." Lord Goddard interrupted Curtis-Bennett to declare that "a man in this state of mind is one of the most dangerous that this country could have within its shores," leaving no doubt where he stood. Fuchs, in a brown suit, with pens and pencils handy in the handkerchief pocket of his jacket, got a final chance to say his piece, and confessed to having "committed certain crimes for which I am charged, and I expect sentence. I have also committed some other crimes which are not crimes in the eyes of the law—crimes against my friends—and when I asked my counsel to put certain facts before you, I did not do it because I wanted to lighten my sentence. I did it in order to atone for those other crimes." The trial, he concluded, had been "fair." (Before his trial, Fuchs had written to Genia Peierls, who had accused him of betrayal, that he had failed to consider the harm he would cause his friends. "I didn't," he agonized, "and that's the greatest horror I had to face when I looked at myself. You don't know what I had done to my own mind. I thought I knew what I was doing, and there was this simple thing, obvious to the simplest decent creature, and I didn't think of it." To another friend he observed, "Some people grow up at fifteen, some at thirty-eight. It is more painful at thirty-eight.")

Expressing fear that Fuchs might "at any other minute allow some curious working in your mind to lead you further to betray secrets of the greatest possible value and importance to this land," the Lord Chief Justice sentenced Klaus Fuchs to the maximum term of fourteen years.

———

Even before the first presidential announcement of the thermonuclear program, Stanislaw Ulam had lost patience with Teller's assaults on Los Alamos's dedication to Super work. "Teller . . . kept insisting on certain special approaches of his own," Ulam writes. "I must admit that I became irritated by his insistence; in collaboration with my friend [Cornelius] Everett one day [in December 1949] I decided to try a schematic pilot calculation which could give an order of magnitude, at least, a 'ballpark' estimate of the promise of [Teller's Super] scheme." Ulam and Everett had prepared calculations for the MANIAC earlier in 1949, but that machine was far from completion, as was John von Neumann's Princeton original. Now they undertook to calculate a simplified version of the problem by hand. At the same time, Los Alamos began preparing a simplified machine version that von Neumann could farm out to the ENIAC.

The problem—whether a D + T burn would ignite a large mass of deuterium—required tracking the progress of thermonuclear reactions using Monte Carlo methods. "Each morning," Ulam recalls, "I would attempt to supply several guesses as to the value of certain coefficients referring to purely geometrical properties of the moving assembly involving the fate of the neutrons and other particles going through it and causing, in turn, more reactions." Ulam supplied his guesses in the form of random numbers which he generated by tossing one die of a pair of dice. Teller certainly understood the mathematical authority of Ulam's random-number generator, but he must have quailed to see his grand design held hostage to a crap shoot.

The calculations were lengthy and tedious, Ulam writes:

> We started work each day for four to six hours with slide rule, pencil and paper, making frequent quantitative guesses. . . . These estimates were interspersed with stepwise calculations of the behavior of the actual motions [of particles]. . . . The real times for the individual computational steps were short . . . and the spatial subdivisions of the material assembly very small. . . . The number of individual computational steps was therefore very large. We filled page upon page with calculations, much of it done by Everett. In the process he almost wore out his own slide rule. . . . I do not know how many man hours were spent on this problem.

At first, writes Françoise Ulam, her husband "worked just with Everett, then with an added bevy of young women who had been hastily recruited to grind manually on electric calculators." Françoise helped out with the effort. It did not favor Teller's design, she notes. "I was well placed to watch how personally Teller took the fact that Stan and Everett were the first to blow the whistle with their crude calculations. Every day Stan would come into the office, look at our computations, and come back with new 'guestimates,' while Teller objected loudly and cajoled every one around into disbelieving the results. What should have been the common examination of difficult problems became an unpleasant confrontation."

Ulam and Everett issued an interim fifty-page report on March 9, 1950, concluding, Carson Mark writes, that the amount of tritium chosen for the calculation "was not nearly enough; so the first calculation was discontinued . . . and a second calculation, with a larger amount of tritium, was started immediately."(Ulam put the conclusion of the first calculation more bluntly; it revealed, he reported, "that the model considered is a fizzle.") As the second calculation proceeded, says Ulam, "it naturally attracted quite a lot of attention among the physicists Teller was trying to interest in the 'super' project. . . ." The results continued to point "to the mediocre progress of the reaction." Teller for his part blamed not his Super design but his opponents for discouraging recruiting. "I feel that the attitude of the members of the GAC has been a serious difficulty in our recruiting efforts, and I continue to wish for a clear-cut change of heart, publicized at least in their closest circle," he wrote William Borden, who was becoming a confidant. "A man like Conant or Oppenheimer can do a great deal in an informal manner which will hurt or further our efforts."

"Teller was not easily reconciled to our results," Ulam observes. "I learned that the bad news drove him once to tears of frustration, and he suffered great disappointment. I never saw him personally in that condition, but he certainly appeared glum in those days, and so were other enthusiasts of the H-bomb project." Ulam visited von Neumann in Princeton that spring and they discussed their ongoing calculation with Fermi and Oppenheimer; Ulam thought Oppenheimer "seemed rather glad to learn of the difficulties." By then Teller suspected that Ulam was deliberately biasing the calculation against the Super; the Polish mathematician told von Neumann on one occasion that Teller "was pale with fury yesterday literally—but I think is calmed down today." Teller wrote von Neumann of his discouragement early in May; von Neumann, preparing to run the ENIAC calculation, responded on May 18 with a letter meant to cheer him up:

> I am sorry to see from your letter that the strain which your work puts on you is exceedingly great. I want to assure you that I certainly don't want to make it any greater, and I regret in particular that I am unable to find a modus procedendi which will reassure you completely regarding the course which our calculations are taking now. . . . This calculation is not and never was intended to be the only one that we shall make. . . . Politically, I would chiefly like to have a tolerable chance of producing a positive result, although I don't see why a negative result should be too much feared, if it produces enough subsidiary information from which to derive better guesses about the plausible limits of a successful arrangement.

But the same day von Neumann heard from Ulam that "the thing gives me the impression of being miles away from going." The Super design that Ulam and Everett were calculating was not only miles away from going. Their calculations also indicated that the one hundred grams of tritium that Teller had told Washington each Super would require—the basis for the Special Committee's March 9 recommendation of ten thermonuclear weapons per year using a total of one kilogram of tritium—would be (the GAC reported a few months later) "quite inadequate" and pointed to "a lower limit for this model in the range of 3 to 5 kilograms." If the AEC succeeded in producing one kilogram of tritium per year, one Super would need at least three to five years' production and would consume neutrons sufficient to make plutonium for about one hundred atomic bombs. No wonder Teller was pale with fury.

Searching for the Unknown Subject—Unsub, "Raymond," Klaus Fuchs's American cut-out—FBI agents interviewed Kristel and Robert Heineman in Cambridge, Massachusetts, in mid-February 1950. Kristel did not remember the name of the man who had visited her home in 1944 and 1945, but she supplied a general description and two significant clues: that he had identified himself as a chemist and had mentioned having a wife and twin children about the age of her own. She or her husband supplied several other details of great significance: that the man had spoken of a partner or colleague whose partners had cheated him in business; that the colleague's firm was known as the Chemurgy Design Corporation; that the colleague had been working on developing an aerosol container and a process for making the pesticide DDT; that Unsub was interested in starting a laboratory of his own. Fuchs had supplied another important point of identification in his confessions: that Raymond was interested in and had done some work on thermal diffusion.

Robert Lamphere and his partner Ernest Van Loon combed the FBI's files for chemists and engineers and came up with several hundred names. They showed photographs of these suspects to both Kristel and Robert Heineman and had them shown to Fuchs in London on March 13. Fuchs paused at a photograph of a heavyset Brooklyn-born civil engineer and said he "might be the man." The Heinemans did not corroborate Fuchs's identification, but Lamphere and Van Loon thoroughly investigated the hapless engineer anyway.

At some point after the Heineman interviews, Van Loon identified the principal of the Chemurgy Design Corporation: Abe Brothman, whom Elizabeth Bentley had named as a source of espionage information in her 1945 confession to the FBI and 1947 grand jury testimony. "Brothman and Gold had been among the first people we'd looked at when I first uncovered Fuchs' probable espionage," Lamphere reveals, "back in September 1949." Fuchs had not identified their photos, but Hoover now was raging for Unsub's capture. The agents went fishing. Hoover evidently authorized a bag job on Abe Brothman's offices sometime in February or early March. The burglary netted what looked like a prize: a "typewritten document," as an FBI summary describes it, "the title of which was obliterated, but the contents of which referred to the industrial application of a process of thermal diffusion. This document was considered of extreme significance. . . . [It] did not bear the name of its author. . . ." Whoever wrote the thermal diffusion document was very likely Fuchs's cut-out.

Gradually, Lamphere and Van Loon's suspicions came to focus on Gold. His 1947 grand jury story of innocent consultation with known Soviet agents was patently phony. He fit the general physical descriptions the Heinemans and Fuchs had supplied. "We instructed the Philadelphia office to open a very active investigation into Harry Gold," Lamphere writes, "and as the reports from Philadelphia started coming in, our interest mounted." Lamphere and Van Loon decided that Gold's file photograph might have misled Fuchs. They asked the Philadelphia office to take still and motion-picture photographs of Gold surreptitiously during surveillance.

Early in May, when Fuchs's appeals process had been completed, Lamphere learned that he would be sent to London to interview the physicist, something Hoover had been trying to arrange for months. The tall Idahoan collected surveillance photographs and motion-picture footage of Gold from Philadelphia agents on his way up from Washington to New York, whence he departed on May 15 with Assistant Director Hugh Clegg, one of Hoover's cronies. That same day, FBI agents began interviewing Abe Brothman and Miriam Moskowitz in New York and Gold in Philadelphia. Moskowitz mentioned Gold's two fantasy children and Gimbel's-model wife. Gold freely discussed his interest in thermal diffusion for the industrial recovery of flue gases and described a paper he had written on the subject; his description matched the paper the FBI had in hand. He also confirmed his knowledge of Brothman's troubles with his business partners, Brothman's work on DDT and an aerosol container and his own interest in starting a laboratory, but he denied ever having traveled west of the Mississippi. The agents showed him photographs, relevant and irrelevant, including a photograph of Fuchs. "When Gold looked at Fuchs' picture," they reported, "he stated, 'That is a very unusual picture—that is the English spy.'" He had never met Fuchs, he told them; he merely recognized the man from the newspapers. After three and a half hours of questioning, Gold begged off to return to Philadelphia General to continue work.

Gold understood immediately that he was under serious suspicion. From May 15 onward, he wrote later, "I was simply fighting desperately for time." He wanted time "to figure out how I was going to tell my father and brother. . . . I wanted to try to warn Tom Black to run. . . . And there was my frenzied effort to get the research work at the Heart Station in good enough shape so that someone else could carry on." The Philadelphia agents interviewed him again on May 19 for six hours. This time they asked him if he had ever been to Boston or Buffalo or Santa Fe. They collected handwriting specimens. They repeatedly showed him photographs of Fuchs and the Heinemans. They confronted him with Moskowitz's story of his wife and children, which he "emphatically denied." They asked for permission to search his house. May 19 was a Friday; Gold put them off until Monday morning, when his father and brother would be away at work. They agreed to the delay, they explained in their report, because they felt "that their showing consideration for the father and brother would help break Gold, inasmuch as it was apparent that he was very much devoted to both of them." Sunday, Gold sat through three and a half more hours of questioning and submitted to still and motion pictures, which went off immediately to Lamphere in London.

Saturday in London, Lamphere and Clegg had arrived at Wormwood Scrubs, the old prison where Fuchs was housed, which Lamphere found "dreary, bare and cold." London itself was a ragged place, still suffering from the war five years past. "There was a great deal of bomb damage, and whole blocks had not been cleared of debris. Meat, butter, other foodstuffs and coal were rationed, along with nearly every other commodity. It was late in May, but the city was still cold, and there was no heat in our hotel room nor in many of the offices we visited during our stay." Lamphere nevertheless "considered the chance to interrogate Klaus Fuchs one of the great opportunities of my life."

Fuchs appeared much as Lamphere had expected: "thin-faced, intelligent and colorless." They sparred at first, the convicted spy wondering why he should cooperate with the FBI; Lamphere quickly realized Fuchs thought of the agency as a sort of Gestapo and was maneuvering to protect his sister, and found a way to imply to Fuchs that his sister's security depended on his cooperation. Out came the photographs then. Fuchs sorted them down to three new surveillance photographs of Harry Gold. "I cannot reject them," he said. Lamphere was ecstatic. But Fuchs was not willing to make a positive identification; he told Lamphere the photographs were not clear enough.

Monday, May 22, 1950, at Wormwood Scrubs, Fuchs viewed the surveillance motion pictures of Harry Gold, shot through the window of a car. After a first showing, he said, "I cannot be absolutely positive, but I think it is very likely him." After a second showing, he still would not confirm an identification. A third showing wedged his identification at "very likely."

In Philadelphia on Monday morning, the agents who had been questioning Gold knocked on his door shortly after eight A.M. The chemist recalls that he had made "only an abortive attempt to 'frisk' my room and elsewhere for incriminating evidence . . . [because] I couldn't do so till after Pop and Joe had left for work . . . [and] over the weekend I was afraid of arousing my family's suspicions." "The search began in the bedroom," the agents reported, ". . . which had considerable papers, books, chemical journals, and a vast amount of personal papers and effects." Several documents the agents turned up unnerved Gold. They were two hours into their search when they finally found something directly incriminating:

The next and most important item located was a Chamber of Commerce map of the City of Santa Fe, New Mexico. This was located behind some books in a bookcase. Gold was shown this and told, "You forgot you had this, didn't you, Harry?" to which Gold replied, "My God, where did that come from?" He then said, "I don't know how that thing got in there." The Agents quickly told Gold that the whole thing was through and that the "jig was up" and he had better explain the whole matter. Gold was obviously very shaken and said that he would like to have a few minutes to think.

Gold explained many years afterward what he thought:

> So when that map of Santa Fe turned up, I had a decision to make. In itself it wasn't too incriminating—along with a couple of other question-arousing items that had been found—but suppose I continued to protest my innocence and cry persecution? All my family and friends and associates at the Philadelphia General Hospital would rally to my aid. Now there existed very real evidence against me: Fuchs might identify me; I had visited his sister Kristel's home in Cambridge a number of times. . . . Plus, who knew if the FBI might not come across Al Slack? And Abe Brothman had panicked once before on interrogation and had given the FBI my name in 1947. The structure of lies was bound to collapse and so I sat on my bed and asked for a cigarette (come to think of it, significant—I dislike them). . . .

"After about one minute and at 10:15 A.M.," the agents' report continues, "Gold stated, 'I am the man to whom Klaus Fuchs gave his information.' "

Gold went into voluntary custody. His brother came to see him at the FBI office that evening and Gold gave him the devastating news. "And the following night, Pop," Gold wrote from prison in 1965: "Judge McGranery was a bit late when he pronounced his sentence; I really got mine when I saw Pop's eyes that night of May 23."

Lamphere did not tell Fuchs that Gold had confessed. The physicist nevertheless identified Gold positively on May 24 from the still photographs the Philadelphia agents had taken the previous Sunday, possibly because the photographs were taken under better lighting conditions, possibly because they were posed frontal and profile shots that obviously had been obtained with Gold's cooperation, implying that identifying him would no longer give him away. As soon as Fuchs saw them he said, "Yes, that is my American contact." "An unbelievably great weight seemed to lift from my shoulders," Lamphere recalls.

Harry Gold was arraigned in Philadelphia at 10:45 P.M. on May 23, 1950; his arrest made headlines the following day. Ruth Greenglass, still convalescing from her burns, had just returned from another and happier hospitalization, her husband recalled. "It was the same day, the day after my wife came home from the hospital after giving birth." Julius Rosenberg found the family at home. "I remember he knocked on the door," David Greenglass continues, "I got up out of the living room chair, opened the door, there he was. It was in the morning, I hadn't gone to work yet. He had the [New York] *Herald-Tribune* or *Times,* anyway there was a picture of Gold on the front page. And he said, That's your man, look at the picture. I said, You're silly, that's not the fellow; my wife said it was not him. He said, That's the man."

Rosenberg handed them a thousand dollars in old bills, tens and twenties.

David: "He said . . . you have to go out of the country. . . . He was excited. . . . He feared he would be arrested; they would pick me up, I would lead to him." David at another time: "He said that he had to leave the country himself and he was making plans for it, and I said, 'Why you?' He said that . . . he knew Jacob Golos . . . and probably [Elizabeth] Bentley knew him." Ruth: "He says, You have a month to spend this; I'll give you more and get what you need. . . . He said you have got to go to the dashers [sic: dacha]; I said, What is that? He said the Soviet Union. I said, Are you going too? How does Ethel feel about it? He said she is disturbed, but she realizes she has got to go. . . . [I] said, we can't go anywhere, we have an infant here; we can't just up and leave. . . . He said your baby won't die; babies are born in the air and on trains, and she will survive." It was a "golden opportunity" to go to the Soviet Union, Ruth says Rosenberg added. Rosenberg and David Greenglass went for a walk then and Rosenberg outlined an escape route through Mexico City and either Stockholm or Berne to Czechoslovakia, thence to the USSR.

Ruth Greenglass claimed later that they "never intended to leave the United States, because this is our country and we want to stay here and live here and raise our children," but said they "accepted the money . . . because David said that if Julius suspected that we would not leave the United States that some physical harm might come to us or our children." What in fact they intended to do is not clear in the record and may not have been clear even to them at the time. On May 28, they had six sets of photographs taken, larger than standard US passport photos, five sets of which they turned over to Julius Rosenberg. The photographs were probably for KGB use, to identify the Greenglasses along their escape route.

Since February, when Robert Lamphere had passed along information from a cable decode, the FBI Albuquerque field office had been investigating who might have been a second spy at Los Alamos during the war. One serious candidate was William Spindel, Greenglass's fellow SED, in whose wife Sarah's apartment Ruth Greenglass had recuperated from her 1945 miscarriage. But Albuquerque had also looked into the possibility that the second spy was a scientist. Stanislaw Ulam was a suspect. So was Victor Weisskopf. But the "most logical suspect for [the] Soviet agent," Albuquerque proposed, was Edward Teller. Albuquerque gave several reasons for its conclusions. Teller was a "close associate of . . . Fuchs at Los Alamos." Mici Teller had traveled to Mexico City with Fuchs and Rudolf and Genia Peierls "in the latter part of 1945." "The Tellers had Fuchs at their home for dinner when Fuchs returned to this country in 1947." "Dr. Teller had considerable contact with Fuchs in England in the summer of 1949." Besides his affiliation with Fuchs, Teller also had recommended for postwar graduate study at the University of Chicago a man with whom he had worked at Los Alamos who "has been identified as a Soviet espionage agent while at Los Alamos."

Teller's name had appeared on a list of possible espionage recruits that the man had compiled. Teller had traveled to New York during the time periods bracketed in the NKVD cable decode and "made frequent trips away from the Los Alamos Project and could have furnished information to the Russians on a regular basis." And, oddly, "Dr. Teller is outspoken against furnishing atomic energy information to Russia, which appears strange in view of the fact that his parents and other relatives are in Hungary under Communist domination."

Had its agents investigated further, the FBI could have learned much more about Edward Teller that might have appeared suspicious. Teller had refused to work on important implosion calculations at Los Alamos in the spring and summer of 1944 and his refusal had led directly to the decision to bring British scientists, including Klaus Fuchs, to Los Alamos. Teller had left Los Alamos to return to private life in 1946 even though he was the leading theoretician responsible for thermonuclear work; his departure had undoubtedly delayed the progress of that work. Teller had insisted on the development of a particular design of thermonuclear weapon, the Super, which had not been determined to be feasible on basic physical principles, when another design, his Alarm Clock, was unquestionably feasible on basic physical principles. The Super design Teller had insisted upon Los Alamos pursuing had recently been shown to be almost certainly inadequate. He had continued to insist on its development, and had encouraged a major commitment of people and funds which the President himself had endorsed, even though the Super was at best a marginal design and even though its development would deprive the country of a large number of atomic bombs which might otherwise be produced. Adding hypothetical charges such as these to the evidence it had already assembled of Teller's associations with Fuchs, the FBI might have built a powerful case that the brooding, volatile Hungarian-born physicist was a Soviet spy. Teller and like-minded patriots such as Lewis Strauss and William Borden would not hesitate to compile similarly hypothetical charges against Robert Oppenheimer in the years to come.

Harry Gold saved Edward Teller from further investigation. The evening Gold saw his father and felt sentenced, Sam Gold asked his son "to try to make up for the damage." Harry decided that the way to do so was to cooperate fully with the FBI. On June 2, he identified his Albuquerque contact as a "US Army" man, "twenty-five years of age, perhaps even younger," from New York, whose wife's name "may have been Ruth." That eliminated Teller. Gold also remembered approximately where the couple had lived in Albuquerque. Greenglass's name was already at hand along with dozens of Los Alamos soldiers, selected on the basis of their furlough records; his Albuquerque apartment fit Gold's description. Lamphere asked New York to acquire surveillance photographs of Greenglass, and on June

4, Gold tentatively identified the young machinist from such a photograph and recalled the Greenglasses' story about receiving packages of kosher food from New York City. He also fingered Al Slack and Tom Black. "Interviewing Gold was like squeezing a lemon," one FBI agent quipped at the time; "there was always a drop or two left."

Julius Rosenberg visited the Greenglasses on June 4, Ruth recalled:

> He came to the house and called David into another room and gave him ... four thousand dollars.... I was in the house when he came.... At that time David and I had already discussed it and decided not to tell Julius that we weren't going to leave because David felt that if he knew of our intentions, some physical harm might come to us, that it would be best to let him believe we were going.... [Julius] was very melodramatic, discussed everything in whispers, he was under the impression there were ears all over the house, he took David down for a walk.... We took the four thousand dollars and David taped the package with Scotch tape and placed it in the fireplace in the flue. The money remained there. It was only there for a few days and David took it out and gave it to my brother-in-law, Louis Abel.... [Julius] promised [David] two thousand dollars additional.... He told him he would be back with it. David told him to keep away and leave us alone; that we did not want the money.

David Greenglass earned $107 per week working as a machinist for the Arma Corporation in Brooklyn; four thousand dollars was most of a year's wages.

The next day, June 5, Greenglass appeared at Arma and requested a month and a half of leave, claiming his wife had relapsed from her burns. The company denied him leave and he committed to returning to work on June 12. By then he knew he was under surveillance, Ruth would testify; when Rosenberg again came calling, "David said he was being followed and to please leave us alone and not to come back any more. Julius said ... he hadn't noticed anyone watching our house and he was sure David was imagining it."

Greenglass decided to run, he told Ronald Radosh in 1979:

> What I did is, I got on a bus at the 50th Street Terminal [on June 11] and I went up to the Catskill Mountains and the FBI followed me all the way. What I had intended to do was to get a place up there and bring Ruthie and the kids and then stay there all summer and then disappear into the hinterlands. Not in any way ever going to Russia or anything like that—but I told Julius I was going to follow what he asked me to do because I wanted to get the money to do something. We never really believed that the FBI would arrest us.

Ruth called the Brooks Farm and Bungalow Colony in Ellenville, New York, the day after David looked it over and reserved a bungalow for the season for $350. Greenglass returned to work that afternoon, June 12.

Two FBI agents knocked on the door of the Greenglass apartment at 265 Rivington Street in lower Manhattan on Thursday, June 15, 1950, at 1:46 P.M. David let them in. "What happened was," he told Ronald Radosh, "I said, nah, I won't say anything to them. They asked me if it was okay to search the house and I said, sure, why not? It's like if you take the Fifth Amendment you're guilty. Searching the house didn't really have anything for them to see except innocent books by Marx, by Lenin." One of the agents immediately found a stack of photographs in a bedroom night table drawer, including a photograph of David and Ruth in front of the Albuquerque house, David in his Army uniform. Greenglass gave up the photographs; the agent ran them to another agent in a waiting car; yet another agent carried them to Pennsylvania Station and boarded a Philadelphia train. In the meantime the second agent at the Greenglasses' turned up a footlocker full of letters, including the couple's wartime correspondence. David let the footlocker go. "The letters?" Greenglass asked Radosh rhetorically. "The letters had been censored, those letters, any letter you write you forget about. Who remembers what was in them? There's nothing incriminating in them unless something happens." The agents took David in for questioning.

Harry Gold was in Holmesburg Prison. An FBI agent mixed the Greenglass photographs freshly arrived from New York into a larger batch of unrelated photographs and asked Gold to look through them. "I made the positive identification of the Greenglasses at about 10 P.M.," Gold recalled. "... (I said 'Bingo!' on seeing a photograph of David and Ruth in front of that house in Albuquerque—Dave was so much younger and thinner then)." The agent in Philadelphia immediately called New York. Confronted with Gold's identification, Greenglass dictated a voluntary statement that implicated Ruth and Julius Rosenberg but not his sister Ethel. After his statement, perhaps giddy with shock, he laughed and said, "I expect to have my day in court, at which time I will plead innocent, repudiate this statement and claim I never saw you guys." Three decades later, Greenglass remembered that he made a more serious threat: "I told ... the FBI right from the start that if my wife was indicted I would not testify. I told [them] I would commit suicide and [they] would have no case."

Greenglass was placed under arrest at 1:32 A.M. on June 16. Concerned that Rosenberg might flee, the FBI picked up the engineer at home and brought him in for questioning just after nine that morning. But another month passed of legal maneuvering to move David Greenglass's venue to New York from New Mexico, where he committed his crimes, before the Greenglasses really opened up to federal prosecutors. During that period, Joel Barr defected from his Paris apartment, leaving behind a new motorcycle as well as his clothes and books; Morton Sobell fled with his wife and children to Mexico; and Ethel Rosenberg, according to Ruth, asked Ruth to convince David not to talk, arguing that "it [i.e., David's eventual sentence] would only be a couple of years and in the long run we would be better off," and that "she would take care of me and my kids." (Ethel Rosenberg would deny Ruth's version of this conversation.) Ruth implicated Ethel in a statement to the FBI on July 17, placing Ethel at the dinner table when Julius proposed that Ruth convince David to pass information and in the kitchen when Julius cut the cardboard identification signal, which Ruth was the first to identify as a "Jello box." Perhaps following his wife's lead, David implicated his sister in an additional statement two days later.

Julius Rosenberg was arrested on July 17. Ethel Rosenberg was arrested on August 11. Ruth Greenglass was questioned extensively but not indicted. Harry Gold would be brought up to New York City's Tombs prison in October and would meet several times with David Greenglass. In the Tombs, Greenglass told Gold—whether proudly or ruefully, the record does not reveal—that both he and Julius Rosenberg had been awarded the Order of the Red Star.

In October 1939, nine months after the discovery of nuclear fission, the colorful Russian-born economist Alexander Sachs, a vice-president of the Lehman Corporation and an informal adviser to Franklin Roosevelt, had carried to Roosevelt a letter from Albert Einstein warning the President that the Germans might be working on an atomic bomb. In spring 1950, Sachs kept another appointment with Paul Nitze, George Kennan's successor as head of the State Department Policy Planning Staff. "He brought three papers with him," Nitze writes: "the first was an analysis of Soviet doctrine on the correlation of forces [i.e., matching Soviet forces against US to see which had advantage]; the second argued that the Soviets would view their successful atomic test and events in China as a favorable change in the correlation; and the third analyzed where and when they might exploit this [change]. Sachs thought that Moscow was naturally cautious, and would try to minimize risks by acting through a satellite. He predicted a North Korean attack upon South Korea sometime late in the summer of 1950."*

Sachs came to the right conclusion on the wrong grounds. Stalin had not felt buoyed by his acquisition of the atomic bomb; to the contrary, it had made him more insecure. He was seventy years old in December 1949 and increasingly suspicious—"weakening mentally as well as physically," says

---

* There was not much that Nitze could do about it. He "managed to put together a $10 million program to give the South Koreans some additional fast patrol boats. That was the limit our aid program in those days would support." Gaddis and Nitze (1980), p. 174.

Nikita Khrushchev, ". . . declining fast." Khrushchev points out that "America had a powerful air force and, most important, America had atomic bombs, while we had only just developed the mechanism and had a negligible number of finished bombs. Under Stalin we had no means of delivery. . . . This situation weighed heavily on Stalin. He understood that he had to be careful not to be dragged into a war." The Soviet Union signed a treaty of alliance and mutual assistance with the new People's Republic of China on February 14, 1950, but only with great reluctance did Stalin agree to "render assistance with all means at our disposal" now that those means included atomic bombs—he was afraid that the Chinese might drag him into atomic war.

If there was room for the Soviet Union to maneuver anywhere, it was in Asia, not in Europe. North and South Korea had been divided by no ancient enmity; they emerged as separate governments exactly as East and West Germany had emerged, because they were divided at the end of the Second World War between Soviet and US occupations. The US began to think about "a firm 'holding of the line' in Korea," (in the language of a 1947 report that Dean Acheson approved) at the same time Truman declared the Truman Doctrine of aid to Greece and Turkey. "Please have [a] plan drafted of [a] policy to organize a definite government of So[uth] Korea," George Marshall had requested casually earlier that year. By the end of 1949 there were two Koreas, the southern Republic of Korea, founded in August 1948 under Syngman Rhee, a seventy-three-year-old Princeton Ph.D. who had fought for Korean independence from the Japanese since before the turn of the century, and the northern Democratic People's Republic of Korea, founded one month later under the young Soviet-trained infantry officer and Communist revolutionary Kim Il Sung. Each leader was rabid to invade the other's country and unite the Korean Peninsula under a common flag. North Korea was much better armed, however; besides captured Japanese equipment, the Soviets had left it all their weapons when they withdrew after independence, and would supply it with more military assistance during the late 1940s and early 1950s than they supplied even the People's Republic of China.

Kim first proposed "liberating" South Korea to Stalin in December 1949. The Korean leader seems to have done the correlating of forces that Alexander Sachs would ascribe to the Soviets; according to a senior Soviet diplomat, "the Koreans were inspired by the Chinese victory and by the fact that the Americans had fled from mainland China completely; they were sure that the same could be accomplished in Korea." Stalin was wary. He doubted, the diplomat writes, that the US would "agree to be thrown out of there and because of that, lose their reputation as a great power." Stalin discussed the question with Mao Zedong that winter when Mao lingered for months in Moscow negotiating the treaty of alliance; the Chinese leader was even more cautious than Stalin about encouraging a war, possibly with the United States, on his northeastern flank.

Kim continued to push. To bolster his contention that the US would not defend little South Korea when it had not come to the rescue of the Nationalist Chinese in the Chinese civil war, he probably cited a speech Dean Acheson delivered to the Washington Press Club on January 12, 1950, which included the curious statement, "So far as the military security of other areas in the Pacific [outside the Aleutians, Japan, Okinawa and the Philippines] is concerned, it must be clear that no person can guarantee these areas against military attack. . . . Should such an attack occur . . . the initial reliance must be on the people attacked to resist it and then upon the commitments of the entire civilized world under the Charter of the United Nations. . . ." Acheson later described this statement as "the warning which I gave . . . which the aggressor disregarded." The Secretary of State was not encouraging a North Korean invasion of the South by writing off the peninsula, as Republican partisans later accused; rather, he was discouraging a South Korean invasion of the North by warning Syngman Rhee that he could not count on immediate US support. Kim Il Sung found a third meaning in Acheson's statement: that a quick invasion could be decisive.

Stalin for his part wanted to isolate China from the West. He suspected correctly that China still scouted an American alliance; US intervention in a Korean war, if it came, would drive the Chinese deeper into the Soviet camp. In the winter and spring of 1950, Kim prepared to attack. By March, three former North Korean high officials testify, "there was a 100,000-man army, tanks, airplanes, artillery—everything was ready." Kim returned to Moscow that month. According to a Korean official who was present at the meeting between the two heads of state, Kim "made four points to persuade Stalin that the United States would not participate in the war: (1) it would be a decisive surprise attack and the war would be won in three days; (2) there would be an uprising of 200,000 Party members in South Korea; (3) there were [Communist] guerrillas in the southern provinces of South Korea; and (4) the United States would not have time to participate. Stalin bought the plan." The Soviet leader did not immediately offer assistance, however: he directed Kim to China to win Mao's endorsement.

Kim saw Mao in May and began masterfully to play off the two Communist leaders each against the other. Mao at that time was planning an invasion of Taiwan, for which he had a promise of Soviet support. (Truman had announced on January 5 that the US would not intervene in Taiwan, a point Mao had taken to heart.) If Mao expressed fear that the US would defend South Korea, he would have to admit the possibility that the US would also defend Taiwan, in which case the Soviets would certainly back away from their promise. Rather than take that risk, Mao tepidly endorsed Kim's adventure. The Korean leader probably exaggerated the temperature of Mao's endorsement when he communicated it to Moscow. Stalin had already started sending Kim weapons; they "began to arrive in huge numbers at the port of Chongjin" in April, says a North Korean general. Now Stalin signaled his approval of Kim's war of liberation by dispatching a team of Soviet advisers to Pyongyang. The Chinese leadership felt betrayed by Stalin's alliance with Kim Il Sung. Dogged by delays, Mao was forced to put his Taiwan invasion on hold—permanently, as it turned out. If there had not been a Korean war in the summer of 1950 there might well have been a Chinese war between the United States and the People's Republic of China.

At four A.M. on Sunday, June 25, 1950, the North Korean army attacked on the Ongjin Peninsula on the western coast of Korea and then generally along the 38th parallel that divided North and South. Fourteen thousand of the invaders were ethnic Koreans from China—Chinese soldiers in Korean uniforms. Kim's Soviet advisers had proposed the day and time; it matched Hitler's invasion of the USSR on the morning of Sunday, June 22, 1941. US Ambassador to South Korea John Muccio, tired from a late night playing strip poker with several of the embassy secretaries, cabled the news to the State Department. Apprised by phone at his Maryland farm, Acheson immediately ordered the department "to take the steps which were necessary" to convene the United Nations Security Council the next day. Only after making that first crucial decision did Acheson call Truman, who was spending the weekend at his wife's family home in Independence, Missouri. "In succeeding days," writes historian Bruce Cumings after authoritative research, "Acheson dominated the decision-making which soon committed American air and ground forces to the fight." Acheson saw Korea as an opportunity to compel a lethargic United States to confront the Communist challenge, "to remove many things," as he later noted, "from the realm of theory," to begin the work of containment if not rollback. "Korea is not a local situation," Acheson would observe a year later in a secret government discussion. ". . . It was the spearpoint of a drive made by the whole Communist control group on the entire power position of the West. . . . [Korea is a] testing ground." "The Greece of the Far East," Truman agreed, alluding to the Truman Doctrine. Of his decision to commit US ground troops to Korea, the President bragged, "We have met the challenge of the pagan wolves."

A Soviet veto in the UN Security Council might have blocked UN commitment (in which case the US would certainly have gone into Korea alone). Andrei Gromyko, who had been the chief Soviet delegate to the United Nations until 1948, advised Stalin to use the veto. Stalin demurred. "In my opinion," he told Gromyko, "the Soviet representative must not take part in the Security Council meeting." Stalin may have believed Kim Il Sung's representation that the war would be over in a matter of days, too quickly for UN intervention, or he may have been maneuvering to forestall a US declaration of war—not likely if the US was fighting under the UN flag—

that might eventually involve China and trigger Soviet participation under the Soviet-Chinese treaty of alliance. Dmitri Volkogonov confirms that "Stalin took an extremely cautious view of events in Korea and from the outset made every attempt to avoid direct confrontation between the USSR and the USA."

US troop strength worldwide, Army and Marines both, was only 669,000 in June 1950, the largest contingent—four divisions, about 100,000 men—positioned in Japan. North Korea could field a force of two hundred thousand if required, China millions more. The Joint Chiefs expressed themselves to be "extremely reluctant" to commit ground troops. Acheson and Truman overruled them when General Douglas MacArthur reported from Japan that the South Koreans had ceased to fight; US forces went in on June 30. By then the North Koreans had already taken Seoul. They would push south almost to the end of the peninsula before MacArthur would contrive to cut them off and drive them back.

Curtis LeMay believed he knew a way to end the invasion as quickly as it had begun. "We slipped a note kind of under the door into the Pentagon," he recalled in retirement, "and said, 'Look, let us go up there . . . and burn down five of the biggest towns in North Korea—and they're not very big—and that ought to stop it.' Well, the answer to that was four or five screams —'You'll kill a lot of noncombatants,' and 'It's too horrible.'" The only lesson LeMay thought he learned in Korea was "how not to use the strategic air weapon." By the time of the Korean War, that weapon was becoming formidable.

22

Lessons of Limited War

CURTIS LEMAY had been building and training the Strategic Air Command with unflagging determination since he took charge in October 1948. "I see no other way of being ready when the whistle blows," he told an audience of officers at the Armed Forces Staff College several years later, "unless you do day after day exactly what you are going to do when the fighting starts. And that's what we're doing now and have been doing for some time."

At the beginning of the new decade of the 1950s, in a little more than a year, SAC had grown by more than one-third, from 52,000 people to more than 71,000. They flew and supported 868 aircraft. Most of SAC's bombers were still B-29s, but the proportion of B-50s was increasing and B-36s were beginning to come in. By January 1950, eighteen atomic-bomb assembly crews had been qualified, with four more to be added by June, when LeMay would have more than 250 operational atomic-modified aircraft.

His primary mission, as a SAC historian describes it, was to "lay down an atomic attack on Russia in the event of war." LeMay interpreted that mission to mean an all-out attack at the beginning of war, a strategy he called a "Sunday punch." To see how well his crews were trained, he staged a major maneuver on June 6, 1950, involving half the aircraft under his command. Without leaving the United States, crews picked up unarmed atomic bombs assembled for the occasion (lacking their nuclear capsules, that is, as fissle core modules had come to be called), flew equivalent distances to forward bases and bombed equivalent targets. Deployment, as planned, took three days, the strike phase three more. Eglin Air Force Base in Florida stood in for Moscow and radar confirmed that ten of the eleven atomic bombs assigned to the Soviet capital in current targeting plans would have found their target had they been dropped. "Screening aircraft preceded the bomb carriers over the target," a report of the exercise notes, "with chaff* and

* Strips of metal foil dropped to confuse enemy radar.—RR

spot jamming being utilized by all aircraft. Bombing was by squadrons, and night withdrawal was to the west." Fifty-eight bomb carriers simulated dropping bombs; some of them arrived back at their bases with as little as seventy-five gallons of fuel left in their tanks. Unlike the debacle over Dayton, Ohio, a year and a half earlier, this exercise was a success; all seventeen designated target areas—on the books, Soviet urban areas—were hit. "This," concludes the report, "was the first realistic test of Strategic Air Command's ability to deploy and execute the initial strike."

Three weeks later, on June 25, North Korea invaded South Korea. LeMay's response was swift. He knew he was prepared to execute his war plan, but he had no atomic bombs. "The military services didn't own a single one," he recalled bitterly in retirement. "These bombs were too horrible and too dangerous to entrust to the military. They were under lock and key of the Atomic Energy Commission. . . . That worried me a little bit. . . . So I finally sent somebody to see the guy who had the key. We were guarding them. Our troops guarded them. But we didn't own them. . . . [On June 27] I sent somebody out to have a talk with this guy with the key. I felt that under certain conditions—say we woke up some morning and there wasn't any Washington or something—I was going to take the bombs." The "guy who had the key" was General Robert M. Montague, the commanding general of Sandia Base in Albuquerque where the atomic bombs were stored. LeMay's chief of staff described LeMay's side bet with Montague in July, when LeMay sought approval for it from USAF Chief of Staff Hoyt Vandenberg, as "in essence that in the event that Washington was destroyed before we had received word to execute our war plan, and in the absence of communications with the alternate headquarters USAF . . . General LeMay would contact General Montague, identifying himself by a system of code words and attain release of the bombs to our bombardment crews." LeMay later explained his thinking to a team of Air Force historians:

> If we got into a position where the President was out of action or something else turned up, I was going to at least get the bombs and get them to my outfits and get them loaded and ready to go—at least do that much. . . . I would have [released them] under certain circumstances, yes. . . . If I were on my own and half the country was destroyed and I could get no orders and so forth, I wasn't going to sit there fat, dumb and happy and do nothing. . . . I may not have waited until half the country was destroyed, but I felt I had to do something in case no one else was capable of doing anything. . . . If we were under attack and I hadn't received orders for some reason, or any other information. . . . I was going to take some action at least to get ready to do something. . . .
>
> What I am trying to say is that SAC was the only force we had that could react quickly to a nuclear attack. It did not make much sense to me to be in a position of not being able to act because I had no weapons. We had

no idea of what confusion might exist, or who the president might be, or where, if a bomb hit Washington. . . . I doubt if I would have retaliated if Washington were the only target hit. But I certainly would not have waited until half the country were destroyed. The main thing is that by making agreements to get the weapons we had some options rather than having none at all.

A Vandenberg staff officer told LeMay in response to LeMay's request that he "could see the desirability of [Vandenberg's] approval" of LeMay's *carpe diem* plan but saw no reason why the other Joint Chiefs needed to hear about or approve it. Appropriating the President's authority under the Atomic Energy Act to decide whether and when to transfer atomic weapons to the military was apparently to be exclusively an Air Force prerogative.

Targeting was another area where LeMay fought for control. Kenneth Nichols recalled in his memoirs that targeting "developed considerable controversy on the Joint [Chiefs] Staff and even within the Air Force. Should cities be the main targets, or should targeting be confined strictly to military and industrial targets?" (Nichols also noted that "One study I saw showed little difference in overall casualties in either case.") LeMay was a realist and understood that bombing at night by radar in a strange country over which you have never previously flown meant unavoidable inaccuracy; he stood for industrial bombing—meaning city bombing—first and last. The US Strategic Bombing Survey had concluded that knocking out the German electric power grid during the Second World War would have crippled that nation's war effort; the target panel in Washington in the early 1950s therefore favored a major effort against Soviet electrical power stations even though such stations would be hard to spot from the air and were essentially local targets, not connected together into a national grid as those in Germany had been. LeMay appeared before the target panel and told the generals that "any target system picked that failed to reap the benefits derived from urban area bombing . . . was wasteful." Rather, LeMay argued, "we should concentrate on industry itself which is located in urban areas. . . ." If a plane missed its aiming point, then "a bonus will be derived from use of the bomb." LeMay won the panel's agreement to send target lists to SAC for comment before seeking JCS approval. "Bonus damage" and "catastrophe bonus" became terms of art at SAC; another popular euphemism was "precision attacks with an area weapon." In the years ahead, target planning served as a royal road to increasing SAC's share of the national defense budget, since target requirements dictated bomb requirements, which dictated in turn how large the air force should be.

But in June 1950, the US faced real war in Korea. General Douglas MacArthur had some five hundred aircraft available in the Far East to throw into the fighting (including one B-29 bomber group), of which about three hundred were combat-operational; the North Koreans had more aircraft than expected, about two hundred. MacArthur wanted more B-29s; the order came through to LeMay on July 1 to send ten aircraft to shore up MacArthur's group—"ten of your standard mediums," General Roger Ramey told LeMay, "that are not especially . . . modified [for atomic bombing], that will hurt you the least." LeMay immediately howled to Vandenberg's deputy Lauris Norstad. "I just wondered if Van[denberg] knows that we can fritter away this [strategic] force pretty quickly," he asked. "Is this going to be the last [requisition of aircraft and crews] or not?" Norstad, trying to save a war that looked as if it might be over in Kim Il Sung's predicted three days, communicated his distress:

> GEN. NORSTAD: . . . Unless the situation changes. God only knows what will
>   happen, but that group [in Korea] has got to be brought up, otherwise,
>   Curt . . . we are not going to operate effectively with it.
> GEN. LEMAY: I know it's not worth a shit. I can tell that from here.
> GEN. NORSTAD: That is exactly right. . . .

Ten B-29s were not to be the last requisition. The next day, July 2, LeMay received orders to send two full groups of bombers; "their job," his aide noted, "will be to destroy the necessary targets north of the 38th parallel and not to directly support tactical operations." LeMay proposed to go to Korea himself and command the operation, but Vandenberg overruled him and he stayed in Omaha. At the same time as it moved two groups of conventional B-29s to the Far East, the Air Force shifted two atomic-capable SAC wings to forward bases in England with Acheson's approval, to encourage the British in their alliance with the US. Truman authorized sending along a supply of unarmed atomic bombs to England with the planes.

LeMay's resistance to supporting the war in Korea was not merely the whim of a strategic-bomber man. The commander told an NBC correspondent off the record a few weeks after the war began that "SAC was the USA's Sunday punch and . . . every effort must be made to make sure that it stayed intact and able to strike and not be pissed away in the Korean war." In April, the CIA had published a top secret "Estimate of the effects of the Soviet possession of the atomic bomb upon the security of the United States and upon the probabilities of direct Soviet military action" in which the armed services had participated. The study estimated that two hundred atomic bombs exploded over the major cities of the United States "might well knock the US out of [a] war" and concluded that the Soviets might have that many bombs "some time between mid-1954 and the end of 1955." Such a historic capability would mean that "the continental US will be for the first time liable to devastating attack."

For the US military, 1954 thus became the year of maximum danger.

LeMay identified it in a May 1950 memorandum to Thomas K. Finletter, the Secretary of the Air Force, as "the critical year at which time we must be prepared to meet, and effectively counter, the full military force of the USSR. . . ." SAC would not have long-range jet bombers by then; the B-52 was not scheduled to begin coming on line until 1955. LeMay would have to make do with B-50s, capacious but lumbering B-36 propeller-jet hybrids and light, medium-range all-jet B-47s. That was why he was ferociously reluctant to piss away his Sunday punch.

Whether the Soviet Union would be able to deliver its presumed arsenal was another matter. The CIA was skeptical, particularly since the only bomber of any range that the Soviets had built since the war was the TU-4, a copy of the B-29 made to measure from the three American bombers lost in Siberia in 1944. "If there are doubts about the ability of the B-36 to deliver the atom bomb against the USSR," the April CIA estimate argued, "how much greater the doubts that the Soviet B-29 could deliver it successfully against an effective and alert US defense." One important condition that the Soviet leadership would probably require before launching atomic war, however, was "virtual certainty that effective US retaliation could be prevented." The CIA concluded that the Soviets would only initiate a war if they could do so with "a successful surprise attack that would seriously cripple or virtually eliminate US retaliatory capabilities." Retaliation was the least that LeMay expected SAC to accomplish.

In the meantime, there was a war on, even if it was officially only a "police action," and Stalin had been right: the United States did not believe it could afford to lose. The Joint Chiefs were even willing to postpone *Greenhouse* if necessary to free Navy vessels to blockade North Korea; the H-bomb could wait. LeMay's first two groups of older, conventional B-29s flew to Yokota Air Force Base, outside Tokyo, and Kadena Air Force Base on Okinawa early in July 1950. (Japan, a senior Japanese diplomat comments dryly, was becoming "America's warehouse," a circumstance which contributed significantly to its economic recovery.) Two more such groups followed in early August. They found crowded emergency conditions, lived through the late-summer monsoons in tents and even loaded their own bombs, some of them labeled "Look out Commies!"

In their first month of operations, the detached SAC groups dropped four thousand tons of bombs in tactical support of UN ground operations. They began as well a little-known strategic bombing campaign that would quickly devastate North Korea, dropping three thousand tons of bombs on strategic targets by mid-August. Their mission directive, cut on July 11 at the order of General George E. Stratemeyer, the commanding general of the Far East Air Forces, listed destroying the "enemy communication system" first, including "highway, railroad and port facilities," ordering the aircraft to keep " 'well clear' of the Manchurian border." Second in priority would be "North Korea industrial targets." The directive disguised an Air Force campaign to bomb out North Korean cities with the standard qualification that strategic bombing enthusiasts had invented during the Second World War: "The Commanding General, Far East Air Forces Bomber Command will not attack urban areas as targets. The attack within urban areas of specific military targets as set forth above is authorized." Urban-area bombing had not been precision during the Second World War and it would not be precision in Korea.

The results were devastating. The FEAF commander announced in October that the strategic air war was over as far as North Korea was concerned; "eighteen major strategic targets had been neutralized." LeMay in retirement told the whole truth, following up his complaint about the Pentagon refusing to "turn SAC loose with incendiaries" at the beginning of the war:

> So we went over there and fought the war and eventually burned down every town in North Korea anyway, some way or another, and some in South Korea too. We even burned down [the South Korean port city of] Pusan—an accident, but we burned it down anyway.... Over a period of three years or so, we killed off—what—twenty percent of the population of Korea as direct casualties of war, or from starvation and exposure? Over a period of three years, this seemed to be acceptable to everybody, but to kill a few people at the start right away, no, we can't seem to stomach that.

Besides high explosives, the weapon of choice was napalm, superior in its combustibility to jellied gasoline and used at major scale for the first time in Korea. A mixture of naphthenic and palmitic acids ignited with phosphorous, it continues to burn inside wounds for up to fifteen days. Military dispatches from Korea characterized the bombed areas as "remov[ed] from off the map," "burn[ed] out," "wilderness of scorched earth."

Historian Bruce Cumings describes the horrific outcome:

> By 1952 just about everything in northern and central Korea was completely leveled. What was left of the population survived in caves, the North Koreans creating an entire underground society in complexes of dwellings, schools, hospitals and factories.... In the final act of this barbaric air war [the US] hit huge irrigation dams that provided water for 75 percent of the North's food production. Agriculture was the only major element of the economy still functioning; the attacks came just after the laborious, back-breaking work of rice transplantation had been done. The Air Force was proud of the destruction created: "The subsequent flash flood scooped clean 27 miles of valley below, and the plunging flood waters wiped out [supply routes, etc.].... The Westerner can little conceive the awesome meaning which the loss of [rice] has for the Asian—starvation and slow death." Many villages were inundated, "washed downstream," and even

Pyongyang, some twenty-seven miles south of one dam, was badly flooded. Untold numbers of peasants died, but they were assumed to be "loyal" to the enemy, providing "direct support to the Communist armed forces." That is, they were feeding the northern population. The "lessons" adduced from this experience "gave the enemy a sample of the totality of war... embracing the whole of a nation's economy and people." This was Korea, "the limited war."

The war dragged on for three years despite all the burning, blasting and killing—and ended, of course, in stalemate, the line dividing the two devastated countries remaining precisely where it had been before the North Korean invasion, at the 38th parallel. Besides at least 1.5 million military dead, wounded or missing on the UN and South Korean side (including 158,000 Americans) and two million North Korean and Chinese casualties, more than two million civilians died in North Korea, almost as many as fire- and atomic-bombing killed in Japan during the Second World War.

Conventional bombing in Korea was destructive enough. The United States also came much closer than most Americans realize to using the atomic bomb in the Korean War. At the first meeting he attended with the Joint Chiefs and Acheson after the North Korean invasion, Truman ordered a study of atomic bombing in case the Soviet Union entered the war. The JCS Plans Division complied with a study of "utilization of atomic bombardment" in Korea. Two weeks later, on July 9, 1950, a message from MacArthur led the Joint Chiefs "to consider whether or not A-bombs should be made available to" the Far East commander; the JCS thought some ten to twenty bombs could be spared from the strategic arsenal. MacArthur had grandiose plans to block a Chinese or Soviet intervention if it came. The inner North Korean border abuts both China and the USSR; MacArthur pointed out to an emissary from the Joint Chiefs that the routes that led into the country from Manchuria and Vladivostok had "many tunnels and bridges. I see here a unique use for the atomic bomb—to strike a blocking blow—which would require a six-month repair job."

To prepare for the possibility that the JCS might send atomic bombs to Korea, a teletype reached SAC headquarters from Washington on July 30 requesting transfer of "ten planes and crews with bombs less nuclear component," as LeMay's aide recorded it. LeMay had no intention of giving up control of atomic-capable aircraft or of atomic bombs to anyone without a fight; he called Roger Ramey and chewed on him. "What ten, Roger?" LeMay asked the Pentagon staff officer. "Don't you know that the two groups that are going haven't got any [atomic] capability?" Ramey demurred that "the top is running it without letting anyone else know." "The 509th [is the only wing that]... has the stinger," LeMay explained ingenuously, "and I don't think they ought to be roaming around. . . . So I talked to Norstad [at the time of the original request for Far East bomber reinforcements] and he said don't break [the 509th] up." Ramey: "The thing is, he didn't tell us that." LeMay: "Do you see any drawback to just cancelling that ten job?" Ramey: "No, not at all." LeMay, passing along what Norstad had told him: "Well, they haven't gone to the President, and [Norstad] isn't expecting that to happen now. . . ." Ramey, correcting: "That decision [to send atomic-capable aircraft and atomic bombs] went to the President originally. If Norstad told you that [i.e., that it had not gone to the President], he's evidently decided against it and just hasn't let us know."

But the President had not decided against moving an atomic delivery capability to the Korean theater of war, and Lauris Norstad was working on LeMay to pry the bombers loose within minutes after the SAC commander finished with Roger Ramey. "The Chief [i.e., Vandenberg] wants the number that was indicated the other day out there," Norstad told LeMay, speaking cryptically over the telephone, "and they want that [atomic] capability." LeMay argued against relinquishing ten atomic-capable bombers to MacArthur's control: "I think we ought to divorce that completely with this other move. . . ." Norstad saw LeMay's objection and accepted it: "Well, let's divorce this. Let's have a separate item on this goddamn thing." LeMay clinched the deal: "Okay, Larry. Have Hank send a wire out cancelling that last one he sent on these ten. . . . We'll make the necessary advance preparations and be all set on it." Norstad was happy to have the dispute resolved: "Yes, that's the best way of doing it, Curt." The next day, July 31, LeMay heard from Ramey that "the ten aircraft are to follow the two Groups now on the way to the Pacific theater"; they would "be under the control of the JCS, not the Theater"—that is, under LeMay's control (SAC reported directly to the Joint Chiefs, which gave LeMay a legal basis for his unusual independence of command). LeMay still wanted to hear the order directly from Vandenberg. He flew to Washington on August 1, where the "decision was made to send the 9th Bomb Wing to Guam as an atomic task force immediately."

Ten B-29s loaded with unarmed atomic bombs and heavy with fuel prepared to leave Fairfield-Suisun Air Force Base, east of San Francisco above Suisun Bay, late on the night of August 5. General Robert Travis, an old friend of LeMay's who was commanding the mission, flew as a passenger in one of the bombers. As Travis's B-29 reached 130 miles per hour on takeoff, still on the runway, its number three propeller ran away to 3,500 rpm. The pilot made a quick decision not to abort the takeoff, feathered the propeller and got off the ground at 150 mph. He began a ninety-degree turn to land the plane, discovered he could not retract his landing gear, saw his number two engine start to run away and crashed approaching the landing runway with forty-five degrees still remaining in his turn. "The crash was not hard," LeMay's aide reports, "and all but twelve people survived. General Travis was hit on the head or hit his head on some object and died on the way to the hospital. His aide, the pilot and co-pilot were not injured." Twenty minutes after the crash, the B-29 exploded, "killing several additional people, firefighters, and injuring approximately sixty others." The explosion—fuel plus the high-explosive shell of the Mark IV atomic bomb aboard—destroyed the bomb and scattered tamper uranium, mildly radioactive, around the airfield. Not ten but nine unarmed atomic bombs went out to Guam that August. The aircraft of the 9th Bomb Group were recalled to the US on September 13, but the bombs and their maintenance teams remained on Guam.

A National Security Council study issued on August 25 argued that Korea might be "the first phase of a general Soviet plan for global war." Despite its contingency preparations, however, the Pentagon did not judge Korea to be a place where atomic bombs would be effective. Frank Pace, the Secretary of the Army at the time, remembered the military's reasoning in an oral history interview:

> [Atomic-bombing] was always discarded on [three] grounds: The first, that it would not be productive, that this was not the kind of war in which the use of the atomic bomb would be effective.... Second, was the concern about the moral use of weapons of this nature against a smaller country in this kind of war, and the third was that if it proved ineffective then its function as a shield for Europe would be either minimized or lost. So you had really three quite compelling reasons against its use....

William Borden expressed similar reservations in late November, when MacArthur was agitating for attacks on North Korean supply lines and bases in Manchuria; Borden told Brien McMahon he opposed atomic bombing in Korea "for three reasons: (1) propaganda capital could be made of their use in the Far East; (2) the fact that each weapon used in Korea would leave one less to be used if necessary against Russia; and, (3), most important of all, I fear that use of these weapons in Korea would not be effective in quickly ending the Korean war and would, therefore, result in an enormous psychological down-grading of the value now attached to atoms."

In August, when LeMay's atomic-capable aircraft were ferrying unarmed atomic bombs to the Far East, the Chinese were already preparing to enter the war. North Korean forces were then pushing the beleaguered United Nations defenders nearly to the end of the Korean peninsula, where they would make a last stand at Pusan. The Chinese had war-gamed MacArthur's options and foreseen that he would launch a counteroffensive by sea on the west coast just below Seoul at Inchon, cutting the overextended North Korean supply lines and trapping Kim Il Sung's armies. (The Chinese warned Kim of the danger, but he chose not to believe them.) "If we do not intervene in the Korean War," Mao told his advisers, "the Soviet Union will not intervene either . . . once China faces a disaster." The Chinese Politburo had concluded that the US would not use atomic bombs if China intervened in Korea because doing so would risk possible Soviet retaliation and because bombing into what the Chinese called "jigsaw pattern warfare" in the Korean mountains would put UN forces at risk. The Chinese also believed that the US was so heavily committed in Europe that it had only limited resources to devote to Korea. (What resources the United States had to spare was indicated at the beginning of October, when Truman approved a joint recommendation of the Atomic Energy Commission and the Department of Defense to spend $1.4 billion—at a time when the entire Defense budget was under $15 billion—to expand uranium and plutonium production facilities, doubling the US capital investment in military atomic energy. In June, Truman had approved construction of two heavy-water reactors at a site on the Savannah River near Aiken, South Carolina, for tritium production at an estimated cost of $250 million. In October, he approved plans for three more.)

US intelligence found some indication of Chinese movement as early as August 30, when LeMay's diary reports news from the Pentagon "that the Chinese have decided to aid the North Koreans, and that four Chinese armies have crossed the border into NK. Many airfields in North Korea are being readied for aircraft which could only be coming from either the Chinese or Russians." The information on army movement at least was garbled and inaccurate (the Chinese New 4th Army was drilling and maneuvering at the time along the Korean border), and no further information came to SAC about it. MacArthur landed his forces at Inchon on September 15. After fighting their way inland, they crossed into North Korea on October 1 and began pushing north toward the Yalu River, which divided Korea from China. The Inchon counteroffensive had shocked Stalin; Chinese sources claim he promised them air cover in response if they intervened on North Korea's behalf.

Mao's armies—twelve, not four—were supposed to begin moving across the Yalu into North Korea in mid-October. "We have decided to send part of the armed forces into Korea," Mao telegraphed Stalin on October 2, "to do combat with the forces of America and its running dog Syngman Rhee and to assist our Korean comrades." Mao's order officially entering the war was issued on October 8. That night, when Kim Il Sung heard the news, he clapped his hands and shouted, "Well done! Excellent!"

But Stalin immediately reneged on his promise of air cover and stalled the Chinese assault. "When the threat emerged [after Inchon]," Nikita Khrushchev recalled, "Stalin became resigned to the idea that North Korea would be annihilated, and that the Americans would reach our border. I remember quite well that in connection with the exchange of opinions on the Korean question, Stalin said, 'So what? Let the United States of America be our neighbors in the Far East. They will come there, but we shall not fight them now. We are not ready to fight.'" Mao and his advisers agonized for most of a week before deciding to go ahead with or without Soviet support, but when that decision reached Moscow, Stalin changed his mind once more and again offered support. The Chinese began crossing the Yalu on October 19 without Soviet air cover, but between October and December, Stalin sent thirteen air divisions to the Korean area. Ten Soviet tank regiments moved into Chinese cities for rear-area defense.

LeMay's diary records a briefing on November 2 confirming "increasing reports of Chinese intervention," including "attacks by Red jet fighters . . . against our troops and fighter [aircraft]." By November 28 "the Korean battle" was "going badly" and "the Chinese" had "made mincemeat out of our latest attack. Gen. MacArthur stated that conditions were out of control." A call to SAC from Vandenberg's office "indicates a feeling of pessimism there on our progress . . . and a growing feeling that we may have to send more [bomber] units to the Far East to bolster the AF there."

Truman confirmed at a press conference on November 30 what he had denied the previous summer: that he had been actively considering using atomic bombs in Korea since the beginning of the war. He also said the military commander in the field—meaning Douglas MacArthur—would have charge of the weapons if their use came to be authorized. "Every New York afternoon paper," writes John Hersey, who reported the story in *The New Yorker* a few weeks later, "[carried] immense front-page headlines [that day] saying that Truman might use the A-bomb—as if it were to happen at any moment." The world was alarmed. The White House "clarified" Truman's statement, but not before British Prime Minister Clement Atlee was compelled by his own party's threat of no confidence to announce in Parliament, "Then I shall have to go to Washington to see the President." Truman met with Atlee reluctantly, with advice from the Joint Chiefs to tell the Prime Minister that the US had "no intention" of using atomic bombs in Korea except to prevent a "major military disaster."

On the Saturday after Truman's Thursday morning announcement, LeMay reported to Vandenberg that "our analysis of available targets together with obvious considerations of possible adverse psychological reaction have led us to conclude that the employment of atomic weapons in the Far East would probably not be advisable at this time unless this action is undertaken as part of an overall atomic campaign against Red China." LeMay proposed to go personally to the Far East to direct any such campaign.

The following week—early December 1950—the JCS sent word to all military commanders that "the situation in Korea has caused the possibility of a general war to increase greatly." Truman declared a national emergency on December 16, announcing a decision to enlarge the armed forces by 3.5 million men and install price controls. LeMay wrote the general commanding SAC operations in Korea, Emmett "Rosie" O'Donnell, Jr., at that time about his atomic-bombing preparations. "As you can imagine," LeMay told O'Donnell, "there is quite a bit of agitation throughout the country to use the A-bomb to support our troops in Korea. In spite of the pressure, I do not feel that the chances of such action materializing are very great. However, until such a time as the Korean campaign has been concluded, there is always the chance that we will be directed to do the job. . . . [So] we are developing a small Shoran bombing capability in SAC, since we are convinced that this is the best method of doing the job [of atomic bombing]. . . ." LeMay's US crews had recently tried bombing using Shoran, a short-range radio navigation system, and in 147 bomb releases from 25,000 feet had achieved an average circular error of only 452 feet. The SAC commander asked O'Donnell to set up Shoran units in Korea "in such a way as to cover the most probable areas where critical situations could develop"; he told O'Donnell he was "relying on your good judgment to handle this matter quietly."

Both the State Department and the Joint Chiefs backed away from atomic bombing in December when MacArthur advised postponing a decision. But the issue reemerged again in early April 1951, a time of great confusion and doubt. MacArthur in particular was increasingly vocal in his opposition to Truman's policy of moving toward a cease-fire, which the Joint Chiefs also resisted. On March 10, MacArthur had asked for a " 'D' Day atomic capability" to use attacking airfields in Manchuria. Vandenberg discussed the problem with Secretary of Defense Robert Lovett and Air Force Secretary Thomas Finletter on March 14, noting "believe everything is set." Then MacArthur went too far, writing House minority leader Joseph Martin that there was "no substitute for victory." Martin read the letter into the *Congressional Record*, making MacArthur's insubordination public.

Early in April, some two hundred Soviet bombers moved onto air bases in Manchuria, a move which put them within striking distance of Japan, and the Soviets communicated through the Indian ambassador to China, Sandar Pannikar, that they would initiate such an attack if the US attacked beyond North Korea. There was movement of Soviet submarines as well. The Soviet threat aroused the Joint Chiefs to prepare to escalate to atomic war. On April 5, they ordered atomic retaliation against the air bases in the event of "a major attack" on UN forces. Gordon Dean, the new chairman of the Atomic Energy Commission, had been alerted the night before to the possibility that the JCS might issue such an order. While the Chiefs deliberated on April 5, Dean looked into the procedure for transferring nuclear capsules to the military—the question had never arisen before. Ramey wired LeMay, putting SAC "on the alert for some atomic operations based out of Okinawa. Apparently the aircraft of the 9th [Bomb Group] are being considered for this job." LeMay arranged to fly to Washington to meet with Vandenberg the next morning.

Omar Bradley, chairman of the JCS, took the decision to Truman for approval that morning of April 6, 1951. Truman saw an opportunity in the Joint Chiefs' conviction that they needed nuclear weapons in Korea to negotiate MacArthur's removal from command, an action the military leaders had been resisting. How could he release nuclear weapons to a commander in whom he had no confidence, who had shown himself to be a loose cannon? Evidently Bradley took the point and carried it back to the JCS.

Truman called Dean to the White House that afternoon, the AEC chairman recorded in his diary:

> I . . . was told by [the President] that the situation in the Far East is extremely serious; that there is a heavy concentration of men just above the Yalu River . . . ; that there is a very heavy concentration of air forces on several fields and the planes are tip-to-tip and extremely vulnerable; that there is a concentration of some seventy Russian submarines at Vladivostok and a heavy concentration on Southern Sakhalin—all of which indicates that not only are the Reds and the Russians ready to push us out of Korea, but may attempt to take the Japanese Islands and with the submarines cut our supply lines to Japan and Korea.
>
> He told me he had a request from the Joint Chiefs of Staff [deleted; probably "to transfer nine atomic bombs to military custody;"] that no decision had been made to use these weapons and he hoped very much that there would be no necessity for using them; that before there was any decision to use them the matter would be fully explored . . . ; that in no event would the bomb be used in Northern Korea where he appreciated, as I pointed out to him, that they would be completely ineffective and psychological "duds.". . .
>
> He then said that if I saw no objection he would sign the order to me directing me to release to the custody of General Vandenberg, Chief of Staff, USAF, nine nuclears. . . .

The same day, Dean wrote, he "called General Vandenberg and told him the President had signed the order; that we were prepared to discuss details with him." Two days later, after a Sunday afternoon discussion, the Joint Chiefs endorsed Truman's decision to fire MacArthur. He did so on Wednesday, April 11, to a firestorm of national hostility. By then the nine Mark IV nuclear capsules had been transferred (on April 10) and were in USAF custody.

Fairfield-Suisun AFB had been renamed Travis AFB in honor of the only US general ever killed by an atomic bomb. The 9th Bomb Group there was assigned once again to carry the bombs, deploying to Guam "on a normal training mission," as LeMay's aide noted deadpan in the general's diary, "and then maneuver[ing] to Okinawa where they will stay for possible action."

(This decision was quickly rescinded in Washington and the 9th was ordered to hold on Guam, probably because Okinawa was within range of the Soviet bombers.) LeMay assigned his deputy commander at SAC, Thomas Power, to direct the mission. Power had led the first and most destructive firebombing of Tokyo; he was "a man so cold, hard and demanding," writes a LeMay biographer, "that several of his colleagues and subordinates have flatly described him as sadistic. LeMay himself, when asked if Power was actually a sadist, has said, 'He was. He was sort of an autocratic bastard. But he was the best wing commander I had on Guam [during the Second World War]. He got things done." Power left for Tokyo to meet with MacArthur's successor, General Matthew Ridgway, on April 24, by which time the 9th Bomb Group appears to have deployed to Guam. LeMay wrote the USAF director of plans on May 7 that Power had negotiated an arrangement with Ridgway that "the use of A-bombs . . . must be approved by the Joint Chiefs of Staff upon request of the theater commander."

Whether or when the nine nuclear capsules were mated up with their HE assemblies is not reported in the declassified record of this first US military nuclear deployment since Hiroshima and Nagasaki. The bomber squadron and its replacements continued to be based on Guam, as did the nuclear and nonnuclear bomb components. As of May 18, Vandenberg and LeMay "determined that General Power's status in the Far East would be unchanged for the time being." Ridgway asked the JCS on August 28, according to LeMay's diary, "that SAC be prepared to deliver on twelve-hour notice an atomic attack if necessity arrives," but the Korean commander may simply have been preparing for future contingencies; he also asked for an aircraft carrier that could "lay . . . down an atomic attack if needed" and wanted "atomic artillery . . . prepared and forwarded as soon as a capability exists." The Navy had only begun to develop such an atomic capability, and atomic artillery was more than a year away. Far from nuclear plenty, LeMay was still working to build up SAC's capability to deliver its Sunday punch; he wrote Vandenberg on August 3, 1951, that his ability to put "bombs on [Soviet] targets" within six days after being ordered to do so had increased from 140 bombs to 146 since April. In a lecture the previous year he had called this order of magnitude "small-scale."

Four months later, on November 27, 1951, LeMay proposed to Vandenberg "discontinuing the Guam deployment," based on the judgment of the interservice watch committee "that the Communists in Korea are not preparing for a general ground offensive in the near future"—SAC could, LeMay pointed out, "quickly redeploy the squadron." Since LeMay's reason for discontinuing the deployment was nothing more urgent than its disruption of his training program—implying that he judged unit training to have higher priority than atomic-bombing China at that time—the bombs and bombers presumably soon came home.

The weapons on which the United States had gambled its security in the years immediately after the Second World War had turned out to be notably useless when war tested the new conditions of the atomic age. Robert Oppenheimer had foreseen that they might. "I have the impression that the general glow which surrounds the atom," he told an audience of officers at the National War College a few months after the beginning of the Korean War, "the excitement about atomic propulsion and super-super bombs and radioactive poisons, have made it hard to concentrate on the question of how many bombs we can use." The experience in Japan had set no precedents, he thought:

Are [atomic bombs] useful in ground combat? Are they useful in preventing the delivery of atomic bombs? What can we do with them? . . . It seems that it is a very hard job, when all the experience you have is of Hiroshima and Nagasaki, of an essentially defeated enemy looking for a chance to get out of the war; of undefended targets with no air cover, of air superiority all on our side; of an army decimated; of a fleet with its shipping all gone— when you have that as your only experience, it is a job that calls for a great deal of imagination to think what is the atom good for in war.

The JCS, who had been fighting for custody for years, never returned the nine bombs to the Atomic Energy Commission; they remained in Curtis LeMay's arsenal. He would have cherished them; Vandenberg had informed him early in 1951 that his side bet with "the guy who had the key" was off:

I agree with you [Vandenberg wrote] as to the confusion that a major attack against Washington would produce with respect to orders that should emanate from the Seat of Government. However, under existing law, authority to initiate the atomic offensive cannot be formally delegated either to field agencies of the Atomic Energy Commission or major commands of the Air Force.

The prospect of atomic-bombing Asia for the second time in less than a decade made Harry Truman sufficiently uncomfortable that he never again deployed armed atomic bombs abroad. His forbearance during a difficult war did not prevent him from indulging the bloody fantasy that atomic bombs might cut the Gordian knot of US-Soviet relations. He wrote a note to himself early in 1952 imagining a brutal ultimatum:

The situation in the Far East is becoming more and more difficult. Dealing with Communist Governments is like an honest man trying to deal with a numbers racket king or the head of a dope ring. The Communist governments, the heads of numbers and dope rackets have no sense of honor and no moral code.

It now looks as if all that the Chinese wanted when they asked for a cease fire was a chance to import war materials and resupply their front lines.

It seems to me that the proper approach now, would be an ultimatum with a ten-day expiration limit, informing Moscow that we intend to blockade the China coast from the Korean border to Indo-China, and that we intend to destroy every military base in Manchuria, including submarine bases, by means now in our control, and if there is further interference we shall eliminate any ports or cities necessary to accomplish our peaceful purposes.

That this situation can be avoided by the withdrawal of all Chinese troops from Korea and the stoppage of all supplies of war materials by Russia to Communist China. We mean business. We did not start this Korean affair but we intend to end it for the benefit of the Korean people, the authority of the United Nations and the peace of the world.

We are tired of these phony calls for peace when there is no intention to make an honest approach to peace. There are events in the immediate past which make it perfectly plain that the Soviet Government does not want peace.

It has broken every agreement made at Tehran, Yalta and Potsdam.

It raped Poland, Rumania, Czechoslovakia, Hungary, Estonia and Latvia and Lithuania.

The citizens of these countries who believe in self government have either been murdered or are in state labor camps.

Prisoners of World War II to the number of some 3,000,000 are still held at state labor contrary to cease fire terms.

Children have been kidnapped in every country occupied by Russia by the thousand [sic] and never again heard from.

This program is evidently a continuing one.

It must stop and stop now. We of the free world have suffered long enough.

Get the Chinamen out of Korea.

Give Poland, Estonia, Latvia, Lithuania, Rumania and Hungary their freedom.

Stop supplying war materials to the thugs who are attacking the free world and settle down to an honorable policy of keeping agreements which have already been made.

This means all out war. It means that Moscow, St. Petersburg [sic: Leningrad], Mukden, Vladivostock, Peking, Shanghai, Port Arthur, Darien, Odessa, Stalingrad and every manufacturing plant in China and the Soviet Union will be eliminated.

This is the final chance for the Soviet Government to decide whether it desires to survive or not.

Curtis LeMay felt similarly frustrated by a difficult political conflict that was building with weapons inventories to higher and higher levels of tension without release. "I, for one," the SAC commander told a Scripps-Howard reporter early in the Korean War, "don't want to stay in this state of tension for fifty years or more." LeMay thought instead that "we should vigorously go after our aims short of war, but at all times be ready for war." He clarified what he meant by "ready" when the reporter asked him about preventive war: "Any responsible Air Force officer will not advocate a preventive war, but you do have to risk it and be ready to fight." LeMay knew the time was coming when the Soviet Union would have the capability to knock out SAC on the ground; the USAF estimated that a minimum Soviet capability to do so would exist by the middle of 1951, whatever the CIA's doubts. Curtis LeMay, who loved his country, intended to find a way to forestall such a disaster even if it meant risking preventive war.

23

Hydrodynamic Lenses and Radiation Mirrors

STANISLAW ULAM and Cornelius Everett finished their second round of Super ignition calculations to discouraging results on June 16, 1950, shortly before the beginning of the Korean War. They established tentatively—pending the results of more detailed computer calculations—that igniting a tank of liquid deuterium with the heat from an atomic bomb would require impractical amounts of tritium, if the process worked at all. Enrico Fermi had returned to Los Alamos for the summer to help out with thermonuclear research. He and Ulam decided to take up calculating the next question raised by Edward Teller's Super design: whether, if the deuterium tank could be ignited at one end, burning would become self-sustaining and progress through the material.

"We did this again in time-step stages," Ulam recalls, "with intuitive estimates and marvelous simplifications introduced by Fermi. . . . I believe this work with Fermi to have been even more important than the calculations made with Everett." Fermi had brought along with him from Chicago a young protégé named Richard Garwin, whose desk abutted Fermi's in the small office they shared; Garwin remembers watching the Super calculations proceeding across the summer:

> Fermi and Ulam would prepare an accounting spread sheet. You write some differential equations calculating the burning of a large cylinder of liquid deuterium. You write a temperature over here and ask what happens as a function of time as neutrons and radiation and so on go down the cylinder. This [column on the spreadsheet] would be time, this [column] would be different radiation and you would have temperatures [in yet another column]. Fermi would do a couple of columns and give them to

[an assistant with a large adding machine] and [the assistant] would bring them back the next day. The problem was that as time went on, this thing would decay; it would be burn versus radius. You can't get [the cylinder of deuterium] to burn because the energy escapes faster than it reproduces itself. With this calculation, they established for their own purposes that the classical Super would not work. . . . All the time [spent developing the classical Super] was wasted. There had been miscalculation, because Teller was optimistic.

In the final report that the two men wrote, Fermi puckishly defined the difficulty as one that would vanish if only the basic physical constants could be changed: "If the cross sections for the nuclear reactions could somehow be two or three times larger than what was measured and assumed, the reaction could behave more successfully." (In fact, the actual cross sections, as remeasured experimentally at Los Alamos during 1951, proved to be even smaller and less promising than the less accurate measurements that Fermi and Ulam used.)

While Fermi and Ulam proceeded with their calculations, work also continued at Los Alamos toward the *Greenhouse* series now scheduled for spring 1951. As plans evolved across the summer, two tests emerged of devices that involved thermonuclear reactions. The first of these was a fission bomb that used DT-gas core boosting to increase its yield—a device called the Booster, designated *Greenhouse* Item. The second and more significant was a device called the Cylinder to be tested as *Greenhouse* George. "Edward Teller was trying to get support for the H-bomb project," Princeton physicist Robert Jastrow describes the genesis of the Cylinder, "and since he could not figure out how to build an H-bomb, he thought up the [George] project instead, as a demonstration piece for the people back in Washington." Carson Mark confirms this objective:

It was not a thing that we were busily working at except as a sequel to the Russian test. So okay, it was agreed we must do something that relates to the thermonuclear business; what would be a good thing to turn attention to? There were at least two different proposals, one mainly talked over and introduced by [George] Gamow and one which Teller favored. Both had to do with notions that fitted into the picture of a classical Super, as things we might be in a position to do and might be useful in that connection—things that would have to be done anyway, if one went that route. And it was from that rather vague scrambling of ideas that the detailed form of the George shot emerged. It was the pattern that Edward had favored. It turned out that it was very much the pattern of the 1945 Fuchs–von Neumann patent. . . . It wasn't by any means a copy from that but it did about the same thing and related to the Super in about the same way of their patent—namely, a possible way of getting the first step from a fission bomb towards heating deuterium to a point where it might burn. And that's what the George shot was about. It was doing things which we had never taken seriously to do before which had been sort of sketched out in an arm-waving fashion.

Theoretical physicist Marshall Rosenbluth, who worked on the Cylinder, explains the strategy of the design: "The idea was to try to look at some burning thermonuclear fuel and to do it in a way where you could clearly diagnose it. It wasn't so very obvious where this was going to get you, other than as an experiment with the relevant material—DT—in which you could look in detail at how the temperature would change, density was changing, all sorts of neutrons were being produced—the detailed diagnostics. Teller's idea was that we needed to put this DT where we could see it, not in the middle of an explosion, so we'll pipe some radiation out to it." One of the test planners described the Cylinder at the time as "an experiment in which [a] nuclear explosion is [to be] used to send material down a tube and cause a thermonuclear reaction of small magnitude in deuterium." Jastrow says the Cylinder used "the energy of a 500-kiloton atomic bomb to ignite a fraction of an ounce of deuterium and tritium placed in a small adjoining chamber. . . . Everyone knew beforehand that it was pretty certain to work; using a huge atomic bomb to ignite the little vial of deuterium and tritium was like using a blast furnace to light a match."

When a mass of uranium or plutonium fissions in an atomic bomb, nearly all of the energy released takes the form of electromagnetic radiation, the class of radiation that includes visible light. Such radiation consists of weight-less packets of waves called photons, which travel at the speed of light: 186,400 miles per second or 1 foot per nanosecond.* Photons have wave-lengths that differ according to their energy—the more energetic the photon, the shorter the wavelength. The coils of an electric heater demonstrate the connection between energy and wavelength as they warm and begin to glow, heating from invisible longer waves of infrared (which humans perceive as warmth) to dark red to orange to yellow, each color produced by the effect on the human eye of photons of progressively shorter wavelength. An even hotter acetylene torch burns blue-white. Farther up the same continuum, a hot sunlamp filament radiates invisible ultraviolet light. Beyond ultraviolet come soft X rays, hard X rays and gamma rays, all a function of the temperature of the radiating material (that is, of the energetic motion of the material's atoms), each type of radiation consisting of photons of shorter wavelength than the last. Graphs of the radiation coming from the electric heater, the acetylene torch, the sunlamp filament or an atomic fireball would

* A nanosecond is one-billionth of a second.

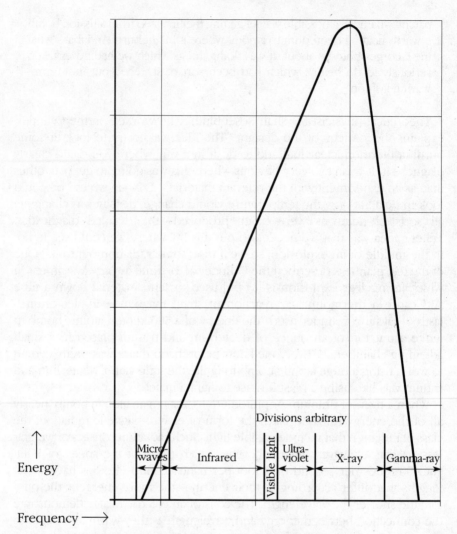

**Heat radiation as a function of frequency for
nuclear fireball (400 million°K).**

all show the same basic sharply peaked curve, with the peak shifted in each
case to whichever region of radiation predominates. The Big Bang that
started the universe created a fireball with a similar curve that shifted its
peak across the aeons downward through progressively longer wavelengths
as the universe cooled until it now peaks in the cold microwave region
below infrared—the famous 2.7° Kelvin cosmic microwave background.
Indeed, the universe itself is a cooling fireball similar to a nuclear-weapon
fireball, an artifact of the Big Bang.

Radiation from an atomic fireball peaks in the soft-X-ray region. Even a small atomic bomb produces an enormous flux of radiant energy; if a cube of that massless flux could somehow be cut out and weighed it would reveal itself to be nearly as heavy as an equivalent cube of air. Clearly, that many extremely hot photons moving densely together can penetrate and heat material objects to high temperatures. As Rosenbluth emphasizes, the idea of the Cylinder was to make some deuterium fuse with some tritium away from the immediate vicinity of an exploding fission bomb, where the reaction could be instrumented and studied. The Booster, in contrast, was designed to compress and ignite some DT *inside* an exploding fission bomb. To arrive at the Cylinder configuration, Teller asked what component of an exploding fission bomb could be used to ignite DT *outside* a bomb. Neutrons pour out of an exploding fission bomb in weighty quantities and could certainly serve that purpose, but because they have mass they move slowly compared to massless photons. In fact, the first component of a fission bomb to move out from the core, well ahead of any material particles such as neutrons or fission fragments, is radiation. An X-ray photon travels ten feet in the time it takes for a uranium nucleus to fission; in the same brief time, fission fragments move only about four inches, neutrons (because they are lighter) somewhat more. "The choice [of using radiation] is forced on you," Mark emphasizes. "It's the most certain, faster-moving thing that progresses from where the fission bomb is to where you might have your DT pellet. Even if you decided you didn't want the radiation, you couldn't really strain it out. It would force itself on your attention." Teller proposed to use the X-radiation to convey energy through a pipe—a radiation channel—to a small capsule of DT outside the fission system, with the intention of studying experimentally rather than only theoretically one small part of his system for the classical Super. His Family Committee concluded design development and froze the designs for the Booster and the Cylinder—for the *Greenhouse* Item and George shots—on October 26, 1950.

The following week, the AEC's General Advisory Committee came to Los Alamos to review the thermonuclear program. The GAC had several new members now, appointed in part because of their enthusiasm for the H-bomb. The most accomplished among them was chemist Willard Libby, whose 1946 development of radiocarbon dating would eventually win him a Nobel Prize in Chemistry. Libby's enthusiasm for the H-bomb translated into suspicion of Robert Oppenheimer as well; Libby believed that Oppenheimer had been "more or less the head" of "a strong Communist contingent at Berkeley" during the Second World War.

The GAC majority had reelected Oppenheimer to chair the committee despite the debacle of the H-bomb debate. Aware that he had become controversial, he had recently offered Gordon Dean his resignation. "[Oppenheimer] said he knew that we had had quite a disagreement on the H-bomb program back in 1949," the AEC chairman recalled. ". . . He told me that he thought that this had perhaps hurt his effectiveness on the . . . Committee, and that he was prepared to get off if for one moment I thought . . . that he could not serve. I thought about it for a few moments . . . and I told him that . . . the . . . Committee would definitely lose, and so would the Commission, if we lost him from it at that time." Oppenheimer convened the GAC at Los Alamos at the end of October 1950 to give the committee's new members "some feeling for the sort of place [Los Alamos] is," he told Norris Bradbury, because they "kept suggesting that a second Los Alamos be set up in order to relieve the work of the first." Oppenheimer's explanation makes it clear that Teller was agitating for a second laboratory before the end of 1950; Bethe traces Teller's vision of a separate laboratory devoted primarily to thermonuclear research even further back, to as early as 1947.

At Los Alamos, the GAC reviewed the Ulam-Everett and Ulam-Fermi calculations, a new model of the Alarm Clock that Teller and John Wheeler had proposed and the plans for the classical Super itself. The advisers were particularly enthusiastic about the Cylinder. "New and elaborate instrumentation forms an essential part of this test," Oppenheimer summarized their understanding. "If the tests and the instrumentation are reasonably successful, radically new information will be obtained. This information bears on the non-equilibrium burning of tritium-deuterium mixtures and on the phenomena associated with the flow of radiation from fission weapons into materials of varying density, and will be relevant to many thermonuclear models."

Despite their praise, the committee members understood that the Cylinder was a demonstration as much as an experiment and was not a potential breakthrough to a thermonuclear. "We wish to make it clear, however, that the test, whether successful or not, is neither a proof firing of a possible thermonuclear weapon nor a test of the feasibility of such a weapon. The test is not addressed to resolving the paramount uncertainties which are decisive in evaluating the feasibility of the Super."

"Teller took the floor to summarize the Super," the official AEC history reports of the Los Alamos GAC meeting. "In his briefing he could offer little more than determination. . . . He had no new ideas. In some way success would be grasped—how, he did not know." Bankrupt of ideas though he admitted himself to be, Teller could still insult his Los Alamos colleagues. "Even the victory might be dangerous to Los Alamos," the AEC history continues its paraphrase of his presentation. "If the [George] test showed the Super impossible, Teller believed the laboratory was strong enough to continue its work, but if the reverse were true—if the test showed the Super was possible—the laboratory might not be strong enough to exploit the triumph." ("I sensed the tension in the corridors," Françoise Ulam remembers of that time, "the bafflement and sometimes annoyance at Teller's autocratic behavior and temperamental outbursts. I had the feeling that nobody quite knew how to handle his demands, and sort of caved under, except [her husband] Stan. . . .")

In fact the failure was Teller's, Hans Bethe observes, not the laboratory's:

That Ulam's calculations had to be done at all was proof that the H-bomb project was not ready for a "crash" program when Teller first advocated such a program in the fall of 1949. Nobody will blame Teller because the calculations of 1946 were wrong, especially because adequate computing machines were not then available. But he was blamed at Los Alamos for leading the Laboratory, and indeed the whole country, into an adventurous program on the basis of calculations which he himself must have known to have been very incomplete.

George Gamow found a way to dramatize how unpromising Teller's Super had proven to be. John McPhee reports the story as Los Alamos physicist Theodore Taylor remembered it. "One day, at a meeting of people who were working on the problem of the fusion bomb . . . Gamow placed a ball of cotton next to a piece of wood. He soaked the cotton with lighter fuel. He struck a match and ignited the cotton. It flashed and burned, a little fireball. The flame failed completely to ignite the wood, which looked just as it had before—unscorched, unaffected. Gamow passed it around. It was petrified wood. He said, 'That is where we are just now in the development of the hydrogen bomb.'"

From October on, Bethe continues, while the President of the United States warned the world that he was prepared to use atomic bombs in Korea and the Chinese armies pushed down the Korean peninsula, Teller "was desperate":

He proposed a number of complicated schemes to save [the classical Super], none of which seemed to show much promise. It was evident that he did not know of any solution. In spite of this, he urged that the Laboratory be put essentially at his disposal for another year or more after the *Greenhouse* test, at which time there should then be another test on some device or other. After the failure of the major part of his program in 1950, it would have been folly of the Los Alamos Laboratory to trust Teller's judgment, at least until he could present a definite idea which showed practical promise.

Teller viewed his position far more grandly. He had stuck by his guns, he wrote Ernest Lawrence at the beginning of December; the criticism was unfair. All he was trying to do was to make sure that a valuable addition to the nation's security was not neglected. "Dr. Teller was never one to keep his candles hidden under bushels," AEC commissioner Sumner Pike would testify. "He was kind of a missionary. I might say that perhaps John the Baptist is a little overexaggeration. He always felt that [his] program had not had enough consideration. . . . I would guess that it would have suited him completely if we had taken all the resources we had and devoted [them] to fusion bombs." Los Alamos theoretician Charles Critchfield, a levelheaded man, was equally blunt. "Teller has a messianic complex," he concluded.

Motivated partly by distaste for Teller's posturing, Stanislaw Ulam had set out systematically to determine if the design upon which the Hungarian-born physicist had insisted with such vehemence for more than five years was simply a fantasy or had some useful connection to physical reality. Ulam and his colleagues demonstrated that the Super's connection was probably tenuous at best. Before the end of the year, John von Neumann's ENIAC calculations came in to add additional weight to that conclusion. "Why did we still follow the classical Super?" Bethe asks of that depressing autumn. "Well, for one thing, it had not been proved that it was dead. For another thing, nobody had an alternative and better idea." Then, Bethe concludes, Ulam's work plus the von Neumann calculations "proved that the Super was dead."

Teller might cling to his fantasy of a thermonuclear weapon of unlimited power; for Ulam, the long effort of calculation had been a necessary clearing away. When the Polish-born mathematician finished his Super calculation work, he began thinking about how fission bombs could use the still-scarce US supply of U235 and plutonium more efficiently. Implosion, by compressing a subcritical core, made it possible to get a bigger bang from a smaller amount of fissile metal than was possible with a gun design, which simply assembled a critical mass without compression, but the degree of compression available for implosion was limited by the force that could be generated from high explosives. In December 1950, Ulam thought of a way to increase implosion compression by orders of magnitude. He called the arrangement "hydrodynamic lensing." The fluid ("hydro") Ulam had in mind to power ("dynamic") his implosion system was the shock wave from an atomic bomb —primarily, he writes, the "enormous flux of neutrons." For a short time during a fission bomb explosion, Ulam pointed out, "one has . . . a neutron 'gas' . . . present together with the other [fission fragment] nuclei at densities comparable to those of ordinary solids." He took his new idea to Carson Mark. "The first thing that Ulam described to me," Mark remembers, "at the beginning of 1951, was using fission for compression—much greater compression than any other method we were aware of allowed. You could compress a little piece of fissionable material, for example, and make it explode."

To use fission for compression, Ulam realized, would require staging: one bomb, which came to be called a "primary," would set off a second, physically separate bomb, a "secondary." Theoretically any number of bomb assemblies could be set off this way in succession. The idea was elegant but premature; Los Alamos had no immediate use for it. It stayed with Ulam into the new year. Sometime in January, he received a memorandum from Darol Froman asking "various people what should be done with the whole 'Super' program. While expressing doubts about the validity of Teller's insistence on his own particular scheme, I wrote to Froman that one should continue at all costs the theoretical work, that a way had to be found. . . ." Shortly after responding to Froman's memorandum, Ulam writes, he saw a way to apply his "iterative" scheme to a thermonuclear. He says he then "put his thoughts in order and made a semi-concrete sketch." Françoise Ulam never forgot the moment she heard:

> Engraved on my memory is the day when I found him at noon staring intensely out of a window in our living room with a very strange expression on his face. Peering unseeing into the garden, he said, "I found a way to make it work." "What work?" I asked. "The Super," he replied. "It is a totally different scheme, and it will change the course of history."

Françoise, who "had rejoiced that the 'Super' had not seemed feasible," was "appalled by this news." She asked her husband what he intended to do:

> He replied that he would have to tell Edward. Knowing how unpleasant Teller had been, I ventured that maybe he ought to test his idea on someone else first. He did. That very afternoon, I think, he went to see Carson [Mark]. But the meeting was not very satisfactory, for when, in the evening, I asked him how it had gone, he only shrugged his shoulders and said that Carson was awfully busy. Too busy, I surmised, to pay much attention to what he was trying to tell him. (Stan could be oblique or Pythic sometimes, especially when particularly preoccupied with a topic he assumed his interlocutor was on the same wavelength [with]. . . .) At any rate, he had to go to Teller after all.

Ulam had gone on to see Norris Bradbury after he talked to Mark that afternoon, he writes, "and mentioned this scheme." Bradbury had been more welcoming. "He quickly grasped its possibilities and at once showed great interest in pursuing it. The next morning, I spoke to Teller." Ulam probably spoke to Teller sometime during the last week in January 1951.

To apply his idea of staging to the thermonuclear, Ulam had to work through the question of compressing the thermonuclear fuel, which is what his shock-implosion system would do. Teller's classical Super was essentially a system for heating uncompressed liquid deuterium to the point, Teller hoped, when it would sustain thermonuclear burning. The problem with the classical Super was that radiation carried heat away from the fuel faster than thermonuclear reactions could replace it. Compressing the fuel would not obviously improve that situation. The question had come up so often, says Carson Mark, that Teller had worked it into his standard briefing on Super design. "Teller used to say, 'Compression makes no difference.' That was number two on his list. He instilled that notion in many people."

Teller himself explains his attitude toward compression by referring to what he calls a similarity relation. Suppose, he says, you have a unit of deuterium in its normal state and another unit that is compressed to one thousand times the first unit's density. Both, he believed, would react the same way if other variables—temperature, hydrodynamic velocity and so on —were the same. The compressed fuel is a thousand times more dense, so its reaction rate is a thousand times faster. But since everything is a thousand times closer together, the amount of time required for the reactions to propagate from one atom to the next is a thousand times shorter. "You can see," Teller concludes, "that all the reactions that depend on a collision between a pair of particles will proceed in a similar way because in the comparison between reaction time and expansion time, both have been equally reduced. The course of the reaction is the same, with the 'insignificant' change of a thousandfold contraction of time. . . ." That is, everything would go faster, but the outcome—including, in the case of thermonuclear burning, the fatal cooling of the fuel mass by radiation loss—would otherwise be similar.

"The theorem is not quite precise," however, Teller continues. A three-body reaction comes into play with compression, he says—the collision of an electron, a nucleus and a photon with absorption of the photon, which prevents it from escaping. At normal pressure, this three-body reaction occurs at a marginal rate. But compression increases the rate, Teller says he eventually realized, "and the greater the density, the more the improvement. Therefore, when we make estimates on the basis of the similarity theorem, we really have been too pessimistic. [Photons] may be absorbed in . . . three-body collisions." Absorbing such radiation would deposit its energy in the compressed fuel as heat which otherwise would have been lost, improving the prospects for sustained thermonuclear burning.

"I don't understand this three-body crap," Carson Mark responds impatiently to Teller's explanation. "Compression reduces the volume and changes the energy needed to overcome the loss due to radiation." According to standard explanations of thermonuclear fusion processes, compression works in thermonuclear fuels in much the same way it works in fission fuels, squeezing nuclei closer together and therefore improving their chances of interacting. With compression, fusion reactions proceed faster and in a smaller volume, which makes that smaller volume correspondingly hotter. Four-fifths of the energy from the fusion reactions goes into heating neutrons, which are electrically neutral and which therefore escape from the burning fuel, carrying away energy. But the remaining 20 percent of the energy released in the fusion reactions goes into heating alpha particles— the helium nuclei formed from the hydrogen nuclei when they fuse—and since alpha particles are positively charged, they interact electrically with the surrounding fuel, colliding and giving up their heat. This process of alpha-particle heating of the fuel increases proportionally as the fuel density increases. With sufficient compression, the alphas can even be stopped from escaping entirely, more than compensating for the loss of heat from radiation and from neutron loss. Whatever Teller's reason for rejecting compression, Mark comments, "eventually he came to understand that it was wrong."

Paradoxically, Soviet research may have contributed significantly to Teller's education in the virtues of compression. It may also have informed Ulam's thinking. Arnold Kramish had carried an intelligence report to Los Alamos in April 1950 that concerned deuterium compression. An Austrian scientist named Schintlemeister, who had recently been repatriated from internship in the Soviet Union, had reported that Peter Kapitza was experimenting with magnetic compression of deuterium cylinders. The report may have been accurate or it may have been garbled. Though Peter Kapitza had been placed under virtual house arrest after challenging Lavrenti Beria's authority, he had not been barred from physics research. Kapitza was a specialist not only in cryogenics but also in the generation of powerful magnetic fields. It may well have occurred to him in the late 1940s that such magnetic fields might be used to compress cylinders of deuterium to produce thermonuclear fusion in a controllable rather than an explosive form —that is, to make a fusion reactor. Andrei Sakharov, however, has reported pursuing magnetic-confinement fusion beginning in the summer of 1950 and describes a different (perhaps independent) origin for the idea: a letter from a young sailor, Oleg Lavrentiev, that Beria passed on to Igor Tamm. "My first vague thoughts on magnetic rather than electrostatic confinement occurred to me," Sakharov writes, "as I read Lavrentiev's letter and wrote my reply." Sakharov's ideas, developed with Tamm, led to the invention of the magnetic-confinement fusion system known as the tokamak.

However the idea originated, Kramish carried it to Los Alamos, where he briefed Teller, Emil Konopinski, John Wheeler, Teller protégé Frederic de Hoffmann and possibly Ulam on April 18, 1950. "I traveled to Los Alamos several times after that," he recalls, "to bring Teller and de Hoffmann up to date on intelligence indicators. Each time, Edward returned to refinements of the Schintlemeister report. Towards the end of January 1951, at Los Alamos, the idea came up again, Edward having made more calculations of temperature and other parameters. And he had remained infatuated with the relationships of compression, temperature and radiation." In Kramish's judgment, Schintlemeister's information about Kapitza's work prepared Teller to understand Ulam's innovation.

If the three-body reaction is "crap," why did Mark and others not pursue compression before Ulam's breakthrough? They did not do so, Mark explains, because chemical explosives were the only known materials with which to generate the necessary pressures. Chemical explosives were inadequate on two grounds: they blew material into the thermonuclear fuel that would probably quench the thermonuclear burn entirely, and they could not generate sufficient compression to make a useful difference in the thermonuclear reaction rates. Ulam's "iterative scheme," says Mark, "changed all that."

Ulam understood that compression was the issue by the time he proposed his staged system to Teller late in January 1951; he described his idea a few years later as "an implosion of the main body of the device . . . [in order to] obtain very high compressions of the thermonuclear part, which then might be made to give a considerable energy yield." Teller was not immediately convinced, Ulam recalled. "I don't think he had any real animosity toward me for the negative results of the work with Everett [that had been] so damaging to his plans, but our relationship seemed definitely strained." (Teller and Ulam "knew each other quite well," Mark comments, "and I knew them both. Each was aware that the other was a pretty bright person —sharp, intelligent. Ulam used to make witty, pointed, scornful, shamefully disreputable remarks about Teller when Teller wasn't there. Once in a while his feelings about Teller couldn't have escaped Edward's notice. Edward reciprocated those feelings generously, so each was talking down the other and that went on for years.")

"For the first half an hour or so during our conversation," Ulam continues, "[Teller] did not want to accept this new possibility. . . ." But "after a few hours," Teller "took up [Ulam's] suggestions, hesitantly at first," then "enthusiastically." Ulam mentions two reasons why Teller warmed to the staging idea: "He had seen . . . the novel elements" and he had "found a parallel version, an alternative to what I had said, perhaps more convenient and generalized." The "novel elements" were presumably compression as a way to improve the reaction rates and staging as a way to achieve it. Teller's "parallel version" was a brilliant adaptation of his own; he proposed to use the *radiation* coming off the fission primary, rather than the neutrons, to compress the thermonuclear secondary. In atomic bombs of low efficiency, such as the wartime Fat Man, radiation is a comparatively minor product: the bomb is relatively opaque. In highly efficient atomic bombs, such as the smaller, higher-yield weapons Los Alamos was then building, energy comes out predominantly as radiation; the bomb is more transparent and the fireball hotter. "From the point of view of the fission [process]," Teller explains, "the escape of radiation was not very important, as Oppenheimer and I estimated [during the war] and as [theoretical physicist] Maria Mayer's group then very explicitly proved. But they calculated opacities and I knew about that in detail, and therefore, when I was forced into thinking about the effects of compression, then I at once knew how to do it." The advantage of using radiation instead of material shock to implode the thermonuclear secondary might be faster and longer-sustained compression of the fusion fuel to greater density.

"From then on," Ulam concludes, "pessimism gave way to hope. In the following days I saw Edward several times. We discussed the problem for about half an hour each time. I wrote a first sketch of the proposal. Teller made some changes and additions, and we wrote a joint report quickly. It contained the first engineering sketches of the new possibilities of starting thermonuclear explosions. We wrote about two parallel schemes based on these principles."

Ulam may have collapsed the time frame in memory. The joint report he and Teller wrote was issued on March 9, 1951. Earlier, Kramish recalls— during the last week in January—"Teller came along and said Ulam has an idea but I have a better one." The "better idea" was radiation implosion:

> Teller assembled de Hoffmann, Max Goldstein and myself to discuss Ulam's new idea. Edward said that Ulam was on track but hadn't gotten it right. Edward suggested a geometry and the four of us began to work out the equations—which turned out to be quite complex. At the end of the afternoon, Edward said he had to go to dinner with family and play the piano. In his polite and forceful way, he said, "That's it. Why don't you three work a little while and we will discuss the meaning of the solution in the morning."
>
> The "little while" was all night. . . . Working with our Frieden and Marchand calculators, we came up with an approximate solution in the morning. Teller, refreshed, became absolutely joyous, while we three were on the point of collapse.

If Kramish's recollection is accurate (and the all-night session eventuated in a paper, "An Estimate of [deleted] Temperatures," issued on February 4), then Ulam's "following days" comprised most of a month. He wrote von Neumann about the breakthrough on February 23. "Had the following couple of thoughts (ideas) about bombs," he noted laconically. He outlined the ideas in a few words, then caricatured Teller's response: "Edward is full of enthusiasm about these possibilities; this is perhaps an indication they will not work." Teller might think the remark cruel; Ulam had earned the right to make it if anyone had.

The joint report of March 9, crediting "work done by" E. Teller and S. Ulam, was titled "On Heterocatalytic* Detonations I: Hydrodynamic

---

* "Hetero" from a Greek word meaning "the other of two," "other," "different"; "catalytic" meaning facilitating a reaction.—RR

Lenses and Radiation Mirrors." It described staging and emphasized compression:

> In this discussion the following general scheme is considered. By an explosion of [deleted; probably "one or more"] auxiliary fission bombs, one hopes to establish conditions for the explosion of a 'principal' bomb. [Deleted]. We proposed to discuss certain general features of such an arrangement. The main purpose of the "auxiliary" system is to induce very high compressions in the principal assembly.
> [Five pages deleted.]
> The scheme then depends on concentrating, as much as possible, the energy released by the explosion of a fission bomb in the mass of the principal assembly and *doing it so as to achieve a high compression* in this mass.

How original were Ulam's and Teller's ideas? Who should receive more credit for the breakthrough? Opinion among knowledgeable participants and observers ranges far and wide. Bethe assessed the Teller-Ulam invention of staging and compression most generously in a nearly contemporary review:

> It is difficult to describe to a non-scientist the novelty of the new concept. It was entirely unexpected from the previous development. It was also not anticipated by Teller, as witness his despair immediately preceding the new concept. I believe that this very despair stimulated him to an invention that even he might not have made under calmer conditions. The new concept was to me, who had been rather closely associated with the program, about as surprising as the discovery of fission had been to physicists in 1939.... [It] created an entirely new technical situation. Such miracles incidentally do happen occasionally in scientific history but it would be folly to count on their occurrence.

Herbert York weighs the difference between concept and design:

> What Ulam did was not a thermonuclear device. It was a general idea. What Teller did was convert that into something which was a sketch of a Super that would work. Teller sketched out a super bomb. Ulam simply presented a fairly general idea in dealing with that topic. I think Teller has slighted Ulam, but I think also Teller does deserve fifty-one percent of the credit.

Ulam discounted the originality of the invention late in life, but by then Teller had worked for years to deny his colleague's contribution. "It was a different arrangement," Ulam said. "It was not new physics. It's not to my mind any such very great intellectual feat. It was partly chance. It could have come a year earlier or two years earlier. It might have . . . shortened the time between the [H-bomb] decision and the actual production of a physical object."

Norris Bradbury in retirement discounted not the originality of the invention but its serendipity:

> Don't ask me who's the father of the H-bomb, because nobody is. . . . The whole thing was a matter of people putting ideas into a pot around a coffee table. . . . Somebody had an idea and that doesn't work, but they give me an idea, but that doesn't work, but you can give me an idea. Because they were beating one idea against another around a table, in front of a blackboard and so on. It wasn't anybody's real invention; it was just a lot of people working on it. . . .
>
> Eventually an idea appeared, which looked as if it might work, and for which the technology for the [primary] was reasonably available. Not completely, but could be gotten together pretty fast if you put some more work on that. . . . It's a very complicated technical question.

Marshall Rosenbluth sets the Teller-Ulam invention in the context of the work Teller in particular had done on the Cylinder:

> It's sort of obvious, once you start doing detailed calculations on radiation flowing out from a bomb to this little test experiment, then you think, why not use a radiation implosion for the secondary? And similarly on the Booster, doing the calculations of how it was going to get imploded and how it would interact with the fissionable material around it—that would be very much related to what you would want to see in the secondary of a thermonuclear explosion. So at least my feeling was that this is kind of a paradigm of the way physics is done. You reach a certain point in your theoretical thinking and then you get stuck for one reason or another, can't go any further, but you try to think what experiments you might do that would shed some light on the situation, even if it's not very obvious—how you go from here to there to actually where you're trying to go, which in this case was the H-bomb. I can see a very clear path. The planning of the *Greenhouse* experiments, at least in my opinion, led Teller to the idea of radiation implosion.

Carson Mark also connects the invention to *Greenhouse:*

> It's true that the Teller-Ulam scheme involved radiation. It's true that Ulam thought that you could use . . . hydrodynamic shock. That's certainly there too. It doesn't move as rapidly as radiation does. It's harder to control and direct. So if you were trying to exercise Ulam's idea of using hydrodynamic

shock you'd find, by God, radiation gets there first. Teller then is supposed to have proposed that it would be better to use radiation than to rely on [material shock]. Well, he was right, it was simpler, but you couldn't have avoided it. Had you sat down to design the thing, asking, now, what's the material shock doing, where is it, how fast does it move? You'd say, dear God, the radiation is going faster, it's there, so let's concentrate on that. So it was hardly an important circumstance that Teller thought of radiation whereas Ulam thought of . . . material [shock]. The fact that Edward thought of radiation was natural because he had been involved in much more detailed work on the George shot than had Ulam.

Niels Bohr, disappointed that a sense of common danger had not prevented such a threatening development, assessed the Teller-Ulam invention most critically in a conversation a decade later with his friend J. Rud Nielson:

[Bohr] told me what Teller's contribution to the design of the hydrogen bomb was. When I asked him how great a contribution this was, he answered: "Old physicists who have turned administrators might not think of this solution. However, if you had asked a good class of physics students, two or three of them would have suggested this solution."

Which was perhaps unfairly sharp; neither the Soviets, the British nor the French found the Teller-Ulam configuration the first time around.

But whatever the technical originality of the ideas Ulam and Teller developed together in February 1951, John Manley concluded, the political effect was electrifying. "Teller and Ulam really won [the argument about building the thermonuclear] by figuring out how to do it. . . . You don't really want to work on something that you don't know how to do." Los Alamos radiochemist George Cowan enlarges shrewdly on Manley's point:

Knowing that it was going to work sure encouraged those people to go ahead and do it. Particularly because now, whether you believed that it was a good idea or not, the fact that that power was available—anyone who has a realistic view of the way power gets used by governments would know that you had to do it. There was no longer an option. Despite all the debates, no responsible government would ever voluntarily forgo developing a very powerful new weapon if it knew how to do it. That is something that you can talk about if you're not in the government, but if you are in the government it is not an option.

Ulam understood the historic fatality of his breakthrough very well, however much he discounted its originality, and laid claim to its power. There is an old Jewish legend of a Frankenstein's monster, the Golem, created out of clay and animated with holy words to defend the Jews from anti-Semitic assaults. The creator of the Golem was reputed to be the great medieval scholar Rabbi Loew of Prague. In some versions of the myth, the Golem goes out of control and becomes destructive, like a robot run amok. Ulam's Aunt Caro, he writes in his autobiography, "was directly related to the famous Rabbi Loew of sixteenth-century Prague, who, the legend says, made the Golem. . . ." Ulam recalls mentioning this exotic connection once to MIT mathematician Norbert Wiener. "[Wiener] said, alluding to my involvement with Los Alamos and with the H-bomb, 'It is still in the family!' " Ulam's implicit claim to have called forth the Golem compares in scale to Robert Oppenheimer's famous characterization of the atomic bomb as a manifestation of Vishnu: "Now I am become Death, the destroyer of worlds." The difference in the two claims is style: Ulam's cool and understated, as he was; Oppenheimer's more urgent and committed but psychically dissonant. Oppenheimer's allusion had amused Ulam when he first heard it; he thought Oppenheimer was referring to himself and considered the claim pretentious. People called the rabbi's Golem "Dumb Yossel," one commentator notes. It sat in the room where the Rabbi held court, its head resting in its hands, without any mind or thought of anything at all, waiting to be summoned.

Edward Teller seems to have found it intolerable that someone might share credit for the historic invention on which he had been working single-mindedly for almost ten years; he moved immediately to take over the technical breakthrough and make it his own. After he and Ulam issued their joint report, Françoise Ulam observes, "my impression is that from then on Teller pushed Stan aside and refused to deal with him any longer. He never met or talked with Stan meaningfully ever again. Stan was, I felt, more wounded than he knew by this unfriendly reception, although I never heard him express ill feelings toward Teller. (He rather pitied him instead.) Secure in his own mind that his input had been useful, he withdrew." (Carson Mark confirms Françoise Ulam's impression: "Ulam felt that he invented the new approach to the hydrogen bomb. Teller didn't wish to recognize that. He couldn't bring himself to recognize it. He's taken occasion, almost every occasion he could, not every one, to deny that Ulam contributed anything. I think I know exactly what happened in the interaction of those two. Edward would violently disagree with what I would say. It would be much closer to Ulam's view of how it happened.")

Later in March, Teller added a crucial additional stage to the Teller-Ulam configuration: a second fission component positioned within the thermonuclear second stage to increase the efficiency of thermonuclear burning. A symmetrical shock wave moving inward through a cylinder of deuterium converges in the middle on itself, at which point the decelerating motion of the imploding material is converted to heat. The small region at the center of the long axis of the cylindrical mass of thermonuclear material, where the heat is thus confined, came to be called the "sparkplug"; it was in this region that thermonuclear burning would initiate. Teller realized that a subcritical stick of U235 or plutonium, positioned where the sparkplug would form at the center axis of the deuterium cylinder, would be compressed to supercriticality by the leading edge of the imploding shock wave. This second fission explosion would then push outward against the implosion that was pushing inward; with careful design, the main implosion and the sparkplug explosion might be made to come to equilibrium, stabilizing in a hot, highly compressed critical layer that would advance outward through the deuterium fuel mass and burn it much more efficiently and completely than could an unboosted sparkplug alone. Teller called this design "an equilibrium thermonuclear gadget" in a report he signed on April 4, 1951; he claimed it in the report's subtitle as "a new thermonuclear device."

The fission sparkplug piggybacked on Ulam's previous breakthrough to staging and compression. It may also have been suggested to Teller, probably subliminally, by Ulam's original idea of using a fission bomb to ignite a second fission bomb, the idea that had led to the breakthrough of staging. Bethe would characterize Teller's invention as representing "the very important second half of the new concept." Carson Mark discounts its originality. "The sparkplug is an obvious idea," he comments, "that occurred to anyone who looked at the system. You ask what do I do if I have it compressed and you think of the sparkplug." However obvious, when physicists refer to the "true" hydrogen bomb as it came to be developed, they customarily cite the Teller-Ulam invention—staging, implosive fuel compression before ignition and a fission-boosted sparkplug—as its ingenious principal mechanism.

Its virtues were not immediately celebrated, as Ulam's remark to von Neumann about the kiss-of-death effect of Teller's enthusiasm makes clear. "I'm not sure that it's fair to say that everybody immediately said, 'Ah ha!' " Marshall Rosenbluth observes. "Everybody realized that the Teller-Ulam invention was a new way of looking at the problem and of course a number of questions immediately arose and people did calculations. After a couple of months the calculations looked pretty good, but they were still pretty crude and it was probably another year before really detailed calculations had been done." Ulam told a meeting of laboratory division leaders on March 6 that they might need years to evaluate his and Teller's new ideas. Mark notes on the other hand that the Teller-Ulam proposals "immediately put everything in a new focus; immediately gave jobs for the cryogenists, metallurgists, the mechanical engineers, the test people, everybody to focus on and proceed. Here was a plan which came out of the theoretical work. They could all see the drawings and say, 'Well, I can do this and you can do that.' That's what was lacking in the earlier time. There wasn't anything to say, 'Well, it's now clear that we really put all our efforts on building this object and getting all the things we need. . . .'"

The fact that the lab was not immediately sold on the equilibrium thermo-nuclear frustrated Teller in the extreme. He debated with Darol Froman organizing a new laboratory division, but increased his lobbying in Washington for a second laboratory, to be sited perhaps north of Denver in Boulder, Colorado, where the National Bureau of Standards was breaking ground in May for a new AEC-funded Cryogenic Engineering Laboratory to produce liquid deuterium. Froman, fronting for Norris Bradbury, resisted setting up a thermonuclear division under Teller; so did Carson Mark.

Greenhouse delayed further discussion of these difficult issues. The Cylinder, which before the Teller-Ulam invention had been a physics experiment at best, now fortuitously promised solid data on radiation implosion. Since the previous summer, a task force had been on station at Eniwetok, a coral atoll at the northwest end of the Marshall Islands chain, building the towers and structures needed for the test. Physicists John Allred and Louis Rosen went out in the winter of 1951 to set up an instrument they designed for measuring the production of fusion neutrons in the Cylinder; they thought the atoll "a monotonous place with a monotonous climate." A gag gift popular among new arrivals was a blank book titled *Sex Life on Eniwetok,* and when Allred and Rosen attended a "special entertainment" of nude movies one night—"(of pristine innocence compared to today's adult fare)" they write parenthetically—they were shocked when the audience "began to cheer and whistle at every slightly suggestive movement." From sheer boredom, they played poker with Teller, who usually forgot to bring money, borrowed a stake and lost that. Ernest Lawrence came out for the tests; so did Gordon Dean, who noted "the unreclaimed [Second World War] planes, tanks, ships etc. which have been bulldozed off the Islands . . . indicating the terrible wastes of most war operations," the "color of the water, the coral formations," a shark attacking a manta ray, the 85-degree heat and 85-percent humidity, "the fine bolting-up of the Cylinder." Herbert York and a Berkeley colleague, Hugh Bradner, prepared another thermonuclear diagnostic experiment for George, determining neutron energy by measuring time of flight; their detectors, mounted close to the Cylinder, would be destroyed in microseconds after the device was fired. "I was at Eniwetok in April when Edward came out and described the Teller-Ulam idea," York recalls. "I remember it well because it was the most exciting time in my life. Just Edward and I in an old aluminum building." ("Teller's new concept was so convincing to any of the informed scientists that it was accepted very quickly," Bethe generalizes.)

George would be a tower shot; after two tests of new, more efficient and compact fission designs, the task force fired the Cylinder on May 9, 1951. It yielded 225 kilotons, its 1,800-foot fireball engulfing the shot tower and melting a crater deep into the white Eniwetok coral. Dean noted his impressions:

> The first daylight shot since Bikini and the amazing destructiveness of [it] as indicated by the complete disintegration and disappearance of the block house used for the X-ray experiments. The vaporization of the 200-foot steel tower, together with 283 tons of equipment on top of the tower. The complete disappearance of the 6 cast iron, 6 feet tall, sample catchers and the crater filled with water.

For Teller, whose last nuclear test had been Trinity, the shot evoked déjà vu. "Rising early that May morning," he recalled, "we walked through the tropical heat to the beach of Eniwetok's placid lagoon. We put on dark glasses, as had been done for the test in Alamogordo. Again we saw the brilliance of another nuclear explosion. Again we felt the heat of the blast on our faces, but still we did not know if the experiment had been a success. We did not know whether or not the heavy hydrogen had been ignited." They would have to wait for the results of the Allred-Rosen experiment. Lawrence invited Teller for a swim that afternoon. The brooding Hungarian found gloom even in the bright day. "When I came out of the water to stand on the white sands of the beach, I told Lawrence that I thought the experiment had been a failure. He thought otherwise, and bet me five dollars. . . ." Rosen sought Teller out the next morning and passed along a first result but swore him to silence; Teller came up with a work-around that conformed to the letter of his vow, caught Lawrence on the airstrip later that morning preparing to leave and silently handed the Berkeley physicist a five-dollar bill. Allred and Rosen measured neutrons with energies of 14 million electron volts (MeV), one unique signature of the reaction of D + T.* Of George's 225 KT, its fission component, the largest yet exploded, probably accounted for 200 KT. The small DT capsule—less than an ounce of deuterium and tritium—yielded the remaining 25 KT, twice the destructive force released over Hiroshima. Los Alamos physicist Jane Hall telexed Robert Oppenheimer on May 10 that "the interesting mixture [i.e., the fusion capsule] certainly reacted well." Teller's gloom veered to optimism; he remarked to Dean "that Eniwetok would not be large enough for the next one."

Greenhouse Item, fired on May 25, proved the principle of DT-gas boosting, yielding 45.5 KT from an all-U235 implosion design that would otherwise have produced no more than half that yield. Los Alamos would need another five years to make a rugged weapon that incorporated boosting.

---

* The reaction is $D + T \rightarrow {}^4He + n + 17.59$ MeV; the ${}^4He$ nucleus carries 3.5 MeV, the neutron 14 MeV.

When the task force returned to the United States, Dean decided "it was high time that we got together all the people who had any kind of a view on H weapons. . . . I talked . . . to two or three of the [AEC] commissioners and said wouldn't it be good if we could get them all around a table and make them all face each other and get the blackboard out and agree on some priorities." Oppenheimer issued invitations to the meeting, to be held at the Institute for Advanced Study in Princeton in mid-June, as chairman of the Weapons Subcommittee of the GAC, and chaired it when it convened on Saturday and Sunday, June 16–17, 1951.

Teller had "very considerable misgivings" about attending the Princeton meeting in the first place, he testified in 1954, ". . . because I expected that the General Advisory Committee, and particularly Dr. Oppenheimer, would further oppose the development" of the H-bomb. He maintained for the rest of his life that he was snubbed at the meeting and the Teller-Ulam invention slighted until he insisted on being heard. "The report on the hydrogen bomb [presented at the conference] did not mention [radiation implosion]," he wrote incorrectly in 1982. "When I asked to speak, Bradbury denied me the opportunity. Although there were several people present who knew about [radiation implosion], including Bethe and Oppenheimer, none chose to speak about it. However, AEC commissioner Smyth, who believed that the other side of the argument should be heard, made my presentation possible. In the developments that immediately followed, this proved to be decisive."

No one else who attended the meeting—the AEC commissioners and managers, the members of the GAC, Bethe, von Neumann, Robert Bacher and the management of Los Alamos—shared Teller's melodramatic recollection. Darol Froman had prepared and circulated an agenda to guide the discussion which Teller had seen beforehand. It began with the *Greenhouse* results and moved systematically through the classical Super and the Alarm Clock (now revivified by the possibility of using radiation implosion to compress it) before arriving at Teller's new equilibrium thermonuclear. With his latest messiah previewed at Eniwetok, Teller was contemptuous of such methodical analysis and evidently assumed it was designed with Oppenheimer's and Bethe's collusion to mislead the AEC. "It was certainly intended," Mark recalls to the contrary, "that I would be reporting what we got from *Greenhouse* and what we thought we knew about the new proposal and what work we had in progress on it. Edward got tremendously impatient and made it impossible for anybody else to complete their job. It was expected that Edward would appear there *after* this presentation, but he more or less short-circuited the thing by starting to talk before."

Teller spoke with passion and eloquence about the promise of the new design. Bethe contributed to the presentation. Oppenheimer remembered that Fermi, who "knew nothing of these developments," was "quite amazed." According to Teller, Oppenheimer "warmly supported this new approach." The GAC chairman thought the meeting accomplished three things. First, "We agreed that the new ideas took top place and that although the old ones should be kept on the back burner, the new ones should be pushed. . . ." Second, the AEC agreed to begin producing lithium deuteride as a possible fuel for both the equilibrium thermonuclear and a radiation-imploded Alarm Clock. Third, said Oppenheimer, they debated "the construction and test schedules for these things. . . ." Teller, among others, argued for moving swiftly and directly to a full-scale test of the equilibrium thermonuclear; an opposing point of view favored testing "components," meaning the various individual inventions that made up the Teller-Ulam breakthrough. Oppenheimer remembered as the consensus of the meeting "that unless the studies of the summer passed out on the feasibility of it, one should aim directly at the large-scale explosion. . . ."

Gordon Dean had wanted a meeting of the minds at Princeton and thought he got it:

At the end of those two days we were all convinced, everyone in the room, that at last we had something for the first time that looked feasible in the way of an idea.

I remember leaving that meeting impressed with this fact, that everyone around that table without exception, and this included Dr. Oppenheimer, was enthusiastic now that you had something foreseeable. I remember going out and in four days making a commitment for a new plant [to produce lithium]. . . . The bickering was gone.

No one had debated morality this time around. Though it should deliver megaton yields, the equilibrium thermonuclear was evidently not "an evil thing considered in any light," as Fermi and Rabi had condemned the classical Super before the President had ordered work on an H-bomb to proceed. Oppenheimer thought the difference lay in the differing technical promise of the two conceptions:

It is my judgment in these things that when you see something that is technically sweet, you go ahead and do it and you argue about what to do about it only after you have had your technical success. That is the way it was with the atomic bomb. I do not think anybody opposed making it; there were some debates about what to do with it after it was made. I cannot very well imagine if we had known in late 1949 what we got to know by early 1951 that the tone of our [October 1949 GAC] report would have been the same.

Why technical promise should decide questions of politics and morality, Oppenheimer did not explain, but he said on the same occasion that the world would be a safer place if the development of thermonuclear weapons could have been avoided, and he knew very well that Soviet scientists were as capable of brewing something technically sweet as US scientists had been. Another and important difference was Korea, which seemed more than a regional threat; Bethe returned to Los Alamos full-time that summer, still heavy with misgivings, because he was concerned that the conflict might spread to Europe. "I was totally opposed to the H-bomb," he recalls. "But the Teller-Ulam invention made it very likely that it could be done. In fact, it was so good that it seemed that the Russians could also do it . . . and I figured that if they could, they probably would. Then I figured if we did it first and did it better—that was really the argument. And then there was the Korean war. Europe seemed threatened."

"Princeton settled that we should go ahead on the hydrogen bomb," Teller concludes. Yet he continued to believe that shadowy forces worked behind the scenes to thwart his dream of a thermonuclear weapon, forces allied with Robert Oppenheimer. Despite Oppenheimer's enthusiasm at the Princeton meeting, Teller would claim, the GAC chairman's recommendations in general "were more frequently . . . a hindrance than a help. . . ." Nor was Teller alone in his suspicions. William Borden spent an evening with Lewis Strauss later that summer and heard the former AEC commissioner express "fear and concern over Oppenheimer. He agreed that it would probably be impossible to confirm or deny these fears through the use of any intelligence methods." Borden "pointed out the parallel thinking which had taken place elsewhere on the same subject," meaning Borden's own growing suspicion that Oppenheimer might be a Soviet spy, "and the feeling of utter frustration about the possibility of any definite conclusions." Both men agreed that they found the recent testimony of a former Communist Party official in California that Oppenheimer had attended Party meetings "inherently believable." Strauss, Borden noted, "is quite sure that another Fuchs may be turned up (if not in the US, then in Britain) but he had no more ideas than we ourselves have been able to generate as to what to do about it. He did like the notion of [using] the lie detector at Los Alamos. . . ." (Borden at this time was prompting Brien McMahon to urge Truman to increase the AEC's industrial capacity sufficiently to build tactical nuclear weapons "numbered in the tens of thousands" in addition to a full strategic arsenal.)

Teller's suspicions found seeming confirmation immediately after the Princeton meeting, back at Los Alamos, when he and Bradbury negotiated a thermonuclear program. "Teller made an offer to stay [at Los Alamos] if he had administrative responsibility over that part of the program only . . . and could actually help it along," Gordon Dean learned from de Hoffmann, but "that led to a deadlock." The negotiation deadlocked because Bradbury had no intention of appointing Teller to head thermonuclear development and offered the physicist either an assistant directorship or a consultancy. Teller waffled through the rest of the summer, threatening to leave, traveling to Washington at one point with de Hoffmann to lobby his cause. Borden invited the scientists to dinner at the Metropolitan Club and convinced McMahon to join them "to make the senator's shoulder available for crying and talk [Teller] into staying [at Los Alamos]." On Henry Smyth's authority, Borden told Dean that "Teller requires a great deal of 'crying on the shoulder' time."

The final straw for Teller was Bradbury's decision to appoint experimental physicist Marshall Holloway, the slim, sometimes imperious director of weapons development at Los Alamos, to head the thermonuclear program. "[Bradbury] was too smart to let [Teller] have any administrative authority," Raemer Schreiber comments. "Edward was a brilliant but erratic theoretical physicist who had magnificent concepts, some of which were virtually impossible to translate into actual hardware. . . . Edward could argue vehemently in support of wild and impractical ideas. He also had more than his share of the traditional wild Hungarian temperament: things were either very black or very white—no gray areas." Bradbury gave Teller an ultimatum, Los Alamos engineer Jacob Wechsler remembers:

> Here's Edward marching along and every time you turned around he wanted to change things. It kept getting worse and worse and finally Marshall said, we've got to get him out of here. You can't tell Edward not to be there. Norris went to do it and Edward really blew his stack. That was when he said, well, either Holloway is going to run this or I'm going to. He wasn't running it anyhow, but he felt like he was. He wasn't in charge of it; he was just blowing out his ideas. Norris said, Edward, if you can't fit the mold, leave.

Appointing Holloway, de Hoffmann told Dean, was "just like waving a red flag before a bull." Teller and Holloway had clashed more than once; Teller also believed Holloway "was a consistent opponent of the Super." Oppenheimer thought the appointment put Teller "in a bad mental way," and agreed to sound out Bradbury on returning to Los Alamos himself, but was not surprised to find that the Los Alamos director "gave no signs of wanting to have the ex-director back and said that he had full confidence in [Holloway], and that was the end of that." A week after Bradbury announced Holloway's appointment on September 17, 1951, Teller resigned. He left "in a huff," says Schreiber. On his way out, Wechsler reports, "he said to Marshall [Holloway], I bet you a dollar it won't work." Carson Mark thought the dispute turned on the question of authority, not of differences over design. Teller had similarly refused to work under Bethe at Los Alamos in 1944, Mark would observe, had "cut out and started working on something else." Now Teller was cutting out again.

54

54–55. Sending mixed signals, the US Navy tested two atomic bombs against shipping at Bikini in 1946 while experts tried to devise a system of international control. When the new General Advisory Committee visited Los Alamos in 1947, it found few bombs and no teams trained to assemble them. Left to right: James B. Conant; J. Robert Oppenheimer; General James McCormack, Jr.; Hartley Rowe; John Manley; I. I. Rabi; Roger Werner.

55

56. At Harvard in 1947, Secretary of State George Marshall (here with J. B. Conant and General Omar Bradley) proposed the Marshall Plan to restore a war-ruined Europe.

57. Secret Anglo-American discussions of rearming West Germany, passed to Stalin by Donald Maclean in April 1948, probably precipitated the blockade of Berlin in June that the US surmounted with a year-long airlift.

58. Edward Teller (right) quit Los Alamos in 1946 to work with Enrico Fermi at the University of Chicago. Later he complained that H-bomb research was neglected.

59. With the Cold War advancing, the new Atomic Energy Commission set to work developing a US atomic arsenal. Left to right: Ernest O. Lawrence (not a member); Lewis L. Strauss; Robert Bacher; Chairman David Lilienthal; Sumner Pike; William Waymack.

61

60

60. Los Alamos had to design "GI-proof" bombs to replace the handmade wartime units: physicist Raemer Schreiber and postwar lab director Norris Bradbury.

61. William Borden, who expected atomic war, became executive director of the new Joint Committee on Atomic Energy. He thought Oppenheimer was a spy.

62

62. In 1947, the US threatened to shut off Marshall Plan aid unless the British turned over their stockpiles of uranium ore. American leaders sought further British concessions in summer 1949, only weeks before Joe 1 exploded. Left to right: David Lilienthal; Dwight Eisenhower; Brien McMahon; Dean Acheson; Louis Johnson.

63. Computer power paced H-bomb progress; H-bomb work pushed early computer development. Mathematician Nicholas Metropolis (right, with Paul Stein) developed the MANIAC at Los Alamos.

64. Edward Teller's unworkable "Super" H-bomb design stalled invention. Physicist George Gamow assembled a mischievous caricature of the protagonists: Stalin with a Soviet bomb; Oppenheimer on the side of the angels; Stanislaw Ulam, Teller and Gamow acting out design ideas. Ulam and Teller made the breakthrough to a staged, radiation-imploded thermonuclear early in 1951.

65. Programming the earlier ENIAC required plugging and unplugging cables.

66

66. Stanislaw Ulam, Richard Feynman and John von Neumann at Los Alamos, late 1940s. Von Neumann invented stored programming; Ulam proved Teller's Super unworkable with hand calculations that the new computers confirmed.

67

69. Theoretician J. Carson Mark worked out the mechanics of the new H-bomb design.

69

67–68. Engineer Jacob Wechsler developed transport dewars to move liquid deuterium for the first staged thermonuclear, the Mike device. A helium engine (left) kept the bomb fuel cold inside a cylindrical thermos bottle. Mike used a similar system.

68

70

70, 72. Two views of the Mike device (above and opposite page), the first megaton-yield thermonuclear, on Eniwetok atoll. Pipes carried early bomblight to distant streak cameras. Physicist Marshall Holloway (opposite center, among construction officials) directed Mike development.

71. The Krause-Ogle box, a nine-thousand-foot plywood tunnel containing helium-filled polyethylene ballonets, carried Mike neutrons and gamma rays to a distant blockhouse for diagnostic measurements before the Mike fireball vaporized it.

71

73–74. Mike yielded 10.4 megatons, one thousand times the yield of the Hiroshima bomb, vaporizing the island of Elugelab. The crater the explosion left behind—two hundred feet deep and more than one mile across—is starkly visible in these before-and-after aerial photographs.

74a   74b

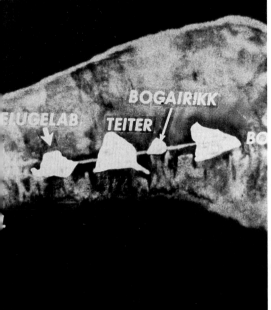

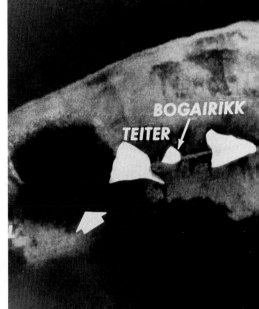

75-76. Bikini Baker, a Nagasaki-scale atomic bomb, versus Mike over Manhattan. Super-imposing the two explosions on the Manhattan skyline dramatizes the thousand-fold dif-ference in destructiveness. The Baker fireball was the diameter of the neck of the mush-room cloud, about eight hundred yards. The Mike fireball alone expanded to more than three miles in diameter; its blast would have obliterated all five New York City boroughs.

76

77–79. Frightened by Soviet advances, military-supremacy dogmatists sought a scapegoat. They found Robert Oppenheimer, who had counseled arms restraint. Left to right: Lewis Strauss led the pack, revoking Oppenheimer's clearance. Kenneth Nichols endorsed the accusations. Attorney Roger Robb brutally cross-examined.

80. Teller testified against Oppenheimer at the 1954 security hearing. In 1963, he shook Oppenheimer's hand when the physicist received the AEC's Enrico Fermi Award. Oppenheimer was already ill with the throat cancer that killed him. At left: Kitty Oppenheimer. Between Oppenheimer and Teller: Glenn Seaborg.

81

81–82. *Castle* Bravo, the first US "dry" (lithium-deuteride-fueled) thermonuclear, exploded at Bikini on March 1, 1954. It ran away to fifteen megatons, endangering scientists at Bikini and irradiating Japanese fishermen. The fireball expanded to nearly four miles in diameter.

82

CASTLE BRAVO

83. Curtis LeMay shaped the Strategic Air Command into a weapon capable of "killing a nation."

84. When the Joint Chiefs requested atomic bombs to use in Korea, Truman traded nine for Douglas MacArthur's dismissal as Far East commander.

85–86. The US quickly weaponized *Castle* Romeo, a lithium-deuteride-fueled thermonuclear tested after Bravo, as the 41,000-pound, eleven-megaton Mark-17. Only the big B-36 bomber (here dwarfing a B-29 half the size of a football field) could carry it.

85

86

87. Soviet premier Nikita Khrushchev took young John F. Kennedy's measure in Vienna in 1961 and decided to install nuclear missiles secretly in Cuba.

88. US aerial reconnaissance spotted missiles and nuclear warhead bunkers in Cuba in October 1962. During the missile crisis, the Strategic Air Command put seven thousand megatons aloft and tried to provoke a Soviet alert that would justify a US preemptive first strike.

89. LeMay challenged Kennedy to invade Cuba. Kennedy cautiously refused. Not until 1989 did the Soviets reveal that there were two dozen nuclear warheads on hand in Cuba during the crisis: invasion would have started nuclear war.

90. Arms race dunce caps: a US Minuteman III missile warhead bus. Each Mk 12A reentry vehicle is 5.9 feet long and 21 inches in diameter, yielding 350 kilotons. Since 1945, no national leader has dared to risk his country by using nuclear weapons.

91-92. The arms race led eventually to economic collapse and the breakup of the Soviet state. Both the US and Russia began chopping up missiles and dismantling nuclear warheads. The US legacy was a $4 trillion national debt.

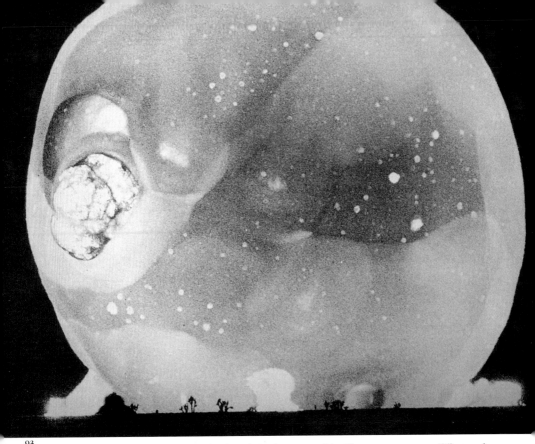

92

93. A nuclear fireball in Nevada dwarfs Joshua trees it will soon consume. "The problem which is posed by . . . atomic energy is a problem of the ability of the human race to govern itself without war," a State Department panel concluded in 1953. Nuclear deterrence banished world war, but halting the arms race took fifty years.

94. The author signing the guest register at the "forester's cabin"—the Kurchatov house, now a museum, on the grounds of the Kurchatov Institute in Moscow— in 1992.

"Edward's tragedy was that he left at the wrong time," Bradbury said later:

Just as the going got tough—in spite of all Edward's protestations about national need and so on—Edward quits, because I wouldn't give him control of the program.

If I'd given him control of the program, I'd have [had] half my division leaders quit, two-thirds of them quit, wouldn't work for him. I couldn't either; I knew him too well. Edward couldn't follow one course of action for two consecutive days—jump here, jump there. . . . I couldn't put him in charge of the program here. I had to tell him so. . . .

I wished he would stay. I tried to persuade him to stay, but I couldn't put him in charge . . . and I wouldn't put him in charge. Oppie hadn't put him in charge. Oppie knew him just as well as I did, perhaps better. . . .

"A lot of us were really teed-off at Edward," Mark concludes, "because if he would have sat down and applied himself to the job, it would of course have gone faster." As it turned out, the job went fast enough, faster certainly than Edward Teller expected. "Had he imagined that Los Alamos could really do this," Mark quotes one of Teller's friends, "he would never have left."

In March 1951, while Stanislaw Ulam and Edward Teller invented the hydrogen bomb and the Korean War escalated, the world had followed the trial of Julius and Ethel Rosenberg and Morton Sobell. Young Roy Cohn assisted prosecutor Irving Saypol in presenting the government case, contriving to introduce a Jello box panel into evidence to represent the Jello box signal Harry Gold had carried with him to Albuquerque. Gold with his remarkable memory made an effective witness. David Greenglass had pled guilty to conspiracy to commit espionage on October 18, 1950, and appeared as a witness as well, as did Ruth Greenglass, who was never charged. The Greenglasses, who assumed that their testimony would merely corroborate other information that the FBI had already acquired, provided the only evidence against Ethel Rosenberg. They were horrified, many years later, to learn that they were responsible for bringing Ethel to trial, but reaffirmed her complicity. "I know she had to know," Ruth told Ronald Radosh in 1979. "I know there are husbands and wives where they never know what they are doing. But this was not the case with them." David had not hesitated to sacrifice his sister to protect his wife. "I got two children," he told Radosh. "If the choice was between [Ruth] and my sister, I'll take [Ruth] any day. That was the choice that I thought I had. In my mind, all [his sister and brother-in-law] had to do was have a conversation [with the FBI], the same as I had a conversation." The Rosenbergs steadfastly maintained their innocence, frequently taking the Fifth.

Judge Irving R. Kaufman was ambitious for appointment to the US Supreme Court and favored the government during the trial, communicating
improperly *ex parte* with the Justice Department and the FBI. After the jury
found the three defendants guilty, Saypol told Kaufman privately that he
wanted the death penalty for the Rosenbergs and thirty years for Sobell;
Kaufman sent the prosecutor to Washington to sound out the Justice Department and the FBI. Neither the Justice Department nor J. Edgar Hoover
favored the death penalty, at least not for Ethel Rosenberg, a mother with
two small children. Kaufman determined to set an example, "to make people
realize that this country is engaged in a life and death struggle with a
completely different system." He told the Rosenbergs at their sentencing on
April 5—the day after Edward Teller issued his report at Los Alamos on the
equilibrium thermonuclear and the day before Gordon Dean met with
Harry Truman to arrange the transfer of nine nuclear cores to Hoyt Vandenberg for possible use against the Chinese—that their crime was "worse than
murder":

> I believe your conduct in putting into the hands of the Russians the
> A-bomb, years before our best scientists predicted Russia would perfect
> the bomb, has already caused, in my opinion, the Communist aggression
> in Korea, with the resultant casualties exceeding fifty thousand and who
> knows but what that millions more innocent people may pay the price of
> your treason. Indeed, by your betrayal, you undoubtedly have altered the
> course of history to the disadvantage of our country.

Espionage, not treason, was the crime of which the defendants had been
convicted, but Kaufman sentenced Julius and Ethel Rosenberg to death.
Morton Sobell got Saypol's requested thirty years. On April 6, Kaufman gave
a shocked David Greenglass fifteen years, a term Saypol had also requested.
The previous fall, when Kaufman had presided at the trial of Abe Brothman
and Miriam Moskowitz, he had commented that his sentencing powers were
"almost Godlike" (he gave Brothman and Moskowitz the maximum sentence
permitted by law for their crimes, seven years). Harry Gold, sentenced
before another judge in December 1950 after pleading guilty, got thirty
years.

Morris and Lona Cohen disappeared from New York sometime late in
1950, after the Rosenberg arrests. (A decade later they turned up in England,
living under assumed names, still working at espionage.) Donald Maclean
disappeared from London with Guy Burgess on May 25, 1951; on June 7 the
press reported that they had defected to the Soviet Union.

Gold wrote a chronicle of his life in the first months after he went to
prison. In it, he reported having "a horrible sense of shame and disgust,
which I can never ever lose, concerning my deeds. . . . I am aware of the

hard fact that, before anything else can transpire, I must be punished, and punished well, for the terribly frightening things I have done. I am ready to accept this penalty. There shall be no quivering, trembling or further pleas for mercy. What was, was, and now I am prepared to pay the price." Fourteen years later, paying the price, he responded from prison to a charge that the judge who appointed his counsel (a distinguished Republican) was indulging a sardonic joke. "Having met Judge McGranery in his professional capacity, as one might say," Harry Gold wrote his attorney, "... I've never been able to find anything funny in the thirty-year sentence he gave me." When he got out, Harry hoped to devote the rest of his life to medical research.

24

Mike

THE TEAM that Marshall Holloway assembled to design and build the first megaton-scale thermonuclear—the Panda Committee, also known as the Theoretical Megaton Group—met for the first time on October 5, 1951, two days after the White House announced the detection of a second Soviet atomic-bomb explosion.* Edward Teller's final argument with Holloway before the volatile Hungarian resigned from Los Alamos had concerned how quickly the laboratory could mount a test of the equilibrium thermonuclear. Teller insisted on a target date of July 1952; Holloway held out for late October, partly because he knew how much fabrication and engineering the shot would take, partly because summer was monsoon time in the Marshall Islands. The Panda Committee had a little more than a year to design and deliver a first experimental H-bomb.

One early and important decision concerned which thermonuclear fuel to use. Lithium deuteride was one choice. Deuterated ammonia was another. Liquid deuterium was a third. Each had its advantages and disadvantages. Lithium deuteride—LiD—would be the simplest material to engineer because it was a solid at room temperature, but breeding tritium within a bomb from lithium required a complex chain of thermonuclear reactions that involved only one of lithium's several isotopes, Li^6. "We were very much aware of lithium deuteride," Hans Bethe comments. "We were not totally sure how well it would work." It was clearly a material that might be tried once the principles of the equilibrium thermonuclear had been proved and

* Joe 2, tested at Semipalatinsk on September 24, 1951, was an advanced and improved implosion device with an all-U235, probably levitated core. It was smaller and lighter than Joe 1, Sarov's Fat Man copy: at half the diameter, it produced twice the yield. Joe 3, an implosion device with a composite core, was air-dropped on October 18. These two tests led to the design of the first Soviet production atomic bomb, which was not released to the Soviet military until 1953. Yuri Smirnov, personal communication, 2.ix.93.

the laboratory was proceeding to weaponize the device; Teller and Frederic de Hoffmann had published a technical report, "Effectiveness of Li^6 in an 'equilibrium Super,' " the previous June.

Deuterated ammonia was a liquid at room temperature, but like LiD, its physical properties were not well known. Los Alamos knew a great deal about the physics of pure deuterium, having measured the relevant cross sections and even having observed D fusing with D and with tritium in the *Greenhouse* George shot. The disadvantage of pure deuterium was that it would have to be maintained below its boiling point of 23.5 degrees Kelvin* to remain liquid; that meant the test device would have to incorporate sophisticated insulation and a cryogenic cooling system. (Cryogenics is the branch of physics that deals with very low temperatures.) Against the possibility that liquid D would be the thermonuclear fuel of choice, the National Bureau of Standards had already begun building its Boulder liquefaction facility. (In those environmentally more innocent times, the citizens and merchants of Boulder had raised seventy thousand dollars by public subscription to buy the land for the plant and deeded it permanently to the federal government.) The technology for handling liquid hydrogen in bulk —storing it, pumping it, transferring it across the country and across the Pacific Ocean to a tropical island—remained to be developed.

Holloway's team soon settled on liquid deuterium despite its engineering challenges, Carson Mark reports, primarily because it would give the cleanest physics:

In the cryogenic design you had nothing but deuterium—essentially an infinite medium of deuterons [i.e., deuterium nuclei]. With lithium deuteride there were as many lithium atoms as there were deuterons, but we were interested in the deuteron reaction, D + D. There wasn't a prominent reaction of deuterium with lithium. The lithium was inert in that respect. In fact it was a diluent. And certainly a complication. And the cryogenic pattern avoided that complication. It introduced some physical complications in construction, in handling, but those one knew how to deal with. They had nothing to do with any of the thermonuclear behavior. . . . The great virtue of lithium, of course, is that it provides you with a free source of tritons [i.e., tritium nuclei]. That's only really true of the isotope Li^6. We didn't have large quantities of separated lithium isotopes. We set out to get them and by 1954 we had them. We could have had them earlier if we had known enough to go after them. The description of the [thermonuclear] burning process of pure deuterium is much simpler than the description of the burning process with either Li^6 or normal lithium deuteride. The

---

*A temperature scale with intervals corresponding to those of the Celsius scale but beginning at absolute zero: 0° Kelvin = −273.15° Celsius = −459.67° Fahrenheit.

description of the compression of liquid deuterium is simpler than of the compound, the deuteride. To avoid discussing the lithium seemed like a virtue. Every departure from the simplest picture seemed like something to avoid.

Jacob Wechsler, a creative, indefatigable engineer from New Jersey trained at Cornell and Ohio State, worked as one of Holloway's deputies. Wechsler had first come to Los Alamos as a Special Engineering Detachment enlisted man during the war, had gone away to Ohio State for his master's degree in 1947 and then had returned to the lab and gained experience at *Sandstone* and *Greenhouse*. He vividly recalls the discussions that shaped the crucial decisions Holloway made that first month:

> One key was setting a time scale early in the game. They said, it may be unrealistic but let's set it and not keep saying, Oh, we've got to slip it. Let's target and try to make it go. That required some real good thinking—namely, what is it you're trying to do in this test? Is it going to be weapon? Is it "proof of principle"? What's it going to be? That was the big discussion. Should it be something that merely demonstrated the principle, so that once we understood the geometry we could try to imagine where we might go from there? That could be a trap. It could be just a wild experiment. How far should you go, how conservative should you be? It's a tough decision because it's a brand new ball game. We had meetings weekly of people from the theoretical division, from the explosives division, from the applied weapons division. The key people had to show up because communications weren't good in those days. You didn't have Xeroxes, you didn't have computers, you couldn't talk over the phone [because phones were not secure].

Devising a system for getting the minutes of Panda Committee meetings out and around the lab in an era when reproducing documents meant using either carbon paper or wax-stencil mimeograph was an important early achievement.

It was clear from the beginning that the test device would be large, since it would have to accommodate a fission primary at one end: the smallest fission bomb available of sufficient yield was forty-five inches in diameter, almost four feet. The complicated device needed thick walls of dense metal to hold it together long enough for a good burn to proceed. Steel was the metal of choice. Who could fabricate thick pieces of steel more than four feet in diameter? Whoever it was would have to be security-checked and cleared. The biggest heavy-equipment manufacturer in the United States was American Car and Foundry of Buffalo, New York, which had built block-buster bomb casings for the US Air Force for use in Korea. Los Alamos began negotiations with ACF in October aimed at initiating engineering design; ACF started fabrication before the end of the month.

The same problem that plagued Panda meetings plagued engineering design: how to communicate effectively, how to prepare and distribute design changes as the test device evolved. "Marshall finally got the idea that we would make a full-scale drawing of this huge beast," Wechsler recalls. "Not a working drawing. A schematic. You wouldn't be able to fabricate anything from it, but it would have the kind of dimensions you'd use when you were doing calculations and wanted to know the spacing and thicknesses of the various materials. The idea was to have something where people could come and look and really see what was going on as we added things in. And everybody said, you're nuts. I mean, this gadget is huge. It's six feet in diameter and twenty feet long." Holloway proposed setting up a giant drafting board in a secure area of the laboratory and putting ACF in charge of maintaining it. For a location, he chose the old Los Alamos Tech Area S building, at the east end of Trinity Drive across from the local dry cleaners. "So ACF built a whole bunch of long sawhorses," Wechsler remembers, "and we put down sheets of plywood to make an area big enough to put the whole drawing on. The guys who were doing the drawing crawled around in their socks filling things in."

The drawing was too big to see from the working floor. Holloway had a balcony built inside S building. "You could go up on the balcony and stand there," says Wechsler, "and look down at this huge, complex drawing. It was neat, especially for the theoreticians—there it was at full scale and they could say, 'Oh, what's that thing over there?' 'Well, don't you remember, that's part of what we've got to hold it up with.' 'Oh my God. Smart man, that Holloway.' "

A crucial decision concerned the internal shape of the test device. (Externally, it would be shaped like a giant capsule—a hollow steel cylinder with rounded ends.) Carson Mark's division worked overtime that autumn and winter to devise the best possible theoretical design. When the fission primary at one end of the test device fired, it would radiate equally in every direction. How much of that radiation could be channeled into the secondary? Originally, Panda thought of building a massively shielded end above the fission primary, thickly coated with lead, to try to contain as much radiation as possible. The radiation would flow down the interior of the big cylinder, which was essentially one big radiation channel—a big pipe. How would the radiation flow? Mark wondered if it flowed like a bucket of water thrown into a big tank. Where did it go? Did it splash along the walls? They would have to substitute calculation for experiment to decide, but the MANIAC would not be finished before spring 1952; hand calculation would have to suffice—hand calculation and instinct. Wechsler says it was Mark more than anyone whose instinct for the physical behavior of the radiation and whose judgment of the corresponding risks they could take with channel design carried the day. "Carson is par excellence a physicist and a theoretician, especially his appreciation for the mechanical aspects of a problem and his ability to assess whether something is really going to make any difference. I just thought that he was fabulous." When Norris Bradbury argued that nobody is the father of the hydrogen bomb, that its development was a group effort, he was probably thinking of this collective work of imagining abstract concepts into physical reality, a work at which steady, reliable men like Mark, Wechsler and Holloway shone. (Edward Teller spent only two weeks at Los Alamos in the crucial six months between October 1951 and April 1952 when the equilibrium thermonuclear was designed; his main contribution seems to have been to kibitz. "Once Teller left Los Alamos," Hans Bethe observes, "even though they were working on 'his' weapon, he found all sorts of reasons why it wouldn't work. He hated the project director, Marshall Holloway.... So he had every reason, he tried to criticize it wherever possible.")

The thermonuclear secondary, which would essentially be a bottle of liquid deuterium with a stick of plutonium mounted inside for a sparkplug, would hang in the center of the big steel casing below the fission primary, behind a heavy blast shield. The flux of soft X rays from the primary would flow down the inside walls of the casing several microseconds ahead of the material shock wave from the primary. The X-ray flux would be dense as solid metal, but it would not have time to exert much pressure directly to implode the secondary. The X rays were hot, however, so hot that they would ionize solid materials instantly and turn such materials into a plasma hot enough to radiate further X rays. (Materials are said to ionize when they are heated sufficiently to break up their atoms into electrons and nuclei— negative and positive ions. Plasma is a fourth state of matter—solids, liquids and gases are three more familiar states of matter—consisting of hot, ionized gas; the sun is a ball of plasma maintained by thermonuclear burning.) So the steel casing would need to be lined with some material that would absorb the radiation and ionize to a hot plasma which could radiate X rays to implode the secondary.

As the radiation flowed from the primary end of the casing around and past the secondary, it would start generating radiation pressure at the end nearer the primary sooner than at the end farther away. "If the pressure is real high at this end and real low at the other end," Wechsler recreates the Panda discussions, "how is it going to work? That was one of the big questions." They first thought to taper the channel, Wechsler says, "leave it as open as it could be near the primary with a minimum amount of material to generate pressure and then put more and more material in to narrow it" down the channel. From the spherical channel of their idealized calculations they evolved to a channel shaped like an inverted bowling pin they called the Schmoo after an imaginary creature of that shape in the *Li'l Abner* comic strip. Apparently they adopted the Schmoo channel design first, then rejected it after further calculation. The Panda Committee froze the basic design of the test device on January 18, 1952, but Hans Bethe notes that "in March 1952, unforeseen difficulties appeared.... These difficulties could only be minimized by a very major redesign.... This redesign came at the latest moment compatible with meeting the test date of November 1952...." Had Los Alamos accepted Teller's proposed July 1952 test date, Bethe observes, there would not have been time to redesign the device and it probably would have failed.

According to Wechsler, Mark led the way to a successful radiation-channel design. "Carson's feeling was that if the system is going to work at all, these pressures are going to be so high that the differences from one end to the other aren't going to be very much in this brief length of time. If it *isn't* going to work, having goofed this up a little bit [i.e., shaped the channel] isn't going to be the reason. The thing about it was, it had a big, big, big channel. It was a huge beast and there was a huge amount of room in there." Panda redesigned the test device with a cylindrical channel, straight-walled from one end to the other. Nor did either end need to be massively reinforced. In the few millionths of a second before the developing explosion vaporized the entire gadget, material shock would do very little damage to the ends. "The far end away from the trigger system," says Wechsler, "nobody knows what the hell is going to happen so far away. So we said, Think of the thing in principle as a long cylinder and ignore the end." The test device acquired a name: the Sausage. By then it was scheduled as one of two large-yield tests for the November 1952 series designated *Ivy*. The other test would try out a Theodore Taylor design, an all-U235 weapon expected to yield four to six hundred kilotons, *Ivy* King—K for kiloton—a big backup fission bomb in case the thermonuclear should fail. The expected megaton-yield Sausage shot was designated *Ivy* Mike.

Teller continued to find fault with the Mike design, Bethe remembers. "At one point he said, well, it may all work perfectly well, except that the radiation will go into the casing and then there will be Taylor instability. Now, I know a lot about Taylor instability, and I worked on this radiation penetration, and then came out with the conclusion that there would not be Taylor instability, and wrote it down and sent it to Teller. That was my main contribution. Certainly my work was not critical to success."

Through late 1951 and early 1952, while the physicists pursued design questions, Wechsler and the other project managers and engineers organized the production of deuterium and the exotic equipment for storing and transporting it in liquid form.

Cryogenics had its substantial start in the work of the Scottish physicist James Dewar, who first produced liquid oxygen in quantity and in 1898 was the first person to liquefy hydrogen. (He was also the co-inventor of cordite, the smokeless explosive that propelled the deadly artillery of the First World War.) At Cambridge University, one of the two institutions where Dewar maintained joint appointments, his low-temperature achievements won celebration in raffish verse:

> *Sir James Dewar*
> *Is a better man than you are*
> *None of you asses*
> *Can liquify gases*

Dewar's most enduring invention—in 1892—was a double-walled flask with the space between the walls pumped out to a good vacuum. A vacuum is an extremely effective insulator against heat convection; the double-walled vacuum flask, with the walls usually silvered to reduce the transport of radiant energy as well across the vacuum, became a standard container for the insulated storage of liquids. In its larger scientific and technical versions such a container is called a dewar; one smaller version adapted for home use is the familiar thermos bottle.

Liquefying gases requires more than straightforward refrigeration, but one important step in the process is incorporated into all home refrigerators: compressing a gas and then allowing it to expand, work which reduces its temperature. In a home refrigerator, compressing and expanding a coolant gas within an arrangement of coiled piping cools the piping, which takes up heat from the food being stored; for liquefying gases, the gas being liquefied is compressed and expanded through a nozzle or in a small piston engine or turbine to cool itself.

Until after the Second World War, the most ingenious and efficient liquefaction system was one developed in the 1930s by Peter Kapitza when he worked at Cambridge University. With John Cockcroft, Kapitza first developed a hydrogen liquefier (in 1932); then, on his own, devised a machine for liquefying helium, which has the lowest boiling point of all gases, only $4.2°K$. Kapitza's helium liquefier first precooled the helium gas with liquid nitrogen, which boils at $77°K$, then cycled it through compression and expansion with a piston and cylinder arrangement. It was because of Kapitza's pioneering cryogenic work that Teller and others had worried, after Joe 1, that the Soviet Union might steal a march on the thermonuclear.

In 1946, an MIT physicist named Samuel Collins had developed a helium liquefier superior to Kapitza's, the Collins Helium Cryostat. Collins, writes the physicist who directed the work on deuterium production at the National Bureau of Standards laboratory in Boulder, "using the same principle as Kapitza, redesigned, modified and then superbly engineered his liquefier to provide a complete, relatively inexpensive, reliable and simple-to-operate [liquid helium] facility. Indeed, Collins's contribution to low-temperature

research has been aptly compared with that of Ford to the automobile."
Collins commercialized his helium liquefier through Arthur D. Little, Inc.,
of Cambridge, Massachusetts, and it was to ADL that Wechsler applied late in
1951 to organize the production of big transport dewars by ADL's Cambridge
Corporation that could store hundreds of gallons of liquid deuterium and
other liquefied gases at Eniwetok. "ADL talked to Sam Collins," Wechsler
remembers. "They asked him, Hey, can we use your helium cryostat to
liquefy hydrogen? You wouldn't do that for commercial production, but
their idea was, if you could build a big dewar for storage, you could have a
little Collins cryostat attached to the big dewar and by keeping on cooling
the hydrogen, recirculating it, you'd never lose any. All you had to do was
to keep the cryostat running. That's a really unique idea."

Coordinating the work at the Cambridge Corporation, Wechsler learned
to live on the road. "We had to have dewars by late spring at Los Alamos,
starting from January 1952. Built from scratch. I got to where I knew all the
Pullman conductors, because air travel was not that good." He would take
the train to Boulder to check out the NBS laboratory, fly a redeye to Washington for a meeting there, sleep on the overnight train from Washington to
Boston, work with the dewar development team, take the New England
States from Boston to Chicago and then change to the Santa Fe, which gave
him a compartment where he could sleep on the way back to Los Alamos as
well as write up his notes. "That was a standard route for me, one little
aspect of what went on in that period of months." Other managers were
circulating around the country on similar routes. American Car and Foundry
eventually opened a plant in Albuquerque to make bomb casings, but before
that, someone cycled back and forth to Buffalo. Carrier Corporation supplied
compressors. Distillation Products supplied pumps. "We had stuff being
shipped from all over the country," Wechsler reminisces. "Not just to Los
Alamos. Some of the things never came here. Some of them had to go
straight overseas."

The big storage dewars that Wechsler developed were considerably more
sophisticated than James Dewar's original double-walled flask. Wechsler's
dewars incorporated nitrogen-cooled shields, Styrofoam and aluminum-foil
insulation and other tricks of the cryogenic trade. That was the first time
Wechsler saw complex shapes cut from Styrofoam with a hot wire. Each
stainless-steel dewar had a capacity of two thousand liters of liquid, about
530 gallons; they were mounted on diesel truck flatbeds with a motor-
generator built onto the flatbed ahead of the dewar housing to supply electrical power. Wechsler ran one cross-country from Boston to Boulder fitted
with recording accelerometers to measure how well the shock-absorber
system worked. All the tubes for transferring the liquids in and out of the
dewars were themselves dewars: vacuum-jacketed stainless steel.

Deuterium production started early, Wechsler emphasizes:

Where do you get enough deuterium to do this job? You start with heavy water. Now you've got to electrolyze the heavy water to break it down to oxygen and deuterium. Do you know any easy way to do that? It's tough. We had a little set of electrolytic cells up there at Boulder. They used 55-gallon drums of heavy water, dumping them in, electrolyzing, collecting the deuterium, pumping it at low pressure into big standard gas holders like you use for natural gas. The liquefiers weren't up and running yet and we didn't have the storage dewars yet but we had to get started because making the deuterium was a long, slow deal. So we had to store the deuterium in gaseous form, and that was the way it went out to Eniwetok. They ordered a bunch of tube banks, big 2,000-pound tube trailer banks pulled with semi's that could be loaded into ships. You could use the tubes for nitrogen, oxygen, you could get them for hydrogen. How many tube trailers did you need? How many places in the country make new, clean, safe tube trailers? They had to be specially modified because we didn't want any standard safety valves on them that might vent the deuterium because we didn't want to lose any—those tubes filled with deuterium were worth more than gold.*

A plant at Eniwetok operated by Ohio State University would liquefy the gases after their delivery by ship. The Cambridge Corporation began shipping storage dewars for Los Alamos in early April 1952.

The cryogenic system for the Sausage was similar to the system in Wechsler's storage dewars but simpler. Liquid hydrogen boils at 20°K, liquid deuterium at 23.5°K, which means liquid hydrogen stores a few degrees colder than liquid deuterium. Taking advantage of that difference, Panda cryogenicists designed the double-walled stainless-steel dewar in the Sausage to connect through pipes in its upper end to a reflux condenser set in a big tank of liquid hydrogen. Vaporizing deuterium would flow by convection through a pipe to the reflux condenser, where it would cool to a liquid again and flow through another pipe back into the Sausage dewar. With this system of continual circulation and cooling dumping any heat the deuterium picked up, the cryogenicists were able to dispense with Styrofoam and aluminum foil. The Sausage dewar, suspended within the big Mike casing—a cylindrical double-walled stainless steel tank with a rounded top and bottom and with the sparkplug assembly mounted inside on a central column that ran the length of the tank—would contain several hundred liters of liquid deuterium. A second evacuated assembly would surround it. Between the second, outer assembly and the dewar, the cryogenicists ingeniously interposed a single floating thermal-radiation shield—another thin-walled tank, probably made of copper, a good reflector of radiant heat. The radiation shield "floated" in the sense that it touched the dewar it contained

---

* Deuterium in 1952 cost about seventy-five cents per gram.—RR.

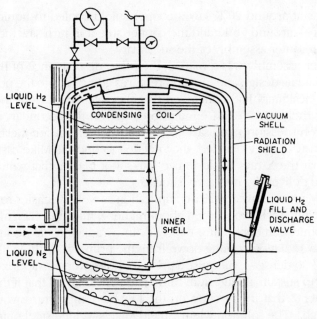

Schematic section of the helium-refrigerated transport dewar.

and the outer assembly that surrounded it in as few places as possible, because any contact would allow heat to flow into the dewar by convection. The contact necessary to hold the components in position was probably made with laminated stacks of thin metal disks, which conduct heat as much as two hundred times less efficiently than would a solid metal bolt.

A thermal-radiation shield floating in a vacuum can significantly reduce radiant-heat transport from a warm exterior to a cold interior. Without it, Wechsler observes, "you're talking a cold surface and a warm surface, and the temperature difference is a couple hundred degrees Kelvin. I don't care if you've got a vacuum between them, the heat leak into the cold surface is serious. But there's a neat little trick. If you can put in a surface with an *intermediate* temperature and float it—isolate it thermally—then the out-side surface sees the intermediate, and the inside one sees the other side of the intermediate, and that cuts the loss way down."

The outer assembly of the Sausage secondary would be warm as the Eniwetok air and the Panda cryogenicists wanted more heat-loss reduction than a single floating shield could accomplish. Rather than clutter the sec-ondary with multiple floating shields, they borrowed another trick, one that Kapitza had used in his helium liquefier. "If you can make a shield float at a temperature lower than it would normally," Wechsler says, "then you can pick up the reflectivity of fifty thermal-radiation shields." To lower the tem-perature of the copper shield, they welded a pan onto its bottom that they kept filled with liquid nitrogen. That cooled the shield. So the liquid deuterium dewar at around 20°K saw a copper shield cooled to liquid-nitrogen temperatures, around 76°K, and the copper shield in turn saw the ambient-temperature outer assembly of the secondary.

The outer assembly was a marvel, the pièce de résistance of the system. To appreciate its design and function requires going back to the point where the Panda designers had concluded that the soft X rays from the primary would not themselves exert enough pressure on the secondary to deliver the high compression necessary to prepare deuterium for thermonuclear burning. Instead, they decided, they needed to line the Mike casing with a material that the X rays could ionize into a hot plasma that would expand rapidly and deliver the necessary shock.

Carson Mark foresaw another complication as well, Wechsler remembers. "He was really, really concerned about higher-Z materials [i.e., materials of higher atomic number] being exposed. In a radiation environment, with high-energy radiation coming down from the primary, anything like steel—because it's so dense—will cause a pressure pulse when it vaporizes. Carson was trying to sustain a deuterium burn, and he was afraid that if things blew off at higher Z that might chop up the fuel, its temperature wouldn't stay high enough." The solution was to shield Mike's welded-steel outer casing:

> You could cover the steel with lead. That would make it more opaque, so it couldn't give a pressure pulse that quickly. That at least would keep you from getting down to the steel. But even the lead, the surface of the lead, would blow off. So then we covered the lead with plastic—with polyethylene. That was low-Z, just CH_2 [i.e., carbon and hydrogen]. From a time point of view, the radiation ionization of the heavy materials with all that shielding would be so late that it would have zilch effect on the overall system.

The polyethylene would function as their plasma generator as well.

A fission implosion bomb used a tamper to smooth out its high-explosive shock wave and to hold the core assembly together by inertia a few microseconds longer to allow a few more chain-reaction generations to occur, increasing the efficiency of the explosion. The Sausage secondary also needed a tamper. It would function not only as a tamper, to hold the secondary together, but also as a pusher, to transfer the energy from the hot, ionized polyethylene plasma into the liquid deuterium. The cylindrical pusher would be cast of thick, heavy pieces of U238, the largest uranium castings made up to that time. X rays from the fission primary would heat the plastic that lined the outer Sausage casing. The resulting hot plasma would reradiate longer-wavelength X rays inward from all sides toward the thick uranium pusher. These X rays would heat the surface of the pusher so hot that it would ablate: boil vaporized uranium off its outer surface. To every action there is an equal and opposite reaction: the ablating vapor would function as burning fuel ejecting from the nozzle of a rocket functions, accelerating the pusher shell inward and rapidly compressing the liquid deuterium to fusion-ignition temperatures. But the Sausage pusher would serve another important function as well. It would serve as an additional source of fuel, soaking up the high-energy neutrons that the thermonuclear reactions would generate that would otherwise escape the explosion and go to waste, neutrons energetic enough to fission U238 and contribute significantly to the overall yield.

It was this thick, heavy U238 assembly that surrounded the deuterium dewar suspended within the Mike casing; the floating, nitrogen-cooled intermediate shield looked at its inner surface. That inner surface posed a problem cryogenically, Wechsler explains. "Oxidized uranium is pretty damned black, about the worst thing you could use in a cryogenic environment. We made the whole actual secondary right here, including the [thermal-]radiation shield parts, the sparkplug assembly and the big uranium pusher parts. Those castings were all made at the old Sigma Building" at the opposite end of the Los Alamos Tech Area from the S building where the Sausage's Brobdingnagian schematic drawing was laid out. "Cast here, machined here, inspected here, everything." To deal with uranium's high emissivity, Wechsler says, they decided to cover the inner surface of the uranium pusher with a coating that would serve as an additional radiation shield, like the silvering on a thermos bottle. The Boulder NBS laboratory had measured the absorptivities of sixty different metallic surfaces. The least absorptive, it found, was gold.

"In the old days," Wechsler comments, "sign painters used gold leaf for signs. They'd buy this really, really thin leaf, so thin that when you hold it, it almost feels as if it will float away." Gold leaf is made by layering gold foil squares between sheets of parchment, wrapping the stack in a sheepskin and beating the sheepskin with a hammer. The finished leaves, less than a thousandth of a millimeter thick—thin enough to see through—are then trimmed into three-inch squares and repacked between sheets of tissue paper in books of twenty-five sheets each. "We got a sign painter," Wechsler continues. "We brought this guy over to Sigma and he glued gold leaf on the inner surface of every one of those uranium pusher sections. It was bubble-free. It was smooth, like a gold mirror."

The fission sparkplug, mounted on a column in the middle of the deuterium dewar, was a plutonium device, cylindrically imploded. It included a chamber for tritium gas to boost the sparkplug yield; a line ran out through the end of the tank for loading the tritium, a few grams. "The problem of plutonium behavior at liquid hydrogen temperatures" had never been faced before, Mark comments, "and there were plenty of problems with plutonium even at room temperature."

The Panda Committee estimated that the Mike device would yield one to ten megatons, with the remote possibility that it might go as high as fifty to ninety megatons. The likeliest yield estimated was five megatons, the equivalent of ten billion pounds of TNT. That was as much as all the explosives used during the Second World War. Steel, lead, waxy polyethylene, purple-black uranium, gold leaf, copper, stainless steel, plutonium, a breath of tritium, silvery deuterium effervescent as sea-wake: Mike was a temple, tragically Solomonic, evoking the powers that fire the sun.

Wechsler looks over an old photograph from June 1952, when the Mike team assembled the Sausage secondary at Los Alamos. "That's a convoy. That was a convoy coming down from Boulder with three dewars loaded with liquid hydrogen. There's an old '49 Pontiac in front. We still had old cars then, we didn't buy new cars for the government every year. This was 1952 and our lead car was a '49 Pontiac." They trucked liquid hydrogen to Los Alamos to do a full cryogenic secondary assembly; the ordinary hydrogen would substitute for the more valuable deuterium. The transport dewars were so good that the cooling systems did not have to be turned on even once between Boulder and Los Alamos, but the liquid hydrogen stratified within the tanks—cooler at the bottom, warmer at the top—with a worrisome build-up of pressure. "One of the engineers said, Drive the damned thing around the block and it will slosh. So, hell, we drove around the block and the pressure dropped right down. We kept those things around for about a week and a half and ran the equipment a little bit and never had to vent anything."

The whole secondary assembly would be suspended within the Mike casing with stainless-steel cables and supported on springs. "The main tubes leading into the assembly were set up with bellows," Wechsler notes, "little bellows in the lines so that things could move a little bit. The secondary didn't have to be perfectly centered. It's not a true implosion system like the primary, where convergence is so critical. If the thing is a little off, you're talking about tremendous pressure. What we needed to make sure was that, in the period of time before it went off, we had everything stable. We did a lot of monitoring. We had a lot of thermocouples and we had thermistors and we had strain gauges. We did physical measurements here when we did the trial fill." They did the trial fill at DP Site, where Los Alamos produced plutonium metal, bomb cores and initiators, away from the main Tech Area. "We did the full cool-down to liquid hydrogen, ran it for about a day and a half and measured everything so we could calibrate all the gauges that we were going to use." The first cool-down revealed problems with the cryogenic assembly. After last-minute modifications, Wechsler's team did a second, successful cool-down in July.

The week of July 14, American Car and Foundry assembled the six-by-twenty-foot Mike casing in Buffalo with a dummy primary and secondary inside a mock-up of the building that would house it—known as a shot cab —at Eniwetok. Mike was never put together in its entirety in the United States. The casing, disassembled again into a set of big rings, and the primary

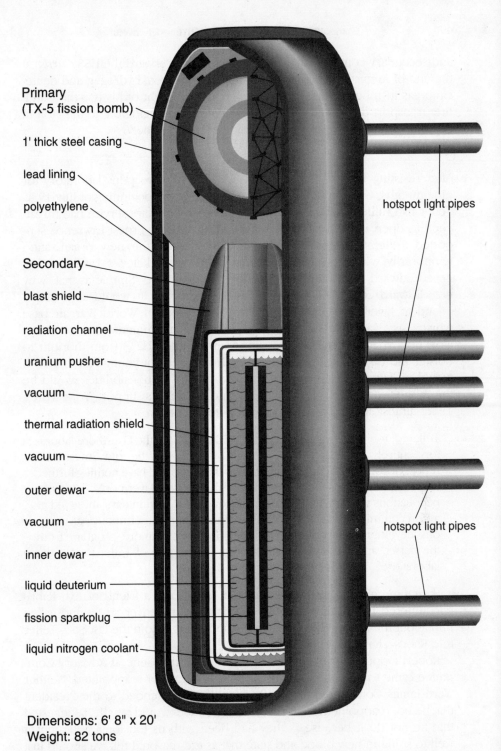

Primary
(TX-5 fission bomb)

1' thick steel casing

lead lining

polyethylene

Secondary

blast shield

radiation channel

uranium pusher

vacuum

thermal radiation shield

vacuum

outer dewar

vacuum

inner dewar

liquid deuterium

fission sparkplug

liquid nitrogen coolant

hotspot light pipes

hotspot light pipes

Dimensions: 6' 8" x 20'
Weight: 82 tons

Mike device (Sausage), tested at Eniwetok, November 1, 1952.
Yield: 10.4 megatons.

and secondary components went off to Eniwetok aboard the USS *Curtiss* at the end of August 1952. Other ships carried nitrogen, hydrogen and deuterium gas in tube trailers. Wechsler shipped out eight of his two-thousand-liter transport dewars. Scientific personnel began staging out to the atoll in September. The Mike shot was scheduled for November 1.

After resisting Edward Teller's drive for a second weapons laboratory for nine months, Gordon Dean finally capitulated at the beginning of July 1952. Teller had rallied the Air Force to his cause; the military service had threatened to open its own laboratory if the AEC refused. Ernest Lawrence supported Teller's proposal and arranged to house the new organization temporarily within his Radiation Laboratory at Berkeley. ("Lawrence believed Edward," Bradbury explains the collaboration. "Simple as that. Why not? Edward could sell refrigerators to Eskimos.") It would soon move inland to Livermore, California, to a former Second World War air base which Luis Alvarez had converted to develop the monumental linear accelerator, now abandoned, that was supposed to breed U233 from thorium to bolster the US atomic-bomb stockpile. Not Teller but young Herbert York would be Livermore's first director, and its first responsibilities would be thermonuclear diagnostic studies. Despite Lawrence's support, York writes, Teller almost aborted the new project before it began:

> Teller ... found the vagueness of the AEC's plans for the Livermore laboratory entirely unsatisfactory. As a result, in early July he told Ernest Lawrence, Gordon Dean, myself, and others that he would have nothing further to do with the plans for establishing a laboratory at Livermore. ... Intense negotiations were resumed among all concerned. Within days, these led to a firm commitment on the part of Gordon Dean that thermonuclear weapons development would be included in the Livermore program from the outset, and a renewed commitment on the part of Teller to join the laboratory.

Sometime after Teller agreed to join Livermore, a friend of I. I. Rabi's encountered the Hungarian-born physicist on a Denver street and asked him about the move. "I am leaving the appeasers to join the fascists," Teller told Rabi's friend sardonically.

Robert Oppenheimer's term as a member of the General Advisory Committee came to an end on August 8, 1952; he was not reappointed. Neither were James Conant and Lee DuBridge. "I recommended to the President that he not reappoint those three," Sidney Souers would recall. "I suggested that he not do it because ... they had been with us too long. I thought it well to bring in new blood and state that as the reason. I felt we should just drop all three and appoint three others who believed in the policy of the President with respect to the H-bomb. At that stage a number of people were talking disloyalty about Oppenheimer to [J. Edgar] Hoover, who passed the information on to me to pass to the President, but I dismissed it."

One H-bomb enthusiast left the field forever that summer. On July 28, after a brief illness, Brien McMahon died of cancer. He was not yet forty-nine years old.

Oppenheimer had written to Niels Bohr a year earlier—in June 1951—that "in spite of all the disappointments and tragedies that have occurred since last we talked, I am still one who does not take a wholly melancholy view. . . . It may seem curious to you that we in this country have been so slow to recognize where lay our true hope and our great danger. I have not despaired that we shall yet have learned in time." With the approach of the Mike shot in 1952, the senior scientists who had opposed the development of the H-bomb in 1949 saw one more opportunity to negotiate a moratorium on thermonuclear weapons with the USSR.

Vannevar Bush went directly to Dean Acheson in the spring of the year to propose postponing Mike. Nineteen fifty-two was a presidential election year. After losing the New Hampshire primary, Truman had decided not to run and had endorsed Adlai Stevenson, the governor of Illinois, as the Democratic candidate; the Republicans reluctantly chose Dwight Eisenhower. The November election would be held only three days after the Mike shot. "I felt that it was utterly improper," Bush testified, ". . . for that test to be [conducted] just before [the] election, to confront an incoming President with an accomplished test for which he would carry the full responsibility thereafter. For that test marked our entry into a very disagreeable type of world." Bush's second reason for proposing that Mike be postponed was even more compelling:

> I felt strongly that that test ended the possibility of the only type of agreement that I thought was possible with Russia at that time, namely, an agreement to make no more tests. For that kind of an agreement would have been self-policing in the sense that if it was violated, the violation would be immediately known. . . . I think history will show that was a turning point . . . [and] that those who pushed that thing through to a conclusion without making that attempt have a great deal to answer for.

Oppenheimer advanced similar concerns; so did Conant. Bethe wrote Gordon Dean that a test would "undoubtedly give food to the Communist propaganda machine" and that "there may be rapid and unpredictable political repercussions, especially in Europe"; he proposed that someone like Oppenheimer be delegated to brief the two presidential candidates and win their approval to postpone Mike until November 15. Bethe had checked with Bradbury at Oppenheimer's request to confirm that a postponement would not threaten the test series; Bradbury was agreeable, but hoped he could get the men home for Christmas.

The AEC commissioners, Lewis Strauss no longer among them, were sympathetic to the idea of delaying the test. Gordon Dean pursued the idea through the National Security Council in mid-August; the word came back from Truman, says the official AEC history, "that the President would not change the date, but he would certainly be pleased if technical reasons cause a postponement." In October, the commissioners decided to send one of their number, Eugene Zuckert, a former assistant secretary of the Air Force, out to Eniwetok to see if "technical reasons" might turn up. Dean solicited the approval of both the President and the Secretary of Defense, Robert Lovett, for his fellow commissioner's mission.

Zuckert remembered playing a larger role; in his recollection it was he who had alerted the President to the problem in the first place:

> Gordon came back and he said, "The President agrees with you." He said, "You go out there and see if you can get that shot stopped." So I got my fanny in an airplane, out to the West Coast, which in those days was prop airplanes. . . .
>
> Then I was given some admiral's plane—I ranked some admiral out of his airplane in Honolulu and churned out to Kwajalein and then to Eniwetok, and I spent many days there and tried to see if we could get that shot postponed. And finally they gave me the responsibility of determining what should happen. Well, finally I decided—I guess the day before—that because of the weather predictions, we should permit them to go ahead.

The President decided, not Zuckert, but Dean only reached Truman on the campaign trail in Chicago on October 29, two days before the scheduled test date on the other side of the International Date Line. A nuclear test requires the explosion of nuclear material, which only a President can approve. Zuckert's memory of the date of that decision at least was correct; approval of the *Ivy* series only passed through the National Security Council on October 30, 1952.

The atoll of Eniwetok, in the northwestern quadrant of the Marshalls about three thousand miles west of Hawaii, is an oval ring twenty miles long and ten miles wide of forty small islands. Like all such ocean structures, it was built up of coral around a submerging volcanic seamount. The United States captured Eniwetok from the Japanese in February 1944 along the way to invading the Marianas, a thousand miles closer to Japan, from which Curtis LeMay's B-29s launched the firebombing of Japanese cities.

After the war, the US had designated Eniwetok along with Bikini (two hundred miles east) as a testing ground for atomic weapons and had removed the atoll's native inhabitants. It had already served as a site for the 1948 *Sandstone* and 1951 *Greenhouse* tests, and a base of mothballed structures built for those earlier series had been restored for *Ivy*. Joint Task Force

ENIWETOK ATOLL

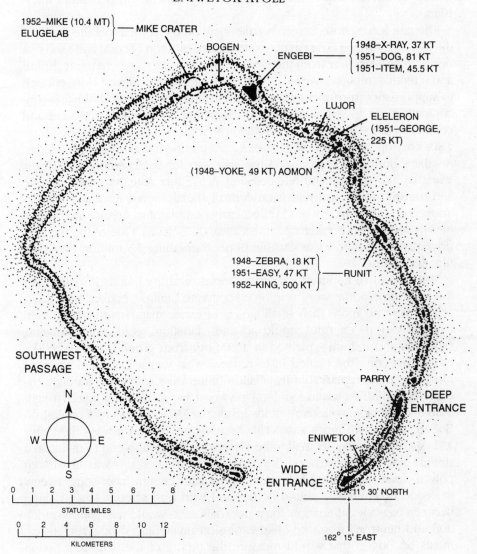

1952–MIKE (10.4 MT)
ELUGELAB — MIKE CRATER

BOGEN

ENGEBI — 1948–X-RAY, 37 KT
1951–DOG, 81 KT
1951–ITEM, 45.5 KT

LUJOR

ELELERON
(1951–GEORGE,
225 KT)

(1948–YOKE, 49 KT) AOMON

1948–ZEBRA, 18 KT
1951–EASY, 47 KT } RUNIT
1952–KING, 500 KT

SOUTHWEST
PASSAGE

N

W   E

S

PARRY

DEEP
ENTRANCE

ENIWETOK

0 1 2 3 4 5 6 7 8
STATUTE MILES

0   2   4   6   8   10   12
KILOMETERS

WIDE
ENTRANCE

11° 30' NORTH

162° 15' EAST

Eniwetok Atoll.

132, with Army, Navy and Air Force components, had begun staging out to
Eniwetok in March 1952; by October, more than nine thousand military and
two thousand civilians lived aboard ships in the Eniwetok area or in tent
encampments on the islands and had to be supplied with food, water, electricity and other necessities. A full task force of ships, including an aircraft
carrier and four destroyers, plied the sea around the atoll; the Air Force
operated more than eighty aircraft for transport and aerial sampling, including twenty-six B-29s, two B-36s and a B-47; a fleet of barges, motorboats,

passenger aircraft and helicopters moved equipment and personnel inter-island.

The task force chose Elugelab, true north on the compass of the atoll, for the Mike shot island. The small, spade-shaped outcrop of coral and sand was opposite and far away from the large atoll island, Parry, where the technical and scientific missions were based. It had islands east and west close enough to support shot instrumentation, and prevailing winds favored blowing contamination out to sea. Task force construction crews bulldozed sand and coral onto Elugelab to raise its elevation to improve lines of sight, then built a six-story open-air shot cab—Mike's zero point—big as an aircraft hangar. Auxiliary buildings surrounded the shot cab for subassembly and repair work, tritium storage and telemetering. A 375-foot antenna would receive and transmit radio and television control signals; seven enclosed mirrors arrayed in a semicircle would reflect early bomblight to streak cameras in a heavily reinforced bunker two miles away on Bogalua. One of the turbine-driven streak cameras was capable of photographing 3.5 million frames of film per second.

More than five hundred scientific stations on thirty islands instrumented the Mike shot. Some were reinforced-concrete bunkers banked with sand; others were no more than small targets of exotic materials that Mike neutrons of various energies would activate—tantalum, gold, sulfur, arsenic, cadmium and indium—planted in a line out from ground zero linked by cables that could be hauled in to retrieve what was left of them after the blast. The most remarkable installation (after Mike itself) was an eight-by-eight-foot, nine-thousand-foot-long plywood tunnel like an idled freight train that ran from concrete shields inside the Mike cab east to a bunker on Bogon, almost two miles away. This Krause-Ogle box, as it was called after its Los Alamos inventors, had to be elevated at the Bogon end to maintain a straight line of sight around the curvature of the earth. It was lined with polyethylene bags called ballonetts. Filled with helium from some twenty thousand two-hundred-pound bottles and with lead screens spaced along between bags for collimation, the tunnel ballonetts would pass gamma radiation and neutrons from the Mike explosion unattenuated by air to instruments on Bogon that would measure the timing of the Sausage's fission phase and the rise of the fusion reaction. The portholes that allowed the gammas and neutrons to pass into the Krause-Ogle box "probably diluted the shot," comments Raemer Schreiber philosophically. "We wanted lots of data, but the portholes permitted energy to leak out that would otherwise propagate [the fusion reaction]." Since Los Alamos expected the Bogon bunker to be damaged, the instruments there were rigged to transmit their findings by radio in real time to another bunker a safe distance beyond; the box and its instrumentation would function even as the Mike fireball roared out from ground zero eating the long tunnel away.

The Sausage itself was instrumented heavily with built-in radio transmitters and exotic materials placed inside and near the device casing that would be radiation-activated and vaporized by the explosion and could be recovered as fallout by sniffer planes downwind. Seven large diagnostic pipes were welded to the Mike casing to isolate the moments when bomblight broke through the casing wall from successive points along the way down the cylinder, signaling the successful propagation of the reaction.

Mike assembly began in September 1952; by September 25 the complex secondary was complete and in position. Vacuum piping went on then and ACF engineers began installing the heavy outer casing. "I remember seeing the guys hammer the big, thick polyethylene plastic pieces inside the casing," Harold Agnew recalls. "They hammered the plastic into the lead with copper nails." Agnew, a tall, raw-boned Westerner, Colorado-born, had helped Fermi build the CP-1 reactor at the University of Chicago during the Second World War, had flown the Hiroshima mission with Luis Alvarez and would be Norris Bradbury's successor as director of Los Alamos. Fishing in his off hours at Eniwetok, Agnew caught a five-foot nurse shark. "We didn't know what to do with it, but Marshall Holloway hadn't been very collegial in managing the program. He tended to give orders without explaining why. So I put the shark in his bed. He never said anything, but after that he was much more collegial."

Agnew was a member of the crew that would load the tritium into the Sausage's fission sparkplug, "a quite small amount," says Carson Mark, "to strengthen the design, make things more certain, push them further in a favorable direction." Agnew recalls that they carried the tritium out to Eniwetok ingeniously hydrided onto uranium:

We had steel buckets, maybe about three-eighths of an inch thick, with heaters built onto the outside. What you do is you cut uranium cubes, pure uranium metal, you put it in this thing, seal it all up and then you heat it real good. Then you cool it down and when you cool it down you activate it. We activated it first with just deuterium gas so that you get powdered uranium deuteride. The cubes turn into powder, probably black powder —I never looked at it, it's pyrophoric as hell.* Then you heat it again and drive off all the deuterium. Now you have a very active bed of uranium powder. Then when we put our tritium in it, it gobbled up the tritium and that's the way we transported it out.

To release the tritium from the uranium they plugged in the heaters on their uranium pots and drove off the gas into a gas bottle, just as they had done with the deuterium before.

---

* That is, it ignites spontaneously upon exposure to air.—RR.

If the deuterium for Mike was worth more than gold, the few grams of tritium were irreplaceable. Loading the exotic gas into the sparkplug, in the center of the complex secondary assembly, would be a delicate business. "You think about it," Wechsler explains, "here's this thing down in the inside. You can't put a float in there or anything else to measure how much tritium you've bled in. So you set up to do a complete mass balance on how much tritium you have; by reading pressures each timed to a volume, you can tell how much mass you've shifted over. Then, as the pressure drops because the gas is condensing, you can add more. They knew just how much and they had so many uranium pots filled with tritium and they kept valving them in one at a time."

Before loading the tritium, Agnew remembers, while the Mike secondary was still warm, "we decided we'd better do a dry run." It almost came to disaster:

We couldn't use tritium because we didn't have very much, so we said, we'll use deuterium because that wouldn't contaminate our lines. So we put the stuff in and watched to make sure we didn't have any leaks. Slowly —My God, there it was, how can we have a leak? Thought, thought and thought and then somebody said, maybe it's hydriding. Because we were putting this deuterium onto warm uranium and it was a slow leak but it was a leak. Well, what are we going to do? Of course we talked with Carson [Mark] and Marshall [Holloway]. We were worried we'd used that amount of deuterium and hydrided it, but hell, that's not going to bother anything. So we said now we'd use an inert gas. I don't remember whether we used argon or helium, but it didn't leak. So the deuterium was hydriding. But we were really upset—there was no way we could have fixed a leak because [the sparkplug] was way in the middle of the [secondary], which had been built up piece by piece by piece.

Agnew also worried about soldiers standing around Elugelab guarding the Mike shot cab with guns. "Sort of drove us bananas, all this liquid deuterium around. I was worried that some nut would shoot and actually hit the bomb. So a couple of us raised hell and got them out. Who's going to invade the island?"

On October 3, the British saluted *Ivy* from the Monte Bello Islands, off the northwest coast of Australia, with their first nuclear test: a fission implosion device named Hurricane suspended in a watertight caisson ninety feet below the frigate HMS *Plym* that yielded twenty-five kilotons and obliterated the *Plym*. The curious detonation scheme was instrumented to inform concerns that the Soviet Union might smuggle a bomb aboard ship into a British harbor.

Practice cryogenic cool-downs with liquid hydrogen started on Mike on October 10. The Navy barged Wechsler's dewars over to Elugelab to deliver the cryogenic liquids and serve for storage. Final filling with liquid deuterium began in the evening on Sunday, October 26. Air froze solid in one of the deuterium lines and had to be removed; otherwise the filling was routine. A stretch of bad weather began the next day, worrying everyone up to the end of the month and grounding the inter-island fixed-wing airlift service. Fortunately, the task force had mustered enough helicopters to substitute for the grounded aircraft.

A new core for the TX-V fission primary arrived by C-124 cargo plane from Los Alamos a few days before the target date. As Agnew remembers it, Marshall Rosenbluth initiated the idea of changing the core of the Sausage primary:

> Marshall may have saved the Mike shot. The nice thing about being overseas was that they fed you very well. They understood that if you want to have happy guys, you ration the booze—a fifth a week or something like that—but you serve very good food: shrimp, steak. Just really good food because you've got all these construction guys. You have movies and food. That's it. Ice cream, lots of ice cream. One night they had shrimp and Marshall ate too much. He couldn't sleep, and he got to thinking about what was going to happen and he decided the core we were using was prone to preinitiate and that we should change the core. He talked with Carson, I guess, and indeed there were rapid renegotiations.

The bellyache that saved the shot must have happened in August, when Mark says discussions of the core change began, "to decrease the chance that the core would predetonate and fail to deliver the proper yield." The substitute core, a levitated composite model built at Los Alamos, contained more uranium and less plutonium than the one it replaced. It would be less likely to chain-react before it reached maximum compression and would therefore react more efficiently, delivering more X-radiation to the secondary than the suspect core might have done.

Estimates of Mike's yield ranged so wide that the entire land task force had to be evacuated from the atoll onto ships, a tour de force of logistics. People packed up and went to sea whenever they finished their assigned preparation work. The scientists would be the last to go because they wanted to leave the Mike instrumentation unattended as briefly as possible. The last to leave would be the arming team, part of the firing party on the USS *Estes* that would monitor Mike by television and trigger the device by radio signal. Before then, Wechsler's group had to dump the excess liquid hydrogen in the transport dewars:

> How do you get rid of liquid hydrogen? You burn it. How do you burn it? You've got to flare it off into vapor first. You open a nozzle and you flare it off at a given pressure and you light it. We used a broom, a regular straw

broom. Light the broom, reach up and light the gas stream, flaring and burning thousands of liters of liquid hydrogen. That's a tremendous amount of energy. Once it starts burning you don't see any vapor trails of water vapor condensing out of the air. There's nothing. There's a roaring noise that sounds like a huge blow torch and it's invisible. If there's any dust in the air you can see a little waviness, but the flame is invisible. When we flared off the extra, the day we were going to leave to get on a ship out there on the Mike site, it was funny. We had two of these dewars sitting there flaring off, couldn't see a thing, just this roaring noise and all these terns flying around. They'd fly along about a hundred feet in the air above you and they'd hit that spot where this invisible hot air was going and whoa, talk about getting your tail feathers singed. That was a real hotfoot.

After the Mike secondary had been filled with liquid deuterium, Wechsler recalls, "we waited a few days to see that everything was stable." The last step in the assembly process was inserting the new primary core. Schreiber was in charge of the pit crew. "They knew what they were doing," he says lightly; "I provided moral support." The Mike casing had a manhole near the top end. "You could use the manhole for loading." The primary pit and core went in on the afternoon of October 31. "Then they buttoned the Mike gadget up." A final dewar of liquid hydrogen to top off the reflux cooler came over at nine-thirty that night. The arming team completed its checklist shortly after midnight on the morning of November 1, 1952, and boarded the *Estes,* which sailed from Eniwetok lagoon at 3:15 A.M. to a point about ten miles beyond the southern rim of the atoll, thirty miles from ground zero. The wind, which had been only marginally favorable on October 31, shifted to southerly at midnight, a direction that would blow the fallout away from the atoll into the unpopulated Pacific north of ground zero, ideal for the shot.

Two primitive but state-of-the-art television cameras broadcast images of the gauges monitoring Mike's systems—monitor dials and timing-signal and go-no-go indicators—to the firing room aboard the *Estes.* "I sat there all that night watching those damned things," Wechsler recalls, "taking notes. We made tables ahead of time of what the pressure balances would be and what this meant in terms of temperature and how full things were. So we knew what we wanted it to be and we knew when a deviation might be excessive. Nothing was moving. Your eyes play tricks on you after a while. We'd had lots of discussion about whether we might get a little bubble. If we did, we needed to know how big it might be, because it might affect the yield. But as near as we could tell that night, [the secondary] was full and it stayed full. Everything worked just the way it was supposed to."

H-hour for the Mike shot was 7:15 A.M., November 1, local time (October

31 in the United States). Before then, B-29 canister-drop, C-54 photo and B-47 and B-36 effects aircraft began orbiting at altitudes from ten to forty thousand feet at prescribed distances and compass headings from ground zero. Three 250-watt Motorola independent radio links communicated manual timing signals, automatic-sequence-timer start and emergency stop signals between the *Estes* and Elugelab. Automatic countdown sequencing began at H − 15 minutes. Two sniffer F-84 jets flew into position at forty thousand feet two minutes prior to H-hour, ready to flank the Mike cloud and take samples. At H − 1 minutes, loudspeakers aboard the ships of the task force instructed the thousands of military and civilian personnel to put on high-density goggles or turn away and cover their eyes. A momentary power failure aboard the *Estes* threw off the timing sequence by half a second, an unnerving stutter; Mike fired at 0714:59.4 ± 0.2, November 1, 1952.

When the radio signal from the *Estes* control room reached Mike, the capacitors in the Mike primary, already charged by the primary battery, discharged into a harness of electrical cables around the primary that carried the high-voltage current simultaneously to the ninety-two electric detonators inserted into the primary's high-explosive shell. (The increased number of detonators in the Mike primary made it possible to shape an implosion without using bulky high-explosive lenses, one way the TX-V device was made smaller and more transparent to radiation.) All ninety-two detonators fired with microsecond simultaneity; a detonation wave spread from each detonator, met other spreading donation waves moving inward and concentrating, emerged from the explosives as a shock wave, crossed to the aluminum pusher shell vaporizing as it passed, rocketed the pusher inward, crossed next to the primary's heavy uranium tamper, liquefied and vaporized the tamper, moved the material to the uranium shell of the core, hammered the uranium shell inward across an air gap to the plutonium ball levitated within, hammered the plutonium ball and crushed the Urchin initiator levitated at the center of the assembly. At that moment of maximum compression, with the vaporizing mass of uranium and plutonium supercritical, the shock wave shaped by the Munroe-effect grooves in the beryllium shell of the Urchin sliced through the shell and mixed beryllium with the polonium plated onto the ball of beryllium inside; alpha particles from the radioactive polonium knocked half a dozen neutrons from the beryllium; the neutrons ejected into the surrounding supercritical mass of uranium and plutonium and a chain reaction began.

Eighty generations later—a few millionths of a second—X-radiation from the furiously heating fission fireball hotter than the center of the sun escaped the primary mass entirely, began to ablate the blast shield over the Mike secondary and flooded down the cylindrical radiation channel inside the Mike casing. Instantly the radiation penetrated the thick polyethylene lining

1. Primary fissions. X rays from primary pass through primary fireball ahead of blast and flow down radiation channel.

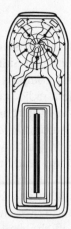

2. X rays from primary vaporize polyethylene lining of Mike casing and heat it to a plasma. Plasma reradiates longer-wavelength X rays that ablate surface of secondary pusher, causing rocket effect that implodes secondary, compressing and heating deuterium to fusion temperature and pressure and imploding fission sparkplug.

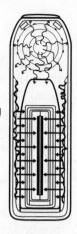

3. Sparkplug fissions, further compressing and heating deuterium from within. Full-scale thermonuclear fusion follows. Neutrons from fusion start fission reactions in U238 pusher shell, generating most of Mike's yield.

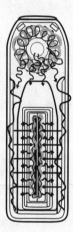

4. Fireball breaks through casing. In microseconds before entire casing vaporizes, light pipes (not shown) carry hotspot breakthrough light ("Teller light") to streak cameras to measure progress of explosions.

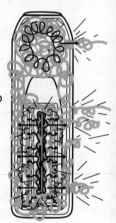

5. Fireball completely vaporizes Mike and quickly expands to more than 3 miles in diameter. Yield: 10.4 megatons.

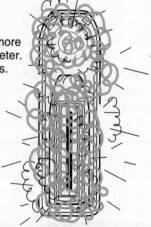

Sequence of events in two-stage radiation implosion (Mike device).

of the casing and heated it to a plasma. The plasma reradiated X rays that shone simultaneously from all sides inward onto the surface of the heavy uranium pusher, heating it instantly to ablation. The ablating surface of the pusher drove it explosively inward even as it liquefied and vaporized. The intense pulse of pressure concentrated as it moved inward, closed the first vacuum gap, compressed the floating thermal shield, closed the next vacuum gap, compressed the outer and inner dewars, encountered the deep, cold mass of liquid deuterium, compressed the deuterium inward and started to heat it. As the pressure pulse that was heating the deuterium to thermonuclear temperatures converged upon itself down the long axis of the secondary, it encountered the fission sparkplug, imploded that cylindrical system and activated a second fission explosion boosted with high-energy neutrons from fusion reactions in the tritium gas the sparkplug compressed.

All these processes, proceeding through microseconds, prepared Mike for thermonuclear burning. Now the escaping X-radiation of the fissioning sparkplug heated the compressed deuterium at its boundaries; the increasing thermal motion of the deuterium nuclei pushed them together until they passed the barrier of electrostatic repulsion between them and came within range of the nuclear strong force, at which point they began to fuse. Some fused to form a helium nucleus—an alpha particle—with the release of a neutron, the alpha and the neutron sharing an energy of 3.27 MeV. The neutron passed through the electrified mass of fusing deuterons and escaped, but the positively charged alpha dumped its energy into the heating deuterium mass and helped heat it further.

Other deuterium nuclei fused to form a tritium nucleus with the release of a proton, the triton and the proton sharing 4.03 MeV. The positively charged proton dumped more energy into the deuterium mass. The tritium nucleus fused in turn with another deuterium nucleus to form an alpha particle and a high-energy neutron that shared 17.59 MeV. The 14-MeV neutrons from this reaction began to escape the hot, compressed deuterium plasma and encountered the U238 nuclei of the vaporized uranium pusher. U238 fissions when it captures neutrons with energies above 1 MeV; so the U238 of the uranium pusher began to fission then under the intense neutron bombardment, flooding more X rays back into the deuterium mass from the outside just as the sparkplug fission reaction was radiating them from the inside, trapping the deuterium between two violent walls of heat and pressure. Deuterium-bred tritium fused with tritium as well, producing a helium nucleus and two neutrons that shared 11.27 MeV of energy. At lower orders of probability, deuterium captured a neutron and bred tritium; deuterium-bred helium fused with deuterium and made heavy helium plus a highly energetic proton, or captured a neutron and bred tritium plus a proton. All these reactions contributed to the force of the Mike explosion.

PRIMARY REACTIONS IN THE HYDROGEN BOMB

$$D + D \longrightarrow {}^3He + n + \ 3.27 \text{ MeV} \tag{1}$$

$$D + D \longrightarrow T + p + 4.03 \text{ MeV} \tag{2}$$

$$D + T \longrightarrow {}^4He + n + 17.59 \text{ MeV} \tag{3}$$

$$T + T \longrightarrow {}^4He + n + n + 11.27 \text{ MeV} \tag{4}$$

$${}^6Li + n \longrightarrow {}^4He + T + 4.78 \text{ MeV} \tag{5}$$

$${}^3He + D \longrightarrow {}^4He + p + 18.35 \text{ MeV} \tag{6}$$

$$D = {}^2H \text{ (deuterium)}; \ T = {}^3H \text{ (tritium)}$$

Moving outward from the cauldron of the secondary as gamma and X-radiation and as escaping high-energy neutrons, that explosion swelled back across the path the radiation-driven implosion had taken. Just as the big uranium pusher had served as a tamper for the secondary, so the thick, lead-lined Mike casing served as a tamper for the entire complex explosion, holding it together a few microseconds longer to give the fuel more time to react, but massive as the casing was, bomblight from its outer surface revealed the breakthrough of the developing explosion before the mass had time even to swell, much less to move.

Once the explosion broke through the casing, it expanded in seconds to a blinding white fireball more than three miles across (the Hiroshima fireball had measured little more than one-tenth of a mile) and rose over the horizon like a dark sun; the crews of the task force, thirty miles away, felt a swell of heat as if someone had opened a hot oven, heat that persisted long enough to seem menacing. "You would swear that the whole world was on fire," one sailor wrote home who turned around like Lot's wife to look. For a moment the fireball seemed to hover; then it began to rise. Los Alamos radiochemist George Cowan, a precise man whose ingenious tests would help measure Mike's yield, was there that day:

> I was stunned. I mean, it was big. I'd been trying to visualize what it was going to be like, and I'd worked out a way to calibrate the shot. The initial fireball I guess I calibrated by holding up a quarter. If the quarter would cover the fireball then the yield would be less than something; if the

fireball were bigger than the quarter, then it would be more than something. The question was, looking through my dark glasses, could I cover the fireball with a quarter. And I couldn't, so I knew it was big. As soon as I dared, I whipped off my dark glasses and the thing was enormous, bigger than I'd ever imagined it would be. It looked as though it blotted out the whole horizon, and I was standing on the deck of the *Estes,* thirty miles away.

Momentarily, the huge Mike fireball created every element that the universe had ever assembled and bred artificial elements as well. "In nanoseconds," writes the physicist Philip Morrison, "uranium nuclei captured neutron upon neutron to form isotopes in measurable amounts all the way from ^{239}U up to mass number 255. Those quickly decayed, to produce a swath of transuranic species from uranium up to element 100, first isolated from that bomb debris and named fermium."

Swirling and boiling, glowing purplish with gamma-ionized light, the expanding fireball began to rise, becoming a burning mushroom cloud balanced on a wide, dirty stem with a curtain of water around its base that slowly fell back into the sea. The wings of the B-36 orbiting fifteen miles from ground zero at forty thousand feet heated ninety-three degrees almost instantly. In a minute and a half, the enlarging fireball cloud reached 57,000 feet; in two and a half minutes, when the shock wave arrived at the *Estes,* the cloud passed 100,000 feet. The shock wave announced itself with a sharp report followed by a long thunder of broken rumbling. After five minutes, the cloud splashed against the stratopause and began to spread out, its top cresting at twenty-seven miles, its stem eight miles across. "It really filled up the sky," notes Raemer Schreiber, who had seen shots before and was not easily impressed. "It was awesome. It just went on and on." At its farthest extent, the Mike cloud billowed out above a thirty-mile stem to form a huge canopy more than one hundred miles wide that loomed over the atoll. Radioactive mud fell out, followed by heavy rain.

Down below, Elugelab had vanished. The fireball had vaporized the entire island, leaving behind a circular crater two hundred feet deep and more than a mile across filled with seawater, a dark blue hole punched into the paler blue of the shallow atoll lagoon. The explosion vaporized and lifted into the air some eighty million tons of solid material that would fall out around the world. It obliterated the Krause-Ogle box and burned and damaged the Bogon bunker. It stripped animals and vegetation from the surrounding islands and flashed birds to cinders in midair. A survey team afterward discovered general ruin:

Rigili is fourteen miles south-southwestward down the lagoon from the Mike shot crater. Yet there the survey team found that the trees and brush facing the test site had been scorched and wilted by the thermonuclear heat. Many of the terns there were sick, some grounded and reluctant to fly and some with singed feathers, particularly the noddy terns and the sooty terns, whose feathers are dark in color. . . .

At Engebi [three miles from ground zero] the group went ashore on an island where the sense of desolation was deepened by the presence of a reinforced concrete building, ruptured and shaken but still standing, on the island flat that had been swept by the blast and the succeeding surge of water. The body of a bird was seen, but no living animals and only the stumps of vegetation. . . . Among the specimens collected were fish which seemed to have been burned. On each of these fish, the skin was missing from one side, as if, the field notes said at the time, the animal "had been dropped in[to] a hot pan."

Red Leader, an F-84 sampler aircraft piloted by a colonel named Meroney, flew into the stem of the mushroom cloud at 42,000 feet almost two hours after the explosion, *Ivy* historians report:

Immediately upon entering the cloud, Red Leader was struck with its intense color. It cast a red glow over the cockpit and his radiological instruments indicated maximum readings. . . . The hand on the Integron, which showed the rate at which radioactivity was being accumulated, "went around like the sweep second hand on a watch. . . . And I had thought it would barely move!" The combination of most instruments indicating maximum readings and the red glow like the inside of a red-hot furnace was "staggering" and Colonel Meroney quickly made a ninety-degree turn to leave the cloud.

Fireball measurements and subsequent radiochemistry put the Mike yield at 10.4 megatons—the first megaton-yield thermonuclear explosion on earth. Its neutron density was ten million times greater than a supernova, Cowan remarks, making it "more impressive in that respect than a star." The Little Boy uranium gun that destroyed Hiroshima was a thousand times less powerful. Mike's fireball alone would have engulfed Manhattan; its blast would have obliterated all New York City's five boroughs. More than 75 percent of Mike's yield, about eight megatons, came from the fission of the big U238 pusher around the secondary; in that sense it was less a thermonuclear than a big, dirty fission bomb. Fission-fusion-fission, the staging arrangement came to be called. If Los Alamos had devised a way to burn unlimited quantities of thermonuclear fuel, it had also devised a way to burn unlimited quantities of cheap ordinary uranium.

Edward Teller had not traveled out to Eniwetok to watch his former colleagues explode the thermonuclear device that he and Enrico Fermi had conceived in the early days of the Manhattan Project, that he had fought for and helped invent. He claimed he was too busy starting the new weapons laboratory at Livermore, but no one doubted that bitterness and hostility, perhaps also jealousy, kept him away. He had not expected Los Alamos to do the job. When he understood that Mike would probably work, he and his colleagues devised a way to observe the explosion from California. Herbert York at Livermore monitored the radio frequency of the Mike firing-signal telemetry on a shortwave radio; Teller in Berkeley with Ernest Lawrence and Luis Alvarez monitored a seismograph. The physicists had calculated the time a seismic wave from a successful shot would need to travel under the Pacific basin to northern California and had calibrated seismic magnitude with yield. When York heard the Mike firing signal, he called Teller. Teller got busy:

> I went down into the basement of the University of California geology building in Berkeley, to a seismograph that had a little light-point marking on photographic film. A tremor of that point would show when the shock wave, generated thousands of miles away on Eniwetok Island, reached Berkeley. I watched the light point but it would not stand still. Try to look at a point of light in the dark; it will dance before your eyes because your eyes are moving. I took a pencil and steadied it against the side of the apparatus; then I could see that the point of light, relative to the pencil tip, was steady.
>
> At exactly the scheduled time I saw the light point move. It moved so slightly that I was not sure whether I just thought it moved or whether it actually had moved. So I stayed around for another ten minutes, lest I miss the real event; then I took the whole film and had it developed. There was the signal, just as predicted. . . . The sound waves took twenty minutes to carry the message under the Pacific and arrive in Berkeley.

The seismic record indicated a big explosion. Teller passed the news to York at Livermore. York says he called Los Alamos. Marshall Rosenbluth, who had returned to Los Alamos before the test, remembers a telegram from Teller dropping into the frustrating silence after H-hour; he and a dozen others had collected in Norris Bradbury's office and were waiting impatiently for the Eniwetok security officers to clear a first report. However the good news traveled, Teller's message claimed paternity: "It's a boy."

The first successful test of a staged thermonuclear, York understood immediately, marked "a moment when the course of the world suddenly shifted, from the path it had been on to a more dangerous one. Fission bombs, destructive as they might have been, were thought of [as] being limited in power. Now, it seemed, we had learned how to brush even these limits aside and to build bombs whose power was boundless." Stanislaw Ulam had been right, that day at the turn of the year when his wife had found him staring out into their garden with a strange expression on his face. The totally different scheme that Ulam had first conceived, that Edward Teller had improved, that Carson Mark and Marshall Holloway's Panda Committee had elaborated into inspired mechanism, would change the course of history—but not in the direction of decisive US advantage that the H-bomb enthusiasts had fantasized.

25

Powers of Retaliation

LAVRENTI BERIA'S NUCLEAR ARCHIPELAGO was thriving. With success had come a measure of indulgence for physicists, who were evidently a valuable commodity worth humoring. At Sarov, Igor Tamm was listening to forbidden BBC programs and reporting them to Andrei Sakharov over breakfast. Yakov Zeldovich had begun a love affair with a *zek* artist named Shiryaeva who had been arrested on charges of anti-Soviet slander and whose husband had denounced her; she had brightened the Sarov theater, VIP dining room and management houses with her murals. "One evening on my way home from work," Sakharov recalled, "I caught sight of Zeldovich. The moon was out, and the bell tower [of the old Sarov monastery] cast a long shadow on the square in front of the hotel. Zeldovich was walking deep in thought, his face somehow radiant. Catching sight of me, he exclaimed: 'Who would believe how much love lies hidden in this heart?' "

Victor Adamsky, then a thoughtful young experimental physicist, was assigned to Sarov in 1950 after he graduated from Moscow University. "I didn't know what the place was," he recalls, "but there were rumors it was a good place to work. I understood that I would go there to work on the hydrogen bomb. There was a good hotel. We would share rooms, two to a room. I came in January, and in March an apartment was finished and allocated to me. Then Igor Tamm, Yuri Romanov, Sakharov—my friends—were given cottages. They took me in. So in our part of the cottage lived Tamm, Romanov and I. Since Sakharov came with a family, he got his own cottage." Adamsky found a collegial spirit at the secret installation. "There was enthusiasm. We worked well. We felt we were doing our duty. It was very interesting physics." The Sarov library had a subscription to the *Bulletin of the Atomic Scientists,* the Chicago-based journal of news and opinion that Leo Szilard had helped found. "We would read it with great interest. What intrigued us were the discussions between scientists, moral and social issues in particular. It gave us a picture of our American counterparts, an image, a character. We could visualize them as human beings—Glenn Seaborg, say, or Szilard himself. We thought Szilard was a leading conscience of humanity."

When theoretician V. I. Ritus arrived at Sarov in 1951, Sakharov was there to greet him. "Andrei Dmitrievich came out of his office with a wide grin and energetically shook my hand.... [He] led me to a blackboard, took a piece of chalk into his left hand, and drew a large circle around the words 'The facility is organized in the following manner.' He is an original, I thought. He wants to familiarize me with the layout of the town, and for economy's sake he is drawing his diagram centro-symmetrically, even though I knew that it was not that way in reality—I had already walked the streets. Andrei Dmitrievich drew a smaller circle concentric with the first and uttered a few more phrases.... Only after awhile did I finally begin to understand that he was talking about something else entirely: the hydrogen bomb."

Although they were still using electronics salvaged from Second World War radio sets, experimentalist Alexander I. Pavlovsky reports, and it was difficult to get permission to leave Sarov for vacation, "we were young, and life seemed full of wonders.... We managed to read a lot, go out with friends, study, go in for sports and do many other things although we worked twelve hours a day and sometimes even round the clock.... On summer Sundays, when we had time, most of us went to the stadium, where we held [sports] competitions between departments." "We skied and went hiking," Sakharov recalled, "and in the summertime we swam." Sarov, the young theoretician concluded, "was a big village."

Eastward at Chelyabinsk, the conscript soldier was allowed access to Techa now—"the academic compound," the conscripts called it. "I got acquainted with the local youth," the soldier reminisces, "mostly with young women. Age is age." He was also allowed access to the neighboring civil settlement in the zone, where there was a cinema. Beria motivated his nuclear *zeks* by commuting two years of their sentences for every one they worked ("providing the work was good," the soldier qualifies), but maintained secrecy by allowing no one to leave the zones to return to private life; the civil settlement within the barbed wire at Chelyabinsk grew up to accommodate released prisoners, military pensioners and civilians formerly attached to the enterprise. "In the summer, naturally, we used to swim in Lake Karachai and in the Techa River," the soldier recalls bitterly, "although there were warning signs on the banks: 'Swimming forbidden!' But why? Nobody told us that the river was contaminated with radiation. We also drank this water and used it for cooking. Once a group of three even risked a swim across the lake, to see if there was barbed wire on the opposite bank. Naive guys!" The conscript soldier was released from duty, sworn to secrecy for life and allowed to go home to Leningrad in 1952;

years later, when he developed multiple chronic illnesses from radiation exposure, his government disallowed compensation because it could find no record that his Chelyabinsk detachment had ever existed.

Beria's humoring had its limits. Zeldovich's love, the muralist Shiryaeva, was shipped off to the east for resettlement in internal exile and gave birth to Zeldovich's daughter in a building with an inch of ice on the floor. A commission arrived in Sarov one day to make sure everyone agreed with Soviet agronomist Trofim Lysenko's Marxian notions of heredity, which Stalin had endorsed. Sakharov expressed his belief in Mendelian genetics instead. The commission let the heresy pass, he writes, because of his "position and reputation at the Installation," but the outspoken experimentalist Lev Altshuler, who similarly repudiated Lysenko, did not fare so well. Sakharov and a colleague had to intercede on Altshuler's behalf with Boris Vannikov's deputy Avrami Zavenyagin—"a man of great intelligence," writes Sakharov, "and an uncompromising Stalinist." Zavenyagin attached a second political commissar to Altshuler's department at Sarov to keep an eye on the miscreant.

What Zavenyagin called Altshuler's "hooliganism" prompted a second confrontation in 1952, this time between Yuli Khariton and Beria himself. Altshuler had spoken out on cultural matters. Vannikov ordered the physicist flown to Moscow and examined him personally, Altshuler remembers:

> He explained to me how bad a man I was. He put my dossier on the desk. "You're working in an installation so secret even the party secretaries don't know about it and you're proposing your own party line in music, literature and biology. If we let just anyone say whatever he wants, we'd be crushed." The Soviet Union was a fortress and everybody around—Europe, the United States, China—was preparing for war. One of the managers at Sarov would declare in a meeting, "One day war will come, and with our bombs we will hit our enemies, the United States." Even after Stalin was gone we lived in this atmosphere.

Altshuler was threatened with exile. Khariton called Beria to save him. The Sarov scientific director "telephoned Beria directly and told him that the project needed Altshuler," Khariton and Yuri Smirnov report. "After a long pause, Beria asked a single question: 'Do you need him very much?' After receiving a positive response, Beria said 'All right,' and hung up. The incident was closed."

If anyone needed reminding of Beria's brutality, an incident Khariton recalls from that period must have served:

> Beria met with about thirty people in his Kremlin office to discuss the preparation of the test site for the first thermonuclear explosion. Those making reports were trying to say where the equipment should be located,

what kind of structures should be erected and how, and what kind of experimental animals should be placed at the site to study the impact of the blast effects. Suddenly, Beria became incensed. He interrupted angrily, moving from one briefer to another, asking strange questions which were not easy to answer.

Finally, Beria completely lost his temper. He screamed: "I will tell you myself!" Then he started to talk nonsense. It gradually became clear from his stormy monologue that he wanted everything at the test site to be totally destroyed in order to provide the maximum terror.

"The participants left the meeting in a gloomy mood," Khariton concludes.

The Soviet Union had a gaseous-diffusion plant now, producing kilogram quantities of U235 per day, and a heavy-water pile for tritium. After problems with the chemical separation of lithium isotopes that drew further threats from Beria ("We have plenty of room in our prisons," he told the responsible official, an MVD general), a physicist working with Lev Artsimovich developed an electromagnetic separation system and produced enough lithium6 to make lithium^6 deuteride for Sakharov's layer-cake thermonuclear. Altshuler's team conducted the first model experimental tests of the design's hydrodynamic parameters in the summer of 1952.

Since the Soviet scientists were building an Alarm Clock, which was limited by its awkward massiveness to around a megaton in yield, Beria's espionage apparatus had evidently not encountered the megaton-range Teller-Ulam configuration. The Mike shot came as a surprise to the Soviets; Pavlovsky reports that "the procedure of sampling radioactive explosion products in the Pacific zone"—sniffer aircraft, surface ships or submarines —"was not available." American sailors on leave from the Mike task force who wrote home or called home from Hawaii quickly broke the story of the *Ivy* test to the world, however, and Sarov attempted to analyze Mike fallout in snow samples collected "in the central belt of Russia." Sakharov remembers an emotionally upset chemist absentmindedly pouring the concentrated snowmelt down the drain. Pavlovsky notes that the fission products which they might have been able to trace—beryllium7 and U237—would have decayed in any case to background levels by the time they had circled three-quarters of the globe and come around to the USSR. More to the point, he says, attempts to analyze the Mike fallout "failed due to one simple reason: we just could not perform such analysis at that time." On the other hand, public accounts of the Mike shot made it obvious that the device had produced a multi-megaton yield; the Soviet scientists would have understood that the United States had achieved a breakthrough beyond layer cakes and alarm clocks. That knowledge was not immediately useful to them because they had no way of knowing what the breakthrough might be, but it spurred them to push completion of the layer-cake device as a counterweight to the new US monopoly.

On the last evening of his life, February 28, 1953, Joseph Stalin had dinner at Kuntsevo, his dacha just outside Moscow, with Beria, Georgi Malenkov, Nikita Khrushchev and Nikolai Bulganin. The seventy-three-year-old Soviet dictator had been suffering from dizzy spells and had recently returned from a Crimean vacation. Bulganin, who was Minister of Defense, briefed him that evening on the status of the Korean War, advising that the war had reached a point of stalemate. Stalin decided then, according to Dmitri Volkogonov, that "he would tell Molotov next day to advise the Chinese and North Koreans to 'try to get the best deal they could in talks,' but in any event to try to bring the armed conflict to a halt." Beria reported on the continuing interrogation of a group of Jewish doctors whom Stalin suspected of plotting against him; in Beria's cellars most of them had already confessed to working for the American Jewish Joint Distribution Committee —"the Joint," Beria called it contemptuously. "The threads run deep," the Minister of Internal Security warned his master, "and are linked to party and military officials." Stalin began a long diatribe, Volkogonov reports, against "people in the leadership who thought they could get by on their past merits. 'They are mistaken.' " At four A.M. on the morning of March 1, Stalin finally stood and left the room. The people in the leadership, with their Kremlin pallors, took their merits home.

When Svetlana Stalin was called out of French class the following day, March 2, and driven to Kuntsevo, she found her father surrounded by doctors, nurses and the ministers of the Presidium. The Soviet dictator was unconscious by then; he had suffered a stroke. The doctors worked him over, his daughter reports, "making a tremendous fuss, applying leeches to his neck and the back of his head, making cardiograms and taking X rays of his lungs. A nurse kept giving him injections.... Everyone was rushing around trying to save a life that could no longer be saved." Alone in his chambers after he had left the room on March 1, Stalin had collapsed and had lain unattended through the day and the evening until almost midnight because the servants were afraid to enter his private rooms unsummoned. They had alerted the leadership then; Volkogonov says Beria had arrived drunk in the wolf hours before dawn on March 2 and had thrown everyone out again, shouting at them that "Comrade Stalin is sound asleep"; the other members of the Presidium had timidly followed his lead and abandoned their leader once more, only returning later that morning with the doctors in tow. Observing the crowd around her dying father, Svetlana noticed that Beria "was behaving in a way that was nearly obscene.... His face, repulsive enough at the best of times, now was twisted by his passions.... He was trying so hard at this moment of crisis to strike exactly the right balance, to be cunning, yet not too cunning." It took Stalin several days to die. When he did so, Beria wasted not a second, Svetlana accuses. "He darted into the hallway ahead of anybody else. The silence of the room where everyone was gathered around the deathbed was shattered by the sound of his loud voice, the ring of triumph unconcealed, as he shouted, 'Khrustalyof! My car!' "

Malenkov and Beria had maneuvered to take over the Central Committee Presidium, with Malenkov as chairman, while the old monster still lay dying. In the months after Stalin's death, Beria consolidated his security organizations into a new MVD with a substantial army as well as police and security forces. He proposed abandoning some of Stalin's vast construction projects, repudiated the Doctors' Plot and requested amnesty for a million *zeks,* not including political prisoners. "The most astonishing thing that happened after Stalin died," correspondent Harrison Salisbury reported from Moscow, "was the quickness with which symptoms of a thaw appeared."

It was Beria's thaw, a strategy for taking power. He might have succeeded, but he made the mistake of pushing through the Presidium a plan to liberalize the regime in East Germany to stop the bleed of East Germans fleeing to the West—half a million since 1951. In mid-June 1953, the East German Politburo rubber-stamped much of the "new course" but resisted lowering labor norms. Beria's liberalization in East Germany had the effect Mikhail Gorbachev's liberalization would have again thirty years later throughout Eastern Europe and the Soviet Union: workers rioted in the streets and had to be put down with Soviet tanks.

Khrushchev rallied Malenkov, Molotov, Bulganin, Pervukhin and other members of the Presidium to oust Beria after the East German debacle. He probably also won support within the Soviet military. He organized a small force of officers led by the commander of the Moscow air defense system, K. S. Moskalenko, that included Marshal Zhukov and young Leonid Brezhnev and secreted them in a room adjoining Malenkov's office before a Presidium meeting there on the afternoon of June 26. Beria arrived open-collared and casual, unsuspecting, his guards and assistants left behind in the reception area. Well into the meeting, Khrushchev sprung his trap: the officers burst into Malenkov's office, Moskalenko announced Beria's arrest and Zhukov searched him. They held him in a nearby room until midnight to evade the MVD forces that guarded the Kremlin—removed and crushed his pince-nez, cut the buttons off his pants so that he would have to hold them up if he tried to run—then slipped him out in a convoy of cars and delivered him to the garrison guardhouse at Lefortovo Prison.

Tanks, armored personnel carriers, self-propelled guns and motorcycles patrolled the streets of Moscow for the next two days. Khrushchev's people swept through the MVD making arrests. The leadership moved Beria from Lefortovo to a little-known bunker under an apple orchard near the Moscow River. He was interrogated there, with how much Berian brutality no one has yet revealed. (One Soviet source claims he staged an eleven-day hunger strike, commenting that "This did not do him much harm, given his health and his build.") The Presidium organized a summer of Central Committee hearings and denunciations in preparation for a December trial.

The new collective leadership was shocked to discover that Beria's nuclear archipelago was working on a hydrogen bomb. When Beria was arrested, Malenkov had appointed Vyacheslav Malyshev head of the atomic-bomb program, renaming it the Ministry of Medium Machine Building. Malyshev—"a short, ruddy-faced man," Sakharov describes him—had been a senior arms-production administrator during the war. "We began to dig into the archives," the new minister told the Central Committee that summer, "and we found that [Beria] had signed a whole number of important decisions without the knowledge of the Central Committee or the government." Zavenyagin confirmed Malyshev's account before the several hundred Central Committee members gathered to hear Beria's crimes:

> I was witness to this story. [After Stalin's death] we prepared a draft decision for the government [authorizing a hydrogen bomb]. For some time it lay on Beria's desk, then he took it with him to read. We had the idea that perhaps he wanted to speak with comrade Malenkov. About two weeks later, he asked us to come by and he began to look at the document. He read it, entered a number of corrections. He got to the end. The signature [block] was [labeled] Chairman of the Council of Ministers G. Malenkov. He crosses it out and says—'This isn't necessary.' And he signs his own name.... The hydrogen bomb ... would be a most important event in worldwide policy. And that scoundrel Beria allowed himself to make this decision outside of the Central Committee.

Evidently Beria had been confident enough of his ascent to power to assume that he would command sole authority by the time the thermonuclear design was ready to be tested, in August 1953.

If there was turmoil in the Soviet Union that spring and summer, there was turmoil in the United States as well as the scheduled day of execution approached of Julius and Ethel Rosenberg. From a first mass meeting of supporters at the Pythian Temple in New York City in March 1952, protests had multiplied in the US and in Europe until the Rosenberg case became an international cause célèbre. The couple had begun to believe that their fellow citizens would save them. "We must soberly realize that our only hope rests with the people," Julius had written Ethel that March. A series of articles in an influential left-wing weekly, the *National Guardian,* arguing their innocence had brought them to public attention and furnished a rallying cry: that they were "the first victims of American fascism." (The American Communist Party, preoccupied with its own trials at that time, took a different view; CP officials privately told a writer who was organizing support for the Rosenbergs, "They're expendable.")

The Rosenbergs began writing public letters to encourage their supporters, letters curious in their claims to martyrdom since the couple had denied political activism at their trial. "Like others we spoke for peace," they wrote from Sing Sing, where they were both separately incarcerated, "because we did not want our two little sons to live in the shadow of war and death. . . . That is why we are in the death house today, as warning to all ordinary men and women." Ethel, more stridently, sometimes preached retribution:

> Wait, wait and tremble, yet mad masters, this barbarism, this infamy you practice upon us, and with which you regale yourselves presently, will not go unanswered, unavenged, forever! The whirlwind gathers, before which you must fly like the chaff!

As the case moved through the superior courts to the Supreme Court, Judge Kaufman shifted the Rosenbergs' execution date forward until it settled on June 19, 1953. The Supreme Court refused to grant certiorari; other late appeals failed as well. A great swell of Soviet-orchestrated protest in Western Europe—Jean-Paul Sartre declared the approaching executions "a legal lynching which smears with blood a whole nation"—drowned a wave of anti-Semitic purges in Eastern Europe and the Soviet Union, including the Doctors' Plot, in that last year of Stalin's life. Europeans argued that Klaus Fuchs, clearly the more serious offender, had received only fourteen years, to which J. Edgar Hoover responded that the US had better business than emulating the "weaknesses" of British security.

Pickets at the White House did not prevent Dwight Eisenhower from refusing to grant the Rosenbergs clemency. The statement he released on February 11, 1953, noted that the couple had been tried by jury and convicted and that their appeals had been rejected. "The nature of the crime for which [the Rosenbergs] have been found guilty," the new President concluded, ". . . involves the betrayal of the whole nation. . . . By their act these two individuals have in fact betrayed the cause of freedom for which free men are fighting and dying at this very hour." Privately, Eisenhower saw the Rosenberg capital sentence less as punishment than as a way of communicating US Cold War resolve. Just three days before the scheduled execution, he explained his position at length in a letter to his son John, then on active duty in Korea:

> The Rosenberg case continues to cause a very considerable amount of furor. Involved in the effort to have the sentence commuted are not only Communists. In addition there are people who honestly believe that there is a doubt as to the Rosenbergs' guilt; others who have conscientious scruples against capital punishment. . . .
>
> I must say that it goes against the grain to avoid interfering in the case where a woman is to receive capital punishment. [But] . . . if there would

be any commuting of the woman's sentence without the man's, then from here on the Soviets would simply recruit their spies from among women. . . .

We know that the Rosenbergs were part of [a spy] ring. If the Soviets can convince prospective recruits that the worst possible penalty they would ever have to pay for exposure as spies would be a relatively short term in prison, then their blandishments and bribes would be much more effective. . . .

If it were possible to assure that these people would be imprisoned for the rest of their natural lives, there would be no question that the vast bulk of the argument would rest on the side of commutation. But the fact is that, if they do not go to the chair, they will be released in fifteen years under federal law.

The White House received tens of thousands of letters in the final weeks before the scheduled execution. Even David Greenglass protested from prison. "I have to live with it," he wrote. "I was never told that my sister would be killed. Or I might not have testified."

Responding clandestinely to the protest, Attorney General Herbert Brownell, Jr., sent Bureau of Prisons Director John V. Bennett to Sing Sing early in June to convince the Rosenbergs to cooperate. Bennett spoke to Julius first, then separately to Ethel, then to both of them together. He offered to see their sentences commuted in exchange for information. He told them that Gordon Dean would appeal to Eisenhower on their behalf if they told the government what they knew. To Julius, the prisons director seemed desperate: "You mean to tell me, Mr. Bennett," he recreated their conversation afterward, "that a great government like ours is coming to two insignificant people like us and saying 'cooperate or die'? It isn't necessary to beat me with clubs, but such a proposal is like what took place during the Middle Ages. It is equivalent to the screw and rack." Ethel, reporting the conversation to their lawyer, offered a rare glimpse of how she may have felt about her share of responsibility for her husband's espionage work, quoting Julius telling Bennett: "Just imagine! Even if it were true, and it is not, my wife is awaiting a horrible end for having typed a few notes! A heinous crime, 'worse than murder,' no doubt, and deserving of the supreme penalty. . . ." The Rosenbergs steadfastly refused to cooperate.

At the last moment, June 18, a lawyer with the Rosenberg defense team pointed out to Judge Kaufman that an execution at eleven P.M. the following evening, Friday, would fall on the Jewish Sabbath. The lawyer thought she had bought the Rosenbergs at least another twenty-four hours, but Kaufman coldly moved the executions up to eight o'clock. FBI agents had to track down the executioner, a Cairo, New York, electrician, and speed him to his task.

The FBI hoped that one or both of the Rosenbergs might decide at the last moment to recant and set up a room on the second floor of the death house to which they might be taken for questioning, with an open line to Washington. Robert Lamphere waited among the FBI officials at the Washington end of the line. "I wanted very much for the Rosenbergs to confess," he writes—"we all did—but I was fairly well convinced by this time that they wished to become martyrs, and that the KGB knew damned well that the USSR would be better off if their lips were sealed tight."

Julius Rosenberg died at 8:06 P.M. Ethel Greenglass Rosenberg, serene enough as she took her place in the electric chair to unnerve the three press-pool reporters on hand to watch her electrocution, withstood a first round of three shocks and had to be shocked twice more. She needed four minutes and fifty seconds to die; a doctor pronounced her dead at 8:15. The double execution was broadcast over the radio like a sporting event. Alger Hiss, in prison in Lewisburg, Pennsylvania, was not listening to the radio but walking outdoors in the yard:

> The June evening of the executions was calm and cloudless. We were all aware that . . . the executions had been scheduled for just before sunset. As the sun sank, silence spread over the recreation yard. Men stopped their games of baseball, boccie, handball, their exercises with weights, their trotting about the cinder track, their endless conversation. We sat or stood in an eerie quiet until after the sun had disappeared. . . .
>
> We felt we were honoring the very moments of death. . . . We had all been aware of the worldwide demonstrations and protests, which at those moments were proved to be futile. In all the months I spent at Lewisburg this occasion was unique—the inmates transcending their own unhappiness and self-involvement and joining in a mood of universal sadness at an act of inhumanity.

There were mass demonstrations in Paris and Rome, a mournful crowd in Manhattan's Union Square. Lamphere says that though he knew the Rosenbergs were guilty, he felt "not satisfaction, but defeat," at their deaths; he thought that the controversy that surrounded and has continued to surround their case represented a major "propaganda victory for the KGB."

Confirmation from the former Soviet Union that Klaus Fuchs was an important Soviet agent, Fuchs's positive identification of Harry Gold as his courier, Gold's early reports of contact with an enlisted man at Los Alamos and subsequent positive identification of David Greenglass, David and Ruth Greenglass's independent testimony to Julius Rosenberg's recruiting, the defections to the Soviet Union of Joel Barr and Alfred Sarant, all add weight to the abundance of evidence in FBI records that Julius Rosenberg was an active Soviet agent. Ethel Rosenberg was convicted of complicity solely on the testimony of the Greenglasses, and they themselves could cite no specific criminal acts. Ruth Greenglass addressed that discrepancy in the interview she and her husband gave Ronald Radosh in 1979. The Rosenbergs' deaths left their sons orphaned, she pointed out: "They had two children. Do you mean that she was going to . . . die for something she knew nothing about, that she had no involvement [in]? That's impossible."

Ethel Rosenberg had framed her choice more starkly: to betray her husband and live or to maintain their mutual innocence and die. "So now, my life is to be bargained off against my husband's," she had written when hints reached her that she might be spared. She could not imagine living with such betrayal, leaving Julius "to drown without a backward glance," nor bequeathing such a legacy to her sons: "And what of our children, noble testament to our sacred union, fruit of our deep and enduring love; what manner of 'mercy' is it that would slay their adored father, and deliver up their devoted mother to everlasting emptiness? Know then, you warped, gross, eaters of dust, you abominations upon this beauteous earth, I should far rather embrace my husband in death than live on ingloriously upon your execrable bounty."

(Lavrenti Beria was accorded a more private execution. After his trial before a special judicial body of the Soviet Supreme Court late in December 1953, he was found guilty and returned to the underground bunker where he had been incarcerated since the previous summer. A Soviet general, Pavel Batitsky, carried out the sentence of death in Beria's cell. Batitsky's widow reports that the great man who had enjoyed personally torturing his victims crawled on his knees and begged for mercy. A bullet ended Beria's brutal life; they burned his body where it lay.)

When the scientists arrived at the Semipalatinsk test site in July 1953 to prepare to test the layer-cake thermonuclear, Andrei Sakharov recalls, they "were confronted by an unexpected complication." The device would be tested on a tower, but no one had given a thought to local fallout. "An explosion of the power we anticipated would spread fallout far beyond the test site," Sakharov explains, "and jeopardize the health and lives of thousands of innocent people." With the help of what they called the Black Book, which seems to have been a copy of the US government publication *The Effects of Atomic Weapons* by Samuel Glasstone, they concluded that they would either have to evacuate tens of thousands of people from the Semipalatinsk area or redesign the bomb for an air drop, which would occasion a six-month delay. (This conclusion refutes the subsequent Soviet and Russian claim that the 1953 design was ready for deployment as a bomb; evidently it had not yet been weaponized.) After insisting they recheck their calculations and demanding that each of them, up to and including Igor Kurchatov himself, publicly concur, Minister of Medium Machine Building Malyshev unhappily arranged the vast evacuation. It continued up to the evening before the test on August 12.

Before then, on August 8, Georgi Malenkov announced in a major speech before the Supreme Soviet that "the United States of America has long since ceased to have a monopoly in the matter of the production of atomic bombs" and added spectacularly, "The United States has no monopoly in the production of the hydrogen bomb either."

"The announcement caught Washington flatfooted," Gordon Arneson writes. "Our detection system . . . had picked up nothing. We were baffled. Was it a propaganda hoax?" Eisenhower had recently appointed Lewis Strauss to the chairmanship of the AEC. Strauss, noting that it had been two years since the Soviets had tested even an atomic weapon, had written the President the day before Malenkov's speech suggesting that the Communists might be leapfrogging to a thermonuclear.

Sakharov lay on the slope of a hill overlooking the test site in the predawn darkness on the morning of August 12, 1953; the scientist who lay beside him, writes Yuri Smirnov, "heard his friend's heart beating fast a few seconds before the explosion. Later, after the explosion, when its success was obvious, the two of them walked to the place where I. V. Kurchatov was standing surrounded by civil and army officials. As he saw Andrei Dmitrievich, Igor Vasilievich bowed to him from the waist, saying, 'Thanks to you, the savior of Russia!' " The layer-cake design, a device of about the same diameter as a Fat Man, had yielded four hundred kilotons, ten times the yield of the previous Soviet fission test.

Joe 4 might have encouraged H-bomb enthusiasts in the United States to believe that the accelerated US effort to build a thermonuclear which they had championed had saved the nation from disaster. The yield of the Soviet device, boosted with scarce tritium, was less than the yield of the *Ivy* King five-hundred-kiloton fission bomb tested after Mike in November 1952. Los Alamos analyzed the fallout from Joe 4 and found it revealing. "I remember our being very intensely involved in trying to reconstruct Joe 4," Carson Mark says, "to figure out what they had done on the basis of the debris evidence that we had. That was quite an intensive effort involving [Hans] Bethe, [Enrico] Fermi, both over a considerable period and a bunch of the rest of us sitting around a table. We managed to speak of an object physically quite similar to what Joe 4 must have been. It didn't lead us to want to emulate it." George Cowan recalls thinking that the device was "a table thumper," something the Soviets had cobbled together to answer Mike. Bethe remembers the analysis revealing that Joe 4 "was compressed by high explosives. It was alternating layers of uranium and lithium deuteride, like our Alarm Clock design. All that we figured out just from seeing the debris. We also figured out from the debris that it was a single-stage device."

"The first thing the Russians did," says Cowan, "the first thing the British did, the first thing the French did were all quite similar and they were the way that people thought at that time. They were HE-initiated, and in a sense they were all failures." "We never tried it," Herbert York adds—"never tried the [HE-initiated] Alarm Clock—because it was a dead end." The crucial difference between Joe 4 and Mike was not the presence in both of thermonuclear fuel or even the fact that both derived about the same percentage of yield from thermonuclear reactions.* The crucial difference was their different methods of compression—Joe 4 with high explosives, Mike with radiation. Only radiation compression made high-megaton yields possible in a device of any reasonable size. (Whether such yields were necessary for national defense or deterrence is another question.) Joe 4 was more than a table thumper, but it was not a weapon that could be much more than doubled in yield.

The fact that the Soviet Union tested such a limited thermonuclear system should have reassured the H-bomb enthusiasts that Soviet scientists had not yet discovered the Teller-Ulam configuration. Until the USSR made that breakthrough, the United States would continue to hold a preponderant advantage in nuclear firepower, which the enthusiasts believed to be a meaningful measure of national security. By August 1953, Los Alamos was actively preparing to test (in 1954) a lighter, lithium-deuteride-fueled successor to Mike that could be weaponized quickly for delivery by air, as well as an emergency-capability, cryogenic, weaponized, air-deliverable version of Mike. The laboratory had proven its ability to produce results. It should have been evident that the US nuclear-weapons program was "ahead"—ahead in tested thermonuclear invention and in numbers of stockpiled atomic bombs.

Mississippi Senator John Stennis of the Senate Armed Services Committee confirmed the US advantage to Stuart Symington (now a senator from Missouri) in October 1953 after touring US Strategic Air Command bases in Europe, North Africa and the Middle East:

I was tremendously impressed and encouraged at the enormous striking power that we could put into action on many fronts in a matter of hours should we be attacked. This is not power on paper; it is actual, real, and to an extent, ready.... Russia is rimmed by lines of bases three deep....

Recent statements of possible atomic or hydrogen bombs on us have emphasized the power of our potential enemies. Some speak as though other powers had the bombs and we had none. I do not discount one bit the terrible destruction involved should such an attack come on us and I

---

* According to Yuli Khariton, Joe 4 derived 15–20 percent of its yield from fusion reactions. Mike derived about 24 percent of its yield from fusion reactions.

know no absolute defense is possible. At the same time, we now have tremendous striking power on our own, which is growing daily, and any nation that commits an atomic attack on us, in my opinion, is committing a suicidal act unless it should wipe us out at the first blow. This is, of course, impossible. Great as our problems of defense may be, Russia's problems are far greater and our own striking power is far greater than hers. She is bound to recognize our overwhelming power of retaliation.

Symington sent Stennis's letter to Curtis LeMay for comment. "I am in general agreement with the comments made by Senator Stennis throughout his letter," LeMay responded, adding, "The balance appears to be tipped in our favor today."

The H-bomb enthusiasts were not convinced by such arguments, however authoritative. They continued to perceive the nation to be endangered and to suspect sabotage. The scientists among them, at least, were men trained in assessing evidence carefully and in discarding hypotheses such evidence falsified. What threatened them so dreadfully that they retreated from that training to unsupported convictions?

Republican US Representative Sterling Cole of New York, Brien McMahon's successor as chairman of the Joint Committee on Atomic Energy, wrote a number of scientists in November 1953 asking them to assess the relative status of the US and Soviet thermonuclear weapons programs. Despite the evidence of the Joe 4 test and the robust success of Mike, John von Neumann responded anxiously to Cole's question about how much the Soviets were behind the US in thermonuclear development. "I . . . no longer think that the time lag . . . is as much as two years in our favor," the distinguished mathematician wrote. "Actually, I would think that it is more probable that it is about a year, and it may very well be zero. Indeed, in some parts of the field the Soviets may be ahead of us." Von Neumann thought that "from 1945 to 1949 there was a uniform time lag of about four years between us and the Soviets in our favor. . . . This time lag seems to me to be now hardly more than one year. . . . So the Russians would seem to have made up in four years at least three years' time. Yet, by all evidence available to us, their technical and scientific manpower does not *yet* exceed ours." Their industrial capacity averaged only 30 percent of ours, von Neumann estimated. He concluded that the Soviets had made up the time because they started earlier on their thermonuclear and because "the level of ability on the part of Soviet scientists, engineers and technicians is very high. *This* must have been the really decisive factor." Yet von Neumann, with his Hungarian background, was in a position to know better than many native-born Americans that Soviet abilities and education were not superior to Western and that a police state is not conducive to sustained creative work or even to exceptional industrial productivity. The very fact that the Soviets had needed espionage to accelerate their program argued against his assessment of their abilities.

Worse, von Neumann believed that SAC was "*highly* vulnerable to an enemy's surprise attack. . . . It is most depressing to contemplate how badly the efficiency of this operation might deteriorate in the case of surprise attack. It might easily endanger the functioning of SAC *altogether.*" Von Neumann thought SAC ought to be dispersed and even go underground. Indeed, he believed that urban and industrial America ought to decentralize and disperse. "I think that the only choice that we have is whether this necessary readjustment will occur after a general disaster, under the conditions of shock, or in an orderly and planned fashion."

John Archibald Wheeler, a distinguished American theoretical physicist who had worked with Niels Bohr in 1939 to elucidate the theory of nuclear fission, had contributed importantly to the design and successful operation of the plutonium production reactors at Hanford, had recently helped Los Alamos calculate the hydrodynamics of the equilibrium thermonuclear and was now a member of the General Advisory Committee, responded even more stridently to Cole's questions. "I know of no evidence," he began, "that would exclude [the Soviets] being substantially ahead of us in production of TN [i.e., thermonuclear] weapons." Wheeler was clear about whom to blame:

> If we had started our effort in 1946 instead of 1950, I see no good reason why we could not have been four years ahead of where we are now. The professional hand-wringers who kept us from getting under way for those four years have much to answer for. But I am even more concerned about the great inertia of those who right now fail to recognize that we are engaged in the most deadly and important armaments race in human history. Our secrecy keeps secret how little we are doing, not how much we are doing. Responsible men like you are all too few. . . . We need to rouse this country to our danger of falling way behind. . . . Is our own hydrogen effort adequate? In my opinion it is shamefully inadequate. . . .

In contrast, I. I. Rabi, Robert Oppenheimer's successor as chairman of the GAC, judged that there was no real time lag between the two nations' nuclear-weapons programs any longer, "no more than I would say that there is a time lag in tanks or aircraft. We are each pursuing an independent program as laid out by our military planners." He did not feel unduly threatened by that parallelism. It was clear that the Soviets were "still stressing their fission program" and Rabi believed "our thermonuclear effort is in excellent shape." He believed "that we could not have undertaken a major thermonuclear program until we had some good ideas of how to proceed. In my judgment, these ideas did not appear before about early 1950. . . . One of the most important factors in our advance was the demonstration by Dr.

Ulam of Los Alamos that the plans of 1946 were scientifically unsound." Rabi thought Soviet thermonuclear progress was "based on the fission bomb techniques," confirming that informed government officials were aware that Joe 4 was an HE-imploded rather than a radiation-imploded design.

All these men had reason to know, deeply and personally, the ferocious destructiveness of nuclear weapons. The difference between Stennis's, Rabi's —and Oppenheimer's—confidence and von Neumann's, Wheeler's—and Teller's, William Borden's and Lewis Strauss's—dread and foreboding would appear to be their different weighing of the deterrent value of those weapons. In the summer of 1953, Oppenheimer had published a memorable essay in the journal *Foreign Affairs,* "Atomic Weapons and American Policy," in which he had ridiculed the proponents of ever-larger and more powerful nuclear arsenals. "The very least we can say," the former GAC chairman wrote, "is that, looking ten years ahead, it is likely to be small comfort that the Soviet Union is four years behind us, and small comfort that they are only about half as big as we are. The very least we can conclude is that our twenty-thousandth bomb, useful as it may be in filling the vast munitions pipelines of a great war, will not in any deep strategic sense offset their two-thousandth." The JCAE had in its files a recent and chilling assessment by a member of the committee staff of how many ten-megaton thermonuclear weapons it would take to destroy totally the eight most productive industrial states in the US. Fortuitously, the estimate confirms Oppenheimer's numbers: 2,010 bombs for New York, Pennsylvania, Ohio, Illinois, California, New Jersey, Indiana and Massachusetts. Double that number to allow for duds and total misses and it is still closer to two thousand than to twenty thousand. John Wheeler, however, had Oppenheimer's essay very much in mind when he wrote to Sterling Cole, "Anybody who says 20,000 weapons are no better than 2,000 ought to read the history of wars." But nuclear weapons are not cannonballs; how many times could either country be destroyed? And who would venture war in the face of total and redundant destruction?

President Eisenhower examined the same issues in a memorandum for the Secretary of State, John Foster Dulles, in September 1953, discussing how "to educate our people in the fundamentals of these problems":

We should patiently point out that any group of people, such as the men in the Kremlin, who are aware of the great destructiveness of these weapons —and who still decline to make any honest effort toward international control by collective action—must be fairly assumed to be contemplating their aggressive use. It would follow that our own preparation could no longer be geared to a policy that attempts only to avert disaster during the early "surprise" stages of a war, and so gain time for full mobilization. Rather, we would have to be constantly ready, on an instantaneous basis, to inflict greater loss upon the enemy than he could reasonably hope to inflict on us. This would be a deterrent. . . .

It was to that point, of course, that the United States had come by 1953, and Stennis's and LeMay's comments testify to the robustness of the deterrent. But deterrence seemed to bring no resolution to the conflict, and Teller, Borden and the others who supported larger and more powerful stockpiles evidently did so because they feared that lack of realism, complacency, the dangerous advice of "professional hand-wringers" or, more insidiously, treasonable deception might allow the Soviet Union to pull ahead. To say it another way, they did not believe that the United States was alert to the danger that its arsenal might be overwhelmed, nor did they believe that a cruel and secretive competitor would accept stalemate.

Eisenhower, instructing Dulles, followed that line of thought out to its logical conclusion:

This would be a deterrent—but if the contest to maintain this relative position should have to continue indefinitely, the cost would either drive us to war—or into some form of dictatorial government. In such circumstances, we would be forced to consider whether or not our duty to future generations did not require us to *initiate* war at the most propitious moment that we could designate.

Such intense anxiety demanded alleviation. Conveniently, there was a scapegoat at hand to slaughter.

26

In the Matter of J. Robert Oppenheimer

When President Dwight Eisenhower offered Lewis Strauss the chairmanship of the Atomic Energy Commission, in May 1953, the financier told the President that he would accept the appointment on one condition: that Robert Oppenheimer not be "connected in any way" to the agency. Strauss explained to the President that he distrusted Oppenheimer because the physicist had failed to report fully Haakon Chevalier's wartime espionage approaches and because he had continued to oppose the hydrogen bomb after President Truman authorized it. Strauss chose not to tell Eisenhower that he believed Oppenheimer might be "another Fuchs," as he had implied to William Borden as long ago as August 1951. His discretion evidently represented no change of heart; he had recently informed the FBI of Oppenheimer activities he considered suspicious, and around this time he promised J. Edgar Hoover that he would purge Oppenheimer when he took up the chairmanship of the AEC.

Oppenheimer had already been separated from the General Advisory Committee. If Sidney Souers had recommended that separation to Truman the previous year, as he later recalled, Strauss had probably influenced Souers's recommendation; not long before Oppenheimer was dropped from the GAC, Borden had reported to Brien McMahon as "late gossip" that "Louie Strauss went to the President and urged him not to reappoint Oppie." The Air Force had also sought to disconnect Oppenheimer from government, suspecting his loyalty and opposing his advice favoring continental defense and the development of tactical atomic weapons to balance strategic bombing. "In 1951," Herbert York writes, "[USAF] Secretary [Thomas] Finletter and General [Hoyt] Vandenberg gave direct orders to . . . the two top civilian scientists in Air Force headquarters not to use Oppenheimer as a consultant . . . and to keep classified Air Force information away from him." In 1952, Borden reported to McMahon that "the Air Force feels that the removal of Dr. Oppenheimer is an urgent and immediate necessity." Early in 1953, Stuart Symington encouraged his Red-baiting colleague Joseph Mc-Carthy to investigate the charismatic physicist. Hostility to Oppenheimer had crossed the gulf between Democratic and Republican administrations; during the transition from Truman to Eisenhower, the incoming Department of Defense under Secretary Charles E. Wilson had abolished the Research and Development Board of which Oppenheimer was a member in order to ease him out of office. "We dropped the whole board," Wilson would brag at a 1954 press conference. "That was a real smooth way of doing that one as far as the Defense Department was concerned." From nearly full-time participation as a government adviser when he was GAC chairman, Oppenheimer worked only two days as a consultant to the AEC in 1952; in 1953, only four.

Yet the physicist's influence on US policy continued to frighten those who believed the nation was in mortal danger from Soviet Communism. Oppenheimer's stature was such that as recently as February 1953 he had briefed the National Security Council with Eisenhower in attendance on the conclusions of a State Department panel he had chaired on disarmament, conclusions favoring increased candor about the US nuclear arsenal that Strauss vehemently opposed. Worse, Eisenhower appeared to be sympathetic to Oppenheimer's long-standing conviction that the nuclear arms race had to be halted. Immediately after Strauss was sworn in as AEC chairman, the President had taken him aside and told him, "Lewis, let us be certain about *this,* my chief concern and your first assignment is to find some new approach to the *dis*arming of atomic energy. . . . The world simply must not go on living in fear of the terrible consequences of nuclear war." If the AEC and the military rejected Oppenheimer's advice, other agencies and even the President continued to seek it. How could the man be severed completely from government?

Contributing to nuclear-weapons policy depended crucially on access to classified information. "You had to be inside the government if you wanted to have an influence, especially on these military matters," I. I. Rabi once noted. "Since there was all that secrecy, you couldn't know what you were talking about unless you were a part of it." Oppenheimer's top secret Q clearance allowed him to know what he was talking about. It followed that withdrawing Oppenheimer's clearance would eliminate him as effectively as if someone had him shot. Strauss began his purge campaign immediately; on July 7, five days after he was sworn in, he ordered all classified AEC documents removed from the security safe in Oppenheimer's office at the Institute for Advanced Study—ostensibly to save the expense of a security guard.

Harold Green, a young attorney at the AEC at that time, remembers that Strauss typically purged people from government positions by "us[ing] his contacts in industry, foundations and educational institutions to produce career opportunities that the objects of his purges could not turn down." Oppenheimer was barely a part-time consultant and was already at the top of his profession, in a position Strauss had arranged. Destroying his influence in any case would require shaming him publicly as well as withdrawing his clearance. Strauss turned aside the investigation that McCarthy was preparing by warning Republican Party leader Robert Taft that such an attack would be "ill-advised and impolitic." In a draft of his letter to Taft he explained that "the McCarthy Committee is not the place for such an investigation, and the present is not the time." Hoover heard him say that "inquiry into Oppenheimer's activities might be well worthwhile, [but] he hoped it would not be done prematurely or by a group that did not thoroughly prepare itself for the investigation." Hoover and Vice-President Richard Nixon helped Strauss convince McCarthy to lay off.

In the meantime, William Borden was preparing a brief summarizing what he believed to be Oppenheimer's crimes. Borden's fortunes had declined since Brien McMahon's death. Republicans controlled the Joint Committee on Atomic Energy. Under Borden's auspices, a member of the committee staff, John Walker, had worked with Princeton physicist John A. Wheeler to prepare a lengthy chronology of H-bomb policy and progress that the JCAE had distributed within the government on January 1, 1953. Walker had continued work during January on a short but detailed review of how Klaus Fuchs might have learned about thermonuclear design principles before he left Los Alamos in 1946. The Walker document revealed highly classified weapons-design information, including references to radiation implosion— the most important secret the United States protected. Walker mailed the document to Princeton for Wheeler to review.

Wheeler promptly lost it. He believed he lost it on the overnight train between Princeton and Washington, and Borden had the Pullman car thoroughly searched and even partly dismantled, but the envelope in which Wheeler had kept the document later turned up empty in his office in Princeton. Borden contrived to avoid allowing the AEC to see a copy of the lost document until the agency appealed to Hoover. When Eisenhower found out about the loss, he lined up the AEC commissioners like errant schoolboys and blasted them with a full measure of his considerable rage. They blamed the Joint Committee. Eisenhower suspected espionage, as did Nixon, who proposed that Borden and his entire staff be investigated. AEC attorney Harold Green writes that Eisenhower approached Bourke Hickenlooper and Sterling Cole and "demanded that the Joint Committee's staff be reorganized so that such a thing could not happen again." Borden left the JCAE under a cloud in May 1953. "Borden's departure," concludes Green, "was a direct consequence of [Eisenhower's] demand." ("Borden is the most indiscreet person I ever met," an AEC security official would characterize him. "He is a greenhorn in the business of atomic energy. . . . He doesn't know anything about the subject of security.")

Borden testified later that he had given "increasing consideration over a period of years" to his doubts about Robert Oppenheimer's loyalty. Before he left the JCAE, he had gone over Oppenheimer's security files once more, listing his questions in the form of investigative leads: "Why is no signed PSQ [i.e., personal security questionnaire] of Oppenheimer available in AEC files?" "What were Oppenheimer's activities in Germany [as a graduate student]?" "What were Oppenheimer's activities during the period 1939 to 1942?" His final list extended to some five hundred questions, which he hoped would serve to alert his successor to the Oppenheimer problem. He went off then to a backwoods retreat near the St. Lawrence River for a month's vacation, but his obsession with Oppenheimer persisted.

Borden was well aware of Strauss's suspicions that Oppenheimer might be a spy. At some point before he left Washington, the thirty-three-year-old lawyer had approached Strauss, hoping that Eisenhower's chief adviser on atomic energy would find his services useful, but Strauss had not offered support. Borden was scheduled to take up work in July as assistant to the manager of the atomic power division of Westinghouse in Pittsburgh, a position Admiral Hyman Rickover had arranged for him. Rickover and Strauss had feuded after Strauss had advised Eisenhower to cancel the Large Ship Reactor that Rickover was developing for aircraft carriers. Evidently these various complications coalesced for Borden, away on his retreat, into a vision of another "Inflammatory Document" like the letter he had written Brien McMahon in 1947 that had won him appointment to McMahon's staff. He told historian Gregg Herken many years later that he hoped as a result of his efforts to be "prosecutor at Oppenheimer's trial for treason." He spoke to Strauss only once between the time he left Washington and the end of the year, but he evidently believed an indictment of Oppenheimer would win the AEC chairman's support, since Strauss would obviously have a strong voice in choosing who might prosecute Robert Oppenheimer if such a trial were held. In Pittsburgh in October, Borden "crystallized [his] thinking" and drafted a letter to Hoover.

The Rosenbergs were executed while Borden was vacationing in the backwoods; while he mulled over Oppenheimer's perfidy, the Soviet Union tested Joe 4. "I couldn't live with myself," he wrote a colleague a few months later, "until I finally got this thing off my chest." He got it off his chest on November 7, 1953, when he finished his letter of denunciation and sent it by regular mail to the FBI for Hoover. "More probably than not," it argued at length, "J. Robert Oppenheimer is an agent of the Soviet Union." Borden listed a number of "factors" that led him to this shocking conclusion, including Oppenheimer's contributions to the Communist Party during the 1930s and early 1940s; the fact that his wife, his younger brother and his "mistress" —Oppenheimer's former fiancée Jean Tatlock, with whom he had spent the night in Berkeley in 1943, after he became director of Los Alamos—had been Communists; his contradictory information about the Haakon Chevalier espionage contacts; and at length, his "tireless" work "to retard the United States H-bomb program." Borden concluded:

1. Between 1929 and mid-1942, more probably than not, J. Robert Oppenheimer was a sufficiently hardened Communist that he either volunteered espionage information to the Soviets or complied with a request for such information. (This includes the possibility that when he singled out the weapons aspect of atomic development as his personal specialty, he was acting under Soviet instructions.)

2. More probably than not, he has since been functioning as an espionage agent; and

3. More probably than not, he has since acted under a Soviet directive in influencing United States military, atomic energy, intelligence and diplomatic policy.

Hoover had heard such charges before. "Many of them are distorted and restated in his own words," the Bureau immediately evaluated them, "in order to make them appear more forceful than the true facts indicate." To AEC commissioner Thomas Murray, Hoover appeared to be "a little at a loss as to why Borden wrote such a letter." Even Strauss, when he learned of Borden's accusations at the end of November, decided to review the Chevalier episode before proceeding further. But when Charles Wilson, the Secretary of Defense, saw the FBI report reviewing Borden's letter on December 1, he was shocked. He wondered if Oppenheimer might be in collusion with John Wheeler (of the lost document) and wanted Oppenheimer's security clearance suspended. That evening Wilson called Eisenhower, the President noted in his diary:

Charlie Wilson states that he has a report from the FBI that carries the gravest implications that Dr. Robert Oppenheimer is a security risk of the worst kind. In fact, some of the accusers seem to go so far as to accuse him of having been an actual agent of the Communists. . . .

The sad fact is that if this charge is true, we have a man who has been right in the middle of our whole atomic development from the very earliest days. . . . Dr. Oppenheimer was, of course, one of the men who has strongly urged the giving of more atomic information to the world. . . .

On December 3, Eisenhower ordered Attorney General Herbert Brownell "to place a blank wall between [Oppenheimer] and all areas of our government operations" and requested advice on "whether further action, prosecutive or otherwise," should be taken. In his diary that day, the President noted that "the so-called 'new' charges . . . consist of nothing more than . . . a letter from a man named Borden," that the letter presented "little new evidence" and that "this same information . . . has been constantly reviewed and re-examined over a number of years, and that the overall conclusion has always been that there is no evidence that implies disloyalty on the part of Dr. Oppenheimer," which did not however mean "that he might not be a security risk." If he were, Eisenhower feared the worst:

Actually, of course, the truth is that no matter now what could or should be done, if this man is really a disloyal citizen, then the damage he can do now as compared to what he has done in the past is like comparing a grain of sand to an ocean beach. It would not be a case of merely locking the stable door after the horse is gone; it would be more like trying to find a door for a burned-down stable.

Eisenhower had reason as well to fear what McCarthy might make of Borden's charges, and since Borden had sent a copy to the Joint Committee, the President could assume someone would leak the document to the Wisconsin senator or his chief counsel, Roy Cohn. Brownell had attacked the Truman administration as recently as November 6 for nominating Harry Dexter White to the International Monetary Fund after Elizabeth Bentley had accused White of espionage; in Oppenheimer, the new Eisenhower administration might have a traitor of its own in its midst.

The immediate problem was how to build Eisenhower's "blank wall." Hoover hoped Oppenheimer could be eased out of government quietly: the FBI files contained "a lot of information which could not be publicly disclosed" (that is, information obtained illegally) and if the case went public Oppenheimer "might get some very clever lawyer and end up by becoming a martyr." The FBI director and Brownell told Strauss "that it was all right to go ahead with the suspension of [Oppenheimer's security clearance] but not to send the notifications around." Oppenheimer was in England at the time, delivering the BBC's Reith Lectures; Strauss and Hoover both feared that if he learned of Borden's charges and the suspension of his Q clearance, "then it was very possible he would depart for the Iron Curtain, which would be most embarrassing, or he might issue a statement and fly back and create quite a furor." If the accused himself could not be notified, Strauss felt no compunction at briefing Teller the same day Eisenhower ordered up his blank wall. "Strauss told me with real fervor of his hope that the President's decision would be reversed or at least modified," Teller writes. "He foresaw disastrous consequences should Oppenheimer's clearance be called into question." Eisenhower's National Security Adviser, Robert Cutler, actually proposed an appeal to the presumptive traitor's patriotism: "If you love your country enough, you will accept this situation and not plunge our world and our national secrets into a bitter, dirty fight."

The solution that Strauss finally came to (after seeking "divine guidance," his general counsel would recall) was to follow AEC security procedures: bring a formal list of charges, offer Oppenheimer the choice of resigning or requesting a security hearing and hope he chose to resign. The AEC general counsel, William Mitchell, set to work drafting the charges on Thursday, December 10. Two AEC members—Eugene Zuckert and Henry Smyth— criticized an early draft Mitchell prepared for questioning Oppenheimer's H-bomb advice. Without the H-bomb issue, Mitchell could find nothing in Oppenheimer's FBI files that had not already been reviewed and cleared in 1947—by Lewis Strauss, among others. In some desperation, on Friday afternoon, the AEC counsel called in his young assistant Harold Green, swore him to secrecy and asked him to take over the work, cautioning him that the commissioners wanted to avoid including charges related to the H-bomb controversy, since dissenting on policy was not a crime.

Saturday morning, Green got to work. Oppenheimer's files surprised and shocked him. So did Kenneth Nichols, now the AEC general manager, when Nichols called the young attorney into his office twice that day to attack Oppenheimer and gloat that they had caught the "slippery sonuvabitch" at last; Nichols would sit in supposedly impartial judgment of Oppenheimer if a security board heard the case. By Sunday noon, Green had a list of thirty-one charges. He called Mitchell, who agreed to meet him at two that afternoon. Green thought over the H-bomb issue while he waited. Secret FBI interviews with Edward Teller dating from May 1952 offered a bounty of new allegations. Green decided he could use Teller's allegations if he framed them to raise questions about Oppenheimer's veracity rather than his judgment. (That is, Green proposed to compare various positions Oppenheimer had taken or actions he was alleged to have performed so as to emphasize their apparent inconsistency.) He set to work.

The Albuquerque FBI office had interviewed Teller on May 10, 1952, and again on May 27, following up on public allegations against Oppenheimer that Kenneth Pitzer had made shortly after he resigned as AEC director of research. Although the Sausage design incorporating the Teller-Ulam breakthrough was well in hand by then and would be tested the following November, Teller's grievances against Oppenheimer had continued to fester. He told the FBI agent that Oppenheimer had opposed the development of the H-bomb since 1945; that the H-bomb would have been a reality by 1951 or earlier if Oppenheimer had not opposed it; that Oppenheimer wrote the October 1949 GAC majority opinion and was the "dominating influence" (in FBI paraphrase) on the committee. Teller alleged a cascade of devious tactics:

Teller claims [Oppenheimer] delayed or hindered [the] development of [the] H-bomb from 1945 to 1950 by opposing it on moral grounds. After [the] President announced [the] H-bomb [was] to be made, [Oppenheimer] opposed it on [the] ground that it was not feasible. . . . After this, [Oppenheimer] changed his approach and opposed [the] H-bomb on [the] basis that there were insufficient facilities and scientific personnel to develop [it], which according to Teller is incorrect.

Teller accused Oppenheimer of convincing Hans Bethe not to join the H-bomb project. He said that Oppenheimer would not make any direct attempt to influence people not to work on the bomb "but would use psychology in the approach." Oppenheimer's opposition, Teller thought, was not due to any subversive intent "but rather to [a] combination of reasons including personal vanity in not desiring to see his work on [the] A-bomb done better on [the] H-bomb, and also because he does not feel [the] H-bomb is politically desirable. Teller also feels [Oppenheimer] [has] never gotten over the shock of [the] first A-bomb being dropped."

Teller then offered a lay psychiatric profile of his former boss:

Teller also said that he has found Oppenheimer to be a very complicated person, even though an outstanding man. He also said that he understands that in his youth Oppenheimer was troubled with some sort of physical or mental attacks which may have permanently affected him. He has also had great ambitions in science and realizes that he is not as great a physicist as he would like to be.

Teller asked the FBI agent not to disseminate these profundities, since "the fact that he, Teller, was repeating such information could prove very embarrassing to him personally."

Teller found a way to affirm Oppenheimer's loyalty to the FBI while implicitly calling it into question. He told the agent "that in all of his dealings with Oppenheimer he has never had the slightest reason or indication to believe that Oppenheimer is in any way disloyal to the United States." But he followed that declaration by asking the agent to hold it in confidence "because he felt that he could be subject to considerable cross-examination on this point when people brought up certain instances like the fact that Oppenheimer's brother, Frank, is an admitted former member of the Communist Party."

In conclusion, according to the FBI report, "Teller states he would do most anything to see [Oppenheimer] separated from [the] General Advisory Committee because of his poor advice and policies regarding national preparedness and because of his delaying of the development of [the] H-bomb."

(The Albuquerque interviews were not the first time Teller had informed on Oppenheimer to the FBI. A Bureau summary document prepared in April 1952 noted that at some earlier date, when asked about the physicist Philip Morrison, Teller had offered the information that "Morrison has the reputation among physicists of being extremely far to the left." Teller then added gratuitously that "Oppenheimer, Robert Serber and Morrison are considered the three most extreme leftists among physicists. [Teller] stated that most of Oppenheimer's students at Berkeley had absorbed Oppenheimer's leftist views." Since Philip Morrison had been revealed by then to have been a member of the Communist Party in his student days, Teller's linkage was especially damaging.)

By the time Mitchell arrived, Green had added seven more charges to his master list. All of them concerned Oppenheimer's position on the hydrogen bomb and all were based on Teller's allegations.

While Strauss and Nichols were preparing his fate, Oppenheimer was blithely visiting Haakon Chevalier and his new wife in Paris, where Chevalier worked as a translator. The Oppenheimers had dinner at the Chevaliers' apartment; the next day Chevalier took them to meet André Malraux. Oppenheimer had identified Chevalier as the cut-out in an espionage contact; his continued relationship with the man looked highly suspect to Strauss and others in the government. When Eisenhower heard about it, some months later, he asked angrily, "How can any individual report a treasonable act [sic] on the part of another man and then go and stay at his home ...?" It was a reasonable question, but Oppenheimer by 1953, after years of high-level government service, had either convinced himself that his loyalty was no longer in doubt or become fatalistic.

He was shocked, on Monday, December 21, 1953, meeting in Strauss's office with the AEC chairman and Nichols, when Strauss handed him a draft copy of the list of charges. Strauss fished for Oppenheimer's resignation; Oppenheimer fished for a request from Strauss that he resign. Neither bit. When Oppenheimer proposed to visit his attorney, Herbert Marks, before making up his mind what to do, Strauss volunteered his car. The shaken physicist went to see another attorney instead, Joseph Volpe, then moved on to Georgetown to have a drink with the Markses. Marks's wife Anne had been Oppenheimer's secretary at Los Alamos. "I can't believe this is happening to me!" he told her. "It was like Pearl Harbor—on a small scale," he reflected later. "Given the circumstances and the spirit of the times, one knew that something like this was possible and even probable, but still it was a shock when it came." Oppenheimer informed Strauss the next day that he wanted a security hearing. "I have thought most earnestly of the alternative suggested," he wrote Strauss, still jockeying the resignation issue. "Under the circumstances, this course of action would mean that I accept and concur in the view that I am not fit to serve this Government that I have now served for some twelve years. This I cannot do." Before Nichols signed the formal notification letter that contained the list of charges, he quipped, "Do we really have to go through with this? Why don't we just turn the files over to McCarthy?"

For the next three months, both sides marshaled their forces. The FBI tapped Oppenheimer's home and office phones at Strauss's specific request and followed the physicist whenever he left Princeton. When the phone taps began to pick up discussions between Oppenheimer and his attorneys, the supervising agent in Newark contacted Bureau headquarters "in view of the fact that [the taps] might disclose attorney-client relations." Washington responded that the taps were "warranted" because Oppenheimer might try to defect. An FBI document notes that the Bureau was "furnishing Strauss [and others] information bearing on the relationship between Oppenheimer, his attorneys and potential witnesses for Oppenheimer." Strauss for his part expressed his appreciation to the Bureau; the surveillance was "most helpful to the AEC," he explained, "in that they were aware beforehand of the moves [Oppenheimer] was contemplating."

Strauss would have extensive access to Oppenheimer's attorney-client discussions in the months ahead. He justified such illegal intrusion on the grounds that the case was of definitive importance to national security. "He felt that if this case is lost," the FBI paraphrases him, "the atomic energy program and all research and development connected thereto will fall into the hands of 'left-wingers.' If this occurs, it will mean another 'Pearl Harbor' as far as atomic energy is concerned. Strauss feels that the scientists will then take over the entire program. Strauss stated that if Oppenheimer is cleared, then 'anyone' can be cleared regardless of the information against them." A decade later, Strauss would reveal to Edward Teller the paranoid depths of his dread:

> [Oppenheimer] had been instrumental in bringing to Los Alamos a number of men known to him to be Communists. It would be reasonable to suppose that they were doing what Fuchs and others did, viz., passing on to the Soviets everything they could discover. Oppenheimer's later decision, therefore, to do what he could to prevent the United States from developing the Super was a decision reached in the knowledge that such weapon data as we then had were in the hands of men whose leaning to the Soviets he knew. Consequently, if he had been able to block the development of the weapon by the United States, its denial to the Russians was beyond his control. It is hardly conceivable that the consequences of such a condition could have been overlooked by a mind as agile as his.

To Strauss, not only Oppenheimer was subversive; here he extends the slander to include the men and women who developed the thermonuclear at Los Alamos as well. If Strauss feared another Pearl Harbor, Oppenheimer's had already roared in.

The physicist chose a Lincolnesque attorney to represent him, Lloyd Garrison, a leader in the American Civil Liberties Union and a great-grandson of abolitionist William Lloyd Garrison. Faced with the possibility that at least one member of the defense team might not be clearable without extensive investigation, Garrison made the serious error of deciding not to seek a security clearance himself; without clearance he would be barred from reviewing many of the documents on which Oppenheimer would be examined and would even sometimes have to vacate the hearing room and leave his client undefended.* Strauss found a seasoned trial lawyer, Roger Robb, an expert at cross-examination, who would turn what was supposed to be an inquiry into a prosecution. Strauss hand-picked the three-man security board as well: Gordon Gray, former Assistant Secretary of the Army, a wealthy North Carolina Democrat who had supported Eisenhower over Adlai Stevenson in 1952 because he considered Stevenson's opposition to Communism insufficiently militant; Thomas Morgan, former president of Sperry Gyroscope, a nonentity who would ask not one question in a month of hearing witnesses; and Ward Evans, a Loyola University chemistry professor who had served on security boards before and who remarked to Gray (as Gray noted) "that in his experience . . . almost without exception those [accused] who turned up with subversive backgrounds were Jewish." Evans told Gray before the hearing began that he thought Oppenheimer was guilty. "I was concerned at this note of clear prejudice," Gray adds, but the security board chairman did nothing about it.

Rabi carried a draft resolution to Strauss from the General Advisory Committee proposing that the entire GAC membership would testify on Oppenheimer's behalf. "Strauss told Rabi he considered this blackmail," an FBI document reports, "and that he could not be swayed by such action." Princeton physicist Eugene Wigner, although politically conservative, tried to intercede on Oppenheimer's behalf as well, as did Hans Bethe. Victor Weisskopf, whom Oppenheimer would credit with having helped open his eyes in the late 1930s to the brutality and incompetence of Soviet Communism, wrote the beleaguered physicist "that I and everybody who feels as I do are fully aware of the fact that you are fighting here our own fight. . . . Please think of us when you are feeling low. Think of all your friends who are going to remain your friends and who rely upon you." One friend Oppenheimer did not hear from was Robert Serber, whom Teller and other informants had denounced as a leftist along with his wife Charlotte, and whose name Roger Robb would invoke often at the security hearing. (Charlotte Serber was active in Spanish relief in the late 1930s and came from a prominent socialist family, but neither she nor her husband had been mem-

---

* He changed his mind and requested security clearance on March 26, 1954, three weeks before Oppenheimer's hearing was scheduled to begin; the AEC chose not to expedite the request. Stern (1969), p. 247.

bers of the Communist Party.) "One morning about three A.M. I got a phone call," Serber remembers sadly. "The guy said he was a lawyer in Garrison's office. Oppie is going to have this loyalty hearing, he told me, and wanted me not to communicate with him because he was sure the phones were going to be tapped and the letters checked. So I didn't communicate with him during that period. Later on, Kitty [Oppenheimer] told me that it wasn't true that Oppie made that request. It all came from Garrison. Oppie didn't know anything about it." Which meant two old and loyal friends each thought the other had abandoned him.

That winter of 1954, Secretary of State John Foster Dulles announced the Eisenhower administration's new policy of "massive retaliation," a Senate subcommittee prepared to hold hearings on charges that McCarthy had abused his authority investigating the Army and a secret Defense Department committee headed by John von Neumann proposed that the US begin building strategic missiles fitted with nuclear warheads. On March 1, Los Alamos and Livermore initiated a new thermonuclear test series at Bikini, *Castle,* exploding the first lithium-deuteride-fueled US thermonuclear, a Los Alamos device called Shrimp tested as *Castle* Bravo.

The room-temperature Shrimp device used lithium enriched to 40 percent lithium6; it weighed a relatively portable 23,500 pounds and had been designed to fit the bomb bay of a B-47 when it was weaponized. It was expected to yield about five megatons, but the group at Los Alamos that had measured lithium fusion cross sections had used a technique that missed an important fusion reaction in lithium7, the other 60 percent of the Shrimp lithium fuel component. "They really didn't know," Harold Agnew explains, "that with lithium7 there was an n, 2n reaction [i.e., one neutron entering a lithium nucleus knocked two neutrons out]. They missed it entirely. That's why Shrimp went like gangbusters." Bravo exploded with a yield of fifteen megatons, the largest-yield thermonuclear device the US ever tested. "When the two neutrons come out," says Agnew, "then you have lithium6 and it went like regular lithium6. Shrimp was so much bigger than it was supposed to be because we were wrong about the cross section."

This time the fireball expanded to nearly four miles in diameter. It engulfed its 7,500-foot diagnostic pipe array all the way out to the earth-banked instrument bunker, which barely survived. It trapped people in experiment bunkers well outside the expected limits of its effects and menaced task force ships far out at sea. "I was on a ship that was thirty miles away," Marshall Rosenbluth remembers, "and we had this horrible white stuff raining out on us. I got 10 rads* of radiation from it. It was pretty frightening. There was a huge fireball with these turbulent rolls going in and out. The thing was glowing. It looked to me like a diseased brain up in

* A chest-X-ray series is equivalent to about 1 rad.—RR.

the sky. It spread until the edge of it looked as if it was almost directly overhead. It was a much more awesome sight than a puny little atomic bomb. It was a pretty sobering and shattering experience." Bravo vaporized a crater 250 feet deep and 6,500 feet in diameter out of the atoll rock; Rosenbluth's "horrible white stuff" was calcium precipitated from vaporized coral.

A Japanese fishing boat, the *Fukuryu Maru*—the *Lucky Dragon*—docked at its home port in mid-March with a sick crew. Trolling some eighty-two nautical miles eastward of Bikini, all twenty-three crewmen had been heavily exposed to Bravo fallout. (So had native peoples on Rongelap, Ailinginae and Utirik.) Once again, Japanese had been poisoned by what one Tokyo newspaper called "ashes of death"; the country erupted. The US offered radiation specialists to treat the fishermen but refused to reveal fallout content for fear the Soviets would learn that the Shrimp had been fueled with lithium deuteride. One of the Japanese sailors died of secondary infection. Lewis Strauss had been en route to Bikini when the *Fukuryu Maru* turned up. At the end of March, back in Washington, he issued a cold disclaimer of responsibility. Bravo, he said, "was a very large blast, but at no time was the testing out of control." He noted that the Soviets had tested a device the previous year "which derived part of its force from the fusion of light elements," and asserted that there was "good reason to believe that they had begun work on this weapon substantially before we did"—a jab at Oppenheimer two weeks before the security hearing was scheduled to begin. Strauss maintained that the *Lucky Dragon* "must have been well within the danger area" even though there was ample evidence to the contrary; privately, the AEC chairman told Eisenhower's press secretary dismissively that the boat was probably a "Red spy ship."

The *Castle* series continued with tests of an unenriched lithium-deuteride device—Runt, *Castle* Romeo—which ran away to eleven megatons, three times its predicted yield, for the same reason Bravo had; of Koon, the first thermonuclear out of Teller's new Livermore lab, a device called Morgenstern with a predicted one-megaton yield that produced only 110 kilotons —a dud; of a radiation-imploded Alarm Clock, Union, that yielded 6.9 megatons; of Yankee, another version of the Runt design that yielded 13.5 megatons; and of Nectar, a thermonuclear weighing only 6,520 pounds that yielded 1.69 megatons. The Runt was Harold Agnew's project. Jacob Wechsler had supervised development of a weaponized version of Mike—Jughead —in case the dry bombs failed; it was supposed to be tested at *Castle*. After the Romeo success, Wechsler reminisces, "Harold said, Got to send a wire to Norris [Bradbury]. I said, Sure. He said, To kill your Jughead. I said, Yeah? He said, Here's the wire: 'Why buy a cow when powdered milk is so cheap?' " "The results of Operation *Castle*," Raemer Schreiber writes, "left me with the unpleasant job of negotiating the closeout of a sizable cryogenic hardware contract." Future US thermonuclear weapons would be fueled with lithium deuteride.

After the Bravo shot, Charles Critchfield recalled, Oppenheimer had a first bitter taste of what life would be like outside the circle of secrecy:

> Robert had lost his clearance and I remember being in my office for some reason. I got a call from Robert and he had heard about the Bravo shot. All he would say on the phone was, Charles, can you give me a number. I said fifteen. He said thank you. He knew what it meant, of course. I knew I was breaking the law but Robert was an old friend of mine and I wasn't about to tell him, "I can't tell you."

McCarthy made headlines early in April claiming Communists in government had delayed "our research on the hydrogen bomb" by eighteen months, effectively pressing Strauss and his prosecution team to produce a culprit. The AEC's inquiry "in the matter of J. Robert Oppenheimer" finally came on for hearing on April 12, 1954, a cool and sunny day.

AEC Building T-3, near the Washington Monument, was a modified barracks structure left over from the Second World War. An executive office on the second floor had been converted into a small hearing room. The three security board members sat at a table along one wall that formed the lintel of a T. Two tables butted together with the opposing lawyers on either side made the T's upright. A witness chair that faced the security board anchored the upright. Whenever Oppenheimer was not testifying, he chain-smoked cigarettes or a pipe on a leather couch against the wall behind the witness chair. It must have been frustrating not to be able to study the witnesses' faces as they spoke.

For the first two days of the hearing, Garrison led Oppenheimer through his life, interrupting to interview a friendly witness and to read an affidavit from John Manley. Near the end of Garrison's direct examination, Oppenheimer read a denunciation of Communism from one of his Reith Lectures: "Perhaps only a malignant end can follow the systematic belief that all communities are one community; that all truth is one truth; that all experience is compatible with all other; that total knowledge is possible; that all that is potential can exist as actual." The statement was eloquent, but the reading must have fallen flat in that unfriendly courtroom. "From the beginning," Garrison would recall many years later of the man he was defending, "[Oppenheimer] had a quality of desperation about him. . . . I think we all felt oppressed by the atmosphere of the time but Oppenheimer particularly so. . . . I found him enigmatic, fascinating of course, with those most beautiful blue eyes, but he was hard to be intimate with. . . . Cold is too strong a word, he wasn't cold but he kept his distance." Roger Robb, fleshy and intimidating as Joe McCarthy but analytical of mind, believed he already knew the answer to the enigma: "There were so many things in those files that didn't add up unless you applied a theory to them which was that Oppenheimer was a Communist and a Russian sympathizer, and that's the only way I could add it up."

Garrison finished presenting his client in direct examination on April 14. Then it was Robb's turn to cross-examine, and he had planned exactly how to proceed:

> Having begun to pull all these strings together, I had been told that you can't get anywhere cross-examining Oppenheimer, he's too smart. He's too fast and he's too slippery. So I said, "Maybe so, but then he's not been cross-examined by me before." Anyway, I sat down and planned my cross-examination most carefully, the sequences to it and the references to the FBI reports and so on, and my theory was that if I could shake Oppenheimer at the beginning, he would be apt to be more communicative thereafter.

Neither Oppenheimer nor his defense team were allowed to see the FBI reports, nor were they aware that Boris Pash and John Lansdale had secretly recorded Oppenheimer's revelations about espionage approaches in 1943. With that privileged information and with documents Oppenheimer had not reviewed for more than ten years, Robb was able to drive the physicist to contradict himself repeatedly on his first morning of cross-examination.

In the last hour of the morning, the AEC counsel closed in. He asked Oppenheimer to describe the Chevalier approach. Oppenheimer repeated his 1946 version:

> One day ... Haakon Chevalier came to our home. It was, I believe, for dinner, but possibly for a drink. When I went out into the pantry, Chevalier followed me or came with me to help me. He said, "I saw George Eltenton recently." Maybe he asked me if I remembered him. That Eltenton had told him that he had a method, he had means of getting technical information to Soviet scientists. He didn't describe the means. I thought I said "But that is treason," but I am not sure. I said anyway something. "This is a terrible thing to do." Chevalier said or expressed complete agreement. That was the end of it. It was a very brief conversation.

Robb then led Oppenheimer through an incriminating catechism:

> Q. Did Chevalier in that conversation say anything to you about the use of microfilm as a means of transmitting this information?
> A. No.
> Q. You are sure of that?

A. Sure.

Q. Did he say anything about the possibility that the information would be transmitted through a man at the Soviet consulate?

A. No; he did not.

Q. You are sure about that?

A. I am sure about that.

Q. Did he tell you or indicate to you in any way that he had talked to anyone but you about this matter?

A. No.

Q. You are sure about that?

A. Yes.

So Robb guided Oppenheimer to contradict his 1943 revelations to Boris Pash that "a man attached to the Soviet consul" had approached, "through other people"—meaning at least Chevalier—"two or three people" about transmitting "information" through "a very reliable guy . . . who had a lot of experience in microfilm." A few more questions and the hearing adjourned for lunch.

Did Oppenheimer understand by then that Robb had trapped him in criminal contradiction? On the evidence of his testimony that afternoon, and of other information that a volume of the official AEC history brought to light for the first time in 1989, it is clear that he did, although he was probably less concerned about the legal conflict in which he had tangled himself than the personal. The 1946 version of his story—that Chevalier had approached him at his home not to solicit espionage but merely to report a contact and that there was only one such approach—was false and Oppenheimer knew that it was false. He had chosen to affirm it anyway under oath at his security hearing. That choice made the version Oppenheimer first told Pash in 1943 appear to be a lie that would have been a felony—falsely informing a federal officer—had the statute of limitations not protected him. But the alternative choice—to affirm the 1943 version and repudiate the 1946 version—would have moved the felony within range of prosecution. More painfully, it would also have implicated a friend in espionage and have exposed Oppenheimer himself to further investigation to discover who the "two or three people" were and how, when, where and by whom they were approached. Oppenheimer had trapped himself, and as he came to see that he had done so, it tore him apart. "Hunched over," Robb remembered the physicist's agony, "wringing his hands, white as a sheet," Oppenheimer bent and broke under Robb's relentless cross-examination:

Q. When did you first mention your conversation with Chevalier to any security officer?

A. I didn't do it that way. I first mentioned Eltenton. . . . I think I said little more than that Eltenton was somebody to worry about.

Q. Yes.

A. Then I was asked why did I say this. Then I invented a cock-and-bull story.

Q. Then you were interviewed the next day by Colonel Pash, were you not?

A. That is right.

"Until the end," Garrison recalled, "[Oppenheimer] never did stop to think 'How could I best put this.' Whether this was because he felt he was so right that he didn't need to pause or that he was contemptuously lording it over the other fellow by spitting the answer out immediately I don't know, but this endless cross-examination simply wore him down. . . ."

Q . . . . Did you tell Pash the truth about this thing?

A. No.

Q. You lied to him?

A. Yes.

Q. What did you tell Pash that was not true?

A. That Eltenton had attempted to approach members of the Project— three members of the project—through intermediaries.

Q. What else did you tell him that wasn't true?

A. That is all I really remember.

Q. That is all? Did you tell Pash that Eltenton had attempted to approach three members of the project—

A. Through intermediaries.

Q. Intermediaries?

A. Through an intermediary.

Q. So that we may be clear, did you discuss with or disclose to Pash the identity of Chevalier?

A. No.

Q. Let us refer, then, for the time being, to Chevalier as X.

A. All right.

Q. Did you tell Pash that X had approached three persons on the Project?

A. I am not clear whether I said there were three X's or that X approached three people.

Q. Didn't you say that X had approached three people?

A. Probably.

Q. Why did you do that, Doctor?

A. Because I was an idiot.

Q. Is that your only explanation, Doctor?

A. I was reluctant to mention Chevalier.

Q. Yes.

A. No doubt somewhat reluctant to mention myself.

Q. Yes. But why would you tell him that Chevalier had gone to three people?

A. I have no explanation for that except the one already offered.

Q. Didn't that make it all the worse for Chevalier?

A. I didn't mention Chevalier.

Q. No; but X.

A. It would have.

Q. Certainly. In other words, if X had gone to three people that would have shown, would it not—

A. That he was deeply involved.

Q. That he was deeply involved. That it was not just a casual conversation.

A. Right.

Q. And you knew that, didn't you?

A. Yes.

Q. Did you tell Colonel Pash that X had spoken to you about the use of microfilm?

A. It seems unlikely. You have a record, and I will abide by it.

Q. Did you?

A. I don't remember.

Q. Did you tell Colonel Pash that X had told you that the information would be transmitted through someone at the Russian consulate? (There was no response.)

Q. Did you?

A. I would have said not, but I clearly see that I must have.

Q. If X had said that, that would have shown conclusively that it was a criminal conspiracy, would it not?

A. That is right.

More cross-examination followed on the episode, Robb reading from transcripts of the 1943 recordings. Significantly, the AEC attorney produced a 1943 telegram revealing that Oppenheimer had told not only Pash that X had made three approaches; when General Groves had insisted Oppenheimer identify X, and Oppenheimer had given Groves Chevalier's name, the physicist had repeated his original charge: "Haakon Chevalier to be reported by Oppenheimer to be professor at Rad Lab who made three contacts for Eltenton. . . . Oppenheimer believed Chevalier engaged in no further activity other than three original attempts." Robb drove the point home:

Q. Why did you go into such great circumstantial detail about this thing if you were telling a cock-and-bull story?

A. I fear that this whole thing is a piece of idiocy.

"I felt sick," Robb recalled. "That night when I came home I told my wife, 'I've just seen a man destroy himself.' "

But the true story of what happened in 1943 between Oppenheimer and Chevalier, and Oppenheimer and Groves, is even more remarkable than the story Oppenheimer told Pash. According to George Eltenton in his 1946 testimony to the FBI, Peter Ivanov, secretary to the Soviet consulate-general in San Francisco, had asked him in 1942 to help collect information on atomic-bomb research at the University of California Radiation Laboratory. Ivanov suggested that Eltenton contact Oppenheimer, Ernest Lawrence and Luis Alvarez. (Evidently Ivanov was unaware that both Lawrence and Alvarez were staunch anti-Communists.) At this point what happened becomes speculative, since only Oppenheimer's various versions of the story remain. Eltenton may have tapped Chevalier to approach Oppenheimer. The approach may have taken place much as Oppenheimer recounted in the security hearing, except that Chevalier obviously told Oppenheimer the whole story: Ivanov, Eltenton, microfilm, support for a beleaguered nation, Lawrence and Alvarez. Oppenheimer rebuffed the approach. Then or later, however, Oppenheimer may have learned another awkward fact: that Chevalier had also approached his brother Frank. Alternatively, Chevalier may have approached Frank *rather than* Robert, in which case Frank filled his brother in and Robert presumably warned Frank off. There the matter rested for six months while Oppenheimer took up his duties as director of Los Alamos.

At Los Alamos, Oppenheimer became more aware of security. Eltenton's activity worried him. He decided to report the British engineer so that security could watch him, but when he did so he tried to avoid implicating anyone else. Eventually he had to tell Groves the truth. Then, remarkably, the official AEC history reports, Groves joined in the cover-up:

> On December 12, 1943, [Groves] learned that Oppenheimer had family concerns as well: apparently Chevalier had also talked to his brother, Frank. As the plot thickened, the truth was irretrievably lost. Had Chevalier actually approached both Oppenheimer brothers, or had he spoken only to Frank, who then turned to his older brother for advice? Was Oppenheimer trying to shoulder the entire burden for his brother and friends? Obviously, a great deal was at stake, including the [Los Alamos] project. Thus, whatever his motives, Oppenheimer secured Groves's pledge not to report his brother's name to the FBI, thereby incredibly implicating the head of the Manhattan Project in his story.

Groves testified to the security board on the morning after Oppenheimer admitted to a "cock-and-bull story." The retired general told the board that his conclusion about the Chevalier affair "was that there was an approach made, that Dr. Oppenheimer knew of this approach, that at some point he was involved, in that the approach was made to him—I don't mean involved in the sense that he gave anything—I mean he just knew about it personally from the fact that he was in the chain, and that he didn't report it in its entirety as he should have done." Groves felt Oppenheimer "was doing what he thought was essential, which was to disclose to me the dangers of this particular attempt to enter the project. . . ." The general testified that he thought "that Dr. Oppenheimer wanted to protect his friends of long stand-ing, possibly his brother. It was always my impression that he wanted to protect his brother, and that his brother might be involved in having been in this chain, and that his brother didn't behave quite as he should have, or if he did, [Robert Oppenheimer] didn't even want to have the finger of suspicion pointed at his brother, because he always felt a natural loyalty to him, and had [a] protective attitude toward him." But Groves did not admit having heard *directly from Oppenheimer* that Frank was involved. Roger Robb, evidently unaware of the general's complicity in this evasion, did not explore how Groves had arrived at his "impression." In any case, the AEC attorney got everything from Groves he could have wanted: when he asked the general if he would have cleared Oppenheimer for the Manhattan Proj-ect "if you had not believed him to be essential to the project," Groves answered reluctantly, "I would not have cleared him if I had not felt that he was essential and if he had not already been so thoroughly steeped in the project. . . ."

Nor do these various explanations, contemporary and historical, com-pletely explain Oppenheimer's 1943 statement to Boris Pash that "a man . . . attached to the Soviet consul"—presumably Peter Ivanov—had indicated indirectly "through people concerned with the project" that he could trans-mit information. Oppenheimer told Pash in 1943 that he knew of "two or three cases," that "two of the men" were "with me at Los Alamos," were "men who are closely associated with me" and had been "contacted for that purpose. . . ." Lawrence and Alvarez had *not* been contacted. Presumably one of the "two or three men" was Frank Oppenheimer. Who else had been contacted? Robb did not pursue these contradictions; they are probably unresolvable.

If the security hearing had stopped at that early point, its inquiry might have been justified; Oppenheimer's contradictory statements had long fes-tered on the record unexamined and unresolved and an espionage ap-proach had indeed occurred. The security board would have had to weigh the physicist's contradictions against the evidence that whoever approached him got nothing from him—the evidence, that is, of his basic loyalty and discretion, to which his successful direction of Los Alamos in wartime and his years of government service after the war (whatever his opinions) gave powerful support. But the AEC's letter of charges raised questions about his conduct during and after the divisive H-bomb debate as well and had to be addressed. Answering those questions occupied much of the rest of the hearing.

One by one, that April spring of 1954, the men of the nation's scientific elite entered the close room and sat down in the witness chair with Oppenheimer behind them on his couch, staring at their backs through the blue smoke of his misery and his indignation. Most were his friends. Seven were his enemies and testified against him.

His friends came first: Gordon Dean, Hans Bethe, David Lilienthal; George Kennan, who called him "one of the great minds of our generation" and insisted that security lay not in "the mathematics of whatever power of destruction we could evolve" but in "our ability to address ourselves to the positive and constructive problems of world affairs. . . ." Rabi, confident in himself and confident in his judgment that the hearing was a travesty, had no difficulty handling Robb's cross-examination:

Q. . . . Perhaps the board may be in possession of information which is not now available to you about the [Chevalier] incident.

A. It may be. On the other hand, I am in possession of a long experience with this man, going back to 1929, which is twenty-five years, and there is a kind of seat of the pants feeling [upon] which I myself lay great weight. In other words, I might even venture to differ from the judgment of the board without impugning their integrity at all.

"You have to take the whole story," Rabi went on. ". . . That is what novels are about. There is a dramatic moment in the history of the man, what made him act, what he did, and what sort of person he was. That is what you are really doing here. You are writing a man's life."

After Rabi came Norris Bradbury, Hartley Rowe, Lee DuBridge. An infuriated Vannevar Bush: "I feel that this board has made a mistake and that it is a serious one. I feel that the letter of General Nichols which I read, this bill of particulars, is quite capable of being interpreted as placing a man on trial because he held opinions, which is quite contrary to the American system, which is a terrible thing." (After that assault, Gordon Gray asked Robb privately if there was a way to end the hearing. No, Robb told him, there was not.) Briefly, Kitty Oppenheimer, the only woman to testify, on her marriage to the Communist Joe Dallet and his death in Spain during the Spanish Civil War. Oppenheimer's wife answered succinctly and to the point as one of the defense attorneys led her through the story. Robb had no questions. Gray wanted her help with the mechanics of Communist Party membership, gumshoe hocus-pocus; she withered him with her answers:

MR. GRAY. Mrs. Oppenheimer, how did you leave the Communist Party?
THE WITNESS. By walking away.
MR. GRAY. Did you have a card?
THE WITNESS. While I was in Youngstown; yes.
MR. GRAY. Did you turn this in or tear it up?
THE WITNESS. I have no idea.

Robb managed to find error in most of the witnesses. Bethe's division at Los Alamos had housed Klaus Fuchs. Senior physicist Charles Lauritsen of Caltech was not aware that Frank Oppenheimer had been a Communist. Robert Bacher had hired Philip Morrison. No one could know, Robb reminded Oppenheimer's friends sharply, what evidence the board had seen that the witnesses had not.

Then came an unprecedented parade of "government" witnesses through what was supposed to be an impartial inquiry, not a trial. It had become a trial and an ordeal, Garrison remembered:

A man's life was at stake. It was like a murder trial and a murder trial in which the evidence was murky and half-known. We spent most evenings back at the Georgetown house [of an Oppenheimer friend, Randolph Paul, where the Oppenheimers were staying]. All we had the energy for was preparation. We were too weary to do much post-morteming.

Of course, Robert was in the most overwrought state imaginable—so was Kitty—but Robert even more so. He would pace his bedroom floor at night, so Randolph Paul told me, and he was just an anguished man. Then his anxieties were added to our own and it was a great torture really.

To Wendell Latimer, the Berkeley chemistry professor who gingered up Ernest Lawrence to push for the Super, Oppenheimer was a Svengali:

You know, he is one of the most amazing men that the country has ever produced in his ability to influence people. It is just astounding the influence that he has upon a group. It is an amazing thing. His domination of the General Advisory Committee was so complete that he always carried the majority with him, and I don't think any views came out of that Committee that weren't essentially his views. . . . Many of our boys [at Berkeley] came back from [wartime Los Alamos] pacifists. I judged that was due very largely to his influence.

Air Force Major General Roscoe Wilson wanted it understood that he was "a dedicated airman" and that "the USSR in the airman's view is a land power." Yet not only was Oppenheimer "interested in what I call the internationalizing of atomic energy, this at a time when the United States had a monopoly"; he also favored the Navy over the Air Force:

> Dr. Oppenheimer . . . opposed the nuclear-powered aircraft. His opposition was based on technical judgment. I don't challenge his technical judgment, but at the same time he felt less strongly opposed to nuclear-powered ships. The Air Force feeling was that at least the same energy should be devoted to both projects.

Wilson's response to these and other differences of opinion with Oppenheimer, the general testified, had been "to go to the Director of Intelligence to express my concern over what I felt was a pattern of action that was simply not helpful to national defense."

At four in the afternoon—by now it was April 28—after Kenneth Pitzer, Edward Teller was sworn in. He had brooded on his testimony for weeks. He had talked to Ernest Lawrence, who was also supposed to testify, and had found him "furious to explain how dangerous Oppenheimer is." Both Lawrence and Alvarez, Teller recalls, "emphasized Oppenheimer's Communist associations, things that were clearly on the record: his wife, his brother, the Tatlock story—and made statements to the effect that a person of that kind cannot be cleared." On a visit to New York, Teller had stopped in to see Lloyd Garrison, the attorney wrote some years later:

> At my request he came to see me in my law office, not very long before the hearings were to begin. He was reluctant to come, and insisted on seeing me alone. I asked him about his associations with Robert and his opinion of Robert's loyalty. His feelings toward Robert were not warm, but he did not challenge his loyalty. He expressed lack of confidence in Robert's wisdom and judgment and for that reason felt that the government would be better off without him. His feelings on this subject and his intense dislike of Robert were so intense that I finally concluded not to call him as a witness. . . . When Dr. Teller did take the stand, his testimony did not depart in any substantial respect from what he had said to me in our interview.

An AEC public information officer, Charter Heslep, had approached Teller innocently at Livermore on April 22 and gotten an earful. The startled Heslep wrote afterward in a confidential memorandum to Strauss that "Teller was interested only in discussing the Oppenheimer case." Heslep explained to Teller that he was merely a speechwriter. He thought to himself that Teller's spill of confidences was none of his business. Teller told him to listen anyway. The lecture lasted more than an hour.

Teller told Heslep he regretted that the case was "on a security basis" because he felt that basis was "untenable." Teller "has difficulty phrasing his assessment of Oppie's loyalty except a conviction that Oppie is not disloyal but rather . . . more of a 'pacifist.' " But since the case was being heard on a security basis, Teller said, he wondered if some way could be found to "deepen the charges" to include documentation of the "consistently bad advice" that Oppenheimer had offered since the end of the war. Very few scientists knew the real situation, Teller went on; Oppenheimer was powerful "politically" among scientists. There was an "Oppie machine." Teller discussed its mechanisms at length and named a long list of names—Oppenheimer's cohorts. Heslep's most revealing point in the light of Teller's subsequent testimony and recollections was his sixth:

> Teller feels deeply that [Oppenheimer's] "unfrocking" must be done or else—regardless of the outcome of the current hearings—scientists may lose their enthusiasm for the [nuclear weapons] program.

In such a mood of determination, Teller proceeded to Washington to testify. The evening before he was scheduled to appear, he met with Robb. "Robb asked, 'How will you testify? Should he be cleared?' I said, 'I will testify that he should.' And then [Robb] said, 'I want you to see a part of his testimony.' . . . He showed me that part where the implication of Chevalier became obvious. . . . And then he asked, 'Would you still testify that he should be cleared?' And I said, 'I don't know.' " Teller's story of his meeting with Robb has the great virtue from his point of view of blaming Oppenheimer for Teller's testimony against him, which is probably why he invented it. Garrison's recollection that Teller's "testimony did not depart in any substantial respect from what he had said to me in our interview," and Teller's diatribe to Heslep are compelling evidence to the contrary that the physicist had planned all along to testify that Oppenheimer's security clearance should be revoked.

Hans and Rose Bethe happened to be in Washington at the time for a meeting of the American Physical Society. They spent the evening trying to persuade Teller to testify in favor of Oppenheimer. "There was not a chance to do that," Bethe recalls. "He said, 'Oppenheimer has made so many mistakes.' " It was a desperate discussion, says Bethe, filled with irrelevancies; the Cornell physicist concluded that Teller was absolutely set in his opinion that Oppenheimer had to be eliminated from his role as a government adviser. Princeton theoretician Freeman Dyson circumstantially corroborates the Bethe-Teller meeting in a memoir:

> I met Hans by chance in a hotel lobby. He was looking grimmer than I had ever seen him. . . . "Are the hearings going badly?" I asked. "Yes," said Hans, "but that is not the worst. I have just now had the most unpleasant

conversation of my whole life. With Edward Teller." He did not say more, but the implications were clear. Teller had decided to testify against Oppenheimer. Hans had tried to dissuade him and failed.

Teller, Dyson said on another occasion, "thought Oppenheimer was somehow a Machiavelli who had far more influence than he really had in the real world. And Teller must have had, somehow, the feeling that if he could once destroy Oppenheimer's political power that somehow things would be all right. And at that time, Oppenheimer had hardly any political power to destroy."

Roger Robb had a powerful weapon at his disposal with which to persuade Teller to testify that Oppenheimer was a security risk had Teller resisted doing so: Teller's secret 1952 FBI interviews. Harold Green had based the charges Robb was pursuing against Oppenheimer on Teller's statements in those interviews. In Teller's testimony, they became the dog that didn't bark.

Robb dispensed with the security-risk issue as soon as Teller was sworn in on the afternoon of April 28, 1954, first egregiously attributing to the Hungarian-born physicist a sense of fair play that Teller's secret FBI testimony decisively contradicts:

Q. I believe, sir, that you stated to me some time ago that anything you had to say, you wished to say in the presence of Dr. Oppenheimer?
A. That is correct.

After briefly reviewing Teller's curriculum vitae, Robb came directly to the point:

Q. To simplify the issues here, perhaps, let me ask you this question: Is it your intention in anything that you are about to testify to, to suggest that Dr. Oppenheimer is disloyal to the United States?
A. I do not want to suggest anything of the kind. I know Oppenheimer as an intellectually most alert and a very complicated person, and I think it would be presumptuous and wrong on my part if I would try in any way to analyze his motives. But I have always assumed, and I now assume that he is loyal to the United States. . . .
Q. Now, a question which is the corollary of that. Do you or do you not believe that Dr. Oppenheimer is a security risk?
A. In a great number of cases I have seen Dr. Oppenheimer act—I understood that Dr. Oppenheimer acted—in a way which for me was exceedingly hard to understand. I thoroughly disagreed with him in numerous issues and his actions frankly appeared to me confused and complicated. To this extent I feel that I would like to see the vital

interests of this country in hands which I understand better, and therefore trust more. In this very limited sense I would like to express a feeling that I would feel personally more secure if public matters would rest in other hands.

In that distinction between loyalty and security, evidently, was the bargain struck.

For the next two hours, Robb elicited from Teller what Harold Green has called a "tame" version of "what [Teller] told the FBI off the record." The allegations that Robert Oppenheimer delayed and subverted the H-bomb program were central among the AEC's charges. Most of the "government" witnesses were called to testify to the substance of those allegations. The allegations originated in Teller's FBI interviews. Yet Robb consistently allowed Teller implicitly to qualify and even to disavow his earlier statements. For Green, Robb's and the security board's handling of the Teller FBI interviews was mysterious:

> Inexplicably, Robb was willing to accept this watering down of his case. If he were interested in establishing the truth, however that might cut, as was his obligation under both the AEC's procedures and [the Eisenhower administration's] Executive Order 10450, he should have attempted to impeach Teller's credibility. Alternatively, if his objective was to make the case against Oppenheimer as effectively as possible, Robb should have pressed Teller to repeat under oath his earlier allegations.
>
> More importantly, the Gray Board itself, which had before it Teller's statements to the FBI, was apparently not interested in ascertaining why Teller had so changed his opinions.

But the reason why Teller changed his opinions is self-evident: he was speaking in the presence of the accused. He barely mustered the bravado to challenge Oppenheimer even then. Gordon Gray, apparently dissatisfied with Teller's craven phrase "personally more secure," decided in the final minutes of the physicist's testimony to force him to put himself more forthrightly on the record:

> MR. GRAY. . . . I would then like to ask you this question: Do you feel that it would endanger the common defense and security to grant clearance to Dr. Oppenheimer?
> THE WITNESS. I believe, and that is merely a question of belief and there is no expertness, no real information behind it, that Dr. Oppenheimer's character is such that he would not knowingly and willingly do anything that is designed to endanger the safety of this country. To the extent, therefore, that your question is directed toward intent, I would say I do not see any reason to deny clearance.

If it is a question of wisdom and judgment, as demonstrated by actions since 1945, then I would say one would be wiser not to grant clearance.

Oppenheimer, listening to Teller testify, made cryptic notes on a yellow legal pad:

> Teller—aggressive
> had conscience
> hysterical
> two sides on H-bomb

Oppenheimer thought more about his Hungarian colleague later in the hearing, remembered Teller's valediction when he left Los Alamos for Livermore and reminded himself to find out who had passed it along. "Since I cannot work with the appeasers," Oppenheimer quoted it from memory, "I will work with the fascists." Revealing though it was, the remark measured nothing of use to Robert Oppenheimer. It measured the extent to which Teller approached American politics with Hungarian values. Working with the fascists of Admiral Nicholas Horthy to overthrow the Communists was precisely what many prosperous, frightened Hungarian Jews had done when Teller was a boy in post-First World War Hungary, his own father among them.

Teller offered Oppenheimer his hand as he was leaving and Oppenheimer shook it, however reluctantly. "I'm sorry," Teller told him. Oppenheimer, disbelieving, responded, "After what you've just said, I don't know what you mean."

Other witnesses continued the testimony against Oppenheimer for six more days: Air Force scientist David Griggs, Luis Alvarez, Boris Pash. William Borden appeared as a surprise witness, shocking Garrison, who had heard only rumors of the young former executive director's letter. Garrison fought to have the document excluded. Gray wanted it read into the record and Borden read it. It was a Pandora's box; after a weekend of deliberation, the defense chose not to question Borden about it. He left the hearing room and disappeared into history.

The hearing ended May 6. On May 27, the security board majority—Gray and Morgan—found that the nation owed Oppenheimer "a great debt of gratitude for loyal and magnificent service," that Oppenheimer was "a loyal citizen" and that "no man should be tried for the expression of his opinions," but did not recommend reinstating his security clearance. His "continuing conduct and associations," wrote the majority, had "reflected a serious disregard for the requirements of the security system"; he had "a susceptibility to influence which could have serious implications for the security interests of the country"; his "conduct in the hydrogen-bomb program" had been "sufficiently disturbing as to raise a doubt as to whether his future participation . . . would be clearly consistent with the best interests of security"; and he had been "less than candid in several instances in his testimony before this Board." Ward Evans, the chemistry professor, had decided to dissent but had produced such an incoherent statement that Robb had to fix it up. It alleged double jeopardy on the grounds that most of the charges against Oppenheimer had been reviewed and cleared by the AEC in 1947. It was not even a stopgap in the flood.

Oppenheimer appealed the security board majority findings to the AEC commissioners. Nichols, in his recommendation to the commissioners on June 12, 1953, emphasized the felonious inconsistencies in the physicist's testimony about the Chevalier incident. In the meantime, Strauss learned from the FBI's phone taps that Oppenheimer and Garrison were worried that the confidential hearing transcript might be published and that publication would be damaging; Oppenheimer, the FBI reported to Strauss, wanted to leak excerpts from the transcript favorable to his case to James Reston of the *New York Times*. Strauss moved to outflank the physicist. Commissioner Henry Smyth had requested a summary of the hearing transcript. Commissioner Eugene Zuckert had left his copy of the summary on a train. The FBI recovered Zuckert's copy, but Strauss used the temporary misplacement as a pretext to publish the entire 992-page transcript on June 15.

Oppenheimer's conflicting statements thus became public, but so did Edward Teller's testimony. Many of the Livermore physicist's scientific colleagues were appalled. Visiting Los Alamos later in June, Teller rose from lunch on the sunny eastern terrace of the main lodge to greet a former student, Robert Christy. Christy cut him, refusing to shake his hand. Shocked, Teller and his wife retreated to their guest room. When Alvarez heard of the incident, he called Strauss, who dictated a record:

> [Alvarez] said that Dr. Teller was now in Los Alamos with his wife and was being given very rough treatment there. . . . I called Dr. Teller and confirmed what I had heard. . . . He besought me to do nothing. . . . "I have made up my mind that I can take the gaff. I am no longer interested in anything except truth. I have complete confidence in the fact that my friends will eventually come to the conclusion that in telling what I believe to be the truth, I have done the cause of science in the service of my country the best that I could."

A majority of the AEC commissioners, led by Lewis Strauss, found on June 29 that "Dr. Oppenheimer is not entitled to the continued confidence of the Government and of this Commission because of the proof of fundamental defects in his 'character.'" (Since Oppenheimer's consultant contract with the AEC was due to expire on June 30, Strauss had to rush the decision through.) By then, Teller had returned to Livermore and had second thoughts. He drafted a brief conciliatory statement. His testimony had been misunderstood, he wrote; it was his duty to testify fully; he in no way meant to imply he believed the right to disagree should be limited; to the contrary, he believed that right is essential in a democracy; he was pleased that the AEC had not allowed the issue of the quality of Oppenheimer's advice to influence its decision to revoke his security clearance.

Teller sent the draft statement to Lewis Strauss, saying he felt now that his testimony was seriously in error. He had not testified that a man's opinion could make him a security risk, he told Strauss, but he had clearly implied it and had come much too close to saying it. The issue was crucial to his colleagues and he would be crushed if they turned away from him. Hence, he wrote, his proposed statement. What did Strauss think of it?

Strauss was more than willing to serve as Teller's conscience. The last thing he wanted was a public recantation by his star witness. He told Teller that the right word for his colleagues' reaction was "misinterpretation," not "misunderstanding." He proposed that Teller consult with Roger Robb. To make sure that Teller did so, Strauss sent the physicist's statement and letter on to the attorney. Robb immediately advised Teller to stand by his testimony, which had required "courage and character" and had performed "a public service of great value." Teller complied.

Robert Oppenheimer was predictably devastated by the withdrawal of his clearance. "I think it broke his spirit, really," Serber reflects. "He had spent the years after the war being an adviser, being in high places, knowing what was going on. To be in on things gave him a sense of importance. That became his whole life. As Rabi said, he could run the Institute with his left hand. And now he really didn't have anything to do." Bethe felt "he was not the same person afterward."

Of the hearing and its consequences for Robert Oppenheimer, Rabi had much to say. During the hearing itself, near the end of his hearing testimony, angrily, he put the government's narrow focus on Oppenheimer's early associations and contradictions in sane perspective:

> I never hid my opinion from Mr. Strauss that I thought that . . . the suspension of the clearance of Dr. Oppenheimer was a very unfortunate thing and should not have been done. In other words, there he was; he is a consultant, and if you don't want to consult the guy, you don't consult him, period. Why you have to then proceed to suspend clearance and go through all this sort of thing—he is only there when called, and that is all there was to it. So it didn't seem to me the sort of thing that called for this kind of proceeding at all against a man who had accomplished what Dr. Oppenheimer has accomplished. There is a real positive record, the way I

expressed it to a friend of mine. We have an A-bomb and a whole series of it, [deleted] and what more do you want, mermaids?

Rabi elaborated on this testimony many years later in a conversation with Bill Moyers that ought to stand as the last word on the destruction of Robert Oppenheimer after he helped the United States end a war and build an unsurpassed arsenal of nuclear weapons:

I was indignant. Here was a man who had done so greatly for his country. A wonderful representative. He was forgiven the atomic bomb. Crowds followed him. He was a man of peace. And they destroyed this man. A small, mean group. There were scientists among them. One reason for doing it might be envy. Another might be personal dislike. A third, a genuine fear of communism. He was an aesthete. I don't think he was a security risk. I do think he walked along the edge of a precipice. He didn't pay enough attention to the outward symbols. He was a very American person of a certain kind. A certain kind of intellectual, aesthetic person of the upper middle classes. . . . [In 1955] we had this [international] conference on the peaceful uses of atomic energy. And Lewis Strauss asked me, whom should we have for president of this conference? And I said, I guess we killed Cock Robin.

27

Scorpions in a Bottle

THROUGH THE YEARS of the Korean War and the development of the hydrogen bomb, Curtis LeMay honed the Strategic Air Command into a weapon capable of killing a nation overnight. "The idea was to have overwhelming strength so that nobody would dare attack us," he would explain in retirement—"at least that was my idea of it, and what I attempted to accomplish out at SAC. . . ." Deterrence was LeMay's formal strategy. He also prepared darker strategies against the hazard that deterrence might fail.

Parochial Air Force politics facilitated and fortuitously reinforced those darker strategies. The Air Force discovered that it could link its expansion to the enlarging nuclear arsenal, a linkage that the other military services sarcastically derided as "bootstrapping"—the Air Force lifting itself by its own bootstraps. First it gained a veto over the official US list of bombing targets. LeMay then began withholding his Basic War Plans from JCS review, arguing imperative secrecy. (He kept SAC targeting plans secret even within the Air Force. In 1952, for example, when Lauris Norstad had become commander of the USAF in Europe, Norstad's staff asked for a list of SAC's atomic targets in order to coordinate plans. LeMay wrote Norstad directly, waving him off: "I keep our target lists from everyone who does not have a real need to know them.") By 1955, SAC essentially controlled its own target planning.

Simultaneously, the Air Force linked target and stockpile numbers to delivery systems. Air Force Chief of Staff Hoyt Vandenberg argued the linkage to the other Joint Chiefs as early as the autumn of 1952:

It must be pointed out that if we do not provide an air force tactically strong enough to deliver atomic weapons on target with a high degree of reliability (and we thereby run out of delivery capability while appropriate targets and unexpended bombs remain) we will have committed a military blunder which will defy logical explanation to the American people. We will have failed to make provision to exploit our one military advantage over the USSR.

The Joint Chiefs accepted Vandenberg's argument and gave SAC priority in the defense budget. "The Fiscal Year 1953 defense budget included an Air Force objective for June 1954 of 143 wings, 48 more than had been proposed [previously]," writes historian David Alan Rosenberg, "... with no corresponding increases in Army and Navy force objectives." President Truman cut the JCS request to 133 wings, but the Air Force nevertheless won more than 40 percent of defense funds that year, and comparable amounts through the decade.

In the first JCS war plans that incorporated atomic weapons, the number of atomic targets had depended on stockpile numbers because the atomic arsenal was small. As the AEC began to acquire plentiful ore supplies after 1950, and as improved bomb design reduced the amount of fissile metal required per bomb, the logic reversed and targets began to drive stockpile numbers. From sixty-six Soviet cities in 1945, Vandenberg would brief Truman in 1952 that the Air Force had identified "perhaps five or six thousand Soviet targets which would have to be destroyed in the event of war." (These were no longer simply city centers; they included the Soviet nuclear production complex, airfields and military bases, oil production and refining and electrical power systems as well as industry.) Truman responded dutifully by regularly approving major increases in funding for the Atomic Energy Commission's enlarging industrial empire.

Just as the United States began a major expansion of its conventional military forces during the Korean War, so also did it expand its nuclear production capacity. A first AEC expansion was authorized in October 1950, a second, larger program in January 1952. Oak Ridge and Hanford doubled in size. Two vast gaseous-diffusion plants came on line, drawing more power than the Tennessee Valley Authority and Hoover, Grand Coulee and Bonneville dams could have delivered in concert; by 1957, the AEC consumed 6.7 percent of total US electrical power. Heavy-water reactors for tritium production at Savannah River required tens of thousands of gallons of the exotic liquid. Building the new production complexes required more than 11 percent of annual US nickel production, 34 percent of stainless steel, 33 percent of hydrofluoric acid. From $1.4 billion in 1947, AEC capital investment increased to almost $9 billion by 1955, exceeding the capital investment of General Motors, Bethlehem and US Steel, Alcoa, DuPont and Goodyear combined. More production capacity meant more weapons, which diversified from strategic bombs into tactical and strategic warheads attached to everything from depth charges to atomic cannons to anti-aircraft missiles to ballistic missiles of every range from battlefield to intercontinental: 298 bombs in 1950 became 2,422 nuclear weapons in 1955. By 1961 there were 18,638 nuclear weapons in the US arsenal; by 1962, the year of the Cuban missile crisis, 27,100.

LeMay's bomber forces increased accordingly, from 668 at the end of 1951, most of them B-50s and B-29s, to about 500 long-range jet B-52s and more than 2,500 air-refuelable medium-range B-47s in 1959. Another 1,000 SAC jet and propeller aircraft supplied transport, aerial refueling and reconnaissance.

In 1950, LeMay had identified 1954 as the year of maximum danger, "the critical year at which time we must be prepared to meet, and effectively counter, the full military force of the USSR...." The Soviet atomic arsenal, though small, was growing, as was the inventory of Soviet bombers; as 1954 approached, this meager strategic counterpart to SAC began to pose a quantifiable threat. A special subcommittee of the National Security Council reported in June 1953 that continental defense programs were "not now adequate either to prevent, neutralize or seriously deter the military or covert attacks which the USSR is capable of launching...." That inadequacy, the subcommittee found, "constitutes an unacceptable risk to our nation's survival." The subcommittee recommended enlarging SAC while developing early-warning radar systems as well as continental air defenses. Joe 4, tested two months later, increased the kilotonnage a single Soviet bomber could carry and therefore raised the stakes.

The Air Force greatly preferred offense to defense; it was axiomatic to LeMay that the bombers always got through. Both civilian and military strategists, including John von Neumann, had soberly discussed preventive war in the late 1940s, before the Soviet Union tested its first atomic bomb; as the year of maximum danger approached, the question came up again for serious review within the US government. In the spring of 1953, a committee headed by retired Air Force General James Doolittle proposed giving the Soviet Union a two-year deadline to come to terms and attacking it if it failed to do so. Eisenhower promptly rejected this bizarre nuclear ultimatum. In August 1953, the new Air Force Chief of Staff, Nathan Twining, reviewed an air staff study, "The Coming National Crisis," which warned that the US would soon have to choose between submitting to "the whims of a small group of proven barbarians [or] be militarily prepared to support such decision as might involve general war." Retaliation—a second strike as opposed to a first—would mean disaster in a nuclear war, the study argued; such a policy was the diabolic invention of a "pseudo-moralist who insists that we must accept this catastrophe."

The question of preventive war came on to Eisenhower for review in the spring of 1954, when a JCS advance study group briefed the President on a plan proposing that the US "deliberately precipitat[e] war with the USSR in the near future ... before the USSR could achieve a large enough thermonuclear capability to be a real menace to [the] Continental US." Army Chief of Staff Matthew Ridgway forthrightly denounced the proposal, he reported afterward:

> At the end of the briefing the President invited comments and I stated that this presentation left me with but one clear impression, which was that this Group was advocating the deliberate precipitation of aggressive war by the US against the USSR; that I thought this was contrary to every principle upon which our Nation had been founded, and which it continued to profess; and that in my opinion it would be abhorrent to the great mass of the American people.

Eisenhower concluded the debate in late 1954 by issuing an updated Basic National Security Policy statement: "The United States and its allies must reject the concept of preventive war or acts intended to provoke war."

Since *preventive* war was not an available remedy to the enlarging Soviet capacity for a first strike, SAC was authorized to plan for *preemption*—for beating the Soviet forces to the punch if intelligence indicated they were beginning a first strike. The CIA estimated that the Soviet Union would need a month to assemble and deliver all its nuclear weapons. The JCS ordered SAC to assign highest priority to a "blunting mission" that would take out Soviet airfields first upon presidential determination that a Soviet attack had begun, followed by attacks on advancing Soviet troops, followed finally by attacks on cities ("urban industry" so called) and government control centers.

LeMay had no interest in dribbling out his forces on three disparate missions. The Soviets might need a month in 1954 to deliver their arsenal of about 150 atomic bombs; his 1,008 bomber crews, once deployed, could deliver as many as 750 bombs in a few hours. The SAC commander continued to believe obstinately that the most effective attack would be his "Sunday punch": simultaneous assault from all sides with everything in the stockpile. A Navy officer, Captain William Brigham Moore, attended a SAC standard briefing on March 15, 1954, kept notes and came away appalled:

> SAC considers that the optimum situation would be to have adequate tankers deployed to overseas bases and also that the bombers would be similarly deployed prior to the major attack. It was estimated that SAC could lay down an attack under these conditions of 600–750 bombs by approaching Russia from many directions so as to hit their [radar] early warning system simultaneously. It would require about two hours from this moment until bombs had been dropped by using the bomb-as-you-go system in which both [airfields] and [urban] targets would be hit as they reached them. This part of the briefing was skillfully done by showing successive charts of

Europe based on one-half-hour time intervals after SAC bombers first hit the Russian early warning screen. Many heavy lines, one representing each wing, were shown progressively converging on the heart of Russia with pretty stars to indicate the many bombs dropped on DGZs [i.e., designated ground zeros]. The final impression was that virtually all of Russia would be nothing but a smoking, radiating ruin at the end of two hours.

During the post-briefing question period, someone asked LeMay what course he would advocate if hostilities were renewed in Korea—by then at truce. He answered that he would drop a few bombs in China, Manchuria and southeastern Russia. "In those 'poker games,'" the Navy captain paraphrases him, "such as Korea and Indo-China [where the French were then engaged], we . . . have never raised the ante—we have always just called the bet. We ought to try raising sometime."

By 1954, Curtis LeMay had apparently begun raising the ante with the Soviet Union on his own, covertly and extralegally. LeMay's crews needed target information to carry out their mission as well as information about Soviet defenses, particularly fighter bases and radar frequencies. In the late 1940s, SAC routinely flew reconnaissance missions in stripped-down B-29s around the perimeter of the USSR; it was from these "weather" squadrons that the first long-range atomic detection aircraft were drawn. But LeMay wanted more than sideward glances into the Soviet Union; he wanted over-flights. Those began no later than early 1950.

They quickly drew a response from Soviet defense forces. A Navy long-range PB4Y-2 engaged in electronic-intelligence gathering was shot down by Soviet fighters on April 8, 1950, probably over Soviet territory; ten men were killed. Such intrusive and provocative flights—legally acts of war—might justify a Soviet attack on Europe; to forestall that eventuality, Truman ordered overflights banned. But LeMay still required reconnaissance, particularly photographs of the images of targets on aircraft radar scopes. He intended to turn such photographs into etched Lucite plates that his crews could overlay onto their radar scopes to practice atomic-bombing. The determined SAC commander thought up a way to work around Truman's ban. With the approval of the JCS he made a deal with the British: the US would supply the Royal Air Force with B-45 medium jet bombers, the newest, fastest high-altitude aircraft available; the RAF would fly photo- and radar-reconnaissance missions over the Soviet Union and share the intelligence with SAC. RAF crews began overflights in March 1952, after Winston Churchill became Prime Minister and approved the risky venture. One of the pilots remembers the Soviet Union distinctly from populous Europe as "one large black hole with odd lights here and there. . . . There were big areas we were supposed to be photographing; most of them were installations out of their radar range, armed installations which are not lit, and once we came up south of Moscow itself you [could] see all the lights. Moscow's a big place and lit up so you do get a good reference point from that."

Soviet interceptors were unable to locate the night-flying B-45s, but the planes were tracked on Soviet radar and sometimes encountered anti-aircraft flak. The Soviets believed the flights were American. By 1954 they were, and US overflights continued in a variety of aircraft, including the notorious U-2, until reconnaissance satellites took their place beginning in 1960. The Soviet Union shot down at least twenty planes during overflights with the loss of an estimated one to two hundred US airmen, some of whom went to the gulag.

LeMay used these reconnaissance flights not only to gather electronic and photographic intelligence; he also used them to probe Soviet air defenses, knowing as he did so that he might be provoking war. There is testimony that he may have meant to do just that. If he could not initiate preventive war, he seems to have concluded, he might be able to push the Soviets to high enough levels of alert to justify launching a full preemptive attack. He linked reconnaissance with provocation in an interview after he retired:

There was a time in the 1950s when we could have won a war against Russia. It would have cost us essentially the accident rate of the flying time, because their defenses were pretty weak. One time in the 1950s we flew all of the reconnaissance aircraft that SAC possessed over Vladivostok at high noon. Two reconnaissance airplanes saw MiGs, but there were no interceptions made. It was well planned, too—crisscrossing paths of all the reconnaissance airplanes. Each target was hit by at least two, and usually three, reconnaissance airplanes to make sure we got pictures of it. We practically mapped the place up there with no resistance at all. We could have launched bombing attacks, planned and executed just as well, at that time.

Soviet defense forces had no way of knowing if LeMay's crisscrossing reconnaissance aircraft carried nuclear weapons or not. If Soviet aircraft had crisscrossed US cities under similar circumstances, SAC would certainly have preempted. The Soviets hunkered down because they had no adequate response, but their lack of defenses predictably emboldened LeMay.

One of LeMay's US reconnaissance crews remembered flying a B-47 deep into the USSR on May 8, 1954, and taking damage from a MiG-17. The mission made it back to England leaking fuel. LeMay ordered the crew to the US, the pilot, Hal Austin, recalled many years later:

[LeMay] said, "I tried to get you guys a Silver Star," but he said "you gotta explain that to Congress and everybody else in Washington . . . so here's a couple of [Distinguished Flying Crosses] we'll give you for that mission."

There wasn't anybody in the room except the wing commander and us three guys, General LeMay and his intelligence officer. . . .

Then General LeMay said, "Well, maybe if we do this overflight right, we can get World War III started."

I think that was just a loose comment for his staff guys, because General Tommy Power, his hatchet man in those days, chuckled and he never laughed very much. So I always figured that was a joke between them. But we thought maybe that was serious.

Austin raised the question with LeMay after the SAC commander retired. "I brought up the subject of the mission we had flown. And he remembered it like it was yesterday. We chatted about it a little bit. His comment again was, 'Well, we'd have been a hell of a lot better off if we'd got World War III started in those days.' "

Was LeMay joking? The best evidence that he was not is his own testimony, in a lecture he delivered to the National War College in April 1956. Decisive victory in a nuclear war, he emphasized on that occasion, would "have been reached in the first few days" of battle. The Soviet Union was not yet capable of achieving such a decisive nuclear victory, but it was "building a global bombing force with aircraft and nuclear weapons of satisfactory quality" to make it capable in time "of devastating the heartland of the United States." The US did have such decisive capability, however, LeMay asserted. He went on to describe "in cold terms what the United States is capable of doing to the Soviet Union today," a description as chilling as any in the literature of war:

Let us assume the order had been received this morning to unleash the full weight of our nuclear force. (I hope, of course, this will never happen.) Between sunset tonight and sunrise tomorrow morning the Soviet Union would likely cease to be a major military power or even a major nation: the bulk of its long-range air power would be shattered, its centers of industry and control devastated. Communications would have been disrupted and much of their economic strength depleted. Dawn might break over a nation infinitely poorer than China—less populated than the United States and condemned to an agrarian existence perhaps for generations to come.

Everything depended on "the forces in being at the outset," LeMay emphasized. ". . . Today, shooting wars are won or lost before they start. If they are fought at all, they would be fought principally to confirm which side had won at the outset. . . . The most radical effect of the changes in warfare is not upon how wars are won or lost, but upon how they will start. . . . The dominant fact is that no nation can arrive at a deliberate decision to wage war today unless it is clear, beyond any doubt, that victory is assured." What those facts meant, LeMay went on significantly, was that "we are at war *now*." By defining the state of affairs between the US and the USSR as war *in progress,* LeMay thus blurred the difference between preventive attack and preemption.

LeMay sketched three phases of "today's war." The first was a decision phase—that is, deciding to wage war. "This decision is reached during what we used to call peace, and now call 'cold war.' We are in the decision phase today." The second was the "proof phase," "the application of . . . power to the enemy to confirm the decision. . . ." The third was the "exploitation phase," which commenced "when the level of radiation in a vanquished belligerent has lowered sufficiently for imposition of the national will of the victor upon the survivors of the vanquished." LeMay saw evidence that the Soviet leaders were "reasonable men," but they had "as their primary goal the perpetuation of their own regime, the retention of power inside the USSR in the few hands in which it now resides," and only secondarily did they desire "to insure the continued security of the Soviet homeland and its people. . . . More than once, millions of Soviet citizens have, in time of peace, had their security and their very lives taken from them for the greater glorification of the all-powerful Party. More than once, the heads of that party have gambled with the security of the USSR, and subjected it to the risk of war, not for national survival, but for the consolidation of their own strength." The ultimate threat, LeMay argued, was "the growing Soviet capability to launch a massive nuclear assault against the free world." And SAC was the answer to that threat. "It can react not in weeks or days, but in hours and minutes, from its present position. . . . SAC is fighting the decisive phase, the IN PLACE WAR [sic], today."

If the Soviet capability was growing, and the bombers always got through, then the time would come when SAC would no longer be able to deliver a victory. The US and the USSR would then be mutually deterred. Robert Oppenheimer had predicted that consequence in his 1953 *Foreign Affairs* analysis: "We may anticipate a state of affairs in which two Great Powers will each be in a position to put an end to the civilization and life of the other, though not without risking its own. We may be likened to two scorpions in a bottle, each capable of killing the other, but only at the risk of his own life."

LeMay evidently found the prospect of such stalemate intolerable. It made the force that he had built with such prodigious energy a wasting asset. "While we had all this superiority," he said in retirement, "we invaded no one; we didn't launch any conquest for loot or territory. We just sat there with the strength. . . . We didn't threaten to use it when it might have brought advantages to the country." If the politicians were craven, he would seize the initiative. The United States, the SAC commander believed, was already at war with a ruthless enemy. He readied his nuclear armada while he prodded the Soviet bear with reconnaissance overflights.

SAC was subject to presidential authority. The Constitution authorized the President, not the SAC commander, to determine when to order the use of military force. But LeMay had decided at the beginning of the Korean War, if not before, that there were circumstances under which he would override the Commander-in-Chief's prerogative. In 1950, LeMay had attempted to arrange privately with the officer commanding Sandia Base "to take the bombs" if "we woke up some morning and there wasn't any Washington or something." By 1957 he no longer needed to take the bombs. He had them, and they would not be fitted with electronic Permissive Action Link (PAL) locks until early in the Kennedy era. (Even then, SAC had the codes.) All that constrained him from delivering them was his soldier's oath. In 1957, a committee Eisenhower appointed to study civil and continental defense sent a delegation to SAC to review the command's defenses against a Soviet surprise attack. The delegation included Robert Sprague, president of the Sprague Electric Company of Massachusetts, and Jerome Wiesner of MIT. LeMay dismissed the delegation with a superficial tour. Sprague arranged for the President to order LeMay to cooperate. From air defense headquarters in Colorado Springs, Sprague had LeMay stage an alert. SAC needed more than six hours to take to the air. To Sprague that performance meant the command was vulnerable to surprise attack: Soviet bombers could make the flight over the North Pole in less than six hours.

At SAC headquarters in Omaha, Sprague challenged LeMay. The general dismissed Sprague's concerns contemptuously. SAC had reconnaissance aircraft flying secret missions over the Soviet Union twenty-four hours a day, he explained. "If I see that the Russians are amassing their planes for an attack, I'm going to knock the shit out of them before they take off the ground." Sprague was shocked. "But General," he countered, "that's not national policy." Sprague remembered LeMay responding, "I don't care. It's my policy. That's what I'm going to do." Wiesner says LeMay responded, "It's my job to make it possible for the President to change his policy"—a less insubordinate answer, but only barely. Sprague chose not to report the renegade commander to the President and buried the incident for thirty years. At least, he reasoned, US strategic bombers would not be destroyed on the ground.

Nineteen fifty-four, the year of maximum danger, passed without preventive war or preemption when the National Security Council determined that "the USSR, by a maximum effort, could launch about 300 aircraft from the Chukotski and Kola areas, 200 to 250 of which might reach their targets." LeMay would not have agreed with this assessment, but it was evidently sufficient to give pause at the Pentagon. A gap had opened between SAC and Washington, SAC believing that it could still deliver a preemptive first strike, Washington already to a considerable degree deterred.

The Soviet Union tested a two-stage, lithium-deuteride-fueled thermonuclear device on November 22, 1955, dropping it from a Tu-16 bomber to minimize fallout. It yielded 1.6 megatons, a yield deliberately reduced for the Semipalatinsk test from its design yield of 3 MT. According to Yuri Romanov, Andrei Sakharov and Yakov Zeldovich worked out the Teller-Ulam configuration in conversations together in early spring 1954, independently of the US development. "I recall how Andrei Dmitrievich gathered the young associates in his tiny office," Romanov writes, "... and began talking about the amazing ability of materials with a high atomic number to be an excellent reflector of high-intensity, short-pulse radiation." The Sarov designers had to fight Vyacheslav Malyshev for resources to develop the new design; the Minister of Medium Machine Building, conservative as his predecessor, wanted them to stick to weaponized layer-cake thermonuclears, and such a device was tested on November 6, 1955, three weeks before the two-stage design, as a backup in case the new system should fail.

Victor Adamsky remembers the shock wave from the new thermonuclear racing across the steppe toward the observers. "It was a front of moving air that you could see that differed in quality from the air before and after. It came, it was really terrible; the grass was covered with frost and the moving front thawed it, you felt it melting as it approached you." Igor Kurchatov walked in to ground zero with Yuli Khariton after the test and was horrified to see the earth cratered even though the bomb had detonated above ten thousand feet. "That was such a terrible, monstrous sight," he told Anatoli Alexandrov when he returned to Moscow. "That weapon must not be allowed ever to be used."

The increasing Soviet capability did not prevent the United States from practicing diplomacy by nuclear threat through the second half of the 1950s. An emboldened Nikita Khrushchev followed suit. Eisenhower had passed an implicit nuclear threat to the Chinese in 1953 that may have contributed to the Korean armistice signed that July, although Stalin's death was a far more important factor in the Chinese decision. Khrushchev, through Bulganin, threatened Britain and France during the Anglo-French-Israeli invasion of Egypt at Suez in 1956, the first such threat from the Soviet Union; Eisenhower checked it by warning that attacks on France or Britain would force a US response and by putting SAC on a modified alert. The President alerted SAC again during the US invasion of Lebanon in 1958, deploying SAC tanker aircraft forward deliberately, he wrote afterward, to show "readiness and determination without implying any threat of aggression." Of that movement Khrushchev commented to Egyptian President Gamal Abdel Nasser, according to a Nasser confidant, that "he thought the Americans had gone off their heads. 'Frankly,' [Khrushchev] said, 'we are not ready for confrontation. We are not ready for World War III.'" In 1958, John Foster Dulles's State Department authorized the Secretary of the Air Force to acknowledge publicly that the US was prepared to use nuclear weapons in the conflict with the People's Republic of China over the islands of Quemoy and Matsu; in response, Khrushchev wrote Eisenhower to warn that "an attack on the People's Republic of China . . . is an attack on the Soviet Union."

Nuclear blackmail continued after 1958, but the stakes increased because the superpowers confronted each other directly, because the conflicts concerned what historian Richard K. Betts calls "the core geographic security zones of the superpowers"—Berlin and Cuba—and because the Soviet nuclear arsenal significantly increased (to 1,050 by 1959 and to 3,100 by 1962). Sputnik, the first earth satellite, which the Soviets had orbited in 1957, had panicked the US, much to Eisenhower's disgust; the US was well along in intercontinental ballistic missile development when Sputnik began beeping and quickly surpassed Soviet deployment. The hemorrhage of East Germans defecting to the West through Berlin in the late 1950s challenged Khrushchev sufficiently that he threatened to sign a separate peace treaty with East Germany and permanently close off Western access. The Cuban revolution, culminating in 1959, brought what turned out to be a Communist government to power ninety miles off US shores. John F. Kennedy, succeeding Eisenhower as President in 1961, moved SAC to 50 percent ground-alert status beginning in July 1961, when Berlin was fulminating: half SAC's bomber fleet held ready for a fifteen-minute scramble. SAC also maintained at least twelve B-52 heavy jet bombers airborne at all times. The Soviet military made its corresponding point by inviting foreign attachés to observe Soviet military maneuvers, including maneuvers of tactical nuclear weapons, four days before the East Germans began throwing up the Berlin Wall.

In the autumn of 1961, newly orbited US reconnaissance satellites revealed that the Soviet Union had fewer strategic delivery systems than US intelligence had previously estimated—only 44 ICBMs and 155 heavy bombers (compared to the US's 156 ICBMs, 144 Polaris submarine-launched ballistic missiles and 1,300 strategic bombers). Paul Nitze warned the Soviet ambassador over lunch one day that the missile gap favored the US; for good measure, the government leaked the story to journalist Joseph Alsop, who reported the gap in a column. Kennedy may also have warned Soviet Foreign Minister Andrei Gromyko directly in a private meeting at the White House.

Partly as a result of these threatening warnings, Khrushchev decided to install nuclear missiles secretly in Cuba, to match fifteen Jupiter intermediate-range ballistic missiles that the US had deployed in Turkey, on the southern border of the USSR. He wanted to protect Cuba from US invasion, he claimed in retirement, and to "[equalize] what the West likes to call 'the balance of power.' . . . We had no desire to start a war. On the contrary, our principal aim was only to deter America from starting a war. We were well aware that a war which started over Cuba would quickly expand into a world war." The Soviets would not need many missiles in Cuba to right the balance. "Before the Soviets put missiles in Cuba," Kennedy Treasury Secretary Douglas Dillon recalls, "it was doubtful whether they could deliver any warheads from Soviet territory at all. So while the Cuban installation didn't add very much to their numbers or didn't change the overall balance very much, my impression at the time was that they radically altered the numbers of *deliverable* warheads, and in that sense, they significantly increased Soviet capability."

The Soviets began a military build-up in Cuba early in 1962. By August, the CIA was reporting that medium-range ballistic missiles might be part of the expansion. A U-2 overflight found MRBM sites in western Cuba first on October 14. Kennedy condemned the secret installation and announced an impending naval "quarantine" on October 22, Monday night of a harrowing week when the world feared (justifiably) that the superpowers verged on full-scale nuclear war.

Thomas Power was SAC commander by then; LeMay had moved up to USAF Chief of Staff, which put him in the Pentagon and at the White House during the Cuban missile crisis. ("Those were ten days when neither Curt nor I went home," General David Burchinal, LeMay's Deputy Chief for Plans and Operations, recalled. "We slept in the Pentagon right around the clock.") Power was at least as eager to "get World War III started" as LeMay. He had been instructed by LeMay's predecessor Chief, Thomas D. White, that he had the authority; White wrote Power in 1957 that "authority to order retaliatory attack may be exercised by CINCSAC [i.e., Commander-in-Chief, SAC] if time or circumstances would not permit a decision by the President." McGeorge Bundy, Kennedy's National Security Adviser, had warned the young President of just such a possibility; "a subordinate commander," Bundy alerted Kennedy in January 1961, "faced with a substantial Russian military action could start the thermonuclear holocaust on his own initiative if he could not reach you (by failure of communication at either end of the line)." LeMay would acknowledge in retirement that Power was "a sadist"; one of Power's subordinate commanders confirms that view:

General Power . . . was demanding; he was mean; he was cruel, unforgiving, and he didn't have the time of day to pass with anyone. A hard, cruel individual. . . . I would like to say this. I used to worry about General Power. I used to worry that General Power was not stable. I used to worry about the fact that he had control over so many weapons and weapon systems and could, under certain conditions, launch the force. Back in the days before we had real positive control [i.e., PAL locks], SAC had the power to do a lot of things, and it was in his hands, and he knew it.

By presidential order, the US military went from DefCon (Defense Condition) 5 to DefCon 3 during Kennedy's Monday night speech. DefCon 5 was business as usual; DefCon 3 was halfway up the scale. When Kennedy began speaking on national television, fifty-four SAC bombers each carrying as many as four thermonuclear weapons thundered off from continental bases to join the twelve-plane around-the-clock airborne alert. Some of the sixty-six bombers orbited the Mediterranean; others circumnavigated North America; others flew an Arctic route across Greenland, north of Canada, across Alaska and down the US Pacific Coast. One orbited above Thule, Greenland, to observe and report any pre-attack Soviet assault on the crucial US early-warning radar there. Polaris submarines put to sea. SAC armed its bomber force, dispersed it to military and civilian airfields and prepared 136 Atlas and Titan ICBMs for firing. Kennedy and Khrushchev began an exchange of belligerent messages. Kennedy had convened an executive committee of government officials to manage the crisis; options discussed by the ExCom ranged from blockade to air strike to Cuban invasion. The President said afterward that the purpose of the alert was to deter a Soviet military response to whatever Caribbean action the US decided to carry out: "The airborne alert," he congratulated SAC, "provided a strategic posture under which every United States force could operate with relative freedom of action." General Power saw a more threatening purpose, however; from his point of view, "This action by the nation's primary war deterrent force gave added meaning to the President's declaration that the US would react to any nuclear missile launched from Cuba with a full retaliatory response upon the Soviet Union itself." Kennedy, that is, was thinking regional engagement under a nuclear umbrella; Power and LeMay were thinking global war.

Wednesday, October 24, when the naval quarantine took effect, SAC ratcheted from DefCon 3 to DefCon 2, the first and only time it ever did so. SAC alerted nuclear weapons increased to 2,952; with 112 Polaris SLBMs, their total destructive force exceeded seven thousand megatons. "We got everything we had in the strategic forces . . . counted down and ready and aimed," General Burchinal said afterward, "and we made damn sure they saw it without anybody saying a word about it." In fact, Power said several words about it, unauthorized and publicly, when he broadcast in the clear—in English rather than in code—to all SAC wings immediately after the move to DefCon 2 was announced:

This is General Power speaking. I am addressing you for the purpose of reemphasizing the seriousness of the situation the nation faces. We are in an advanced state of readiness to meet any emergencies and I feel that we are well prepared. I expect each of you to maintain strict security and use calm judgment during this tense period. Our plans are well prepared and are being executed smoothly. . . . Review your plans for further action to insure that there will be no mistakes or confusion. . . .

SAC routinely transmitted DefCon increases as unclassified messages until 1972, and Power was clearly emphasizing control. His broadcast was nevertheless a warning to the Soviets, whom Power knew monitored such transmissions, that the US had gone to full alert and might be planning "further action."

Equally unsanctioned, and potentially catastrophic, was the launch of an Atlas ICBM from Vandenberg Air Force Base across the Pacific to the Kwajalein test range at four A.M. on October 26, the height of the Cuban missile crisis. SAC had taken over the test missiles at Vandenberg at the time of the DefCon 3 alert, programmed them with Soviet targets and begun attaching nuclear warheads. The Atlas had been scheduled for testing; it was launched on its precrisis schedule with SAC concurrence, a deliberate provocation.

The US's first squadron of Minutemen I solid-fuel missiles was undergoing testing and certification prior to deployment at Malmstrom Air Force Base in Montana when the missile crisis began. SAC, the Air Force Systems Command and contractor personnel worked nonstop to ready the Minutemen for launch. A declassified history of the missile wing reports that "lack of equipment, both standard and test, required many work-arounds." The first Minuteman was up on October 26; five had been made operational by October 30. But miswiring, wire shorts and other problems left the missiles capable of being accidentally armed; one had to be shut down and restarted five times because its guidance and control systems failed, and all ten Minutemen at Malmstrom had to be taken off alert repeatedly for repair in the course of the crisis. Missiles did not have PAL locks in 1962; for safety and control, launch required redundant, coordinated keying by four officers in two physically separate launch control centers. The Malmstrom work-around overrode that safety system. One officer who controlled the Minutemen during the missile crisis told nuclear safety expert Scott Sagan, "We didn't literally 'hot wire' the launch command system—that would be the wrong analogy—but we did have a second key.... I could have launched it on my own, if I had wanted to." An Air Force safety inspection report noted after the crisis that "possible malfunctions of automated equipment...posed serious hazards [including] accidental launch...." Another possibility that the inspectors chose not to mention was *unauthorized* launch.

Air Defense Command F-106s armed with nuclear air-to-air missiles scrambled at Volk Field in Wisconsin on October 25 when a launch klaxon went off in the middle of the night. With practice alert drills canceled at DefCon 3, the interceptor crews assumed they were going to war. Since they had not been briefed that SAC bombers were aloft dispersing and did not know SAC airborne alert routes, nuclear friendly fire was a real possibility. The launch klaxon sounding was a mistake; an Air Force guard at the Duluth

Sector Direction Center had sounded a sabotage alarm that somehow keyed the klaxon at Volk Field. The guard had seen someone climbing the base security fence and had fired at the figure. An officer flashing his car lights managed to stand down the F-106s; on closer inspection, the saboteur had turned out to be a bear.

There were other serious command-and-control snafus during the Cuban missile crisis as well: a U-2 strayed over Siberia, leading Khrushchev to complain to Kennedy "that an intruding American plane could be easily taken for a nuclear bomber, which might push us to a fateful step"; air defense interceptors flew fully armed with nuclear rockets with all safety devices removed; US radar picked up an apparent missile launch from Cuba with a near-Tampa trajectory on Sunday morning, October 28, that was determined only after predicted impact to be a computer test tape. The US Navy tracked Soviet submarines aggressively throughout the world—forcing them to surface and reveal their positions, a serious provocation—when it had been ordered to do so only in the area of quarantine.

More dangerous by far than all these incidents was Curtis LeMay's overconfident and belligerent advice to President Kennedy, whom he believed to be a coward. Knowing that the US and the USSR were approaching mutual deterrence and that SAC was therefore a wasting asset, LeMay pushed Kennedy to up the ante, bomb Cuba and take out the missile sites. "The Kennedy administration thought that being strong as we were was provocative to the Russians and likely to start a war," the SAC general said with disgust in retirement. "We in the Air Force, and I personally, believed the exact opposite. . . . We could have gotten not only the missiles out of Cuba, we could have gotten the Communists out of Cuba at that time. . . . During that very critical time, in my mind there wasn't a chance that we would have gone to war with Russia because we had overwhelming strategic capability and the Russians knew it." Believing the crisis a poker game, LeMay imagined that the US held the best cards. "The Russian bear has always been eager to stick his paw in Latin American waters," he taunted during the crisis. "Now we've got him in a trap, let's take his leg off right up to his testicles. On second thought, let's take off his testicles, too." As LeMay's castration imagery implies, he may have been goading Kennedy to attack Cuba as an excuse to launch full strategic preemption; discussing the missile crisis twenty years later with historian Ernest May, he said "that it was his belief that at any point the Soviet Union could have been obliterated without more than normal expectable SAC losses on our side. . . ." Kennedy administration Secretary of Defense Robert McNamara remembers that "LeMay talked openly about a first strike against the Soviet Union if the Russians ever backed us into a corner."

Kennedy fortunately resisted LeMay's goading, Robert Kennedy writes:

When the President questioned what the response of the Russians might be, General LeMay assured him that there would be no reaction. President Kennedy was skeptical.... "They, no more than we, can let these things go by without doing something. They can't, after all their statements, permit us to take out their missiles, kill a lot of Russians and then do nothing. If they don't take action in Cuba, they certainly will in Berlin."

The President's instincts were sharper than the general's. The blockade worked; the crisis passed; Khrushchev capitulated. LeMay was outraged. Daniel Ellsberg, then a member of the ExCom staff under McNamara, says LeMay chewed Kennedy out. McNamara confirms the story: "After Khrushchev had agreed to remove the missiles, President Kennedy invited the Chiefs to the White House so that he could thank them for their support during the crisis, and there was one hell of a scene. LeMay came out saying, 'We lost! We ought to just go in there today and knock 'em off!' "

. At the height of the crisis, according to a retired SAC wing commander, SAC airborne alert bombers deliberately flew past their turnaround points toward Soviet airspace, an unambiguous threat which Soviet radar operators would certainly have recognized and reported. "I knew what my target was," the SAC general adds: "Leningrad." The bombers only turned around when the Soviet freighters carrying missiles to Cuba stopped dead in the Atlantic.

Nuclear crises are not poker games. What Curtis LeMay and Thomas Power did not know—what no one in the US government knew until it was revealed at a conference between Soviet and US missile crisis participants in Moscow in 1989—was that, contrary to CIA estimates, the Soviet forces in Cuba during the missile crisis possessed twenty nuclear warheads for medium-range R-12 ballistic missiles that could be targeted on US cities as far north as Washington, DC, as well as nine tactical nuclear missiles which the Soviet field commanders in Cuba were delegated authority to use—the only time such authority was ever delegated by the Soviet leadership. The medium-range missiles would probably have been launched as well, McNamara believes. "If they'd been NATO missiles without PALs, then the NATO officers, acting without Presidential authorization, would have been likely to use them rather than lose them. The fear that Soviet or Cuban officers might have reacted the way NATO officers might have was one reason I was extremely reluctant to risk the air strike."

In 1954, when LeMay calculated that he could deliver a Sunday punch of 750 atomic bombs to targets in the Soviet Union overnight, the Defense Department Weapons Systems Evaluation Group estimated that Soviet and Soviet bloc casualties would total seventeen million injured and sixty million dead. In 1962, Power was prepared to deliver almost three thousand strategic nuclear weapons, many of them thermonuclear bombs, with yields totaling seven thousand megatons. Under such a rain of destruction, the United

States would have killed at least 100 million human beings* in pursuit of the small group of Soviet leaders, as LeMay said, "[who] have as their primary goal the . . . retention of power inside the USSR in the few hands in which it now resides." If the Soviet field commanders in Cuba had launched their missiles as well, more millions of Americans would have been killed. Seven thousand megatons was also more than enough fire and brimstone to initiate a lethal nuclear winter over at least the Northern Hemisphere, freezing and starving yet more millions in Europe, Asia and North America—a phenomenon that scientists had not yet identified and that neither SAC nor Washington had yet assessed. How extraordinary that Curtis LeMay believed for the rest of his life that the United States "lost" the Cuban missile crisis and the Cold War. If John Kennedy had followed LeMay's advice, history would have forgotten the Nazis and their terrible Holocaust. Ours would have been the historic omnicide.

The Soviet Union never went to full nuclear alert in all the years of the Cold War. After the Cuban missile crisis, the United States never did again. Nor did the two nations ever again directly confront each other.

---

*One hundred million is an extremely conservative estimate; in 1984, the World Health Organization estimated that a ten-thousand-megaton nuclear exchange would account for 1.15 billion dead and 1.1 billion injured. International Committee (1984).

Epilogue: 'The Gradual Removal of Prejudices'

ROBERT OPPENHEIMER turned fifty during his security hearing in 1954; he went home to Princeton visibly aged. A former student who saw him there remarked that he had always looked younger than his years, but now looked older. He continued to direct the Institute for Advanced Study for another decade, "eating out his heart in frustration," George Kennan would say at his funeral, "over the consciousness that the talents he knew himself to possess, once welcomed and used by the official establishment of his country to develop the destructive possibilities of nuclear science, were rejected when it came to the development of the great positive ones he believed that science [offered]."

From Princeton, in 1959, Oppenheimer could watch with the rest of the country the public humiliation of Lewis Strauss, two grueling months of Senate hearings reviewing Strauss's fitness to serve as Secretary of Commerce. The manipulative financier was pilloried for his arrogance and rigidity and caught in a lie under oath; at one point he feared his phones had been tapped. With Clinton Anderson of New Mexico leading the fight, the Senate rejected Strauss 49–46. His brother said he never got over it. In that despondency, Lewis Strauss and Robert Oppenheimer finally had something in common.

Oppenheimer was called to speak to the world from Paris, from South America, from England and Japan and finally from within the US. John Kennedy invited him to dine at the White House with forty-nine Nobel laureates in 1961 and planned to present him the Enrico Fermi Award, the AEC's highest honor, on December 2, 1963; Lyndon Johnson, in a time of mourning, made the presentation to Oppenheimer in the White House Cabinet Room—a medal and fifty thousand dollars to take home from an agency that still denied him clearance as a security risk.

Edward Teller sought personal reconciliation with Oppenheimer then. Teller had received the Fermi Award the previous year. He nominated

Hyman Rickover, Leo Szilard and Oppenheimer as possible successors. When he heard that Oppenheimer had been chosen, he sent him a note of congratulation. Teller wrote that he was happy remembering their 1942 Berkeley summer. He still thought the Acheson-Lilienthal Report was the only "honest and effective" arms control proposal anyone had ever made. He had wanted to speak to Oppenheimer before the occasion of the award, he confessed, but had not been sure that he would be doing the right thing. He thought it might have been better if Oppenheimer had received the Fermi Award first. He wished his old colleague good luck.

Oppenheimer had been wounded too deeply for reconciliation. At the Institute for Advanced Study he drafted a friendly note and thought of enclosing a recent lecture. He thought again, refiled the lecture and reduced the note to formality:

Dear Edward:

Thank you for writing me. I am very glad that you did.
With good wishes.

Robert Oppenheimer

Oppenheimer retired from the institute in 1966, when illness weakened him. He died of cancer of the throat on February 18, 1967, at the age of sixty-two. His ashes were scattered on the ocean off the Virgin Islands, where he often vacationed. Among his last published words were these: "Science is not everything, but science is very beautiful."

He left behind an evocative self-portrait. In 1963, *The Christian Century* had asked him, "What books did most to shape your vocational attitude and your philosophy of life?" He had responded with a list which the magazine transcribed exactly as he wrote it—his personal ranking of the competing claims of head and heart:

Les Fleurs du Mal, by Charles Baudelaire
The Bhagavad-Gita
Collected Works, by Bernhard Riemann
Theaetetus, by Plato
L'education sentimentale, by Gustave Flaubert
The Divine Comedy, by Dante Alighieri
The Three Centuries, by Bhartrihari
The Waste Land, by T. S. Eliot
The notebooks of Michael Faraday
Hamlet, by William Shakespeare

Edward Teller became the Richard Nixon of American science—dark, brooding, indefatigable. As he predicted, many of his colleagues turned away from him. Their rejection hurt him deeply, he told his biographers one bitter day in 1974:

> If a person leaves his country, leaves his continent, leaves his relatives, leaves his friends, the only people he knows are his professional colleagues. If more than ninety percent of these then come around to consider him an enemy, an outcast, it is bound to have an effect. The truth is it had a profound effect. It affected me, it affected [my wife] Mici, it even affected her health.

Some of Teller's colleagues rejected him, perhaps, out of personal animosity. But many others, especially at Los Alamos, did so because they knew it was Edward Teller, not Robert Oppenheimer, who had delayed the development of the hydrogen bomb. He did so by defining that device from the beginning as a mechanism for achieving megaton-range yields. His characteristic grandiosity blinded him to the more modest possibilities of his Alarm Clock, which was inherently physically feasible and practical at high-kiloton yields. I. I. Rabi made the point in his response to Sterling Cole's 1953 questionnaire that asked for comparisons between US and Soviet hydrogen-bomb progress. While Teller, John Wheeler and John von Neumann responded by blaming US delay, Rabi saw the essential difference: "One of the faults of our program is that we set our sights too high in aiming for [megaton yields]. The Russians were apparently much more modest in their goals." From a longer perspective, and with information direct from Russian sources about the Soviet layer-cake design, Carson Mark reached the same conclusion in 1994:

> The idea of the Alarm Clock, which was similar to Sakharov's layer cake, came up here at the end of the summer of 1946. It came on the scene—Teller's excited and enthusiastic proposal—as a way to build a hydrogen bomb. The hydrogen bomb, however, carried with it the irrelevant trapping that it wasn't a hydrogen bomb unless it gave a megaton or was multi-megaton. It wasn't a hydrogen bomb unless it had the potential to increase the yield indefinitely. The Russians seem to have approached Joe 4 by asking, What can you do with this new idea in a Trinity-size system? They did that and they got half a megaton. What got asked here was, how big would it have to be to give a megaton, or ten megatons. The answer that came back was that it would have to be too big. You needed to start with a fission bomb of possibly a megaton size all by itself—and at that time we had a perfectly clear way of making twenty kilotons, period. Now, of course, it didn't *have* to be bigger than the Trinity bomb, but it did if it had to carry these trappings. Semantics. So we dropped it and never put it to a test in the form which, had we happened to think of it, could have been interesting in

the sense that Joe 4 was. You say, Why didn't you realize that you were screwing things up by taking that approach? I think we realize now—or at least it's my conviction—that we did screw up by taking that approach.

If, starting in late 1946, the US had pursued building a simple Alarm Clock with a single layer of fusion material rather than the unwieldy multi-layered object Teller proposed, then it almost certainly could have tested a half-megaton version by 1949, when the other weapons in its arsenal were still under one hundred kilotons, and a megaton version within another year or two. When the Soviet Union tested Joe 1, in August 1949, US political, scientific and military leaders would have enjoyed the security of knowing that the US still controlled a unique capability. Ulam and Teller might still have invented a two-stage system by 1951. An ugly division in the nation's political and scientific communities might have been avoided.

The nation went another way, more expensive, divisive and dangerous. It did so primarily because Edward Teller, an ambitious and ruthless but also a charismatic man, defined the terms of US hydrogen-bomb research, and Teller's *idée fixe* blinded him to the promise of incremental achievement that Andrei Sakharov immediately recognized.

Teller had believed since childhood that only the technological application of science could save the human world from doom. In little Hungary that doom seemed Russian—"a totalitarian regime for seven hundred years," Teller characterized Russian governing once. In the United States, despite the nation's awesome military and economic power, that doom seemed to Teller Russian still. Stanislaw Ulam recalled speculating with John von Neumann about the reasons why Hungary produced so many brilliant scientists in the early years of the twentieth century. "Johnny used to say that it was a coincidence of some cultural factors which he could not make precise: an external pressure on the whole society of this part of Central Europe, a feeling of extreme insecurity in the individuals, and the necessity to produce the unusual or else face extinction." They came to America and devoted themselves gratefully to inventing weapons, thinking to make their adopted country more secure; instead, as in old tragedy, they extended the conditions of their critically unstable Central European past across the earth: the pressure became universal, the insecurity general, the dark unusual spilled forth at every hand, the human world faced extinction. They did not, of course, work alone.

Hans Bethe was the principal scientific adviser to the US government during negotiations on the 1963 Nuclear Test Ban Treaty, which moved most nuclear testing underground. He received the 1967 Nobel Prize in Physics for his work on energy production in stars. He was still publishing original scientific papers in 1995.

Enrico Fermi died prematurely of stomach cancer in the autumn of 1954. John von Neumann died prematurely of brain cancer in 1957.

Klaus Fuchs was released from Wakefield Prison in 1959 after serving nine years of his fourteen-year sentence. England had revoked his citizenship, a punishment more painful to him than prison; he flew immediately to East Germany, where he became deputy director and then director of the Institute for Nuclear Research in Rossdorf, near Dresden. He died in 1988.

Harry Gold was paroled in 1966 after serving sixteen years of his thirty-year sentence. He found work in medical research at a Philadelphia hospital, as he had hoped to do. He patented a device for the office testing of blood, but his time ran out before he could turn his patent to account: he died during open-heart surgery in 1972.

David and Ruth Greenglass changed their family name after David was paroled but continue to live in the New York area. David has prospered as an inventor. When Ronald Radosh interviewed him in 1979, he declared himself to be a Debsian socialist.

Morris and Lona Cohen continued espionage work for the KGB in Britain as Peter and Helen Kroger, antiquarian booksellers, until 1961, when they were arrested, convicted and sentenced to a twenty-year term. The KGB negotiated the exchange of their Soviet control but abandoned them to serve out their time. After their release from prison they emigrated to the Soviet Union and lived to see it collapse.

Kim Philby defected to the Soviet Union in 1963. The KGB suspected him of being a double agent and limited his work to lectures at its espionage school. He died in Moscow in 1988.

Curtis LeMay teamed with George Wallace to run unsuccessfully in 1968 as an independent candidate for the Vice-Presidency of the United States. He died in 1990.

On the last morning of his life, February 7, 1960, a Sunday, Igor Kurchatov drove to a spa near Moscow to visit Maria and Yuli Khariton, who were guests there. Playfully the Beard tuned the Kharitons' radio to a waltz and danced with his old friend Maria Nikolaevna. He told her how he had recently wangled a ticket to a wonderful performance of Mozart's *Requiem* and asked her to help him choose floor tiles for a new lab. All the time, write the Zukermans, the clock was ticking a countdown as it had so many times in his work:

Kurchatov puts on his coat and takes Khariton by the arm. "Let's go for a little walk, Yuli Borisovich, and talk some shop . . ." And the clock is still silently ticking away the remaining minutes.

They head for the park. It's below freezing out, sunny. The bare trees are powdered with snow. Igor Vasilievich picks out a park bench and brushes off the snow to make a place for the two of them.

"Let's sit for a while right here."

Yuli Borisovich starts telling Kurchatov about his latest experimental results. Kurchatov, who is always responsive in conversation, isn't answering. Yuli Borisovich is gripped with sudden alarm. He turns quickly and sees that Igor Vasilevich's eyes are glassing over. He shouts at the top of his voice, "There's something wrong with Kurchatov!" Secretaries and doctors come running, but it's too late. A small blood clot had occluded the coronary artery. The countdown clock read zero. The heart had stopped, and the mind had ceased all functioning.

Kurchatov was fifty-seven. Russians say that the strain of working for Lavrenti Beria shortened his life.

Yuli Khariton served as scientific director of Sarov until 1992, his eighty-eighth year, traveling back and forth to Moscow in his private railroad car. His colleagues call him "the Biological Phenomenon."

Andrei Sakharov stopped by Victor Adamsky's office at Sarov one day in 1961 to show him a story. It was Leo Szilard's short fiction "My Trial as a War Criminal," one chapter of his book *The Voice of the Dolphins,* published that year in the US. "I'm not strong in English," Adamsky says, "but I tried to read it through. A number of us discussed it. It was about a war between the USSR and the USA, a very devastating one, which brought victory to the USSR. Szilard and a number of other physicists are put under arrest and then face the court as war criminals for having created weapons of mass destruction. Neither they nor their lawyers could make up a cogent proof of their innocence. We were amazed by this paradox. You can't get away from the fact that we were developing weapons of mass destruction. We thought it was necessary. Such was our inner conviction. But still the moral aspect of it would not let Andrei Dmitrievich and some of us live in peace." So the visionary Hungarian physicist Leo Szilard, who first conceived of a nuclear chain reaction crossing a London street on a gray Depression morning in 1933, delivered a note in a bottle to a secret Soviet laboratory that contributed to Andrei Sakharov's courageous work of protest that helped bring the US-Soviet nuclear arms race to an end.

Was that arms race necessary? By one estimate that properly counts delivery systems as well as weapons, it cost the United States $4 trillion—roughly the US national debt in 1994. Soviet costs were comparable and were decisive in the decline of the Soviet economy that triggered the USSR's collapse. Cold warriors have argued from that fact that spending the Soviet Union into bankruptcy itself justifies the arms race. Their argument overlooks the inconvenient reality that the expense of the arms race contributed to US decline as well, decline evident in an oppressive national debt, in decaying infrastructure and social and educational neglect. The potlatch theory of the arms race also overlooks the unconscionable risk both superpowers took of omnicidal war. "One should always remember," the British scientific adviser Solly Zuckerman wrote in 1988, "that thirty years ago Eisenhower, the one

President who could challenge the Joint Chiefs on their own ground, saw the advantage of a comprehensive test ban without any provisions for verification."

What nuclear strategists quaintly called "existential deterrence"—deterrence at the level of personal dread—set in almost from the beginning. Stalin was deterred from August 6, 1945, onward, which is why he moved so expeditiously to acquire nuclear arms of his own. Harry Truman was deterred by a troubled conscience as well as by a sense that to use nuclear weapons again, making their use credible, would be to reap the whirlwind someday. He said as much in his last State of the Union address shortly before leaving office in 1953:

> For now we have entered the atomic age, and war has undergone a techno-logical change which makes it a very different thing from what it used to be. War today between the Soviet empire and the free nations might dig the grave not only of our Stalinist opponents, but of our own society, our world as well as theirs. . . .
> The war of the future would be one in which man could extinguish millions of lives at one blow, demolish the great cities of the world, wipe out the cultural achievements of the past—and destroy the very structure of a civilization that has been slowly and painfully built up through hun-dreds of generations. Such a war is not a possible policy for rational men.

Four months later, the Soviet leaders who took power after Stalin's death discovered peaceful coexistence; Georgi Malenkov, in a Moscow address, declared that "a third world war would mean the destruction of world civilization. . . ." Nikita Khrushchev never forgot his initial encounter with existential deterrence at his first full briefing on nuclear weapons in Septem-ber 1953:

> When I was appointed First Secretary of the Central Committee and learned all the facts about nuclear power, I couldn't sleep for several days. Then I became convinced that we could never possibly use these weapons, and when I realized that, I was able to sleep again.

Syngman Rhee, the South Korean leader, traveled to Washington to see Dwight Eisenhower one day in July 1954, two years after the Korean War had been fought to armistice at great cost in lives and matériel. Rhee told Eisenhower he wanted to "unify" his country. "He said," the notes of the conference report, "that his nation might propose to start some positive action at the front so that the United Nations forces would not have to remain there for a long time." He taunted Eisenhower: "People may say that England, France and Italy are presently free. But that is not so. They are afraid. Now [the Communists] have won in Indochina, Vietnam is parti-tioned. Pretty soon Thailand will be gone and South America will come next.

. . . How can you say that . . . we must sit still and let the Communists conquer and conquer and conquer? If we still believe that people amount to anything, we must never be afraid. If we are afraid, democracy will be conquered. Your efforts to save the world at peace will suddenly end." John Foster Dulles intervened to mollify the Korean leader. By then Eisenhower had heard enough and answered Rhee with furious conviction:

> There is no disposition in America at any time to belittle the Republic of Korea. But when you say that we should deliberately plunge into war, let me tell you that if war comes, it will be horrible. Atomic war will destroy civilization. It will destroy our cities. There will be millions of people dead. War today is unthinkable with the weapons which we have at our command. If the Kremlin and Washington ever lock up in a war, the results are too horrible to contemplate. I can't even imagine them. But we must keep strong. . . . I assure you that we think about these things continuously and as seriously as you do. The kind of war that I am talking about, if carried out, would not save democracy. Civilization would be ruined, and those nations and persons that survived would have to have strong dictators over them just to feed the people who were left. That is why we are opposed to war.

(To which Rhee countered slyly: "Suppose we had a plan that would not risk world war but would provide the unification of Korea?")

Niels Bohr had anticipated this new limit to war when he went to see Franklin Roosevelt and Winston Churchill in 1944 to propose that they begin arms-control discussions with Stalin before the bomb could be finished and used. A few years later, he summarized his understanding in a single sentence in a conversation with a friend: "We are in a completely new situation that cannot be resolved by war." By 1954, Eisenhower had come independently to an identical conclusion, reported in the notes of a National Security Council meeting that year: "The President commented that, as so often, we had again gone around in a circle and come back to the same place. The problem of the Soviet Union was a new kind of problem, and the old rules simply didn't apply to our present situation."

If all sides were deterred, why did the arms race continue? Business as usual, certainly. Khrushchev's memory of his first encounter with nuclear energy goes on: "But all the same we must be prepared. Our understanding is not sufficient answer to the arrogance of the imperialists." The hawks on both sides bear much responsibility; their fears, fanatic and often paranoid, helped drive the accumulation of overkill. Even realists, believing they were erring on the side of caution, erred instead on the side of risk, arguing that the US had to arm not against what the Soviets *might* do but against what they *could* do—against the capability represented by Soviet weapons, that is, not against the fact of redundant targeting and the annihilating US capacity for certain retaliation. But weapons that would only make the rubble bounce do not count as capability.

The truth is, the US was deterred throughout the Cold War at the lowest possible level of potential retaliation and so, evidently, was the Soviet Union. Robert McNamara, speaking in 1987 of the official US war plan at the time of the Cuban missile crisis, called it "totally unreasonable *before* Cuba and . . . totally unreal *after* Cuba. . . . Does anyone believe that a President or a Secretary of Defense would be willing to permit thirty warheads to fall on the United States? No way!" Solly Zuckerman, after castigating "the Pentagon's 'defense intellectuals' " for "weaving for their paymasters a network of abstract reasons why the nuclear arms race had to continue," identified the political reality in Europe as of 1988:

> Helmut Schmidt, the former West German chancellor, is credited with the critical step that led to the deployment of American intermediate-range missiles in Europe. He has now written that were nuclear war ever to erupt, he would expect German forces to break off the battle were even two warheads to explode on German territory. That—not the hundreds of weapons detonated in the prolonged abstract nuclear battles of the academic theorists—is the more likely nuclear reality. . . .

McGeorge Bundy reported the same reality as early as 1969:

> In light of the certain prospect of retaliation, there has been literally no chance at all that any sane political authority, in either the United States or the Soviet Union, would consciously choose to start a nuclear war. This proposition is true for the past, the present and the foreseeable future. . . .
> In the real world of real political leaders . . . a decision that would bring even one hydrogen bomb on one city of one's own country would be recognized in advance as a catastrophic blunder; ten bombs on ten cities would be a disaster beyond history; and a hundred bombs on a hundred cities are unthinkable.

These various statements reflect what the historical record demonstrates: that the US deterrent was credible from 1945 onward and the Soviet deterrent from 1949 onward. How many cities would a political leader be ready to lose? US leaders were prepared to lose not one, whatever patriotic gore their advisers gushed. If Soviet leaders were prepared to lose one or ten, as self-righteous cold warriors liked to allege, the least deterrent the US marshaled after 1949 never allowed them such a monstrous choice.

If real political leaders understood from one end of the Cold War to the other that even one hydrogen bomb was sufficient deterrence, why did they allow the arms race to devour the wealth of the nation while it increased the risk of an accidental Armageddon? In 1982, political scientist Miroslav Nincic examined the economics of the arms race and discovered that it was hardly a race at all; US and Soviet levels of defense spending were only weakly coupled at best. Far more influential on the US side were such domestic political phenomena as competition among the military services, coalitions of scientific and industrial organizations promoting new technologies, the pressure of "defense" as a political issue and defense spending to prime the economic pump, particularly in election years. Similar patterns obtained along somewhat different lines for the Soviet command economy. "The arms race," Nincic summarized, "is imbedded in circumstances proper to the domestic political and economic systems of the superpowers *in addition* to dynamics inherent in the interaction between the two nations." Having worked the numbers, Nincic concluded that all the high claptrap of arms strategy was essentially decorative: "Strategic doctrines are designed, in large part, to justify the weaponry that the arms race has imposed on both the United States and the Soviet Union." Which is independent confirmation of John Manley's dictum that in Washington (and in Moscow), "You don't do staff work and then make a decision. You make a decision and then do the staff work." Since defense spending reduces civilian investment and consumption (each dollar spent on defense between 1938 and 1969, for example, implied a loss of forty-two cents from consumption—that is, went into building weapons rather than cars or public schools), any more than the minimum is evidently parasitic.

Politicians found it possible to rationalize encouraging such parasitism for political gain because they believed they had reliable command and control of the nuclear arsenal. The Cuban missile crisis demonstrated how much less reliable was that command and control than they believed, to such an extent that it scared both superpowers off escalation and direct confrontation for the duration of the conflict. Potentially catastrophic false warnings and accidents recurred down through the years.

Efforts at arms limitation foundered not only on Soviet refusal to admit inspection, as cold warriors claim. It also foundered on the resistance of US hawks, Edward Teller prominent among them, to any reduction in the pace of technological advance even when that advance—attaching multiple warheads to missiles, for example—actually gave away advantage. Minimal deterrence was political suicide so long as the Soviet Union existed, as Jimmy Carter learned when he proposed it shortly after his election in 1976.

A charitable view of what the US and Soviet leaderships thought they were doing when they continued to add to bloated arsenals must take into account the necessity for extreme prudence where such vast destruction is threatened. After criticizing his own predecessors for "aiming to spread socialism throughout the world," Mikhail Gorbachev bluntly blamed the United States for the arms race when he spoke at Westminster College in Fulton, Missouri, in 1992, forty-six years after Winston Churchill defined the Iron Curtain there:

But the West, and the United States in particular, also committed an error. Its conclusion about the probability of open Soviet military aggression was unrealistic and dangerous. This could never have happened, not only because Stalin, as in 1939–1941, was afraid of war, did not want war, and never would have engaged in a major war. But primarily because the country was exhausted and destroyed; it had lost tens of millions of people, and the public hated war. Having won a victory, the army and the soldiers were dying to get home and get back to a normal life.

By including the "nuclear component" in world politics, and on this basis unleashing a monstrous arms race—and here the initiator was the United States, the West—"defense sufficiency was exceeded," as the lawyers say. This was a fateful error.

Historical evidence supports Gorbachev's accusations, however defensive Americans may feel about them. Gorbachev ignores at least two important facts, however: that the Soviet Union was demonstrating open and covert aggression in Eastern Europe and that its opacity as a closed society made a realistic assessment of its intentions difficult.

From a longer perspective, the arms race was a process of communication and of learning, paced as all progress is paced by the limitations of human intellect. The discovery in 1938 of how to release nuclear energy introduced a singularity into the human world—a deep new reality, a region where the old rules of war no longer applied. The region of nuclear singularity enlarged across the decades, sweeping war away at its shock front until today it excludes all but civil wars and limited conventional wars.

Science once claimed power over nature as its goal, as if science and power were one thing and nature another. Niels Bohr observed to the contrary that the more modest but relentless goal of science is "the gradual removal of prejudices." One of those prejudices, which has accounted for immense human suffering, was the belief that in an anarchic world there are no limits to national sovereignty except those that conflict might determine. Knowledge of how to release nuclear energy, knowledge that only science was structured to perceive, came to define a natural limit. The authority of the institution called science, that is, has taken precedence, at least in this extreme arena, over the authority of the nation-state. Science has fielded no armies in order to do so and is indeed pacifist; rather, it has gradually removed the prejudice that there is a limited amount of energy available in the world to concentrate into explosives, that it is possible to accumulate more of such energy than one's enemies and thereby militarily to prevail. Science has revealed at least world war to be historical, not universal, a manifestation of destructive technologies of limited scale. In the long history of human slaughter, that is no small achievement.

Even as they reduce international violence by limiting national sovereignty, nuclear weapons also simultaneously and paradoxically threaten and protect such sovereignty. The technology has proliferated by a process of political homeostasis: the United States racing, as it believed, to beat Nazi Germany; the Soviet Union racing to catch up with the United States; Britain and France, unwilling to credit US assurances that it would sacrifice itself to save them, developing independent minimum deterrents against the Soviet Union; China balancing the US and the USSR; India balancing China; Pakistan balancing India. Iraq and Iran have pursued a nuclear capability against each other and the world, but especially to balance Israel. At the margins of this homeostatic activity, the process becomes degenerate: North Korea builds a weapon or two to send a message to the West that its neglect could be dangerous; the old South African government of apartheid builds half a dozen uranium guns against the world and dismantles them when it democratizes itself. These complexities imply that the chance of a terrorist weapon is vanishingly small. Nuclear weapons will never be easy to make, and they are uniquely destabilizing, invoking the deterrent forces of all the major nuclear powers against even local and regional threats. What nation— whether or not it possessed nuclear weapons of its own—would allow a subnational group to build them on its territory? Why build a nuclear weapon to blow up the World Trade Center or a federal building in Oklahoma City when you can blow it up with nitrate fertilizer and fuel oil? Why bomb Tokyo when you can poison it with nerve gas?

Robert Oppenheimer chaired a panel on disarmament for Dean Acheson in 1952. Members included Vannevar Bush and Allen Dulles, who was soon to become the director of the CIA. McGeorge Bundy served as secretary and rapporteur. The panel came to a conclusion about the new knowledge of how to release nuclear energy that is as valid today as it was then, and as final:

> Fundamentally, and in the long run, the problem which is posed by the release of atomic energy is a problem of the ability of the human race to govern itself without war. There is no permanent method of excising atomic energy from our affairs, now that men know how it can be released. Even if some reasonably complete international control of atomic energy should be established, knowledge would persist, and it is hard to see how there could be any major war in which one side or another would not eventually make and use atomic bombs. In this respect the problem of armaments was permanently and drastically altered in 1945.

The world will not soon be free of nuclear weapons, because they serve so many purposes. But as instruments of destruction, they have long been obsolete.

Glade
February 1990–January 1995

Acknowledgments

My wife, Ginger Rhodes, contributed her time, skill and good sense to this book: planned our extensive travel for research, recorded and supervised Helen Haversat's transcription of interviews, tracked down books and documents, reviewed every chapter along the way.

Charles Till, at Argonne, besides investigating the 305 reactor, connected me with Yuri Orechwa, who connected me with Elena Bonner in Moscow, who introduced me to the Russian physics community and endorsed Alexander Goldin, who proved to be a superb research associate. Victor Adamsky, Lev Altshuler, Susan Eisenhower, Scott Horton, Yuli Khariton, Victor Mikhailov, Evgenii Negin, Tim and Jen Sergay, Tatiana Yakelevich and particularly Yuri Smirnov all helped with interviews, information or advice. Alan Schriesheim at Argonne always contributes.

Chuck Hansen shared his years of research on weapons history and technology. Ronald Radosh shared his unique files on the Greenglasses and Harry Gold. Robert Lamphere was kind enough to read the espionage episodes in the manuscript for accuracy and to sit for an interview.

Jay Wechsler at Los Alamos supplied a unique perspective. Roger Meade guided me through the LANL Archives. Sig Heckler, Kay Manley and Paula Dransfield offered moral support. Edwardo de Los Alamos sent along a helpful document. George Cowan, the late Charles Critchfield, Carson Mark, Nicholas Metropolis and Raemer Schreiber gave valuable interviews. So, at various times and places, did Philip Abelson, Harold Agnew, the late Luis Alvarez, Hans Bethe, Robert Cornog, Rudolf Peierls, Marshall Rosenbluth, Glenn Seaborg, the late Emilio Segrè, Robert Serber, Rubby Sherr, the late Cyril Smith, Bill Spindel, Ted Taylor, Edward Teller, Al Weinberg, Victor Weisskopf, John A. Wheeler and Herbert York. Françoise Ulam and William Arnold shared photographs.

Historians contributed documents and insights: Tom Cochran, Bruce Cumings, Stanley Goldberg, Dick Hallion, Gregg Herken, David Holloway, Amy Knight, Arnold Kramish, Priscilla McMillan, Mike Neufeld, Stan Norris, Tom Powers, Jonathan Weisgall, Herman Wolk and Steven Zaloga. My New Haven research associate, Steve Rice, did vital work. Stephen Kim also helped. Personal thanks to Augustus Ballard, Harry Bayne, Louis Brown, Gil Elliot, Dan Ellsberg, Rachel Fermi, Eric Markusen, Esther Samra, Frank Shelton, Jann Wenner and Don Wille.

Libraries and museums made available their collections. Thanks to The Bancroft Library; Fred Bauman in the Manuscript Room at the Library of Congress; the St. Petersburg public library; Penny Abel at Yale's Sterling Memorial Library; Ben Zobrist and Liz Safly at the Harry S. Truman Library; Dan Holt and David Haight at the Dwight D. Eisenhower Library. Thanks also to Jonathan Brent at Yale University Press, Alexander Shlyakhter at Harvard and Rick Ray at the National Atomic Museum.

The Alfred P. Sloan Foundation provided significant support. I acknowledge that grant elsewhere. Here I'd like to thank my colleagues on the Sloan Technology Book advisory committee: John Armstrong, Mike Bessie, Vic McElheny, the late Elting Morison, Ralph Gomory, Hirsh Cohen, Sam Gibbon, Jr., Frank Mayadas and especially Art Singer.

Kazunari Fujita, Ryukichi Imai, Sakae Shimizu and Fumihiko Yoshida contributed in Japan.

Michael Korda, best of editors, saw this long work through, and the one before it and the several between. Special thanks at Simon & Schuster to Rebecca Head, Eve and Frank Metz and Victoria Meyer.

Last is also a place of honor: Mort Janklow and Anne Sibbald kept the ship afloat through five expensive years. Bless them.

Notes

SOURCES

The most significant Soviet sources to which I refer are the espionage documents given at Visgin (1992). These were supplied to the Russian Institute for the History of Science and Technology by the KGB, in what was evidently an attempt by that discredited institution to demonstrate to the new Russian government the historic importance of its work. The Institute published the documents in its journal *Voprosy istorii estestvoznaniia i tekhniki (Problems in the History of Science and Technology)*. Before that issue had been completely distributed, the Russian government required that it be withdrawn from circulation on the grounds that two of the documents violated the Nuclear Nonproliferation Treaty. Lengthy excerpts from all but those two documents appeared as an appendix in Sudoplatov and Sudoplatov (1994), and David Holloway (1994) notes (p. 372) that all the documents have been quoted in the Russian press. My source is Visgin; I have quoted from the two prohibited documents—Numbers 12 and 13, both descriptions of the Fat Man bomb —only that information which has been declassified in the United States (excluding, for example, the exact dimensions of the components of the Urchin initiator). Internal evidence, particularly discussions of bomb, reactor and isotope-separation physics and engineering, substantiates the authenticity of the Visgin documents.

They make it possible to match up espionage information with the confessions of Klaus Fuchs, Harry Gold, David and Ruth Greenglass and Alan Nunn May. I have done so at length: first, because doing so reveals an important mechanism of technology transfer in this ominous field; second, because Gold, the Greenglasses and the FBI at least have been accused of fabricating testimony and evidence implicating Julius and Ethel Rosenberg in espionage. The matches are extensive and convincing. So is the FBI record, which is far too diverse, detailed and circumstantially corroborated to have been fabricated. I interviewed Robert Lamphere; his revelation of decoded wartime cable intercepts fills important gaps in the chronology of the FBI investigation.

Other Soviet and Russian sources are largely anecdotal; I have relied on eyewitnesses wherever possible. I interviewed Russian atomic scientists and government officials and personally inspected the F-1 reactor at the Kurchatov Institute on two visits to Russia in 1992. Another corroboration of the information I collected has been David Holloway's history *Stalin and the Bomb,* which I reviewed in bound galleys after I had substantially finished drafting my narrative of the Soviet story. I made minor changes based on Holloway's evidence and noticed areas where my sources pointed to conclusions different from his, but our narratives basically agree.

Many Soviet and Russian documents came to me as unpaginated translations via E-mail; hence the absence of page numbers for the corresponding references in the endnotes. Russian readers should have no trouble finding the citations in the original periodicals.

The invaluable daily diary maintained by Curtis LeMay's various aides was only recently declassified; it corroborates LeMay's efforts to gain control of US nuclear weapons and his part in the transfer of nine atomic bombs to the Far East during the Korean War. The other major source for that transfer is Gordon Dean's diary—Anders (1987). Sources for SAC's Cuban missile crisis excesses are sketchier—necessarily, since those excesses were illegal and nearly fatal—but reliable wherever they can be checked. The story of SAC efforts to force a nuclear confrontation from 1950 onward is probably even more frightening than I have been able to document. Nor have likely Soviet (and, in the case of the missile crisis, Cuban) provocations yet seen the light of day.

George Racey Jordan's account of wholesale espionage shipments through Great Falls under cover of Lend-Lease was widely discounted when it first appeared. I discuss its several corroborations in my text; based on those corroborations, it appears to me to be largely credible. Soviet espionage was indeed wholesale, and obviously successful; Jordan's black suitcases explain how the ten thousand pages of secret documents that shocked Yakov Terletsky might have been transported.

Independent scholar Chuck Hansen has assembled the best extant collection in private hands of declassified documents relating to weapons design, development and testing. Chuck's collection, which he generously made available, was invaluable in reconstructing significant events and in filling out the testimony of the scientists and engineers whom I interviewed. To the best of my knowledge, none of the information on weapons design in *Dark Sun* is restricted; all such information came to me from properly cleared persons and/or declassified sources. Where there was doubt—in the case of the KGB sources, for example—I opted for exclusion.

ABBREVIATIONS

DDEL	Dwight D. Eisenhower Library, Abilene, Kansas
FBI	US Federal Bureau of Investigation
FOIA	US Freedom of Information Act
HHL	Herbert Hoover Library, West Branch, Iowa
HSTL	Harry S. Truman Library, Independence, Missouri
LANL, LASL	Los Alamos National Laboratory, Los Alamos, New Mexico, formerly Los Alamos Scientific Laboratory
LC	US Library of Congress, Washington, DC
MED	Manhattan Engineer District (informally, the Manhattan Project) of the US Army Corps of Engineers
RG	Record Group (at the National Archives, Washington, DC)

Other abbreviations are explained in the text or in corresponding entries in the Bibliography.

PROLOGUE

PAGE

17 "We were . . . work": Alvarez (1987), p. 4.

18 "I looked . . . population": ibid., p. 7.

18 "This is . . . dreams": Luis to Walter Alvarez, 6.viii.45. Luis Alvarez, personal communication. Most of this text is reproduced in Alvarez (1987), p. 8.

19 "really harrowing . . . in": Robert Serber, 1994 Pegram Lectures, Brookhaven National Laboratory, unpub. MS. (Courtesy Robert Serber.)

19 "We came . . . nothing": interview with Bruce C. Hopper, 8th Air Force historian, 7.ix.43, p. 3, LeMay Papers, LC.

19 "We tottered . . . future": speech by Maj. Gen. Curtis E. LeMay before Ohio Society of New York, 19.xi.45, pp. 1–2, ibid.

20 "Hit it . . . back": file biography, "Colonel Curtis E. LeMay as bombardment wing and division commanding officer," c. 1942, ibid.

20 "Iron Ass"; "absolutely the . . . Army": B. W. Crandell, " 'Iron Ass' was the name." From June 1944 history, 3rd Bomber Division, ibid.

20 "We couldn't . . . defense": LeMay (1965), p. 347.

20 Radar bombsights, air power: Hansell (1986), p. 228.

21 "was *not* . . . concentrated": Foreword, Tactical mission report, Mission No. 40, 10 March 1945, LeMay Papers, LC.

21 "twenty-two industrial . . . industries": Tactical mission report, ibid.

21 "The physical . . . property": quoted in LeMay (1965), p. 353.

21 "485 B-29s . . . personnel": Curtis LeMay, lecture headed "General Bull, Gentlemen:", p. 16, filed between 8.iii.52 and 15.iii.52 documents, Box 200, LeMay Papers, LC.

21 "how much . . . soldier": quoted in Hurley and Ehrhart (1979), pp. 200–201.

22 "that if . . . crews": LeMay (1965), p. 390.

22 "I think . . . all": quoted in Hurley and Ehrhart (1979), p. 197.

22 "Like many . . . tired": LeMay (1965), p. 390.

22 "General LeMay . . . fourth": Theodore E. Beckemeier, "Resume of events while aide-de-camp to Major General Curtis E. LeMay, June 25, 1943 to Sept. 25, 1945," entry for 3.ix.45, LeMay Papers, LC.

22 "Offhand . . . at large": LeMay (1965), p. 393.

22 "The trip . . . made": James Doolittle to Carl Spaatz, 31.viii.45, Doolittle Papers, LC.

22 "there are . . . feasible": Doolittle from Spaatz, message NR 4021, 5.ix.45, ibid.

22 "We got . . . runways": LeMay (1965), p. 393.

23 B-29 flight schedule and events: "Log of the flight Japan-Washington," *New York Times,* 20.ix.45; "Brass band greets crews on arrival at capital," *Chicago Tribune,* 20.ix.45.

23 "That night . . . sweat": LeMay (1965), p. 394.

23 "I went . . . back": ibid., p. 395.

23 "the only . . . airliner": "The B-29
Flight," *Chicago Tribune,* 21.ix.45,
p. 14.

23 *A Strategic Chart of Certain
Russian and Manchurian Urban
Areas,* 30.viii.45: RG 77, MED,
Stockpile Storage and Military
Characteristics file, National
Archives; for estimated bomb
requirements cf. "Atomic bomb
production," Lauris Norstad to L. R.
Groves, 15.ix.45, "Tab C," p. 1.
Chuck Hansen collection.

24 18.x.45 Fat Man plans: Visgin
(1992), p. 127ff.

**CHAPTER ONE: 'A SMELL OF
NUCLEAR POWDER'**

27 "got a frenzied going-over":
Golovin (1968), p. 31. David
Holloway questions the existence
of this letter (Holloway, 1981, p.
191, fn. 17; Holloway, 1994, p. 384,
n. 5), arguing in 1981 that a letter
from Joliot-Curie to Ioffe "cannot
have been so because of the
timing," noting that "other Soviet
sources say that Soviet scientists
learned of the discovery when the
foreign journals arrived in
February," and asserting in 1994 on
the basis of a 1992 interview with
Igor Golovin that "the story of the
letter from Joliot-Curie is a myth."
When Yuri Smirnov interviewed
Golovin on my behalf in 1993,
however, Golovin specifically
remembered speaking with
participants in the seminar that
resulted from the letter, including
G. N. Flerov, V. A. Davidenko, M. I.
Pevsner, I. I. Gurevich and others,
although he no longer remembers
who told him specifically about the
letter. (Yuri Smirnov, personal

communication, 13.ix.93.)
Holloway quotes Flerov as writing,
"we first learned of the new
phenomenon from the work of
Joliot-Curie," which tends to
corroborate Golovin. A letter could
not have been written before early
January 1939, since Lise Meitner
and Otto Robert Frisch did not
work out the physics of fission
until early in the new year. I
conclude, a) that Joliot-Curie did
write Ioffe and b) that the letter
reached the Soviet physicists
before the February journals.

27 "The first . . . air": Flerov (1989),
p. 54.

27 Nuclear fission confirmed: cf.
Frisch (1939), Meitner and Frisch
(1939).

27 "new sources . . . envisaged":
quoted in Holloway (1994),
p. 29.

28 Radium extraction: ibid., pp. 30–
31.

28 "hunger and . . . 1921": quoted in
Zukerman and Azarkh, unpub. MS,
p. 130.

28 "The Institute . . . him": Frish
(1992).

28 "the kindergarten": Golovin
(1968), p. 17.

28 Lenin on electrical power: cf.
Badash (1985), p. 39, n. 11.

28 "Stalin's realism . . . us": Snow
(1966), p. 258.

29 "it will . . . pleases": quoted in
Kramish (1959), p. 6.

29 "I went . . . Institute": quoted ibid.,
p. 7.

29 "I was . . . General": quoted in
Zukerman and Azarkh, unpub. MS,
p. 123f.

30 Kurchatov's cyclotron: Cochran and
Norris (1993), p. 5.

30 24 papers: Golovin (1968), p. 27.

PAGE

30 "the liveliest . . . joke": quoted in Zukerman and Azarkh, unpub. MS, p. 125.

30 "lanky stripling . . . cheeks": Golovin (1968), p. 21.

30 "Such a . . . him": Eddie Sinelnikov, writing home in 1930, in Street (1947), p. 29.

30 "a young . . . forehead": Frish (1992).

30 "worked harder . . . head": Golovin (1968), p. 21.

31 "This is . . . it": quoted ibid., p. 11.

31 "It was . . . trip": quoted ibid., p. 20.

32 "remarkably clear . . . neutron": Peierls (1985), p. 110.

32 "compact, ascetically . . . sprightly": quoted in Zukerman and Azarkh, unpub. MS, p. 137.

32 "Stalin killed . . . perished": Kravchenko (1946), p. 470.

32 "From 1 . . . camp": quoted in Conquest (1993).

32 "the full . . . thousands": Medvedev (1978), p. 34.

33 "for the . . . people": quoted in Badash (1985), p. 98.

33 "by his . . . endeavors": Bohr report dated 28.vi.44, in Oppenheimer Papers, LC.

33 "After a . . . released": Medvedev (1978), p. 37.

33 "Lev Landau . . . Communist": Edward Teller, personal communication, vi.93.

33 "Although the . . . practice' ": Alexandrov (1988).

33 "In those . . . night": Alliluyeva (1967), p. 140.

33 "Stalin personally . . . annihilated": Conquest (1991), p. 206.

34 Truckloads of bodies: cf. Remnick (1993), p. 135.

34 "The Stalin Epigram": Forché (1993), p. 122.

PAGE

35 Spontaneous fission: Petrzhak and Flerov (1940).

35 "Yuli Borisovich . . . late": Zeldovich (1992, 1993), II, p. 637.

35 "not a . . . degree": Sakharov (1990), p. 132f.

35 "We immediately . . . possible": quoted in Golovin and Smirnov (1989).

35 1939 Fiztekh seminar: Khariton and Smirnov (1993). One atomic bomb: cf. Igor Tamm's comment in Golovin and Smirnov (1989).

35 "These possibilities . . . physicist": Serber (1992), p. xxvii.

36 "In order . . . hydrogen": Zeldovich (1993), II, p. 7.

37 "another possibility . . . 235": ibid., p. 14.

37 "completely safe," "significantly increase": ibid., p. 16.

37 "hasty conclusions . . . reality": ibid., p. 19.

37 "It would . . . process": ibid., p. 16.

38 2 to 3 kilograms: according to Igor Kurchatov, this amount constituted the entire country's supply as late as April 1943; cf. Visgin (1992), Doc. # 4.

38 Borst-Harkins letter: Borst and Harkins (1940).

38 "Thus . . . heavy water": Visgin (1992), Doc. # 4.

38 "Contrary to . . . years": Khariton and Smirnov (1993).

39 "In the . . . material": Yuli Khariton, personal communication, 25.x.93.

39 Khariton studied centrifuge: Khariton and Smirnov (1993), p. 23.

39 Kvasnikov initiative: Leskov (1993), p. 38. Yatzkov (1992) also discusses this early initiative.

39 28,000 NKVD agents purged: Volkogonov (1988, 1991), p. 332.

39 "the various . . . bearable": Gold (1951), p. 24.

PAGE

39 "It seemed . . . succeeded": quoted in Golovin and Smirnov (1989).

40 *New York Times* article and George Vernadsky: Holloway (1994), p. 60.

40 "Uranium has . . . waters": quoted ibid., p. 62.

40 Kurchatov at 5th conference: Zaloga (1993), p. 8.

41 "would be . . . war": Wilson (1975), p. 55.

41 "The situation . . . room]": Golovin (1989), p. 198.

41 "The conflict . . . year": quoted in Volkogonov (1988, 1991), p. 393.

41 "if the . . . century": quoted in Holloway (1981), p. 168.

41 "A quarter . . . disappointed": Golovin (1989), p. 198.

42 "The tension . . . result": Altshuler et al. (1991), p. 26 (translation edited).

42 Stalin's 22.vi.41 meetings: Knight (1993), p. 111, citing Stalin's diary of visitors.

42 "simply lost . . . up!' ": Volkogonov (1988, 1991), p. 409f.

43 "We got . . . face": quoted ibid., p. 411.

43 "Soviet losses . . . fighting": ibid., pp. 418–419.

43 Stalin's July 3 speech: Werth (1971), p. 2.

43 "Stalin spoke . . . stood": quoted in Werth (1964) p. 166f.

43 "It was . . . before": ibid., p. 162.

43 "In every . . . war": Kravchenko (1946), pp. 354–356.

44 "all the . . . State": quoted in Werth (1964), p. 165.

44 "an utterly . . . own": Volkogonov (1988, 1991), p. 250.

44 "By early . . . apparatus": Knight (1993), p. 90.

44 "A magnificent . . . courtier": Alliluyeva (1967), p. 8.

PAGE

44 "somewhat plump . . . solicitude": Djilas (1962), p. 108.

44 "Beria was . . . sleep": quoted in Knight (1993), p. 195.

44 "Beria was . . . whip": Kravchenko (1946), p. 404.

45 "agreeing to . . . it": quoted in Volkogonov (1988, 1991), p. 412.

45 "How long . . . solutions": Kaftanov (1985).

46 "Do you . . . calamity?": quoted in Werth (1964), p. 272.

46 "The street . . . rhythm": Gouzenko (1948), p. 70f.

46 "as office . . . himself!' ": Sakharov (1990), p. 43.

46 "The situation . . . gates": Gouzenko (1948), p. 71.

46 "the most . . . II": ibid.

46 2 million evacuated: Werth (1964), p. 241.

46 "West of . . . folklore": ibid., p. 264.

47 "So crammed . . . ground": Rosenberg (1988), p. 86f.

47 Flerov report: Khariton and Smirnov (1993), p. 24; Golovin (1968), p. 39.

47 "Flerov's report . . . two": quoted in Golovin and Smirnov (1989).

47 "He suggested . . . material' ": Khariton and Smirnov (1993), p. 43.

48 "which . . . victory": Golovin (1968), p. 38.

48 Kurchatov saved report: Khariton and Smirnov (1993), p. 24.

48 "Kurchatov knew . . . Murmansk": Golovin (1968), p. 39.

48 "Scientific work . . . required": quoted in Kramish (1959), p. 49.

CHAPTER TWO: DIFFUSION

49 "I think . . . it": Smith and Weiner (1980), p. 143.

PAGE

49 "solution": Chambers (1952), p. 193.

49 "In the . . . crises": ibid., p. 191.

49 "The same . . . influence": ibid., p. 192f.

49 "his decision . . . birth": ibid., p. 193.

50 "The Communist . . . repudiates": ibid.

50 "We were . . . country": quoted in Williams (1987), p. 51. Ruth Kuczynski's real name was Ursula, but she preferred Ruth, and that was what her parents called her: Pincher (1984), p. 9.

50 "We didn't . . . Stalin": quoted in Moss (1993), p. 11.

50 "In only . . . opportunity": Gold (1951), p. 27.

50 "That wonderful . . . so": ibid., p. 16.

50 "[numerous] . . . way": Taschereau and Kellock (1946), p. 44f.

51 "This object . . . standards": ibid., p. 71 (italics in original).

51 "Construction": for details of the school's operation, cf. Costello and Tsarev (1993), p. 275ff.

51 "In April . . . Soviets": quoted ibid., p. 277.

51 Cohen: Lamphere and Shachtman (1986), p. 276ff; Chikov (1991b), p. 36ff.

51 "the Socialist . . . cells": Straight (1983), p. 60.

52 "At one . . . inhibitions": the classical scholar Maurice Bowra, quoted in Andrew and Gordievsky (1990), p. 206.

52 "fascinating, charming . . . ruthless": quoted ibid., p. 217.

52 Burgess recruited Cairncross: according to Andrew and Gordievsky (1990), p. 217. Chapman Pincher (1987), p. 66, says James Klugmann recruited Cairncross.

52 "It's like . . . it": quoted in Cecil (1989), p. 77.

52 "Cairncross was . . . bore": quoted in Andrew and Gordievsky (1990), p. 219.

52 September 1941: according to Anatoli Yatzkov; Yatzkov (1992). Cairncross denied ever having passed atomic secrets. He claimed his position with Lord Hankey amounted to nothing more than that of a scheduler and that Donald Maclean had access to atomic documents. But "List" was not Maclean's code name, and Oleg Gordievsky, a former KGB *rezident* in London, reports having heard from the head of the KGB British desk that Cairncross provided "literally tons of documents." (Andrew and Gordievsky, 1990, pp. 262–263.) Nor was Maclean anywhere near the British uranium committee work at that time—after returning from the British Embassy in Paris following the fall of France he was stuck in the Foreign Office General Department until he left for the United States in spring 1944.

52 "a short . . . eyes": one of Gorsky's wartime agents, quoted in Andrew and Gordievsky (1990), p. 293.

53 "#6881/1025 . . . London": Visgin (1992), Doc. # 1.

53 MAUD report: Office of Scientific Research and Development records, S-1 Bush-Conant file, RG 227, National Archives.

53 a second transmission: Visgin (1992), Doc. # 2.

54 An American physicist: Lona Cohen confirmed the existence of this unidentified American-born physicist in a telephone interview with Walter Schneir, reported in

Dobbs (1992), p. A37. Yatzkov (1992) similarly discusses a "physicist" who contacted "our source" in September 1941. Other details about the Cohens and the physicist are from Chikov (1991a, 1991b). I use Chikov cautiously; much of his information is obviously inflated, if not actually invented (i.e., the physicist, if there were such, could not have been recruited for Los Alamos in 1941; the location had not yet been proposed).

54 Yatzkov New York *rezident:* Robert Lamphere thinks it probable: Robert Lamphere, personal communication, vi.94.

54 "from the . . . future": Yatzkov (1992).

54 "I . . . Fuchs": Peierls (1985), pp. 162–163.

54 "a very happy childhood": Williams (1987), p. 180. For details of Fuchs's background cf. also Moss (1987), Pilat (1952) and Pincher (1984).

54 "My father . . . convention": quoted in Williams (1987), p. 180.

55 "At this . . . candidate": quoted ibid., p. 181.

55 "In spite . . . escaped": quoted ibid., p. 182.

55 "I was . . . lapel": quoted ibid.

55 "I was . . . decision": quoted ibid.

55 "arrogance and naiveté": Peierls (1985), p. 223.

55 "You must . . . system": quoted ibid.

55 "I was . . . Germany": quoted in Williams (1987), p. 183.

55 "noble . . . admiration": Gold (1951), p. 78.

56 "shy and reserved"; "accusing the . . . man": quoted in Williams (1987), p. 23.

56 "did some . . . solids": Peierls (1985), p. 163.

56 "a very . . . eyes": quoted in Moss (1987), p. 20.

56 Propaganda leaflets: Fuchs FBI FOIA files, 65-58805-1412, p. 7.

56 Details of internment: Williams (1987), p. 34.

56 Fuchs billeted: Peierls (1985), p. 163.

56 "I felt . . . people": quoted in Williams (1987), p. 184.

56 "had complete . . . death": quoted ibid.

57 "If we . . . word": quoted in *New York Times,* 24.vi.41, p. 7.

57 Description of Fuchs: Fuchs FBI FOIA files, 65-58805-1412, pp. 49–50.

57 "Fuchs . . . Theoretic": quoted from the weekly Harwell newspaper *AERA News* in Moss (1987), p. 98.

57 "[Fuchs] was . . . slot": Peierls (1985), p. 163.

57 "When I . . . Party": quoted in Williams (1987), p. 184.

57 "On his . . . Union": Fuchs FBI FOIA files, 65-58805-1412, p. 8.

58 Fuchs's reports: cf. ibid.; Williams (1987), p. 45. Cf. also itemized list attached to Fuchs FBI FOIA files, 65-58805-1412.

58 KZ-4: Visqin (1992), Doc. # 3.

59 "If every . . . expedite"; "I think . . . essence": quoted in Rhodes (1986), p. 406.

59 The file to Pervukhin: Holloway (1983), p. 17.

59 "might have . . . did": Pervukhin (1978). Pervukhin places this discussion in September or October 1942. Other sources place it in April. The date of the Beria report, March 1942, makes April the most probable month.

PAGE

59 "Ukrainian partisans . . . uranium": Kaftanov (1985).

60 "In three . . . war": ibid.

60 Five telegrams: reported by Flerov in his letter to Stalin, Appendix 2, Cochran and Norris (1990), p. 27.

60 Flerov and Stalin Prize: Holloway (1981), p. 173.

60 "Dear Josef . . . mistake": Flerov, letter to Stalin in Cochran and Norris (1992), p. 90.

61 "I see . . . through": ibid.

61 "To choose . . . world": quoted in Conquest (1991), p. 107.

61 "asked them . . . work": Pais (1990), p. 13, quoting an Igor Golovin interview in *Moscow News,* 8.x.89.

61 Expense of bomb program: Holloway (1983), p. 18.

61 "I said . . . unarmed": Kaftanov (1985).

61 "Stalin said . . . it' ": ibid.

61 "The Stalingrad . . . evacuation": Golovin and Smirnov (1989).

62 "Civilians were . . . all' ": Werth (1964), p. 367.

62 Combat casualties: ibid., pp. 401–403.

62 "We now . . . him": quoted in Golovin and Smirnov (1989).

62 "Top secret . . . leadership": quoted in Chikov (1991a), p. 39.

63 Fuchs knew Ruth Kuczynski: after he was arrested in 1950, he denied knowing her until he learned, in prison, that she was safely in East Germany; then he identified her to MI5. MI5 chose not to tell the FBI. Pincher (1984), p. 145.

63 "controlled schizophrenia": quoted in Williams (1987), p. 184.

63 "It was . . . man": quoted in Moss (1993), p. 10.

63 "no hesitation . . . had": quoted in Williams (1987), p. 184.

PAGE

63 "He said . . . work": Golovin (1989), p. 200.

63 "I got . . . Kurchatov": Kaftanov (1985).

63 "After the . . . well": quoted in Zukerman and Azarkh, unpub. MS, p. 116f.

64 "Alikhanov . . . well-known": Kaftanov (1985).

64 "It was . . . needed": ibid. Kaftanov goes on to say he chose Kurchatov after a meeting with him in May 1943. The record does not support so early a date of appointment.

64 "We invited . . . 'Yes' ": quoted in Alexandrov (1988).

64 "The outcome . . . too' ": Kaftanov (1985).

64 "was in . . . me": Resis (1993), p. 56.

65 "The work . . . immediately": quoted in Alexandrov (1988).

CHAPTER THREE: 'MATERIAL OF IMMENSE VALUE'

66 "Stalin's shadow," "harsh man": Volkogonov (1988, 1991), p. 244, p. 245.

66 "generating the . . . stupid": quoted ibid., p. 244.

66 "His leadership . . . effective": Khariton and Smirnov (1993), p. 26.

66 "modest, precise and thrifty": quoted in Volkogonov (1988, 1991), p. 244.

66 "It was . . . enemy": Pervukhin (1985).

66 Pervukhin on start-up problems: ibid.

67 "Our suggestion . . . done": ibid.

67 "They were . . . wrong": ibid.

67 "It was . . . away!": ibid.

67 "Until 1945 . . . resources": Khariton and Smirnov (1993), p. 26.

PAGE

67 "Stalingrad was . . . Siberia": Werth (1971), p. 9.

67 "with 1,200 . . . tanks": quoted in Werth (1964), p. 406.

68 40,000 civilians killed at Stalingrad: ibid., p. 442.

68 "whole columns . . . pavements": quoted ibid., p. 453.

68 "whole, huge . . . ruins": Ehrenburg and Simonov (1985), pp. 203–204.

68 "the other . . . Germans": quoted in Werth (1964), p. 456.

68 "a battle . . . fighting": quoted ibid., p. 464.

68 "The sky . . . underfoot": Ehrenburg and Simonov (1985), p. 204.

68 German high command withheld clothing: Werth (1964), p. 550.

69 " 'Funny blokes . . . shoes' ": ibid., p. 554.

69 "All the . . . east": ibid., pp. 552–553.

69 "At that . . . specialists": Kaftanov (1985).

70 "In no . . . plutonium": Golovin (1968), pp. 42–43.

71 "The Devil . . . case": quoted in Alexandrov (1988).

71 "He said . . . missing' ": Resis (1993), p. 56.

71 14-page review: Visgin (1992), Doc. # 4.

72 Fuchs passed diffusion theory reports: cf. itemized list attached to Fuchs FBI FOIA files, 65-58805-1412.

73 "Kurchatov was . . . data": Khariton and Smirnov (1993), p. 25.

74 "In the . . . reactor": ibid.

74 March 22, 1943, Kurchatov letter: Visgin (1992), Doc. # 5.

74 "very important . . . 238": ibid., Doc. # 4.

75 element 93: McMillan and Abelson (1940).

PAGE

75 the German physicist: cf. Rhodes (1986), p. 350.

76 "daughter product 94^{239}": McMillan and Abelson (1940).

77 *"Conclusion* . . . abroad": Visgin (1992), Doc. # 4.

77 "one should . . . program": Khariton and Smirnov (1993), p. 25.

77 "The world . . . Nagasaki": Glenn Seaborg, personal communication, 1.ix.93.

78 "a neat . . . trees": Golovin and Smirnov (1989).

78 Laboratory for Thermal Engineering: Kaftanov (1985).

78 "Most of . . . privations": Golovin (1968), p. 43.

78 "We used . . . flowers": Golovin and Smirnov (1989).

78 "On Kaluzhskaya . . . entrances": Golovin (1968), p. 44.

78 "We had . . . wrong": quoted in Golovin and Smirnov (1989).

78 "examined many . . . necessary": Kaftanov (1985).

79 "Igor Vasilievich . . . compound": Pervukhin (1985).

79 "Kurchatov . . . USSR": Golovin (1968), p. 45.

79 "Kira has . . . beginning": Street (1947), p. 321.

80 An unidentified person with access to the National Research Council: the documents Kurchatov reviewed include work from Columbia, Chicago, the University of Iowa, UC Berkeley and elsewhere. Since compartmentalization obtained in the Manhattan Project at that time, the documents could not have come from any one of these sources alone. The NRC reference committee, which served as a clearing house for reactor- and bomb-directed work, is the likeliest point of leakage.

PAGE

80 Kurchatov July 3, 1943, analysis: Visgin (1992), Doc. # 6.

80 "Of 237 . . . uranium.": V. P. Visgin, in Visgin (1992).

81 Morris Cohen drafted: Dobbs (1992).

81 94[239] cross section for fission: Kennedy et al. (1946).

82 "It was . . . problem": Khariton and Smirnov (1993), p. 25.

CHAPTER FOUR: A RUSSIAN CONNECTION

83 Golodnitsky odyssey: Gold FBI FOIA files, 65-57449-185, p. 5.

83 "fertile soil . . . beaten": Gold (1951), p. 4.

83 "under extremely . . . Jews": ibid., p. 5.

84 "were crudely . . . encountered": ibid., p. 6.

84 "So Sam . . . boys": ibid., p. 7.

84 "Many other . . . anti-Semitic": ibid.

84 "also espoused . . . horrified": ibid., pp. 7–8.

84 Gold told his draft board: Gold FBI FOIA files, 65-57449-185, p. 49.

84 "Here . . . weeks": Gold (1951), pp. 10–11.

84 "somewhat . . . hand": Gold FBI FOIA files, 65-57449-185, p. 43.

85 "Mom hurriedly . . . hand": Gold (1951), p. 14.

85 "nothing was . . . children": ibid., p. 17, p. 21.

85 "My family . . . them": ibid., p. 22.

85 "with his . . . peasant": ibid., p. 23.

85 "but then suddenly stopped": ibid., p. 22.

85 Black volunteered for USSR: Lamphere and Shachtman (1986), p. 165.

85 Tom Black's espionage: Gold FBI FOIA files, 65-57449-667, p. 3.

85 "I said . . . to": Gold (1951), p. 25.

85 Gold's list of reasons: ibid., pp. 26–40.

PAGE

86 "It might . . . business": ibid., p. 27.

86 "almost suicidal . . . it": ibid., p. 33.

86 "there must . . . myself": ibid., p. 34.

86 "letting down . . . years": ibid., p. 42.

86 "the lies . . . affairs).": ibid., p. 43.

86 "The planning . . . years": ibid., pp. 42–43.

87 "Best of . . . me": ibid., p. 41.

87 "It had . . . work": ibid., pp. 53–54.

87 "with a . . . martinet": ibid., pp. 65–66.

87 Xavier College; Ben Smilg: Gold FBI FOIA files, 65-57449-229, p. 1.

88 "brown complexion . . . first": Gold (1951), p. 13. Draft: Gold FBI FOIA files, 65-57449-185, p. 47.

88 "a hard . . . individual": Gold FBI FOIA files, 65-57449-185, p. 26.

88 "nervous . . . do": Mrs. Charles Mahoney (Claire Bleyman), ibid., p. 27.

88 "had a . . . smile": Gold (1951), p. 66.

88 "and especially . . . do": ibid., p. 67.

88 "I was . . . anymore": Gold FBI FOIA files, 65-57449-591, p. 22.

88 "Sam called . . . Union": ibid., 65-57449-68, p. 2.

88 Golos and Brothman: ibid., 65-57449-184, p. 6, quoting Elizabeth Bentley.

88 Elizabeth Bentley: cf. Lamphere and Shachtman (1986), p. 36ff.

89 A contact with a technical background: Gold FBI FOIA files, 65-57449-184, p. 22, quoting Elizabeth Bentley.

89 "an important . . . engineer": ibid., 65-57449-59, p. 30.

89 The two chemists connected: cf Gold's narrative at ibid., 65-57449-591, p. 29ff.

89 "Starting in . . . $50,000": ibid., p. 66.

89 "a bunch . . . him": ibid., p. 36

PAGE

89 "He told . . . elated": ibid., p. 38.
89 "I remember . . . morning": ibid.
90 "He said . . . Brothman' ": ibid., pp. 39–40.
90 "because . . . Buna-S": ibid., p. 43.
90 "Once, in . . . epithets": Gold (1951), pp. 57–58.
90 "But as . . . hopeless": ibid., pp. 58–59.
91 "just before . . . store": Gold (1965b).
91 Slack at Oak Ridge: Gold FBI FOIA files, 65-57449-591, p. 13.
91 "The purpose . . . meeting": ibid., p. 51.
91 "Sam was . . . morning": ibid., p. 52.
91 "A good . . . States": ibid., p. 52f.
92 Harry buttered up/Order of the Red Star: cf. Gold FBI FOIA files, 65-57449-584, p. 33; Gold (1951), p. 73ff.
92 "His greatest . . . making)' ": Gold (1951), p. 71.
92 "And, Sam . . . time' ": ibid., pp. 72–73.
92 "I am . . . know": ibid., p. 73.
92 "ulterior motives . . . done": ibid.
93 "an affair . . . seal": Gold (1965a).
93 "free trolley . . . Moscow": Gold FBI FOIA files, 65-57449-584, p. 33.
93 Gold told others about Red Star: ibid.
93 "I was . . . agreed": Gold FBI FOIA files, 65-57449-591, p. 67.
93 "work of . . . move": ibid., 65-57449-68, p. 33 (also at Williams, 1987, p. 197).
93 "Sam then . . . importance": ibid., 65-57449-591, p. 67.
93 "didn't elaborate . . . was": ibid., 65-57449-185, p. 60.
93 "In any . . . Fuchs": ibid., 65-57449-68, p. 33 (also at Williams, 1987, p. 197).

PAGE

93 Henry Street Settlement: ibid., 65-57449-551, Harry Gold statement of 7-10-50, p. 1.

CHAPTER FIVE: 'SUPER LEND-LEASE'

94 Gore Field: for the Gore Field story cf. Jordan (1952) and JCAE (1951), p. 184ff.
94 "I believe . . . world": Vandenberg (1952), p. 11.
95 Lend-Lease totals: JCAE (1951), p. 185.
95 "Just imagine . . . trucks]": quoted in Keegan (1989), p. 218.
95 "complete alcohol . . . documents": JCAE (1951), p. 185.
95 "to put . . . pay": quoted in Harriman and Abel (1975), p. 108.
95 4.5 million: Werth (1964), p. 401. Other numbers from Elliot (1972).
95 "We've lost . . . Spam": quoted in Werth (1964), p. 628.
96 "One can . . . happiness": quoted ibid., p. 414.
96 "made the . . . Lindbergh": Jordan (1952), p. 21.
96 "Capt. Jordan . . . here": quoted ibid., p. 68.
96 "the unusual . . . Moscow": ibid., pp. 68–69.
97 "But the . . . immunity' ": ibid., p. 69.
97 "Highest diplomatic character"; "I am . . . containers": ibid.
97 "answered yes . . . go": ibid., p. 71.
97 "As we . . . any": ibid., p. 73.
98 "Always just . . . apart": ibid., p. 82.
98 "I had . . . reports": ibid., p. 78.
98 "Oak Ridge . . . was"; "Uranium 92 . . . deuterons": ibid., p. 83.
99 Blueprints and patents: ibid., pp. 135–136.
99 "runs into . . . thousands": HUAC, 1949, quoted ibid., p. 136.

PAGE

99 "Another 'diplomatic' . . . Russia": ibid., p. 137.

99 "I began . . . there": ibid., p. 66.

99 "atomic materials"; table: extracts from Jordan's compilation of Soviet lists at Jordan (1952), p. 142, p. 181.

99 1.2 quarts heavy water: JCAE (1951), p. 189.

100 March 1943 uranium orders: JCAE (1951), p. 186.

100 "Where that . . . 1942": quoted in Jordan (1952), p. 85f.

100 Black-market transaction: JCAE (1951), p. 187.

100 "urged the . . . cent": Jordan (1952), p. 107.

100 New York Times, 31.viii.51: quoted ibid., p. 66n.

101 "On the . . . Lend-Lease": quoted ibid., p. 265f.

101 "diplomatic mail": quoted ibid., p. 267.

101 "mass production": Gouzenko (1948), p. 123.

101 "There were . . . world": ibid., p. 65.

101 "When I . . . sent": ibid., pp. 129–130.

102 "The persistence . . . England": ibid., p. 123.

102 "This one . . . Moscow": quoted ibid., p. 211.

102 "What the . . . material": Bentley (1951), pp. 169–170.

102 Bentley microfilm: ibid., p. 175.

102 "I know . . . consignment": quoted in Jordan (1952), p. 249.

CHAPTER SIX: RENDEZVOUS

103 "hysterical laughter": Frisch (1979), p. 148.

103 Fuchs's arrival: Fuchs FBI FOIA files, 65-58805-1412, p.10ff.

103 Kristel Fuchs Heineman details: cf. ibid., 65-58805-1202, p. 12;

PAGE

Williams (1987), p. 17ff; Gold FBI FOIA files, 65-57449-549, p. 2ff.

104 "around Christmas 1943": Fuchs FBI FOIA files, 65-58805-1412, Fuchs's confession (following p. 51), p. 3.

104 "late January . . . 1944": Gold FBI FOIA files, 65-57449-551, p. 2. Fuchs later agreed that January 1944 was probable; cf. Fuchs FBI FOIA files, 65-58805-1412, p. 11.

104 "to carry . . . ball": quoted in Williams (1987), p. 197.

104 Fuchs was apprehensive: Fuchs FBI FOIA files, 65-58805-1412, p. 11. That Fuchs first rendezvoused with Gold from the Barbizon Plaza fixes the month of this first meeting as January, before Fuchs moved to an apartment on February 1.

105 "was wearing . . . hand": ibid., Fuchs's confession (following p. 51), p. 3.

105 Gold's gloves: "Gold recalled he had to stop in a store at New York City to buy a pair of gloves." Gold FBI FOIA files, 65-57449-520, p. 13f.

105 "indicated he . . . assignment": Fuchs FBI FOIA files, 65-58805-1412, p. 12.

105 "We went . . . meetings": quoted in Williams (1987), p. 197.

105 Manny Wolfe's: Gold FBI FOIA files, 65-57449-520, p. 14.

105 "[Fuchs] told . . . meeting": Fuchs FBI FOIA files, 65-58805-1412, p. 12.

105 Fuchs remembered no dinner: ibid.

105 "one of . . . physicists": Gold (1951), p. 78.

105 "I liked . . . caution)": ibid., p. 79.

106 "Intelligence information . . . rezidents": Yatzkov (1992).

PAGE

106 "Our present . . . high": Street (1947), p. 322f.

106 "A turning . . . salvoes": Golovin (1968), p. 46.

106 *Perelom* year: Werth (1964), p. 759.

106 "None of . . . resistance": ibid., p. 766f.

106 "The Black Death": ibid., p. 835.

106 "The reason . . . so": Golovin (1989), p. 198.

107 "the joke . . . disinformation": Lev Altshuler interview, vi.92.

107 "He was . . . motion": Gold (1951), p. 75.

107 "John"/Yatzkov duck-like walk: Gold (1965b), p. 18.

107 "I failed . . . like": Chikov (1991c).

107 "about being . . . homesick": Walter Carl Neunson, at Yakovlev FBI FOIA files, 100-346193-26, p. 3.

107 "spoke with . . . ones": Gold (1951), p. 52.

107 "which led . . . thing": ibid., p. 63.

107 "the intention . . . Queens": Gold FBI FOIA files, 65-57449-551, pp. 1–2.

108 "anything but . . . area": FBI paraphrase at Fuchs FBI FOIA files, 65-58805-1412, p. 13.

108 "several passages . . . streets": Gold FBI FOIA files, 65-57449-551, p. 2.

108 "That is . . . dream": ibid., 65-57449-185, p. 86.

108 "Klaus knew . . . process": ibid., 65-57449-551, p. 27.

108 "I . . . aside": ibid.

108 Philip Abelson and thermal diffusion: cf. Rhodes (1986), p. 550ff.

108 "worked in . . . Ridge": quoted in Williams (1987), p. 198.

108 "made good . . . John": Gold FBI FOIA files, 65-57449-551, p. 27.

108 "Rest": Pincher (1984), p. 140.

109 "Drop copy" program: Robert

PAGE

Lamphere, personal communication, vi.94.

109 One-time pads: Lamphere and Shachtman (1986), p. 78ff.

109 "It was . . . John": Gold FBI FOIA files, 65-57449-551, p. 2.

109 "Two or . . . meeting": Fuchs FBI FOIA files, 65-58805-1412, p. 14.

109 "I, with . . . 1944": ibid., Fuchs's confession (following page 51), p. 5.

110 "a deviation . . . rules": Gold FBI FOIA files, 65-57449-551, p. 3.

110 "[Fuchs] advised . . . him": Fuchs FBI FOIA files, 65-58805-1412, p. 15.

110 "the manpower . . . orally": ibid., p. 16.

110 "we went . . . chess": Gold FBI FOIA files, 65-57449-551, pp. 2–3.

111 "*May 8 . . . B[ritain]*": Fuchs FBI FOIA files, 65-58805-1202, "Director of Security" handwritten notes appended at end of serial, p. 8. (Hoover's underlining.)

111 Hoover's notes: ibid., pp. 8–9.

111 "should either . . . intermission"; dinner details: Gold FBI FOIA files, 65-57449-551, p. 3.

111 "some 25 . . . detail": ibid., p. 4.

112 "During this . . . inquiry": ibid., p. 5.

112 "several typewritten . . . them": ibid., p. 28.

112 "I did . . . so": ibid.

112 "near an . . . Southwest": ibid., p. 6.

112 "that his . . . O.K.": ibid., pp. 6–7.

113 "On this . . . prey": ibid., p. 7.

113 "Our principal . . . York": ibid., p. 8.

113 "talked at . . . tight' ": ibid., p. 10.

113 "that Fuchs . . . inquiry": ibid.

113 "highly pleased": ibid., p. 11.

113 John dictated a message: ibid., pp. 11–12.

114 Kristel Heineman positively identified a photograph of Harry Gold: ibid., 65-57449-401. The

identification, on 16.vi.50, took place under unusual circumstances, as the FBI teletype explains: "Kristel Heineman . . . identified Gold without hesitation and positively as chemist who visited her home in Jan dash Feb dash March, fortyfive. She had just finished a shock treatment prior to identification. She did not know Gold by any name whatsoever. Doctors at hospital feel that she is as good mentally now as she ever will be. . . ." Fuchs's sister was hospitalized for "schizophrenia"; she eventually recovered.

114 "the approximate . . . home": ibid., 65-57449-549, p. 4.

114 "Mrs. Heineman . . . children": ibid., 65-57449-551, p. 13.

115 "Dear Kristel . . . Klaus": ibid., 65-57449-549, pp. 5–6.

115 "provisionally until . . . December": quoted in Williams (1987), p. 74. July 14: Fuchs FBI FOIA files, 65-58805-1202, "Director of Security" handwritten notes appended at end of serial, p. 4.

117 "blow in . . . beer": quoted in Rhodes (1986), p. 479.

117 "[Bethe] wanted . . . subjects": quoted in Blumberg and Owens (1976), p. 131.

118 "If Fuchs . . . damage": quoted in Williams (1987), p. 74.

118 "One of . . . division": quoted ibid., p. 76.

118 "Whenever I . . . hours": Nicholas Metropolis interview, vi.93.

118 Fuchs's papers: Report No. R88004, LASL Authors Shared Database, 4/22/87, Author Index, p. 771.

119 February 11, 1945: Gold FBI FOIA files, 65-57449-549, p. 4.

119 "seemed surprised . . . right' ": ibid., p. 7.

119 Fuchs called Manhattan: J. Edgar Hoover reached the same conclusion and ordered a search for the unknown contact; cf. Fuchs FBI FOIA files, 65-58805-1268. For Fuchs's denial cf. 65-58805-1395.

120 "With some . . . Friday": Gold FBI FOIA files, 65-57449-551, pp. 13–14.

120 "I went . . . me": ibid., p. 14.

CHAPTER SEVEN: 'MASS PRODUCTION'

121 "mass production": Gouzenko (1948), p. 123.

121 "new, very . . . text)": quoted in Holloway (1994), p. 102.

121 "about 10,000 . . . on": Terletsky (1973).

122 "a fellow traveler": USAEC (1954b), p. 113.

122 Steve Nelson background: JCAE (1951), p. 173.

122 Steve Nelson at Barcelona Intelligence School: "One graduate [in addition to Morris Cohen] was a United States citizen who went on to become a member of the espionage network that helped the Soviets steal from the United States the production secrets for nuclear weapons." Costello and Tsarev (1993), p. 276.

122 "assigned . . . bomb": JCAE (1951), p. 174.

122 "Late one . . . denomination": HUAC (1950), p. 4. HUAC identified "Joe" as physicist Joseph Weinberg. Cf. also Groves's testimony at the Oppenheimer security hearing: "On [Joseph] Weinberg, I would like to emphasize that the information he passed was probably with respect to the electromagnetic process. . . ." USAEC (1954b), p. 176.

PAGE

123 "Nelson later . . . communism": HUAC (1950), p. 4.

123 "a chemical . . . Union": USAEC (1954b), p. 135.

123 "A man . . . supply": ibid., p. 144, quoting a transcript of a conversation on 26.viii.43 between Oppenheimer and Boris Pash.

123 "if you . . . ago": ibid., p. 145.

123 "was introduced . . . separation": Sagdeev (1993), p. 33.

124 "a pure . . . idiocy": USAEC (1954b), pp. 146–149.

124 "I might . . . information]": ibid., pp. 145–146.

124 "Let me . . . treasonable": ibid., p. 146.

124 "To put . . . door": ibid., p. 144.

124 "But it . . . whatever": ibid., p. 146.

125 "varied approaches . . . Intelligence": Gouzenko (1948), p. 67.

125 "wholly false": USAEC (1954b), p. 149.

125 "One day . . . conversation": ibid., p. 130.

125 "[Eltenton] admitted . . . information": Fuchs FBI FOIA files, 65-58805-1202, p. 5.

125 Chevalier's version: "On June 26, 1946, Haakon M. Chevalier was interviewed by Bureau agents. He furnished a signed statement admitting that some time prior to March 1, 1943, he was approached by George Charles Eltenton regarding the possibility of getting information regarding work being done at the Radiation Laboratory; that Eltenton stated that any information concerning the research being conducted would be of use to the Soviet Scientists [sic] and that they could benefit from it. Chevalier advised that Eltenton stated that he had been approached by someone connected with the Soviet Union in an effort to obtain this information. Chevalier stated that he mentioned to J. Robert Oppenheimer a matter concerning an approach having been made to him in which an inquiry was made if any part of the secret of the project should be made available to Russian Scientists [sic]." Ibid., p. 5. Oppenheimer's 5.ix.46 version: ibid., p. 4.

126 "Q. Had you . . . suggestions": USAEC (1954b), p. 135.

126 Gouzenko's arrival: cf. Taschereau and Kellock (1946), p. 11.

126 "tall, handsome, personable," Gouzenko (1948), p. 182.

126 "magnetic personality attracted contacts": ibid., p. 188.

127 Israel Halperin: Taschereau and Kellock (1946), p. 131ff. Halperin's address book: cf. facsimile attached to Fuchs FBI FOIA files.

127 "I did . . . reciprocal": Taschereau and Kellock (1946), p. 74.

127 "a charming . . . humor": in Leslie Groves's paraphrase: JCAE (1951), p. 52.

127 "Before coming . . . Moscow": Taschereau and Kellock (1946), p. 449.

127 "The whole . . . gain": ibid., p. 456.

128 "he worked . . . developments": L. R. Groves to B. B. Hickenlooper, 12.iii.46, quoted in JCAE (1951), p. 52.

128 Nunn May and reactor poisoning: cf. ibid.: "He also, at this time [i.e., October 1944], probably acquired knowledge of some technical problems which we encountered in the operation of the first Hanford pile."

129 Donald Maclean details: cf. in particular Cecil (1989).

PAGE

129 "He is . . . man": quoted ibid., p. 60.

129 "drinking orgy . . . bed": quoted ibid.

129 Kim Philby meetings: Cecil (1989), p. 64.

129 Combined Policy Committee: cf. Hewlett and Anderson (1962), p. 277ff.

130 CDT and ore: cf. ibid., p. 285f; Cave Brown and MacDonald (1977), pp. 191–199.

130 "medium-grade . . . great": quoted in Cave Brown and MacDonald (1977), p. 198.

130 Maclean and Yatzkov: Cecil (1989), p. 71.

130 Maclean's travel to New York: Robert Lamphere, personal communication, vi.94.

130 "the question . . . deposits": Visgin (1992), Doc. # 7.

130 "As a . . . purification": Hewlett and Anderson (1962), p. 283.

131 John Anderson: Williams (1987), p. 92.

131 "We simply . . . development": quoted ibid., p. 93.

131 Churchill told Roosevelt: Williams (1987), p. 92.

131 "I therefore . . . amused": Peierls (1985), p. 201.

132 "a physicist": quoted in Dobbs (1992), p. A37.

132 "On at . . . her": Yatzkov/Yakovlev FBI FOIA files, 100-346193-64, p. 3.

132 " 'No' . . . position' ": Bentley (1951), p. 180.

133 "Another group . . . identify": quoted in Radosh and Milton (1983), pp. 229–230. Cf also Lamphere and Shachtman (1986), p. 38.

133 Knickerbocker Village; "they always . . . Julius' ": quoted in Nizer (1973), pp. 188–189.

133 JR told Morton Sobell: according to Max Elitcher; cf. Lamphere and Shachtman (1986), p. 196; Radosh and Milton (1983), p. 176.

133 "that . . . him": JCAE (1951), p. 104.

133 Julius Rosenberg and Ethel Greenglass details: Greenglass FBI FOIA files, 65-59028-187, p. 2ff; Radosh and Milton (1983), p. 48ff.

134 "violent Communists . . . cause": Greenglass FBI FOIA files, 65-59028-345, p. 6.

134 Converting David Greenglass: cf. DG testimony at JCAE (1951), p. 68ff; Samuel Greenglass: Greenglass FBI FOIA files, 65-59028-345, p. 6.

134 "Samuel Greenglass . . . States": ibid.

134 Ethel Rosenberg CP membership: cf. ibid., p. 495, n. 53.

134 Branch 16B; Joel Barr and Alfred Sarant: ibid., p. 496.

135 "He told . . . proposition": quoted from Rosenberg trial transcript, ibid., p. 176.

135 "I've got . . . espionage": David and Ruth Greenglass interview with Ronald Radosh and Sol Stern, 1979; Ronald Radosh, personal communication.

135 "damned fool nonsense": Gold FBI FOIA files, 65-57449-591, p. 67.

135 Montreal drugstore: Gouzenko (1948), p. 221.

135 "Although I'd . . . socialism's": Greenglass FBI FOIA files, 65-59028-193, p. 17.

136 "Well darling . . . future": ibid., p. 18.

136 Fort Ord: ibid., p. 32.

136 "the people . . . reforms": ibid., p. 22.

136 "terribly let . . . time": ibid., pp. 22–23.

136 "Rosenberg told . . . missed": FBI

PAGE

FOIA file, Rosenberg Case
Summary, pp. 51–52.

137 Nine attempts: Gold FBI FOIA files,
65-57449-667, p. 8.

137 Elitcher cable: Lamphere and
Shachtman (1986), p. 191.

137 "Darling, I . . . peoples": Greenglass
FBI FOIA files, 65-59028-193, p. 25.

137 Greenglass's transfer to Oak Ridge:
FBI FOIA file, Rosenberg Case
Summary, p. 12.

137 "I had . . . 'Gee' ": David and Ruth
Greenglass interview with Ronald
Radosh and Sol Stern, 1979; Ronald
Radosh, personal communication.

137 "Julie was . . . yourself": quoted in
Radosh and Milton (1983), p. 65.

137 "Dear, I . . . comrade": Greenglass
FBI FOIA files, 65-59028-193, p. 25.

138 "I don't . . . him": ibid., 65-59028-
332, p. 61.

138 High-speed cameras: ibid., 65-
59028-193, p. 32.

138 "About a . . . employees": ibid., 65-
59028-149, p. 33.

138 "The group . . . arrangements":
Hoddeson et al. (1993), p. 279.

138 "I am . . . in": quoted in Radosh and
Milton (1983), p. 66. The authors
found this contemporary
corroboration of the Greenglasses'
accusations against the Rosenbergs
among a file of handwritten
original letters which the FBI
retrieved from the Greenglass
apartment during a consent search
on 15.vi.50. The FBI had neglected
to transcribe it and it had gone
unnoticed for almost thirty years.
None of the letters had been used
as evidence in the Rosenberg-
Greenglass trials; their existence
was unknown to historians until
they came to light in an FOIA
action in 1975 brought by the
Rosenbergs' sons.

PAGE

138 "that business . . . Russians": David
and Ruth Greenglass interview with
Ronald Radosh and Sol Stern, 1979;
Ronald Radosh, personal
communication.

138 "I got . . . things": Greenglass FBI
FOIA files, 65-59028-193, p. 51.

139 "Julius Rosenberg . . . made": Gold
FBI FOIA files, 65-57449-614, pp.
9–10.

139 "I asked . . . secret": Greenglass FBI
FOIA files, 65-59028-332, p. 7.

139 "He also . . . bomb": Gold FBI FOIA
files, 65-57449-614, p. 10.

139 "He felt . . . participate": Greenglass
FBI FOIA files, 65-59028-332, p. 7.

139 "I didn't . . . it": ibid., pp. 7–8.

140 "about $150 . . . trip": Gold FBI
FOIA files, 65-57449-614, p. 10.

140 RG's travel problems: Greenglass
FBI FOIA files, 65-59028-193, pp.
28–29.

140 "accumulated plenty . . . back":
ibid., p. 29.

140 "We went . . . conversation": JCAE
(1951), p. 70.

140 Ruth Greenglass began with atomic
bomb: Greenglass FBI FOIA files,
65-59028-332, p. 11.

140 "I was very surprised": Gold FBI
FOIA files, 65-57449-614, p. 2.

140 "David asked . . . me": Greenglass
FBI FOIA files, 65-59028-332, p. 11.

140 "She said . . . information": Gold
FBI FOIA files, 65-57449-614, p. 2.

140 "that she . . . it": JCAE (1951), p. 71.

140 "I felt . . . understanding": quoted
in Nizer (1973), p. 132.

140 Ruth asked David what he thought:
JCAE (1951), p. 71.

140 "you're jumping . . . water": David
and Ruth Greenglass interview with
Ronald Radosh and Sol Stern, 1979;
Ronald Radosh, personal
communication.

140 "At first . . . it": JCAE (1951), p. 71.

PAGE

140 "memories and . . . mind": ibid., p. 108.

140 "I felt . . . doubts": ibid., p. 113.

140 "She asked . . . her": ibid., p. 71.

140 David Greenglass mentioned Oppenheimer et al.: ibid.

141 "how it . . . out": Greenglass FBI FOIA files, 65-59028-332, p. 13.

141 More RG train trouble: ibid., 65-59028-193, p. 29.

141 "alone . . . alone": ibid., 65-59028-332, p. 12.

141 "I didn't . . . made": ibid., p. 66.

141 "The scientists . . . on": ibid., p. 60.

141 "We were . . . interrupting": ibid., p. 13.

141 Julius Rosenberg's morning visit: In various testimony David Greenglass placed this transfer of information elsewhere, in particular at the subsequent dinner where the Jello box was cut. I follow his testimony to FBI agents after he pled guilty (and therefore had little reason to lie), paraphrased ibid., p. 5. This was the story DG told at the Rosenberg trial; cf. JCAE (1951), p. 72ff. Their first meeting after DG arrived on furlough is also the most plausible time for this first exchange of information.

142 "Rosenberg described . . . functions": Gold FBI FOIA files, 65-57449-614, p. 3; "He said . . . question]": Greenglass FBI FOIA files, 65-59028-332, pp. 67–68; "He said . . . described": JCAE (1951), p. 93.

142 "He told . . . up": JCAE (1951), p. 72.

142 Rosenberg's lists: ibid., pp. 72–73.

142 "a number . . . molds": ibid., p. 73.

142 the type he drew for Rosenberg: cf reproduction at Langer (1966), p. 1501. "the flat . . . mold": JCAE (1951), p. 141.

PAGE

142 "It has . . . lens": ibid., p. 75.

142 Initiator design: cf. Hoddeson et al. (1993), p. 280.

143 Rosenberg apartment: JCAE (1951), p. 117.

143 Chappaqua: according to Ruth Greenglass, not (as usually cited) Cleveland, where they lived in 1950; Gold FBI FOIA files, 65-57449-614, p. 11.

143 Ruth had seen Ann before: Greenglass FBI FOIA files, 65-59028-332, p. 16; David had never met her: ibid., p. 69.

143 "she was . . . wallet": ibid., p. 14.

143 Los Alamos authorized families: according to Ruth Greenglass at ibid., p. 108.

143 "the Rosenbergs . . . forthcoming": JCAE (1951), p. 78.

143 "[Julius] said . . . lenses": ibid.

143 "I recall . . . him": Greenglass FBI FOIA files, 65-59028-332, p. 16.

143 "Rosenberg told . . . Secret": ibid., 65-59028-422, p. 3. NB: "Greenglass stated [this] information was given to him in January 1945 by Rosenberg.": ibid.

143 "For some . . . gift": Bentley (1951), p. 209. Order of the Red Star: ibid., p. 254.

144 "entitles you . . . free": ibid., p. 255; Gorsky: Andrew and Gordievsky (1990), p. 320.

144 "[Julius] stated . . . Russia": JCAE (1951), pp. 101–102.

144 "asked to . . . around": Greenglass FBI FOIA files, 65-59028-149, p. 37.

144 "all over . . . answer": JCAE (1951), p. 79.

144 "the high-explosive . . . him": Greenglass FBI FOIA files, 65-59028-361, p. 6.

144 Greenglass's news of implosion: based on my analysis of the NKVD's

PAGE

response and of Soviet documents, below.

144 " 'Go home . . . him": JCAE (1951), p. 79.

145 Semenov to Vladivostok: Gold FBI FOIA files, 65-57449-491, p. 51.

145 "I knew . . . for": Greenglass FBI FOIA files, 65-59028-332, p. 66.

CHAPTER EIGHT: EXPLOSIONS

146 Alexandrov looking for Lab. No. 2: Nikolai Ivanov, personal communication, v. 92.

146 "These talented . . . substance": Pervukhin (1985).

146 "very small . . . plutonium": Knyazkaya (1986).

146 "It looks . . . gardening": Street (1947), pp. 323–324.

147 "The first . . . Prize": Zukerman and Azarkh, unpub. MS, p. 44.

147 "intensely studying . . . shock-mounts": ibid., p. 22.

147 "For a . . . torpedoes": ibid.

148 "Just look . . . camps": Knyazkaya (1986).

148 "At that . . . deposits": Kaftanov (1985).

148 "not received . . . ago": quoted in Holloway (1994), p. 102.

148 Donkeys: Alexandrov (1988).

148 Uranium production: Golovin (1989), p. 201; graphite production: Pervukhin (1985).

148 "The roads . . . roads": Ehrenburg and Simonov (1985), pp. 378–379.

149 Lend-Lease feeding Russians: Werth (1971), p. 19.

149 Stalin confirmed two-thirds: Herring (1973), p. 116.

149 "is the . . . Nazis": quoted in Andrew and Gordievsky (1990), p. 290.

149 "Stalin then . . . never!' ": Djilas (1962), p. 74.

PAGE

149 "What frightens . . . logical": quoted in Herring (1973), p. 135.

149 "settle about . . . down": Churchill (1959, 1987), pp. 885–886.

150 Nunn May in February 1945: Hyde (1980), citing Nunn May's defense barrister, Gerald Gardiner. Note, however, that Doc. # 11, below, which may have been Nunn May's "technical" Part One, is said to have been furnished to Kurchatov under a cover letter dated 25.xii.44, a plausible earlier date: Visgin (1992).

150 Swail Avenue: Taschereau and Kellock (1946), p. 455.

150 "a man . . . trapped": quoted in Gouzenko (1948), p. 238.

150 "I told . . . States": quoted ibid.

150 "The report . . . Washington": Gouzenko (1948), pp. 238–239.

151 "costly errors": ibid., p. 239.

151 "two methods . . . implosion": Visgin (1992), Doc. # 7.

151 "unlimited control . . . Congo": ibid.

151 "The material . . . considered": ibid., Doc. # 8.

151 Hydride gun work at Los Alamos: cf. Hoddeson et al. (1993), p. 181.

151 "that would . . . material": Visgin (1992), Doc. # 8.

152 "The 'implosion' . . . study": ibid.

152 "I went . . . there": Gold FBI FOIA files, 65-57449-185, p. 62.

152 *Mrs. Palmer's Honey,* candy: ibid., 65-57449-184, pp. 4–5.

152 Gold's wife and children: ibid., 65-57449-42, pp. 7–8.

152 "Mrs. Heineman . . . sitting": ibid., 65-57449-491, p. 34.

153 "saying 'I . . . minutes": ibid., 65-57449-551, p. 14

153 "he . . . bomb"; tremendous progress: JCAE (1951), p. 150.

153 "that he . . . Cambridge": Gold FBI FOIA files, 65-57449-551, p. 14.

PAGE

153 "[Klaus] told . . . Fe": ibid., p. 15.

153 "that I . . . April": ibid.

153 "a yellow . . . Fe": ibid., 65-57449-520, p. 27.

153 "a quite . . . information": ibid., 65-57449-551, p. 16.

153 "Fuchs wrote . . . core": quoted in Williams (1987), p. 190.

154 Los Alamos technical studies: Fuchs FBI FOIA files, 65-58805-1270, p. 6.

154 "Mrs. Heineman . . . back": Gold FBI FOIA files, 65-57449-551, p. 16.

154 "that I . . . man": ibid.

154 "Fuchs held . . . it": Gold (1965b).

155 "He turned . . . thing": Fuchs FBI FOIA files, 65-58805-1412, p. 31.

155 "I left . . . York": Gold FBI FOIA files, 65-57449-551, p. 16.

155 "[He] told . . . lens": JCAE (1951), p. 151.

155 February 28, 1945, Los Alamos meeting: Hoddeson et al. (1993), p. 308, p. 312.

155 "Now we . . . bomb": quoted ibid., p. 271.

155 uranium gun: ibid., p. 249.

155 "Since there . . . attack": Groves (1962), p. 230.

156 "We did . . . there": Putney (1987), p. 55.

156 "In a . . . destroyed": Groves (1962), p. 231.

156 "It was . . . earlier": Werth (1971), p. 16.

156 "Russia was . . . regime": Werth (1964), p. 982.

156 "When the . . . world": Ehrenburg and Simonov (1985), pp. 446–447.

157 "The Russians . . . Russians": quoted in Herring (1973), p. 140f.

157 Rosenberg feared his espionage exposed: according to Max Elitcher: Rosenberg FBI FOIA file, 94-3-4-317-348X, p. 52.

157 "I am . . . falsehood": quoted in Nizer (1973), p. 228.

PAGE

157 Julius Rosenberg CP card: Radosh and Milton (1983), p. 496, citing a photostat in U.S. Army intelligence files.

157 Julius Rosenberg meeting with Ruth Greenglass in February: Radosh and Milton (1983), p. 198, citing RG's trial testimony.

157 Denver: David Greenglass: ibid.; Ruth Greenglass: ibid., p. 16.

158 "in front . . . Albuquerque": Greenglass FBI FOIA files, 65-59028-332, p. 68.

158 "I think . . . live": ibid., pp. 18–19.

158 Sara Spindel: ibid., p. 33; March 19: ibid., p. 34.

158 "a very . . . area": ibid., p. 3.

158 "on the . . . apartment": William Spindel, personal communication, ix.93.

158 "that she . . . Saturdays": quoted in Nizer (1973), p. 124.

158 April 7, 1945 Kurchatov report: Visgin (1992), Doc. # 10.

159 "a slightly . . . oxide": quoted in Taschereau and Kellock (1946), p. 455. Mid-April: "A handwritten Russian entry in one of the [GRU] notebooks, signed by [Angelov] referring to a meeting where a sample of uranium 235 was delivered by Dr. May, states: —200 dollars Alek and 2 bottles of whiskey handed over 12.4.45." (Ibid., p. 66.) In his confession, Nunn May merges the delivery of the U235 sample with a later delivery of U233; the U233 delivery occurred August 9, as attested by one of the documents Igor Gouzenko passed to Canadian authorities in the course of his defection; cf. ibid., p. 450. (NB: the document is misdated July 9—9.7.45—but its reference to "the bomb dropped on Japan" dates it

precisely between the Hiroshima and Nagasaki bombings, as the Royal Commission recognizes in its discussion on p. 450.)

160 "I attacked . . . through": quoted in Werth (1964), p. 995.

160 Hitler's suicide: Bullock (1992), p. 892.

160 Soviet and German Berlin casualties: according to Zhukov: Werth (1964), pp. 995–996.

160 Berlin-Grünau: CIA (1957), p. 7.

160 "A remnant . . . disposal": quoted in Zukerman and Azarkh, unpub. MS, p. 139.

160 "The plant . . . started": quoted in Rhodes (1986), pp. 608–609.

161 "about 1200 . . . war": quoted ibid., p. 613.

161 "Through our . . . year": quoted in Zukerman and Azarkh, unpub. MS, pp. 139–141.

162 Kaiser Wilhelm Institute: Walker (1989), p. 183.

162 Austrian uranium and heavy water: HQ EUCOM Frankfurt to WDGID, INFO: Gen. Groves, 31.vii.47, National Archives.

162 German scientists; Sinop; Agudzeri: cf. Walker (1989), p. 183ff; CIA (1957) *passim.* "We had to manufacture metallic uranium. The Germans had solved that problem. Nikolaus Riehl, a Baltic German, replicated German technology for making metallic uranium." Nikolai Ivanov, personal communication, v.92.

162 "The world . . . soil": Ehrenburg and Simonov (1985), p. 481.

162 "was an . . . seen": Werth (1964), p. 969.

163 "unfortunate and even brutal": quoted in Herring (1973), p. 207.

163 "completely unsatisfactory . . .

significance": quoted in Holloway (1994), pp. 102–103.

163 "From the . . . cellar' ": Yatzkov (1992).

CHAPTER NINE: 'PROVIDE THE BOMB'

165 "so that . . . Fe": Greenglass FBI FOIA files, 65-59028-332, p. 36.

165 "one meeting . . . Fuchs": JCAE (1951), p. 151.

165 "a circular . . . drink: ibid.

165 "I told . . . trip": ibid.

165 "I have . . . is": quoted from Rosenberg trial transcript in Radosh and Milton (1983), p. 211.

166 "And that . . . go": JCAE (1951), p. 152.

166 Gold's written instructions: Gold (1965b), p. 15.

166 "Then a . . . Julius' ": JCAE (1951), p. 152.

166 "Frank Kessler," "Frank Martin": cited in Radosh and Milton (1983), p. 158f; "Ben from Brooklyn": Greenglass FBI FOIA files, 65-59028-332, p. 36.

166 "John told . . . them": ibid.

166 "I was . . . have": ibid.

166 Jello boxtop: note that it was Ruth Greenglass who first specified that the recognition device was a "Jello boxtop"; it was not an invention of Roy Cohn's, though the one displayed to such notorious effect at the Rosenberg trial was of course not the original. Greenglass FBI FOIA files, 65-59028-332, p. 14.

166 Gold's route to Santa Fe: Gold FBI FOIA files, 65-57449-551, p. 17.

166 "extremely short . . . carefully": ibid.

167 $400: Greenglass FBI FOIA files, 65-59028-332, p. 37.

PAGE

167 "the only . . . midst": Gold FBI FOIA files, 65-57449-551, p. 17.

167 "I had . . . Fe": ibid., p. 18.

167 Fuchs's car: Anthony French, personal communication, x.93.

167 "Klaus arrived . . . settled": Gold FBI FOIA files, 65-57449-551, p. 18.

167 "I . . . meeting": Fuchs FBI FOIA files, 65-58805-1412, Fuchs's confession (following p. 51), p. 7.

167 "Klaus told . . . day": Gold FBI FOIA files, 65-57449-551, pp. 18–19.

167 "the names . . . TNT": Fuchs FBI FOIA files, 65-58805-1412, Fuchs's confession (following p. 51), p. 7.

168 "Baratol" and "Composition B": quoted in Williams (1987), p. 191.

168 Gun bomb: Fuchs FBI FOIA files, 65-58805-1412, p. 32.

168 "due to . . . 1945": Gold FBI FOIA files, 65-57449-551, p. 19.

168 "a considerable . . . information": ibid.

168 "that among . . . itself": Fuchs FBI FOIA file, 65-58805-1455, p. 29.

168 "I delivered . . . efficiency": ibid., 65-58805-1412, Fuchs's confession (following p. 51), p. 7.

168 "He reported . . . system": quoted in Williams (1987), p. 191.

169 "I went . . . Yakovlev": Greenglass FBI FOIA files, 65-59028-332, p. 38.

169 "I was . . . morning": JCAE (1951), p. 153.

169 P. M. Sherer: the FBI established in 1950 that Sherer, 75, occupied an apartment in the Greenglasses' building from Christmas 1944 until June 1945. Sherer remembered the Greenglasses, but he was unable to identify Harry Gold five years after the fact. "That really would have been remarkable," Gold comments, "considering that our meeting took place at night on a

darkened porch." Gold (1965b), p. 25. Cf. Greenglass FBI FOIA files, 65-59028-78, p. 41, and 65-59028-93, p. 71.

169 "So . . . Albuquerque": Gold (1965b), p. 25.

169 "Finally, about . . . there": Gold FBI FOIA files, 65-57449-551, pp. 19–20.

169 "Now, with . . . experience": Gold (1965b), p. 35.

169 "I clearly . . . going": ibid., p. 26.

169 "We had . . . box": JCAE (1951), p. 81.

170 "The whole . . . stripes": Gold (1965b), p. 27.

170 Gold offered food: JCAE (1951), p. 81.

170 "I didn't . . . friendly": Greenglass FBI FOIA files, 65-59028-332, p. 20.

170 "He just . . . work": JCAE (1951), p. 81.

170 "cut him . . . Union": ibid., p. 153.

170 "Greenglass was . . . after": Gold (1965b), p. 27.

170 "He agreed . . . elsewhere": Greenglass FBI FOIA files, 65-59028-332, p. 40.

170 "Mrs. Greenglass . . . Julius": ibid., pp. 40–42. David Greenglass denied having made any such statement, telling an FBI agent in 1950 "that on the one occasion he met Harry Gold, he did not furnish Gold with a telephone number where he could be contacted in New York City; and that he never arranged with Gold for any future contact at any place." (Ibid., p. 90.) Greenglass received an early furlough in September 1945, not at Christmastime, though he could not have known in June that the war would end in August. But Gold remembered the arrangement for a

possible second meeting with Greenglass *before* he remembered the GI any more specifically than as "a man in Albuquerque" whom "John" wanted him to contact and before he knew about any "Julius" —he recalled the contact as David's father-in-law, as living in the Bronx and whose name "may have been Philip." (Ibid., 65-59028-78, p. 3.) And Greenglass, in his first confession, before he spoke to anyone other that the FBI, said, "Gold told me that he would come back to see me again and I agreed to see him; however, he never contacted me again." (Ibid., 65-59028-149, p. 33.)

171 "With all . . . mission": Gold (1965b), p. 28.

171 "most anxious . . . available": ibid., p. 35.

171 "waiting for . . . jostling": ibid., p. 21.

171 "I got . . . experiment": JCAE (1951), pp. 81–83.

171 "the growth . . . there": Greenglass FBI FOIA files, 65-59028-332, p. 73.

171 "David and . . . cameras": Gold FBI FOIA files, 65-57449-614, p. 13.

172 "Gold told . . . currency": ibid., p. 4.

172 Gold remembered Greenglass looking disappointed: Greenglass FBI FOIA files, 65-59028-332, p. 38.

172 "[Gold] said . . . it' ": ibid., p. 75.

172 "I said . . . while": JCAE (1951), p. 82.

172 "that they . . . bread": Greenglass FBI FOIA files, 65-59028-332, p. 41.

172 "We went . . . money": JCAE (1951), p. 82.

172 "The taking . . . worse": Gold FBI FOIA files, 65-57449-614, p. 13.

172 "I was . . . paid": quoted in Nizer (1973), p. 133.

172 "I furnished . . . ally": Greenglass FBI FOIA files, 65-59028-193, p. 4.

172 "to see . . . street]": Gold (1965b). The FBI eventually confirmed that such a parade took place on Sunday, 3.vi.45, in Albuquerque. Radosh and Milton (1983), p. 469ff.

172 "On the . . . 'Other' ": JCAE (1951), p. 154.

173 Ruth Greenglass savings account: Greenglass FBI FOIA files, 65-59028-78, p. 39.

173 "to save . . . evening": ibid., 65-59028-332, p. 43.

173 "I met . . . envelopes": JCAE (1951), p. 155.

173 "at the . . . Albuquerque": ibid.

173 Fuchs's knowledge of production rates: he knew them in September; cf. Chapter 10 below.

173 Perseus; Lona Cohen in Albuquerque: Yatzkov (1992).

174 "Top Secret . . . Kurchatov": Visgin (1992), Doc. # 12.

174 Maclean, Yatzkov and the CPC: cf. Cecil (1989), pp. 70–71.

175 Yatzkov and Brothman's espionage: according to Harry Gold; cf. JCAE (1951), p. 160.

175 "whereby . . . me"; details of Gold recontact: JCAE (1951), p. 156.

175 "Suddenly, there . . . nature": Rabi (1970), p. 138.

176 "Believe Japs . . . homeland": Ferrell (1980), p. 42.

176 "I casually . . . Japanese' ": Truman (1955), p. 416.

176 "Stalin . . . up' ": Zhukov (1971), pp. 674–675, with interpolations from David Holloway's translation of the same passage from the Russian edition at Holloway (1994), p. 117.

176 "Truman didn't . . . role": Resis (1993), p. 56.

PAGE

176 "try to ... uranium": Taschereau and Kellock (1946), p. 450.

176 "handed over ... lamina": ibid.

176 "Herb said ... went": Alvin Weinberg, personal communication, 2.xi.93.

177 "uranium samples ... it!' ": Gouzenko (1948), p. 241.

177 "I didn't ... me": Alliluyeva (1967), p. 188.

177 "after the ... Kurchatov": Terletsky (1973).

177 "Stalin summoned ... refusals' ": Alexandrov (1988).

178 Soviet press announcement: Werth (1964), p. 1037.

178 "On my ... revered": Sakharov (1990), p. 92.

178 "I declared ... note": Resis (1993), p. 21.

178 "This is ... whatsoever": quoted in Kramish (1959), p. 87.

178 "Yet the ... wasted' ": Werth (1964), p. 1037.

179 "France recently ... ground": Ehrenburg and Simonov (1985), p. 485.

179 Soviet statistics: Werth (1964), p. 103ff; Werth (1971), p. 232ff. War deaths: Rummel (1990), p. 151ff.

179 "In the ... women:" Werth (1971), p. 24.

179 "Ninety-eight ... vanished": quoted ibid., p. 232f.

179 "A single ... us": quoted in Holloway (1981), p. 183.

CHAPTER TEN: A PRETTY GOOD DESCRIPTION

180 "Where will they fight?": Emilio Segré, personal communication, vi.83.

181 Soviet force reductions: Werth (1971), p. 63ff; US force reductions:

Kohn and Harahan (1988), p. 74, n. 78.

181 " 'Give [the ... it' ": Djilas (1962), p. 114f.

181 "he regarded ... enemy": ibid., p. 82.

181 "A lot ... problem": Golovin (1989), p. 201.

181 "Until 1945 ... combines": Khariton and Smirnov (1993), p. 26.

182 Special Committee: Stickle (1992), p. 203, n. 14.

182 "Stalin's word ... power": Alexandrov (1988).

182 "Once the ... it": Khariton and Smirnov (1993), p. 26.

182 "I also ... published": Taschereau and Kellock (1946), p. 455.

183 Gouzenko preparing to defect for a year: since September 1944; cf. Gouzenko (1948), p. 215.

183 Oranges: Sawatsky (1984), p. 7.

183 "You take ... things": quoted ibid., p. 5.

183 "a very ... gentleman": A. Clare Anderson, quoted ibid., p. 1.

183 109 documents: Gouzenko (1948), p. 264.

183 "You still ... happened": ibid., p. 267.

184 "short, with ... fright": quoted in Sawatsky (1984), pp. 22–23.

184 "could see ... crazy": Gouzenko (1948), p. 268.

184 "They were ... do": quoted in Sawatsky (1984), p. 26.

184 "[Gouzenko] was ... fantastic": quoted ibid., pp. 28–29.

184 "I am ... back": Gouzenko (1948), p. 270.

184 "The day ... office": ibid., p. 271.

185 "It's two ... government": ibid., p. 272.

185 RCMP inspector: Sawatsky (1984), p. 37.

PAGE

185 "I said . . . tomorrow' ": quoted ibid., p. 38.

185 Main a corporal: ibid., p. 40.

185 "He was . . . him": quoted ibid., p. 41.

185 "three or . . . them": John MacDonald, quoted ibid., p. 43.

186 "You weren't . . . thought": Gouzenko (1948), p. 277.

186 Robertson decided to send Gouzenko back: Hyde (1980), p. 23.

186 Stephenson and protective custody: Sawatsky (1984), p. 47ff. Robert Lamphere doubts that Stephenson would have exposed himself by personally visiting Gouzenko's apartment. Robert Lamphere, personal communication, vi.94.

186 "It was . . . go": quoted in Hyde (1980), pp. 26–27.

186 "Narrated the . . . implicated": quoted ibid., pp. 46–47.

187 "above all . . . direction": quoted ibid., p. 46.

187 Ruth Greenglass to New York with David: JCAE (1951), p. 92.

187 "He came . . . so": ibid., pp. 92–95.

187 "I did . . . again": quoted (from Ruth Greenglass's trial testimony) in Nizer (1973), p. 125.

187 "lens molds . . . beryllium": extract of David Greenglass's confession typed 19.vii.50, quoted in Anders (1978), p. 390ff.

188 Greenglass and Munroe effect: cf. ibid., p. 394.

188 Cone initiator an advanced design: Rubby Sherr, personal communication, xi.93.

188 Cone initiator patent: ibid.

188 "which was . . . Alamos": Greenglass FBI FOIA files, 65-59028-332, p. 87.

188 "The solid . . . mixtures": Hans Bethe, personal communication, v.93.

189 "The way . . . push": McPhee (1974), p. 218.

189 "She couldn't make it": Greenglass FBI FOIA files, 65-59028-345, p. 7.

189 Max Elitcher and Julius Rosenberg: Rosenberg FBI FOIA files, 94-3-4-317-348X, p. 52.

189 "told me . . . Albuquerque": JCAE (1951), p. 157.

189 Tom Black and wire transfer: Gold FBI FOIA files, 65-57449-560.

190 Gold registered in his own name, a fact the FBI discovered on May 24, 1950, *before* it knew anything about the Rosenbergs: Fuchs FBI FOIA files, 65-58805-1239.

190 "very late . . . tardy": quoted in Williams (1987), p. 215.

190 "friends with . . . weapon": quoted ibid.

190 British Mission party: cf. invitation reproduced at Fakley (1983), p. 189; Brode (1960), XI, p. 8.

190 "En route . . . deliver": Fuchs FBI FOIA files, 65-58805-1412, Fuchs's confession (following p. 51), p. 8.

190 Fuchs-Gold 19.ix.45 Santa Fe meeting: for quotations and paraphrases cf. JCAE (1951), p. 157ff; Williams (1987), p. 215ff; Gold FBI FOIA files, 65-57449-401, pp. 14–15.

191 Gold's itinerary: Williams (1987), p. 217ff; Gold FBI FOIA files, 65-57449-401, pp. 16–17.

191 Meetings with Yakovlev: Rosenberg All-Case Summary (FBI FOIA 94-3-4-317-348), p. 41.

192 "very touchy . . . up": Gold FBI FOIA files, 65-57449-401, p. 17.

192 "Plutonium is . . . elements": quoted in Groueff (1967), pp. 151–152.

192 "During our . . . ends": Smith (1954), p. 88.

192 Fuchs's September report: cf. Fuchs

PAGE

FBI FOIA files, 65-58805-1412, pp. 32–33, and 65-58805-1246, p. 3; Williams (1987), p. 191.

192 Fuchs and levitation: Fuchs FBI FOIA files 65-58805-1246, p. 3. I surmise that Fuchs reported to Yatzkov on levitation; why would he not?

193 18.x.45 Fat Man plans: Visgin (1992), p. 127ff.

193 Rubby Sherr named the Urchin: Rubby Sherr, personal communication, xi. 93.

195 "barium plastic sphere": JCAE (1951), p. 97.

195 "a plastic . . . explosives": quoted in Anders (1978), p. 390, from Greenglass's 19.vii.50, confession.

196 "With all . . . Stalin": Gubarev (1989), p. 11.

196 "At first . . . us": quoted ibid.

196 Soviet film of Nagasaki: Kazunari Fujita, Chogoku Broadcasting Company (Hiroshima), personal communication, iv.94.

196 "Is the . . . scientists": quoted in Holloway (1990).

196 Beria and Tula shotgun: Knight (1993), p. 136. Pavel Sudoplatov, who claims to have made the presentation, says "an inlaid Belgian shotgun." (Sudoplatov 1994, p. 202.)

196 "Peter Leonidovich's . . . later": Alexandrov (1988).

196 "All the . . . position": quoted in Badash (1985), pp. 62–63.

197 "coarse imitati[on]": quoted ibid., p. 63.

197 "[Comrade Beria] . . . weak": quoted in Andrew and Gordievsky (1990), p. 376.

197 "I told . . . people": quoted in Knight (1993), p. 137.

197 "I will . . . him": quoted in Holloway (1990), p. 24.

PAGE

198 "an atomic . . . it": quoted in Herken (1980), p. 48.

198 "At one . . . bomb' ": RG 77, MED, 20 (miscellaneous), National Archives. Cf. also Herken (1980), pp. 48–49.

CHAPTER ELEVEN: TRANSITIONS

201 "essentially came . . . period": Norberg (1980f), p. 26ff.

201 "a new . . . nature": Rabi (1970), p. 138.

201 "It kind . . . on": Norberg (1980f), p. 25.

202 "We all . . . peacetime": Bethe (1982a), p. 45.

202 "And Fermi . . . cold' ": Robert Oppenheimer, "The Atomic Age," 1.ix.50, p. 7, Oppenheimer Papers, LC.

202 "No monopoly . . . it!": quoted in Gleick (1992), p. 204.

202 Feynman in Manhattan bar: Richard Feynman, personal communication, x.76.

202 "We used . . . in": Feynman (1985), p. 115.

203 "an extraordinary . . . breakdown": Chester Barnard, paraphrased in Oppenheimer FBI FOIA files, "Julius Robert Oppenheimer," 18.iv.52, p. 37.

203 "You will . . . despair": Smith and Weiner (1980), p. 297.

203 "We were . . . future": Oppenheimer (1946a), p. 265.

203 "our only . . . thing": Edward Teller to Leo Szilard, 2.vii.45. Box 71, Oppenheimer Papers, LC.

203 "not very . . . strongly": Asahi Shimbun interview with Edward Teller, 10.vi.91, tape 4, p. 5.

203 Teller's cover note: "Dear Oppy, You may have guessed that one of the men 'near Pa Frank' whom I

PAGE

have seen in Chicago was Szilard. His moral objections to what we are doing are in my opinion honest. After what he told me I should feel better if I could explain to him my point of view. This I am doing in the enclosed letter. What I say is, I believe, in agreement with your views. At least in the main points. I hope you will find it correct to send my letter to Szilard." Box 71, Oppenheimer Papers, LC.

203 Ernest Lawrence's impression of Oppenheimer: Childs (1968), p. 366.

203 "We are . . . impossible: Interim Committee Scientific Panel to the Secretary of War, 17.viii.45, Oppenheimer Papers, LC.

204 "I . . . following": Smith and Weiner (1980), p. 301.

204 "in the . . . ahead": quoted ibid.

204 "the smartest . . . lot": *Time,* 29.x.45, p. 30.

204 "Dr. Oppenheimer . . . this": George L. Harrison, Memorandum for the files, 25.ix.45, National Archives.

204 "and obtain . . . scientists": Memo, "Mr. Harrison and Dr. Oppenheimer," 25.ix.45, National Archives.

205 "In the . . . wash' ": quoted in Davis (1968), pp. 257–258.

205 " 'cry baby' . . . energy": Harry Truman to Dean Acheson, 7.v.46, HSTL.

205 "Truman said . . . kids' ": quoted in Herken (1980), p. 11.

205 "There are . . . used": Harold D. Smith diary, 5.x.45, quoted in Gaddis (1987), p. 106.

205 "today that . . . humanity": Smith and Weiner (1980), pp. 310–311.

PAGE

206 "There was . . . cease": Bradbury (1948), p. 10.

206 "continual uncertainty . . . time": Norberg (1980d), p. 60.

206 "Oppenheimer thought . . . advice": ibid., p. 94.

206 "Oppenheimer made . . . avoided": P. B. Moon and R. E. Peierls to James Chadwick, 22.viii.45. RG 77, National Archives.

206 "In this . . . been": quoted in Blumberg and Owens (1976), p. 185.

206 *Darkness at Noon:* Edward Teller, personal communication, vi.93.

207 "quite favorable . . . superbomb": Scientific Panel to the Secretary of War, 17.viii.45. RG 77, National Archives.

207 "the scientists . . . policy": George L. Harrison memorandum for the record, 18.vii.45. RG 77, National Archives.

207 "that no . . . maintained": quoted in USAEC (1954a), p. 10.

207 "We feel . . . consideration": quoted in Galison and Bernstein (1989), p. 276.

207 "General Groves . . . Super": quoted in Sawyer (1954), p. 288.

207 Edward Teller 31.x.45 letter on thermonuclear: quoted in USAEC (1954a), pp. 7–9.

207 "probably at . . . enterprise": "Possibilities of a Super Bomb," James Bryant Conant to Vannevar Bush, Oct. 20, 1944, p. 2. OSRD files, National Archives.

208 "I said . . . year": Teller (1962), p. 22.

209 "I was . . . way": ibid., p. 23.

209 "The development . . . years": Robert R. Wilson for the Committee to Robert Oppenheimer, 7.ix.45, p. 3. RG 77, National Archives.

PAGE

209 "almost certainly ... time": Henry Stimson, "Memorandum for the President," 29.viii.45, pp. 1–2. RG 77, National Archives.

209 JIS "Soviet Capabilities report": JIS 80/15, 9.xi.45.

210 Farrell memorandum: "Time for Russia to make an Atomic Bomb," T. F. Farrell to L. R. Groves, 12.x.45. RG 77, National Archives.

210 Groves's 20-year estimate: Groves biographer Stanley Goldberg, personal communication, iv.94.

210 "that not ... expensive": "Fission Technology: Retrospect and Prospect," in Behrens and Carlson (1989), p. 103.

210 Groves to Lawrence: 21.viii.45. RG 77, National Archives.

211 "simply does ... situation": Groves (1948).

211 "the complete ... map": ibid.

211 Suitcase joke: Herbert York, personal communication.

211 "He was ... confidence": Schreiber (1991), p. 45.

212 "The use ... is": Truslow and Smith (1947), p. 362.

212 "The project ... suicidal": ibid., p. 358.

212 "We had ... empire": Norberg (1980a), p. 43.

212 "engineering ... established": Truslow and Smith (1947), p. 362.

212 "stockpile the ... model": "Notes on a talk given by Comdr. N. E. Bradbury at Coordinating Council, 1.x.45." 8.x.45. RG 77, National Archives.

212 "that the ... feasible?": Truslow and Smith (1947), p. 363.

213 60 bombs: anon. to Brig. Gen. T. F. Farrell, x.45, LANL Archives A-84-019, 19-4.

213 "It was ... people": Pervukhin (1985).

PAGE

213 "every institute ... metallurgy)": Golovin (1968), p. 49.

213 "Yesterday I ... production": quoted in Holloway (1994), pp. 136–137.

213 "possible indications ... A-bomb": Harriman to Sec. of State, 16.xi.45, State Dept. ALH-1910-H. RG 77, National Archives.

213 North Korean mining sites: Zaloga (1993), p. 40, citing US Far East Command sources.

214 "the Czechoslovakian ... government": Steinhardt to Sec. of State, State Dept. DVF-886-K. RG 77, National Archives.

214 Jáchymov: Proctor (1993), citing Czech sources.

214 Czech forced labor: ibid. Soviet needs: Zaloga (1993), p. 46, citing CIA sources. Cf. also CIA (1951).

214 Soviet domestic uranium sources: ibid., p. 73:10.

214 Tankograd: Werth (1964), p. 218.

214 Chelyabinsk power station: ibid., p. 622.

214 12 labor camps: Cochran and Norris (1993), p. 9.

214 Chelyabinsk details: Cochran and Norris (1991) and (1993); name withheld by request, personal communication, St. Petersburg, vi.92.

215 "S for Sudoplatov": "Special Tasks" transcript of videotaped interview with Pavel Sudoplatov.

215 "had no ... [1946]": quoted in "Russians Deny US Scientists Gave Atom Data," *New York Times,* 6.v.94, p. A5. Sudoplatov disconnected from atomic information x.46, according to the Russian agency.

215 *Special Tasks:* Sudoplatov et al. (1994).

215 Smyth Report discrepancy: Arnold

Kramish, personal communication, 28.vii.94.

215 Report to Soviet intelligence: Arnold Kramish interviewed a young physicist in 1949 who had been an assistant to Kikoin who reported reading the Smyth Report in a top secret translation in Kurchatov's Laboratory No. 2 soon after it was published in the US. Ibid.

215 "General Groves . . . edition": ibid.

215 "In spite . . . later": Lithoprint edition of Smyth Report, pp. VIII-4–VIII-5. (Copy courtesy Arnold Kramish.)

216 "Xenon . . . control rod": Wheeler (1962), p. 35.

217 "As examples . . . decay": Ostriker (1993), II, p. 19.

217 Department S translating Smyth Report: Terletsky (1973), p. 48. Department S noticing discrepancy: Arnold Kramish: "The Russian Smyth Report corresponds sentence by sentence to the September 1, 1945, Princeton edition except for [a] few exceptions. . . . [It may be that] the [pile-poisoning] discrepancy was noted by the technical editor who deliberately included it to make the Russian edition as complete as possible." Arnold Kramish to H. A. Fidler, "Russian Smyth Report," 17.ix.48. (Memorandum courtesy Arnold Kramish.)

217 "A pivotal . . . problem": Sudoplatov et al. (1994), p. 205. In the "Special Tasks" transcript of Anatoli Sudoplatov's videotaped interview with his father, Pavel Sudoplatov is less specific: "Our people ran into a particular difficulty. Our specialists who were working in a certain direction ran into a dead

end." Presumably Sudoplatov's co-authors questioned him further or otherwise filled in the blanks; Anatoli coaches his father with the phrase "To launch the atomic reactor" on the videotape.

217 "We decided . . . warm": "Special Tasks" transcript of videotaped interview with Pavel Sudoplatov.

217 "were photocopies . . . secret": Terletsky (1973), p. 13.

218 "What physicist . . . anyway?": ibid., p. 19.

218 "was inclined . . . us": ibid.

218 "starting with . . . shop": ibid., p. 21.

218 "filled with . . . egg-shaped": ibid.

218 "When we . . . it": ibid., pp. 22–23.

219 "was completely . . . Bohr": ibid., p. 23.

219 "He would . . . whom' ": ibid., pp. 24–25.

219 "the people . . . abundance": ibid., p. 32.

219 Bohr notified . . . Groves: Stanley Goldberg, "Observations on P. Sudoplatov, *Special Tasks,*" unpub. MS, v. 94, p. 7.

220 "Terletsky brought . . . Report)": Aage Bohr, e-mail to Kurt Gottfried, Cornell University, 28.iv.94, p. 2.

220 "that in . . . future": Terletsky (1973), p. 44.

220 "only international . . . independently": "Reconstructed account of conversations between Niels Bohr and Y. P. Terletsky in Copenhagen, 14 & 16 Nov 1945. Trans. from Russian original by Roald Sagdeev." (Translation slightly amended.) (Facsimile courtesy Thomas Powers.)

220 "I already . . . orders": Terletsky (1973), p. 46.

221 "the Niels Bohr Interrogation": ibid.

221 "at his . . . it": ibid., p. 47.

PAGE

221 "the decision . . . taken": Niels Bohr
to Robert Oppenheimer, 9.xi.45,
Oppenheimer Papers, LC.

221 "Question 15 . . . plutonium":
"Reconstructed account of
conversations between Niels Bohr
and Y. P. Terletsky in Copenhagen,
14 & 16 Nov 1945. Trans. from
Russian original by Roald Sagdeev."
(Translation slightly amended.)
(Facsimile courtesy Thomas
Powers.)

221 "Bohr told . . . Report": Terletsky
(1973), p. 53.

221 "crude curses . . . Americans":
quoted in AIP (1994), p. 5.

221 *The British* . . . minutely": Sakharov
(1990), p. 92.

222 "The institute's . . . handy":
Altshuler et al. (1991),
pp. 477–479.

222 "at least . . . compromised": Arnold
Kramish to H. A. Fidler, "Russian
Smyth Report," 17.ix.48.
(Memorandum courtesy Arnold
Kramish.)

222 "With the . . . upon": Snow (1981),
p. 89.

222 "a question . . . not?": quoted in
Holloway (1994), p. 148.

222 "In the . . . enough' ": Khariton and
Smirnov (1993), p. 27. I have
amended this text using a different
translation by A. Goldin, however,
of the original text of the
Kurchatov Institute talk of 12.xii.92
on which this article was based.

223 "in respect . . . need": ibid.

223 "Financial resources . . . 1948–
1949": Medvedev (1978), p. 44f.

223 "Our state . . . cars": quoted in
Holloway (1994), p. 148.

223 Kurchatov house: *Academician I. V.
Kurchatov's Memorial House* (n.d.).
Brochure distributed by the
Kurchatov Museum, Moscow.

PAGE

223 "the forester's cabin": Alexandrov
(1988).

CHAPTER TWELVE: PECULIAR SOVEREIGNTIES

224 "during this . . . requirements":
"Atomic bomb production," Lauris
Norstad to L. R. Groves, 15.ix.45.
RG 77, National Archives.

224 "plans for . . . program": Carl Spaatz
memorandum, 8.viii.45, A-84-019,
19-4, LANL Archives.

225 509th: Rosenberg (1983), p. 14.

225 "striking the first blow": quoted in
Kaku and Axelrod (1987), p. 29.

225 JCS planning document (JCS 1496):
Kaku and Axelrod (1987), p. 30.
According to Kaku and Axelrod this
was a revised version of a
document drafted 19.vii.45.

225 "Offense . . . defense": quoted from
JCS 169 1/7 (30.vi.47) in ibid.

225 October 1945 plan: JIC329/1,
"Strategic Vulnerability of the USSR
to a Limited Air Attack," cited in
Kaku and Axelrod (1987), p. 31.

225 "If we . . . us": quoted ibid.

225 "Such a . . . States": quoted in Sagan
(1994), p. 78, from *New York
Times,* 2.ix.50, p. 4.

226 "the whims . . . barbarians": Nathan
Twining in viii.53, quoted in Sagan
(1994), p. 80. "Official Air Force
doctrine manuals continued to
support preventive war ideas,"
Sagan notes (p. 79).

226 Norstad study: "Atomic bomb
production," Lauris Norstad to L. R.
Groves, 15.ix.45. RG 77, National
Archives.

226 "My general . . . excessive": quoted
in Rosenberg (1983), p. 77.

226 "the effect . . . Forces": quoted ibid.

226 "enormously expensive . . . Force":
quoted ibid., p. 78.

PAGE

226 Oak Ridge and Hanford production: cf. ibid., p. 11ff.

227 "spending [a] . . . seen": these and following quotations from Curtis LeMay, Ohio Society of New York speech, 19.xi.45. Box 41, LeMay Papers, LC.

228 "but it . . . interested": LeMay (1965), p. 396.

228 "The reason . . . program": Herbert York interview, 27.vi.83.

228 "the development . . . missile": *Preliminary Design of an Experimental World-Circling Spaceship,* Rand report SM-11827, 2.iv.46, Chapter 2. (Courtesy Gary Dorsey.)

228 LeMay and long-range detection: in August 1946: Borowski (1982), p. 188.

228 "test the . . . Navy": Strauss (1962), p. 209.

229 "nothing that . . . world-union": Teller (1946c), p. 13.

229 "to prevent . . . ends": ¶ 2, Agreed Declaration, 15.xi.45, quoted in Truman (1955), p. 542.

229 "Oppenheimer and . . . Report": quoted in Bernstein (1975), II.

230 "Once [Oppenheimer] . . . it": quoted ibid.

230 "the chief . . . group": Johnson (1989), p. 47.

230 "one of . . . life": Lilienthal (1964), p. 13.

230 Lilienthal and newspaper clipping: Lang (1959), p. 69ff.

230 "a luxurious . . . plane": Lilienthal (1964), p. 17.

230 "with the . . . *real*": ibid., p. 20.

230 "It wasn't . . . sovereignties": quoted in Lang (1959), pp. 79–80.

231 "deep pleasure . . . faith": Niels Bohr to Robert Oppenheimer, 17.iv.46. Box 31, Oppenheimer Papers, LC.

PAGE

231 "Any system . . . system": Barnard et al. (1946), p. 8.

231 "every stage . . . control": ibid., p. 6.

231 "if the . . . peoples": ibid., p. 21.

231 "then . . . misused": ibid., p. 22 (italics in original).

232 "it may . . . bomb": quoted in Lang (1959), p. 80, who reports Winne's authorship.

232 "a systematic . . . nations": Barnard et al. (1946), p. 47.

232 "This will . . . disadvantage": ibid.

233 Stalin, 9.ii.46 speech excerpts: quoted in Thomas (1986), pp. 7–17.

234 30%; 25 million: Volkogonov (1988, 1991), p. 504.

234 "Stalin used . . . system": ibid., p. 503.

234 "The Declaration . . . III": Millis (1951), p. 134.

234 "with cold . . . stone": Kennan (1967), pp. 292–293.

234 "horrified amusement . . . conspiracy": ibid., p. 294.

234 Excerpts from Kennan telegram: ibid., p. 547ff.

235 "utter ruthlessness . . . clique": quoted in Millis (1951), p. 140.

235 "had it . . . services": Kennan (1967), pp. 294–295.

236 "splendid analysis": quoted in Yergin (1977), p. 170.

236 "it came . . . President": quoted in Shlaim (1983), p. 52.

236 "unless Russia . . . Soviets": Harry Truman to James F. Byrnes, 2.i.46 (discussed but probably not sent), quoted in Yergin (1977), p. 161.

236 "official loneliness . . . end": Kennan (1967), p. 295.

236 "the dire . . . circles": Churchill (1959), p. 996.

236 Byrnes and Truman read Churchill

PAGE

speech, Truman denied doing so:
Rossi (1986), p. 114, p. 118; Clifford
(1991), p. 102.

237 Byrnes Overseas Press Club
speech: Graybar (1986), p. 890.

237 "The President . . . ahead":
Churchill (1959), p. 996.

237 Churchill Iron Curtain speech:
Vital Speeches of the Day 12,
15.iii.46, p. 329ff.

238 "Mr. Churchill . . . world": quoted
in Ingram (1955), pp. 31–32.

238 "I found . . . was' ": Smith (1950),
pp. 28–29.

239 "reputation was . . . spell": Acheson
(1969), p. 154.

239 "I [could . . . start now": Baruch
(1960), p. 361.

239 "the only . . . way": Truman (1956),
p. 10.

239 "was most . . . are!' ": Baruch
(1960), p. 363.

239 "did not . . . crucial": ibid., p. 361.

239 "swift and . . . rules": ibid., p. 367.

240 "The 'swift . . . occurring": Acheson
(1969), p. 155.

240 "There . . . died": ibid., p. 156.

240 "was the . . . meetings": quoted in
Davis (1968), p. 260.

240 "Once I . . . encountered": Baruch
(1960), p. 365.

240 "I found . . . focussed": quoted in
Bernstein (1975), p. 84f.

241 "I think . . . within": Else (1981),
p. 26.

241 "Whether we . . . it": I. I. Rabi,
interviewed by Bill Moyers in *A
Walk Through the Twentieth
Century*.

241 "doubts [about . . . minds":
Oppenheimer (1948a), p. 246.

241 "more reality . . . peoples": ibid.,
p. 250.

241 "When will . . . never": Davis
(1968), p. 260.

242 "Khariton arrived . . . consent":

PAGE

Zukerman and Azarkh, unpub. MS,
p. 50.

242 "This time . . . that": ibid., p. 51.

242 "It was . . . points": quoted ibid.,
p. 141.

243 Vannikov; "We immediately . . .
realized": ibid., p. 169.

243 Sarov names: Khariton and
Smirnov (1993), p. 20.

CHAPTER THIRTEEN: CHANGING HISTORY

244 "several further . . . 1946": quoted
in Williams (1987), p. 191. Perrin
paraphrases Fuchs to say the
meetings took place with *"the*
Russian agent," i.e., Gold, but that
appears to be an assumption.

244 Fuchs denied "Sonia": "Not long
after he had been in prison . . . he
learned from a visitor . . . that Sonia
was safely in East Germany and he
then told MI5 officers that his
courier had been Kuczynski's
sister." Pincher (1984), p. 145.

244 Heisenberg approach: cf. Walker
(1989), p. 184ff.

245 "a physicist": quoted in Dobbs
(1992), p. A37.

245 "Yakovlev . . . information": JCAE
(1951), p. 158.

245 Maclean and Attlee visit: Cecil
(1989), p. 72.

245 "was very . . . appointments": Gold
FBI FOIA files, 65-57449-486,
p. 17.

245 Halperin's address book:
reproduced at Fuchs FBI FOIA
files, 65-58805-1202, p. 26. The
Canadian roundup took place on
16.ii.46.

246 Nunn May arrest: cf. Hyde (1980),
pp. 38–39 and ff.

246 "On previous . . . Greek": Gold
(1965b), p. 28.

246 Lafazanos: cf. Fuchs FBI FOIA files,

65-58805-1202, p. 10; Gold FBI
FOIA files, 65-57449-185, p. 39;
-491, p. 39ff; -486, p. 18ff; -549, pp.
13–14.

246 Gold vitamin laboratory loan: Gold
FBI FOIA files, 65-57449-185, p. 39.

246 "soon after . . . Kingdom": quoted
in Williams (1987), p. 191;
information Fuchs passed: Fuchs
FBI FOIA files, 65-58805-1412,
p. 25ff.

246 Fuchs heard reports: chronicled at
ibid., 65-58805-1246, pp. 5–6.

246 Fuchs passing information about
the Super: Perrin reports of his
interview with Fuchs: "During 1947
Fuchs was asked on one occasion
by the Russian agent for any
information he could give about
'the tritium bomb.' He said that he
was very surprised to have the
question put in these particular
terms and it suggested to him . . .
that the Russians were getting
information from other sources."
(Quoted in Williams (1987), p.
192.) Norman Moss, however, in
his biography of Fuchs, quotes the
actual exchange between Perrin
and Fuchs based on a
contemporary transcript of the
interview, which was recorded. It
tells a significantly different story:
"Fuchs agreed with Perrin's
suggestion that the most important
information he gave the Russians
came from Los Alamos. When he
was talking about his meetings with
Raymond [i.e., Harry Gold] in Santa
Fe, he said at one point: 'They
asked me what I knew about the
tritium bomb, the super. I was very
surprised because I hadn't told
them anything about it.'
"Perrin said: 'Let me get this

clear. *They* asked *you* what you
knew?'
"Fuchs: 'Yes . . . [sic] I hadn't told
them anything about it. I was
surprised.'
" 'Did you tell them anything?'
" 'I gave them some simple
information. I couldn't explain it to
Raymond because he wouldn't
understand a thing. All I could give
them was something on paper.' "
(Moss, 1987, p. 144.)
If Raymond—Harry Gold—
asked Fuchs about the Super, then
the exchange could not have taken
place in 1947; it had to have taken
place during one or another of
Fuchs's meetings with Gold after
Fuchs took up residence at Los
Alamos and learned about the
thermonuclear work—that is,
sometime between February 1945
and June 1946. Perrin's further
discussion of what information
Fuchs passed along, which
included "the current ideas in Los
Alamos *when he left* on the
design," suggests that Fuchs passed
information on the thermonuclear
in Britain as well; apparently Perrin
conflated the two locations and
assigned the events the later date.
Bruno Pontecorvo and Alan Nunn
May are both possible independent
Soviet sources for information
about a "tritium bomb"—both
were aware of, and involved in,
heavy-water research and
development during the war.
"Perseus," of course, is another
possible source. Perrin's misdating
of Fuchs's contacts has been a
pervasive source of confusion in
the literature, most recently to
David Holloway, who almost

PAGE

caught the discrepancy. ("If Fuchs is correct about the date, the information received by Kurchatov in 1946 must have come from another source." Holloway, 1994, p. 296. It was not Fuchs but Perrin who was wrong about the date.) Lewis Strauss was also confused; his belief that thermonuclear spying continued at Los Alamos after Fuchs left—one important reason why he suspected Robert Oppenheimer—appears to have originated in Perrin's conflation.

246 "Only in . . . energy": quoted in Lang (1959), p. 69.

247 "has a . . . hydrogen": quoted in "Concerning uranium. Tonizo Laboratory, April 43." (Document copy and translation in the private collection of P. Wayne Reagan, Kansas City MO.) Cf. Rhodes (1986), p. 375.

247 "an enormous . . . helium": Oliphant et al. (1934), p. 694.

247 "This was . . . nucleus": Irving (1967), p. 45.

247 "In the . . . reach": York (1976), p. 21.

248 Fermi, Teller and the thermonuclear: cf. Rhodes (1986), p. 374ff.

248 Konopinski and tritium: JCAE (1953), p. 3.

248 D + T cross section: USAEC (1954a), p. 15.

248 Teller Los Alamos studies: JCAE (1953), p. 3.

248 Ulam's first assignment: Ulam (1976), p. 148ff.

249 "All the . . . theoreticians": Ulam (1966), pp. 595–596.

250 Von Neumann photographic memory: Goldstine (1972), p. 167.

250 "The story . . . humor": ibid., p. 176.

PAGE

250 "Sometime in . . . ENIAC": ibid., p. 182.

250 "the most . . . computers": ibid., p. 191.

251 "The logical . . . out": quoted ibid., p. 193.

251 "In early . . . trial": Metropolis and Nelson (1982), p. 352.

251 ½ million cards, 100 man-years: Aspray (1990), p. 47.

251 "It seemed . . . problems": quoted ibid., p. 47.

251 Super Handbook, 5.x.45 technical report: JCAE (1953), p. 10.

252 Teller boosting disclosure: Fuchs FBI FOIA files 65-58805-1246, p. 10.

252 "present knowledge . . . made": Frankel (1946), abstract.

252 "a large-scale . . . justified": Frankel (1946), p. 47.

252 "concomitant with this program": JCAE (1953), p. 11.

252 Super Conference chronology: Fuchs FBI FOIA files, 65-58805-1246, pp. 6–8.

252 "The classical . . . happen": J. Carson Mark interview, 3.vi.94.

253 "scale is . . . attained": Bretscher et al. (1946, 1950), p. 4.

253 "on Edward . . . elsewhere": Serber (1992), p. 4, n. 2.

253 Cylindrical implosion: Fuchs FBI FOIA files, 65-58805-1412, p. 34.

253 "Dr. von . . . process": ibid., 65-58805-1246, p. 6.

253 Fuchs's claim; "laughingly": ibid., 65-58805-1412, p. 28.

253 "Edward first . . . recovered": Serber (1992), p. xxxi.

254 "I can . . . system": Bradbury press conference, 24.ix.54, LANL Archives.

254 "tritium plus . . . reaction": Fuchs FBI FOIA files, 65-58805-1246, pp. 7–8.

PAGE

254 Fuchs could not remember: ibid., 65-58805-1412, p. 34.

254 "several times . . . material": Bretscher et al. (1946, 1950), p. 4.

254 "indicated that . . . ignite": ibid., p. 24.

254 "conclusively demonstrated": ibid., p. 25.

255 "It is . . . feasible": ibid., p. 44.

255 "that further . . . policy": ibid., p. 46.

255 Tritium production: Hansen (1994a), p. 34.

255 Serber on optimistic Super Conference report: Robert Serber, personal communication.

255 "the promising . . . continuation": Ulam (1976), p. 184.

255 "essentially the . . . conference": Bretscher et al. (1946, 1950), Foreword, p. 1.

255 "We are . . . dead": Bernard Baruch, "Atomic Energy Control," *Vital Speeches of the Day* XII:18, 1.vii.46, p. 546.

255 Mark and theoreticians' time: "There was about as much time devoted in the Theoretical Division during this period [1946–1950] to studies of thermonuclear problems as to studies of fission weapons." Mark (1954, 1974), p. 12.

256 Gurevich et al. report: Gurevich et al. (1946, 1991).

256 "I think . . . realized": quoted in Gershtein (1991).

256 "an American . . . country": Visgin (1992), Doc. # 14, 31.xii.46.

256 Golovin says: in a talk at the Woodrow Wilson Center on 7.x.92; Gregg Herken, personal communication.

256 "Zeldovich looked . . . work": Doran (1994), III.

256 "In order . . . material]": Gurevich et al. (1946, 1991).

PAGE

256 "Alarm Clock"; "wake up . . . world": JCAE (1953), p. 1.

256 "was probably . . . mission": ibid., p. 12f.

256 "research on . . . Sciences": Romanov (1990).

257 "the status . . . identical": ibid.

257 "We had . . . over": Stickle (1992), p. 133.

257 "One day . . . loyal": Conquest (1991), p. 274.

257 "the drought . . . USSR": quoted in Werth (1971), p. 219.

258 "the housekeeper . . . ruins": Alliluyeva (1967), p. 189.

258 "Does the . . . States?": quoted in Ingram (1955), p. 57.

258 "Merciless beatings . . . grinder": quoted in Andrew and Gordievsky (1990), p. 273.

258 "Plants were . . . reactors": Pervukhin (1985).

258 CIA estimate: Zaloga (1993), p. 282, n. 20.

258 Greenglass discharge: Greenglass FBI FOIA files, 65-59028-78, p. 11.

258 G & R Engineering; April 1946: ibid., pp. 5–6.

259 "I got . . . told"; "He wanted . . . stall": Greenglass FBI FOIA files, 65-59028-332, pp. 79–80. Cf. also JCAE (1951), p. 99.

259 "I told . . . it": quoted in Nizer (1973), p. 132.

259 Fuchs's last act at Los Alamos: Brien McMahon and General Omar Bradley discussed this point on 23.ii.50: *"The Chairman:* I am also advised that before Fuchs left Los Alamos, he withdrew from the archives everything that we had on hydrogen bombs, kept it for an inordinate period of time. They also contained diagrams for the proposed construction of them,

PAGE

and there is no question, but what they went to the Soviets in—I won't be certain of the date, but I think it was 1946. *General Bradley:* That is my understanding." JCAE (1953), p. 42.

259 "to give . . . experience": Fuchs FBI FOIA files, 65-58805-1412, p. 36.

259 "He is . . . history": Hans Bethe interview, 3.v.93.

259 Feklisov: cf. Feklisov (1990); Nags Head pub: Doran (1994), III.

259 "At his . . . come": Feklisov (1990).

260 "a scheme . . . it": Gold FBI FOIA files, 65-57449-503, p. 5.

260 "he had . . . home": ibid., 65-57449-184, p. 18.

260 "Gold thereupon . . . family": ibid., 65-57449-42, pp. 7–8.

260 "In 1946 . . . offers": ibid., 65-57449-184, p. 13.

260 Canadian report: Taschereau and Kellock (1946).

260 "defenseless and . . . explosion": Altshuler et al. (1991), p. 40.

261 Fat Man scale model: Zaloga (1993), p. 53, citing a Soviet document; about 14″: 350 millimeters: Lev Altshuler, personal communication, 27.iv.93.

261 One of Beria's aides; RDS: Khariton and Smirnov (1993), p. 20.

261 Soviet Bikini observers: Weisgall (1994), p. 144; Holloway (1994), p. 163.

261 US stockpile numbers: Darol Froman to Morris Kolodney, 13.vi.46; Darol Froman to James Taub, 13.vi.46, LANL Archives.

261 "The damned . . . again": quoted in Weisgall (1994), p. 186.

261 "Well, it . . . overrated": Bradley (1948), p. 58.

262 "Not so much": quoted in Weisgall (1994), p. 187.

PAGE

262 "obvious miscalculation . . . results": quoted ibid., p. 204.

262 "common blackmail . . . disarmament": *New York Times,* 4.vii.46, p. 4.

262 "the Bikini . . . intimidation": quoted in Graybar (1986), p. 900.

262 "not a . . . threat": quoted ibid., p. 902.

262 "possible by . . . less": "Memorandum for General Groves, Remote Air Sampling," Philip G. Krueger, 18.ix.46, LANL Archives.

263 "(1) Atomic bombs . . . possible": quoted in Rosenberg (1983), pp. 93–94. For evaluation board report cf. Ross and Rosenberg (1989), "The Final Report of the Joint Chiefs of Staff Evaluation Board for Operation Crossroads," 30.vi.47.

263 "Experience, experience . . . experience" : LeMay (1965), p. 400.

263 "I wish . . . defense": Commanding General of the MED to JCS, 7.viii.46, quoted in Graybar (1986), p. 904.

CHAPTER FOURTEEN: F-1

264 F-1 begun in July: according to Golovin (1968), p. 54, the first of four exponential piles was completed at Laboratory No. 2 on August 1.

264 "The brand-new . . . war": ibid., p. 46.

264 "Usually about . . . minister' ": Zukerman and Azarkh, unpub. MS, p. 117f.

265 Kurchatov details: Raisa Kuznetsova, Kurchatov Museum, personal communication, vi.92.

265 "But administrative . . . him": Golovin (1968), p. 46.

PAGE

265 "was taking . . . existed": Alexandrov (1988).

265 "We see . . . safety": Knyazkaya (1986).

266 "We needed . . . them": Alexandrov (1988).

266 "The Germans . . . it": Nikolai Ivanov interview, vi.92.

266 Nikolaus Riehl and uranium processing: Nikolai Ivanov interview, vi.92.

266 xi.46: "In November 1946, new specifications for calcium of higher purity (comparable to American purity requirements) were received at Bitterfeld." CIA (1951), Sec. 73, p. 16.

266 "The specifications . . . piles": CIA (1951), Sec. 73, p. 16.

266 "Before the . . . bomb": Frish (1992), ch. 12.

267 "the ability . . . technology": ibid.

267 "305": A. White to Dr. C. E. Till, Argonne National Laboratory memorandum "Re: the 305 Reactor," 14.iv.94. (Courtesy Charles Till.)

267 F-1–305 comparison table: adapted from Kramish (1959), p. 112.

267 "A single . . . emergency": Arnold Kramish, personal communication, 23.v.94.

268 305 parameters: A. White to Dr. C. E. Till, Argonne National Laboratory memorandum "Re: The 305 Reactor," 14.iv.94. (Courtesy Charles Till.)

268 "The first . . . espionage": Kramish (1959), p. 113.

268 Alvin Weinberg on F-1 coincidence: personal communication, 2.xi.93.

268 "To come . . . man": Charles Till, personal communication, 25.iv.94.

268 Several other Met Lab scientists: cf. e.g. Pilat (1952), p. 132ff.

PAGE

269 "rather than . . . self-confidence": Kramish (1959), p. 113.

269 "there was . . . undertaking": quoted ibid., p. 119.

269 "We paid . . . later": Pervukhin (1985).

269 "On a . . . graphite": Golovin (1968), p. 47. For details of the construction of F-1 cf. also Knyazkaya (1986) and Panasyuk (1967).

270 "The tests . . . cm^2": Panasyuk (1967).

270 "Kurchatov advised . . . available": ibid.

270 "He was . . . didn't": Yatzkov (1992).

270 "Kurchatov used . . . again": Sagdeev (1993), p. 33.

270 "Data of . . . obtained": Panasyuk (1967).

270 "The method . . . elements": ibid.

271 "Measurements and . . . cm^2": ibid.

272 "Kurchatov . . . then": Golovin (1968), p. 54.

272 "Layer by . . . clock": Knyazkaya (1986).

272 German aircraft electronics: ibid.

272 "alarmed everyone . . . rest": Golovin (1968), p. 54.

272 "This discovery . . . industry": Panasyuk (1967).

272 "reassured everyone . . . imminent": Golovin (1968), p. 54.

272 November 10: Nikolai Ivanov interview, vi.92. Golovin (1968), p. 54, says "December."

273 F-1 used all available uranium metal: Nikolai Ivanov interview, vi.92.

273 90 kg, 218 kg: Panasyuk (1967).

273 Briquettes: Kramish (1959), p. 119ff; periphery of lattice: Golovin (1968), p. 54.

273 Lend-Lease graphite: Nikolai Ivanov interview, vi.92.

PAGE

273 "The density . . . exaggerated":
Panasyuk (1967).

273 "It became . . . reaction": Golovin
(1968), p. 55. For reactor start-up
cf. also Panasyuk (1967) and
Knyazkaya (1986).

273 Criticality layer: Holloway (1994),
p. 181, has layer 55, citing a 1947
report by Kurchatov and
Panasyuk.

273 "Kurchatov was . . . ready":
Panasyuk (1967).

273 "All radiation-measuring . . .
orders": Knyazkaya (1986).

274 "Everybody got . . . steadied":
Panasyuk (1967).

274 "Ever since . . . activate": quoted in
Zukerman and Azarkh, unpub. MS,
p. 97.

274 "Kurchatov quickly . . . beginning":
Panasyuk (1967).

275 "In thirty . . . rods": ibid.

275 "Well, we . . . it": according to
Dubovsky in Knyazkaya (1986).

275 "this level . . . users": Panasyuk
(1967).

275 12 seconds: ibid.

275 Beria's visit: Golovin and Smirnov
(1989).

275 "the idea . . . scale": Yatzkov (1992).

276 "L. R. Kvasnikov . . . punishment":
ibid.

276 "security precautions . . . loyalty":
Fuchs FBI FOIA files, 65-58805-
1412, p. 38.

276 "Every day . . . design": Pervukhin
(1985).

277 "The successful . . . efforts": ibid.

277 "Everyone understood . . . trouble":
Golovin and Smirnov (1989).

277 Emergency Po capability: Hansen
(1994f), p. 109.

277 "that this . . . week": Darol Froman
to M. Kolodney, "Fabrication of 49
cores," 15.vi.46, LANL Archives.

PAGE

277 "an American . . . country": Visgin
(1992), Doc. # 14, 31.xii.46.

278 "the presence . . . relearned': Norris
Bradbury to L. R. Groves, 29.viii.46.
A-84-019, LANL Archives.

278 "The uranium . . . kilotons?": Jacob
Wechsler interview, 3.vi.94.

278 Initiator production: cf. Norris
Bradbury to L. R. Groves, 29.viii.46.
A-84-019, LANL Archives.

278 "authorization of . . . vacancy":
Schreiber (1991), p. 5.

278 "What was . . . personnel": ibid.

279 Officer training; "It was . . .
laboratory": ibid., p. 6.

279 "Since the . . . end": extract from
N. E. Bradbury to AEC, 14.xi.46, in
App. 9 to supplement to Manhattan
District History, Book VIII, Los
Alamos Project, vol. 2, Technical.
Chuck Hansen Collection.

279 "not very promising": quoted in
JCAE (1953), p. 15.

279 "I remember . . . it": Edward
Condon interview with Charles
Weiner, p. 24, American Institute of
Physics.

279 McMahon, atomic bomb and Christ:
according to John Manley; Manley
(1985), II, p. 2.

279 "an island . . . economy": quoted in
Hewlett and Anderson (1962),
p. 4.

280 "He's a . . . disaster": quoted in
Lilienthal (1964), p. 562.

280 "Reds, phonies . . . heavenly":
quoted in Gaddis (1987), p. 33.

280 "too much . . . acre": quoted ibid.

280 Clifford-Elsey report: Krock (1968),
p. 417ff.

281 "the simultaneous . . . confronted":
ibid., p. 419.

281 "a direct . . . reserves": ibid., p. 468.

281 "main deterrent . . . Union": ibid.,
pp. 477–478.

PAGE

281 "I read . . . Kremlin": quoted in Clifford (1991), p. 123.

281 "The same . . . it": Curtis LeMay to Sol Rosenblatt, 21.xi.47. Box A4, LeMay Papers, LC.

281 "Simple arithmetic . . . future": quoted in Borowski (1982), pp. 94–95, p. 107.

282 Silverplate B-29s: Wainstein et al. (1975), p. 71.

282 "I remember . . . third": quoted in Shlaim (1983), p. 94.

282 "symbolic reference . . . bomb": quoted in Clifford (1991), p. 62.

282 "This . . . thinking": ibid.

282 "The President . . . war": Ferrell (1991), p. 161.

282 "Groves had . . . world]": Norberg (1980a), p. 57.

282 "I was . . . had": quoted in Evans (1953), pp. 292–293.

283 "I knew . . . did": Norberg (1980a), p. 57ff.

283 "Probably one . . . operable": quoted in Herken (1980), p. 196n.

283 "We walked . . . document": Lilienthal (1964), p. 165.

283 "the President . . . come": Lilienthal (1980), p. 1.

284 "This news . . . paper": ibid., p. 2.

284 "He turned . . . difficulties": Lilienthal (1964), p. 165.

284 "We had . . . weapons' ": Jacob Wechsler interview, 3.vi.94.

CHAPTER FIFTEEN: *MODUS VIVENDI*

285 "Our future . . . answer": Altshuler et al. (1991), p. 41.

285 "one of . . . Archipelago' ": ibid., p. 40.

285 "the harsh . . . night": ibid., p. 41.

285 "It was . . . somewhere": Victor Adamsky interview, vi.92.

286 "Leaving the . . . them": Altshuler et al. (1991), p. 46.

286 Altshuler biographical details: Lev Altshuler interview, vi.92.

286 "an event . . . come": Zukerman and Azarkh, unpub. MS, p. 36.

286 "Levka the Dynamite Man": ibid., p. 88.

286 "Everything we . . . them": ibid., p. 53.

287 Sarov details: ibid., p. 52ff.

287 "We have . . . work": quoted ibid., p. 54.

287 "I came . . . concern": Lev Altshuler interview, vi.92.

288 "No one . . . outside": Zukerman and Azarkh, unpub. MS, p. 71.

288 "there was . . . lives": quoted in Altshuler et al. (1991), p. 586.

288 "depressing": Lev Altshuler interview, vi.92.

288 "The Bearded . . . guarded": quoted in Zukerman and Azarkh, unpub. MS, p. 71.

288 "limitless desolation . . . sorrow": Djilas (1962), p. 141f.

288 "blond, with . . . steel": Zukerman and Azarkh, unpub. MS, p. 93.

289 "Physicists in . . . vaseline": ibid., p. 59.

289 Vacuum cleaner: ibid., p. 61.

289 Castor oil: ibid., p. 70.

289 Barbershop mirror: ibid., p. 61ff.

289 "The summer . . . strength": ibid., p. 57.

289 26.xii.46 Gold/Yatzkov meeting: cf. JCAE (1951), p. 159ff; Yakovlev FBI FOIA files, 100-346193-64; Gold (1965b).

290 "He was . . . Yakovlev": Yakovlev FBI FOIA files, 100-346193-64, p. 3.

290 "but said . . . paper": JCAE (1951), p. 160.

290 Paris Metro, Gold to England: Yakovlev FBI FOIA files, 100-346193-64, p. 3.

PAGE

290 Gold to write French chemists: ibid.

291 "I told ... me": JCAE (1951), p. 160.

291 S. S. *America,* Cherbourg: Yakovlev FBI FOIA files, 100-346193-18, p. 1.

291 29.v.47: Gold FBI FOIA files, 65-57449-180, p. 6; pp. 6–9 report the 1947 episode.

291 "Brothman was ... met' ": ibid., 65-57449-591, p. 70; p. 69f report the 1947 incident as Gold confessed to it on 11.vii.50.

291 "[In] about ... it": ibid., p. 71.

292 "Harry ... not familiar": ibid., pp. 75–76.

292 Visit to Gold's home: Fuchs FBI FOIA files, 65-58805-1239, p. 4.

292 "everything ... spying": Gold FBI FOIA files, 65-57449-591, pp. 79–80.

293 "Abe kept ... spy]": ibid., pp. 81–82.

293 "repressed longing ... family": Gold (1951), p. 72.

293 "I had ... mistake": ibid., p. 48.

294 Maclean had already had access: Cecil (1989), p. 70.

294 "that we ... consent": quoted in Gowing (1964), p. 439f.

294 "full collaboration ... Japan": quoted in Acheson (1969), p. 165.

294 "full and effective cooperation": ibid.

294 3.iv.47 AEC report: "Report to the President of the United States from the Atomic Energy Commission, January 1–April 1, 1947," HSTL. Not enough uranium for Hanford: Hewlett and Duncan (1969), p. 274.

295 "Immense Russian ... territory": Werth (1971), p. 332f.

295 "to support ... pressures": quoted in Clifford (1991), p. 130.

295 "During our ... Europe": Bohlen (1973), pp. 262–263.

PAGE

296 "some action ... needed": quoted in Hewlett and Duncan (1969), p. 274.

296 "The somewhat ... agreement": Acheson (1969), p. 167.

296 "There was ... do": Lilienthal (1964), pp. 175–176.

296 "This was ... materials": ibid., p. 182.

296 "When he ... weapons": Johnson (1989), pp. 27–28.

296 "that I ... States": Vandenberg (1952), p. 354.

296 "The United ... so": *Address of Secretary of State George C. Marshall at Harvard University, June 5, 1947,* Pogue (1987), pp. 525–528.

297 "It is ... it": Lilienthal (1964), pp. 215–216.

297 "the present ... intolerable": quoted in Hewlett and Duncan (1969), p. 275.

297 "I shall ... Britain": quoted in Newton (1984), p. 56.

297 "some important ... way": Lilienthal (1964), p. 236.

298 "Got me ... complaint": Lilienthal (1964), p. 248.

298 Hickenlooper queries to JCS: JCS 1745/7, p. 34, in Ross and Rosenberg (1989), n.p.; "Nagasaki type ... 100": JCS 1745/7, Enclosure "C"; JCS 1745/15, 27.vii.48, pp. 52–53, referring to JCS 1745/5: ibid.

298 "to determine ... declassified": Cecil (1989), p. 81f. The conference was held 14–17.xi.47.

298 "that Fuchs ... clever' ": Fuchs FBI FOIA files, 65-58805-1321, p. 1.

299 "astounding ... program": Vandenberg (1952), p. 361.

299 "something that ... matter": Lilienthal (1964), p. 259.

299 "a young ... journal)": ibid., p. 258.

PAGE

299 "that before . . . uranium": ibid., p. 260.

299 "it has . . . essential": ibid., p. 260. Terms: cf. Hewlett and Duncan (1969), p. 279.

300 "accepted the . . . dispute": Lilienthal (1964), pp. 265–266.

300 "Vandenberg and . . . materials": quoted in Newton (1984), p. 64.

300 "preemption of . . . security": JCS 1745/7, 17.xii.47, Decision on JCS 1745/7, a memorandum by the Director, Joint Staff, on production of fissionable material, n.p., in Ross and Rosenberg (1989), n.p.

300 "A *modus* . . . way' ": Johnson (1989), p. 36.

300 "undramatic and . . . organization": Lilienthal (1964), p. 282.

301 " 'You know . . . right' ": Johnson (1989), p. 36. Arneson attributes the comment to "some British fellow"; Gullion, at Newton (1984), p. 180, identifies Maclean.

301 "had to . . . wraps": Cecil (1989), p. 82.

301 "left the . . . treated": Acheson (1969), p. 168.

301 "the estimates . . . period": USAEC to J. Edgar Hoover, 10.vii.51, p. 2, at Philby, Maclean, Burgess FBI FOIA files, Set I, Referrals.

CHAPTER SIXTEEN: SAILING NEAR THE WIND

302 "adding . . . justified": Ulam (1991), p. 44.

302 "tremendously long . . . anything": ibid., p. 146.

302 "I found . . . problems": ibid., p. 149.

302 "When I . . . ingenuity": ibid., p. 151.

303 "helpful, willing . . . Project": ibid.

303 "I used . . . apart": ibid., p. 174.

303 "fantastic headache . . . endured": ibid.

303 "The surgeon . . . liberally": ibid., p. 176.

303 "One morning . . . know' ": ibid., p. 177.

303 "the security . . . secrets": ibid.

303 "just like before": quoted ibid., p. 179.

304 "It occurred . . . events": ibid., p. 197.

304 "At each . . . on": ibid.

304 "statistical approach . . . treatment": Taub (1963), V, p. 751.

304 Neutrons transfer energy: "Excerpts from Supplement to Manhattan District History, Book VIII, Los Alamos Project (Y) Vol. 2, Technical," 15.x.47, section 5.5. Chuck Hansen collection.

304 Implosion as unsuitable geometry: "The explosion of a fission bomb of the ordinary design probably will be inadequate to ignite a thermonuclear reaction, both on account of the relatively small mass of active material used in such bombs and of the unfavorable geometrical conditions. New methods of assembly should be designed and tested. . . ." USAEC (1954a), p. 16.

304 "The radiation . . . explosive": Hans Bethe interview, v.93.

305 "The main . . . tricks": Norberg (1980d), p. 80.

305 "One might . . . problem": Norberg (1980c), p. 30.

305 "probably feasible . . . years": JCAE (1953), pp. 15–16.

305 "some adverse . . . account"; tritium doubling: Mark (1954, 1974), p. 8.

PAGE

305 Teller re Alarm Clock feasibility: JCAE (1953), pp. 15–16.

306 "most pessimistic": Bethe (1982a), p. 47.

306 "that may . . . probable": JCAE (1953), p. 16.

306 Lithium deuteride physics: cf. York (1976), p. 28, n. 11.

306 Lithium disadvantage: Hansen (1994a), p. 7.

306 Li^{6}D, number of kilograms needed: JCAE (1953), p. 16; Hansen (1994a), p. 50.

306 "I think . . . out": quoted in Mark (1954, 1974), p. 9.

306 "The very . . . met": ibid., p. 10.

307 Richtmyer calculation and thermonuclear: cf. ibid., p. 9.

307 "was tall . . . intelligence": Ulam (1991), p. 192.

307 "Richtmyer took . . . Clock": Edward Teller interview, vi.93.

307 "a fully . . . explosion": Mark (1954, 1974), p. 5.

307 Stockpile numbers: "Outline for stockpile reports to be received by MLC . . . as of 31 December 1947," Defense Nuclear Agency. Chuck Hansen collection.

307 Bomb assembly and delivery, end of 1947: Wainstein (1975), p. 72.

308 "In the . . . undertake": USAEC (1954b), p. 67.

308 "The problem . . . 1947": ibid.

308 "I suppose . . . weapons": ibid., p. 69.

308 "an intellectual hotel": quoted in Regis (1987), p. 5.

308 "I regard . . . benefit"; news over car radio: ibid., pp. 138–139.

309 "The new . . . spot": *Life,* 29.xii.47, p. 58.

309 "about his . . . space": ibid., p. 59.

309 "that he . . . information": FBI FOIA

files, "Julius Robert Oppenheimer," 18.iv.52., pp. 25–26.

309 "natty, energetic . . . badly": Alsop and Lapp (1954), p. 35.

310 "magnificent . . . magnificent' ": Lewis Strauss to Edward Teller, 6.vi.61, Strauss Papers, HHL.

310 "[Strauss] was . . . obstinate": Bromberg (1978), p. 6.

310 "whether to . . . Columbus": quoted in Bernstein (1975), p. 84f. For Strauss's religious doubts about Oppenheimer, cf. Pfau (1984), pp. 98–99.

310 "Strauss spoke . . . him": Emilio Segrè interview, vi.83.

310 "the only . . . security": AEC minutes of meeting No. 95, 19. viii.47. Microfilm.

310 "The debate . . . resigned": L. R. Groves, "Memorandum for personal file," 30.i.48, National Archives.

311 "a bad . . . it": Lilienthal (1964), pp. 238–240.

311 "It is . . . once": quoted in Strauss (1962), pp. 201–202.

311 William Golden, AEC, USAF and Long-Range Detection Committee: Strauss (1962), p. 201ff; Ziegler (1988).

312 "still in . . . dead": LeMay (1965), p. 401.

312 "At a . . . them": ibid., p. 411.

312 "away out . . . troops": ibid.

312 "I told . . . Belgium": ibid., (1965), p. 412.

312 "We zigzagged . . . existed": ibid., pp. 412–413.

313 "We are . . . war": Millis (1951), pp. 350–351.

313 "convinced that . . . future": quoted in Kofsky (1993), p. 82.

313 "Stalin said . . . stronger": quoted in Millis (1951), p. 327.

PAGE

313 "his . . . collar": Djilas (1962), p. 147f.

313 "Stalin spoke . . . state": ibid., p. 153.

313 Attacks on Byrnes, Forrestal, Truman: Werth (1971), p. 332.

313 "In the . . . existed": quoted ibid., pp. 335–336.

314 "vigilance campaign": quoted ibid., p. 332.

314 "There is . . . vigilant": quoted ibid., p. 336.

314 "soon developed . . . spy-mania": ibid., p. 332.

314 "[The MVD . . . stone": Rosenberg (1988), p. 181.

314 Boris Kurchatov, Khlopin and Pu extraction: Golovin (1968), p. 57.

314 "A large . . . complaint": ibid., p. 59.

315 Herbert Hoover and Kyshtym: CIA (1951), p. 73-19n.

315 "The town . . . built": name withheld by request, unpub. MS.

315 "a serious . . . electrodes": CIA (1951), p. 73-18.

315 5,500 tons of graphite; "With the . . . tons": ibid., p. 73-17.

315 "was cut . . . away": name withheld by request, interview, vi.92.

315 March 1948: Holloway (1994), p. 186.

315 "You and . . . for!": quoted ibid.

316 "When reactor . . . manhole' ": Pervukhin (1985).

316 "During this . . . later": Zukerman and Azarkh, unpub. MS, pp. 63–64.

316 "Suddenly there . . . barrel' ": ibid., p. 65.

316 "in favor . . . Truman": ibid., p. 71.

316 "charge for . . . month": Lev Altshuler, personal communication, 27.iv.93.

317 "they were . . . incident": Zukerman and Azarkh, unpub. MS, p. 66.

317 "the formation . . . Dominions": quoted in Wiebes and Zeeman (1983), p. 352.

317 "America may . . . now": quoted in Zubok and Pleshakov (1994), p. 67.

317 "the advance . . . position": quoted in Shlaim (1983), p. 94.

317 "We know . . . chance": quoted in Werth (1971), p. 268.

317 "Full information . . . Force": quoted in Kofsky (1993), p. 86.

317 US aviation industry troubles: cf. Kofsky (1993).

318 War scare: cf. ibid.

318 "Marshall talked . . . Russia": quoted ibid., p. 88.

318 Lucius Clay prediction: Shlaim (1983), p. 31.

318 "For many . . . advisable": quoted in Kofsky (1993), p. 104.

318 "as secure . . . home": quoted ibid., p. 106.

318 "that the . . . testimony": quoted ibid.

319 "Bevin judged . . . place": Cecil (1989), p. 85.

319 "did sail . . . wind": quoted in Wiebes and Zeeman (1983), p. 363.

319 "among others . . . join": quoted ibid., p. 361.

319 "during the . . . costs": Goncharov et al. (1993), p. 58.

319 "at that . . . 1948": Cecil (1989), p. 86.

320 "Clay had . . . teams": Nichols (1987), pp. 260–261.

320 *Sandstone* devices and yields: Announced U.S. nuclear detonations & tests, 1945–1962, Hansen (1994h), p. 1.

320 Soviet warship: Lilienthal (1964), p. 296; submarine: ibid., p. 301.

320 "The most . . . US": Hansen (1994f), pp. 74–75.

PAGE

321 "marked the . . . hand": Memorial Committee (1994), p. 178.

321 "During one . . . tons": LeMay (1965), p. 415.

321 "immediate . . . industry": quoted in Rosenberg (1983), p. 108.

321 Truman and Leahy: ibid., p. 109; "aggressive purposes": quoted ibid.

321 "the United . . . differences": quoted in Kofsky (1993), p. 218.

322 "a statement . . . record": quoted ibid., p. 219.

322 "in agreement . . . us": quoted ibid.

322 "threw more . . . peace": quoted ibid., p. 220.

322 Soviet response propaganda: cf. Millis (1951), pp. 442–444.

322 24.v. State Department meeting: Shlaim (1983), p. 149.

322 "not to . . . U.S.": quoted ibid., p. 96.

322 "The old . . . areas": Department of State Office of Public Affairs Information Memorandum No. 28, 7.i.49, p. 4. Elsey Papers, HSTL.

323 "Since Sokolovsky . . . Berlin": Clay (1950), p. 364.

323 "When the . . . run": ibid., p. 366.

Chapter Seventeen: Getting Down to Business

324 USAFE and RAF cargo capacity: Tunner (1964), p. 158.

324 "I didn't . . . possible": quoted in Schlaim (1983), p. 206.

324 "and proceeded . . . Department": LeMay Daily Diary, 27.vi.48, Box 47, LeMay Papers, LC.

324 "The President . . . period": Millis (1951), p. 454.

325 "It was . . . altitude": LeMay Daily Diary, 29.vi.48, Box 47, LeMay Papers, LC.

325 "No one . . . negotiations": Tunner (1964), p. 159.

325 "The Russians . . . scale": quoted in Shlaim (1983), p. 211, n. 38.

325 "Nobody regarded . . . first": LeMay (1965), p. 416.

325 "that Berlin . . . population": Bedell Smith (1950), p. 238.

325 "that the . . . Berlin": Truman (1956), p. 123.

325 "would have . . . Germany": ibid., p. 124.

325 160 C-54s: Clay (1950), p. 368.

326 "an emergency . . . afford": Truman (1956), p. 125. Truman may be merging two meetings; the debate with Vandenberg may have occurred at a second Clay visit in October: cf. Clay (1950), p. 384. Clay told Richard McKinzie that the debate took place "on my second trip." McKinzie (1974), p. 38.

326 "Truman realized . . . operation": McKinzie (1974), p. 40.

326 60 B-29s to East Anglia: Shlaim (1983), p. 237.

326 "bringing nuclear . . . regulated": Walter Millis in a 1957 book, quoted ibid., p. 238.

326 Silverplate B-29s not sent abroad: Borowski (1982), p. 128; p. 135, n. 51.

326 "the question . . . one' ": Millis (1951), p. 458.

326 "[Weapon] storage . . . base?": K. D. Nichols, "Organization for military application of atomic energy," 9.ix.48, p. 14, RG 77, National Archives.

326 "one of . . . business" ': Lilienthal (1964), pp. 388–389.

327 "Our fellas . . . anyway": quoted ibid., pp. 390–391.

327 "I don't . . . around": quoted ibid., p. 391.

PAGE

327 Forrestal and HALFMOON: Rosenberg (1983), p. 110, citing an entry in Forrestal's manuscript diary and JCS documents.

327 "The President . . . so": Millis (1950), p. 487.

327 "unanimous agreement . . . used": ibid., p. 488.

327 "Forrestal, [Omar] . . . mess": quoted in Shlaim (1983), p. 338.

328 "is blue . . . blue": Lilienthal (1964), p. 406.

328 "I believe . . . situation": quoted in Holloway (1994), p. 260.

328 Wedemeyer reviewing airlift: Tunner (1964), p. 161ff.

328 "a real cowboy operation": ibid., p. 167.

328 "Pilots were . . . airfields": ibid., pp. 160–168.

329 "You can . . . fog": ibid., p. 172.

329 "All planes . . . day": ibid.

329 Leahy aide 28.ix.48 report: "Memorandum for the President from Colonel Robert B. Landry," Box 46, LeMay Papers, LC.

330 "Neither Stalin . . . city": Bedell Smith (1950), p. 253.

330 "The Russians . . . it": Tunner (1964), pp. 184–185.

330 James Hill narrative: James Arthur Hill interview, xi.91.

331 "Here comes . . . coal!": quoted in Jackson (1988), p. 128.

331 "We were . . . colleagues": Pervukhin (1985).

331 A reactor start-up: cf. Holloway (1994), p. 186ff.

331 "At the . . . hours": *Nazis and the Russian Bomb,* NOVA #2004 (1988), Journal Graphics transcript, p. 5.

332 "Kurchatov was . . . expected": Golovin (1989), p. 203.

332 A reactor operation delayed to end

PAGE

of year: Igor Golovin, Woodrow Wilson Center talk, 7.x.92; Gregg Herken, personal communication.

332 "Given the . . . frivolous": Khariton and Smirnov (1993), p. 22.

332 "had conducted . . . it": Yuri Smirnov, personal communication, 22.ix.93.

332 Zeldovich group and Teller's design: Sakharov (1990), p. 94. Teller's Super was the design considered at the Super Conference.

332 "Toward the . . . espionage": ibid.

333 "extremely talented . . . students": Sakharov (1990), pp. 95–96.

333 "A few . . . ass!' ": ibid., p. 94.

333 "Guards sat . . . assignments": Altshuler et al. (1991), p. 483.

333 "During the . . . science": Drell and Kapitza (1991), p. 127.

333 "At twenty-seven . . . defense": ibid.

333 "It seemed . . . row": Altshuler et al. (1991), p. 483.

334 "Despite summer's . . . pursuits": Sakharov (1990), p. 96.

334 *"essential* . . . psychology": ibid., p. 97.

334 "I radically . . . released": ibid., p. 102.

334 First Idea = Alarm Clock: Carson Mark: "The first Soviet thermonuclear was identical to the Alarm Clock. Romanov says that. And an article in *Sakharov Remembered* calls it the layer cake and says that the pattern was similar to one by Teller and called the Alarm Clock and that's correct." J. Carson Mark interview, 3.vi.94.

334 "alternating layers . . . (U238)": Drell and Kapitza (1991), p. 127.

334 "This means . . . sakharization' ":

PAGE

Lev Altshuler, personal communication, 16.ix.92.

334 "later·became . . . -fission": ibid.

335 "The great . . . feasibility": J. Carson Mark interview, 3.vi.94.

335 "he'd been . . . knowledge": Sakharov (1990), pp. 102–103.

335 "affectionately named . . . production": Drell and Kapitza (1991), pp. 127–128.

335 "because a . . . time": Sakharov (1990), p. 104.

335 "When in . . . it": Altshuler et al. (1991), p. 551.

336 "The direct . . . say": Sakharov (1990), p. 105.

336 "This, then . . . messages": Lamphere and Schachtman (1986), p. 85.

337 "that someone . . . Ordnance": ibid., p. 91.

337 "Background checks . . . engineering": ibid., p. 92.

337 "He'd been . . . 1938": ibid., p. 94.

337 "acted as . . . (1944)": quoted ibid., p. 93.

337 "Christian name . . . Soviets": quoted ibid., pp. 95–96.

337 "We came . . . 1948": ibid., p. 96.

338 "on the . . . job": FBI FOIA files, Rosenberg Case Summary, p. 55.

338 "On the . . . discontinued": quoted in Lamphere and Schachtman (1986), p. 93.

338 "once talked . . . right": quoted in Schneir (1965), p. 123.

338 "On the . . . threat": Gold FBI FOIA files, 65-57449-591, p. 83.

339 Gold job at Philadelphia General: ibid., 65-57449-185, p. 31, p. 41.

339 "I fell . . . hours": Gold (1951), p. 86.

339 "that his . . . Germantown": Gold FBI FOIA files, 65-57449-185, p. 12.

339 "Even in . . . cause": Gold (1951), pp. 86–87.

339 "She . . . that area": Gold FBI FOIA files, 65-57449-520, p. 12.

340 AEC luncheon for Maclean: Cecil (1989), p. 84.

340 509th training program: Borowski (1980), p. 109.

340 "The personnel . . . mission": quoted in Borowski (1982), p. 146.

340 "Vandenberg asked . . . LeMay": ibid., p. 149.

340 "The first . . . plan' ": Kohn and Harahan (1988), p. 79.

340 "Then I . . . unrealistic": ibid.

341 "We didn't . . . job": LeMay (1965), pp. 429–430.

341 "The day . . . going": Kohn and Harahan (1988), pp. 81–82.

341 "Everybody thought . . . otherwise": ibid., p. 79.

341 "a realistic . . . air": ibid.

341 "Oh, I'll . . . upstairs": LeMay (1965), p. 433.

341 Dayton bombing scores: Borowski (1982), p. 167.

341 "just about . . . *one*": LeMay (1965), p. 433.

341 "There wasn't . . . shacks": ibid., p. 432.

341 "My goal . . . good": Kohn and Harahan (1988), p. 84.

342 "develop a . . . aggression": NSC 20/4, quoted in Borowski (1982), p. 138.

342 "remembered the . . . nothing": Kohn and Harahan (1988), p. 84.

342 "It looked . . . away": quoted in Shlaim (1983), p. 377, n. 191.

342 Counterblockade, 45%: ibid., p. 378, n. 192.

343 "When and . . . yield": quoted ibid., p. 138.

343 "[Army] General . . . 'No' ": Kohn and Harahan (1988), p. 85.

PAGE

344 "This isn't ... it": quoted in Lilienthal (1964), p. 474.

CHAPTER EIGHTEEN: 'THIS BUCK ROGERS UNIVERSE'

345 1946 speech: Curtis LeMay, "Remarks at Cleveland, Ohio," 8.x.46, Box 44, LeMay Papers, LC.

346 91%: Curtis LeMay, untitled lecture headed "General Bull, Gentlemen:"; n.d. (probably 28.iii.50; cf. LeMay Daily Diary this date), p. 16, Box 200, ibid.

346 "Hit it ... back": file biography, "Colonel Curtis E. LeMay as bombardment wing and division commanding officer," c. 1942, ibid. Connection to postwar situation: LeMay (1965), p. 436.

346 "war aim ... submission": quoted in Rosenberg (1983), p. 85.

346 "offensive measures ... us": Final Report of the Joint Chiefs of Staff Evaluation Board for Operation Crossroads, Enclosure C, The Evaluation of the Atomic Bomb as a Military Weapon, p. 111 (JCS 1691/10), Ross and Rosenberg (1989).

347 "the primary ... hours": LeMay Daily Diary, 4.xi.48, "Notes for discussion with General Vandenberg," LeMay Papers, LC.

347 "to such ... attack": quoted in Rosenberg (1983), p. 116f.

347 Most recent JCS war plan: JCS 1952/ 1, 21.xii.48 (rev. 10.ii.49), Ross and Rosenberg (1989). 30 days, 133 bombs: Rosenberg (1979), p. 70. 2.7 million dead, 4 million casualties: estimates from the Harmon Report, a more conservative document than the JCS war plan, cited in Rosenberg (1983), p. 126.

PAGE

347 Air University commitment: Rosenberg (1983), p. 118.

347 "killing a nation": quoted ibid., p. 95, n. 2.

347 Roger Ramey discussion: LeMay Daily Diary, 16.xii.48, LeMay Papers, LC.

347 "My determination ... *then*": LeMay (1965), p. 436.

347 "We took ... did": Kohn and Harahan (1988), pp. 80–81.

348 "This will ... mission": LeMay Daily Diary, 4.xi.48, "Notes for discussion with General Vandenberg," LeMay Papers, LC.

348 B-36 capabilities: Knaack (1988), p. 24.

348 "the first ... weapon": JCS 1745/18, Appendix, p. 61, Ross and Rosenberg (1989).

348 "Soviet antiaircraft ... capability": JCS 1951/1, 21.xii.48, Appendix, pp. 12–16, ibid.

349 "General LeMay ... figure": LeMay Daily Diary, 28.iii.50, LeMay Papers, LC.

349 "over 8,000 ... Honolulu": extracts from speeches and articles on B-36, Box 95, ibid.

349 *Lucky Lady II*: Borowski (1982), p. 153.

349 "Flying ... upstairs": LeMay (1965), p. 436.

349 "more closely ... presentations": Borowski (1982), p. 169.

349 San Francisco 600 times: LeMay (1965), p. 436.

349 Conscript soldier's narrative: name withheld by request, personal communication, vi.92.

350 "during the ... risk": Nikipelov et al. (1990).

350 B installation radiation exposure: ibid., Table I. Dosage relation to clinical signs: Glasstone and Dolan (1977), p. 580, Table 12.108.

PAGE

350 3–11 rem US and British estimated lifetime dose: Marshall (1990), p. 474.

350 "We used . . . bomb": Steve Fetter, personal communication.

351 Fission wastes into Techa and Arctic: Cochran and Norris (1990), p. 15.

351 "[Looking] out . . . future": International Commission Against Concentration Camp Practices (1959), p. 66.

351 "Shop No. 9": Holloway (1994), p. 189.

352 "They asked . . . away": quoted ibid., p. 203.

352 "Is it . . . war": Sakharov (1990), p. 108.

352 "Vannikov appeared . . . work": quoted in Gubarev (1989).

352 "The specialists . . . negatively": Khariton and Smirnov (1993), p. 28.

353 "We may . . . them": quoted in Zaloga (1993), p. 58. Zaloga cites no source for this version of Khariton's meeting with Stalin.

353 Test delay and second core: name withheld by request, personal communication, Moscow, vi. 92. Corroborating this authoritative anonymous information is the obvious fact that the bomb core was ready in June, when the scientists reported to Stalin; why otherwise would they have waited until late August to test? In an overlooked passage in his biography of Kurchatov, Golovin implicitly confirms production of a second core, writing that after criticality tests of the first core, "while preparations were under way [for the test] . . . Kurchatov rushed preparation of a second bomb." Golovin (1968), p. 63.

PAGE

Confirming, Golovin imputes to Zeldovich at the time of the first Soviet test the thought "True, we've got another plutonium charge ready. . . ." Golovin (1991), p. 20.

353 Sarov theater: Zukerman and Azarkh, unpub. MS, p. 73.

353 "Jim calls . . . time": quoted in Hoopes and Brinkley (1992), p. 437.

353 "Jewish or . . . wires": quoted ibid., p. 440.

353 "Bill, something . . . me": quoted ibid.

354 "Bob, they're after me": quoted ibid., p. 451.

354 "of the . . . war": Captain George N. Raines, quoted in Rogow (1963), p. 7.

354 Forrestal's suicide: cf. ibid., p. 18.

354 "a lot . . . globe": USAEC (1954b), p. 601.

354 "Well . . . it": quoted in Lilienthal (1964), p. 525.

355 "own first-hand . . . probed": McLellan and Acheson (1980), pp. 121–122.

355 "The years . . . scientifically": Mark and Fernbach (1969), p. 4.

355 "a bold . . . sound": Teller (1946b), p. 10f.

355 "One is . . . man": ibid.

355 "agreement with . . . peace": Teller (1947b), p. 356.

355 "Edward-a how . . . anything": quoted in Ulam (1991), p. 164.

355 "Due to . . . incomplete": Teller (1948a), p. 5n.

355 "world government . . . Government": Teller (1948b), p. 204.

356 "likely to . . . warfare": "The Russian Atomic Plan," Edward Teller to Norris Bradbury, 3.ix.48, LANL Archives.

PAGE

356 "Norris was . . . Alamos": Ulam (1991), pp. 192–193.

356 Oppenheimer encouraged Teller: USAEC (1954b), p. 77.

356 "Oppenheimer had . . . there": ibid., p. 714.

356 "giving most . . . science": Edward Teller to Norris Bradbury, 30.viii.48, LANL Archives.

356 "Russia was . . . Russia": USAEC (1954b), p. 654.

357 "men who . . . us": Borden (1946), p. x.

357 "galvanic effect . . . world": quoted in Herken (1985), p. 6.

357 "think straight . . . weapons": Borden (1946), p. ix.

357 "War . . . sovereignties": ibid., p. 23.

357 "war has . . . deterrent": ibid., p. 28.

357 "We are . . . truce": ibid., p. 41.

358 "giving away . . . attached": ibid., pp. 111–112.

358 "as wartime . . . all": ibid., p. 218.

358 "rocket Pearl Harbor": ibid., p. 225.

358 "If the . . . warhead": ibid., p. 219.

358 "quixotic . . . control": ibid., pp. 223–224.

358 Details of William Borden's early career: USAEC (1954b), pp. 832–833; Herken (1985), p. 39f.

358 "the Inflammatory . . . war": quoted in Herken, p. 39.

359 "incredible mismanagement": quoted in Hewlett and Duncan (1969), p. 358.

359 Strauss repeated his complaints: "Statement of Lewis L. Strauss . . . to the Joint Committee of the Congress," 9.vi.49, Box 70, Oppenheimer Papers, LC.

359 "No man . . . all": quoted in Goodchild (1980), p. 195.

359 "There was . . . face": quoted ibid., p. 196.

359 "that he . . . 1941": FBI FOIA files,

PAGE

"Julius Robert Oppenheimer," 18.iv.52, p. 56.

359 "tremendously impressed . . . program": quoted in Goodchild (1980), p. 192.

359 Frank Oppenheimer working for EOL during the war: Robert Oppenheimer: "All during the war, my brother worked in the lab. He worked very hard. And the testimony is that he did a good job." CU-369, Herbert Childs interview with Robert Oppenheimer, Box 6, reel 17, side 1, The Bancroft Library.

359 "Lawrence thought . . . visit": quoted in Davis (1968), p. 275.

359 "[Frank] did . . . that": CU-369, Herbert Childs interview with Robert Oppenheimer, Box 6, reel 17, side 1, The Bancroft Library.

360 Rabi thought Lawrence had not forgiven Oppenheimer: "To Lawrence, Oppenheimer's leaving Berkeley seemed treason." Quoted in Davis (1968), p. 254.

360 "I think . . . resolved": CU-369, Herbert Childs interview with Robert Oppenheimer, Box 6, reel 17, side 1, The Bancroft Library.

360 "fragmentary information . . . completion"; 50 and 20 bombs by 1955: "Status of USSR atomic energy project," 1.i.49, attached to R. H. Hillenkoetter to Bourke B. Hickenlooper, 13.xiii.48, JCAE Classified Document No. 129, HSTL.

360 "that it . . . it": USAEC (1954b), p. 658.

360 "As the . . . strong": Sloan Foundation (1982), pp. 67–71.

361 "asked whether . . . bombs": JCAE (1953), p. 23.

361 "On the . . . also": Robert

PAGE

Oppenheimer to Cuthbert Daniel, 30.v.49, Box 30, Oppenheimer Papers, LC.

361 63%, 75%: Rosenberg (1983), p. 121, n. 2; 400 bombs by 1.i.51: ibid., p. 121.

362 JCS expansion reasoning (including quotations): Draft of "Report by the Special Committee appointed by the President to review the proposed acceleration of the atomic energy program," 1.ix.49, DDEL.

362 "the air . . . aggression": quoted (from JCS documents) in Rosenberg (1983), p. 122.

362 "this Buck . . . universe": USAEC (1954b), p. 650.

363 $300 million: Harry Truman to Sidney Souers, 26.vii.49, HSTL.

363 Blair House meeting: cf. Lilienthal (1964), pp. 543–552.

363 "looking like a tired owl": ibid., p. 547.

363 "I am . . . weapons": quoted in Rosenberg (1983), p. 131.

363 NSC Special Committee: cf. Harry Truman to Sidney Souers, 26.vii.49, DDEL.

363 "the detonation . . . bomb": W. J. Sheehy to William Borden, 28.vii.49, JCAE Classified Document No. 515, HSTL.

363 1.vii.49 CIA status report: "Status of the U.S.S.R. Atomic Energy Project," CIA Joint Nuclear Energy Intelligence Committee, HSTL.

CHAPTER NINETEEN: FIRST LIGHTNING

364 "We traveled . . . ground": Zukerman and Azarkh, unpub. MS, pp. 79–80.

364 "Along the . . . lakes": quoted in Holloway (1994), p. 213.

PAGE

365 "A freight . . . staircases": Zukerman and Azarkh, unpub. MS, p. 80.

365 "one-story . . . observed": Holloway (1994), p. 214.

365 Bomb assembly: Golovin (1991), p. 17ff; Holloway (1994), p. 215ff.

365 "holding onto . . . posture": Golovin (1991), p. 17.

366 "It drizzled . . . problems": Zukerman and Azarkh, unpub. MS, pp. 80–81.

366 "Nothing . . . Igor": quoted in Doran (1994), II. "We'll . . . it!": quoted in Golovin (1991), p. 20. For "Nothing . . . Igor," Golovin gives "I don't think you are going to get it, Kurchatov." Ibid.

366 "In front . . . zero' ": quoted in Holloway (1994), p. 217.

366 "It worked . . . worked!": Zukerman and Azarkh, unpub. MS, p. 81.

366 German chronicler: cited in Holloway (1994), p. 215.

366 "Right then . . . relief": quoted in Gubarev (1989), p. 16.

366 "The white . . . avalanche": quoted in Holloway (1994), p. 217.

367 "The shock . . . hugging": Pervukhin (1985).

367 "Did it . . . Good!": Golovin (1991).

367 "It's urgent . . . dust!' ": Golovin and Smirnov (1989).

367 "There were . . . back": V. Vlasov, quoted in Gubarev (1989), p. 14.

368 "losing its . . . fire": quoted in Holloway (1994), p. 217.

368 History of US long-range detection: cf. Ziegler (1988); Strauss (1962), p. 201ff.

368 "We cannot . . . realistic": quoted in Ziegler (1988), p. 202.

369 "Hell! we . . . it": quoted in Strauss (1962), p. 203.

369 "grave doubts . . . actually": quoted in Ziegler (1988), p. 207.

PAGE

369 "was frantic": Roscoe Wilson, USAEC (1954b), p. 695.

369 "was busy . . . pitch": quoted in Ziegler (1988), p. 209.

369 "had come . . . atomized": quoted ibid., p. 210.

370 "not only . . . beautiful": quoted ibid., p. 213.

370 "was confident . . . burst": quoted ibid., p. 214.

370 "the door . . . unguarded": quoted ibid., p. 215.

370 Strauss conclusions re Oppenheimer and long-range detection: ibid., p. 227, n. 49.

370 "That's impossible . . . conclusions": Luis W. Alvarez oral history interview, American Institute of Physics.

371 111 samples: Ziegler (1988), p. 217.

371 Sample measurement: Ziegler explains this procedure at ibid., p. 226, n. 30.

371 "More evidence . . . [US]": John Manley, "Joe One," handwritten MS, 5.vii.85, LANL Archives.

371 "night and . . . clock": quoted in Ziegler (1988), p. 219.

371 Halifax and Mosquitos: Hyde (1980), p. 138.

371 "The 'cloud' . . . English": Wayne Brobeck, "Memorandum for the file: Long-range detection," 19.i.51, JCAE Classified Document No. 272, National Archives.

371 "By September . . . Stalin": John Manley, "Joe One," handwritten MS, 5.vii.85, LANL Archives.

371 Lab estimates of explosion: cf. Ziegler (1988), p. 219; R. W. Spence, "Identification of radioactivity in special samples," 4.x.49, HSTL.

372 plutonium and uranium: Ziegler (1988), p. 220.

372 "I recall . . . firm": Sloan Foundation (1983), p. 82f.

372 "as if . . . Jim": Lilienthal (1964), p. 569.

372 "deeply worried . . . it": ibid., p. 570.

373 "German scientists . . . question": ibid., p. 571.

373 "Admiral Souers . . . unlikely": Arneson (1969), v, p. 28.

373 "those asiatics": quoted in York (1976), p. 34.

373 "feeling badly . . . have": Lilienthal (1964), p. 572.

373 "10:30 P.M. . . . bed": ibid.

374 "a bath . . . Alamos": Ulam (1991), p. 169.

374 "met by . . . people": Sloan Foundation (1983), pp. 40–41.

374 "He saw . . . flat": Greenglass FBI FOIA files, 65-59028-11, p. 56.

374 "I hope . . . tonight": Teller (1962), p. 33. For a more candid version of the Chadwick episode, cf. Edward Teller to Lewis Strauss, 13.xi.51, Strauss Papers, HHL.

374 "to the . . . it": York (1976), p. 34.

374 "that within . . . USSR": quoted ibid.

374 At least 100 atomic bombs: estimating from official numbers of 56 in 1948 and 169 in 1949. "Nuclear Notebook," *Bul. Atom. Sci.* 49(10):57.

375 "Then, in . . . message": Lamphere and Shachtman (1986), p. 133.

375 "And so . . . suspect": ibid., p. 135.

375 22.ix.49 case file: Fuchs FBI FOIA files, 65-58805-1202, p. 22. Abe Brothman: Lamphere and Shachtman (1986), p. 240ff.

375 Gold and Sarytchev: Gold FBI FOIA files, 65-57449-341, -576X, -696; Gold (1965b).

375 "Remember John . . . New York":

PAGE

Gold FBI FOIA files, 65-57449-341, pp. 1–2.

375 Gold promotion: ibid., 65-57449-185, p. 40.

376 "completely natural ... own": Gold (1951), p. 87.

376 "Gold advised ... speechless": Gold FBI FOIA files, 65-57449-696, pp. 5–7.

376 "lay low ... arrested": ibid., p. 11.

377 "Several fragments ... Britain": Lamphere and Shachtman (1986), p. 128.

377 Fuchs's conscience: Williams (1987), pp. 184–185.

377 Philby's August appointment: Cecil (1989), p. 100. Lamphere thinks the mole was Roger Hollis. Lamphere and Shachtman (1986), p. 244.

377 Fuchs's October meeting with Arnold: Williams (1987), pp. 1–2.

377 "deal a ... Harwell": ibid., p. 185.

377 "the British ... informant": Fuchs FBI FOIA files, 65-58805-1202, p. 23. British to FBI, 29.x.49; FBI response, 2.xi.49.

377 Fuchs's promotion and house: Pilat (1952), p. 178.

377 "I happened ... impact": Sloan Foundation (1983), p. 39.

378 "that the ... announcement": "Minutes, 16th meeting of the General Advisory Committee to the U. S. Atomic Energy Commission, September 22–23, 1949," pp. 20–21, LANL Archives.

378 "Keep your ... undertaking": USAEC (1954b), p. 714.

378 "I was ... pouch' ": Sloan Foundation (1983), pp. 67, 97a–100a, 112a. Golden's letter was dated 25.ix.49.

379 23 methods; "possible methods ... all-out": USAEC (1954a), p. 20.

379 "a step ... test": ibid., pp. 20–21.

PAGE

379 *Greenhouse* named, construction begun: mid-August 1949. Hansen (1994c), p. 9.

379 "I think ... us": USAEC (1954a), p. 21.

379 "If all ... order": ibid.

379 "a fission ... weight": Hansen (1994a), p. 61.

379 "Sardines?": USAEC (1954b), p. 429.

380 "It requires ... production": USAEC (1954a), p. 21.

380 "the cost ... gram": USAEC (1954b), p. 432.

380 "If we ... table": USAEC (1954a), p. 21.

380 "that we ... weapon": USAEC (1954b), p. 432.

380 Borden conclusions: Hewlett and Duncan (1969), p. 372.

380 "keen enthusiasm ... me": Lilienthal (1964), p. 364.

381 "larger stockpile ... expedition": Strauss (1962), pp. 216–217.

381 "Strauss came ... busy": Sidney Souers oral history interview, 16.xii.54, p. 1f. HSTL.

CHAPTER TWENTY: 'GUNG-HO FOR THE SUPER'

382 "really got concerned": USAEC (1954b), p. 659.

382 "Latimer and ... impossible": ibid., p. 774. For Alvarez's narrative of this period cf. Alvarez (1987), p. 168ff.

382 "got hold ... campus": USAEC (1954b), p. 659.

382 "Talked with ... years": ibid., p. 775.

383 "a subject ... heart": ibid., p. 776.

383 "They give ... ready": ibid., p. 775.

383 "a conscious ... Contraption": Memorial Committee (1994), p. 85.

PAGE

383 "we really . . . available": Norberg (1980c), p. 36.

383 Need for calculation: cf. Mark (1954, 1974), p. 10.

383 "I was . . . it": Edward Teller interview, Los Alamos, vi.93, p. 43.

383 "Every time . . . backwards": quoted in Hershberg (1993), p. 468.

384 "Now you . . . shirt": Edward Teller interview, Los Alamos, vi.93, p. 45.

384 "We agreed . . . long": USAEC (1954b), p. 775.

384 "Ernest and . . . AEC": Alvarez (1987), p. 170.

384 "had made . . . agents": USAEC (1954b), p. 776.

384 "The two . . . taken": William Borden to JCAE files, 10.x.49, JCAE Classified Document No. 66, National Archives.

385 "Dr. Lawrence . . . us": ibid.

385 "They were . . . thing": Alvarez (1987), p. 170.

385 "We keep . . . area": Lilienthal (1964), p. 577.

386 "I must . . . subject": USAEC (1954b), pp. 777–778.

386 "Ernest Lawrence . . . decisions": Lilienthal (1964), p. 577.

386 "very happy . . . too": USAEC (1954b), p. 778.

386 "It is . . . work": quoted ibid.

386 "Following announcement . . . flexibility": USAEC (1954b), p. 452.

386 "They were . . . optimistic": ibid., pp. 460–461.

387 Lawrence and Nichols: Bundy (1988), p. 205.

387 "I think . . . bigger": Sloan Foundation (1983), p. 104f.

387 "He said . . . first": ibid., pp. 104–107.

388 "one of . . . weapon": JCAE (1953), p. 29; USAEC (1954b), p. 683.

PAGE

388 "quite contrary . . . work": Sloan Foundation (1983), II, pp. 61A-62A.

388 "long and . . . Harvard": quoted in Hershberg (1993), p. 471. Hershberg (1993) is a useful guide to the chronology of Oppenheimer's communications with Conant, but he mistakenly subscribes to the notion that Conant wrote Oppenheimer a letter outlining his views. As my reconstruction will make clear, there was no such letter, at least not as of October 21. When Bethe and Teller came to call, they would remember Oppenheimer waving a letter at them when he reported Conant's views. That was most probably *his* letter to Conant, which is clearly his first substantive communication with Conant since their Cambridge meeting. No one ever testified actually to reading the supposed Conant letter; Conant himself did not even remember seeing Oppenheimer's letter, much less writing one of his own. Robert* Serber, who talked to Oppenheimer after Bethe and Teller, agrees that Oppenheimer may not have shown him a letter, though if Conant wrote in response to Oppenheimer's letter of October 21, there would have been time for Oppenheimer to receive such a letter before Serber arrived. Testimony in the historical record about Conant's supposed letter is certainly confusing; Hershberg's inattention unfortunately only adds to the muddle. Peter Galison and Barton Bernstein also fall victim to the missing-letter syndrome; cf. Galison and Bernstein (1989), p. 288.

PAGE

388 "When we ... relevant": USAEC
(1954b), p. 242.

388 Oppenheimer in doubt: ibid.,
p. 231.

388 "Conant told ... formulated": ibid.

388 "he would ... folly": ibid., p. 76.

389 "on a ... considerations": ibid.,
p. 385.

389 "was sort ... weapons]": ibid.,
p. 387.

389 "the battlefield ... repudiated":
quoted in Hershberg (1993),
p. 476.

389 "is always ... peace": quoted ibid.

389 "I remember ... it": USAEC
(1954b), p. 246.

389 "tacitly assumed ... possible": JCAE
(1953), p. 28.

390 "why it ... program": ibid., pp. 27–
28.

390 Oppenheimer and Super progress:
as he implies in his October 21
letter; cf. USAEC (1954b), p. 242ff.

390 "we did ... bomb": ibid.,
p. 479.

390 Oppenheimer October 21 letter:
ibid., p. 242ff.

391 Oppenheimer meeting with
LeBaron and McCormack: ibid.,
p. 683.

392 "Bethe ... come": Edward Teller,
personal communication, vi.93.

392 "the greatest ... me": Bethe
(1982a), p. 52.

392 "some ... refuse": USAEC (1954b),
p. 328.

392 "to come ... Princeton": ibid.,
p. 715.

392 "equally undecided ... given":
ibid., p. 328.

392 "over my dead body": ibid., p. 715.

392 "did not ... coming' ": ibid.

393 "I remember ... powerful": quoted
in Bernstein (1980), p. 93.

393 "We both ... more": USAEC
(1954b), p. 328.

PAGE

393 missed plane, traded coats: Galison
and Bernstein (1989), p. 289.

393 "A few ... relieved": quoted in
Bernstein (1980), p. 94. At the
Oppenheimer hearing, Bethe
testified that he called Teller from
New York. Since Teller was
probably in transit at that time, I
quote Bethe's later recollection. Cf.
USAEC (1954b), p. 328.

393 "refused out of hand": Edward
Teller, personal communication,
vi.93.

393 "no reaction ... opposed": USAEC
(1954b), p. 782.

393 Pike letter, 21.x.49: USAEC (1954a),
p. 22c.

394 "The Atomic ... do": Glenn
Seaborg, personal communication,
16.xi.92.

394 Bradley's description of Super
design: Hansen (1994a), p. 68.

394 "I warned ... point": Robert Serber
interview, 22.ix.94.

394 Alvarez traveled east: USAEC
(1954b), pp. 783–784.

395 Conant moved to a moral position:
Teller would certainly have
remembered if Oppenheimer had
mentioned such a stance at their
meeting of October 21; moral
condemnation of the
thermonuclear was a position he
never forgot and never forgave. But
of Princeton he only reported
Conant's adamant "over my dead
body."

395 "I was ... opposition": Robert
Serber interview, 22.ix.94.

395 28–30.x.49 GAC meeting: cf.
minutes, A-92-024 17–8, LANL
Archives; Oppenheimer summary
reproduced in Seaborg (1990), III,
p. 317Aff.

395 "Well ... letter": Glenn Seaborg,
personal communication, 16.xi.92.

PAGE

395 "Although I . . . program": Glenn Seaborg to Robert Oppenheimer, 14.x.49, reproduced in Seaborg (1990), III, p. 282.

395 "of the . . . America"; Smith's impression of Kennan talk: Cyril Smith to Richard G. Hewlett, 27.iv.67, reproduced in Norberg (1980e), App. 2, pp. 48–50.

395 "on the . . . said": ibid., p. 49.

396 Bethe spoke of Weisskopf vision: "I transmitted this spirit [of his discussions with Weisskopf] to several members of the AEC [sic: GAC]. . . ." Bethe to Weisskopf, 31.x.49, quoted in Galison and Bernstein (1989), p. 289.

396 "Fermi said . . . it": Robert Serber interview, 22.ix.94.

396 "very solemnly . . . program": USAEC (1954b), p. 453.

396 "asked the . . . two": ibid., pp. 518–519.

396 "there were . . . weapon": ibid., p. 453.

397 "in the . . . discussion": ibid., p. 247.

397 "a result . . . churches": quoted in Hershberg (1993), p. 475.

397 "It was . . . committee": USAEC (1954b), pp. 510–511.

397 "A dramatic . . . gray": Lilienthal (1964), p. 581.

397 "Campbell's . . . Campbell's' ": ibid., p. 580.

397 "the famous . . . pictures": USAEC (1954b), p. 785.

397 "no longer . . . useless' ": Lilienthal (1964), pp. 582–583.

397 "The reaction . . . 'psychological' ": Manley (1987), pp. 12–13.

398 "[the] chief . . . months' ": Lilienthal (1964), p. 581.

398 "That was . . . it": Alvarez (1987), p. 172.

398 "I said . . . discussion": USAEC (1954b), p. 247.

398 "was dead": ibid., p. 786.

398 "Conant flatly . . . time' ": Lilienthal (1964), p. 581. Lilienthal deleted "on moral grounds" from the published text of his diary: cf. Galison and Bernstein (1989), p. 291.

399 Smyth, Dean judged reactions to be visceral: for Smyth, cf. Sloan Foundation (1983), pp. 146–148; for Dean, cf. USAEC (1954a), p. 106.

399 GAC 29-30.x.49 reports: reproduced at Seaborg (1990), III, p. 317Aff.

400 "Edward promised . . . reaction": J. Carson Mark interview, 3.vi.94.

400 "When you . . . impossible": USAEC (1954b), pp. 454–455.

401 "Fermi and . . . world": quoted in Bernstein (1975), II.

402 "a group . . . bloodthirsty' ": Lilienthal (1964), p. 582.

403 "Q. In fact . . . small": USAEC (1954b), pp. 235–236.

403 "The notion . . . with": ibid., p. 80.

403 "We thought . . . undertaking": ibid., p. 249.

404 "there [was] . . . discussion": John Manley diary, 1.xi.49, LANL archives.

404 "pretty discouraging . . . time": Lilienthal (1964), p. 584f.

404 "He was . . . gray": ibid., pp. 583–584.

404 "a frenetic . . . converts": Manley (1987), p. 15.

404 Teller wrote Alvarez: Edward Teller to Luis Alvarez, c. 31.x.49, National Archives.

404 "Robert's views prevailed": Luis Alvarez to Edward Teller, 10.xi.49, National Archives.

PAGE

404 "morose and . . . years!": Manley (1985), "Two Papers and Three Chairman: Another Teller," p. 5.

404 Truman saw GAC reports: USAEC (1954b), p. 403.

404 "that he . . . happy": John Manley diary, 9.xi.49, LANL Archives.

404 Strauss, McMahon and Johnson meetings: Marx Leva, "Afterthoughts on Strauss," *Washington Post,* 24.vi.59.

404 Truman cut off debate: Harry Truman to Sidney Souers, 19.xi.49, HSTL.

404 "The habit . . . suggestions": Bundy (1988), p. 215.

404 "There is . . . deterred": quoted in York (1976), p. 66.

405 "The Dean . . . Oppenheimer": Sloan Foundation (1983), pp. 41a–42a.

405 "Nichols, be . . . score": ibid., pp. 75a–77a.

405 "[Acheson] was . . . example"?' ": Arneson (1969), v, p. 29.

405 "His sense . . . stillborn": ibid.

405 "George was . . . day": Sloan Foundation (1983), pp. 41a–43a.

406 "I say . . . means": Lilienthal (1964), p. 614.

406 "All we . . . them": ibid., pp. 616–617.

406 JCS response to GAC: JCS to Secretary of Defense, "Request for comments on military views of members of the General Advisory Committee," 13.i.50, DDEL.

407 "made a . . . do": Dean Acheson, file memorandum, 19.i.50, HSTL.

407 "What the . . . it": quoted in Sidney Souers oral history interview, 16.xii.54, p. 8, HSTL. Souers places this statement at the time "when the Commission told [Truman] about the H-bomb," but the

context suggests that he is referring to the Special Committee meeting of 31.i.50.

407 "the Atomic . . . super-bomb": quoted in Arneson (1969), v, p. 27.

407 "the White . . . beginning": Sidney Souers oral history interview, 16.xii.54, p. 7, HSTL.

407 "February 4 . . . Russians": Ferrell (1991), p. 340.

408 "We don't . . . again)": quoted in Lilienthal (1964), p. 594. (Lilienthal's parentheses.)

408 "I never . . . done": quoted in Bernstein (1975), II.

CHAPTER TWENTY-ONE: FRESH HORRORS

411 "On January . . . Union": Arneson (1969), vi, p. 26.

411 "sort of . . . story": Lamphere and Shachtman (1986), p. 135.

411 "Were you . . . thing": quoted in Hyde (1980), p. 141. Hyde interviewed Skardon.

412 US awareness of Fuchs investigation: Williams (1987), pp. 115–116.

412 "tie on your hat": quoted in Lilienthal (1964), p. 634.

412 "the recent . . . Alamos": quoted in Williams (1987), p. 116.

412 "It will . . . publicly": quoted in Galison and Bernstein (1989), p. 311.

412 "The roof . . . orgies": Lilienthal (1964), pp. 634–635.

412 "Is it . . . Bethe": quoted in Davis (1968), p. 324.

412 "I don't . . . there": quoted ibid., p. 325.

413 "all hell . . . headquarters": Lamphere and Shachtman (1986), p. 137.

PAGE

413 "have to . . . hot": Greenglass FBI FOIA files, 65-59028-378, p. 4.

413 "I'll . . . again"; how much he loved America: Radosh (1979).

413 "something is . . . States": quoted in Greenglass FBI FOIA files, 65-59028-378, pp. 4–5.

413 "Julius said . . . one' ": ibid., 65-59028-422, p. 2.

413 Greenglass's New Year's Eve thoughts: ibid., 65-59028-378, p. 5.

413 "He came . . . him": JCAE (1951), p. 118.

414 initiator thefts: Greenglass FBI FOIA files, 65-59028-378, p. 6.

414 "Before leaving . . . time": Gold FBI FOIA files, 65-57449-696, p. 16.

414 Gold's last attempt at contact: Gold (1965a).

414 "completely panicked . . . horror-stricken": Gold (1951), p. 83.

414 "As nearly . . . man": quoted in Radosh and Milton (1983), p. 492, n. 37.

414 "did not . . . welfare": Gold FBI FOIA files, 65-57449-621, pp. 1–2.

414 "He said . . . suicide": quoted in Radosh and Milton (1983), p. 492, n. 37.

414 "Black counseled . . . there": Gold FBI FOIA files, 65-57449-402, p. 1.

414 "We walked . . . go": Greenglass FBI FOIA files, 65-59028-332, pp. 81–82.

415 Ethel Rosenberg's suggestion: ibid., 65-59028-378, p. 7.

415 Greenglasses' accident: Radosh and Milton (1983), p. 77; Greenglass FBI FOIA files, 65-59028-104, p. 1.

415 "reason to . . . agent": Lamphere and Shachtman (1986), p. 175.

415 Gamow's cartoon: cf. Ulam (1991), p. 212.

416 "Both Gamow . . . ambition": ibid.

416 "miserable": quoted in Hansen (1994a), p. 84.

PAGE

416 "The idea . . . AEC": Norberg (1980b), p. 66.

416 "for the . . . connection": Edward Teller to Robert Oppenheimer, 17.ii.50, Oppenheimer Papers, LC.

416 "I still . . . position": Hans Bethe to Norris Bradbury, 14.i.50, LANL Archives.

416 "several conversations . . . intellect": Segrè (1993), p. 238.

417 "I thought . . . motivation": Marshall Rosenbluth interview, 26.v.94.

417 "the Laboratory . . . men": Teller's response to a question from Kenneth Pitzer: "Notes of a briefing held at the Los Alamos Scientific Laboratory on February 23, 1950," p. 13. Chuck Hansen collection.

417 "Our scientific . . . themselves": Teller (1950), p. 72.

417 "If things . . . this": Norberg (1980c), p. 35.

417 "a tendency . . . views": Norberg (1980d), p. 92.

417 "It was . . . Edward": Norberg (1980c), p. 35.

417 Teller thermonuclear paper: LA-643, 16.ii.50. Chuck Hansen collection.

418 "nobody could . . . densities": Hans Bethe interview, 3.v.93.

418 "the idea . . . implementation": Altshuler et al. (1991), p. 24.

418 Sakharov's move: Sakharov (1990), p. 101.

419 Bradbury briefing: "Notes of a briefing held at the Los Alamos Scientific Laboratory on February 23, 1950." Chuck Hansen collection.

419 copper shell liner: Hansen (1994c), p. 36.

420 Loper memorandum: H. B. Loper to Robert LeBaron, "A basis for estimating maximum Soviet

PAGE

capabilities for atomic warfare."
16.ii.50, HSTL.

420 "If [the . . . indeed": quoted in
Galison and Bernstein (1989),
p. 311.

420 LeBaron endorsement: Robert
LeBaron to Secretary of Defense,
20.ii.50, attached to H. B. Loper to
Robert LeBaron, "A basis for
estimating maximum Soviet
capabilities for atomic warfare."
16.ii.50, HSTL. CIA estimate: ORE
91-49, 6.iv.50 (based on a draft of
10.ii.50), Declassified Documents
microfilm.

420 "an all-out . . . position": Louis
Johnson to the President, 24.ii.50,
HSTL.

420 "Los Alamos . . . 1943": Norris
Bradbury press conference,
24.ix.54, LANL Archives.

420 10 thermonuclear weapons; 1 kg T;
"all-out . . . bombs": Hansen
(1994b), p. 19.

420 "When I . . . sanity": quoted in
Hershberg (1993), p. 483.

420 Fuchs's trial: cf. especially Moss
(1987), p. 157ff.

421 "would only . . . himself": quoted
ibid., p. 162.

421 "a man . . . shores": quoted ibid.,
p. 161.

421 "committed certain . . . fair": quoted
ibid., p. 163.

421 "I didn't . . . thirty-eight": quoted
ibid., pp. 152–153.

422 "at any . . . land": quoted ibid.,
p. 165.

422 "Teller . . . scheme": Ulam (1991),
p. 213.

423 "Each morning . . . reactions": ibid.,
p. 214.

423 Ulam using die: Nicholas
Metropolis interview, 9.vi.93.

423 "We started . . . problem": Ulam
(1991), pp. 214–215.

PAGE

423 "worked just . . . confrontation":
ibid., p. 310.

423 "was not . . . immediately": Mark
(1954, 1974), p. 8.

423 "that the . . . fizzle": quoted in
Hewlett and Duncan (1969), p. 440.

423 "it naturally . . . reaction": Ulam
(1991), p. 215.

423 "I feel . . . efforts": quoted in
Hewlett and Duncan (1969), p. 440.

424 "Teller was . . . project": Ulam
(1991), p. 216.

424 "seemed rather . . . difficulties":
ibid., p. 217.

424 "was pale . . . today": quoted in
Hewlett and Duncan (1969), p. 440.

424 "I am . . . arrangement": John von
Neumann to Edward Teller,
18.v.50, LANL Archives.

424 "the thing . . . going": quoted in
Hewlett and Duncan (1969), p. 440.

424 "quite inadequate . . . kilograms":
summary of GAC meeting of
1.xi.50, quoted in Hansen (1994b),
pp. 43–45.

425 Kristel Heineman information:
Gold FBI FOIA files, 65-57449-491,
p. 31ff.

425 Unsub significant details: cf. ibid.,
65-57449-185, p. 85. I deduce that
this information came from the
Heinemans from the fact that their
names are handwritten in the
margin of the page, linked by an
asterisk to the text citation
"Chemurgy Design Corporation."

425 "might be the man": Fuchs FBI
FOIA files, 65-58805-1156, p. 13.

425 "Brothman and . . . 1949":
Lamphere and Shachtman (1986),
p. 143.

425 "typewritten document . . . author":
Fuchs FBI FOIA files, 65-58805-
1239, p. 5; FBI bag job in February
or March: ibid., referring to "a
confidential source with access to

PAGE

Brothman's office" which "provided on May 6, 1950 a typewritten document. . . ." The document (cf. below) was Harry Gold's paper on thermal diffusion. But 65-58805-1156, dated 30.iii.50, comparing known facts about Unsub with facts about Joseph Robbins, already refers to Abe Brothman, and on p. 5 notes specifically that "Robbins is not known to have written any articles on the thermal diffusion of gases."

425 "We instructed . . . mounted": Lamphere and Shachtman (1986), p. 143.

426 "When Gold . . . spy' ": Gold FBI FOIA files, 65-57449-185, p. 86.

426 "I was . . . on": Gold (1965b), pp. 48–50.

426 "emphatically denied": Gold FBI FOIA files, 65-57449-185, p. 87.

426 "that their . . . them": ibid., p. 88.

427 "dreary, bare and cold": Lamphere and Shachtman (1986), p. 147.

427 "There was . . . stay": ibid., pp. 146–147.

427 "considered the . . . life": ibid., p. 146.

427 "thin-faced . . . colorless": ibid., p. 148.

427 "I cannot reject them": quoted ibid., p. 149.

427 "I cannot . . . likely": quoted ibid., p. 150.

427 "only an . . . suspicions": Gold (1965b), p. 49.

427 "the search . . . effects": Gold FBI FOIA files, 65-57449-185, p. 89.

427 "The next . . . think": ibid.

428 "So when . . . them)": Gold (1965b), pp. 44–45.

428 "After about . . . information' ": Gold FOIA files, 65-57449-185, p. 89.

428 "And the . . . 23": Gold (1965b), p. 50.

428 "Yes, that . . . shoulders": Lamphere and Shachtman (1986), p. 151.

428 "It was . . . man": Greenglass FBI FOIA files, 65-59028-332, pp. 83–84.

429 "He said . . . him": ibid.

429 "He said . . . Golos . . . him": JCAE (1951), p. 104.

429 "He says . . . survive": Greenglass FBI FOIA files, 65-59028-332, pp. 23–25.

429 "golden opportunity": Gold FBI FOIA files, 65-57449-614, p. 14.

429 "never intended . . . children": ibid.

429 Spindel, Ulam, Weisskopf: Greenglass FBI FOIA files, 65-59028-11, p. 57ff.

429 Albuquerque's reasons for suspecting Teller: ibid., 65-59028-11, pp. 56–57; -38, pp. 4–5.

430 "to try . . . damage": Gold (1965b), p. 50.

430 "US Army . . . Ruth": Greenglass FBI FOIA files, 65-9028-51, p. 2.

430 June 4 photograph and kosher food: ibid., 65-59028-53.

431 "Interviewing Gold . . . left": Gold FBI FOIA files, 65-57449-229, p. 1.

431 "He came . . . money": Greenglass FBI FOIA files, 65-59028-332, pp. 26–27.

431 $107: ibid., 65-59028-57, p. 5.

431 Greenglass and Arma leave: ibid., 65-59028-57, pp. 1–2, -78, p. 1, p. 21.

431 "David said . . . it": ibid., 65-59028-332, p. 27.

431 "What I . . . us": Radosh (1979), p. 1.

432 Brooks Farm rental: Greenglass FBI FOIA files, 65-59028-193, p. 72.

432 Greenglass search: ibid., 65-59028-149, p. 28ff.

PAGE

432 "What happened . . . Lenin": Radosh (1979), p. 1.

432 "The letters? . . . happens": ibid.

432 "I made . . . then)": Gold (1965b), p. 17.

432 "I expect . . . guys": Greenglass FBI FOIA files, 65-59028-193, p. 42.

432 "I told . . . case": Radosh (1979), p. 2.

433 "it would . . . kids": ibid.

433 Ruth Greenglass statement: Gold FBI FOIA files, 65-57449-614, p. 9ff; "Jello box": ibid., p. 12.

433 David implicated Ethel: ibid., p. 7.

433 Gold met Greenglass in the Tombs: Gold denied, however, that they were ever questioned together by the FBI or by Justice Department attorneys or that they rehearsed and concerted their stories: Augustus Ballard to John Hamilton, 17.viii.65. (Courtesy Ronald Radosh.)

433 Order of the Red Star: Greenglass FBI FOIA files, 65-59028-361, p. 9.

433 "He brought . . . 1950": Gaddis and Nitze (1980), p. 174.

433 "weakening mentally . . . fast": Khrushchev (1970), pp. 307–308.

434 "America had . . . war": quoted in Holloway (1994), p. 270.

434 "a firm . . . Korea": quoted in Cumings (1990), p. 46.

434 "Please have . . . Korea": quoted ibid., p. 35.

434 North Korean armaments: Goncharov et al. (1993), p. 133.

434 December 1949: this chronology follows Goncharov et al. (1993).

434 "the Koreans . . . power": quoted ibid., p. 138.

435 "So far . . . disregarded": quoted in Cumings (1990), pp. 421–422.

435 Acheson discouraging South Korea: cf. ibid., p. 428.

435 China wanted US alliance: cf. Zhisui (1994).

435 "there was . . . ready": quoted in Goncharov et al. (1993), p. 143.

435 "made four . . . plan": quoted ibid., p. 144.

435 Mao took Truman 5.i.50 announcement to heart: cf. his statement of betrayal when the US moved the Seventh Fleet into the Taiwan Straits in late June, quoted ibid., p. 157.

436 "began to . . . Chongjin": quoted ibid., p. 147.

436 14,000 Chinese: ibid., pp. 140–141.

436 Soviet advisers proposed invasion day: ibid., p. 154.

436 strip poker: Cumings (1990), p. 545.

436 "to take . . . necessary": according to John Hickerson, quoted ibid., p. 625. Cumings chronicles Acheson's preemption of early decisions on the Korean situation.

436 "In succeeding . . . fight": ibid., p. 625.

436 "to remove . . . theory": quoted ibid., p. 888, n. 5.

436 "Korea is . . . ground": quoted ibid., p. 628.

436 "The Greece . . . East": quoted ibid., p. 629.

436 "We have . . . wolves": quoted ibid., p. 630.

436 "In my . . . meeting": quoted in Goncharov et al. (1993), p. 161.

437 Stalin's UN maneuvers: Goncharov, Lewis and Xue Litai offer these conclusions ibid.

437 "Stalin took . . . USA": Volkogonov (1988, 1991), p. 540.

437 "We slipped . . . horrible' ": quoted in Cumings (1990), p. 756.

PAGE

437 "how not . . . weapon": Kohn and Harahan (1988), p. 88.

CHAPTER TWENTY-TWO: LESSONS OF LIMITED WAR

438 "I see . . . time": Curtis LeMay, "The Strategic Air Command," Armed Forces Staff College, 11.ix.50, Box 93, LeMay Papers, LC.

438 SAC statistics: Kohn and Harahan (1988), p. 6; Borowski (1982), p. 191; 250 atomic-modified aircraft: Rosenberg (1983), p. 119.

438 "lay down . . . war"; June 6 maneuver details: "Maneuver of June 1950," Topical Study Monograph, 1950, Box 196, LeMay Papers, LC.

438 "Sunday punch": LeMay Daily Diary, 22.vii.50, ibid.

439 "The military . . . bombs": Kohn and Harahan (1988), p. 92.

439 "in essence . . . crews": A. W. Kissner to record, 24.vii.50, Box 196, LeMay Papers, LC. Cf. also LeMay Daily Diary, 27.vi.50, confirming meeting with Montague "to Bring Montique up to date on any alternate plans that we may have to mmett the situation in Korea" (sic).

439 "If we . . . all": Kohn and Harahan (1988), pp. 90–95.

440 "could see . . . approval": A. W. Kissner to record, 24.vii.50, Box 196, LeMay Papers, LC.

440 "developed considerable . . . case": Nichols (1987), p. 278.

440 "any target . . . bomb": LeMay Daily Diary, 23.i.51, LeMay Papers, LC.

440 Target lists to SAC: Rosenberg (1983), p. 165.

440 "Bonus damage . . . weapon": cf. Steiner (1991), p. 66.

PAGE

441 "ten of . . . least": LeMay Daily Diary, transcript of telephone conversation, General LeMay and General Ramey, 1.vii.50, LeMay Papers, LC.

441 "I just . . . right": LeMay Daily Diary, transcript of telephone conversation, General LeMay and General Norstad, 1.vi.50, ibid.

441 "their job . . . operations": LeMay Daily Diary, 2.vii.50, ibid.

441 LeMay proposal to go to Korea: ibid.

441 2 wings to England with atomic bombs: ibid., 8.vii.50. and ff.

441 "SAC was . . . war": ibid., 22.vii.50.

441 CIA estimate: ORE 91-49, 6.iv.50, Declassified Documents microfilm.

442 "the critical . . . USSR": draft cover letter to Finletter memorandum, 11.v.50, in LeMay Daily Diary, LeMay Papers, LC.

442 Joint Chiefs and *Greenhouse:* Anders (1987), p. 64.

442 Transfer of SAC groups to Far East; "Look out Commies!": "The Deployment of Strategic Air Command Units to Far East, July-August 1950." LeMay Papers, LC.

442 "America's warehouse": Ryukichi Imai, personal communication.

442 First-month bomb tonnages: "The Deployment of Strategic Air Command Units to Far East, July-August 1950," LeMay Papers, LC.

442 Mission directive: George E. Stratemeyer to Commanding General, FEAFBC, 11.vii.50, in LeMay Daily Diary, ibid.

443 "eighteen major . . . neutralized": quoted from the 3.x.50 *Washington Post* in "The Deployment of Strategic Air Command Units to Far East, July-August 1950," p. 30, ibid.

PAGE

443 "So we . . . that": Kohn and Harahan (1988), p. 88.

443 napalm continues to burn: Cumings (1990), p. 917, n. 146.

443 "remov[ed] from . . . earth": quoted ibid., p. 753.

443 "By 1952 . . . war": ibid., pp. 755–756.

444 Casualties: LeMay (1965), citing *The World Book;* 2 million civilians: Cumings (1990), p. 748.

444 Atomic bombs in Korea: cf. in particular Dingman (1988) and Cumings (1990) as well as specific references below.

444 "utilization of . . . job": quoted in Cumings (1990), p. 749.

444 "ten planes . . . component": LeMay Daily Diary, 30.vii.50, LeMay Papers, LC.

444 "What ten . . . know": "Transcript of telephone conversation, General LeMay and General Ramey," 30.vii.50, ibid.

445 "The Chief . . . Curt": ibid.

445 "the ten . . . Theater": LeMay Daily Diary, 31.vii.50, ibid.

445 "decision was . . . immediately": ibid., 1.viii.50.

445 "The crash . . . others"; other details of crash: ibid., 6.viii.50.

446 Planes recalled, bombs remained on Guam: "A wire was received from USAF ordering the 9th Bomb Group in Guam to return without bombs, leaving the teams there to supervise them.": ibid., 13.ix.50.

446 "the first . . . war": quoted in Trachtenberg (1988), p. 16.

446 "[Atomic-bombing] was . . . use": Hess (1972), p. 137.

446 "for three . . . atoms": William Borden to Brien McMahon, 28.xi.50, JCAE Classified Document No. 1785, National Archives.

PAGE

446 Chinese foresaw Inchon landing: Goncharov et al. (1993), p. 171ff.

446 "If we . . . disaster": quoted ibid., p. 182.

446 Chinese Politburo considerations: ibid., p. 166.

447 "that the . . . Russians": LeMay Daily Diary, 30.viii.50, LeMay Papers, LC.

447 Chinese New 4th Army: Cumings (1990), p. 734.

447 "We have . . . comrades": quoted in *New York Times,* 26.ii.92, p. A4.

447 Official order, "Well done! Excellent!": quoted in Goncharov et al. (1993), pp. 184–185.

447 For a discussion and chronology of Soviet-Chinese negotiations in x–xi.50, cf. ibid., p. 187ff.

447 "When the . . . fight' ": quoted ibid., p. 191.

448 13 air divisions, 10 tank divisions: ibid., p. 200.

448 "increasing reports . . . [aircraft]": LeMay Daily Diary, 2.xi.50, LeMay Papers, LC.

448 "the Korean . . . there": ibid., 28.xi.50.

448 "Every New . . . moment": Hersey (1950).

448 "Then I . . . President": quoted ibid.

448 "no intention . . . disaster": quoted in Dingman (1988), p. 67.

448 "our analysis . . . China": "Personal for Vandenberg from LeMay," Box 196, LeMay Papers, LC.

448 "the situation . . . greatly": quoted in LeMay Daily Diary, 6.xii.50, ibid.

449 "As you . . . job": Curtis LeMay to Emmett O'Donnell, Jr., 16.xii.50, Box 197, ibid.

449 Shoran results: Commanding General's notes, Wing Commanders Conference, 6-7.xii.50, Box 100, ibid.

449 "in such . . . quietly": Curtis LeMay

PAGE

to Emmett O'Donnell, Jr., 16.xii.50, Box 197, ibid.

449 " 'D' Day . . . capability": quoted in Cumings (1990), p. 750.

449 "believe everything is set": quoted ibid., p. 751.

449 "no substitute for victory": quoted in Dingman (1988), p. 72.

449 200 Soviet bombers, Pannikar warning: Cumings (1990), p. 751.

449 "a major attack": quoted in Dingman (1988), p. 72.

449 Dean alerted: Anders (1987), p. 134.

449 "on the . . . job"; Vandenberg: LeMay Daily Diary, 5.iv.51, LeMay Papers, LC.

450 "I . . . with him": Anders (1987), p. 137.

450 Mark IVs: ibid., p. 138; April 10: ibid., p. 142. That these were nuclear capsules and not complete bombs is clear from the LeMay Daily Diary record of the movements of the 9th Bomb Group; but Dean makes the fact explicit in his situation note of 27.iii.51, referring to "9 nuclear cores" and pointing out that "the figure 9 is arrived at . . . because there are * * * * non-nuclear parts. . . ." (Ibid., pp. 127–128.) Hoyt Vandenberg was assigned custody of the bombs as the personal representative of the President, "acting as executive agent of the JCS." Wainstein et al. (1975), p. 32.

450 "on a . . . action": LeMay Daily Diary, 7.iv.51. LeMay Papers, LC.

451 9th holding on Guam: ibid.

451 "a man . . . done": Coffey (1986), p. 276.

451 Power left for Tokyo; 9th to Guam: LeMay Daily Diary, 24.iv.51, LeMay Papers, LC. In his otherwise

PAGE

outstanding analysis of atomic diplomacy during the Korean War, Roger Dingman mistakes this first and only movement of nuclear capsules to the Far East for a "second movement westward of nuclear-configured aircraft." Dingman (1988), p. 75.

451 "the use . . . commander": Curtis LeMay to Thomas White, 7.iv.51, Box 197, LeMay Papers, LC.

451 "determined that . . . being": LeMay Daily Diary, 18.v.51, ibid.

451 "that SAC . . . exists": ibid., 28.viii.51.

451 "bombs on [Soviet] targets": Curtis LeMay to Hoyt Vandenberg, 3.viii.51, Box 198, ibid.

451 "small-scale": Curtis LeMay, "The Strategic Air Command," Armed Forces Staff College, 11.ix.50, Box 93, ibid.

451 "discontinuing the . . . squadron": Curtis LeMay to Hoyt Vandenberg, 27.xi.51, Box 198, ibid. Dingman has "the B-29s and their nuclear cargoes return[ing] home late in June 1951," based on the 9th Bomb Group's squadron history. (Dingman [1988], p. 78.) LeMay's letter makes it clear that the 9th returned home in rotation with other squadrons, but did not bring home the bombs. Cf. also LeMay Daily Diary, 24.v.51: "Gen Everest was in this morning. . . . He was briefed on our plans for rotation of a B-50 outfit to the Far East to replace the 9th Wing there now."

452 "I have . . . war": Robert Oppenheimer, National War College lecture, 1.ix.50, Oppenheimer Papers, LC.

452 "I agree . . . Force": Hoyt Vandenberg to Curtis LeMay, 5.ii.51, Box 197, LeMay Papers, LC.

PAGE

452 "The situation . . . not": "From the desk of Harry S. Truman," 27.i.52, HSTL.

454 "I, for . . . fight": "Interview of General LeMay by Mr. McNeil, Scripps-Howard Papers," 20.x.50, LeMay Daily Diary, LeMay Papers, LC.

454 USAF estimate of minimum Soviet capability in 1951: "Comments re the intelligence briefing," attached to Commanding General's Notes, Wing Commanders Conference, 6-7.xii.50, Box 100, ibid.

CHAPTER TWENTY-THREE:
HYDRODYNAMIC LENSES AND RADIATION MIRRORS

455 "We did . . . Everett": Ulam (1991), p. 219.

455 "Fermi and . . . optimistic": Richard Garwin Asahi Shimbun interview, n.d.

456 "If the . . . successfully": quoted in Ulam (1991), p. 219.

456 Cross-section measurements: Mathews and Hirsch (1991), p. xiii.

456 "Edward Teller . . . Washington": Jastrow (1983), p. 27.

456 "It was . . . fashion": J. Carson Mark interview, 3.vi.94.

457 "The idea . . . it": Marshall Rosenbluth interview, 26.v.94.

457 "an experiment . . . deuterium": Alvin Graves to Norris Bradbury, quoted in Hansen (1994c), p. 9.

457 "the energy . . . match": Jastrow (1983), p. 27.

459 "The choice . . . attention": J. Carson Mark interview, 3.vi.94.

459 "more or . . . Berkeley": Terrall (1983), p. 76.

459 "[Oppenheimer] said . . . time": USAEC (1954b), p. 307.

460 "some feeling . . . first": Robert Oppenheimer to Norris Bradbury, 15.ix.50, VFA 1448, LANL Archives.

460 Bethe traces 2nd lab to 1947: Hans Bethe, personal communication.

460 "New and . . . Super": quoted in Hansen (1994b), pp. 43–45.

460 "Teller took . . . triumph": Hewlett and Duncan (1969), p. 530.

460 "I sensed . . . Stan": Françoise Ulam to John Manley, 10.viii.88, A-92-024, 15-6, LANL Archives.

461 "That Ulam's . . . incomplete": Bethe (1982a), p. 47.

461 "One day . . . bomb' ": McPhee (1974), p. 90.

461 "was desperate . . . promise": Bethe (1982a), p. 48.

461 Teller's comments to Ernest Lawrence: Edward Teller to Ernest Lawrence, 5.xii.50, The Bancroft Library.

461 "Dr. Teller . . . bombs": USAEC (1954b), p. 437.

462 "Teller has . . . complex": Charles Critchfield interview, 11.vi.93.

462 "Why did . . . dead": Hans Bethe interview, 3.v.93.

462 "enormous flux . . . solids": Ulam (1966), p. 599.

462 "The first . . . explode": J. Carson Mark, personal communication, 16.xii.94.

463 "various people . . . found": Ulam (1991), p. 219. Ulam remembers receiving Froman's memo in February; but Arnold Kramish's recollection of calculating LAMS-1210 argues for a January date.

463 "put his . . . sketch": ibid.

463 "Engraved on . . . history": Ulam (1991), p. 311.

463 "had rejoiced . . . all": Françoise Ulam to John Manley, 10.viii.88, A-92-024, 15-6, LANL Archives.

PAGE

463 "and mentioned . . . Teller": Ulam (1991), p. 220.

464 "Teller used . . . people": J. Carson Mark, personal communication, 16.xii.94.

464 Teller explains compression: Edward Teller interview, vi.93.

464 "You can . . . time": Teller (1979), pp. 214–215.

464 "The theorem . . . collisions": ibid.

464 "I don't . . . radiation": J. Carson Mark, personal communication, 16.xii.94.

465 "eventually he . . . wrong": ibid.

465 Arnold Kramish and Schintlemeister report: Arnold Kramish, personal communication, 27.xii.94.

465 "My first . . . reply": Sakharov (1990), p. 139.

465 "I traveled . . . radiation": Arnold Kramish, personal communication, 27.xii.94.

466 Mark's explanation why HE unsuitable; "changed all that": quoted in Hansen (1994i), p. 28.

466 "an implosion . . . yield": Stanislaw Ulam to Glenn Seaborg, 22.iii.62, LANL Archives.

466 "I don't . . . strained": ibid.

466 "knew each . . . years": J. Carson Mark interview, 3.vi.94.

466 "For the . . . possibility": Stanislaw Ulam to Glenn Seaborg, 22.iii.62, LANL Archives.

466 "after a . . . enthusiastically": Ulam (1991), p. 220.

466 "He had . . . generalized": ibid.

466 "From the . . . it": Edward Teller interview, vi.93.

467 "From then . . . principles": Ulam (1991), p. 220.

467 "Teller came . . . one": Arnold Kramish, personal communication, 25.i.94.

467 "Teller assembled . . . collapse": ibid., 27.xii.94.

467 "An Estimate of [deleted] Temperatures": LAMS-1210. Declassified cover sheet courtesy Arnold Kramish.

467 "Had the . . . work": quoted in Hewlett and Duncan (1969), p. 537.

467 Teller-Ulam report: LAMS-1225. Chuck Hansen collection.

468 "In this . . . mass": ibid. (italics in original).

468 "It is . . . occurrence": Bethe (1982a), p. 49.

468 "What Ulam . . . credit": Herbert York interview, 27.vi.83.

468 "It was . . . object": Sloan Foundation (1983), p. 12, p. 29.

469 "Don't ask . . . question": Norberg (1980a), p. 76.

469 "It's sort . . . implosion": Marshall Rosenbluth interview, 26.v.94.

469 "It's true . . . Ulam": J. Carson Mark interview, 3.vi.94.

470 "[Bohr] told . . . solution' ": Nielson (1963), p. 30.

470 "Teller and . . . do": Norberg (1980d), p. 90.

470 "Knowing that . . . option": George Cowan interview, 8.vi.93.

471 "was directly . . . family!' ": Ulam (1991), p. 109.

471 Ulam thought Oppenheimer's reference pretentious: cf. ibid., pp. 170–171.

471 "my impression . . . withdrew": ibid., p. 311.

471 "Ulam felt . . . happened": J. Carson Mark interview, 3.vi.94.

472 "an equilibrium . . . device": LA-1230. Chuck Hansen collection.

472 "the very . . . concept": Bethe (1982a), p. 48.

472 "The sparkplug . . . sparkplug":

PAGE

J. Carson Mark, personal communication, 16.xii.94.

472 "I'm not ... done": Marshall Rosenbluth interview, 26.v.94.

472 Ulam at division leaders meeting: Hewlett and Duncan (1969), p. 540.

472 "immediately put ... need": Norberg (1980c), p. 38.

473 Teller debating Froman: Hewlett and Duncan (1969), pp. 540–541; Cryogenic Engineering Laboratory: Timmerhaus (1960), p. 2.

473 Carson Mark and thermonuclear division: cf. "Comments on plan for setting up separate thermonuclear division," 15.iii.51, LANL Archives.

473 "a monotonous ... movement": Allred and Rosen (1976), p. 50.

473 *Sex Life on Eniwetok:* Anders (1987), p. 144.

473 "the unreclaimed ... Cylinder": ibid., pp. 143–144.

473 "I was ... building": Herbert York, personal communication, 20.v.94.

473 "Teller's new ... quickly": Bethe (1982a), p. 48.

474 "The first ... water": Anders (1987), p. 144.

474 "Rising early ... dollars": Teller (1987), p. 80.

474 200 KT, 25 KT: Hansen (1994c), p. 50ff.

474 "the interesting ... well": quoted ibid., p. 149, n. 113.

474 "that Eniwetok ... one": Anders (1987), p. 144.

474 Item yield data: Hansen (1994h), p. 2; 5 years: J. Carson Mark interview, 3.vi.94.

475 "it was ... priorities": USAEC (1954b), p. 305.

475 Oppenheimer invitations: Robert Oppenheimer memorandum, 29.v.51, Box 175, Oppenheimer Papers, LC.

PAGE

475 "very considerable ... development": USAEC (1954b), p. 720.

475 "The report ... decisive": Teller (1982).

475 Teller's claim incorrect: cf. "Los Alamos Scientific Laboratory Thermonuclear Program," 22.vi.51. Chuck Hansen collection.

475 Froman agenda: Hewlett and Duncan (1969), p. 542; "Los Alamos Scientific Laboratory Thermonuclear Program," 22.vi.51. Chuck Hansen collection.

475 "It was ... before": Norberg (1980c), p. 42.

475 Bethe contributed: Bethe (1982b), p. 1270.

475 "knew nothing ... amazed": USAEC (1954b), p. 84.

475 "warmly supported ... approach": ibid., p. 720.

476 "We agreed ... explosion": ibid., p. 84.

476 "At the ... gone": ibid., p. 305.

476 "It is ... same": ibid., p. 81.

477 "I was ... threatened": Hans Bethe interview, 3.v.93.

477 "Princeton settled ... bomb": Edward Teller interview, vi.93.

477 "were more ... help": USAEC (1954b), p. 721.

477 "fear and ... Alamos": William Borden, "Memorandum for the file," 13.viii.51.

477 "numbered in ... thousands": J. K. Mansfield, "Memorandum for the Chairman," draft edited in William Borden's handwriting, 15.vii.51, JCAE Classified Document No. 2283, National Archives.

477 "Teller made ... deadlock": ibid., pp. 155–156.

478 "to make ... time": Anders (1987), pp. 160–161.

PAGE

478 "[Bradbury] was . . . areas":
Schreiber (1991), p. 12.

478 "Here's Edward . . . leave": Jacob
Wechsler interview, 9.vi.93.

478 "just like . . . bull": Anders (1987),
p. 164.

478 "was a . . . Super": Edward Teller
interview, vi.93.

478 "in a . . . way": Anders (1987),
p. 164.

478 "gave no . . . that": USAEC (1954b),
p. 85.

478 "in a huff": Schreiber (1991),
p. 12.

478 "he said . . . work": Jacob Wechsler
interview, 9.vi.93.

478 "cut out . . . else": Norberg (1980c),
p. 46.

479 "Edward's tragedy . . . better":
Norberg (1980a), p. 84ff.

479 "A lot . . . faster": Norberg (1980c),
p. 45.

479 "Had he . . . left": ibid.

479 "I know . . . them": quoted in
Radosh and Milton (1983), p. 167.

479 "I got . . . conversation": David and
Ruth Greenglass interview, 12.vi.79.
(Text courtesy Ronald Radosh.)

480 Irving Kaufman *ex parte*
communications: cf. Radosh and
Milton (1983), p. 277ff.

480 "to make . . . country": quoted ibid.,
p. 284; Meeropol and Meeropol
(1975), p. 34.

480 "almost Godlike": quoted in
Meeropol and Meeropol (1975),
p. 288.

480 "a horrible . . . price": Gold (1951),
pp. 117–121.

481 "Having met . . . me": Gold (1965b).

CHAPTER TWENTY-FOUR: MIKE

482 Deuterated ammonia: Jacob
Wechsler interview, 9.vi.93.

PAGE

482 "We were . . . work": Hans Bethe
interview, 3.v.93.

483 Teller–de Hoffman report: cited in
JCAE (1953), p. 64.

483 Boulder public land purchase:
Timmerhaus (1960), p. 2.

483 "In the . . . avoid": J. Carson Mark
interview, 3.vi.94.

484 "One key . . . secure]": Jacob
Wechsler interview, 9.vi.93.

484 American Car and Foundry: ibid.;
October negotiations: Hansen
(1994c), p. 81.

485 "Marshall finally . . . in": Jacob
Wechsler interviews, 9.vi.93, 3.vi.94.

485 "You could . . . Holloway' ": Jacob
Wechsler interview, 9.vi.93.

485 Carson Mark wondered: ibid.

486 "Carson is . . . fabulous": ibid.

486 Teller 2 weeks at Los Alamos:
Galison and Bernstein (1989),
p. 324.

486 "Once Teller . . . possible": Hans
Bethe interview, 3.v.93.

486 "If the . . . it": Jacob Wechsler
interview, 3.vi.94.

486 Schmoo: Jacob Wechsler, personal
communication, 27.xii.94; Raemer
Schreiber interview, vi.93.

487 "In March . . . 1952," Teller
discussion: H. A. Bethe to Gordon
Dean, 23.v.52. Chuck Hansen
collection.

487 "Carson's feeling . . . end": Jacob
Wechsler interview, 3.vi.94.

487 King a backup to Mike: Moore and
Bechanan (n.d.), p. 24.

487 "At one . . . success": Hans Bethe
interview, 3.v.93.

488 "Sir James . . . gases": quoted in
Scurlock (1992), p. 255.

488 "using the . . . automobile": F. G.
Brickwedde et al., in ibid., p. 371.

489 "ADL talked . . . overseas": Jacob
Wechsler interview, 9.vi.93.

PAGE

489 Dewar technology: Timmerhaus (1960); Scott (1959); Jacob Wechsler interviews.

490 Sparkplug assembly configuration: J. Carson Mark interview, 3.vi.94. Deuterium volume: Chuck Hansen estimates 150–600 liters based on Mike's fusion yield and a range of estimated efficiencies from 10 to 50 percent. Chuck Hansen, personal communication, 31.xii.94.

492 Polyethylene for energy: Harold Agnew interview, 27.v.94.

492 Largest uranium castings: according to H. D. Smyth at a JCAE meeting on 21.ii.52. Chuck Hansen, personal communication, 31.xii.94.

493 "Oxidized uranium . . . mirror": Jacob Wechsler interview, 9.vi.93.

493 NBS absorptivity measurements: Timmerhaus (1960), p. 12.

493 "The problem . . . temperature": LANL (1983), p. 37.

493 Mike yield estimates: Hansen (1994d), pp. 26–27.

494 "That's a . . . anything": Jacob Wechsler interview, 3.vi.94.

494 "The main . . . use": ibid.

494 Two cool-downs with modifications: WT-608, extracted version, p. 22. Chuck Hansen collection.

494 All Mike shipped on *Curtiss:* Hansen (1994d), pp. 60–62; WT-608, extracted version, p. 26. Chuck Hansen collection.

496 "Lawrence believed . . . Eskimos": Norberg (1980a), p. 90.

496 Thermonuclear diagnostic studies: Herbert York to Norris Bradbury, 3.vi.52, LANL Archives.

496 "Teller . . . laboratory": York (1976), p. 133.

496 "I am . . . fascists": quoted in

PAGE

Blumberg and Owens (1976), p. 290.

496 "I recommended . . . it": Sidney W. Souers oral history, 16.xii.54, pp. 9–10, HSTL.

497 "in spite . . . time": Robert Oppenheimer to Niels Bohr, 27.vi.51, Box 31, Oppenheimer Papers, LC.

497 "I felt . . . for": USAEC (1954b), p. 562.

497 "undoubtedly give . . . Europe": Hans Bethe to Gordon Dean, 9.ix.52. Chuck Hansen collection.

497 Bradbury's postponement concerns: Norris Bradbury to Robert Oppenheimer, 11.vi.52, quoted in Hansen (1994d), pp. 17–19.

498 "that the . . . postponement": ibid., pp. 591–592.

498 Zuckert episode: cf. Hewlett and Duncan (1969), pp. 590–592.

498 "Gordon came . . . ahead": Hess (1971), pp. 60–61.

500 Mike photography: cf. Brixner (n.d.).

500 Mike instrumentation: WT-608, extracted version (n.d.); Moore and Bechanan (n.d.).

500 "probably diluted . . . reaction]": Raemer Schreiber interview, 9.vi.93.

501 secondary complete by September 25: Hansen (1994d), p. 63.

501 "I remember . . . nails": Harold Agnew interview, 27.v.94.

501 "We didn't . . . collegial": Harold Agnew, LANL talk, vi.93.

501 "a quite . . . direction": J. Carson Mark interview, 3.vi.94.

501 "We had . . . out": Harold Agnew interview, 27.v.94.

502 "You think . . . time": Jacob Wechsler interview, 3.vi.94.

PAGE

502 "we decided . . . piece": Harold Agnew interview, 27.v.94.

502 "Sort of . . . island": ibid.

502 Hurricane test: Norris et al. (1994), pp. 25–26.

502 Practice cool-downs on October 10: Hansen (1994d), p. 63.

503 Mike final filling: WT-608 (n.d.), p. 58.

503 Bad weather: ibid., p. 28.

503 "Marshall may . . . renegotiations": Harold Agnew interview, 27.vi.94.

503 "to decrease . . . yield": J. Carson Mark interview, 3.vi.94.

503 Substitute core built at Los Alamos: Jacob Wechsler interview, 9.vi.93; increased U and decreased Pu: Gordon Dean to James S. Lay, "Approval for Operation Ivy," 30.x.52, HSTL.

503 More efficient primary core: Raemer Schreiber interview, 8.vi.93.

503 Evacuation: WT-608 (n.d.), p. 32ff.

503 "How do . . . hotfoot": Jacob Wechsler interview, 9.vi.93.

504 "we waited . . . stable": ibid., 3.vi.94.

504 "They knew . . . up": Raemer Schreiber interview, 8.vi.93.

504 Arming party details: WT-608 (n.d.), pp. 57–59; Moore and Bechanan (n.d.), p. 270.

504 Wind: Moore and Bechanan (n.d.), p. 270.

504 Monitoring systems, control room: WT-608 (n.d.), pp. 47–48.

504 "I sat . . . to": Jacob Wechsler interview, 3.vi.94.

504 H-hour details: Moore and Bechanan (n.d.), p. 272ff.

508 Area effects of Mike explosion: ibid., p. 274ff; Hansen (1994d), p. 66ff.

508 "You would . . . fire": quoted in Hansen (1994d), p. 70.

508 "I was . . . away": George Cowan interview, 8.vi.93.

509 Every element in fireball: ibid.

509 "In nanoseconds . . . fermium": Morrison (1991), p. 133.

509 "It really . . . on": Raemer Schreiber interview, 8.vi.93.

509 Rigili is . . . pan' ": quoted in Hansen (1994d), p. 71.

510 "Immediately upon . . . cloud": Moore and Bechanan (n.d.), p. 277.

510 "more impressive . . . star": George Cowan interview, 8.vi.93.

511 Lawrence and Alvarez at seismograph: Hansen (1994d), p. 131, n. 125.

511 "I went . . . Berkeley": Teller (1979), pp. 150–151.

511 Rosenbluth remembers telegram: Marshall Rosenbluth interview, 26.v.94.

511 "It's a boy": quoted in Herbert York interview, 27.vi.83.

511 "a moment . . . boundless": quoted in Hansen (1994d), p. 74.

Chapter Twenty-five: Powers of Retaliation

513 "One evening . . . heart' ": Sakharov (1990), p. 134.

513 "I didn't . . . humanity": Victor Adamsky interview, vi.92.

514 "Andrei Dmitrievich . . . bomb": Ritus (1990).

514 "we were . . . departments": Altshuler et al. (1991), p. 497.

514 "We skied . . . swam": Sakharov (1990), p. 128.

514 "was a big village": ibid., p. 134.

514 "the academic . . . guys!": name withheld by request, personal communication, St. Petersburg, vi.92.

515 Shiryaeva exiled: Sakharov (1990), p. 134.

PAGE

515 "position and . . . Installation": ibid.,
p. 135.

515 "a man . . . Stalinist": ibid., p. 136.

515 Second commissar: Lev Altshuler
interview, vi.92.

515 "He explained . . . atmosphere":
ibid.

515 "telephoned Beria . . . closed":
Khariton and Smirnov (1993),
p. 27.

515 "Beria met . . . mood": ibid.

516 "We have . . . prisons": quoted in
Holloway (1994), p. 305.

516 Artisimovich and lithium: ibid.

516 Altshuler tests: Lev Altshuler,
personal communication, 16.xi.92.

516 "the procedure . . . Russia":
Altshuler et al. (1991), pp. 499–500.

516 Upset chemist: Sakharov (1990),
p. 158.

516 "failed due . . . time": Altshuler et
al. (1991), p. 499.

517 "he would . . . halt": Volkogonov
(1988, 1991), p. 570.

517 "the Joint . . . mistaken' ": ibid., pp.
570–571.

517 "making a . . . saved": Alliluyeva
(1967), p. 7.

517 "Comrade Stalin is sound asleep":
quoted in Volkogonov (1988,
1991), p. 572.

517 "was behaving . . . car!' ": Alliluyeva
(1967), p. 8.

518 For Beria's "thaw" cf. especially
Knight (1993), p. 180ff.

518 "The most . . . appeared": quoted
ibid., p. 186.

518 Beria's arrest: this is Khrushchev's
version; Beria's son has claimed his
father was arrested at home: ibid.,
p. 197. Buttons and pince-nez:
Hansen (1990), p. 105.

518 "This did . . . build": quoted in
Hansen (1990), p. 107.

519 "a short, ruddy-faced man":
Sakharov (1990), p. 169.

PAGE

519 "We began . . . government": Stickle
(1992), p. 84.

519 "I was . . . Committee": ibid., p. 130.

519 "We must . . . people": quoted in
Radosh and Milton (1983), p. 322.

519 "the first . . . fascism": quoted ibid.,
p. 323.

519 "They're expendable": David
Alman, quoted ibid., p. 327.

520 "Like others . . . women": quoted
ibid., p. 336.

520 "Wait, wait . . . chaff!": quoted in
Meeropol and Meeropol (1975),
p. 150.

520 "a legal . . . nation": quoted in
Radosh and Milton (1983), p. 351.

520 "weaknesses": quoted ibid., p. 376.

520 "The nature . . . hour": quoted in
Lamphere and Shachtman (1986),
p. 262.

520 "The Rosenberg . . . law": Dwight
Eisenhower to John Eisenhower,
16.vi.53, DDEL.

521 "I have . . . testified": Radosh
(1979).

521 "You mean . . . rack": quoted in
Meeropol and Meeropol (1975),
p. 208.

521 "Just imagine! . . . penalty": quoted
ibid., p. 217.

522 "I wanted . . . tight": Lamphere and
Shachtman (1986), p. 265.

522 "The June . . . inhumanity": Hiss
(1988), p. 180f.

522 "not satisfaction . . . KGB":
Lamphere and Shachtman (1986),
p. 286.

523 "They had . . . impossible": Radosh
(1979).

523 "So now . . . bounty": quoted in
Meeropol and Meeropol (1975),
p. 185.

523 Beria's execution: Hansen (1990),
p. 107.

523 "were confronted . . . people":
Sakharov (1990), pp. 170–171.

PAGE

523 Air drop delay: ibid., p. 172.

524 "the United . . . either": quoted in Arneson (1969), vi, p. 27.

524 "The announcement . . . hoax?": ibid., p. 26.

524 Strauss wrote Eisenhower: Galison and Bernstein (1989), p. 329.

524 "heard his . . . Russia' ": Altshuler et al. (1991), p. 613.

524 "I remember . . . it": J. Carson Mark interview, 3.vi.94.

524 "a table thumper": George Cowan interview, 8.vi.93.

524 "was compressed . . . device": Hans Bethe, personal communication, v.93.

525 "The first . . . failures": George Cowan interview, 8.vi.93.

525 "We never . . . end": Herbert York, personal communication, 20.v.94.

525 "I was . . . retaliation": John Stennis to Stuart Symington, 29.x.53, Box A5, LeMay Papers, LC.

526 "I am . . . today": Curtis LeMay to Stuart Symington, 12.xi.53, ibid.

526 John von Neumann on thermonuclear program: John von Neumann to Sterling Cole, 23.xi.53, RG 128, Box 60, National Archives.

527 John Wheeler on thermonuclear program: John Wheeler to Sterling Cole, 1.xii.53. JCAE No. 3797, ibid.

527 I. I. Rabi on thermonuclear program: I. I. Rabi to Sterling Cole, 24.xi.53, JCAE No. 3777, ibid.

528 "The very . . . two-thousandth": Oppenheimer (1953), pp. 527–528.

528 JCAE study: Edward L. Heller, memorandum for the files, "H-bomb destruction potential," 17.xi.53, JCAE No. 3764, Box 62, RG 128, National Archives.

528 Eisenhower to Dulles: Dwight Eisenhower, "memorandum for

PAGE

the Secretary of State," 8.ix.53, DDEL.

CHAPTER TWENTY-SIX: IN THE MATTER OF J. ROBERT OPPENHEIMER

530 "connected in any way": quoted in Pfau (1984), p. 139.

530 "another Fuchs": William Borden, "Memorandum for the file," 13.viii.51, JCAE Classified Document No. 3464. Chuck Hansen collection.

530 Oppenheimer activities suspicious to Strauss: Pfau (1984), p. 140; Strauss promised Hoover: Green (1977), p. 16.

530 "late gossip . . . Oppie": William Borden to Brien McMahon, 28.v.52, JCAE No. 3831, National Archives.

530 "In 1951 . . . him": York (1976), p. 139.

531 "the Air . . . necessity": John Walker and William Borden, "Memorandum to Senator McMahon," 4.iv.52, JCAE Classified Document No. 7490. National Archives.

531 Symington encouraged McCarthy: Bernstein (1984–85), p. 10. In a later study, Bernstein inconsistently dates this activity from May 1953 "during Borden's last weeks on the Joint Committee": Bernstein (1990b), p. 1431.

531 "We dropped . . . concerned": quoted in York (1976), p. 139.

531 Oppenheimer AEC work: USAEC (1954b), p. 1045.

531 "Lewis, let . . . war": quoted in Strauss (1962), p. 336.

531 "You had . . . it": quoted in Bernstein (1975), II.

532 "us[ing] his . . . down": Green (1977), p. 16.

PAGE

532 "ill-advised and . . . time": quoted in Pfau (1984), pp. 140–141.

532 "inquiry into . . . investigation": quoted ibid., p. 141.

532 Nixon: ibid.

532 Wheeler believed he lost document on train: John A. Wheeler, personal communication, 16.xi.92; Walker document envelope in Princeton: W. P. Brobeck, "Weekly conference with Walter J. Williams, deputy general manager, AEC," 6.ii.53, quoted in Hansen (1994d), p. 95.

532 "demanded that . . . demand": Green (1977), p. 14.

533 "Borden is . . . security": Bryan LaPlante, quoted in Bernstein (1990b), p. 1444.

533 "increasing consideration . . . years": USAEC (1954b), p. 833.

533 "Why is . . . 1942?": quoted in Robert Oppenheimer FBI FOIA files, 100-17828-427, p. 2.

533 Borden approached Strauss: Pfau (1984), p. 150.

533 "prosecutor at . . . treason": Gregg Herken, personal communication, 3.x.94.

533 Borden speaking to Strauss: Hewlett and Holl (1989), p. 62.

533 October; "crystallized [his] thinking": USAEC (1954b), p. 839.

533 "I couldn't . . . chest": William Borden to Corbin Allerdice, 6.xii.53, quoted in Bernstein (1990b), p. 1438, n. 262.

533 Borden letter: USAEC (1954b), pp. 837–838.

534 "Many of . . . indicate": quoted in Bernstein (1990b), p. 1440.

534 "a little . . . letter": quoted ibid.

534 Strauss checking Chevalier episode: ibid., p. 1442.

534 Wilson response: cf. ibid., p. 1443.

PAGE

534 "Charlie Wilson . . . world": Dwight Eisenhower, "Note for diary," 2.xii.53, DDEL.

534 "to place . . . otherwise": Dwight Eisenhower, "Memorandum for the Attorney General," 3.xii.53, DDEL.

535 "the so-called . . . stable": Dwight Eisenhower, "P.S. to previous note," 3.xii.53, DDEL.

535 "a lot . . . furor": John Edgar Hoover to Tolson, Ladd and Nichols, 3.xii.53, Robert Oppenheimer FBI FOIA files, 100-17828-[illegible].

535 "Strauss told . . . question": Teller (1987), p. 63.

536 "If you . . . fight": quoted in Bernstein (1990b), p. 1447.

536 "divine guidance": Stern (1969), p. 224.

536 Zuckert and Smyth criticism: Hewlett and Holl (1989), p. 78.

536 "slippery sonuvabitch": quoted in Bernstein (1990b), p. 1449.

536 McCabe, Albuquerque, to Director FBI, received 14.v.52, and SAC Albuquerque to Director FBI, 27.v.52, Robert Oppenheimer FBI FOIA files.

538 "Morrison has . . . views": Teller is identified as the source of this information in "Julius Robert Oppenheimer" summary file, 18.iv.52., p. 14, ibid.

538 "How can . . . home": quoted in Bernstein (1990b), p. 1476.

538 Strauss-Oppenheimer meeting: cf. Nichols's contemporary memorandum at Strauss (1962), p. 443ff; Hewlett and Holl (1989), p. 78ff.

538 "I can't . . . came": quoted in Stern (1969), p. 232.

538 "I have . . . do": quoted ibid., p. 233.

538 "Do we . . . McCarthy?": quoted ibid., p. 229.

PAGE

539 Strauss requested taps: cf. FBI document quoted at Bernstein (1984–85), p. 9: "At the specific request of Admiral Lewis L. Strauss, Chairman, Atomic Energy Commission . . . a technical surveillance was installed at the residence of Dr. J. Robert Oppenheimer at Princeton, New Jersey on 1-1-54." The official AEC history says "Hoover . . . authorized taps on Oppenheimer's home and office telephones; these were installed on January 1, 1954." Hewlett and Holl (1989), p. 80.

539 "in view . . . warranted": quoted in Hewlett and Holl (1989), p. 81.

539 "furnishing Strauss . . . Oppenheimer": quoted in Bernstein (1984–85), p. 9.

539 "most helpful . . . contemplating": quoted in Pfau (1984), p. 162.

539 "He felt . . . them": Robert Oppenheimer FBI FOIA files, 100-17828-704, p. 2.

539 "[Oppenheimer] had . . . his": Lewis Strauss to Edward Teller, 12.vii.61, Strauss Papers, HHL.

540 Garrison's security clearance: the defense team member was Herbert Marks; cf. Hewlett and Holl (1989), p. 81ff.

540 "that in . . . Jewish": quoted in Bernstein (1984–85), p. 13.

540 "I was . . . prejudice": quoted in Bernstein (1990b), p. 1471.

540 "Strauss told . . . action": Robert Oppenheimer FBI FOIA files, 100-17828-704, p. 1.

540 Weisskopf and Communism: cf. Oppenheimer's testimony at USAEC (1954b), p. 10.

540 "that I . . . you": quoted in Stern (1969), p. 255.

541 "One morning . . . it": Robert Serber interview, 22.ix.94.

541 "They really . . . section": Harold Agnew interview, 27.v.94.

541 "I was . . . experience": Marshall Rosenbluth interview, 26.v.94.

542 "ashes of death": quoted in Hewlett and Holl (1989), p. 176.

542 "was a . . . area": "The H-Bomb and World Opinion," *Bul. Atom. Sci.,* 31.iii.54, p. 163.

542 "Red spy ship": quoted in Hewlett and Holl (1989), p. 177.

542 *Castle* series shot statistics: Hansen (1994h), pp. 5–6.

542 "Harold said . . . cheap?' ": Jacob Wechsler interview, 3.vi.94.

542 "The results . . . contract": Schreiber (1991), p. 13.

543 "Robert had . . . you' ": Charles Critchfield interview, 11.vi.93.

543 "our research . . . bomb": quoted in "McCarthy and the H-bomb," *New York Times,* 8.iv.54, p. 26.

543 Security hearing setting: Stern (1969), p. 257ff.

543 "Perhaps only . . . actual": USAEC (1954b), p. 98.

543 "From the . . . distance": quoted in Goodchild (1980), p. 228.

544 "There were . . . up": quoted ibid., p. 230.

544 "Having begun . . . thereafter": quoted ibid., p. 231.

544 "One day . . . conversation": USAEC (1954b), p. 130.

544 "Q. Did . . . yes": ibid.

545 "Hunched over . . . sheet": quoted in Stern (1969), p. 280.

545 "Q. When . . . right": USAEC (1954b), pp. 136–137.

546 "Until the . . . down": quoted in Goodchild (1980), p. 241.

546 "Q. Did . . . right": USAEC (1954b), pp. 137–138.

547 "Haakon Chevalier . . . attempts": Kenneth Nichols to Peer DeSilva, ibid., p. 153.

PAGE

547 "Q. Why . . . idiocy": USAEC (1954b), p. 149.

548 "I felt . . . himself' ": quoted in Stern (1969), p. 280.

548 Oppenheimer espionage contact story: Hewlett and Holl (1989), p. 94ff.

548 "On December . . . story": ibid., p. 96.

548 "was that . . . him": USAEC (1994b), pp. 167–168.

549 "if you . . . project": ibid., p. 170.

550 "one of . . . generation": ibid., p. 357; "the mathematics . . . affairs": ibid., p. 367.

550 "Q . . . Perhaps . . . life": ibid., pp. 469–470.

550 "I feel . . . thing": ibid., p. 565.

550 Gray and Robb privately: Goodchild (1980), p. 251.

551 "MR. GRAY . . . idea": USAEC (1954b), p. 575.

551 "A man's . . . really": quoted in Goodchild (1980), p. 249.

551 "You know . . . influence": USAEC (1954b), p. 660.

551 "a dedicated . . . projects": ibid., p. 684.

552 "to go . . . defense": ibid., p. 684.

552 "furious to . . . is": Edward Teller interview with Herbert Childs, Box 6, reel 15, side 2, n.d., CU-369, The Bancroft Library.

552 "emphasized Oppenheimer's . . . cleared": Edward Teller interview, vi.93.

552 "At my . . . interview": Stern (1969), p. 516.

552 Charter Heslep to Lewis Strauss, "Conversation with Edward Teller at Livermore on April 22, 1954," 3.v.54, Strauss Papers, HHL.

553 "Robb asked . . . know' ": Edward Teller interview, vi.93.

553 "There was . . . mistakes' ": Hans

PAGE

Bethe interview, 3.v.93. Cf. also Hans Bethe oral history interview, American Institute of Physics, p. 38ff.

553 "I met . . . failed: Dyson (1979), p. 90.

554 "thought Oppenheimer . . . destroy": WGBH (1980), p. 10.

554 "Q. I . . . correct": USAEC (1954b), p. 709.

554 "Q. To . . . hands": ibid., p. 710.

555 "tame . . . record": Green (1977), p. 59.

555 "Inexplicably, Robb . . . opinions": ibid.

555 "MR. GRAY. . . . clearance": USAEC (1954b), p. 726.

556 "Teller . . . fascists": quoted in Kunetka (1982), p. 242.

556 "I'm sorry . . . mean": quoted in Stern (1969), p. 340.

556 "a great . . . Board": USAEC (1954b), pp. 1016–1019.

557 Publishing hearing transcript: cf. Hewlett and Holl (1989), p. 105.

557 "[Alvarez] said . . . could' ": Lewis Strauss file memorandum, 23.vi.54, Strauss Papers, HHL.

557 "Dr. Oppenheimer . . . character' ": USAEC (1954b), p. 1049.

558 Teller's second thoughts: Edward Teller to Lewis Strauss, 2.vii.54; Strauss to Teller, 6.vii.54; Roger Robb to Teller, 8.vii.54; various statement drafts interleaved. Teller-Strauss correspondence file, Strauss Papers, HHL. Teller's draft text appears at Blumberg and Owens (1976), p. 368.

558 "I think . . . do": Robert Serber interview, 22.ix.94.

558 "he was . . . afterward": Else (1981), p. 28.

558 "I never . . . mermaids": USAEC (1954b), p. 468.

559 "I was . . . Robin": Bill Moyers,

PAGE

interview with I. I. Rabi, *A Walk Through the Twentieth Century*.

CHAPTER TWENTY-SEVEN: SCORPIONS IN A BOTTLE

560 "The idea . . . SAC": Kohn and Harahan (1988), p. 108.

560 "bootstrapping": quoted in Rosenberg (1983), p. 168.

560 SAC targeting control: ibid., p. 204ff.

560 "I keep . . . them": Curtis LeMay to Lauris Norstad, 9.vii.52, Box 201, LeMay Papers, LC.

560 "It must . . . USSR": quoted in Rosenberg (1983), p. 168.

561 "The Fiscal . . . objectives": ibid.

561 "perhaps five . . . war": quoted ibid., p. 176.

561 AEC expansion statistics: Anders (1987), p. 4.

561 Stockpile numbers: Norris and Arkin (1993); Chuck Hansen, personal communication, 10.ii.5.

562 "the critical . . . USSR": draft cover letter to Finletter memorandum, 11.v.50, in LeMay Daily Diary, LeMay Papers, LC.

562 "not now . . . survival": quoted in Rosenberg (1983), p. 193.

562 Doolittle proposal: ibid., p. 195.

562 "The Coming . . . catastrophe": quoted in Kaku and Axelrod (1987), p. 100; cf. also Rosenberg (1983), p. 196.

562 "deliberately precipitat[e] . . . people": quoted in Kaku and Axelrod (1987), p. 101.

563 "The United . . . war": quoted in Rosenberg (1983), p. 197.

563 preemption: cf. discussion ibid., p. 198ff.

563 1,008 crews, 750 bombs: Rosenberg (1981–82).

PAGE

563 "SAC considers . . . hours": quoted ibid., p. 25.

564 "In those . . . sometime": quoted ibid., p. 27.

564 PB4Y-2 shootdown: Prados (1992), p. 12; Lashmar (1994c), p. 8.

564 Lucite radar plates: Rosenberg (1983), p. 161.

564 US-British reconnaissance deal: cf. Lashmar (1994c).

564 "one large . . . that": quoted ibid., p. 13.

565 20 planes, 100–200 airmen: cf. various estimates at Prados (1992), p. 12.

565 "There was . . . time": Kohn and Harahan (1988), pp. 95–96.

565 "[LeMay] said . . . serious": quoted in Lashmar (1994a), p. 24.

566 "I brought . . . days' ": quoted ibid.

566 LeMay War College speech: "Presentation to National War College by General C. E. LeMay," 18.iv.56, Box 93, LeMay Papers, LC.

567 "We may . . . life": Oppenheimer (1953), p. 529.

567 "While we . . . country": Kohn and Harahan (1988), p. 112.

567 SAC had PAL codes: Jervis (1989), p. 147.

568 Sprague episode: cf. Kaplan (1983), p. 132ff.

568 "If I . . . do": quoted ibid., p. 134.

568 "It's my . . . policy": quoted in Jervis (1989), p. 143, n. 18.

568 "the USSR . . . targets": quoted in Kaku and Axelrod (1987), p. 104.

569 "I recall . . . radiation": Romanov (1990).

569 "It was . . . you": Victor Adamsky interview, vi.92.

569 Kurchatov and Khariton at ground zero: Holloway (1994), p. 317.

569 "That was . . . used": quoted ibid.

569 Nuclear diplomacy: cf. especially Betts (1987).

PAGE

569 Eisenhower warning: cf. ibid., p. 42ff.

569 "readiness and . . . aggression": quoted ibid., p. 67.

569 "he thought . . . III' ": quoted ibid.

570 "an attack . . . Union": quoted ibid., pp. 73–74.

570 "the core . . . superpowers": ibid., p. 82.

570 Soviet stockpile: Norris and Arkin (1993), p. 57.

570 Soviet and US missiles and bombers: Blight and Welch (1989), p. 31.

570 Khrushchev reacted to US warnings: cf. Raymond Garthoff's comments, ibid.

570 "[equalize] what . . . war": Khrushchev (1970), pp. 494–495.

571 "Before the . . . capability": Blight and Welch (1989), pp. 30–31.

571 CIA report: ibid., p. 375.

571 "Those were . . . clock": Kohn and Harahan (1988), p. 116.

571 "authority to . . . President": quoted in Sagan (1993), p. 150.

571 "a subordinate . . . line": quoted ibid.

571 "General Power . . . it": Horace Wade, SAC 8th AF, quoted ibid.

571 DefCon scale: cf. ibid., p. 64.

572 22.x.62 airborne alert: ibid., p. 63ff.

572 Purpose of alert: cf. ibid., p. 66ff.

572 "The airborne . . . action": quoted ibid., p. 67.

572 "This action . . . itself": quoted ibid., p. 66.

572 7,000 MT: Betts (1987), p. 118.

572 "We got . . . it": quoted ibid., p. 120, n. 120.

572 "This is . . . confusion": quoted in Sagan (1993), p. 68. (italics deleted).

573 SAC DefCon transmittal policy: ibid., p. 69, n. 45.

573 Atlas missile launch: cf. ibid., p. 79.

573 "lack of . . . work-arounds": quoted ibid., p. 81.

573 Minutemen problems: cf. ibid., p. 81ff.

573 "We didn't . . . to": quoted ibid., p. 90.

573 "possible malfunctions . . . launch": quoted ibid.

573 Volk Field incident: cf. ibid., p. 99ff.

574 "that an . . . step": quoted ibid., p. 142; armed interceptors: ibid., p. 96; false missile launch: ibid., p. 130ff.

574 Navy tracking subs: Betts (1987), p. 119.

574 LeMay believed Kennedy to be a coward: Fletcher Knebel "said he got the idea for *Seven Days in May* while interviewing Gen. Curtis LeMay, onetime Air Force Chief of Staff, who went off the record to accuse President Kennedy of cowardice in his handling of the Bay of Pigs crisis." Knebel obituary, *New York Times,* 28.ii.93.

574 "The Kennedy . . . it": Kohn and Harahan (1988), pp. 112–116.

574 "The Russian . . . too": quoted without attribution in Brugioni (1991), p. 469.

574 "that it . . . side": Blight and Welch (1989), p. 91.

574 "LeMay talked . . . corner": ibid., p. 29.

575 "When the . . . Berlin' ": quoted ibid., p. 370.

575 LeMay chewed Kennedy out: Daniel Ellsberg, personal communication, iv.94.

575 "After Khrushchev . . . off' ": Blight and Welch (1989), p. 50.

575 SAC wing commander's story: personal communication, xi.91.

575 Soviet missiles in Cuba: McNamara (1992), p. A25; Zaloga (1993), p. 211.

PAGE

575 "If they'd . . . strike": Blight and
Welch (1989), p. 52.

575 60 million dead: Rosenberg (1981–
82), p. 30.

576 "[who] have . . . resides":
"Presentation to National War
College by General C. E. LeMay,"
18.iv.56, p. 20, Box 93, LeMay
Papers, LC.

576 LeMay believed US "lost": "As a
matter of fact," LeMay said in 1988,
"we lost because we didn't
threaten to use [our strength] when
it might have brought advantages
to the country." Kohn and Harahan
(1988), p. 112.

**Epilogue: 'The Gradual Removal of
Prejudices'**

577 "eating out . . . [offered]": George
Kennan, "Contribution to
memorial service for J. Robert
Oppenheimer," 25.ii.67, Box 43,
Oppenheimer Papers, LC.

577 Strauss nomination: cf. Pfau (1984),
p. 228ff.

577 Oppenheimer and Teller exchange
after AEC award: cf. Teller's
handwritten note on Eastern Air
Lines stationery and
Oppenheimer's draft and final
replies, dated 23.iv.63, Box 71,
Oppenheimer Papers, LC.

578 "Science is . . . beautiful": quoted in
Rhodes (1979), p. 116.

578 Oppenheimer booklist: *The
Christian Century,* 15.v.63, p. 647.

579 "If a . . . health": quoted in
Blumberg and Owens (1976),
p. 365.

579 "One of . . . goals": I. I. Rabi to
Sterling Cole, 25.xi.53. Cf. also
Edward Teller to S.C., 18.xii.53;

PAGE

John Wheeler to S.C., 1.xii.53; John
von Neumann to S.C., 25.xi.53,
National Archives.

579 "The idea . . . approach": J. Carson
Mark interview, 3.vi.94.

580 "a totalitarian . . . years": Los Alamos
National Laboratory colloquium,
28.vi.82, videotape, LANL archives.

580 "Johnny used . . . extinction": Ulam
(1976), p. 111.

581 Cohens: cf. Andrew and Gordievsky
(1990), p. 442ff.

581 "Kurchatov puts . . . functioning":
Zukerman and Azarkh, unpub. MS,
pp. 126–127.

582 "I'm not . . . peace": Victor Adamsky
interview, vi.92.

582 $4 trillion: estimate by Defense
Budget Project, Washington, DC,
reported in *Albuquerque Journal,*
24.xii.94, p. B5. (Courtesy Edwardo
de Los Alamos.)

582 Soviet decline: cf. Aleksandr
Yakovlev in *New Yorker* 43(37):6
(2.xi.92).

582 "One should . . . verification":
Zuckerman (1988), p. 33.

583 "For now . . . men": Truman (1966),
pp. 1124–1125.

583 "a third . . . civilization": quoted
in *Bul. Atom. Sci.* X(5) (v.54),
p. 167.

583 "When I . . . again": quoted in
Holloway (1994), p. 339.

583 Syngman Rhee and Eisenhower:
"American-Korean talks," 27.vii.54,
DDE Diary, Box 4, DDEL.

584 "We are . . . war": quoted in Nielson
(1963), p. 30.

584 "The President . . . situation": 129th
NSC, 21.xii.54. DDEL.

584 "But all . . . imperialists": quoted in
Holloway (1994), p. 339.

585 "totally unreasonable . . . way!":
Blight and Welch (1989), p. 33.

PAGE

585 "the Pentagon's . . . reality":
Zuckerman (1988), p. 33.
585 "In light . . . unthinkable": Bundy
(1969), pp. 9–10.
586 "The arms . . . nations": Nincic
(1982), p. 82.
586 "Strategic doctrines . . . Union":
ibid., p. 107.
586 1938–1969 study: quoted ibid.,
p. 50.

PAGE

586 False warnings and accidents: cf.
Sagan (1993).
586 "aiming to . . . world": Gorbachev
(1992), p. 22.
587 "But the . . . error": ibid.
587 "the gradual . . . prejudices": Bohr
(1958), p. 31.
588 "Fundamentally, and . . . 1945":
Bundy (1982), pp. 14–15.

Glossary of Names

Abel, Louis. Ruth Greenglass's brother-in-law.

Abelson, Philip. American experimental physicist; developed thermal diffusion of uranium.

Acheson, Dean. US Secretary of State, under Harry S. Truman.

Adamsky, Victor *(ah-dahm'-ski)*. Soviet physicist; younger colleague of Andrei Sakharov at Sarov.

Agnew, Harold. American experimental physicist; flew Hiroshima mission; directed development of Mark-17 hydrogen bomb; third director of Los Alamos Laboratory.

Alexandrov, Anatoli Petrovich *(ahl-ek-sahn'-droff, ah-nuh-toe'-lee pet'-tro-veetch)*. Soviet physicist; Director of Institute of Physical Problems, 1946–1955.

Alexandrov, Simon. MGB geologist; attended Bikini tests.

Alikhanov, Abram Isaakovich *(ahl-eek-ahn'-off, ahb'-rahm ees-sa-ak'-o-veetch)*. Soviet physicist; directed development of first Soviet heavy-water reactor.

Alliluyeva, Svetlana *(ahl-lay-lew-yave'-uh, svet-lahn'-uh)*. Josef Stalin's daughter.

Allred, John. American physicist who measured fusion neutrons from *Greenhouse* George shot.

Alsop, Joseph. American journalist and columnist.

Altshuler, Lev *(alt'-shoe-lur, leff)*. Soviet experimental physicist; worked on implosion at Sarov.

Alvarez, Luis W. American experimental physicist; Nobel laureate.

Anderson, Clinton. US Senator (Democrat–New Mexico) who brought down Lewis Strauss.

Anderson, Herbert. American experimental physicist; pioneered reactor development.

Anderson, John. Director of British atomic-bomb research ("Tube Alloys").

Angelov (Lieutenant) *(ahn'-jell-off)*. Soviet army intelligence officer in Ottawa; controlled Alan Nunn May.

Arneson, Gordon. US Foreign Service officer specializing in atomic energy issues.

Arnold, Henry. British security officer at Harwell who befriended Klaus Fuchs.

Arnold, Henry "Hap." Wartime commanding general, USAAF.

Artsimovich, Lev *(art-seem-o'-veetch, leff)*. Soviet physicist who developed electromagnetic separation of uranium isotopes.

Attlee, Clement. Prime Minister of England, 1945–1951.

Ayers, Eben. Press secretary to Harry S. Truman.

Bacher, Robert *(bock'-er)*. American physicist; member of first US Atomic Energy Commission.

Balezin, S. A. *(bahl-a'-zeen)*. Senior aide to Sergei Kaftanov.

Barnard, Chester. President, New Jersey Bell; member of the board of consultants that framed the Acheson-Lilienthal Report.

Barr, Joel. American electrical engineer associated with Julius Rosenberg; Soviet espionage agent. Defected to USSR in 1950; as Iozef Veniaminovich Berg, worked as microelectronics expert at Leningrad Design Bureau.

Baruch, Bernard. American financier. Headed US delegation to UN Atomic Energy Commission.

Batitsky, Pavel *(bah-teet'-ski, pa-vell')*. Soviet military officer; executed Lavrenti Beria.

Bedell Smith, Walter. US Army officer; ambassador to the USSR, 1946–1949; Director of the CIA, 1950–1953.

Belenky, Semyon *(bale-yane'-kee, same'-yawn)*. Soviet physicist in Tamm's thermonuclear development group.

Bellet, Samuel. Physician; director of Heart Station at Philadelphia General Hospital where Harry Gold worked.

Bennett, John V. Director of the US Bureau of Prisons under Dwight D. Eisenhower.

Bentley, Elizabeth. Vassar-educated American; Soviet espionage courier.

Beria, Lavrenti Pavlovich *(bear'-e-uh, lah-vren'-tee pahv'-low-veetch)*. Head of Soviet secret police; Deputy Prime Minister of USSR during Second World War; directed Soviet atomic bomb project.

Bethe, Hans *(bate'-uh, hahnz)*. German emigré theoretical physicist; directed Theoretical Division at Los Alamos during Second World War; Nobel laureate.

Bethe, Rose. German emigré. Mrs. Hans Bethe.

Bevin, Ernest. British Foreign Secretary in the postwar Labour government of Clement Attlee.

Black, Tom. American chemist. Soviet courier; recruited Harry Gold.

Blunt, Anthony. British art historian; member of the Cambridge Five; Soviet espionage agent.

Bohlen, Charles. American Foreign Service officer and translator, later ambassador.

Bohr, Aage *(bore, oh'-ah)*. Danish theoretical physicist; Nobel laureate; son of Niels Bohr.

Bohr, Niels. Danish theoretical physicist; Nobel laureate.

Borden, William Liscum. American attorney; author of *There Will Be No Time* (1946); executive director of Joint Committee on Atomic Energy; accused Robert Oppenheimer of espionage.

Born, Max. German emigré theoretical physicist; taught Klaus Fuchs in Edinburgh.

Borst, L. B. American physicist; estimated deuterium capture cross section.

Bradbury, Norris. American experimental physicist; second director of Los Alamos Laboratory.

Bradley, David. Radiation monitor at Bikini; author of *No Place to Hide* (1948).

Bradner, Hugh. American physicist who designed diagnostics for *Greenhouse* George.

Brewer, Len. Ruth Kuczynski's husband.

Brezhnev, Leonid *(brezh'-neff, lay'-oh-need)*. Member of the Central Committee of the Communist Party of the Soviet Union; later Soviet Premier.

Brish, Arkady Admovich *(breesh, are-kay'-dee ahd'-mo-veetch)*. Soviet physicist at Sarov.

Brokovich, Boris *(broke'-oh-veetch, bore'-ees)* Soviet engineer at Chelyabinsk-40; later Director.

Brothman, Abraham. American chemist and entrepreneur; sold industrial espionage to Soviets through Elizabeth Bentley and Harry Gold.

Browder, Earl. Chairman of the American Communist Party.

Brownell, Herbert, Jr. US Attorney General under Dwight D. Eisenhower.

Buckley, Oliver. President of Bell Telephone Laboratories; member of General Advisory Committee.

Bulganin, Nikolai *(bool-gah'-neen, neek'-oh-lie).* Member of the State Defense Committee in wartime; Marshal of the Soviet Union postwar.

Bundy, McGeorge. National Security Adviser to John F. Kennedy.

Burchinal, David. US Air Force officer; Deputy Chief for Plans and Operations under Curtis LeMay when LeMay was USAF Chief of Staff.

Burgess, Guy. British diplomat; leader of Cambridge Five; Soviet espionage agent.

Bush, Vannevar. American science administrator; directed the wartime Office of Scientific Research and Development.

Byrnes, James "Jimmy." US Secretary of State, 1945–1947.

Cairncross, John ("List"). British diplomat; member of the Cambridge Five; Soviet espionage agent.

Carlisle Smith, Ralph. Los Alamos security officer.

Catton, Jack. US Air Force officer; staff officer to Curtis LeMay at SAC.

Chadwick, James. British experimental physicist; discovered the neutron; directed the wartime British Mission to the US for atomic-bomb research; Nobel laureate.

Chambers, Whittaker. *Time* editor and essayist; Soviet espionage courier.

Chernyayev, Ilia Iliich *(chern-yah'-yeff, eel'-ee-ah eel'-ee-eech).* Soviet chemist and Academician.

Chevalier, Haakon *(shev-ahl'-yea, hoe'-kun).* Academic and translator of French literature who solicited Robert Oppenheimer for espionage.

Christy, Robert. American physicist; proposed solid "Christy" core for plutonium implosion.

Churchill, Winston. Prime Minister of Great Britain, 1940–1945, 1951–1955.

Clay, Lucius. US Army officer; military governor of the American zone of occupied Germany.

Clegg, Hugh. FBI Assistant Director assigned with Robert Lamphere to interview Klaus Fuchs.

Clifford, Clark. American attorney; postwar aide to Harry S. Truman.

Cockroft, John. British physicist; headed atomic-bomb research in Canada during Second World War; Nobel laureate.

Cohen, Lona (Leontine Patka). Soviet espionage courier; Mrs. Morris Cohen.

Cohen, Morris. American Communist and Soviet espionage agent.

Cohn, Roy. American attorney; prosecuted Rosenbergs; served as aide to Senator Joseph McCarthy.

Cole, Sterling. US Representative (Republican–New York) who succeeded Brien McMahon as chairman of the Joint Committee on Atomic Energy.

Collins, Samuel. American physicist; invented Collins Helium Cryostat.

Colville, John. British diplomat.

Compton, Arthur H. American experimental physicist; wartime Director of Metallurgical Laboratory of University of Chicago; Nobel laureate.

Compton, Karl. President of MIT; headed Bikini test evaluation board.

Conant, James Bryant. American chemist; President of Harvard University; wartime head of US Office of Scientific Research and Development; postwar member of General Advisory Committee.

Condon, Edward. American theoretical physicist; directed National Bureau of Standards.

Connally, Tom. US Senator (Democrat–Texas).

Conquest, Robert. British journalist and historian.

Corbino, Orso *(core-bean'-o, or'-so).* Italian physicist; mentor to Enrico Fermi.

Coulson, Fernande. Canadian Crown Attorney staff member who tried to help the Gouzenkos.

Cowan, George. American radiochemist at Los Alamos who analyzed fallout from Soviet and US weapons tests.

Critchfield, Charles. American theoretical physicist; protégé of Hans Bethe; worked on Mike device.

Curie, Marie. Polish physicist; co-discoverer of polonium and radium; Nobel laureate.

Curie, Pierre. French physicist. Co-discoverer of polonium and radium; Nobel laureate.

Curtis-Bennett, Derek. English barrister; Klaus Fuchs's defense counsel.

Cutler, Robert. National Security Adviser to Dwight D. Eisenhower.

Dallet, Joe. American Communist activist; member of Abraham Lincoln Brigade killed during siege of Madrid. Kitty Oppenheimer's first husband.

Daniel, Cuthbert. American statistician.

Davidenko, V. A. *(dah-veed-yen'-ko).* Soviet physicist; colleague of Igor Kurchatov.

de Hoffman, Frederic. Austrian theoretical physicist; protégé of Edward Teller at Los Alamos.

Dean, Gordon. Attorney; second chairman of the US Atomic Energy Commission.

Debs, Eugene V. US Socialist Party candidate for President.

Dewar, James *(doo'-er).* British cryogenic pioneer who invented the dewar vacuum bottle.

Dillon, Douglas. Member of Cuban missile crisis Executive Committee (ExCom).

Djilas, Milovan *(jee'-lus, mee'-oh-vahn).* Yugoslavian diplomat; author of *Conversations with Stalin* (1962).

Donovan, William. Founder and director of Office of Strategic Services (OSS).

Doolittle, James. US Air Force officer and air pioneer.

Douglas, Lewis. US ambassador to England during Berlin Airlift.

Douglas, William O. US Supreme Court Associate Justice.

Draper, William. Undersecretary of the Army during Berlin Airlift.

Dubovsky, Boris G. *(due-boff'-ski, bore'-ees).* Soviet physicist active in nuclear reactor development at Laboratory No. 2.

Dulles, Allen. CIA Director under Dwight D. Eisenhower.

Dulles, John Foster. US Secretary of State under Dwight D. Eisenhower.

Dyson, Freeman. British theoretical physicist; member of Institute for Advanced Study.

Eberstadt, Ferdinand. Financier; adviser to James Forrestal.

Ehrenfest, Paul. Austrian theoretical physicist.

Einstein, Albert. Theoretical physicist; formulated mass-energy equivalency *($E = mc^2$);* Nobel laureate.

Eisenhower, Dwight D. US Army Chief of Staff, 1945–1948; thirty-fourth President of the United States, 1953–1961.

Elitcher, Helene. Mrs. Max Elitcher.

Elitcher, Max. American engineer who testified that Julius Rosenberg solicited him for espionage.

Ellsberg, Daniel. Member of ExCom staff during Cuban missile crisis; later leaked Pentagon Papers.

Elsey, George. Assistant to Clark Clifford in Truman administration.

Eltenton, George. British petroleum engineer and Soviet espionage agent.

Erdös, Paul *(air'-dose).* Hungarian mathematician; colleague of Stanislaw Ulam.

Evans, G. Foster. American theoretical physicist; contributed to Super calculations.

Evans, Ward. Member of security board that investigated Robert Oppenheimer.

Everett, Cornelius. American mathematician; calculated Super ignition with Stanislaw Ulam.

Farrell, T. F. US Army Corps of Engineers officer; assistant to General Groves.

Fedosimov, Pavel Ivanovich *(fay-doe'-seem-off, pah'-vell ee-von'-o-veetch).* Anatoli Yatzkov's superior.

Feklisov, Alexander Semonovich *(feck'-lih-soff; seem-yawn'-o-veetch).* MGB agent; Klaus Fuchs's postwar London control.

Fermi, Enrico *(fair'-me, en-reek'-o).* Italian theoretical and experimental physicist; co-inventor of the nuclear reactor; Nobel laureate.

Fermi, Laura. Mrs. Enrico Fermi.

Fersman, A. Y. *(fairs'-mahn).* Soviet geologist.

Feynman, Arlene. Mrs. Richard Feynman. Died of tuberculosis, 1945.

Feynman, Richard *(fine'-mun).* American theoretical physicist; member of Theoretical Division at wartime Los Alamos; Klaus Fuchs's roommate; Nobel laureate.

Finletter, Thomas K. Secretary of the Air Force.

Fitin, P. M. *(feet'-een).* Chief of the Foreign Department of the NKVD.

Flerov, Georgi *(fly'-roff, gay-org'-ee).* Soviet physicist; co-discoverer of spontaneous fission.

Forrestal, James. US Secretary of the Navy during Second World War; first US Secretary of Defense, 1947–1949, under Harry S. Truman; suicide.

Frankel, Stanley. American physicist at Los Alamos.

Fraser, Elizabeth. *Ottawa Journal* reporter who interviewed the Gouzenkos.

French, Anthony. British physicist at Los Alamos; bought Klaus Fuchs's car.

Frenkel, Yakov I. *(frenk'-uhl, yock'-off).* Soviet theoretical physicist.

Frisch, Otto Robert. Austrian physicist; with Lise Meitner, first described the breakup of uranium under neutron bombardment and named the process "fission."

Frish, Sergei E. *(freesh, sare'-gay).* Soviet physicist.

Froman, Darol. American physicist; postwar Associate Director of Los Alamos.

Frowde, Chester. *Ottawa Journal* night city editor.

Fuchs, Elizabeth. Klaus Fuchs's older sister; anti-Nazi activist; suicide.

Fuchs, Emil. Klaus Fuchs's father. German religious leader; anti-Nazi activist.

Fuchs, Gerhard. Klaus Fuchs's brother. Anti-Nazi activist; concentration-camp victim.

Fuchs, Klaus *(fukes)* ("Rest"). German emigré theoretical physicist; member of British Mission to Los Alamos; Soviet espionage agent.

Fursov, V. S. *(foor'-soff)* Soviet physicist on F-1 crew.

Gamow, George *(gahm'-off).* Russian emigré theoretical physicist; contributed to Super design.

Gardner, Meredith. Cryptanalyst at US Army Security Agency; decoded wartime NKVD documents.

Garner, John Nance. Thirty-second Vice President of the United States, 1933–1941.

Garrison, Lloyd. American attorney; chief counsel to Robert Oppenheimer during Oppenheimer security hearing.

Garwin, Richard. American theoretical physicist; protégé of Enrico Fermi; worked on thermonuclear at Los Alamos.

Giles, Barney. US Air Force officer; flew B-29 nonstop trans-Pacific with Curtis LeMay in September 1945.

Ginzburg, Vitaly *(geenz'-burk, fee-tall'-ee).* Soviet physicist; member of Tamm's thermonuclear development group.

Goddard (Lord). Lord Chief Justice of England; Klaus Fuchs's judge.

Gödel, Kurt *(gure'-dell).* Mathematician at the Institute for Advanced Study.

Gold, Celia (Mrs. Samuel Gold). Harry Gold's mother.

Gold, "Essie and David." Harry Gold's imaginary twin children.

Gold, Harry ("Raymond"). American industrial chemist; Soviet espionage courier.

Gold, Samuel (Samuel Golodnitsky). Harry Gold's father.

Gold, Yosef. Harry Gold's younger brother.

Golden, William T. American financier; aide to Lewis Strauss.

Goldstein, Max. Los Alamos physicist.

Goldstine, Herman. American mathematician; pioneered digital computer.

Golos, Jacob *(go'-lows)*. Soviet espionage agent; Elizabeth Bentley's control and lover; Julius Rosenberg's control prior to Anatoli Yatzkov.

Golovin, Igor N. *(goal'-o-veen, ee'-gore)*. Soviet physicist; Igor Kurchatov biographer.

Gorbachev, Mikhail *(gore'-bah-choff, mee-kile')*. First and last President of the USSR.

Gorsky, Anatoli Borisovich *(gore'-ski, ah-nuh-toe'-lee bore-ees'-oh-veetch)* ("Henry," "Vadim") (aka Gromov). Soviet espionage officer; the Cambridge Five's London NKVD control; later Elizabeth Bentley's and Donald Maclean's Washington control.

Gouzenko, Igor *(goo-zenk'-oh, ee'-gore)*. Soviet cipher clerk; defected from Soviet Embassy in Ottawa, September 1945.

Gouzenko, Svetlana "Anna." Mrs. Igor Gouzenko.

Gray, Gordon. Chairman of security board that investigated Robert Oppenheimer.

Green, Harold. AEC attorney who drafted list of charges against Robert Oppenheimer.

Greenglass, David. Machinist at Los Alamos during Second World War; confessed to espionage, naming Julius Rosenberg as his control; convicted of espionage.

Greenglass, Ruth Printz. Mrs. David Greenglass. Confessed to complicity in espionage; never charged.

Greenglass, Samuel. David and Ethel Greenglass's older brother.

Griggs, David. American geophysicist; testified against Robert Oppenheimer.

Gromyko, Andrei *(grow-meek'-oh)*. Soviet diplomat; Foreign Minister, 1957–1985.

Groves, Leslie R. US Army officer; commanding general of Corps of Engineers Manhattan Engineer District during Second World War and of Armed Forces Special Weapons Project (AFSWP) afterward.

Gullion, Edmund. US Foreign Service officer.

Gurevich, I. I. *(goo-ray'-veetch)*. Soviet physicist; co-author of 1946 thermonuclear weapon proposal *Utilization of the nuclear energy of the light elements.*

Hagiwara, Tokutaro *(hah-gee-wahr'-ah, toe-ku-tar'-oh)*. Japanese physicist; first proposed using an atomic bomb to fuse hydrogen.

Hahn, Otto. German radiochemist; with Fritz Straussmann, discovered breakup of uranium nucleus under neutron bombardment (fission); Nobel laureate.

Halban, Hans. Austrian physicist working in France; colleague of Frédéric Joliot-Curie and Lew Kowarski.

Hall, Jane. American physicist and Los Alamos administrator.

Halperin, Israel ("Bacon"). Canadian mathematician and Soviet espionage agent.

Hankey (Lord). Minister in British War Cabinet; chairman of Scientific Advisory Committee.

Harkins, William D. American physicist; estimated deuterium capture cross section.

Harriman, W. Averell. American financier and diplomat; wartime US ambassador to the Soviet Union.

Harrison, George L. Aide to Henry Stimson.

Harrison, Stewart. British physician. Kitty Oppenheimer's second husband.

Harteck, Paul. German physicist; with Ernest Rutherford and Marcus Oliphant, discovered thermonuclear fusion reaction.

Heineman, Kristel Fuchs. Klaus

Fuchs's younger sister. Mrs. Robert Heineman.

Heineman, Robert. American Communist. Kristel Fuchs's husband.

Heisenberg, Werner. German theoretical physicist; Nobel laureate.

Henri, Ernst. Alias of Soviet agent who recruited Guy Burgess.

Hersey, John. Novelist and journalist; author of *Hiroshima* (1946).

Heslep, Charter. AEC public information officer on whom Edward Teller unburdened himself during Oppenheimer security hearing.

Hickenlooper, Bourke. US Senator (Republican–Iowa).

Hill, James Arthur. US Air Force officer; line pilot during Berlin Airlift; later USAF Vice Chief of Staff.

Hillenkoetter, Roscoe. US Navy officer; first Director of the CIA.

Hindenburg, Paul von. German field marshal; second and last President of the Weimar Republic.

Hinshaw, Carl. US Representative (Republican–California); member of Joint Committee on Atomic Energy.

Hirohito. Emperor of Japan, 1926–1989.

Hitler, Adolf. German Führer; Nazi Party leader; dictator of Germany.

Holloway, Marshall. American experimental physicist who directed development of first megaton-range thermonuclear test device (the Sausage, *Ivy* Mike) at Los Alamos.

Hoover, Herbert. Thirty-first President of the United States, 1929–1933.

Hoover, J. Edgar. Director of the Federal Bureau of Investigation, 1924–1972.

Hopkins, Harry. Personal adviser to Franklin D. Roosevelt; wartime director of Lend-Lease.

Horthy, Nicholas. Regent of Hungary, 1919–1944.

Hoyar-Millar, Derek. Counselor of the British Embassy in Washington.

Hull, Cordell. US Secretary of State under Franklin D. Roosevelt.

Ickes, Harold L. *(ick'-eez).* Secretary of the Interior under Franklin D. Roosevelt and Harry S. Truman.

Ioffe, Abram Fedorovich *(ee-yoff'-ee, ahb'-rahm fee-door'-o-veetch).* Soviet senior physicist; founder of the Leningrad Institute of Physics and Technology ("Fiztekh").

Ivanov, G. N. (aka G. N. Kolchenko) *(ee-von'-yoff; coal-chen'-ko).* NKVD translator who translated Smyth Report.

Ivanov, Nikolai. Member of F-1 crew; in 1992, F-1 chief engineer.

Ivanov, Peter. Vice consul at Soviet Consulate in San Francisco and NKVD officer; Steve Nelson's and George Eltenton's Soviet control.

Jackson, Henry "Scoop." US Representative, later Senator (Democrat–Washington), on Joint Committee on Atomic Energy.

Jodl, Alfried. German commander; signed German surrender.

Johnson, Louis. American Legion commander; James Forrestal's successor as Secretary of Defense under Harry S. Truman.

Joliot-Curie, Frédéric. French nuclear physicist; Nobel laureate.

Jordan, George Racey. USAAF officer; expedited Soviet Lend-Lease shipments through Gore Field, Montana.

Kabul, Bogdan *(kah-bool', bohg'-don).* NKVD torture specialist under Lavrenti Beria.

Kaftanov, Sergei *(calf-tahn'-off, sare'-gay).* Soviet Minister of Higher Education during Second World War.

Kapitza, Peter *(kuh-pete'-zuh).* Soviet theoretical physicist; protégé of Ernest Rutherford; Nobel laureate.

Kaufman, Irving R. Federal judge at trial of Rosenbergs and Sobell.

Keith, Dobey. American engineer; head of Kellex Corporation.

Kennan, George. American diplomat; first Director of State Department Policy Planning Staff; "Mr. X."

Kennedy, Joseph W. American chemist; with Glenn Seaborg and Arthur C. Wahl, first isolated plutonium at Berkeley in 1940.

Kennedy, Robert. US Attorney General under John F. Kennedy; chairman of Cuban missile crisis ExCom.

Khariton, Maria Nikolaevna *(har'-ih-tone, mah-ree'-uh nick-oh-lave'-nuh)*. Mrs. Yuli Khariton.

Khariton, Yuli Borisovich *(har'-ih-tone, you'-lee bore-ees'-o-veetch)*. Soviet theoretical physicist; scientific director of Arzamas-16 (Sarov).

Khlopin, Vitali Grigorievich *(klope'-een, vih-tall'-ee grig-ore-ee-a'-veetch)*. Soviet radiochemist; chairman of Soviet uranium commission.

Khrushchev, Nikita *(kroos'-choff, nee-kee'-tah)*. Premier of the Soviet Union, 1958–1964.

Kikoin, Isaak Konstantinovich *(kee-coin', ees-sa-ak' cone-stan-teen'-o-veetch)*. Soviet experimental physicist; directed development of gaseous diffusion and centrifuge methods of isotope separation.

Kim Il Sung. First Premier of the Democratic People's Republic of Korea, 1948–1994.

King, Mackenzie. Prime Minister of Canada, 1935–1948 (and previously).

Kirov, Sergei Mironovich *(keer'-off, sare'-gay meer-own'-oh-veetch)*. Leningrad Communist Party leader whose assassination in 1934 marked the beginning of the Great Terror.

Kistiakowsky, George *(kiss-tee-ah-cough'-ski)*. Ukrainian emigré chemist; developed explosive-lens technology at Los Alamos during the Second World War.

Koestler, Arthur *(kest'-ler)*. Hungarian-born novelist, journalist and critic; author of *Darkness at Noon* (1940).

Komarovsky, A. N. *(kome-ah-rahv'-ski)*. Director of construction at Chelyabinsk-40.

Konopinski, Emil *(ko-no-pin'-ski)*. American theoretical physicist; first proposed using tritium to lower temperature of fusion ignition in hydrogen bomb.

Koski, Walter. Group leader for implosion studies at wartime Los Alamos.

Kotikov, Anatoli N. *(coe'-tih-koff, ah-nuh-toe'-lee)*. Red Army colonel; head of Soviet Alsib Pipeline mission.

Kowarski, Lew *(coe-wahr'-ski)*. Russian physicist working in France; colleague of Frédéric Joliot-Curie and Hans Halban.

Kravchenko, Victor *(krahf-chen'-ko)*. Soviet defector; author of *I Chose Freedom* (1946).

Kremer, Simon Davidovich *(kray'-mer, see-moan' dah-veed'-o-veetch)* ("Alexander"). Soviet Red Army Intelligence (GRU) agent; Klaus Fuchs's first control in England.

Kuczynski, Jurgen *(kyu-zin'-ski, yurg'-en)*. German emigré economist; political leader of German Communist Party in England. Ruth Kuczynski's brother.

Kuczynski, Ruth ("Sonia") (Ruth Brewer; Mrs. Len Brewer). German Communist and Soviet courier.

Kurchatov, Antonina *(kur-chot'-off, ahn-toe-neen'-yah)*. Igor and Boris Kurchatov's older sister; died in childhood of tuberculosis.

Kurchatov, Boris. Soviet chemist. Igor Kurchatov's brother.

Kurchatov, Igor Vasilievich *(kur-chot'-off, ee'-gore vahs-ill-ee-a'-veetch)*. Soviet nuclear physicist; scientific director of Soviet nuclear project, 1943–1960.

Kurchatov, Marina Dmitrievna *(mah-reen'-ah dih-mee-tree-ave'-nuh)* (Marina Sinelnikov). Kirill Sinelnikov's sister. Mrs. Igor Kurchatov.

Kvasnikov, Leonid R. *(kvass'-nee-koff, lay'-oh-need)*. In 1940, head of NKVD science and technology department.

La Guardia, Fiorello. Mayor of New York City. Headed UNRRA postwar.

Lafazanos, Konstantin. Graduate student who lived with Kristel and Robert Heineman.

Lamphere, Robert *(lam'-fear)*. FBI

agent responsible for identifying Klaus Fuchs.

Landau, Lev *(lawn'-dow, leff)*. Soviet theoretical physicist jailed during the Great Terror; Nobel laureate.

Lanning, Mary. Harry Gold's co-worker at Philadelphia General Hospital; the woman he hoped to marry.

Lansdale, John, Jr. Manhattan Engineer District intelligence officer.

Lattimer, Wendell. American chemist at Berkeley; testified against Robert Oppenheimer.

Lauritsen, Charles. Danish physicist; testified for Robert Oppenheimer.

Lavrentiev, Oleg *(lah-vrehn'-tee-ehv, oh'-lek)*. Soviet sailor whose proposal for controlled thermonuclear fusion influenced Andrei Sakharov.

Lawrence, Ernest O. American experimental physicist; cyclotron inventor; founder of the Lawrence Radiation Laboratory; Nobel laureate.

Leahy, William. US Navy officer; Chief of Staff to Harry S. Truman.

LeBaron, Robert. Chairman of Military Liaison Committee.

Leipunski, Alexander I. *(luh-poon'-ski)*. Polish-born Soviet physicist; directed the Kharkov Institute of Physics and Technology, 1932–1937.

LeMay, Curtis. US Air Force officer; commanding general of the Strategic Air Command, 1948–1957; USAF Chief of Staff during the Cuban missile crisis.

Libby, Willard F. American chemist; developed radiocarbon dating; Nobel laureate.

Lilienthal, David E. First chairman of the US Atomic Energy Commission.

Lindbergh, Charles. Pioneer American aviator.

Loper, Herbert B. US Air Force officer; member of Military Liaison Committee.

Lovett, Robert. Under Secretary of State; Under Secretary of Defense; Secretary of Defense under Harry S. Truman.

Lysenko, Trofim *(lie-syen'-kho, troe'-feem)*. Soviet agronomist whose theory of acquired inheritance became official Stalinist dogma.

MacArthur, Douglas. US Army officer; commander of occupied Japan, 1945–1951; commander of UN forces in South Korea, 1950–1951.

Maclean, Donald *(muck-lane')*. British diplomat; member of the Cambridge Five; Soviet espionage agent.

Maclean, Melinda. Donald Maclean's American wife.

Main, Harold. RCAF corporal; the Gouzenkos' neighbor.

Makins, Roger. British diplomat; deputy chairman of Combined Policy Committee.

Malenkov, Georgi *(mall-yen'-koff, gay-org'-ee)*. Member of the Soviet State Defense Committee; chairman of the Council of Ministers, 1953–1955.

Malraux, André. French Resistance leader, novelist and art historian.

Malyshev, Vyacheslav *(mall'-ih-sheff, yatch'-ace-slaff)*. Protégé of Georgi Malenkov; first Minister of Medium Machine Building.

Manakova, Maria Alekseevna *(mahn-ah-cove'-ah, mah-ree'-ah ahl-eck-see-ave'-nah)*. Physicist at Sarov.

Mandelstam, Osip *(mahn'-dell-schtahm, oh'-seep)*. Soviet poet; died in gulag.

Manley, John H. American experimental physicist; Los Alamos associate director, 1947–1951; secretary to General Advisory Committee.

Mao Zedong. First chairman of the People's Republic of China, 1949–1976.

Mark, J. Carson. Canadian theoretical physicist; led theoretical group at Los Alamos postwar; led theoretical work on Mike device.

Marks, Anne. Aide to Robert Oppenheimer in wartime Los Alamos. Mrs. Herbert Marks.

Marks, Herbert. Attorney; State Department liaison with Acheson-Lilienthal board of consultants;

adviser to Robert Oppenheimer during Oppenheimer security hearing.

Marshall, George C. Wartime US Army Chief of Staff; Secretary of State and Secretary of Defense under Harry S. Truman.

Martin, Joseph. US Representative (Republican–Massachusetts); House Minority Leader during Korean War.

Mayer, Maria. German theoretical physicist; Nobel laureate.

Mazerall, Edward Wilfred. Canadian electrical engineer and Soviet espionage agent.

McCarthy, Joseph. US Senator (Republican–Wisconsin) whose Communist witchhunts gave his name to the McCarthy Era.

McCloy, John J. Member of Secretary of State's 1946 Committee on Atomic Energy that supervised preparation of Acheson-Lilienthal Report for international control of atomic energy.

McCormack, James, Jr. US Army officer; US Atomic Energy Commission Director of Military Applications.

McMahon, Brien. US senator (Democrat–Connecticut); chairman of Joint Committee on Atomic Energy, 1948–1952.

McMillan, Edwin. American physicist; co-discoverer with Glenn Seaborg of element 93, neptunium; Nobel laureate.

McNamara, Robert. US Secretary of Defense under John F. Kennedy and Lyndon B. Johnson.

Medvedev, Zhores *(med'-veh-deff, zor'-ace)*. Exiled Soviet geneticist.

Menninger, William. American psychiatrist; treated James Forrestal.

Merkulov, Vsevolod Nikolayevich *(murk'-you-loff, vess'-vo-lod neek-o-lay'-eh-veetch)*. Head of wartime Soviet foreign intelligence (NKGB) under Lavrenti Beria.

Metropolis, Nicholas. Greek-American mathematician; pioneered computer development at Los Alamos postwar.

Mikolajczyk, Stanislaw *(mee-coe-lodge'-zick, stahn'-ees-laff)*. Member of postwar Polish Provisional Government.

Mikoyan, Anastas *(mee-koy-yawn', ahn'-ess-stass)*. Armenian Bolshevik; member of State Defense Committee.

Miller, Byron. Co-author with James R. Newman of the US Atomic Energy Act.

Mitchell, William. AEC general counsel at time of Oppenheimer security hearings.

Molotov, Vyacheslav *(mole'-oh-toff, yatch'-ace-slaff)*. First Deputy Chairman of the Soviet Council of Ministers; member of State Defense Committee.

Montague, Robert M. Commanding general of Sandia Base, Albuquerque.

Morgan, Thomas. Member of security board that investigated Robert Oppenheimer.

Morrison, Philip. American theoretical physicist; with Marshall Holloway, developed pit assembly for Fat Man bomb at wartime Los Alamos.

Moskalenko, K. S. *(mose-kuh-len'-ko)*. Soviet Red Army Marshal.

Moskowitz, Miriam. Abraham Brothman's secretary and alleged mistress.

Motinov (Lieutenant Colonel) *(moh'-teen-off)*. GRU officer at Soviet Embassy in Ottawa.

Mott, Nevill. British physicist; taught Klaus Fuchs; Nobel laureate.

Muccio, John. In 1950, American ambassador to the Republic of Korea.

Murphy, Robert. Under Secretary of State under Harry S. Truman.

Murray, Thomas. AEC commissioner.

Nasser, Gamal Abdel. Prime Minister and President of Egypt, 1954–1970.

Nelson, Steve. Croatian-born American Communist espionage agent.

Newman, James R. Attorney. Aide to US Senator Brien McMahon; co-author with Byron Miller of the US Atomic Energy Act.

Nichols, Kenneth. US Army officer. Manhattan Project deputy to Leslie R.

Groves; commander of Armed Forces Special Weapons Project (AFSWP); general manager, Atomic Energy Commission.

Nitze, Paul. Director of State Department Policy Planning Staff, succeeding George Kennan.

Nixon, Richard. Thirty-seventh President of the United States, 1969–1974; 36th Vice-President under Dwight D. Eisenhower.

Norstad, Lauris. USAF officer; aide to USAF Chief of Staff Hoyt Vandenberg.

Nunn May, Alan ("Alek"). British physicist. Wartime Soviet espionage agent in Canada.

O'Donnell, Emmett "Rosie," Jr. US Air Force officer; flew B-29 nonstop trans-Pacific with Curtis LeMay in September 1945.

Oliphant, Marcus. Australian physicist; with Ernest Rutherford and Paul Harteck, discovered thermonuclear fusion reaction.

Oppenheimer, Frank. American experimental physicist; Robert Oppenheimer's younger brother.

Oppenheimer, J. Robert. American theoretical physicist; wartime Director of Los Alamos Laboratory; first chairman of General Advisory Committee.

Oppenheimer, Jacquenette "Jackie." Mrs. Frank Oppenheimer.

Oppenheimer, Katherine "Kitty." Mrs. J. Robert Oppenheimer.

Ordzhonikidze, Sergei. *(ord-zhawn-uh-kid'-zuh, sare'-gay).* Chairman of the Soviet Supreme Council of National Economy.

Pace, Frank. Secretary of the Army during the Korean War.

Panasyuk, Igor Semenovich *(pahn'-yah-syuk, ee'-gore same-yon'-o-veetch).* Soviet physicist; worked with Igor Kurchatov on F-1 reactor.

Pannikar, Sandar. Indian ambassador to People's Republic of China during Korean War.

Pariyskaya, L. V. *(par-ees-kah'-yah).* Soviet engineer; colleague of Andrei Sakharov at FIAN.

Parsons, William "Deke." US Navy officer; directed wartime Explosives Division at Los Alamos.

Pash, Boris L. Russian-American US Army intelligence officer.

Pasternak, Boris *(pahs'-tur-nahk).* Soviet poet and novelist.

Patterson, Robert. Under Secretary of War under Henry Stimson; Secretary of War under Harry S. Truman.

Paul, Randolph. Friend of the Robert Oppenheimers.

Pearson, Drew. Radio commentator.

Peierls, Genia *(pie-earls', jean-ee-ah).* Russian wife of Rudolf Peierls.

Peierls, Rudolf *(pie-earls').* German emigré theoretical physicist; hired Klaus Fuchs for British atomic-bomb project.

Penney, William. British nuclear physicist, specializing in hydrodynamics; directed development of British atomic bomb.

Pervukhin, Mikhail Georgievich *(purr-view'-kin, mee'-kile gay-or-gee-ave'-veetch).* Soviet electrical engineer; People's Commissar of the Chemical Industry.

Petrzhak, Konstantin A. *(pet'-er-zak, cone'-stan-teen).* Soviet physicist; co-discoverer with Georgi Flerov in 1940 of spontaneous fission of uranium.

Philby, Kim. British intelligence officer; member of the Cambridge Five; Soviet espionage agent.

Pike, Sumner. US Atomic Energy Commission commissioner; sardine czar.

Pitzer, Kenneth. Berkeley chemist and US Atomic Energy Commission Director of Research.

Placzek, George *(plot'-zeck).* Bohemian emigré physicist; advised Hans Bethe against hydrogen-bomb work.

Pomeranchuk, Isaak *(pome-air-ahn'-chook, ees-sah-ahk').* Soviet physicist; co-author of 1946 thermonuclear weapon proposal *Utilization of the nuclear energy of the light elements.*

Pontecorvo, Bruno. Italian physicist;

Soviet espionage agent working in Canada during the Second World War and at Harwell in England afterward; defected to the USSR in 1950.

Poskrebyshev, A. N. *(pose-kreb-bee-choff')* Principal aide to Josef Stalin.

Powell, Lewis F., Jr. Wartime USAAF intelligence officer; later Associate Justice, US Supreme Court.

Power, Thomas. USAF officer; succeeded Curtis LeMay as commanding general of SAC.

Rabi, I. I. *(rah'-bee).* American experimental physicist; Nobel laureate.

Rabinovich, Matvei S. *(rah-been'-o-veetch, maht'-vay).* Soviet physicist; Sakharov classmate at FIAN.

Raines, George. US Navy physician; treated James Forrestal.

Ramey, Roger. US Air Force officer; commanded one of SAC's air forces under Curtis LeMay.

Reston, James. *New York Times* correspondent and columnist.

Rhee, Syngman. First President of the Republic of Korea, 1948–1960.

Richtmyer, Robert. American theoretical physicist; at postwar Los Alamos, planned computer program analyzing fission explosion.

Rickenbacker, Eddie. First World War US air ace.

Rickover, Hyman. US Navy officer; pioneered nuclear submarine.

Ridgway, Matthew. US Army officer; succeeded Douglas MacArthur as Far Eastern commander.

Riehl, Nikolaus *(reel).* German metallurgist; as war prisoner, taught Soviets uranium purification.

Ritus, V. I. *(ree'-toos).* Soviet theoretical physicist at Sarov.

Robb, Roger. Counsel to security board during Oppenheimer security hearing.

Robertson, Norman. Deputy to Canadian Prime Minister Mackenzie King.

Roentgen, Wilhelm. German physicist who discovered X rays; Nobel laureate.

Romanov, Yuri A. *(rho-mahn'-off, you'-ree).* Soviet physicist; colleague of Sakharov at Sarov.

Roosevelt, Franklin D. Thirty-second President of the United States, 1933–1945.

Rose, Fred. Communist member of Canadian parliament; Soviet espionage agent.

Rosen, Louis. American physicist at Los Alamos who measured fusion neutrons from *Greenhouse* George shot.

Rosenberg, Ethel (Ethel Greenglass). David Greenglass's sister; Mrs. Julius Rosenberg; convicted of espionage and executed at Sing Sing in 1953.

Rosenberg, Harry. Julius Rosenberg's father.

Rosenberg, Julius. American engineer and Soviet espionage agent; convicted of espionage and executed at Sing Sing in 1953.

Rosenberg, Sofie. Julius Rosenberg's mother.

Rosenberg, Suzanne. Canadian raised in USSR; author of *A Soviet Odyssey* (1988).

Rosenbluth, Marshall. American theoretical physicist; worked on Mike design.

Rowe, Hartley. Engineer; member of General Advisory Committee.

Rusinov, Lev *(roos'-ee-noff, leff).* Soviet physicist; with Georgi Flerov, measured secondary neutrons from fission in April 1939.

Rutherford, Ernest. New Zealand-born experimental physicist who discovered atomic nucleus; Nobel laureate.

Sachs, Alexander. Economist who predicted Korean War.

Sagdeev, Roald *(sahg-day'-eff, rho'-ald).* Russian astrophysicist.

Sakharov, Andrei Dmitrievich *(sock'-ah-roff, ahn'-drey dih-mee-tree-a'-veetch).* Soviet theoretical physicist; inventor of the "layer-cake" hydrogen bomb; co-inventor with Yakov Zeldovich of the Soviet staged, radiation-imploded hydrogen bomb;

prominent dissident; Nobel Peace Prize laureate.

Sarant, Alfred. American electrical engineer associated with Julius Rosenberg; defected to USSR; as Filpp Staros, directed computer development at Leningrad Design Bureau.

Sartre, Jean-Paul. French existential philosopher and Communist.

Sarytchev, Filipp *(sahr'-eet-cheff, fee-leep').* Soviet espionage agent who visited Harry Gold in September 1949.

Saypol, Irving. US prosecuting attorney who prosecuted the Rosenbergs.

Schied, Beatrice. Acquaintance of Harry Gold at Pennsylvania Sugar.

Schmidt, Helmut. Chancellor of West Germany, 1974–1984.

Schreiber, Raemer. American experimental physicist. Alternate division leader for atomic weapon engineering at Los Alamos, 1947–1951; division leader, 1951–1962. Los Alamos Deputy Director, 1972–1975.

Seaborg, Glenn T. American radiochemist; co-discoverer of plutonium; Nobel laureate.

Segrè, Emilio. Italian emigré physicist; Los Alamos group leader; Nobel laureate.

Semenov, N. N. *(same-yawn'-off).* Soviet physicist and chemist; Nobel laureate.

Semenov, Semen N. *(same-yawn'-off, same'-yawn)* ("Sam"). MIT graduate engineer; Soviet espionage agent who controlled Harry Gold, 1940–1943.

Serber, Charlotte. Los Alamos librarian in wartime. The first Mrs. Robert Serber.

Serber, Robert. American theoretical physicist; directed design of Little Boy uranium gun bomb.

Shawcross, Hartley. British Attorney General; prosecutor at Klaus Fuchs's trial.

Shchelkin, Kirill *(shh-chel'-keen, keer'-eel).* Soviet physicist; deputy to Yuli Khariton.

Sherer, P. M. Father of Greenglasses' Albuquerque landlady.

Sherr, Rubby. American physicist; co-inventor with Klaus Fuchs of advanced initiator.

Shiryaeva *(sheer-ee-yave'-ah).* Zek artist at Sarov; Yakov Zeldovich's lover.

Sidorovich, Ann. Friend of Julius and Ethel Rosenberg whom David Greenglass identified as an espionage cut-out.

Sidorovich, Michael. American engineer. Husband of Ann Sidorovich.

Siegbahn, Manne (Karl M. G. Siegbahn) *(seeg'-bahn, man'-nee).* Swedish physicist; Nobel laureate.

Simon, Franz. German emigré chemist; did pioneer work in England on gaseous diffusion.

Simonov, Konstantin *(seem-yawn'-off, kone'-stan-teen).* Soviet novelist and war correspondent.

Sinelnikov, Eddie (Edna) *(see-nell'-nee-koff).* Kirill Sinelnikov's English wife.

Sinelnikov, Kirill *(see-nell'-nee-koff, keer'-eel).* Soviet physicist; director of Laboratory No. 1 (Kharkov). Igor Kurchatov's brother-in-law.

Skardon, William. British MI5 officer who broke Klaus Fuchs.

Skyrme, Tony *(skurm).* British physicist; member of British Mission to Los Alamos.

Slack, Al. Eastman Kodak employee who sold industrial espionage to Soviets through Harry Gold.

Slavsky, E. P. *(slaff'-ski).* Soviet metallurgist; first Director of Chelyabinsk-40.

Smilg, Ben. American aeronautical engineer whom Harry Gold attempted unsuccessfully to recruit for espionage.

Smirnov, Yuri *(schmeer'-noff, you'-ree).* Russian theoretical physicist.

Smith, Cyril Stanley. British-American metallurgist; division leader for metallurgy at wartime Los Alamos; member of General Advisory Committee postwar.

Smith, Durnford. Canadian scientist exposed as Soviet espionage agent.

Smith, Harold D. Budget director under Harry S. Truman.

Smith, Kingsbury. American journalist whose question to Stalin prompted resolution of Berlin blockade.

Smith, Ralph Carlisle. Cf. **Carlisle Smith,** Ralph.

Smith, Walter Bedell. Cf. **Bedell Smith,** Walter.

Smyth, Henry D. (rhymes with *scythe*). American physicist; author of Smyth Report; member, US Atomic Energy Commission.

Snow, C. P. British physicist, novelist and essayist.

Sobell, Morton *(so-bell')*. American engineer associated with Julius Rosenberg; convicted of espionage.

Sokolovsky, Vassily *(zho-co-loff'-ski, vahs'-ee-lee)*. Military governor of Soviet zone of occupied Germany.

Solzhenitsyn, Alexander *(sole-zhen-eet'-zin)*. Russian writer and gulag survivor; author of *The Gulag Archipelago* (1973); Nobel Literature Prize laureate.

"Sonia." Cf. **Kuczynski,** Ruth.

Souers, Sidney. US Navy officer; directed Central Intelligence Group and National Security Council.

Spaatz, Carl "Tooey." Commander of US Pacific air forces during Second World War.

Spindel, Sara. Mrs. William Spindel.

Spindel, William. Member of Army Special Engineering Detachment at Los Alamos; friend of David Greenglass.

Sprague, Robert. President of Sprague Electric; confronted Curtis LeMay about war plans.

Stalin, Josef Vassarionovich *(yo'-seff vahss-are-ee-yawn'-oh-veetch)*. Secretary-General of the Soviet Communist Party and Soviet Premier; head of state, 1924–1953.

Standley, William H. US Navy officer; early wartime US ambassador to the USSR.

Stephenson, Sir William ("Intrepid"). Director of British intelligence for the Western Hemisphere.

Stettinius, Edward, Jr. US Secretary of State, 1944–1945, under Franklin D. Roosevelt and Harry S. Truman.

Stevenson, Adlai. Governor of Illinois; Democratic candidate for the presidency opposing Dwight Eisenhower, 1952 and 1956.

Stilwell, Joseph W. "Vinegar Joe." US Army officer; served on Bikini evaluation board.

Stimson, Henry. US Secretary of War under Franklin D. Roosevelt and Harry S. Truman.

Straight, Michael. American Communist; author of *After Long Silence* (1983).

Stratemeyer, George E. Commanding general, Far East air forces, during the Korean War.

Strauss, Lewis L. *(straws)*. American financier. Member, US Atomic Energy Commission, 1947–1950; chairman, 1953–1958.

Sudoplatov, Pavel *(sue-doe-plot'-off, pah'-vell)*. NKVD officer; directed assassination of Leon Trotsky; supervised translation of atomic espionage materials for Lavrenti Beria, 1945–1946; co-author of *Special Tasks* (1994).

Symington, Stuart. Secretary of the Air Force; Senator (Democrat–Missouri).

Szilard, Leo *(zil-ard')*. Hungarian emigré theoretical physicist; co-inventor with Enrico Fermi of the nuclear reactor.

Taft, Robert. US Senator (Republican–Ohio).

Tamm, Igor E. *(tom, ee'-gore)*. Soviet theoretical physicist; mentor to Andrei Sakharov; Nobel laureate.

Taylor, Theodore. American physicist; designed nuclear weapons at Los Alamos.

Teller, Edward. Hungarian-born theoretical physicist; co-inventor with Stanislaw Ulam of the staged, radiation-imploded hydrogen bomb.

Terletsky, Yakov Petrovich *(tare-lets'-ski, yock'-off pet'-tro-veetch).* Soviet physicist on the staff of Lavrenti Beria.

Thomas, Charles. Vice President, Monsanto Chemical; member of the board of consultants that framed the Acheson-Lilienthal Report.

Thomas, Norman. American socialist.

Thornton, Robert L. American physicist on 1947 declassification panel with Klaus Fuchs.

Tibbets, Paul. USAAF officer; commanded 509th Composite Group; piloted *Enola Gay* at Hiroshima.

Till, Charles. Canadian physicist; in 1995, Associate Director of the Argonne National Laboratory.

Tolman, Richard. American physicist; dean of graduate studies at Caltech.

Travis, Robert. US Air Force officer; killed in plane crash during transfer of unarmed atomic bombs to the Far East.

Trudeau, Arthur G. US Army officer. Commanded constabulary during Berlin Airlift.

Truman, Harry S. Thirty-third President of the United States, 1945–1953; thirty-fourth Vice President of the United States, 1945, under Franklin D. Roosevelt.

Tuck, James. British physicist; member of British Mission to Los Alamos; proposed using explosive lenses for implosion.

Tunner, William. US Air Force officer; commanded Operation Vittles (Berlin Airlift).

Turkevich, Anthony. American mathematician at Los Alamos who calculated Super hydrodynamics.

Twining, Nathan. USAF officer; USAF Chief of Staff, 1953–1960; chairman JCS, 1957–1960.

Ulam, Françoise. Mrs. Stanislaw Ulam.

Ulam, Stanislaw *(oo'-lahm, stahn'-ees-laff).* Polish emigré mathematician; co-inventor with Edward Teller of the staged, radiation-imploded hydrogen bomb.

Urey, Harold. American chemist who discovered deuterium; Nobel laureate.

Van Loon, Ernest. FBI agent; Robert Lamphere's partner.

Vandenberg, Arthur. US Senator (Republican–Michigan).

Vandenberg, Hoyt. US Air Force officer; USAF Chief of Staff, 1948–1953.

Vannikov, Boris Lvovich *(vahn'-ih-koff, bore'-ees live'-o-veetch).* Wartime Soviet People's Commissar of Munitions; administered Soviet nuclear weapons program, 1945–1953.

Vasilenko, Semen *(vahs-eel-yen'-coe, same'-yon).* NKVD officer; transported black suitcases to USSR during wartime.

Vasilevsky, Lev *(vahs-eel-yeff'-ski, leff).* NKVD officer; deputy to Pavel Sudoplatov.

Vaughan, Harry. Military aide to Harry S. Truman.

Veblen, Oswald. Mathematician at Institute for Advanced Study.

Vernadski, Vladimir I. *(vare-nahd'-ski, vlad'-ih-meer).* Professor of mineralogy at Moscow University; founded State Radium Institute in Leningrad.

Vernadsky, George *(vare-nahd'-ski).* Yale University historian. Son of Vladimir Vernadski.

Volkogonov, Dmitri *(vole-coe'-gone-off, dih-mee'-tree).* Retired Red Army colonel general and head of the Institute of Military History; Stalin biographer.

Volpe, Joseph. American attorney. Counsel to AEC and to Robert Oppenheimer.

von Neumann, John *(fon noy'-mahn).* Hungarian emigré mathematician; developed high-explosive lens theory; pioneered digital computer.

Voroshilov, Kliment *(vor-osh'-eel-off, klee'-ment).* Marshal of the Red Army and member of the State Defense Committee.

Vyshinsky, Andrei *(vih-shin'-ski).*

Soviet prosecutor; chief prosecutor during the purge trials that accompanied the Great Terror.

Wahl, A. C. American chemist. With Glenn Seaborg and Joseph W. Kennedy, first isolated plutonium at Berkeley in 1940.

Walker, John. Member of Joint Committee on Atomic Energy staff.

Wallace, Henry. Thirty-third Vice President of the United States, 1941–1945, under Franklin D. Roosevelt; 1948 Progressive Party candidate for President.

Waymack, William W. Iowa newspaper editor; member of first Atomic Energy Commission.

Webster, William. Chairman of Military Liaison Committee.

Wechsler, Jacob. American engineer at Los Alamos; developed cryogenic system for Mike secondary.

Wedemeyer, Albert. US Army officer; Army Director of Plans and Operations during Berlin Airlift.

Weinberg, Alvin M. American experimental physicist; pioneered development of nuclear reactor.

Weinberg, Joseph. American physicist; suspected of espionage at Berkeley.

Weisskopf, Victor *(vice'-kopf)*. Austrian emigré theoretical physicist; advised Hans Bethe against hydrogen-bomb work.

Weizsäcker, Carl Friedrich von *(fon vite'-seck-ehr)*. German physicist.

Werth, Alexander. Russian-born British war correspondent and historian.

Wheeler, John Archibald. American theoretical physicist; developed fission theory with Niels Bohr; directed thermonuclear research at Princeton, 1950–1951.

White, Harry Dexter. US government official whom Elizabeth Bentley accused of Soviet espionage.

White, Thomas D. US Air Force officer; Curtis LeMay's predecessor as USAF Chief of Staff.

Wiener, Norbert. American mathematician at MIT.

Wiesner, Jerome. MIT physicist; confronted Curtis LeMay about war plans.

Wigner, Eugene. Hungarian emigré theoretical physicist; predicted "Wigner's disease" in nuclear reactors; Nobel laureate.

Wilson, Carroll. First General Manager of US Atomic Energy Commission.

Wilson, Charles E. US Secretary of Defense under Dwight D. Eisenhower.

Wilson, Roscoe. US Air Force officer; testified against Robert Oppenheimer.

Winne, Harry. Vice President, General Electric; member of the board of consultants that framed the Acheson-Lilienthal Report.

Yatzkov, Anatoli Antonovich (aka **Yakovlev**) *(yachts'-koff, ah-nuh-toe'-lee ahn-toe'-no-veetch; yah'-kove-leff)*. NKVD New York *rezident* for espionage; controlled Morris and Lona Cohen, Harry Gold and Julius and Ethel Rosenberg.

York, Herbert. American theoretical physicist; first Director of Lawrence Livermore Laboratory.

Zabotin, Nicolai *(zuh-bo'-teen, nee'-coe-lie)*. Red Army colonel; head of GRU in wartime Canada.

Zacharias, Jerrold. MIT physicist.

Zakharova, Tatyana Vasilievna *(zock'-har-oaf-ah, taught-yon'-ah vaze-eel-ee-ave'-nya)*. Colleague of Yuli Khariton.

Zavenyagin, Avrami Pavlovich *(zha-vehn-yag'-een, ahv-rahm'-ee pahv'-low-veetch)*. Deputy director of the NKVD.

Zeldovich, Yakov B. *(zel-doe'-veetch, yak'-off)*. Soviet theoretical physicist; with Yuli Khariton, developed theory of nuclear chain reactions; co-inventor with Andrei Sakharov of Soviet staged, radiation-imploded hydrogen bomb.

Zernov, Pavel Mikhailovich *(zehrn'-yoff, pah'-vell mee-kile'-o-veetch)*. Director of Arzamas-16 (Sarov).

Zhdanov, Andrei *(zh-don'-yoff).* Political boss of Leningrad during its wartime siege; member of the Politburo.

Zhukov, Georgi Konstantinovich *(zoo'-koff, gay-org'-ee cone-stan-teen'-o-veetch).* Marshal of the Red Army.

Zinn, Walter. American physicist. Protégé of Enrico Fermi; reactor pioneer; first director of Argonne National Laboratory.

Zubilin, Vassili (aka **Zarubin**) *(zoo-beel'-een, vaz'-eel-ee; zah-rube'-een).* Third secretary, Soviet Embassy, Washington, DC; NKVD officer.

Zuckerman, Solly. British security adviser.

Zuckert, Eugene. AEC commissioner.

Zukerman, Veniamin Aronovich *(sook'-air-mahn, fen'-ee-yah-meen air-ron'-o-veetch).* Soviet experimental physicist; developed flash radiography of explosions; author with his wife, Z. M. Azarkh, of *People and Explosions.*

Bibliography

Acheson, Dean. 1969. *Present at the Creation*. W. W. Norton.

Alexandrov, A. P. 1967. The heroic deed. *Bul. Atom. Sci.* xii.

———. 1988. How we made the bomb. *Izvestia* 205. 22.vii.

Alliluyeva, Svetlana. 1967. *Twenty Letters to a Friend*. Harper & Row.

Allred, John, and Louis Rosen. 1976. First fusion neutrons from a thermonuclear weapon device. In Bogdan Maglich, ed., *Adventures in Experimental Physics*. World Science Foundation.

Alsop, Joseph, and Stewart Alsop. 1946. Your flesh *should* creep. *Sat. Even. Post*. 13.vii.

———. 1954. We accuse! *Harper's*. x.

Alsop, Stewart, and Ralph E. Lapp. 1954. The strange death of Louis Slotin. *Sat. Even. Post*. 6.iii.

Altshuler, B. L., et al., eds. 1991. *Andrei Sakharov: Facets of a Life*. France: Editions Frontières.

Alvarez, Luis W. 1987. *Alvarez*. Basic Books.

Ambrose, Stephen E. 1990. *Eisenhower*. Simon & Schuster.

American Institute of Physics (AIP). 1994. Historians, physicists mobilize to refute spy stories. *AIP History Newsletter*. Fall.

Anders, Roger M. 1978. The Rosenberg case revisited: the Greenglass testimony and the protection of atomic secrets. *American Historical Review* 83(2). iv.

———, ed. 1987. *Forging the Atomic Shield*. University of North Carolina Press.

Andrew, Christopher, and Oleg Gordievsky. 1990. *KGB: The Inside Story*. Harper-Perennial.

Anon. 1953. The hidden struggle for the H-bomb. *Fortune*. v.

Anon. 1971. The billion-dollar bomber. *Air Enthusiast*. vii–x.

Arneson, R. Gordon. 1969. The H-bomb decision. *Foreign Service Journal*. v, vi.

Aspray, William. 1990. *John von Neumann and the Origins of Modern Computing*. MIT Press.

Badash, Lawrence. 1985. *Kapitza, Rutherford, and the Kremlin*. Yale University Press.

Bailey, Greg. 1992. Farewell to SAC. *Bul. Atom. Sci*. vi.

Bamford, James. 1982. *The Puzzle Palace*. Penguin.

Barnard, Chester I., et al. 1946. *A Report on the International Control of Atomic Energy* [Acheson-Lilienthal Report]. Department of State.

Baruch, Bernard M. 1960. *The Public Years*. Holt, Rinehart and Winston.

Bedell Smith, Walter. 1950. *My Three Years in Moscow.* J. B. Lippincott.

Behrens, James W., and Allan D. Carlson, eds. 1989. *50 Years with Nuclear Fission.* American Nuclear Society.

Bell, George I. 1965. Production of heavy nuclei in the Par and Barbel devices. *Phys. Rev.* 139(5B):B1207.

Bellman, Richard. 1984. *Eye of the Hurricane.* World Scientific.

Bentley, Elizabeth. 1951. *Out of Bondage.* Devin-Adair.

Bernstein, Barton J. 1982. "In the matter of J. Robert Oppenheimer." *Historical Studies in the Physical Sciences* 12(2):195.

———. 1984–85. The Oppenheimer conspiracy. *Our Right to Know.* Fund for Open Information and Accountability. Fall-Winter. (Rep. from *Discover,* iii.85.)

———. 1990a. Essay review—from the A-bomb to Star Wars: Edward Teller's history. *Technology and Culture* 31(4):846.

———. 1990b. The Oppenheimer loyalty-security case reconsidered. *Stanford Law Review* 42(6):1383.

Bernstein, Jeremy. 1975. Physicist. *New Yorker.* I: 13.x; II: 20.x.

———. 1980. *Hans Bethe: Prophet of Energy.* Basic Books.

Bethe, Hans A. 1950. The hydrogen bomb: II. *Scientific American.* iv.

———. 1964. Theory of the fireball. Los Alamos Scientific Laboratory (LA-3064).

———. 1965. The fireball in air. *J. Quant. Spectrosc. Radiat. Transfer* 5:9.

———. 1982a. Comments on the history of the H-bomb. *Los Alamos Science.* Fall.

———. 1982b. Hydrogen bomb history. (Letter.) *Science* 218:1270.

Betts, Richard K. 1987. *Nuclear Blackmail and the Nuclear Balance.* Brookings Institution.

Birkhoff, Garrett, et al. 1948. Explosives with lined cavities. *J. App. Phys.* 19(6):511.

Blight, James G., and David A. Welch. 1989. *On the Brink.* Hill and Wang.

Blumberg, Stanley A., and Gwinn Owens. 1976. *Energy and Conflict.* G. P. Putnam's Sons.

Blumberg, Stanley A., and Louis G. Panos. 1990. *Edward Teller.* Charles Scribner's Sons.

Bohlen, Charles E. 1973. *Witness to History.* W. W. Norton.

Bohr, Niels. 1958. *Atomic Physics and Human Knowledge.* John Wiley.

Borden, William Liscum. 1946. *There Will Be No Time.* Macmillan.

Borowski, Harry R. 1980. Air Force atomic capability from V-J Day to the Berlin blockade—potential or real? *Military Affairs* XLIV. x.

———. 1982. *A Hollow Threat.* Greenwood Press.

Borst, L. B., and William D. Harkins. 1940. Search for a neutron-deuteron reaction. *Physical Review* 37 (1.iv.40): 659.

Bracken, Paul. 1983. *The Command and Control of Nuclear Forces.* Yale University Press.

Bradbury, Norris E. 1948. An address at Pomona College. *Pomona College Bulletin* XLVI(6). ii.

Bradley, David. 1948. *No Place to Hide.* Little, Brown.

Bretscher, Egon, et al. 1946, 1950. Report of conference on the Super. Los Alamos Scientific Laboratory (LA-575).

Brickwedde, Ferdinand G. 1982. Harold Urey and the discovery of deuterium. *Physics Today.* ix.

Brixner, Berlyn. n.d. High-speed photography of the first hydrogen-bomb explosion. Los Alamos National Laboratory (LA-UR-92-2514).

Brode, Bernice. 1960. Tales of Los Alamos. *LASL Community News.* 2.vi., 22.ix.

Brode, Harold L. 1968. Review of nuclear weapons effects. *Ann. Rev. Nucl. Sci.* 18:153.

Brodie, Bernard. 1948. The atom bomb as policy maker. *Foreign Affairs.* x.

Bromberg, Joan. 1978. Interview with Herbert York. The Bancroft Library.

Brugioni, Dino A. 1991. *Eyeball to Eyeball.* Random House.

Bullock, Alan. 1992. *Hitler and Stalin.* Knopf.

Bundy, McGeorge. 1969. To cap the volcano. *Foreign Affairs* 48(1). x.

———. 1982. Early thoughts on controlling the nuclear arms race. *International Security* 7(2). Fall.

———. 1988. *Danger and Survival.* Random House.

Caldwell, Erskine. 1942. *All-Out on the Road to Smolensk.* Duell, Sloan and Pearce.

Cameron, A. G. W. 1959. Multiple neutron capture in the Mike fusion explosion. *Canadian Journal of Physics* 37(3):322.

Cave Brown, Anthony, and Charles B. MacDonald, eds. 1977. *The Secret History of the Atomic Bomb.* Delta/Dell.

Cecil, Robert. 1989. *A Divided Life.* William Morrow.

Central Intelligence Agency (CIA). 1951. *National Intelligence Survey 26: USSR.* Section 73: Atomic Energy.

———. 1957. The problem of uranium isotope separation by means of ultracentrifuge in the USSR. Report No. EG-1802. 8.x.

Chambers, Whittaker. 1952. *Witness.* Random House.

Chikov, Vladimir. 1991a. How Soviet intelligence service "split" the American atom. *New Times* (Moscow) 16.

———. 1991b. How Soviet intelligence service "split" the American atom. *New Times* (Moscow) 17.

———. 1991c. Recollections of Colonel Yatzkov. *Army* 19.

Childs, Herbert. 1963. Interview with Robert Oppenheimer. The Bancroft Library.

———. 1968. *An American Genius.* E. P. Dutton.

———. n.d. Interview with Edward Teller. The Bancroft Library.

Churchill, Winston. 1959, 1987. *Memoirs of the Second World War (Abridged).* Houghton Mifflin.

Clark, John C. 1957. We were trapped by radioactive fallout. *Sat. Even. Post.* 20.vii.

Clay, Lucius D. 1950. *Decision in Germany.* Doubleday.

Clifford, Clark. 1991. *Counsel to the President.* Random House.

Cochran, Thomas B., and Robert Standish Norris. 1990. *Soviet Nuclear Warhead Production* (NWD 90-3). Natural Resources Defense Council.

———. 1991. A first look at the Soviet bomb complex. *Bul. Atom. Sci.* v.

———. 1993. *Russian/Soviet Nuclear Warhead Production* (NWD 93-1). Natural Resources Defense Council.

Coffey, Thomas M. 1986. *Iron Eagle.* Crown.

Coit, Margaret L. 1957. *Mr. Baruch.* Houghton Mifflin.

Collier, Richard. 1978. *Bridge Across the Sky.* McGraw-Hill.

Conquest, Robert. 1990. *The Great Terror.* Oxford University Press.

———. 1991. *Stalin.* Penguin.

————. 1993. "The evil of this time." *NYRB.* 23.ix.

Costello, John, and Oleg Tsarev. 1993. *Deadly Illusions.* Crown.

Craxton, R. Stephen, Robert L. McCrory and John M. Soures. 1986. Progress in laser fusion. *Scientific American* 255(2):68.

Cumings, Bruce. 1990. *The Origins of the Korean War.* Vol. 2. Princeton University Press.

Dallin, David J. 1955. *Soviet Espionage.* Yale University Press.

Davis, Nuel Pharr. 1968. *Lawrence and Oppenheimer.* Simon and Schuster.

De Geer, Lars-Erik. 1991. The radioactive signature of the hydrogen bomb. *Science & Global Security* II. Gordon and Breach Science.

de Seversky, Alexander P. 1946. Atomic bomb hysteria. *Reader's Digest.* ii.

Diamond, H., et al. 1960. Heavy isotope abundances in Mike thermonuclear device. *Phys. Rev.* 119(6):2000.

Dingman, Roger. 1988. Atomic diplomacy during the Korean War. *International Security* 13(3):50.

Djilas, Milovan. 1962. *Conversations with Stalin.* Harcourt, Brace & World.

Dobbs, Michael. 1992. How Soviets stole U.S. atom secrets. *Washington Post.* 4.x.

Doolittle, James H. With Carroll V. Glines. 1991. *I Could Never Be So Lucky Again.* Bantam.

Doran, Jamie. 1994. *The Red Bomb.* 3 vols. Documentary videotape. Discovery Communications.

Douhet, Giulio. 1942. *The Command of the Air.* Coward-McCann.

Drell, Sidney D., and Sergei P. Kapitza, eds. 1991. *Sakharov Remembered.* American Institute of Physics.

Dyadkin, Iosif G. 1983. *Unnatural Deaths in the USSR, 1928–1954.* Transaction.

Dyson, Freeman. 1979. *Disturbing the Universe.* Harper & Row.

Ehrenburg, Ilya, and Konstantin Simonov. 1985. *In One Newspaper.* Trans. Anatol Kagan. Sphinx Press.

Eisenhower, Dwight D. (Louis Galambos, ed.). 1989. *Nato and the Campaign of 1952. The Papers of Dwight David Eisenhower.* Vol. 13. Johns Hopkins University Press.

Elliot, Gil. 1972. *Twentieth Century Book of the Dead.* Charles Scribner's Sons.

Else, Jon. 1981. *The Day After Trinity.* Transcript. KTEH-TV.

Emmett, John L., John Nuckolls and Lowell Wood. 1974. Fusion power by laser implosion. *Scientific American.* vi.

Evans, Medford. 1953. *The Secret War for the A-Bomb.* Henry Regnery.

Fakley, Dennis C. 1983. The British Mission. *Los Alamos Science.* Winter-Spring.

Feynman, Richard P. 1985. *"Surely You're Joking, Mr. Feynman!"* Bantam.

Feklisov, A. S. 1990. The heroic deed of Klaus Fuchs. *Voyennoi Istorischeski Zhurnal* 12.

Ferrell, Robert H., ed. 1980. Truman at Potsdam. *American Heritage.* vi/vii.

————, ed. 1991. *Truman in the White House: The Diary of Eben A. Ayers.* University of Missouri Press.

Flerov, Georgi N. 1989. Soviet research into nuclear fission before 1942. In James W. Behrens and Allan D. Carlson, eds., *50 Years with Nuclear Fission.* American Nuclear Society.

Forché, Carolyn. 1993. *Against Forgetting.* W. W. Norton.

Frankel, S. 1946. Prima facie proof of the feasibility of the Super. Los Alamos Scientific Laboratory (LA-551) 15.iv.

Fried, Yehuda, and Joseph Agassi. 1976. *Paranoia. Boston Studies in the Philosophy of Science.* Vol. 50. D. Reidel.

Frisch, Otto Robert. 1939. Physical evidence for the division of heavy nuclei under neutron bombardment. *Nature* 143:276.

———. 1978. Lise Meitner, nuclear pioneer. *New Scientist.* 9.xi.

———. 1979. *What Little I Remember.* Cambridge University Press.

Frish, S. E. 1992. *Skvoz' prizmu vremen: Vospominania (Through the Prism of Time: Memoirs).* Politizdat.

Furman, Necah Stewart. 1990. *Sandia National Laboratories.* University of New Mexico Press.

Gaddis, John Lewis. 1987. *The Long Peace.* Oxford University Press.

Gaddis, John Lewis, and Paul Nitze. 1980. NSC 68 and the Soviet threat reconsidered. *International Security* 4:164. Spring.

Galison, Peter, and Barton Bernstein. 1989. In any light: scientists and the decision to build the Superbomb, 1942–1954. *HSPS* 19:2.

Galtung, Johan. 1987. *United States Foreign Policy: As Manifest Theology.* University of California Institute on Global Conflict and Cooperation Policy Paper No. 4.

Gershtein, S. S. 1991. From reminiscences about Ya. B. Zeldovich. *Usp. Fiz. Nauk* 161:170–171 (v.). American Institute of Physics.

Glasstone, Samuel, and Philip J. Dolan. 1977. *The Effects of Nuclear Weapons.* USGPO.

Gleick, James. 1992. *Genius.* Pantheon.

Gold, Harry. 1951. The circumstances surrounding my work as a Soviet agent—a report. Unpub. MS.

———. 1965a. Memorandum to Augustus S. Ballard, August 25, 1965. Holograph MS.

———. 1965b. Memorandum to Augustus S. Ballard, September 24, 1965. Holograph MS.

Goldstine, Herman H. 1972. *The Computer from Pascal to von Neumann.* Princeton University Press.

Golovanov, Yaroslav. 1990. The portrait gallery. *Poisk* 15–21.ii.90. (Trans. JPRS-UST-90-006, 31.v.90, 87ff.)

Golovin, I. N. 1968. *I. V. Kurchatov.* Trans. William H. Dougherty. Selbstverlag Press.

———. 1989. The first steps in the atomic problem in the USSR. In James W. Behrens and Allan D. Carlson, eds., *50 Years with Nuclear Fission.* American Nuclear Society.

———. 1991. A crucial moment. *Science in the USSR* 1:17.

Golovin, I. N., and Yuri N. Smirnov. 1989. *It Began in Zamoskvorechie.* Kurchatov Institute.

Goncharov, Sergei N., John W. Lewis and Xue Litai. 1993. *Uncertain Partners.* Stanford University Press.

Goodchild, Peter. 1980. *J. Robert Oppenheimer.* Houghton Mifflin.

Gorbachev, Mikhail. 1992. The river of time. *Bul. Atom. Sci.* vii/viii.

Gouzenko, Igor. 1948. *The Iron Curtain.* E. P. Dutton.

Gowing, Margaret. 1964. *Britain and Atomic Energy 1939–1945.* Macmillan.

————. 1974. *Independence and Deterrence.* Vol. 2. Macmillan.

Graybar, Lloyd J. 1986. The 1946 atomic bomb tests: atomic diplomacy or bureaucratic infighting? *Jour. Am. Hist.* 72:4. iii.

Green, Harold P. 1977. The Oppenheimer case: a study in the abuse of law. *Bul. Atom. Sci.* ix.

Groueff, Stephane. 1967. *Manhattan Project.* Little, Brown.

Groves, Leslie R. 1948. The atom general answers his critics. *Sat. Even. Post* 22:15 +. 19.vi.

————. 1962. *Now It Can Be Told.* Harper & Brothers.

Gubarev, Vladimir. 1989. Nuclear trace. *Pravda* 25.viii.89:1, 4. (Trans. JPRS-TND-89-021.)

Gurevich, I. I., et al. 1946, 1991. Utilization of the nuclear energy of the light elements. *Sov. Phys. Usp.* 34(5). v. (Trans. G. M. Volkoff; American Institute of Physics.)

Hansell, Haywood S., Jr. 1986. *The Strategic Air War Against Germany and Japan.* Office of Air Force History.

Hansen, Chuck. 1988. *U.S. Nuclear Weapons.* Aerofax/Orion Books.

————. 1994a. Thermonuclear weapons development: overview. Unpub. MS.

————. 1994b. The status of the H-bomb program, January 1950. Unpub. MS.

————. 1994c. Operation GREENHOUSE. Unpub. MS.

————. 1994d. April 1952: the AEC reports to the President. Unpub. MS.

————. 1994e. The "Emergency Capability" program. Unpub. MS.

————. 1994f. Postwar U.S. fission weapons development. Unpub. MS.

————. 1994g. Weapons physics. Unpub. MS.

————. 1994h. U.S. nuclear weapons tests, 1945–1962. Unpub. MS.

————. 1994i. The application of radiation implosion. Unpub. MS.

Hansen, James H. 1990. The Kremlin follies of '53 . . . The demise of Lavrenti Beria. *Intelligence and Counterintelligence* 4(1):101.

Harriman, Averell, and Elie Abel. 1975. *Special Envoy to Churchill and Stalin 1941–1946.* Random House.

Hawkins, David. 1994. (Letter.) *Bul. Atom. Sci.* ix/x.

Herken, Gregg. 1980. *The Winning Weapon.* Knopf.

————. 1983. Mad about the bomb. *Harper's.* xii.

————. 1985. *Counsels of War.* Knopf.

Herring, Jr., George C. 1973. *Aid to Russia 1941–1946.* Columbia University Press.

Hersey, John. 1950. Conference in Room 474. *New Yorker.* 16.xii.

Hershberg, James. 1993. *James B. Conant.* Knopf.

Hess, Jerry N. 1970. Robert G. Nixon oral history interview. HSTL.

————. 1971. Eugene M. Zuckert oral history interview. HSTL.

————. 1972. Frank Pace, Jr., oral history interview. HSTL.

Hewlett, Richard G., and Oscar E. Anderson, Jr. 1962. *The New World.* Pennsylvania State University Press.

Hewlett, Richard G., and Francis Duncan. 1969. *Atomic Shield, 1947/1952.* Pennsylvania State University Press.

Hewlett, Richard G., and Jack M. Holl. 1989. *Atoms for Peace and War, 1953–1961.* University of California Press.

Hines, Neil O. 1962. *Proving Ground.* University of Washington Press.

Hiss, Alger. 1988. *Recollections of a Life.* Arcade.

Hoddeson, Lillian, Paul W. Henriksen, Roger A. Meade and Catherine Westfall. 1993. *Critical Assembly.* Cambridge University Press.

Hogan, William J., Roger Bangerter and Gerald L. Kulcinski. 1992. Energy from inertial fusion. *Physics Today.* ix.

Hogerton, John F. 1948. There is no shortcut to the bomb. *Look.* 16.iii.

Holloway, David. 1979–80. Research note: Soviet thermonuclear development. *International Security* 4.

———. 1981. Entering the nuclear arms race: the Soviet decision to build the atomic bomb, 1939–45. *Social Studies of Science* 11:2, 159. v.

———. 1983. *The Soviet Union and the Arms Race.* Yale University Press.

———. 1990. The scientist and the tyrant. *NYRB.* 1.iii.

———. 1994. *Stalin and the Bomb.* Yale University Press.

Hoopes, Townsend, and Douglas Brinkley. 1992. *Driven Patriot.* Knopf.

Hoover, J. Edgar. 1958. *Masters of Deceit.* Henry Holt.

House Committee on Un-American Activities (HUAC). 1950. *Report on Atomic Espionage.* 81st Congress, 1st Session: Hearings of Sept. 29, 1949. USGPO.

Hughes, Emmet John. 1975. *The Ordeal of Power.* Atheneum.

Hurley, Alfred F., and Robert C. Ehrhart. 1979. *Air Power and Warfare.* Office of Air Force History.

Hyde, H. Montgomery. 1980. *The Atom Bomb Spies.* Atheneum.

Ingram, Kenneth. 1955. *History of the Cold War.* Philosophical Library.

International Commission Against Concentration Camp Practices. 1959. *The Regime of the Concentration Camp in the Post-War World, 1945–1953.* Centre International d'Edition et de Documentation.

International Committee of Experts in Medical Sciences and Public Health to Implement Resolution WHA34.38. 1984. *Effects of Nuclear War on Health and Health Services.* World Health Organization.

Irving, David. 1967. *The Virus House.* William Kimber. (In US: *The German Atomic Bomb,* Simon and Schuster, 1968.)

Jackson, Robert. 1988. *The Berlin Airlift.* Patrick Stephens.

Jastrow, Robert. 1983. Why strategic superiority matters. *Commentary* 75(3). iii.

Jervis, Robert. 1989. *The Meaning of the Nuclear Revolution.* Cornell University Press.

Johnson, Ken. 1970. A quarter century of fun. *The Atom.* Los Alamos Scientific Laboratory. ix.

Johnson, Niel M. 1989. R. Gordon Arneson oral history interview. HSTL.

Joint Committee on Atomic Energy (JCAE). 1951. *Soviet Atomic Espionage.* USGPO.

———. 1953. Policy and progress in the H-bomb program: a chronology of leading events. National Archives.

Jones, Joseph M. 1955. *The Fifteen Weeks.* Viking.

Jordan, George Racey. 1952. *From Major Jordan's Diaries.* Harcourt, Brace.

Kaftanov, S. V. 1985. On alert (interview by V. Stepanov). *Chemistry and Life* 3.

Kaku, Michio, and Daniel Axelrod. 1987. *To Win a Nuclear War.* South End Press.

Kanet, Roger E., and Edward A. Kolodziej. 1991. *The Cold War as Cooperation.* Johns Hopkins University Press.

Kaplan, Fred. 1983. *The Wizards of Armageddon.* Simon and Schuster.

Keegan, John. 1989. *The Second World War.* Penguin.

Kennan, George. 1967. *Memoirs, 1925–1950*. Pantheon.

Kennedy, J. W., G. T. Seaborg, E. Segrè and A. C. Wahl. 1946. Properties of 94(239). *Phys. Rev.* 70:555.

Kerr, Walter. 1944. *The Russian Army*. Knopf.

Kevles, Dan. 1990. Cold war and hot physics: science, security, and the American state, 1945–56. *HSPS* 20:2.

Khariton, Yuli, and Yuri Smirnov. 1993. The Khariton version. *Bul. Atom. Sci.* v.

Khrushchev, Nikita. 1970. *Khrushchev Remembers*. Little, Brown.

King, John Kerry, ed. 1979. *International Political Effects of the Spread of Nuclear Weapons*. USGPO.

Kissinger, Henry. 1994. *Diplomacy*. Simon & Schuster.

Knaack, Marcelle Size. 1988. *Post-World War II Bombers, 1945–73. Encyclopedia of U.S. Air Force Aircraft and Missile Systems*. Vol. 2. Office of Air Force History.

Knight, Amy. 1993. *Beria*. Princeton University Press.

Knyazkaya, N. V. 1986. Starting up: the story, told by a participant (B. G. Dubovksy). *Chemistry and Life* 12.

Kofsky, Frank. 1993. *Harry S. Truman and the War Scare of 1948*. St. Martin's.

Kohn, Richard H., and Joseph P. Harahan, eds. 1988. *Strategic Air Warfare*. Office of Air Force History.

Kramish, Arnold. 1959. *Atomic Energy in the Soviet Union*. Stanford University Press.

———. 1994. Safety in quarks? (Letter.) *Science*. 25.ii.

Kravchenko, Victor. 1946. *I Chose Freedom*. Charles Scribner's Sons.

Krock, Arthur. 1968. *Memoirs*. Funk & Wagnalls.

Kuchment, Mark. 1985. The American connection to Soviet microelectronics. *Physics Today,* ix. Rep. in Hafemeister, David, ed. 1991. *Physics and Nuclear Arms Today*. American Institute of Physics.

Kunetka, James W. 1982. *Oppenheimer*. Prentice-Hall.

Lamphere, Robert J., and Tom Shachtman. 1986. *The FBI-KGB War*. Random House.

Lang, Daniel. 1959. *From Hiroshima to the Moon*. Simon and Schuster.

Langer, Elinor. 1966. The case of Morton Sobell: new queries from the defense. *Science* 153:1501.

Lanouette, William. 1992. *Genius in the Shadows*. Charles Scribner's Sons.

Lashmar, Paul. 1994a. Stranger than "Strangelove." *Washington Post National Weekly Edition*. 11–17.vii.

———. 1994b. Shootdowns. *Aeroplane Monthly*. viii.

———. 1994c. Skulduggery at Sculthorpe. *Aeroplane Monthly*. x.

LeMay, Curtis E. With MacKinlay Kantor. 1965. *Mission with LeMay*. Doubleday.

Leskov, Sergei. 1993. Dividing the glory of the fathers. *Bul. Atom. Sci.* v.

Leva, Marx. 1959. Afterthoughts on Strauss. (Letter.) *Washington Post*. 24.vi.

Lilienthal, David E. 1964. *The Atomic Energy Years, 1945–1950. The Journals of David E. Lilienthal,* Vol. 2. Harper & Row.

———. 1980. *Atomic Energy: A New Start*. Harper & Row.

Lindl, John D., Robert L. McCrory and E. Michael Campbell. 1992. Progress toward ignition and burn propagation in inertial confinement fusion. *Physics Today*. ix.

Los Alamos National Laboratory (LANL). 1983. The evolution of the laboratory. *Los Alamos Science*. Winter-Spring.

Lukacs, John. 1991. Ike, Winston and the Russians. *NYTBR*. 10.ii.

Manley, John H. 1985. Recollections and memories. Unpub. MS. LANL Archives.
———. 1987. Star Wars and the H-bomb. Unpub. MS. LANL Archives.
Mark, Hans, and Sidney Fernbach, eds. 1969. *Properties of Matter Under Unusual Conditions.* Interscience.
Mark, J. Carson. 1954, 1974. A short account of Los Alamos theoretical work on thermonuclear weapons, 1946–1950. Los Alamos Scientific Laboratory (LA-5647-MS).
Markusen, Eric, and David Kopf. 1995. *The Holocaust and Strategic Bombing: Genocide and Total War in the Twentieth Century.* Westview Press.
Marshall, Eliot. 1981. Richard Garwin: defense adviser and critic. *Science* 212:763.
———. 1990. Radiation exposure: hot legacy of the Cold War. *Science* 249:474.
Mathews, William G., and Daniel Hirsch. 1991. Preface to the 1991 edition of Stanislaw Ulam, *Adventures of a Mathematician.* University of California Press.
McCracken, Daniel D. 1955. The Monte Carlo method. *Scientific American.* v.
McKinzie, Richard D. 1974. Lucius D. Clay oral history interview. HSTL.
McLellan, David S., and David C. Acheson, eds. 1980. *Among Friends: Personal Letters of Dean Acheson.* Dodd, Mead.
McMillan, Edwin, and Philip H. Abelson. 1940. Radioactive element 93. *Phys. Rev.* 57:1185.
McNamara, Robert S. 1992. One minute to Doomsday. *New York Times.* 14.x.
McPhee, John. 1974. *The Curve of Binding Energy.* Farrar, Straus and Giroux.
Medvedev, Zhores A. 1978. *Soviet Science.* Oxford University Press.
Meeropol, Robert, and Michael Meeropol. 1975. *We Are Your Sons.* Houghton Mifflin.
Meitner, Lise, and O. R. Frisch. 1939. Disintegration of uranium by neutrons: a new type of nuclear reaction. *Nature* 143:239.
Memorial Committee, J. Robert Oppenheimer. 1994. *Behind Tall Fences.* J. Robert Oppenheimer Memorial Committee.
Metropolis, N., and E. C. Nelson. 1982. Early computing at Los Alamos. *Annals of the History of Computing* 4(4).
Millis, Walter. 1951. *The Forrestal Diaries.* Viking.
Moore, Jr., Frank E., and H. Gordon Bechanan. n.d. *History of Operation Ivy.* Department of Defense.
Moorehead, Alan. 1952. *The Traitors.* Charles Scribner's Sons.
Morrison, Philip. 1991. Review, *The Elements Beyond Uranium. Scientific American.* v.
Morse, Philip M. 1977. *In at the Beginnings: A Physicist's Life.* MIT Press.
Moss, Norman. 1987. *Klaus Fuchs.* St. Martin's.
———. 1993. "Sonya" explains. *Bul. Atom. Sci.* vii/viii.
Newton, Verne. 1984. *The Cambridge Spies.* Madison Books.
Nichols, K. D. 1987. *The Road to Trinity.* Morrow.
Nielson, J. Rud. 1963. Memories of Niels Bohr. *Physics Today.* x.
Nikipelov, Boris V., A. S. Nikiforov, O. L. Kedrovsky, M. V. Strakhov and E. G. Drozhko. n.d. Practical rehabilitation of territories contaminated as a result of implementation of nuclear material production defense programs. Trans. Alexander Shlyakhter.
Nikipelov, Boris V., Andrei F. Lizlov and Nina A. Koshurniknva. 1990. Experience with the first Soviet nuclear installation. *Priroda.* ii. Trans. Alexander Shlyakhter.

Nincic, Miroslav. 1982. *The Arms Race*. Praeger.

Nizer, Louis. 1973. *The Implosion Conspiracy*. Doubleday.

Norberg, Arthur Lawrence. 1980a. Interview with Norris E. Bradbury. The Bancroft Library.

———. 1980b. Interview with Darol K. Froman. The Bancroft Library.

———. 1980c. Interview with J. Carson Mark. The Bancroft Library.

———. 1980d. Interview with John H. Manley. The Bancroft Library.

———. 1980e. Interview with Cyril Stanley Smith. The Bancroft Library.

———. 1980f. Interview with Raemer E. Schreiber. The Bancroft Library.

Norris, Robert S. 1992. *Questions About the British H-Bomb* (NWD 92-2). Natural Resources Defense Council.

Norris, Robert S., and William M. Arkin. 1993. Nuclear notebook: estimated nuclear stockpiles, 1945–1993. *Bul. Atom. Sci.* xii.

Norris, Robert S., Andrew S. Burrows and Richard W. Fieldhouse. 1994. *British, French and Chinese Nuclear Weapons. Nuclear Weapons Databook*. Vol. 5. Westview Press.

O'Keefe, Bernard J. 1983. *Nuclear Hostages*. Houghton Mifflin.

Oliphant, M. L. E., P. Harteck and Lord Rutherford. 1934. Transmutation effects observed with heavy hydrogen. *Proc. Roy. Soc. A* 144.

Oppenheimer, J. Robert. 1945. Atomic weapons and the crisis in science. *Sat. Rev. Lit.* 24.xi.

———. 1946a. The atom bomb and college education. *The General Magazine and Historical Chronicle*. University of Pennsylvania. Summer.

———. 1946b. The atom bomb as a great force for peace. *New York Times Magazine*. 9.vi.

———. 1948a. International control of atomic energy. *Foreign Affairs* 26(2).

———. 1948b. Physics in the contemporary world. *Bul. Atom. Sci.* iii.

———. 1953. Atomic weapons and American policy. *Foreign Affairs* 31(4).

———. 1958. An inward look. *Foreign Affairs* 36(2).

Ostriker, J. P., ed. 1992, 1993. *Selected Works of Yakov Borisovich Zeldovich*. 2 vols. Princeton University Press.

Pais, Abraham. 1990. Stalin, Fuchs and the Soviet bombs. *Physics Today* 43:13.

Panasyuk, I. S. 1967. First Soviet nuclear reactor. *Sovetskaya Atomnaya Nauka i Tekhnika (Soviet Atomic Science and Technology)*. Atomizdat.

Pavlovsky, A. I. 1991. *Usp. Fiz. Nauk* 161:137.

Peattie, Lisa. 1984. Normalizing the unthinkable. *Bul. Atom. Sci.* iii.

Peierls, Rudolf. 1985. *Bird of Passage*. Princeton University Press.

Pervukhin, M. G. 1985. First years of the nuclear project: interview by M. Chernenko (iii.78). *Chemistry and Life* 5.

Petrov, Vladimir. 1955. Mystery of missing diplomats solved. *U.S. News & World Report*. 23.ix.

Petrzhak, Konstantin A., and Georgi N. Flerov. 1940. *Dokladi Akademii Nauk SSSR* 28:6, 500.

Pfau, Richard. 1984. *No Sacrifice Too Great*. University Press of Virginia.

Philby, Kim. 1968. *My Silent War*. Ballantine.

Pilat, Oliver. 1952. *The Atom Spies*. G. P. Putnam's Sons.

Pincher, Chapman. 1981. *Their Trade Is Treachery*. Sidgwick & Jackson.

———. 1984. *Too Secret Too Long*. St. Martin's.

———. 1987. *Traitors.* St. Martin's.

Pogue, Forrest C. 1987. *George C. Marshall: Statesman.* Viking.

Poirier, Bernard W. 1980. W. Averell Harriman oral history interview. HSTL.

Powers, Thomas. 1993. *Heisenberg's War.* Knopf.

Prados, John. 1992. High-flying spies. *Bul. Atom. Sci.* ix.

Proctor, Robert M. 1993. (Letter.) *Science* 259:1676.

Putney, Diane T., ed. 1987. *ULTRA and the Army Air Forces in World War II.* Office of Air Force History.

Rabi, I. I. 1970. *Science: The Center of Culture.* World.

———. 1983. How well we meant. Edited transcript of LANL 40th anniversary talk. LANL Archives.

Radosh, Ronald. 1979. Unpublished interview with David and Ruth Greenglass.

Radosh, Ronald, and Joyce Milton. 1983. *The Rosenberg File.* Holt, Rinehart and Winston.

Ranelagh, John. 1986. *The Agency.* Simon and Schuster.

Raymond, Ellsworth. 1948. Russia is ready for the wrong war. *Look.* 16.iii.

Reeves, Thomas C. 1982. *The Life and Times of Joe McCarthy.* Stein and Day.

Regis, Ed. 1987. *Who Got Einstein's Office?* Addison-Wesley.

Remnick, David. 1993. *Lenin's Tomb.* Random House.

Resis, Albert, ed. 1993. *Molotov Remembers.* Ivan. R. Dee.

Rhodes, Richard. 1979. *Looking for America.* Penguin.

———. 1986. *The Making of the Atomic Bomb.* Simon and Schuster.

Ridenour, Louis N. 1950. The hydrogen bomb. *Scientific American.* iii.

Ritus, V. I. 1990. Who else if not I? *Priroda* 8:10.

Rogow, Arnold A. 1963. *James Forrestal.* Macmillan.

Romanov, Yuri A. 1990. Father of Soviet hydrogen bomb. *Priroda* 8:20-24.

Rosenberg, David Alan. 1979. American atomic strategy and the hydrogen bomb decision. *Journal of American History.* v.

———. 1981–82. "A smoking, radiating ruin at the end of two hours." *International Security* 6(3).

———. 1982. US nuclear stockpile, 1945 to 1950. *Bul. Atom. Sci.* v.

———. 1983. "Toward Armageddon: The Foundations of United States Nuclear Strategy, 1945–1961." Unpub. Ph.D. thesis, University of Chicago.

Rosenberg, J. Philipp. 1982. The belief system of Harry S. Truman and its effect on foreign policy decision-making during his administration. *Presidential Studies Quarterly* XII(2). Spring.

Rosenberg, Suzanne. 1988. *A Soviet Odyssey.* Oxford University Press.

Ross, Steven T., and David Alan Rosenberg. 1989. *The Atomic Bomb and War Planning. America's Plans for War Against the Soviet Union, 1945–1950,* Vol. 9. Garland.

Rossi, John P. 1986. Winston Churchill's Iron Curtain speech: forty years after. *Modern Age* 30. Winter.

Roth, Julius L. 1992. (Letter.) Who won Cold War? *New York Times.* 30.viii.

Rummel, R. J. 1990. *Lethal Politics.* Transaction.

Sagan, Scott. 1993. *The Limits of Safety.* Princeton University Press.

———. 1994. The perils of proliferation: organization theory, deterrence theory, and the spread of nuclear weapons. *International Security* 18(4):66.

Sagdeev, Roald. 1993. Russian scientists save American secrets. *Bul. Atom. Sci.* v.

Sakharov, Andrei. 1990. *Memoirs*. Knopf.

Sanders, Jerry W. 1983. *Peddlers of Crisis*. South End Press.

Sawatsky, John. 1984. *Gouzenko*. Macmillan of Canada.

Sawyer, Roland. 1954. The H-bomb chronology. *Bul. Atom. Sci.* x.

Schilling, Warner R. 1961. The H-bomb decision: how to decide without actually choosing. *Political Science Quarterly* LXXVI(1).

Schneir, Walter, and Miriam Schneir. 1965. *Invitation to an Inquest*. Pantheon.

Schreiber, Raemer. 1991. Reminiscences. Unpub. MS. LANL Archives.

Schumar, James F. 1959. Reactor fuel elements. *Scientific American*. ii.

Scott, Russell B. 1959. *Cryogenic Engineering*. D. Van Nostrand.

Scurlock, Ralph G., ed. 1992. *History and Origin of Cryogenics*. Clarendon Press.

Seaborg, Glenn T. 1958. *The Transuranium Elements*. Yale University Press.

——. 1990. *Journal of Glenn T. Seaborg, 1946–1958*. Vols. 1–4. Lawrence Berkeley Laboratory.

Segrè, Emilio. 1993. *A Mind Always in Motion*. University of California Press.

Serber, Robert. 1992. *The Los Alamos Primer*. University of California Press.

Sherry, Michael S. 1987. *The Rise of American Air Power*. Yale University Press.

Shimizu, Sakae. 1982. Historical sketch of the scientific field survey in Hiroshima several days after the atomic bombing. *Bulletin of the Institute for Chemical Research, Kyoto University* 60(2):39.

Shlaim, Avi. 1983. *The United States and the Berlin Blockade, 1948–1949*. University of California Press.

Simonov, Konstantin. 1958, 1975. *The Living and the Dead*. Trans. Alex Miller. Moscow: Progress Publishers.

Sloan Foundation, Alfred P. 1982. The H-bomb decision. Transcript of videotaped conference at Princeton University.

Smith, Alice Kimball, and Charles Weiner, eds. 1980. *Robert Oppenheimer: Letters and Recollections*. Harvard University Press.

Smith, Bruce L. R. 1966. *The Rand Corporation*. Harvard University Press.

Smith, Cyril Stanley. 1954. Metallurgy at Los Alamos 1943–1945. *Metal Progress* 65(5):81.

Smyth, Henry D. 1945a. *A General Account of the Development of Methods of Using Atomic Energy for Military Purposes Under the Auspices of the United States Government, 1940–1945*. War Department lithoprint.

——. 1945b. *Atomic Energy for Military Purposes*. Princeton University Press.

Snow, C. P. 1966. *Variety of Men*. Charles Scribner's Sons.

——. 1981. *The Physicists*. Little, Brown.

Steiner, Barry H. 1991. *Bernard Brodie and the Foundations of American Nuclear Strategy*. University Press of Kansas.

Stern, Philip M. With Harold P. Green. 1969. *The Oppenheimer Case: Security on Trial*. Harper & Row.

Stettinius, Jr., Edward W. 1949. *Roosevelt and the Russians*. Doubleday.

Stickle, D. M., ed. 1992. *The Beria Affair*. Nova Science Publishers.

Straight, Michael. 1983. *After Long Silence*. W. W. Norton.

Strauss, Lewis L. 1950. A-bomb fallacies are exposed. *Life*. 24.vii.

——. 1962. *Men and Decisions*. Doubleday.

Street, Lucie, ed. 1947. *I Married a Russian*. Emerson Books.

Sudoplatov, Pavel, and Anatoli Sudoplatov. With Jerrold L. Schecter and Leona P. Schecter. 1994. *Special Tasks*. Little, Brown.

Szilard, Leo. 1961. *The Voice of the Dolphins*. Stanford University Press.

Szulc, Tad. 1984. The untold story of how Russia "got the bomb." *Los Angeles Times*. 26.viii.

Taschereau, Robert, and R. L. Kellock. 1946. *The Report of the [Canadian] Royal Commission*. Edmond Cloutier.

Taub, A. H. 1963. *John von Neumann: Collected Works*. Pergamon.

Teller, Edward. 1946a. Scientists in war and peace. *Bul. Atom. Sci.* iii.

———. 1946b. The State Dep't report—"a ray of hope." *Bul. Atom. Sci.* iv.

———. 1946c. Dispersal of cities and industries. *Bul. Atom. Sci.* iv.

———. 1947a. How dangerous are atomic weapons? *Bul. Atom. Sci.* ii.

———. 1947b. Atomic scientists have two responsibilities. *Bul. Atom. Sci.* iii.

———. 1948a. The first year of the Atomic Energy Commission. *Bul. Atom. Sci.* i.

———. 1948b. Comments on the "draft of a world constitution." *Bul. Atom. Sci.* vii.

———. 1950. Back to the labs. *Bul. Atom. Sci.* iii.

———. 1955. The work of many people. *Science* 121:267.

———. 1962. *The Legacy of Hiroshima*. Doubleday.

———. 1979. *Energy from Heaven and Earth*. W. H. Freeman.

———. 1982. Hydrogen bomb history. (Letter.) *Science* 218:1270.

———. 1987. *Better a Shield Than a Sword*. Free Press.

Terletsky, Y. P. 1973. Operation "Niels Bohr Interrogation." Trans. Catherine A. Fitzpatrick. (Courtesy Thomas Powers.)

———. n.d. Reconstructed account of conversations between Niels Bohr and Y. P. Terletsky in Copenhagen, 14 and 16 November 1945. Trans. Roald Sagdeev. (Courtesy Thomas Powers.)

Terrall, Mary. 1983. Willard F. Libby oral history interview. The Bancroft Library.

Thirring, Hans. 1946. *Die Geschichte der Atombombe*. Neues Osterrreich.

Thomas, Hugh. 1986. *Armed Truce*. Atheneum.

Timmerhaus, K. D., ed. 1960. *Advances in Cryogenic Engineering I*. Plenum Press.

Trachtenberg, Marc. 1988. A "wasting asset": American strategy and the shifting nuclear balance, 1949–1954. *International Security* 13(3).

Truman, Harry S. 1955. *Year of Decision*. Doubleday.

———. 1956. *Years of Trial and Hope*. Doubleday.

———. 1966. *Public Messages, Speeches and Statements of the President, 1952–1953*. USGPO.

Truslow, Edith C., and Ralph Carlisle Smith. 1947. Part II: Beyond Trinity. In David Hawkins, Edith C. Truslow and Ralph Carlisle Smith. 1983. *Project Y: The Los Alamos Story*. Tomash Publishers.

Tunner, William H. 1964. *Over the Hump*. Office of Air Force History.

Ulam, Françoise. 1991. Postscript to adventures. In Ulam, Stanislaw M. 1991. *Adventures of a Mathematician*. University of California Press.

Ulam, Stanislaw M. 1966. Thermonuclear devices. In R. Marshak, ed., *Perspectives in Modern Physics*. Interscience.

———. 1986. Mark C. Reynolds, Gian-Carlo Rota, eds. *Science, Computers, and People*. Birkhäuser.

———. 1991. *Adventures of a Mathematician*. University of California Press.

United States Atomic Energy Commission (USAEC). 1954a. Thermonuclear weapons program chronology. LANL Archives. 22.iv.

———. 1954b. *In the Matter of J. Robert Oppenheimer*. MIT Press.

Urey, Harold C., F. G. Brickwedde and G. M. Murphy. 1932. A hydrogen isotope of mass 2 and its concentration. *Phys. Rev.* 40(1):1.

Vance, Robert W., and Harold Weinstock. 1969. *Applications of Cryogenic Technology*. Tinnon-Brown.

Vandenberg, Arthur H., Jr. 1952. *The Private Papers of Senator Vandenberg*. Houghton Mifflin.

Visgin, V. P., ed. 1992. At the source of the Soviet atomic project: the role of espionage, 1941–1946. *Problems in the History of Science and Technology* 3:97.

Volkogonov, Dmitri. 1988, 1991. *Stalin*. Grove Weidenfeld.

Voyetekhov, Boris. 1943. *The Last Days of Sevastopol*. Knopf.

Wainstein, L., C. D. Cremeans, J. K. Moriarty and J. Ponturo. 1975. *The Evolution of US Strategic Command and Control and Warning*. Institute for Defense Analyses.

Walker, Mark. 1989. *German National Socialism and the Quest for Nuclear Power*. Cambridge University Press.

Weiner, Charles. 1967–73. Interviews with Edward Condon. American Institute of Physics.

Weisgall, Jonathan M. 1994. *Operation Crossroads*. Naval Institute Press.

Werth, Alexander. 1944. *Leningrad*. Hamish Hamilton.

———. 1964. *Russia at War: 1941–1945*. E. P. Dutton.

———. 1971. *Russia: The Post-War Years*. Taplinger.

WGBH. 1980. *A Is for Atom, B Is for Bomb*. WGBH Transcripts.

Wheeler, John A. 1962. Fission then and now. *IAEA Bulletin*. 2.xii.

Wiebes, Cees, and Bert Zeeman. 1983. The Pentagon negotiations March 1948: the launching of the North Atlantic Treaty. *International Affairs* (UK) 59. Summer.

Williams, Robert Chadwell. 1987. *Klaus Fuchs: Atom Spy*. Harvard University Press.

Wilson, Jane, ed. 1975. *All in Our Time*. Bulletin of the Atomic Scientists.

Wilson, Robert R. 1958. Books. *Scientific American*. xii.

Winterberg, Friedwardt. 1981. *The Physical Principles of Thermonuclear Explosive Devices*. Fusion Energy Foundation.

Wittlin, Thaddeus. 1972. *Commissar: The Life and Death of Lavrenti Pavlovich Beria*. Macmillan.

WT-608, extracted version. 1952. Operation *Ivy,* report of commander, Task Group 132.1.

X (George Kennan). 1947. The sources of Soviet conduct. *Foreign Affairs*. vii.

Yatzkov, Anatoli A. 1992. The atom and intelligence. *Problems in the History of Science and Technology* 3:103.

Yergin, Daniel. 1977. *Shattered Peace*. Houghton Mifflin.

York, Herbert F. 1975. The debate over the hydrogen bomb. *Scientific American*. x.

———. 1976. *The Advisors*. W. H. Freeman.

Yurechko, John J. 1983. The day Stalin died: American plans for exploiting the Soviet succession crisis of 1953. *Journal of Strategic Studies* 3. v.

Zaloga, Steven J. 1993. *Target America*. Presidio.

Zeldovich, Yakov Borisovich. 1992, 1993. *Selected Works*. 2 vols. Princeton University Press.

Zhisui, Li. 1994. *The Private Life of Chairman Mao.* Random House.

Zhukov, G. K. 1971. *The Memoirs of Marshal Zhukov.* Delacorte.

Ziegler, Charles. 1988. Waiting for Joe-1: Decisions leading to the detection of Russia's first atomic bomb test. *Social Studies of Science* 18:197.

Zimmerman, Carroll L. 1988. *Insider at SAC.* Sunflower University Press.

Zubok, Vladislav, and Constantine Pleshakov. 1994. The Soviet Union. In David Reynolds, ed. *The Origins of the Cold War in Europe.* Yale University Press.

Zuckerman, Solly. 1988. Bomber barons and armchair warriors. *NYTBR.* 22.v.

Zukerman, V. A., and Z. M. Azarkh. n.d. *People and Explosions.* Trans. Timothy D. Sergay. Unpub. MS.

PHOTO CREDITS

Courtesy Robert Serber: 1.

USAF: 2–3, 32.

From A. P. Grinberg and V. Ya. Frenkel, *Igor Kurchatov in the A. F. Ioffe Physical-Technical Institute, 1925–1943,* courtesy V. Ya. Frenkel: 6.

Courtesy David Holloway: 7.

ITAR-TASS/Sovfoto © Fotokhronika TASS: 9.

Sovfoto: 14.

LANL: 15, 25–26, 34, 36, 55, 60, 65, 69, 70–73, 86.

UPI/Bettmann: 17, 46, 61, 77, 88–89.

AP/Wide World: 20–21, 56–57, 79, 84–85, 87.

FBI: 27.

National Archives: 29, 31.

Imperial War Museum, London: 30.

Yoshito Matsushige: 33.

Niels Bohr Institute, Copenhagen: 35.

RIA–Novosti/Sovfoto: 37, 92.

Courtesy Lev Altshuler: 40, 42–45.

Courtesy Yuri Smirnov: 50.

Courtesy Frank Shelton: 54, 59, 78, 82.

Courtesy John Marshall: 58.

Courtesy Françoise Ulam: 63, 66.

Courtesy Jacob Wechsler: 67–68.

Chuck Hansen Collection: 74, 81, 90.

Peter Miller/The Image Bank: 76.

Ralph Morse, *LIFE* Magazine © Time Inc.: 80.

Reuters/Bettmann: 91.

© The Harold E. Edgerton 1992 Trust, courtesy of Palm Press, Inc.: 93.

Ginger Rhodes: 94.

Index